D0164282

Applied Numerical Methods for Engineers

USING MATLAB® AND C

Robert J. Schilling
Sandra L. Harris
Clarkson University

Pacific Grove • Albany • Belmont • Boston • Cincinnati • Johannesburg • London • Madrid • Melbourne
Mexico City • New York • Scottsdale • Singapore • Tokyo • Toronto

Sponsoring Editor: *Bill Stenquist*
Marketing Team: *Nathan Wilbur, Christina De Veto, Samantha Cabaluna*
Editorial Assistants: *Shelley Gesicki, Meg Weist*
Production Coordinator: *Laurel Jackson*
Production Service: *WestWords, Inc.*
Manuscript Editor: *Charles R. Batten*
Media Editor: *Marlene Thom*
Interior Design: *Andrew Herzog*
Cover Design: *Christine Garrigan*
Cover Illustration: *Robert J. Schilling*
Interior Illustration: *Heather Theurer*
Print Buyer: *Vena Dyer*
Typesetting: *WestWords, Inc.*
Printing and Binding: *R. R. Donnelley & Sons Co., Crawfordsville Mfg. Div.*

For more information, contact:
BROOKS/COLE PUBLISHING COMPANY
511 Forest Lodge Road
Pacific Grove, CA 93950 USA
www.brookscole.com

Printed in the United States of America

10 9 8 7 6 5 4 3

Library of Congress Cataloging-in-Publication Data
Schilling, Robert J. (Robert Joseph).
Applied numerical methods for engineers using MATLAB and C / by Robert J. Schilling and Sandra L. Harris.
p. cm.
Includes bibliographical references (p.).
ISBN 0-534-37014-4
1. MATLAB. 2. Engineering mathematics—Data processing. 3. C (Computer program language) I. Harris, Sandra L. II. Title.
TA345.S34 1999
620'.001'51—dc21

99-28210
CIP

Dedicated to our parents:
Edgar and Bette Rose Schilling
and
George and Florence Harris

Contents

CHAPTER 1 Numerical Computation 1

CHAPTER 6 Optimization 250

CHAPTER 7 Differentiation and Integration 307

Preface

Numerical computing is a powerful tool for solving practical mathematical problems that occur throughout engineering. In this book, we focus on the application of numerical methods to solve both analysis and design problems. In today's computing environment, inexpensive hardware and software are available to solve realistic numerical problems very quickly and with modest effort.

TOPICAL COVERAGE

The topics included in this book are summarized in the block diagram shown in Figure 1. The number in the lower right corner of each block indicates the chapter or appendix where the topic is discussed, while the text in the lower left corner indicates the software module associated with the block.

In Chapter 1, we examine potential sources of error in numerical computations. Two of the most common mistakes that casual users of numerical software make are to accept the validity of numerical output at face value and to confuse precision for accuracy. In addition to examining the effects of round-off error and formula truncation error, Chapter 1 also focuses on techniques for efficiently generating random numbers with desired statistical properties. We included random number generation in Chapter 1 because, like round-off error, it can be tied directly to the internal representation of numbers within the computer. In addition, random numbers prove useful for testing

FIGURE 1 Topics*

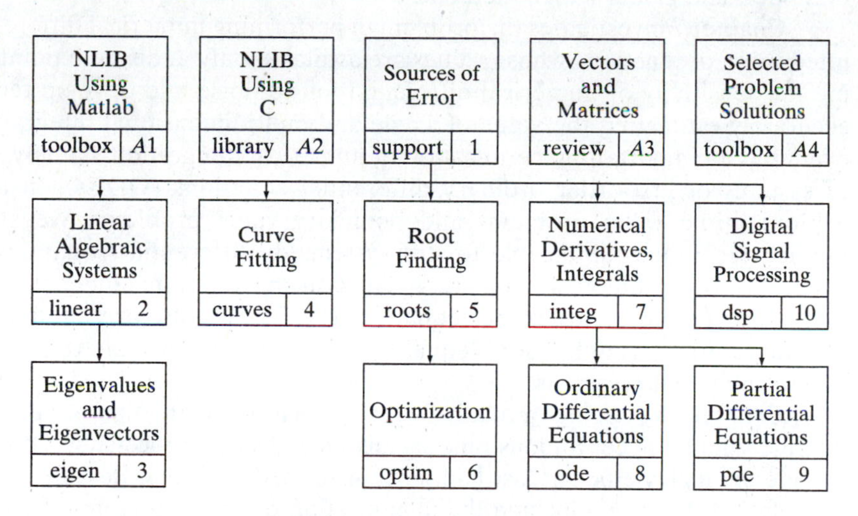

*In each box, the lower left corner lists the appropriate software module; the lower right corner lists the chapter or appendix number.

virtually all the numerical methods that follow. Vectors and matrices are used in engineering applications because they provide an elegant, concise way to describe the essential relationships between problem variables. They are used extensively both in this book and in the MATLAB and C software packages that accompany it. We highly recommend that students whose mathematical background includes little or no exposure to the use of vectors and matrices read the brief review of the topic in Appendix 3 before proceeding beyond Chapter 1. It is *not* necessary for students to have taken a course in linear algebra to use this book. However, it is important that students at least be comfortable with the *notation* of vectors and matrices and be generally familiar with basic operations and fundamental definitions.

Chapter 2 focuses on the problem of solving systems of linear algebraic equations using both direct methods based on elementary row operations and iterative methods that are attractive for large sparse systems. In Chapter 3, we examine the problem of finding the eigenvalues and eigenvectors of a square matrix. Knowing the locations of the eigenvalues allows one to draw conclusions about the stability of linear dynamic systems, and this knowledge is also helpful in analyzing the convergence properties of linear iterative methods.

Chapter 4 addresses the problem of fitting curves and surfaces to experimental data. In this chapter, we have included techniques for interpolation, extrapolation, least-squares curve fitting, and cubic splines. Several of these techniques surface again in later chapters as parts of other algorithms. Chapter 5 is devoted to the problem of solving nonlinear algebraic equations, also called root finding. Iterative techniques are developed and compared on the basis of speed and ease of implementation. We also consider in detail the important special case of finding the roots of polynomials. Chapter 6 focuses on the more general problem of minimizing an objective function subject to equality and inequality constraints. Optimal design problems in engineering can often be formulated as constrained minimizations. In this chapter, we examine both local and global search methods.

Chapter 7 investigates the problem of performing numerical differentiation and integration of functions whose values are available only at discrete points. We examine the sensitivity of numerical differentiation to noise and develop techniques for accurately estimating the value of single and multidimensional integrals. Chapter 8 examines an important generalization of numerical integration, namely, the solution of systems of first-order ordinary differential equations (ODEs). In addition, we explore initial value problems and boundary value problems. We also consider implicit techniques applicable to stiff systems of differential equations. Chapter 9 generalizes the problem still further by introducing additional independent variables in the form of partial differential equations (PDEs). We also present finite difference techniques for solving Poisson's equation, the heat equation, and the wave equation, in one and two dimensions.

Chapter 10 examines a relatively recent topic, the extraction of information from discrete samples of continuous-time signals. Digital signal processing (DSP) techniques include the highly efficient fast Fourier transform, digital filter design, correlation, and convolution. They also include the identification of linear discrete-time systems from input and output measurements using least-squares and adaptive methods.

STYLE OF PRESENTATION

This book is written in an informal style in order to ease the student gradually into each new topic and to smooth the transition between topics. The book contains numerous algorithms and examples, but there are no formal explicit theorems, definitions, or proofs. Terms defined or emphasized within the text are *italicized*. The usage of mathematical notation is summarized in Appendix 3.

Each chapter follows the template shown in Figure 2. Chapters start with a brief introduction to the problem or class of problems to be solved. This is followed by a section on motivation ("Why solve this problem?") and chapter objectives ("What will you learn?"). The core of each chapter is the development of a sequence of increasingly sophisticated numerical methods for solving the problem. We introduce the simplest and most specialized techniques first; then we present the more general ones. Each method is demonstrated with at least one example. For all the computational examples, corresponding example programs are available on the CD that accompanies the text. The example programs follow the naming convention *exyz.m* (for MATLAB) and *exyz.c* (for C), where x is the chapter number, y is the section within the chapter, and z is the example within the section.

The algorithm development sections are followed by an applications section that includes several case study–type examples from different fields of engineering (chemical, civil, electrical, and mechanical). The case study problems are developed in detail, and complete computational solutions are provided in MATLAB and C. Several case study examples reappear throughout subsequent chapters as we investigate different aspects of the problem and apply different mathematical techniques. The applications section is followed by a chapter summary that compares the different methods in terms of speed, accuracy, and applicability. At the end of each chapter is a section with homework problems, including subsections on analysis and computation. Problem solutions are available both in an Instructor's Manual and on an accompanying Solution Disk. In addition, solutions to selected problems are provided in Appendix 4. We encourage students to use these problems, marked with an (S), to check their understanding of the material.

MATHEMATICAL BACKGROUND

An increasingly common problem faced by college and university instructors is the large variation in the mathematical backgrounds of students taking a course such as

FIGURE 2 Chapter Format

Chapter Problem
Motivation and Objectives
Algorithm Development with Examples
Applications
Chapter Summary
Homework Problems

applied numerical methods. This book is targeted for use by undergraduates in all fields of engineering. It is assumed that the students have taken the standard sequence of courses in calculus and differential equations and are in at least the spring semester or quarter of their sophomore year.

There is enough material in the book, and enough flexibility in the order in which topics can be covered, to provide for two distinct ways the book can be used. The choice depends on the academic maturity of the class (sophomore, junior, senior) and the length of time available for the course (one quarter, one semester, two quarters). For students who have a less sophisticated mathematical background, the *light* chapter sequence shown in Figure 3 is probably most appropriate. In the light sequence, the sources-of-error material in Chapter 1 is followed directly by root-finding techniques in Chapter 5, skipping the section on systems of nonlinear equations at the end of the chapter. This way, students can see important problems and solution techniques using only *scalar* mathematics. The notation, basic operations, and fundamental definitions associated with the use of vectors and matrices are then covered in Appendix 3. This is necessary because vectors and matrices are used extensively in both the algorithm development and the accompanying software. Once students are comfortable with problem formulations that use vectors and matrices, they can go on to the fundamental topic of solving linear systems of algebraic equations, covered in Chapter 2. Eigenvalues and eigenvectors (Chapter 3) can be skipped without interrupting the flow, which then leads to curve fitting techniques in Chapter 4. Optimization (covered in Chapter 6) is a more advanced topic that can be skipped without loss of continuity, so students can proceed directly to Chapter 7, in which we discuss numerical differentiation and integration. This leads naturally to the solution of ordinary differential equations in Chapter 8. As an alternative, and with time permitting (one semester), the topic of partial differential equations can be considered after either Chapter 7 or Chapter 8, for courses for which PDEs are important.

For classes that are at least junior level and for courses of longer duration (that is, one semester or two quarters), some variation of the more complete treatment

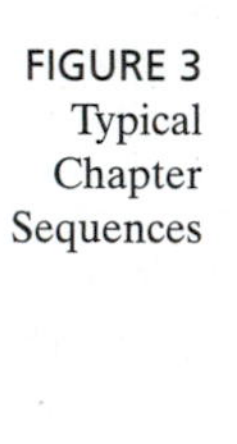
FIGURE 3 Typical Chapter Sequences

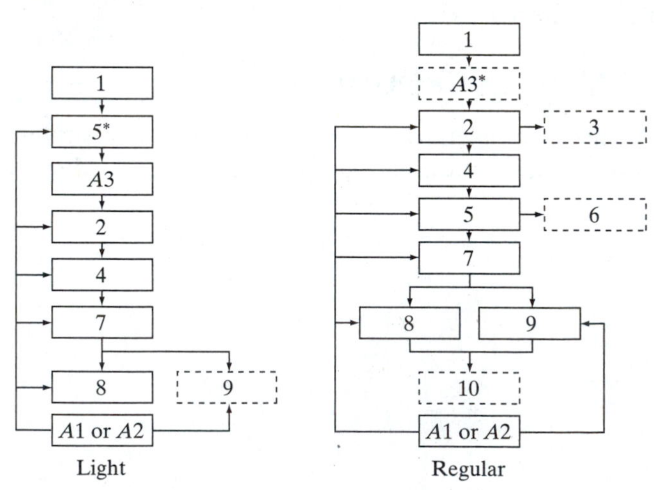

shown in the *regular* sequence in Figure 3 can be used. The main difference here is that the chapters are covered more or less in the order they are presented in the book, with the optional chapters on eigenvalues and eigenvectors (Chapter 3), optimization (Chapter 6), and digital signal processing (Chapter 10) included, depending on the students' interests, the students' backgrounds, and the time available. It should be emphasized that even though inclusion of Appendix 3 is optional in the regular course sequence, most students should read at least the first part of Appendix 3, which reviews vector and matrix notation and basic operations.

SOFTWARE BACKGROUND

The key to the successful application of numerical methods is effective software. The software included assumes that the student is familiar with the fundamentals of the MATLAB programming environment or the fundamentals of the programming language C. For MATLAB users, a numerical toolbox of MATLAB functions (NLIB) is available. The NLIB toolbox is described in detail in Appendix 1. It includes functions that implement the algorithms developed throughout the text. Also included is a set of main-program support functions, which are low-level utility functions designed to ease user interaction and the display of numerical results. At the start of each Applications section, the student should read the accompanying section of the software appendix to see how the algorithms are implemented.

For C users, there is a corresponding library of functions described in detail in Appendix 2. The NLIB library follows the ANSI C standard so as to maximize *portability* between programming platforms. Although the MATLAB toolbox is somewhat easier to use and is therefore recommended, the advantage of the C library is that the corresponding functions execute faster. This is particularly noticeable for computationally intensive applications, such as optimization and solution of partial differential equations. The complete source code for the NLIB library is provided on the distribution CD. In addition, precompiled versions of the library are available for the Microsoft Visual C++ and the Borland C++ compilers.

ACKNOWLEDGMENTS

This project has been several years in the making, and many individuals have contributed to its completion. We would like to thank our colleagues at Clarkson University for early reviews of the draft (Rangaswami Mukundan) and for suggesting some of the case study examples (Goodarz Ahmadi and Hayley Shen).

The reviewers commissioned by Brooks/Cole made numerous thoughtful and insightful suggestions that we incorporated into the final draft. In particular, we would like to express our gratitude to the following individuals for their contributions: G. Donald Allen, Texas A&M University; Ara Arabyan, University of Arizona; Daniel E. Bentil, University of Vermont; Neil E. Berger, University of Illinois at Chicago; Barbara S. Bertram, Michigan Technological University; Daniel Boley, University of Minnesota; Roberto Celi, University of Maryland; Chichia Chu, Michigan State University;

Prodromas Daoutidis, University of Minnesota; Prabir Daripa, Texas A&M University; Robert E. Fulton, Georgia Institute of Technology; Deborah Furey, Texas A&M University; J. Wallace Grant, Virginia Polytechnic Institute and State University; Willian Grimes, Central Missouri State University; Ali Hajjafar, The University of Akron; James G. Hartley, Georgia Institute of Technology; A. Scottedward Hode, Auburn University; Robert Jajcay, Indiana State University; Martin E. Kaliski, California Polytechnic State University; Charles Kenney, University of California, Santa Barbara; William H. Marlow, George Washington University; Tom I.-P. Shih, Carnegie Mellon University; S. V. Sreenivasan, University of Texas at Austin; Jery R. Stedinger, Cornell University; and Alfred G. Striz, University of Oklahoma.

We would also like to thank Brooks/Cole and WestWords, who helped to shepherd the project to completion and mold the final product. Special thanks to Peter Gordon, who got us started; Nancy Hill-Whilton; David Dietz; Bill Barter; Eric Frank; Liz Clayton; Laurie Jackson; and Richard Saunders.

Finally, we would like to acknowledge Clarkson University for supporting a sabbatical leave that helped to accelerate completion of the manuscript and software.

Robert J. Schilling
Sandra L. Harris

CHAPTER 1

Numerical Computation

Personal computers are powerful tools for solving numerical problems in science and engineering. With the aid of high-level programming languages and customized software libraries, realistic analysis and design problems can be solved quickly, with a minimum of effort. The focus of this book is on developing, implementing, and applying numerical algorithms to problems in science and engineering.

This chapter starts with a discussion of some of the pitfalls of numerical computing. It is all too common for users of numerical software to accept, without skepticism, the results of computations. We investigate sources of possible error by first examining how numbers are represented internally within a computer. Binary, decimal, and hexadecimal integers are discussed along with the two's complement representation of negative integers. Next, we discuss floating-point number representation, including the practical problems of overflow and underflow. We then discuss the machine epsilon and explain how it can be used to estimate the spacing between adjacent floating-point values. Next, we explore the notion of round-off error; we discuss propagation of accumulated round-off error through fundamental arithmetic operations, including the case of catastrophic cancellation where an abrupt significant loss of accuracy can occur. In addition, we examine a second major source of error, formula truncation error, and develop bounds on the magnitude of the error produced by truncating a Taylor series. We then examine the problem of generating random sequences of numbers with desired statistical properties. (Random numbers are useful for testing numerical algorithms.) The chapter concludes with a brief discussion of numerical software, including a comparison of compilers and interpreters. Finally, a numerical library called NLIB is introduced and illustrated using MATLAB™ and C.

1.1 MOTIVATION AND OBJECTIVES

1.1.1 A Simple Calculation

Although the computer is an ideal tool for performing complex numerical computations, casual or careless use of the output from a computer program can lead to highly undesirable consequences. Indeed, one of the most common mistakes made by new users is to accept, almost as a matter of faith, the validity of numerical output produced by an operational computer program. A relatively benign manifestation of this phenomenon arises when one attributes more precision to a numerical output than the accuracy of the input data or the underlying mathematical model justifies.

In other instances, a value produced by an operational computer program can be totally meaningless in the sense that it is not accurate to even one digit. This can result from an accumulation of round-off error due to the way a calculation is structured. As a simple illustration, consider the following quadratic equation.

$$ax^2 + bx^2 + c = 0 \tag{1.1.1}$$

Note that this equation is labeled (1.1.1) where the first digit identifies the chapter, the second digit identifies the section within the chapter, and the last digit identifies the equation within the section. A triple numbering convention is also used to refer to figures, tables, examples, and algorithms throughout the text. The parentheses around the three numbers, as in (chapter.section.number), indicate that the item being referenced is an *equation*. All nonequation items are identified explicitly and do not use the parentheses.

Suppose we want to find the two roots of (1.1.1). Using the well-known quadratic formula, the roots are

$$x_{1,2} = \frac{-b \pm \sqrt{b^2 - 4ac}}{2a} \tag{1.1.2}$$

To make the problem more specific, suppose the parameters are $a = 1$, $b = -(10^4 + 10^{-4})$, and $c = 1$. Carrying out the calculation in (1.1.2) yields the roots $x_1 = 10^4$ and $x_2 = 10^{-4}$. Interestingly enough, if we perform this simple calculation on a computer that has a precision of seven decimal digits (not uncommon), then the results are $x_1 \approx 10^4$ and $x_2 = 0$. That is, the first root is easily obtained, but the error in the second root is 100 percent! This is a consequence of accumulated round-off error. As we shall see in Chapter 5, there is a simple way to restructure this particular calculation to eliminate the problem.

In some instances, the computation is simple enough that we can check the validity of a numerical result by comparing it to an approximate solution obtained through problem simplification. In other cases, say, the solution of a system of nonlinear differential equations, the accuracy of the solution produced by the computer may be more difficult to judge. Here, special knowledge of the problem must be brought to bear to make sure the answers produced are at least plausible.

1.1.2 Chapter Objectives

When you finish this introductory chapter, you will understand the potential pitfalls of numerical computing. You will know how the principal sources of error in numerical computations arise, including round-off error and formula truncation error. You will understand how numerical data are represented internally within a computer and the effect this has on the precision and range of values that can be represented. You will learn how to efficiently generate pseudo-random numbers with desired statistical properties, numbers that can be used to test numerical algorithms. You will also learn how to use NLIB support functions, either MATLAB or C, that facilitate the development of user programs. You will achieve these overall goals by mastering the following chapter objectives.

Objectives for Chapter 1

- Understand how integers are represented internally within a computer.
- Know how to convert between binary, decimal, and hexadecimal integers.

- Know how to find the two's complement representation of negative integers.
- Understand how floating-point numbers are represented internally within a computer.
- Know how underflow and overflow conditions arise and what they mean.
- Know how to define, calculate, and use the machine precision.
- Understand how round-off error occurs and when it can lead to catastrophic cancellation.
- Be able to analyze error propagation in basic arithmetic operations.
- Be able to specify bounds on the size of formula truncation error.
- Understand the meaning of formula order, $O(h^n)$.
- Know how to generate uniform and Gaussian random numbers.
- Know how to use the main program support functions (MATLAB or C) in the numerical library NLIB.
- Understand how random numbers can be used in practical applications.

1.1.3 Mathematical Background

This book is targeted mainly at undergraduates specializing in engineering, math, and science. It is assumed that the students are at least in the second semester of the sophomore year and are taking the typical calculus and differential equations sequence. It is not assumed that students have taken a course in linear algebra. Nonetheless, it is important that students be comfortable with at least the *notation* of vectors and matrices. Students should also be generally familiar with basic matrix operations, such as the addition and multiplication of matrices and the notion of the determinant and inverse. A summary of these fundamental concepts is provided in Appendix 3. It is *highly recommended* that students who have not had a course in linear algebra review the material in Appendix 3 before proceeding beyond Chapter 1. Although some of the numerical techniques that follow can be developed without reference to vectors or matrices, the use of vector and matrix notation throughout allows for a *unified* treatment. Consequently, a brief review of the material in Appendix 3 will pay significant dividends, particularly for those students with a relatively modest mathematical background.

1.2 NUMBER REPRESENTATION

There are several potential sources of error in numerical computations. To cultivate an appreciation of how errors arise, it is useful first to examine how numbers are represented in computers.

1.2.1 Binary, Decimal, and Hexadecimal Numbers

Information is stored and processed in a computer with individual transistor elements that are turned on and off. In these two states, the power consumption of the transistor is minimal because in the first state (on), the output voltage is small, while in the second state (off), the output current is small. Because transistor elements are efficiently operated in this on/off manner, the natural way to represent and store numbers is to use a *binary* or base 2 number system, which contains the following two digits.

$$\text{binary digits} = \{0, 1\} \tag{1.2.1}$$

Like the more familiar decimal (base 10) numbers, binary numbers can be evaluated by taking a weighted sum of powers of the base or radix. For example, the equivalent decimal value of the binary number $(1010.01)_2$ can be evaluated as follows.

$$(1010.01)_2 = 1(2)^3 + 0(2)^2 + 1(2)^1 + 0(2)^0 + 0(2)^{-1} + 1(2)^{-2} = 10.25 \tag{1.2.2}$$

Here the notation $()_m$ is used to denote a number expressed in base m, the default being $m = 10$. For binary numbers, each binary digit is referred to as a *bit.* Although there are some computer operations that act on individual bits, for numerical computations it is more common to work with one or more bytes of data where one *byte* consists of eight bits. Since $2^8 = 256$, the range of non-negative decimal integers that can be represented with a single byte of data is 0 to 255.

Even though binary numbers are a natural choice for computations performed with a computer, users find base 2 awkward because long strings of digits, which are not easily interpreted, are required. Consequently, automatic conversion between base 10 and base 2 is performed by modern programming languages. Conversion of an n-bit binary integer $b = b_{n-1} \cdots b_0$ to its decimal equivalent x is accomplished by forming a sum of n powers of 2 as follows.

$$\boxed{x = \sum_{k=0}^{n-1} b_k 2^k} \tag{1.2.3}$$

A positive decimal integer x, in the range 0 to $2^n - 1$, is converted to its n-bit binary equivalent $b = b_{n-1} \cdots b_0$ by performing a sequence of n divisions by decreasing powers of 2. That is, the digits of the binary number are computed starting with the most significant bit, b_{n-1}, and ending with the least significant, b_0, as follows.

Alg. 1.2.1 Decimal to Binary

1. Set $m = 2$ and $y = m^{n-1}$.
2. For $k = n - 1$ down to 0 do

{

$$b_k = \text{fix}(x/y)$$
$$x = x - b_k y$$
$$y = y/m$$

}

Here the notation *fix* in step (2) of Alg. 1.2.1 indicates that the result of the division x/y is truncated to an integer. Note that by changing the value of the integer m in step (1),

TABLE 1.2.1 Decimal-to-Binary Conversion

k	b_k	x	y
7	1	109	64
6	1	45	32
5	1	13	16
4	0	13	8
3	1	5	4
2	0	1	2
1	0	1	1
0	1	0	0.5

conversion from decimal to an arbitrary base in the range $(1 < m < 10)$ can be achieved assuming $0 \le x \le m^n - 1$.

EXAMPLE 1.2.1 Decimal to Binary

Consider the problem of representing the decimal number $x = 237$ with its 8-bit binary equivalent. From step (1) of Alg. 1.2.1, the initial value of y is 128. The sequence of values generated by step (2) is summarized in Table 1.2.1.

It follows that the binary equivalent of x is $b = (11101001)_2$. This can be confirmed by converting from binary back to decimal using (1.2.3).

$$x = \sum_{k=0}^{7} b_k 2^k = 128 + 64 + 32 + 8 + 4 + 1 = 237$$

In some instances, it is useful to represent numerical data with *hexadecimal* or base 16 numbers. Since hexadecimal numbers have a larger base or radix than decimal numbers, some extra symbols are required to obtain 16 distinct digits. The first six letters of the alphabet are used to augment the decimal digits as follows.

$$\text{hexadecimal digits} = \{0, 1, 2, 4, 5, 6, 7, 8, 9, A, B, C, D, E, F\} \tag{1.2.4}$$

Hexadecimal numbers are again evaluated by taking a weighted sum of powers of the base or radix which in this case is 16. For example, the address of the COM1 serial port on an IBM personal computer is $(3F8)_{16}$. The equivalent decimal value in this case is determined as follows.

$$(3F8)_{16} = 3(16)^2 + F(16)^1 + 8(16)^0 = 1016 \tag{1.2.5}$$

The virtue of the hexadecimal or *hex* representation is that a single hex digit can be used to represent exactly four bits. Because there are eight bits in a byte, one byte corresponds to two hex digits with the hex value ranging from 0_{16} to FF_{16}. The hexadecimal representation is much more compact than the binary representation, yet conversion between the two can be done very easily because all one has to do is group

TABLE 1.2.2 Binary, Decimal, and Hexadecimal Numbers

Binary	*Decimal*	*Hexadecimal*	*Binary*	*Decimal*	*Hexadecimal*
0000	00	0	1000	08	8
0001	01	1	1001	09	9
0010	02	2	1010	10	A
0011	03	3	1011	11	B
0100	04	4	1100	12	C
0101	05	5	1101	13	D
0110	06	6	1110	14	E
0111	07	7	1111	15	F

the bits into sets of four and make the corresponding association as summarized in Table 1.2.2.

1.2.2 Integers

The simplest numbers to represent in computers are integers. Although the number of bytes used to represent an integer varies between computers and programming languages, an integer is typically allocated at least two bytes as shown in Figure 1.2.1. An unsigned integer therefore takes on decimal values between 0 and $2^{16} - 1$. However, it is more common to work with both positive and negative integers, in which case the range is -2^{-15} to $2^{15} - 1$, which corresponds to

$$-32768 \le \text{integers} \le 32767 \tag{1.2.6}$$

A reduction in the range of magnitudes arises because the most significant bit (bit 15) is reserved to hold the *sign* while bits 0 through 14 represent the magnitude. Positive numbers have a sign bit of 0, and negative numbers have a sign bit of 1.

Conceptually, the easiest way to represent a negative integer is to set the sign bit to 1 and the remaining bits to the magnitude or absolute value of the integer. However, this *sign-magnitude* representation is rarely used. Instead, a two's complement format is used to represent negative integers because it makes computations more efficient at the hardware level. The *two's complement* of a binary number is obtained by complementing each of the bits (changing 1 to 0 and 0 to 1), and then adding 1. Any carry from the most significant bit is ignored. For a two-byte integer, this is equivalent to performing the following operation:

$$\text{two's complement} = 2^{16} - \text{number}. \tag{1.2.7}$$

Notice from (1.2.7) that taking the two's complement of the two's complement of a number results in the original number, as it should, since two minus signs cancel. The

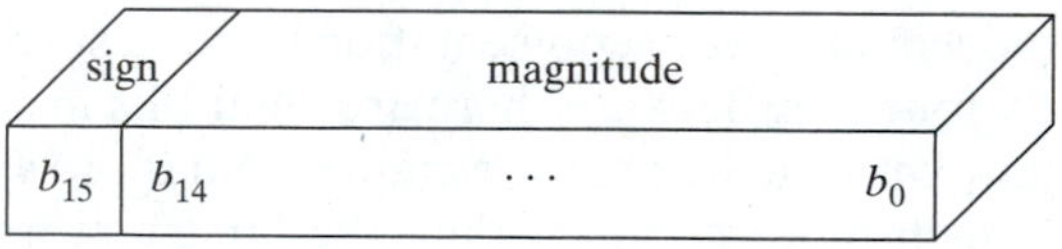

FIGURE 1.2.1 A Two-Byte Integer

sign	magnitude		
b_{15}	b_{14}	$\cdots$	b_0

process of complementing each of the bits is sometimes referred to as taking the one's complement.

EXAMPLE 1.2.2 Two's Complement

As an illustration of the two's complement representation, consider the problem of finding the two's complement representation of the number $x = -49$ and expressing the result in both binary and hexadecimal. We convert $|x|$ from decimal to binary using Alg. 1.2.1, complement the bits, and then add one.

$$\begin{aligned} |x| &= (0011\,0001)_2 \\ x &= (1100\,1110)_2 + (0000\,0001)_2 \\ &= (1100\,1111)_2 \\ &= (CF)_{16} \end{aligned}$$

1.2.3 Floats

The majority of numerical computations are done using real numbers otherwise referred to as floating-point numbers or simply floats. As with integers, the number of bytes allocated to a float depends on the machine and the programming language. Typically, at least four bytes are used to represent a float as shown in Figure 1.2.2. The four bytes are decomposed into three fields of bits, one for the sign, one for the exponent, and one for the mantissa or fractional part. Thus the representation of a float is of the following general form.

$$\boxed{\text{float} = \text{sign} \times \text{mantissa} \times 2^{\text{exponent}}} \tag{1.2.8}$$

The details of the allocation of fields vary from machine to machine and compiler to compiler. In the illustration in Figure 1.2.2, the sign is a one-bit field consisting of the most significant digit, b_{31}. This is followed by an 8-bit field used to represent the exponent, $b_{30} \cdots b_{23}$. In order to represent both very large numbers and very small numbers, positive and negative exponents must be used. Recall that a single byte can represent unsigned integers in the range 0 to 255. Therefore exponents ranging from -127 to 128 can be represented by simply adding 127 to the actual exponent. This is called a *biased* binary representation since the numbers are biased (shifted) by 127.

The last field in Figure 1.2.2 is a 23-bit field, $b_{22} \cdots b_0$, used to represent the magnitude of the mantissa or fractional part of the float. In representing floats, it is assumed that the number has been *normalized* so that the first digit of the mantissa is always 1.

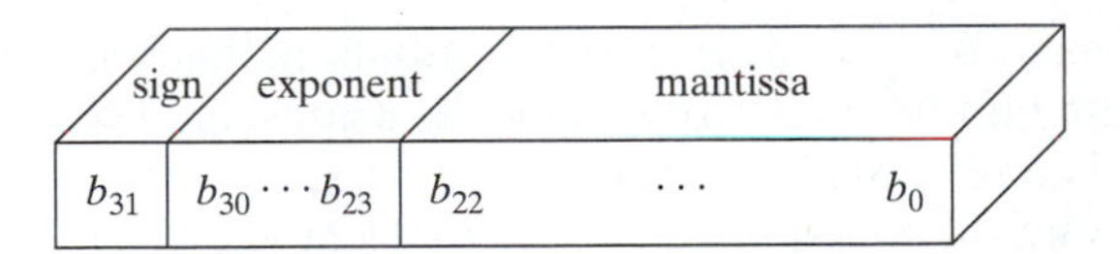

FIGURE 1.2.2 A Four-Byte Float

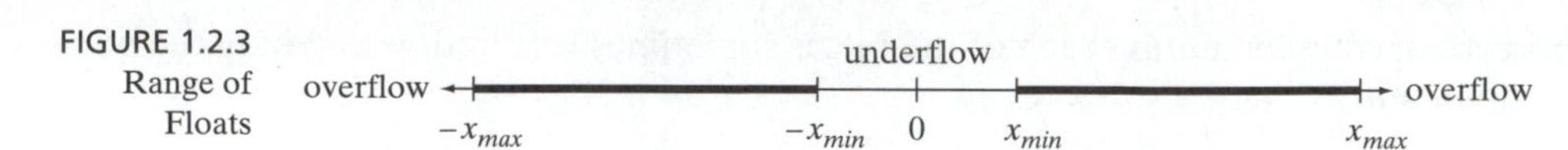

FIGURE 1.2.3 Range of Floats

Normalization is equivalent to selecting the exponent to ensure that the mantissa lies in the following normalized range.

$$\frac{1}{2} \leq \text{normalized mantissa} < 1 \tag{1.2.9}$$

Note that since the mantissa is *always* normalized in this manner, there is no need to explicitly store the first binary digit because it is known to be 1. Consequently, a normalized 24-bit mantissa can be stored using only 23 bits, $b_{22} \cdots b_0$, by adopting the convention that the most significant bit, always 1, is a *hidden* bit.

Given the number of bits of precision with which the mantissa and the exponent are represented, the range of possible values for floats can be determined. If x is a float, then the smallest value for $|x|$ corresponds to a normalized 24-bit mantissa of $(10 \cdots 0)_2$ and an exponent of -127. Thus $|x| \geq x_{min}$ where

$$\begin{aligned} x_{min} &= (.10 \cdots 0)_2 \times 2^{-127} \\ &= 2^{-128} \\ &\approx 2.94 \times 10^{-39} \end{aligned} \tag{1.2.10}$$

Similarly, the largest value of $|x|$ corresponds to a normalized 24-bit mantissa of $(11 \cdots 1)_2$ and an exponent of 128. Thus $|x| \leq x_{max}$ where

$$\begin{aligned} x_{max} &= (.11 \cdots 1)_2 \times 2^{128} \\ &\approx 2^{128} \\ &\approx 3.40 \times 10^{38} \end{aligned} \tag{1.2.11}$$

The lower limit x_{min} represents the smallest positive number that can be represented by a float. Numerical computations which produce nonzero values in the range $|x| < x_{min}$ result in an *underflow* condition which typically generates an error (or warning) message. Similarly, the upper limit x_{max} represents the largest number than can be represented by a float. In this case numerical computations which produce values in the range $|x| > x_{max}$ result in an *overflow* condition which generates an error message. A diagram of the range of realizable values for floats is shown in Figure 1.2.3.

1.3 MACHINE PRECISION

Although real numbers can take on values over a continuum, this is not true of floats, which take on only a finite number of discrete values as a consequence of their limited precision. For example, the only float in the range $|x| < x_{min}$ is $x = 0$, as can be seen in Figure 1.2.3. More generally, there is space between adjacent values of all floats. To see this, it is useful to first examine the smallest positive float ε_M that can be added to one and produce a sum that is greater than one. This is referred to as the *machine epsilon* and is a convenient measure of floating-point precision.

$$1 + \varepsilon_M > 1 \tag{1.3.1}$$

By definition, the float that follows 1 is $1 + \varepsilon_M$. If the details of the internal representation of floats are known, then the machine epsilon can be derived. For example, for the 32-bit representation in Figure 1.2.2, the number 1 has a 24-bit mantissa of $(10 \cdots 0)_2$ and an exponent of 1, while the next larger number has a 24-bit mantissa of $(10 \cdots 01)_2$ and exponent of 1. Thus, the difference between these two numbers is

$$\begin{aligned} \varepsilon_M &= (.00 \cdots 01)_2 \times 2^1 \\ &= 2^{-23} \\ &\approx 1.19 \times 10^{-7} \end{aligned} \tag{1.3.2}$$

Each computer and programming language has its own machine epsilon. If the details of the internal representation of floats are not known, then the machine epsilon can be computed. The following procedure estimates ε_M by starting with $\varepsilon = 1$ and cutting the estimate in half until condition (1.3.1) is violated.

Alg. 1.3.1 Machine Epsilon

1. Set $\varepsilon_M = 1$.
2. Do

   {

   $$\varepsilon_M = \varepsilon_M / 2$$
   $$x = 1 + \varepsilon_M$$

   }
3. While $(x > 1)$
4. $\varepsilon_M = 2\varepsilon_M$

In Alg. 1.3.1, the variables x and ε_M must be declared to be of the proper data type. For example, the machine epsilon for the data types float (four bytes), double (eight bytes), and long double (ten bytes) can be computed in this manner with the results, for an IBM PC, summarized in Table 1.3.1.

TABLE 1.3.1 Machine Epsilon on an IBM Personal Computer

Data Type	*Bytes*	*Visual C++*
Float	4	1.19×10^{-7}
Double	8	2.22×10^{-16}
Long double	10	1.08×10^{-19}

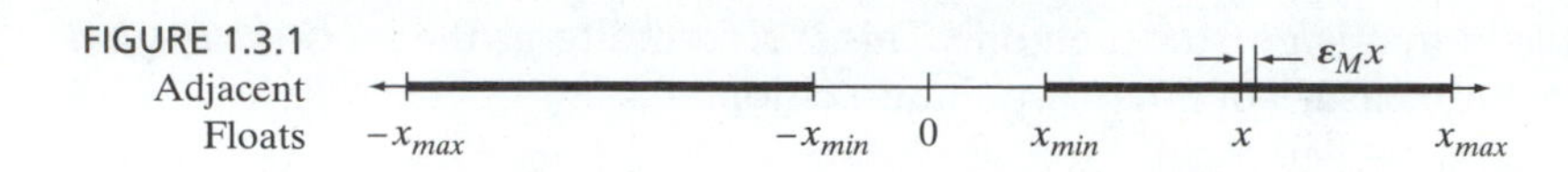

FIGURE 1.3.1 Adjacent Floats

The utility of the machine epsilon is that it allows us to estimate the distance between any two adjacent float values. In particular, the interval between a float x and the next larger float is approximately $\varepsilon_M x$. Thus, if x can be represented exactly, then the next larger float is $(1 + \varepsilon_m)x$ and the next smaller float is $(1 - \varepsilon_M)x$.

$$\text{floats:} \qquad \ldots, (1 - \varepsilon_M)x,\ x,\ (1 + \varepsilon_M)x, \ldots \tag{1.3.3}$$

Using (1.3.3), we can compute the spacing between the smallest positive float x_{min} in (1.2.10) and the next smallest positive float. From (1.3.2) the separation is approximately $\varepsilon_M x_{min} = 1.66 \times 10^{-46}$. Notice that this is much smaller than the separation between 0 and x_{min}, hence the spacing between adjacent floats is not uniform. A diagram of the spacing between floats is shown in Figure 1.3.1.

1.4 ROUND-OFF ERROR

There are several potential sources of error in a numerical calculation. To begin with, the data upon which the calculations are based may contain errors. This is particularly true when the data are generated from measurements of physical variables. In addition, errors are introduced by the computational process itself. Perhaps the most troublesome type of error, because it can accumulate and become significant, is *round-off* error. Round-off error is the general term given to inaccuracies that arise because a finite number of digits of precision are used to represent numbers (Wilkinson, 1963).

To examine the effects of finite precision we begin by defining the *absolute error* of a variable as the approximate value minus the exact value. It is often more meaningful to work with the *relative error,* which is the absolute error normalized by the exact value.

$$\boxed{\text{relative error} \triangleq \frac{\text{approximate value} - \text{exact value}}{\text{exact value}}} \tag{1.4.1}$$

The absolute error has to be interpreted with some care because a relatively small difference between two large numbers can appear to be large, and a relatively large difference between two small numbers can appear to be small. The relative error, by contrast, can be expressed as a percentage of the exact value. This is usually more meaningful as long as the exact value is not too close to zero.

1.4.1 Chopping and Rounding

Real numbers can be approximated with a finite number of digits of precision in two ways. If n digits are used to represent a real number, then the simplest scheme is to keep the first n digits and *chop* off all remaining digits. In terms of hardware implementation, this is the most appealing scheme because the excess digits are simply ignored.

Alternatively, one can *round* to the nth digit by examining the values of the remaining digits. The rounding operation is actually a modified version of chopping. Suppose a nonzero real number x, expressed in base b, is to be rounded to n digits. Then the following two steps can be used to round x.

Alg. 1.4.1 Rounding

1. Add $\text{sgn}(x)b/2$ to digit $n + 1$ of x.
2. Chop x to n digits.

Here, $\text{sgn}(x) = x/|x|$ denotes the *sign* of x with $\text{sgn}(0) \triangleq 0$. The effect of the add and chop method of rounding is to round digit n up (away from zero) if the first digit to be chopped, digit $n + 1$, is greater than or equal to $b/2$, otherwise digit n is left as is. In contrast, the magnitude of a chopped number never increases; that is, chopped numbers are always rounded down by discarding the excess digits. Whether chopping or rounding is used, the errors that result from this process are referred to as *round-off* error. For uniformly distributed random numbers, rounding is more accurate than chopping half of the time, and it is never less accurate.

EXAMPLE 1.4.1 Round-off Error

As an illustration of the round-off error associated with chopping and rounding, consider the base of the natural logarithm, $e = 2.718281828$, expressed as a decimal number using $n = 3$ digits. Using chopping results in $e_1 = 2.71$ and the following relative error.

$$r_1 = \frac{2.71 - e}{e} \approx -0.305\% \left.\vphantom{\frac{1}{1}}\right\} \text{chopping}$$

Thus, the relative error is negative and fairly significant because only $n = 3$ digits of precision are used. To determine the relative error due to rounding, we first add $b/2 = 5$ to digit 4 of e and then chop to $n = 3$ digits.

$$e \approx 2.718 + 0.005 = 2.723 \approx 2.72$$

The relative error associated with the rounded version of e is then as follows.

$$r_2 = \frac{2.72 - e}{e} \approx 0.063\% \left.\vphantom{\frac{1}{1}}\right\} \text{rounding}$$

In this case the sign of the error has changed, and the magnitude of the error is reduced.

1.4.2 Error Propagation

Since chopping always rounds the magnitude down towards zero, it follows from (1.4.1) that the relative error introduced by chopping is never positive. If a long sequence of calculations is performed, the round-off error can *accumulate* and rapidly grow to be significant. Accumulated round-off error associated with rounding can also be a significant problem. For example, if a sum of m numbers is computed, it can be shown

(Ralston and Rabinowitz, 1978) that the accumulated error associated with rounding is proportional to $\sqrt{m}$.

$$\text{rounding error} \sim \sqrt{m} \tag{1.4.2}$$

To examine how round-off errors propagate through calculations and accumulate, suppose x_e and y_e represented the *exact* values of two real numbers whose *approximate* values are x and y.

$$x = x_e + \Delta x \tag{1.4.3}$$

$$y = y_e + \Delta y \tag{1.4.4}$$

Here $\Delta x = x - x_e$ and $\Delta y = y - y_e$ represent the absolute errors of the variables x and y, respectively. The relative errors are obtained by normalizing as in (1.4.1).

$$r_x = \frac{\Delta x}{x_e} \quad , \quad r_y = \frac{\Delta y}{y_e} \tag{1.4.5}$$

There are four basic arithmetic operations. We begin by examining the relative error of the sum r_{x+y}. The objective in this case is to express r_{x+y} in terms of r_x and r_y so that we can determine the relative sizes of the errors and identify conditions under which the error might grow. Using (1.4.1), we have

$$r_{x+y} = \frac{(x + y) - (x_e + y_e)}{x_e + y_e} = \frac{\Delta x + \Delta y}{x_e + y_e} \tag{1.4.6}$$

To reformulate (1.4.6) in terms of r_x and r_y, it is useful to introduce a parameter α, which represents the ratio of y_e to x_e.

$$y_e = \alpha x_e \tag{1.4.7}$$

Note that α is well defined as long as $x_e \neq 0$. By substituting (1.4.7) into (1.4.6) and using (1.4.5) one can show, after simplification, that the relative error of the sum can be expressed in terms of the relative errors of the operands as follows.

$$\boxed{r_{x+y}(\alpha) = \frac{r_x + \alpha r_y}{1 + \alpha}} \tag{1.4.8}$$

It is clear from (1.4.8) that the size of the accumulated round-off error of a sum depends on the relative sizes of the two operands. If $y_e \approx x_e$, then $\alpha \approx 1$ and r_{x+y} is approximately the average of r_x and r_y. On the other hand if $|y_e| \ll |x_e|$, then $|\alpha| \ll 1$ and $r_{x+y} \approx r_x$. Similarly, if $|y_e| \gg |x_e|$, then $|\alpha| \gg 1$ and $r_{x+y} \approx r_y$. Thus the round-off error of the sum is dominated by the round-off error of the operand whose magnitude is largest.

The most interesting case occurs when $y_e \approx -x_e$ which corresponds to subtracting two nearly identical numbers. In this case $\alpha \approx -1$, and from (1.4.8) this can result in a *very large* relative error! This is sometimes referred to as *catastrophic cancellation* because it results in a significant loss of accuracy.

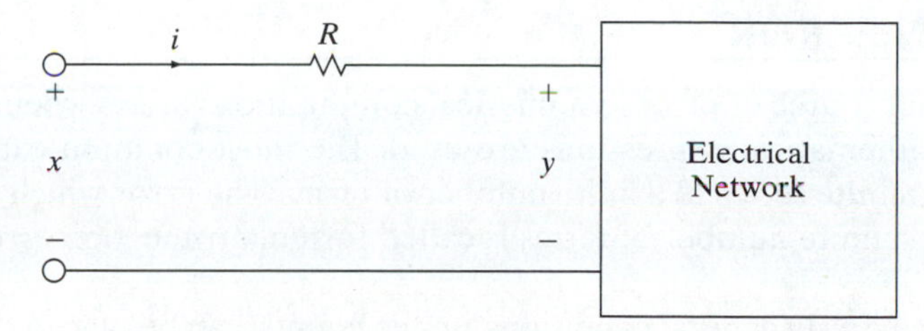

FIGURE 1.4.1 Catastrophic Cancellation

EXAMPLE 1.4.2 Catastrophic Cancellation

To illustrate the case of catastrophic cancellation, suppose some voltage measurements are performed on the electrical network shown in Figure 1.4.1. Here R is a 10-ohm external resistor inserted in order to determine the amount of current drawn by the network. Suppose voltage measurements, using a digital multimeter, yield $x = 5.09$ volts and $y = 5.04$ volts. If the exact values for these variables are $x_e = 5.07$, and $y_e = 5.05$, respectively, then the relative errors are $r_x = 0.39\%$ and $r_y = -0.20\%$.

The objective is to estimate the current flowing through the resistor. Using the voltage drop across the resistor and Ohm's law, the estimated current i and the exact current i_e are as follows.

$$i = \frac{x - y}{R} = 0.500 \text{ mA}$$

$$i_e = \frac{x_e - y_e}{R_e} = 0.200 \text{ mA}$$

Since the relative errors in the operands are less than 1%, one might (casually) expect that the relative error in the current estimate r_i is also on the order of 1%. This is clearly *not* the case.

$$r_i = \frac{i - i_e}{i_e} = 150\%$$

This is a case where subtracting two nearly identical numbers in the numerator of i has caused a catastrophic loss of accuracy, almost three orders of magnitude!

The effects of multiplication and division on the round-off error can be investigated in a similar manner. In this case, if it is assumed that the operand errors are small, $|r_x| \ll 1$ and $|r_y| \ll 1$, then the following approximations to the round-off error can be obtained (see Problems 1.13–1.14).

$$r_{xy} \approx r_x + r_y \tag{1.4.9}$$

$$r_{x/y} \approx r_x - r_y \tag{1.4.10}$$

Thus, the operations of multiplication and division are relatively well behaved in the sense that the error will not grow dramatically with a single operation, instead it accumulates in a predictable manner. Note that when chopping is used, the errors for multiplication grow faster than the errors for division because $r_x \leq 0$ and $r_y \leq 0$ for chopped numbers.

1.5 TRUNCATION ERROR

Another significant source of error in numerical computations arises when approximations to exact mathematical expressions are used. The most common example is the truncation of an infinite series to a finite number of terms. The error which results from terminating after a finite number of terms is called formula truncation error or simply *truncation error.*

Under certain fairly general conditions, upper bounds can be placed on the size of the truncation error. To illustrate, suppose a function $f(x)$ is infinitely differentiable over an interval which includes the point $x = a$. Then the *Taylor series* expansion of $f(x)$ about $x = a$ is as follows.

$$f(x) = \sum_{k=0}^{\infty} \frac{f^{(k)}(a)(x-a)^k}{k!} \tag{1.5.1}$$

Here it is understood that $0! = 1$ and the notation $f^{(k)}(a)$ denotes the kth derivative of $f(x)$ evaluated at $x = a$.

$$f^{(k)}(a) \triangleq \left. \frac{d^k f(x)}{dx^k} \right|_{x=a} \tag{1.5.2}$$

The special case $a = 0$ in (1.5.1) is called the *MacLaurin series.* If the series is truncated after n terms, it is equivalent to approximating $f(x)$ with a polynomial of degree $n - 1$.

$$f_n(x) \triangleq \sum_{k=0}^{n-1} \frac{f^{(k)}(a)(x-a)^k}{k!} \tag{1.5.3}$$

The error in the approximation, $E_n(x)$, is the sum of neglected higher-order terms which is sometimes called the *tail* of the series. The tail can be represented by using the first neglected term (see, e.g., Gerald and Wheatley, 1989).

$$E_n(x) \triangleq f(x) - f_n(x) = \frac{f^{(n)}(\xi)(x-a)^n}{n!} \tag{1.5.4}$$

For convenience, suppose $x > a$. Then ξ is a constant which lies in the interval $[a, x]$. Depending on the nature of the function $f(x)$, it may be possible to place an upper bound on the size of $E_n(x)$. In particular, suppose the *maximum* value of $|f^n(\xi)|$ over the interval $[a, x]$ is known or can be estimated.

$$M_n(x) \triangleq \max_{a \le \xi \le x} \{|f^{(n)}(\xi)|\} \tag{1.5.5}$$

Then from (1.5.4) and (1.5.5) a conservative worst-case bound on the size of the truncation error can be expressed as follows:

$$\boxed{|E_n(x)| \le \frac{M_n(x)|x-a|^n}{n!}} \tag{1.5.6}$$

In discussing truncation error, it is often more meaningful to describe the asymptotic behavior of the error as x approaches a, the point about which the series is expanded. For example, if $h = x - a$, then we can say that the truncation error $E_n(x)$ is of *order* $O(h^n)$, which means that as h approaches zero, $E_n(x)$ goes to zero at the same rate as h^n. Consequently, the order notation, $O(h^n)$, can be interpreted as follows where α is some nonzero constant.

$$\boxed{O(h^n) \approx \alpha h^n \quad , \quad |h| \ll 1} \tag{1.5.7}$$

In general, a function which is of order $O(h^n)$ also includes higher-order terms, but it is the first term that dominates in the limit as h approaches zero. As we shall see, numerical algorithms are often classified based on the order of their formula truncation error. Higher-order algorithms tend to converge faster and be more accurate than lower-order algorithms, but they require more computational effort per iteration.

EXAMPLE 1.5.1 Taylor Series

Suppose $f(x) = \exp(-x)$ is to be expanded about the point $a = 0$ and truncated to $n = 5$ terms.

$$\exp(-x) \approx 1 - x + \frac{x^2}{2} - \frac{x^3}{6} + \frac{x^4}{24}$$

If this approximation is evaluated at $x = 1$, the relative truncation error in this case is

$$r_5(1) = \frac{9/24 - \exp(-1)}{\exp(-1)} \approx 1.94\% \quad \Big\} \text{ truncation error}$$

To compare this with the error bound (1.5.6), first note that $f^{(k)}(\xi) = (-1)^k \exp(-\xi)$. Thus, from (1.5.5),

$$M_5(1) = \max_{0 \le \xi \le 1} \{\exp(-\xi)\} = 1$$

It then follows from (1.5.6) that the upper bound on the magnitude of the absolute error due to truncation is

$$|E_5(1)| \le 0.0083$$

Converting this to a relative error (by dividing by $\exp(-1)$) results in $|r_5(1)| \le 2.26\%$, which is a reasonably good estimate in this case.

1.6 RANDOM NUMBER GENERATION

Random numbers are useful for testing numerical methods. Random sequences with a variety of statistical properties can be generated using many different techniques (Press et al., 1992). In this section, we examine two common distributions of random numbers and methods for generating them.

1.6.1 Uniform Distribution

The simplest type of random sequence is a sequence of numbers in an interval $[a, b]$ where each number in the interval is equally likely to occur. A sequence of this form is said to be *uniformly distributed* over the interval $[a, b]$ as illustrated in Figure 1.6.1. The curve in Figure 1.6.1 is called the *probability density* function $p(x)$ for a uniform distribution.

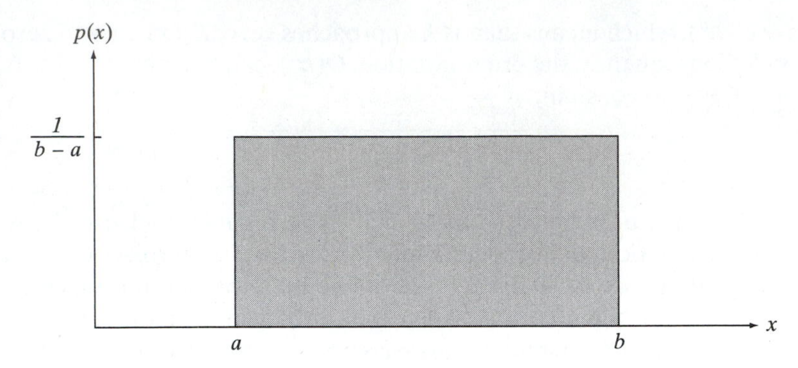

FIGURE 1.6.1 Probability Density of a Uniform Distribution

$$p(x) = \begin{cases} \dfrac{1}{b-a} & , \quad a \le x \le b \\ 0 & , \quad \text{otherwise} \end{cases} \tag{1.6.1}$$

For a general probability density function, $p(x)$, the probability that a random number will lie in the interval $[\alpha, \beta]$ is

$$P_{[\alpha,\beta]} = \int_{\alpha}^{\beta} p(x)\, dx \tag{1.6.2}$$

That is, the area under a section of the probability density function is the probability that a random number will fall in that segment. Since a random number has to fall somewhere, the area under the entire probability density function is always one.

A survey of many random number generators can be found in (Park and Miller, 1988). The most widely used numerical technique for generating a sequence of uniformly distributed random integers, u_k, is the *linear congruential method* which is based on the following iterative formula.

$$u_{k+1} = (\alpha u_k + \beta) \,\%\, \gamma \tag{1.6.3}$$

Here % denotes the *modulo* operator. That is, $a \,\%\, b$ is the remainder after dividing a by b. Therefore random number u_{k+1} is obtained as the remainder of $\alpha u_k + \beta$ divided by γ. The integers α, β, and γ are called the *multiplier, increment* and *modulus,* respectively. The sequence will repeat itself after, at most, γ numbers so it is referred to as a *pseudo-random* sequence. Typically γ is chosen to be very large. If the parameters $\{\alpha, \beta, \gamma\}$ are selected with care, then the period p will equal the maximum value γ. In this case, every integer in the range 0 to $\gamma - 1$ appears once in the sequence. The arbitrary initial value, u_0, is called the *seed* of the random number generator. Each seed generates a different random sequence because each seed starts at a different point in the cycle.

A very efficient random number generator can be produced if the programming language supports unsigned four-byte (32 bit) integer arithmetic. In this case, the modulus can be taken to be $\gamma = 2^{32}$. This eliminates the need to express the random numbers modulo γ because truncation to four bytes occurs automatically. Suggested values for the multiplier and increment in this case (see, e.g., Press et al., 1992) are as follows.

$$u_{k+1} = 1664525u_k + 1013904223 \tag{1.6.4}$$

This is a computationally inexpensive scheme for generating random numbers with unsigned four-byte integer arithmetic because it requires only one multiplication and one addition per random number.

Another approach is to make the increment $\beta = 0$. This results in a *multiplicative congruential* method. Park and Miller (1988) have proposed using a multiplier of $\alpha = 7^5$ and a modulus of $\gamma = 2^{31} - 1$. Thus, the iterative formula is

$$u_{k+1} = 16807u_k \,\%\, 2147483647 \tag{1.6.5}$$

This random number generator has been successfully used for many years. However, note that $\alpha(\gamma - 1)$ is too large to be represented by a four-byte integer, which means that (1.6.5) can not be implemented with four-byte integer arithmetic (Schrage, 1979). Instead, longer integers must be used. The period of the random number generator in (1.6.5) is $p = 2^{31} - 2$. Thus, it generates in excess of 2.147 *billion* random integers before repeating. Note that the seed u_0 must always be *nonzero* because otherwise $u_k \equiv 0$. This is characteristic of all multiplicative congruential generators.

The random number generators discussed thus far produce integers. To convert integers in the range 0 to u_{max} to floats in the range a to b, we simply add a scale factor and an offset as follows.

$$x_k = a + \frac{(b - a)u_k}{u_{max}} \tag{1.6.6}$$

A simple test of the distribution of random numbers produced by the four-byte multiplicative congruential method in (1.6.5) is shown in Figure 1.6.2. The selected

FIGURE 1.6.2 Measured Probability Density of Uniform Random Numbers

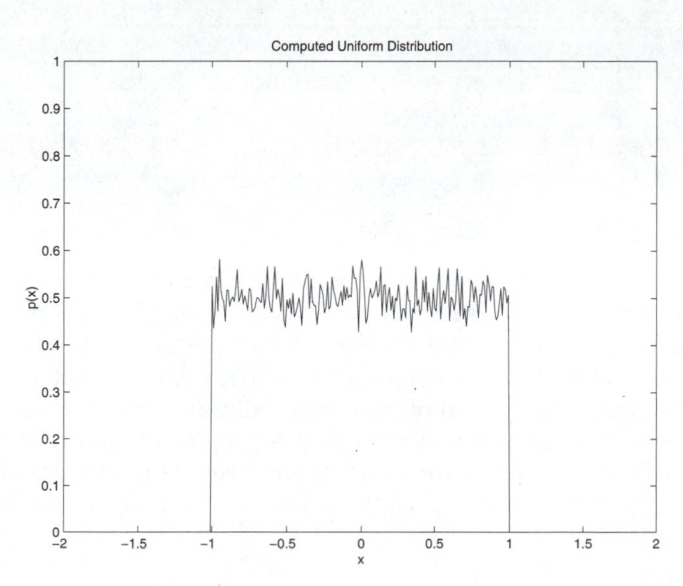

interval is $[a,b] = [-1,1]$, and 50000 random numbers are generated. The plot in Figure 1.6.2 is a histogram which shows the relative frequency of occurrence of each number. The interval $[-1,1]$ is partitioned into subintervals and a normalized version of the number of random numbers in each subinterval is plotted. As the total number of random numbers increases, the measured plot in Figure 1.6.2 begins to look more and more like the ideal plot in Figure 1.6.1.

The simple test of randomness in Figure 1.6.2 does not take into account the possibility of *correlation* between successive random numbers in the sequence. Serial correlation becomes apparent when one takes sets of two successive numbers and uses them as coordinates to fill a square. Even though every integer in the range 0 to u_{max} may appear once in the sequence, this does not mean that every pair of coordinates, i.e., every point in the square, will appear. The effects of serial correlation become even more pronounced when one is filling a cube in three-dimensional space.

A simple technique will reduce the correlation between successive random numbers generated by a linear or multiplicative congruential random number generator. The basic idea is to *shuffle* the order of the random numbers, using a scheme developed by Bays and Durham and described in (Knuth, 1981). The following algorithm is an implementation which generates a sequence of m random numbers in the interval $[a,b]$. It assumes that there is a basic random number generator, RAND, which returns random integers in the range, 0 to u_{max}.

Alg. 1.6.1 Random Shuffle

1. Pick $n \gg 1$, and seed the random number generator, RAND.
2. For $k = 1$ to n set $z_k = \text{RAND}$.
3. For $k = 1$ to m do

{

$$p = 1 + n\text{RAND}/(u_{max} + 1.0)$$
$$x_k = a + (b - a)z_p/u_{max}$$
$$z_p = \text{RAND}$$

}

First, we initialize the system by generating a vector of n random integers z. Then each random number in the sequence $\{x_1, x_2, \ldots, x_m\}$ is generated by two calls to the basic random number generator. The first call generates a random integer p in the range 1 to n which serves as an *index* into the array z. The selected z_p is then output as x_k after suitable scaling and offset to normalize it to the interval $[a,b]$. Finally, the value of the random integer z_p is updated using a second call to the basic random number generator. This provides for a *random ordering* of the random numbers and thereby reduces the effects of any serial correlation inherent in the basic random number generator.

1.6.2 Gaussian Distribution

The uniform distribution of random numbers in Figure 1.6.1 is useful in many applications in engineering. However, there are other instances where it is more appropriate to generate random numbers which are not restricted to an interval of finite length. Nonuniform distributions of random numbers can be constructed from a uniform distribution over the interval [0, 1]. Suppose $p(x)$ is the desired *probability density* function. Let $P(x)$ denote the integral of $p(x)$.

$$P(x) \triangleq \int_{-\infty}^{x} p(\alpha)d\alpha \tag{1.6.7}$$

The function $P(x)$ is called the *cumulative probability distribution* function. Note that $P(x)$ is the probability that a random number will be less than or equal to x. Since the area under a probability density function must be one, it follows that $P(\infty) = 1$.

Let u_k be the kth random number in a sequence that is uniformly distributed over the interval [0, 1]. To generate a random sequence $\{x_0, x_1, \ldots\}$ that has density function $p(x)$ we use the following transformation.

$$x_k = P^{-1}(u_k) \tag{1.6.8}$$

Thus the inverse of the desired cumulative probability distribution function maps a uniform distribution over [0, 1] into the desired distribution. This transformation method is particularly suitable when a closed-form expression for the inverse of $P(x)$ is available.

EXAMPLE 1.6.1 Exponential Distribution

Suppose the desired probability density function is the following one-sided exponential or Boltzmann-type distribution where $a > 0$.

$$p(x) = \begin{cases} a\exp(-ax) & , \quad x \geq 0 \\ 0 & , \quad x < 0 \end{cases}$$

Note that $p(x) \geq 0$ for all x, and the area under $p(x)$ is unity. The desired cumulative distribution function in this case is

$$P(x) = \int_{-\infty}^{x} p(\alpha)d\alpha = \begin{cases} 1 - \exp(-ax) & , \quad x \geq 0 \\ 0 & , \quad x < 0 \end{cases}$$

Setting $P(x) = u$ and solving for x in terms of u then yields the following expression for the inverse of $P(x)$.

$$P^{-1}(u) = \frac{-\ln(1-u)}{a}$$

Thus, to generate a sequence of random numbers with exponential probability density $p(x)$ we start with random numbers u_k uniformly distributed over [0, 1] and then use

$$x_k = \frac{-\ln(1-u_k)}{a}$$

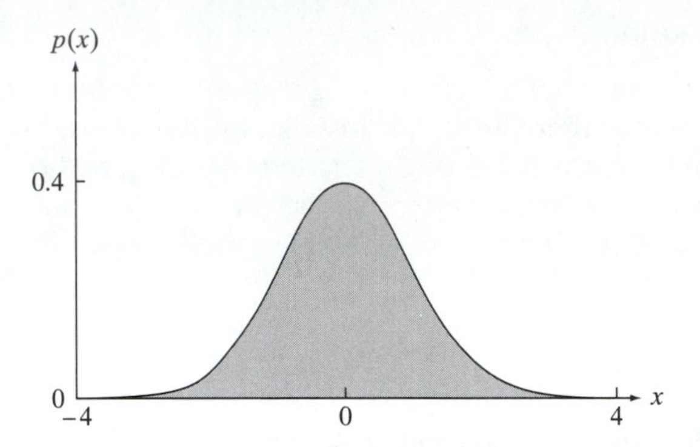

FIGURE 1.6.3 Probability Density of Gaussian Distribution

The most widely used nonuniform distribution of random numbers is the *Gaussian distribution,* also called the normal distribution. The probability density function of the Gaussian distribution is as follows.

$$p(x) = \frac{1}{\sigma\sqrt{2\pi}} exp\left[\frac{-(x-\mu)^2}{2\sigma^2}\right] \tag{1.6.9}$$

Here the parameters μ and σ are the *mean* and the *standard deviation* of the distribution of random numbers. A plot of the probability density function for a Gaussian distribution with mean $\mu = 0$ and standard deviation $\sigma = 1$ is shown in Figure 1.6.3. Note that the mean is the point about which the bell-shaped curve is *centered,* while the standard deviation is a measure of the *spread* of random numbers about the mean.

There is no closed-form expression for the cumulative Gaussian probability distribution function $P(x)$. However, it is possible to transform a uniform distribution into a Gaussian distribution using a pair of random numbers with the *Box-Muller* method. Let u_1 and u_2 be two random numbers uniformly distributed over $[0, 1]$. Then the following random numbers each have a Gaussian distribution.

$$x_1 = \cos(2\pi u_2)\sqrt{-2 \ln u_1} \tag{1.6.10a}$$

$$x_2 = \sin(2\pi u_2)\sqrt{-2 \ln u_1} \tag{1.6.10b}$$

The Gaussian distribution in this case has a mean of $\mu = 0$ and a standard deviation of $\sigma = 1$. To get a more general mean and standard deviation, we use scaling and offset as follows.

$$z = \mu + \sigma x \tag{1.6.11}$$

A plot of the measured probability density of random numbers produced by the transformation in (1.6.10) is shown Figure 1.6.4. The selected mean and standard deviation are $\mu = 0$ and $\sigma = 1$, respectively, and 50000 random numbers are generated. The plot in Figure 1.6.4 is a histogram that shows the relative frequency of occurrence of each number. Again, as the number of random numbers increases, the

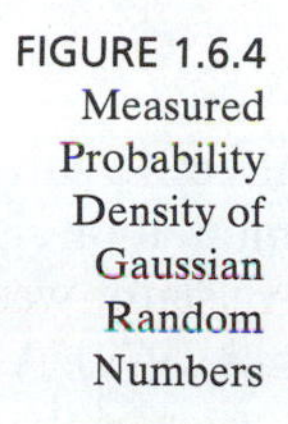

FIGURE 1.6.4 Measured Probability Density of Gaussian Random Numbers

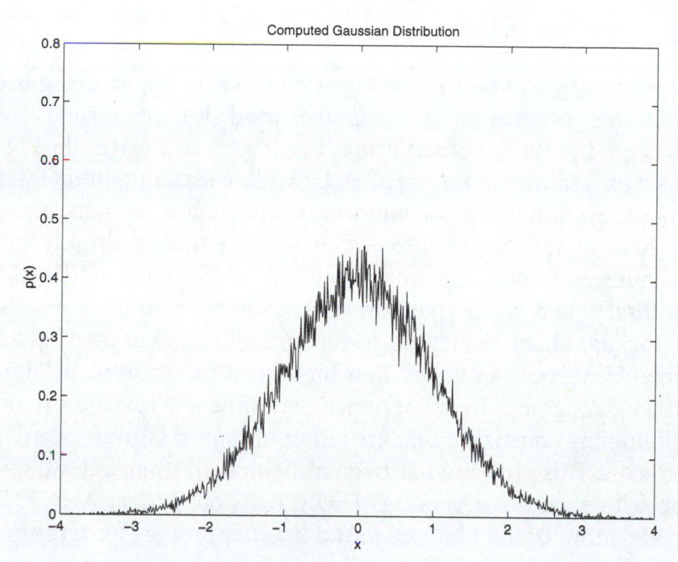

FIGURE 1.6.5 Gaussian Distribution in Two-Dimensional Space

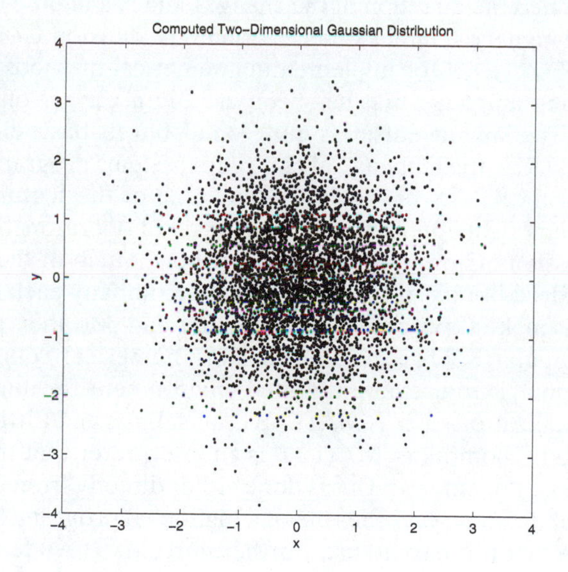

measured plot in Figure 1.6.4 begins to look more and more like the ideal plot in Figure 1.6.3.

Another way to view a Gaussian distribution is to generate a sequence of random 2×1 vectors using pairs of random numbers for the x and y coordinates. Suppose a Gaussian distribution with zero mean and unit standard deviation is used. The resulting distribution of 2500 points in the $x-y$ plane is shown in Figure 1.6.5. Note how the cloud of points is most dense around the mean at the origin. The fact that there are no spokes, rings, or streaks is consistent with the random points being uncorrelated from one another as a result of the shuffling operation.

1.7 NUMERICAL SOFTWARE

The key to successful *application* of numerical methods in engineering is effective software. The traditional programming language used for numerical computation is FORTRAN. Developed in the nineteen fifties, FORTRAN or FORmula TRANslator was the first high-level programming language to be widely disseminated (Backus, 1979). As such, FORTRAN enjoys a number of important advantages over its rivals. One is that there is a large base of software written in FORTRAN that is already out in the field. Examples in the numerical methods area include the IMSL (International Mathematical and Statistical Library) and NAG (Numerical Algorithms Group) libraries (Rice, 1993). These software packages have been field-tested and refined over many years and are efficient and reliable. However, being the first high-level programming language also has some disadvantages. Programming experience over time has revealed that there are certain useful programming constructs that are either absent or implemented in an awkward manner in FORTRAN. This problem has been mitigated, to some extent, by the evolution of the language with successive versions (FORTRAN-66, FORTRAN-77, FORTRAN-90) including more and more of the features found in other modern programming languages (Chapman, 1998). However, the need to maintain upward compatibility with the existing software base has constrained the direction of the changes, and has limited their scope.

The programming language C, and its object-oriented extension C++, provide an attractive alternative to FORTRAN for implementing numerical methods. C is an elegant modern programming language that has been used in a variety of applications (Kelley and Pohl, 1990). The language itself is quite small, but its basic elements have been carefully constructed. C is the language of choice for systems programming, and it is the native language of the UNIX operating system. One of the features of C that makes it appealing for numerical applications is its rich set of operators that allow for compact powerful expressions. C programs tend to be short. Much of the power of C lies in the extensive standard library of functions which accompany each implementation of C. These functions make C programs both modular and portable.

An increasingly popular approach to performing numerical computations, in both industry and academia, is to use an integrated environment for numerical computation and graphic visualization such as MATLAB (Hanselman and Littlefield, 1998). Unlike the FORTRAN and C compilers, MATLAB is an interpreter that translates and executes commands as they are entered. This is done either directly from the keyboard or indirectly from a script file that plays the role of a high-level program. MATLAB has the important advantage that it is easy to use. Furthermore, MATLAB features a powerful set of vectorized operators and an extensive numerical and graphics library. However, MATLAB it is not as flexible as a lower-level programing language, and it tends to require more memory and run slower than equivalent executable files produced by a compiler.

The basic approach taken in this text is to allow the user to choose between the flexibility and efficiency of a programming language and the ease of use of a specialized interpreter. Implementations of the numerical techniques developed in subsequent chapters are provided on the distribution CD in both MATLAB and C. In this way, users can choose the programming environment within which they are most comfortable.

FIGURE 1.7.1 The Numerical Library NLIB

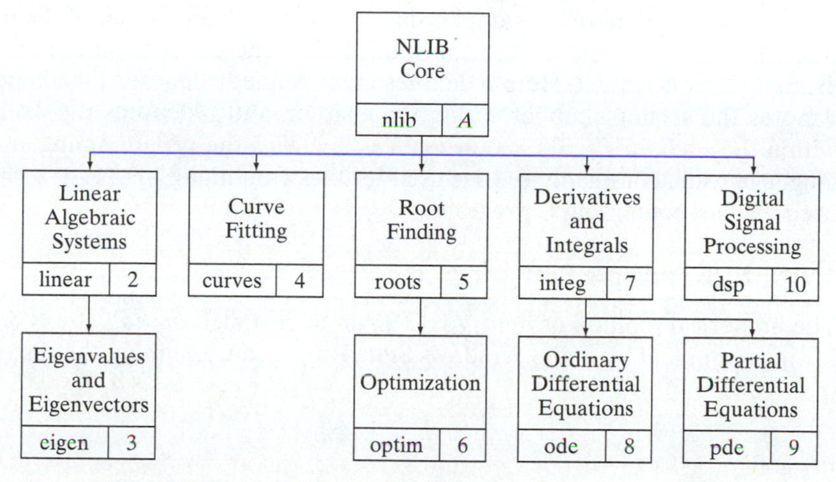

1.7.1 A Numerical Library: NLIB

The distribution CD that accompanies the text contains a **N**umerical **LIB**rary or toolbox of functions called NLIB. The NLIB library is composed of the modules pictured in Figure 1.7.1. The number appearing in the lower right corner of each block indicates the chapter where the algorithms implemented in the module are discussed in detail.

The NLIB Core module on the top level of Figure 1.7.1 contains general-purpose utility functions. More specifically, the module *nlib* contains low-level functions that are used by all of the remaining application modules. It also contains functions that facilitate the input, output, and display of numerical data. The contents of the module *nlib* depend on the programming environment used and they are described in more detail in the appropriate appendix.

The modules below the first level in Figure 1.7.1 contain implementations of the numerical methods developed in subsequent chapters. The Linear Algebraic Systems module in *linear* includes direct and iterative methods for solving linear algebraic systems of equations. The second module in this group is the Eigenvalue and Eigenvector module, *eigen,* which focuses on computing eigenvalues and eigenvectors of square matrices.

Problems of interpolation, extrapolation and least-squares curve fitting to discrete data are covered in the Curve Fitting module in *curves.* The Root Finding module in *roots* contains iterative solution techniques for nonlinear equations. The more general problem of finding the minimum of an objective function subject to equality and inequality constraints is covered in the Optimization module in *optim.*

The topics of numerical differentiation and integration are the focus of the Derivatives and Integrals module in *integ.* This leads naturally to the problem of solving systems of ordinary differential equations subject to initial or boundary conditions, which is treated in the Ordinary Differential Equations module in *ode.* The Partial Differential Equations module in *pde* includes techniques for solving various classes of partial differential equations including hyperbolic, parabolic, and elliptic equations. Finally, the Digital Signal Processing module in *dsp* focuses on numerical techniques for processing discrete-time signals, including spectral analysis, digital filtering, and system identification.

All of the numerical examples that appear in the remainder of the text have solutions that can be found on the distribution CD. The convention used for naming the example files is *exy.ext.* Here *e* denotes an example, *x* denotes the chapter number, *y* denotes the section number within the chapter, and *z* denotes the example number within the section. The file extension *.ext,* specifies the programming environment or language used. For example, *.m* is used for files containing MATLAB scripts, while *.c* is used for files containing C programs.

1.7.2 NLIB Example Browser

The numerical toolbox or library NLIB can be installed on a PC by executing the following command from the Windows Start/Run menu, assuming the distribution CD is in drive D.

```
D:\nlib\setup
```

More detailed instructions for installing the NLIB toolbox for both PC and non PC users can be found by viewing the plain text files *user_mat.txt* for MATLAB or *user_c.txt* for C.

The most convenient way for MATLAB users to view and run all of the NLIB software is to use the *NLIB Example Browser,* which is launched by entering the following command from the MATLAB command prompt:

```
browse
```

The file *browse.m* is an easy to use menu-based program that allows the user to select from a number of options. A typical screen for the MATLAB version of the example browser is shown in Figure 1.7.2.

The *run* option allows the user to execute all of the computational examples discussed in the text as well as solutions to selected problems discussed in Appendix 4. The

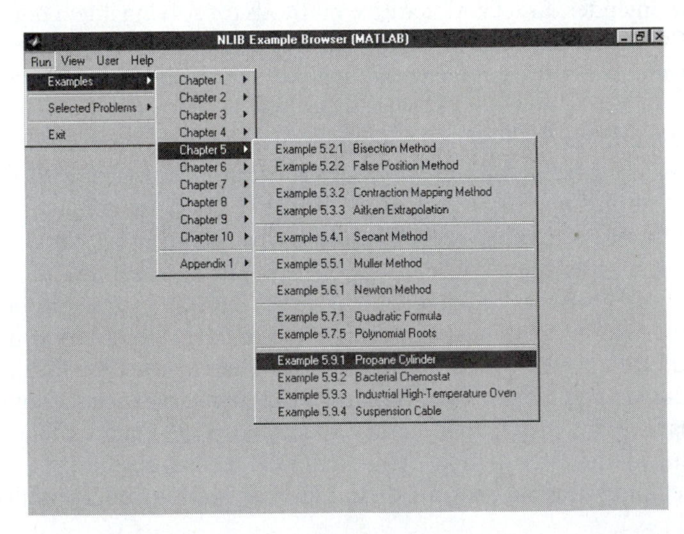

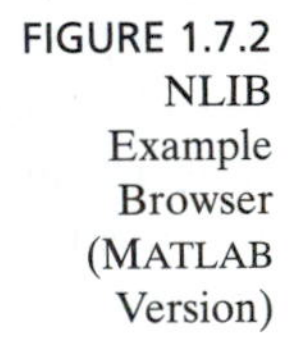
FIGURE 1.7.2 NLIB Example Browser (MATLAB Version)

view option is similar to run, but it allows the user to view the complete source code of the examples and the selected problems including data files. The *user* option provides a simple means of configuring *browse* to include user scripts. Selections are provided to create, run, and view user scripts as well as add them to or remove them from the browser menus. In this way, the user can employ *browse* as an environment for program development. Finally, the *help* option allows the user to view documentation on all of the functions in the NLIB toolbox, and learn how to obtain software upgrades as they become available.

The NLIB Example Browser is a fast and easy way to access all of the MATLAB software available on the distribution CD, and it also serves as an environment for user program development. There is also a C version of the NLIB Example Browser available for PC users. It is launched by entering the command *browse.exe* from the Windows Start/Run menu.

An alternative way to obtain online documentation of the NLIB toolbox functions is to use the MATLAB *help* command. To display a list of the names of all of the functions in the NLIB toolbox, arranged by category, enter the following command from the MATLAB command prompt:

```
help nlib
```

To obtain user documentation on a specific NLIB function, replace the operand *nlib* with the appropriate function name.

```
help funcname
```

Again, a similar help utility called *help.exe* is available for viewing user documentation of the NLIB functions in the C version of the library.

1.7.3 Pseudo-Prototypes

The Problems section at the end of each chapter includes an Analysis subsection and a Computation subsection. Some of the programming tasks in the Computation subsection are described by providing a *pseudo-prototype* of the function to be written. A pseudo-prototype is similar to a function prototype in C in that it specifies the calling arguments and their data types. In particular the following convention is used for a function named *funcname*:

```
[type_1 out_1,...,type_q out_q] = funcname (type_1 in_1,...,type_p
in_p)
```

Here *in_p* denotes the pth input argument, *out_q* denotes the qth output argument, and *type_i* denotes the data type for the ith input or output. The values of input arguments are supplied to the function when it is called, and the values of output arguments are returned to the calling program by the function. The details of how the function is actually implemented and called depend on the programming environment. The

TABLE 1.7.1 Data Types Used with Pseudo-Prototypes

Data Type	*Description*
void	empty argument list
int	integer
float	real number
complex	complex number
string	character string
vector	real one-dimensional array
cvector	complex one-dimensional array
matrix	real two-dimensional array
cmatrix	complex two-dimensional array
function	user-defined function

pseudo-prototype merely provides a convenient mechanism for describing how the function is used in general terms. The list of data types used by the pseudo-prototypes is summarized in Table 1.7.1.

The *void* data type is used to specify a function that has no inputs or no outputs. The subscripts of vectors and matrices are assumed to start at one (as in MATLAB) rather than zero (as in C). Examples of functions described by pseudo-prototypes can be found in problems 1.18 through 1.24 at the end of this chapter.

Two sets of functions accompany the text on the distribution CD. The first is a toolbox of MATLAB M-file functions that are described in Appendix 1, while the second is a library of C functions that are described in Appendix 2. The MATLAB functions are easier to use and are recommended. Users familiar with C/C++ may want to use the library of C functions, which execute faster. The C functions do not produce direct graphical output because they are written in ANSI C in order to maximize portability between programming environments.

1.8 APPLICATIONS

The following examples illustrate some simple applications of random numbers using both MATLAB and C. They also illustrate the use of a number of low-level NLIB support functions. The relevant MATLAB functions in the NLIB toolbox are described in Sections 1.1–1.2 of Appendix 1, while the corresponding C functions in the NLIB library are described in Sections 2.1–2.5 of Appendix 2.

1.8.1 Throwing Darts to Estimate π: MATLAB

It is assumed that the MATLAB user has been exposed to the fundamentals of MATLAB programming, and knows how to write a MATLAB script (see, e.g., Hanselman and Littlefield, 1998). In an effort to ease the task of user program development, a toolbox of functions called NLIB has been developed that parallels the algorithms discussed in this text. The NLIB toolbox also includes a number of low-level utility

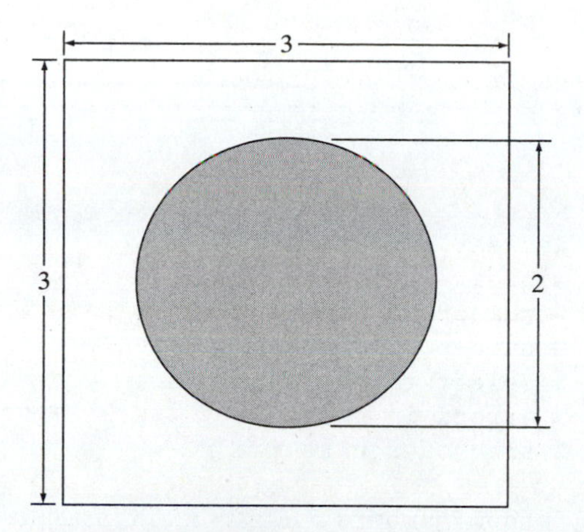

FIGURE 1.8.1 Circular Dart Board on a Square Mat

functions designed to facilitate user program development. These general purpose support functions are described in detail in Sections 1.1–1.2 of Appendix 1. Functions for tabular and graphical display of numerical data, and functions for generating random numbers are included. These functions should be helpful in solving the computational problems at the end of each chapter. The following example is a supplement to the examples in Appendix 1. It features a number of the MATLAB support functions.

Consider the problem of numerically estimating the value π. A rough estimate can be obtained by using random numbers and exploiting the observation that π is the area of the unit circle. In particular, suppose there is a circular dart board of unit radius that is mounted on a 3 by 3 square mat as shown in Figure 1.8.1. The area of the circle is $A_c = \pi$, while the area of the square that encloses the circle is $A_s = 9$. Thus the ratio of the area of the circle to the area of the square is

$$\frac{A_c}{A_s} = \frac{\pi}{9} \tag{1.8.1}$$

Now suppose a series of n random darts hit the square. Of these, let $p \le n$ be the number of darts that hit the circular dart board within the square. If the dart locations are random and uniformly distributed over the square, then for sufficiently large values of n, the ratio p/n should be approximately equal to the ratio of the areas A_c/A_s. Substituting p/n for A_c/A_s in Equation (1.8.1) and multiplying both sides by nine, results in the following rough approximation for π.

$$\pi \approx \frac{9p}{n} \tag{1.8.2}$$

The following MATLAB script uses this approach and a number of NLIB support functions to find a rough estimate for the value of π. This is not to suggest that this is the only way, or even the best way, to approximate π. Rather, script *e181.m* is simply a vehicle employed to illustrate a number of the NLIB support functions for the MATLAB user. Additional NLIB examples can be found in Appendix 1.

```
%--------------------------------------------------------------
% Example 1.8.1: Estimating PI by Throwing Darts
%--------------------------------------------------------------

% Initialize

   clc                         % clear command window
   clear                       % clear variables
   n = 1000;                   % number of darts
   p = 0;                      % number of darts inside unit circle
   seed = 1000;                % select random sequence
   e = zeros(n,1);             % errors in estimate
   s = 3;                      % size of square mat

% Generate estimate with random numbers

   fprintf ('Example 1.8.1: Estimating PI by Throwing Darts\n');
   randinit(seed);
   for i = 1 : n
      z = randu (2,1,-s/2,s/2);
      r = norm(z,2);
      if r <= 1
         p = p + 1;
      end
      q = s^2*p/i;
      e(i) = q - pi;
   end

% Display results

   show ('Number of darts',n)
   show ('Final estimate of PI',q)
   graphmat (e,'Error in Estimate of PI','number of darts',...
             'error')
%--------------------------------------------------------------
```

In this case, a total of $n = 1000$ darts or random points are generated using the NLIB function *randu*, which computes a sequence of random 2×1 vectors z whose elements are uniformly distributed over the interval $[-1.5, 1.5]$. The MATLAB function *norm* is used to determine if the point is inside the circle by computing its radius.

$$r = \sqrt{z_1^2 + z_2^2} \tag{1.8.3}$$

The NLIB function *show* is used to display the number of random points n and the final estimate which is $q = 3.132$. The *show* function can also be used to display vectors and matrices. The NLIB function *graphmat* is used to create a graph that shows how the error in the estimate e changes with the number of darts. The graphical output produced by script *e181.m* is shown in Figure 1.8.2. Notice that the error gradually gets smaller as the number of darts increases, but it does not do so monotonically. The detailed shape

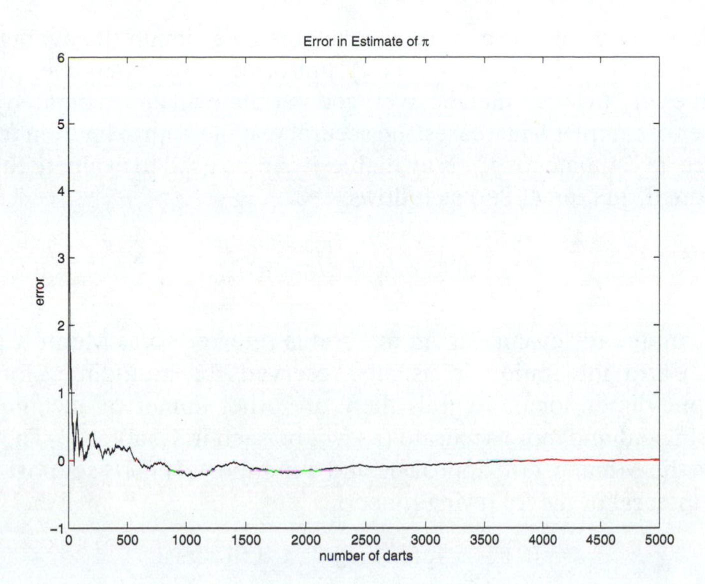

FIGURE 1.8.2 Graphical Output from MATLAB Script *e181.m*

of the curve depends on the random sequence selected, which is specified by the seed chosen for the NLIB function *randinit.*

1.8.2 Monte Carlo Integration: C

We have assumed that users who prefer C over MATLAB have been exposed to the fundamentals of C/C++ programming (see, e.g., Kelley and Pohl, 1990). To ease the task of C program development, a library of functions called NLIB has been developed, which parallels the algorithms discussed in this text. The NLIB library also includes a number of low-level utility functions designed to facilitate user program development. These general purpose support functions are described in detail in Sections 2.1–2.5 of Appendix 2. They include functions for dynamic allocation, input and output, and basic operations with vectors and matrices. They also include functions for tabular and graphical display of numerical data, and functions for generating random numbers. These functions should be helpful in solving the computational problems at the end of each chapter. The C functions are written in ANSI C to maintain portability over a wide variety of programming platforms. As such, the C functions do *not* produce direct graphical output. Instead, they produce script files that contain the data to be graphed and the graphing commands. These script files are plain text files that can be read by a separate plotting program. They are formatted in such a way that they are directly executable from within MATLAB. The following example is a supplement to the examples in Appendix 2. It illustrates the use of a number of NLIB support functions for the C user.

Suppose $f(x)$ is a continuous function defined over an interval $a \leq x \leq b$. Let f_{ave} denote the *average value* of $f(x)$ over $[a, b]$. The average value can be expressed in terms of the integral as follows.

$$f_{ave} = \frac{1}{b-a} \int_a^b f(x)\, dx \tag{1.8.4}$$

Next suppose an independent method is available to estimate the average value f_{ave}. For example, a set of n random numbers, x_k, uniformly distributed over $[a, b]$ might be used. The values of $f(x_k)$ can then be averaged to obtain an approximate value for f_{ave}. As the number of samples n increases, the accuracy of the approximation for f_{ave} should improve. Once an estimate for f_{ave} is available, it can be used to evaluate the integral by simply rewriting Equation (1.8.4) as follows.

$$\int_a^b f(x)\, dx = (b - a) f_{ave} \tag{1.8.5}$$

This technique for evaluating an integral is referred to as Monte Carlo integration. Monte Carlo integration is usually reserved for multidimensional integrals because for one-dimensional integrals there are other numerical techniques that are much more efficient and more accurate (as will be seen in Chapter 7). The following C program uses the Monte Carlo approach, and a number of NLIB support functions, to estimate the integral of the following function.

$$f(x) = x \ln(1 + x) \quad , \quad 0 \le x \le 1 \tag{1.8.6}$$

Again, this is not to suggest that this is the preferred way to compute the integral in (1.8.5). Indeed, for the particular function f in (1.8.6), the integral can be evaluated in closed form to yield 0.25. Instead, program *e182.c* is simply a convenient vehicle employed to illustrate a number of the NLIB support functions for the C user. Additional NLIB examples can be found in Appendix 2.

```
/* ------------------------------------------------------------- */
/* Example 1.8.2: Monte Carlo Integration                        */
/* ------------------------------------------------------------- */

#include "c:\nlib\util.h"

float f (float x)
{
   return x*log(1+x);
}

int main (void)
{
    int     n     = 3000,         /* number of random points */
            a     = 0,            /* lower limit */
            b     = 1,            /* upper limit */
            seed  = 2000,         /* select random sequence */
            i;
    float ye      = 0.25,         /* exact value of integral */
            z     = 0,
            x,y;
    vector e = vec (n,"");        /* errors in estimates */

  /* Estimate average value of f(x) with random numbers */
```

```
      printf ("\nExample 1.8.2: Monte Carlo Integration\n");
      randinit(seed);
      for (i = 1; i <= n; i++)
      {
         x = randu (a,b);
         z += f(x);
         y = (b-a)*z/i;
         e[i] = y - ye;
      }

/* Display results */

   shownum ("Number of random points",n);
   shownum ("Final estimate of integral",y);
   graphfun (a,b,"Integrand","x","f(x)",f,"fig183");
   graphvec (e,n,"Error in Estimate","number of random points",
            "error","fig184");
   return 0;
}
%-----------------------------------------------------------x
```

In this case, a total of $n = 3000$ random samples are generated using the NLIB function *randu*. The NLIB function *shownum* is used to display the number of random points n and the final estimate, which is $y = 0.2502$. There are equivalent *showvec* and *showmat* functions that can be used to display vectors and matrices, respectively. The NLIB function *graphfun* is used to create a graph of the integrand, $f(x)$. The call to *graphfun* creates a script file named *fig183.m* that contains the data to be graphed and the graph commands. When *fig183* is executed from within MATLAB, it produces the graph shown in Figure 1.8.3.

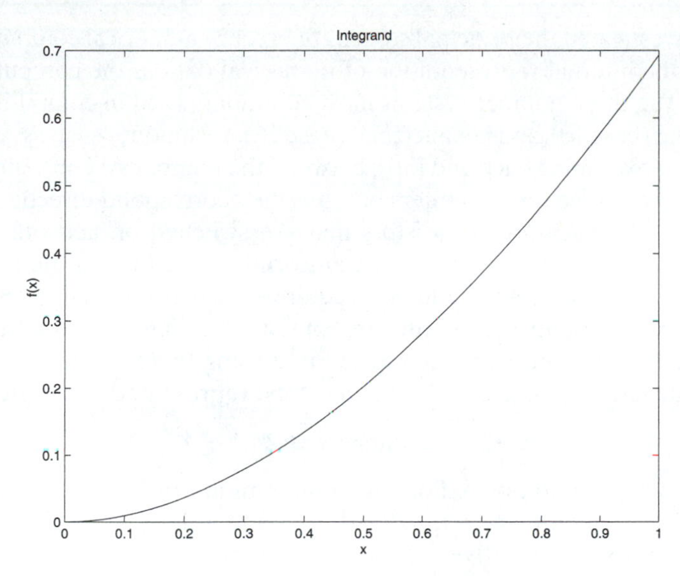

FIGURE 1.8.3 Graphical Output from Script *fig183.m* Produced by *e182.c*

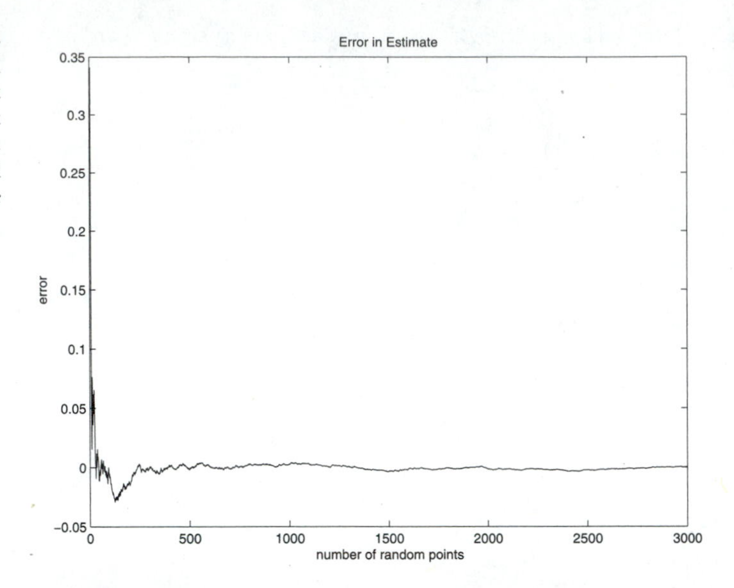

FIGURE 1.8.4 Graphical Output from Script *fig184.m* Produced by *e182.c*

The NLIB function *vec* is used to allocate the space for the error vector *e*, and the NLIB function *graphvec* is used to create a graph of *e* showing how the error in the estimate changes with the number of random points. The call to the *graphvec* function creates a script file *fig184.m*. When this script is executed from the MATLAB command prompt, it produces the output shown in Figure 1.8.4. Notice that the error gradually gets smaller as the number of random points increases, but it does not do so monotonically.

1.9 SUMMARY

This chapter investigated the principal sources of error in numerical computations, and it explored how the internal representation of numerical data in the computer can affect computations. The three number systems most commonly used in computing are binary (base 2), decimal (base 10), and hexadecimal (base 16). Techniques such as Alg. 1.2.1 were developed for transforming back and forth between the number systems. Binary numbers $\{0, 1\}$ are a natural choice for computers because they correspond directly to the underlying hardware, which features transistors that are switched on and off. Hexadecimal numbers are useful and efficient because transforming to and from binary numbers is straightforward, and hexadecimal numbers require exactly two digits per byte.

The binary representation of integers consists of a sign bit and a field of bits for the magnitude. Negative integers are represented using two's complement arithmetic. For an n-bit integer, the range of values that can be represented is as follows.

$$-2^{n-1} \leq \text{integers} \leq 2^{n-1} - 1 \tag{1.9.1}$$

The binary representation of floating-point numbers or floats consists of a sign bit, a field of bits for the mantissa or fractional part, and a field of bits for the exponent, which can be positive or negative. The mantissa is always normalized to the interval

$[0.5, 1)$ by adjusting the value of the exponent as needed. This way the most significant bit of the mantissa is always one, which means it can be dropped and regarded as a hidden bit. For a float that uses r bits for the exponent, the approximate range of values for the magnitude is as follows where $q = 2^{r-1}$.

$$2^{-q} \le |\text{floats}| < 2^q \tag{1.9.2}$$

When a computation is performed that results in a float with a magnitude less than 2^{-q} an underflow condition is generated. A magnitude greater than 2^q causes an overflow condition. The precision of floats can be expressed in terms of the machine epsilon, ε_M, which is the smallest positive number satisfying the following inequality.

$$1 + \varepsilon_M > 1 \tag{1.9.3}$$

Not all floating-point values can be represented. If x is a float, the next larger number that can be represented is $(1 + \varepsilon_M)x$. The machine epsilon can be computed numerically as in Alg. 1.3.1. If the binary representation of floats uses m bits for the mantissa, then

$$\varepsilon_M = 2^{-m} \tag{1.9.4}$$

There are two principal sources of error in numerical computations. The first, called round-off error, is associated with the use of a finite number of digits to represent numbers. Round-off error is caused by either rounding or chopping. Rounding is somewhat more accurate than chopping (ignoring excess digits), but chopping is typically used because it is simpler to implement in hardware. When a series of computations is performed, round-off error can accumulate and become significant. In particular, when two nearly identical large numbers are subtracted, the relative error of the result can be large. This process is called catastrophic cancellation.

The second major source of error in numerical computations is called truncation error. Truncation error is the error that arises when approximations to exact mathematical expressions are used, such as the truncation of an infinite series to a finite number of terms. For the case of a Taylor series, bounds on the magnitude of the truncation error can be expressed in terms of the size of the first neglected term.

The topic of random number generation was also included in this chapter because efficient generation of random numbers is tied directly to how numbers are represented internally in the computer. In addition, random numbers are useful to test the numerical algorithms developed in subsequent chapters. Random numbers are typically generated using either the linear or the multiplicative congruential method. The computation uses the remainder after dividing by a number called the modulus. If the modulus is chosen to be $\gamma = 2^n$ where n is the number of bits used to represent integers, then the division by γ step can be eliminated because it occurs automatically when the numbers are chopped to n bits. The random numbers generated are called pseudo-random sequences because they repeat after a finite, but very long, period. The point in the sequence where the computation starts is called the seed, so each seed generates a different sequence. Random numbers with a Gaussian or normal distribution can be computed in terms of random numbers with a uniform distribution using the Box-Muller method.

The final topic covered in this chapter was the introduction of a numerical toolbox or library called NLIB that implements the numerical algorithms covered in the remainder of the text. The MATLAB version of NLIB is described in detail in Appendix 1, and the C version is described in Appendix 2. It is highly recommended that readers who have little or no familiarity with vectors and matrices also review the material in Appendix 3 *before* proceeding beyond Chapter 1.

PROBLEMS

The problems are divided into Analysis problems, which can be solved by hand, and Computation problems, which require the use of MATLAB or C. Solutions to selected problems can be found in Appendix 4. Students are encouraged to use these selected problems, which are identified with an (S), as a check on their understanding of the material. Problems marked with a (P) are programming problems that require the student to implement one or more of the algorithms discussed in the chapter. The remaining Computation problems require the student to write a script or main program that uses one or more of the NLIB functions discussed in Appendix 1 or Appendix 2.

Analysis

1.1 Find the decimal values of the following numbers:

(a) $x = (10010110)_2$

(b) $x = (51D4)_{16}$

(c) $x = (777)_8$

1.2 Find the binary and hexadecimal values of the following numbers.

(a) $x = 329$

(b) $x = 203$

1.3 Modify Alg. 1.2.1 so that it converts a positive base p integer $c = c_{q-1} \cdots c_0$ into a base m integer $b = b_{n-1} \cdots b_0$ where $1 < p < 10$ and $1 < m < 10$. You can assume that $p^q \leq m^n$. Hint: Convert c to decimal first.

1.4 What is the range of decimal values that can be represented by a signed four-byte integer?

1.5 Find the one-byte two's complement binary representation of the following numbers.

(a) $x = -119$

(b) $x = -(3F)_{16}$

(c) $x = -(00101110)_2$

1.6 Explain how an n-bit mantissa of a floating-point number can be represented using only $n - 1$ bits.

(S)1.7 Suppose six bytes are used to represent a floating-point number with $r = 10$ bits used for the exponent.

(a) Find the largest positive number that can be represented.

(b) Find the smallest positive number that can be represented.

1.8 Explain what is meant by the error messages "floating-point underflow" and "floating-point overflow." You can assume r bits are used to represent the exponent. Be specific.

1.9 Suppose five bytes are used to represent a floating-point number. Alg. 1.3.1 is executed and the machine epsilon returned is $\varepsilon_M \approx 4.657 \times 10^{-10}$. How many bits are used for the exponent?

1.10 Suppose the machine epsilon for a four-byte floating-point number is $\varepsilon_M = 1.19 \times 10^{-7}$.

(a) Can the number $x = 1024$ be represented exactly? Why or why not?

(b) Find the floating-point number that immediately follows x using 6 decimal places.

(c) Find the floating-point number that immediately precedes x using 6 decimal places.

1.11 When is the relative error of a variable $x = x_e + \Delta x$ not well defined? Here x_e is the exact value.

1.12 Suppose the number $\pi = 4 \tan^{-1}(1)$ is approximated using $n = 5$ decimal digits.

(a) Find the relative error due to chopping. Express it as a percent.

(b) Find the relative error due to rounding. Express it as a percent.

(S)1.13 Derive the following expression for the relative error of the product where $x = x_e + \Delta x$ and $y = y_e + \Delta y$. You can assume that $|r_x| \ll 1$ and $|r_y| \ll 1$.

$$r_{xy} \approx r_x + r_y$$

1.14 Derive the following expression for the relative error of the quotient where $x = x_e + \Delta x$ and $y = y_e + \Delta y$. You can assume that $|r_x| \ll 1$ and $|r_y| \ll 1$.

$$r_{x/y} \approx r_x - r_y$$

1.15 Consider the trigonometric function $f(x) = \sin x$.

(a) Find the Taylor series expansion of $f(x)$ about $x = 0$.

(b) Suppose the Taylor series is truncated to $n = 6$ terms. Find the relative error at $x = \pi/4$ due to truncation. Express it as a percent.

(c) Find an upper bound on the magnitude of the relative error at $x = \pi/4$ expressed as a percent.

1.16 Consider a sequence of random numbers with the following two-sided exponential probability density function. What is the probability that a random number lies in the interval $[-1, 1]$?

$$p(x) = \frac{1}{2} \exp(-|x|)$$

1.17 Consider a sequence of random numbers with the following probability density function

$$p(x) = \begin{cases} \dfrac{2}{\pi(1 + x^2)} &, \quad x \geq 0 \\ 0 &, \quad x < 0 \end{cases}$$

(a) Find the cumulative probability distribution function $P(x)$.

(b) Suppose y_k is uniformly distributed over $[0, 1]$. Show how to compute x_k from y_k such that x_k has probability density $p(x)$.

Computation

(SP)1.18 Write a function called *quadroot* that finds the roots of the polynomial, $p(x) = ax^2 + bx + c$, using the quadratic formula in (1.1.2). The pseudo-prototype is

```
[cvector x] = quadroot (float a,float b,float c)
```

On entry to *quadroot*: a, b, and c are the polynomial coefficients. On exit, the 2×1 complex vector x contains the roots. Include a check for the case $a = 0$. Test *quadroot* by writing a main program that prompts the user for the coefficients and then displays the roots.

1.19 Use Alg. 1.2.1 to write a function called *dec_bin* that converts a decimal integer $0 \leq d \leq 32767$ into a binary number. The pseudo-prototype is

```
[string b] = dec_bin (int d)
```

On entry, d is the decimal integer to be converted. On exit, the string b contains the binary representation of d. Modify Alg. 1.2.1, as needed, such that the array b starts with the most significant bit. Test *dec_bin* by writing a main program that prompts the user for an integer and then displays its binary value. Check your function by comparing it with the output from the built-in MATLAB function *dec2bin.*

1.20 Write a function called *twos* that computes the two's complement of a binary number. The pseudo-prototype is

```
[string c] = twos (string b)
```

On entry, b is a string of binary digits. On exit, c is a binary string that contains the two's complement of b. The arrays b and c should both start with the most significant bit. Test *twos* by writing a main program that prompts the user for a binary string and then displays its two's complement.

1.21 Modify the results of problem 1.19 so that it works for decimal numbers in the range $-32678 \leq d \leq 32767$. Use the function *twos* from problem 1.20 to express the negative numbers using the two's complement representation.

1.22 Write a function called *bin_dec* that converts a binary number into a non-negative decimal integer. The pseudo-prototype is

```
[int d] = bin_dec (string b)
```

On entry, b is a string containing the binary number. On exit, $d \geq 0$ is the decimal equivalent of b. Test *bin_dec* by writing a main program that prompts the user for a binary string and displays its decimal value. Check your function by comparing the results with the output from the built-in MATLAB function *bin2dec.*

1.23 Modify the results of problem 1.22 so it works for both positive and negative numbers assuming negative numbers are expressed using the two's complement representation. Use the function *twos* from problem 1.20.

1.24 Use the technique in Equations (1.6.5) and (1.6.6) to write a function called *randf* that generates uniformly distributed random numbers in the interval $[0, 1]$. The pseudo-prototype is

```
[int u] = randf (void)
```

When *randf* is called, it should generate the next random number in the sequence. Start the sequence using a seed of one. Test *randf* by writing a main program that computes and displays the first 20 random numbers.

1.25 Write a program that computes and displays the complex number $u = j^i$ where $j = \sqrt{-1}$. Verify the answer by hand by using the fact that $j = \exp(j\pi/2)$.

1.26 Write a program that uses the function *randf* from problem 1.24 to create a 8×6 matrix A with random elements uniformly distributed over the interval $[-10, 10]$ Display A, and compute and display the minimum, maximum and mean of the elements of A.

CHAPTER 2

Linear Algebraic Systems

One of the most fundamental problems in numerical methods is that of solving a set of simultaneous linear algebraic equations. Equations of this form arise naturally in a wide variety of applications in the fields of science and engineering. A system of n linear equations can be expressed in compact form as follows.

$$\boxed{Ax = b}$$

Here A is an $n \times n$ coefficient matrix, b is an $n \times 1$ right-hand side vector, and x is an $n \times 1$ vector of unknowns. The objective is to find an x which satisfies the equations. To illustrate the diversity of applications of linear algebraic systems, we begin by examining a number of examples where they occur. Vector and matrix notation, used throughout the remainder of this text, is then briefly reviewed. This is followed by a presentation of direct solution methods based on elementary row operations. We show that the number of floating-point operations, or FLOPs, required to obtain a solution is proportional to n^3 where n is the number of unknowns. The direct elimination methods examined include the Gauss-Jordan method, the Gaussian elimination method, and the LU decomposition method. LU factorization of the coefficient matrix A is particularly effective when the system of equations must be solved repeatedly, but with different right-hand side vectors b. Next, we examine the practical issue of ill-conditioned systems, systems whose solutions are highly susceptible to accumulated roundoff error. A simple one-step technique for improving numerical accuracy, called iterative correction, is then presented. This leads naturally to a discussion of multistep iterative methods. They include the Jacobi method, the Gauss-Seidel method, and the successive relaxation method. Iterative techniques are very effective for large sparse systems of equations such as those that arise in solving linear partial differential equations. The chapter concludes with a summary of linear algebraic system techniques and a presentation of engineering applications. The selected examples include a five-stage chemical absorption process, a planar truss, a DC bridge circuit, and a mass-spring-damper system.

2.1 MOTIVATION AND OBJECTIVES

Linear algebraic systems of equations occur in many applications. The simplest case is two equations with two unknowns, but it is not uncommon to encounter systems with tens, hundreds, or even several thousand equations and unknowns.

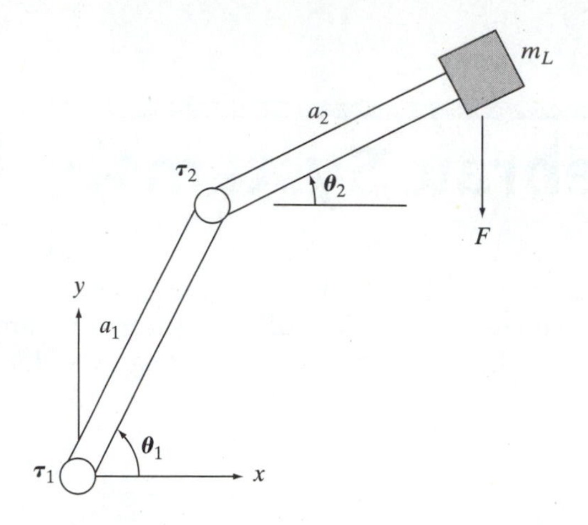

FIGURE 2.1.1 Two-Axis Planar Robot Holding Object of Mass m_L

2.1.1 Robotic Arm

When a human reaches down to pick up an object it is possible, using feel, to estimate roughly how much the object weighs. Consider the problem of having the two-axis planar robotic arm in Figure 2.1.1 perform this task. Here a_k is the length of link k, and θ_k is the angle that link k makes with the x axis for $1 \le k \le 2$.

Suppose the arm is instrumented in such a way that one can measure the torque τ_k induced at joint k as a result of the load mass m_L at the end of the arm. The objective is to use the measurements of τ_1 and τ_2 to estimate the value of the mass m_L. This can be achieved by making use of the following linear algebraic system, which relates the induced torques to the end-of-arm force (see, e.g., Schilling, 1990).

$$(a_1 \cos \theta_1)F_y - (a_1 \sin \theta_1)F_x = \tau_1 \tag{2.1.1a}$$

$$(a_2 \cos \theta_2)F_y - (a_2 \sin \theta_2)F_x = \tau_2 \tag{2.1.1b}$$

Here the unknowns F_x and F_y denote the x and y components, respectively, of the force at the end of the arm as a result of gravity g acting on the mass m_L. For example, if the robot is oriented in such a way that gravity points in the direction opposite to the y axis, then $F_x = 0$ and $F_y = -m_L g$. In this case, one can solve the linear algebraic system for F_x and F_y and then estimate the mass of the object in the robotic hand using

$$m_L = \frac{-F_y}{g} \tag{2.1.2}$$

2.1.2 Converter Circuit

As a second example of a linear algebraic system, consider the problem of finding the voltages and currents in an electrical circuit. Every circuit that contains only voltage sources, current sources, and resistors can be modeled with a system of linear algebraic equations. As an illustration, consider the resistor ladder network shown in Figure 2.1.2.

This type of circuit can be found inside a variety of commercial digital-to-analog converter chips. Applying Kirchhoff's voltage law to each mesh of the ladder and divid-

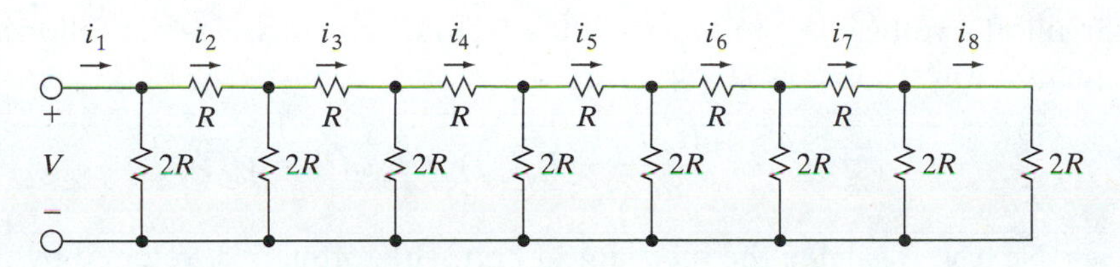

FIGURE 2.1.2 Resistor Ladder Network

ing through by the resistance R, we find the mesh currents $\{i_1,i_2,\ldots,i_8\}$ must satisfy the following system of linear algebraic equations.

$$
\begin{aligned}
2i_1 - 2i_2 \qquad\qquad\qquad\qquad\qquad\qquad\qquad\qquad &= V/R \\
-2i_1 + 5i_2 - 2i_3 \qquad\qquad\qquad\qquad\qquad\qquad &= 0 \\
-2i_2 + 5i_3 - 2i_4 \qquad\qquad\qquad\qquad\qquad &= 0 \\
-2i_3 + 5i_4 - 2i_5 \qquad\qquad\qquad\qquad &= 0 \\
-2i_4 + 5i_5 - 2i_6 \qquad\qquad\qquad &= 0 \\
-2i_5 + 5i_6 - 2i_7 \qquad\qquad &= 0 \\
-2i_6 + 5i_7 - 2i_8 &= 0 \\
-2i_7 + 4i_8 &= 0
\end{aligned}
\qquad (2.1.3)
$$

Given the repetitive structure of the ladder network, the circuit equations also have a special structure. Note how, at most, three of the eight unknowns appear in each equation. As we shall see, this will be called a *tridiagonal* system. Systems of this form are particularly easy to solve.

2.1.3 DC Motor

As another example of a linear algebraic system, consider the diagram of a permanent-magnet DC motor shown in Figure 2.1.3. Literally millions of small electrical motors are in use throughout the industrial world. The electrical part is modeled by the applied armature voltage V_a, the armature-winding current I_a, and the armature-winding resistance R_a and inductance L_a, respectively.

When the motor starts to turn, it builds up a voltage across it, V_b, that is proportional to the motor speed ω. If k_b denotes the constant of proportionality, then applying

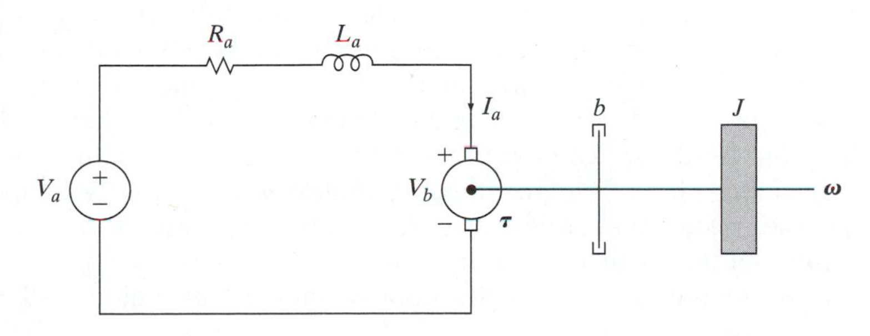

FIGURE 2.1.3 DC Motor

Kirchhoff's voltage law to the armature winding circuit yields the following electrical equation where t denotes time.

$$L_a \frac{dI_a(t)}{dt} + R_a I_a(t) + k_b \omega(t) = V_a(t) \tag{2.1.4}$$

Next we examine the mechanical part. The torque τ developed by the magnetic field turns the shaft whose moment of inertia is J and friction coefficient is b. The developed torque is proportional to the armature-winding current I_a. If k_t denotes the constant of proportionality, then applying Newton's second law to the motor shaft yields the following mechanical equation.

$$J \frac{d\omega(t)}{dt} = k_t I_a(t) - b\omega(t) \tag{2.1.5}$$

Suppose a constant voltage V_a is applied to the motor, say from a battery, and consider the following question. How fast will the motor turn when it reaches its maximum speed? This question can be answered by solving a simple linear algebraic system of equations. To see this, first note that when the motor reaches top speed, the shaft speed will no longer be changing with time which means that $d\omega(t)/\,dt = 0$. But if the motor is running at constant speed, then the current drawn will also be constant which means $dI_a(t)/\,dt = 0$. If the derivatives in (2.1.4) and (2.1.5) are set to zero, this results in the following linear algebraic system in the unknown I_a and ω.

$$R_a I_a + k_b \omega = V_a \tag{2.1.6a}$$

$$k_t I_a - b\omega = 0 \tag{2.1.6b}$$

Solving this system then yields the running speed ω, which is typically expressed in RPMs. The solution of this linear algebraic system can also be used to answer another practical question, namely, how long will the battery last? Suppose the battery is designed to supply E watt-hours of energy to the motor. The rate at which power is dissipated by the motor is $P = V_a I_a$ watts. It follows that the battery will last T hours where

$$T = \frac{E}{V_a I_a} \tag{2.1.7}$$

2.1.4 Chapter Objectives

When you finish this chapter you will be able to efficiently solve linear algebraic systems of equations. You will also be able to compute the determinant and the inverse of a matrix. You will learn how to represent linear systems of equations using compact vector notation, and then solve them using direct methods based on elementary row operations. You will be able to compare the computational effort required for the direct methods including Gauss-Jordan, Gaussian elimination, and LU decomposition. You will know when a system of equations is difficult to accurately solve by using the condition number of the coefficient matrix. You will also be able to efficiently solve very large sparse systems of equations using iterative methods including the Jacobi, Gauss-

Seidel, and successive relaxation methods. These overall goals will be achieved by mastering the chapter objectives summarized below.

Objectives for Chapter 2

- Know how to formulate linear algebraic systems of equations using vector and matrix notation.
- Know how to premultiply and postmultiply one matrix by another.
- Know how to perform elementary row operations and represent them as matrix multiplications.
- Be able to solve a linear algebraic system using Gauss-Jordan elimination.
- Know how to identify square, upper-triangular, lower-triangular, tridiagonal, diagonal, and sparse matrices.
- Be able to solve a linear algebraic system using Gaussian elimination.
- Understand how to define and compute the matrix determinant.
- Understand how to define and compute the matrix inverse.
- Be able to factor a matrix into lower-triangular and upper-triangular parts, and solve a linear algebraic system using LU decomposition.
- Be able to efficiently solve tridiagonal systems of equations.
- Know how to compute the norm and condition number of a matrix.
- Know how many floating-point operations (FLOPs) are needed by the direct solution methods.
- Know how to improve the accuracy of a solution using iterative correction.
- Be able to solve linear algebraic systems using the iterative Jacobi, Gauss-Seidel, and successive relaxation methods.
- Know the types of problems for which iterative methods are most applicable, and why.
- Understand the conditions under which the iterative methods converge.
- Understand how linear algebraic system techniques can be used to solve practical engineering problems.
- Understand the relative strengths and weaknesses of each computational method and know which are most applicable for a given problem.

2.2 GAUSS-JORDAN ELIMINATION

The previous examples of linear algebraic systems are special cases of the following set of simultaneous linear equations.

$$\begin{aligned}
A_{11}x_1 + A_{12}x_2 + \cdots + A_{1n}x_n &= b_1 \\
A_{21}x_1 + A_{22}x_2 + \cdots + A_{2n}x_n &= b_2 \\
\vdots \qquad\qquad & \quad \vdots \\
A_{n1}x_1 + A_{n2}x_2 + \cdots + A_{nn}x_n &= b_n
\end{aligned} \tag{2.2.1}$$

Here the values of $\{A_{kj}\}$ and $\{b_k\}$ are given, and it is the n unknowns $\{x_1, x_2, \ldots, x_n\}$ that must be determined. To facilitate a concise general solution, it is helpful to reformulate the equations using the notation of vectors and matrices as follows.

$$\underbrace{\begin{bmatrix} A_{11} & A_{12} & \cdots & A_{1n} \\ A_{21} & A_{22} & \cdots & A_{2n} \\ \vdots & \vdots & & \vdots \\ A_{n1} & A_{n2} & \cdots & A_{nn} \end{bmatrix}}_{A} \underbrace{\begin{bmatrix} x_1 \\ x_2 \\ \vdots \\ x_n \end{bmatrix}}_{x} = \underbrace{\begin{bmatrix} b_1 \\ b_2 \\ \vdots \\ b_n \end{bmatrix}}_{b} \tag{2.2.2}$$

Here x and b are $n \times 1$ column vectors whose kth elements are x_k and b_k, respectively. Unless specifically noted otherwise, *all* vectors in this text are regarded as column vectors. Similarly, A is an $n \times n$ matrix whose elements are A_{kj} where the first subscript specifies the row and the second the column. (A review of the notation and basic properties of vectors and matrices, including matrix multiplication and the determinant and inverse of a matrix, can be found in Appendix 3.) Given the definitions of A, x, and b in Equation (2.2.2), the set of simultaneous equations in (2.2.1) can be written much more compactly as follows.

$$Ax = b \tag{2.2.3}$$

The beauty of this concise formulation of a system of linear algebraic equations is that it is the same for *all* systems of equations. It does not matter if there are two equations or two thousand equations! Using matrix multiplication (see Appendix 3 for a review if needed), the kth equation in the linear algebraic system can be expressed as follows.

$$\boxed{\sum_{j=1}^{n} A_{kj}x_j = b_k \quad , \quad 1 \le k \le n} \tag{2.2.4}$$

The brute force way to solve the linear algebraic system in (2.2.3) is to use the matrix inverse. From Appendix 3, if the $n \times n$ matrix A is nonsingular, then there exists a unique $n \times n$ matrix A^{-1} called the *inverse* of A such that $A^{-1}A = I$ where I is the $n \times n$ diagonal *identity matrix* with ones along the diagonal. Multiplying both sides of (2.2.3) on the left by A^{-1} and using $Ix = x$, we get the following solution.

$$x = A^{-1}b \tag{2.2.5}$$

The formulation in Equation (2.2.5) is a conceptual way to *express* the solution rather than a practical way to numerically *compute* the solution. This is because there are more efficient methods that find x without first finding A^{-1}.

A number of techniques for solving a system of linear algebraic equations are based on the use of elementary row operations. The basic idea is to *reformulate* the equations, using elementary operations, so as to make the solution transparent. The three elementary row operations that are used are summarized in Table 2.2.1. Operation one simply interchanges rows k and j. Since this is applied to the rows of both A and b, this is equivalent to *reordering* the equations. Operation two multiplies row k by a scalar

TABLE 2.2.1 Elementary Row Operations

Operation	*Description*	*Matrix*
1	Interchange rows k and j	$E_1(k,j)$
2	Multiply row k by $\alpha \neq 0$	$E_2(k,\alpha)$
3	Add α times row j to row k	$E_3(k,j,\alpha)$

$\alpha \neq 0$. Thus this operation *scales* both sides of the kth equation. Finally, operation three adds α times row j to row k. This type of operation can be used to *eliminate* (make zero) a coefficient in row k, thereby simplifying the equations. Each of these elementary row operations leaves the solution of the system unchanged.

If an elementary row operation is applied to the identity matrix I, the resulting matrix is called an *elementary row matrix*. Application of an elementary row operation is equivalent to *premultiplication* (multiplication on the left) by the corresponding elementary row matrix. Therefore, performing a sequence of elementary row operations is equivalent to premultiplying by a sequence of elementary row matrices. Two matrices that are related to one another through a sequence of elementary row operations are said to be *row equivalent*.

EXAMPLE 2.2.1 Elementary Row Operations

To illustrate the three elementary row operations and how they can be implemented with elementary row matrices, consider the case $n = 3$. The first operation, $E_1(2,3)$, interchanges rows 2 and 3 of A. The second operation, $E_2(2,\alpha)$, multiplies row 2 of A by $\alpha \neq 0$. The third operation, $E_3(1,3,\alpha)$, adds α times row 3 to row 1. In each case the elementary row matrix on the left is obtained by applying the corresponding row operation to the 3×3 identity matrix I.

$$\begin{bmatrix} 1 & 0 & 0 \\ 0 & 0 & 1 \\ 0 & 1 & 0 \end{bmatrix} \begin{bmatrix} A_{11} & A_{12} & A_{13} \\ A_{21} & A_{22} & A_{23} \\ A_{31} & A_{32} & A_{33} \end{bmatrix} = \begin{bmatrix} A_{11} & A_{12} & A_{13} \\ A_{31} & A_{32} & A_{33} \\ A_{21} & A_{22} & A_{23} \end{bmatrix}$$

$$\begin{bmatrix} 1 & 0 & 0 \\ 0 & \alpha & 0 \\ 0 & 0 & 1 \end{bmatrix} \begin{bmatrix} A_{11} & A_{12} & A_{13} \\ A_{21} & A_{22} & A_{23} \\ A_{31} & A_{32} & A_{33} \end{bmatrix} = \begin{bmatrix} A_{11} & A_{12} & A_{13} \\ \alpha A_{21} & \alpha A_{22} & \alpha A_{23} \\ A_{31} & A_{32} & A_{33} \end{bmatrix}$$

$$\begin{bmatrix} 1 & 0 & \alpha \\ 0 & 1 & 0 \\ 0 & 0 & 1 \end{bmatrix} \begin{bmatrix} A_{11} & A_{12} & A_{13} \\ A_{21} & A_{22} & A_{23} \\ A_{31} & A_{32} & A_{33} \end{bmatrix} = \begin{bmatrix} A_{11}+\alpha A_{31} & A_{12}+\alpha A_{32} & A_{13}+\alpha A_{33} \\ A_{21} & A_{22} & A_{23} \\ A_{31} & A_{32} & A_{33} \end{bmatrix}$$

By using a sequence of elementary row operations, we can reduce the original linear algebraic system to an equivalent system where the solution is obvious by inspection. First, we form an $n \times (n + 1)$ *augmented* matrix $C = [A, b]$ which consists of the n columns of A augmented by the $n \times 1$ column vector b which corresponds to column $n + 1$. The basic idea is to apply row operations so as to reduce the first n columns of the augmented matrix C to the identity matrix I. This is done by working with one column at a time starting from the left. A search is made for the element whose magnitude is largest. This element, called the *pivot*, is brought into position with a row interchange, normalized to one, and then used to eliminate (make zero) the other elements in the column. The procedure is then repeated for the next column until all n columns are completed. The final row-reduced version of C is then partitioned into $[I, x]$ where the last column is the solution vector x. Conceptually, the process can be summarized as follows, where the arrow indicates row operations.

$$[A, b] \rightarrow [I, x] \tag{2.2.6}$$

The following algorithm, which can be used to solve $Ax = b$, is called the Gauss-Jordan elimination method. It returns an *error code* E that is zero if the coefficient matrix A is singular.

Alg. 2.2.1 Gauss-Jordan Elimination

1. Set $E = 1$ and form the augmented $n \times (n + 1)$ matrix $C = [A, b]$.
2. For $j = 1$ to n do

   {

   (a) Compute the pivot index $j \leq p \leq n$ such that

   $$|C_{pj}| = \max_{i=j}^{n}\{|C_{ij}|\}$$

   (b) If $C_{pj} = 0$, set $E = 0$ and exit.

   (c) If $p > j$, interchange rows p and j.

   (d) Divide row j by the pivot C_{jj}.

   (e) For each $i \neq j$, subtract C_{ij} times row j from row i.

   }
3. Partition C as $C = [I, x]$.

If Alg. 2.2.1 returns with $E = 1$, then the unique solution, x, has been found, and it is located in column $n + 1$ of the augmented matrix C. If the algorithm returns with $E = 0$, then the coefficient matrix A was determined to be singular in which case a unique solution does not exist. Of course, the test for singularity in step (2b) is an *equality* test with floats and therefore must be implemented with some care. In practice, if $|C_{pj}|$ is very small it is regarded as zero. The following example illustrates the application of Alg. 2.2.1.

EXAMPLE 2.2.2 Gauss-Jordan Elimination

Consider the following linear algebraic system.

$$\begin{bmatrix} 1 & -1 & 0 \\ -2 & 2 & -1 \\ 0 & 1 & -2 \end{bmatrix} x = \begin{bmatrix} 2 \\ -1 \\ 6 \end{bmatrix}$$

To streamline the application of Alg. 2.2.1, we combine some of the steps. In the first step we select a pivot (2a), move it into position (2c), and normalize it to unity (2d). In the second step we use the pivot row to eliminate the other elements in the column by subtracting appropriate multiples of the pivot row (2e). The process is then repeated for the remaining columns until the identity matrix is obtained. This yields the following sequence of augmented matrices.

$$C = [A, b] = \left[\begin{array}{ccc|c} 1 & -1 & 0 & 2 \\ -2 & 2 & -1 & -1 \\ 0 & 1 & -2 & 6 \end{array}\right] \rightarrow \begin{bmatrix} 1 & -1 & 0.5 & 0.5 \\ 1 & -1 & 0 & 2 \\ 0 & 1 & -2 & 6 \end{bmatrix}$$

$$\rightarrow \begin{bmatrix} 1 & -1 & 0.5 & 0.5 \\ 0 & 0 & -0.5 & 1.5 \\ 0 & 1 & -2 & 6 \end{bmatrix} \rightarrow \begin{bmatrix} 1 & -1 & 0.5 & 0.5 \\ 0 & 1 & -2 & 6 \\ 0 & 0 & -0.5 & 1.5 \end{bmatrix}$$

$$\rightarrow \begin{bmatrix} 1 & 0 & -1.5 & 6.5 \\ 0 & 1 & -2 & 6 \\ 0 & 0 & -0.5 & 1.5 \end{bmatrix} \rightarrow \begin{bmatrix} 1 & 0 & -1.5 & 6.5 \\ 0 & 1 & -2 & 6 \\ 0 & 0 & 1 & -3 \end{bmatrix}$$

$$\rightarrow \left[\begin{array}{ccc|c} 1 & 0 & 0 & 2 \\ 0 & 1 & 0 & 0 \\ 0 & 0 & 1 & -3 \end{array}\right] = [I, x]$$

Thus the unique solution is $x = [2, 0, -3]^T$ where the superscript T denotes the transpose operation.

Next we consider an example of the Gauss-Jordan elimination method applied to a system for which the coefficient matrix is singular.

EXAMPLE 2.2.3 Singular A

Consider the following linear algebraic system.

$$\begin{bmatrix} 2 & 4 & -8 \\ 0 & 1 & -2 \\ 1 & 3 & -6 \end{bmatrix} x = \begin{bmatrix} 6 \\ 2 \\ 1 \end{bmatrix}$$

Applying Alg 2.2.1 as in Example 2.2.2 yields

$$C = [A, b] = \left[\begin{array}{ccc|c} 2 & 4 & -8 & 6 \\ 0 & 1 & -2 & 2 \\ 1 & 3 & -6 & 1 \end{array}\right] \rightarrow \begin{bmatrix} 1 & 2 & -4 & 3 \\ 0 & 1 & -2 & 2 \\ 1 & 3 & -6 & 1 \end{bmatrix}$$

$$\rightarrow \begin{bmatrix} 1 & 2 & -4 & 3 \\ 0 & 1 & -2 & 2 \\ 0 & 1 & -2 & -2 \end{bmatrix} \rightarrow \left[\begin{array}{ccc|c} 1 & 0 & 0 & -1 \\ 0 & 1 & -2 & 2 \\ 0 & 0 & 0 & -4 \end{array}\right]$$

Since the pivot $C_{33} = 0$, the algorithm terminates with $E = 0$. It follows that A is singular, and consequently a unique solution does not exist. It is clear from inspection that the last column of C cannot be expressed as a linear combination (weighted sum) of the remaining columns, which means that there is no solution at all in this case.

Steps (2a) through (2c) of Alg. 2.2.1 implement the *pivoting* operation that is used to improve the numerical accuracy of the solution. At first glance, it might appear that these steps could be eliminated, thereby saving some operations. Although this simplification works in many instances, it is not a *robust* procedure for two reasons. First, C_{jj} may turn out to be zero for some $1 \le j \le n$. Clearly, this makes the normalization in step (2d) impossible. More subtle is the case when C_{jj} is nonzero, but small. In this case, significant round-off error can occur that might otherwise be avoided by using pivoting. The following example illustrates how problems can occur in the absence of pivoting.

EXAMPLE 2.2.4 No Pivoting

Suppose we reconsider the system from Example 2.2.2. Applying Alg. 2.2.1, but without the pivoting in steps (2a) through (2c), yields

$$C = [A, b] = \left[\begin{array}{ccc|c} 1 & -1 & 0 & 2 \\ -2 & 2 & -1 & -1 \\ 0 & 1 & -2 & 6 \end{array}\right] \rightarrow \begin{bmatrix} 1 & -1 & 0 & 2 \\ 0 & 0 & -1 & 3 \\ 0 & 1 & -2 & 6 \end{bmatrix}$$

At this point, the simplified version of Alg. 2.2.1 would terminate in an ungraceful way because there would be an attempt in step (2d) to divide by $C_{22} = 0$. This occurs in spite of the fact that this system has a well-defined solution, namely, $x = [2, 0, -3]^T$.

The pivoting procedure in Alg. 2.2.1 is called *partial* pivoting because the search for a pivot element is restricted to entries in column j at or below row j. Alternatively, one can use *complete* pivoting, which involves searching for pivots in the submatrix whose upper left corner is C_{jj}. Full pivoting requires more bookkeeping because it uses column interchanges as well as row interchanges. Recall that interchanging rows is equivalent to changing the order of the equations. Interchanging columns is equivalent to interchanging the unknowns, the elements of x. Consequently, column interchanges must be recorded so the solution can be "unscrambled" at the end.

Sometimes in engineering applications the elements of one row of the coefficient matrix have magnitudes that are much larger than those of the other rows. Typically, this large row will produce a pivot even if the pivot element is the smallest element in its row. When this happens, the normalization of the pivot can result in very large entries in the remainder of the row, a process that can lead to an increase in round-off error.

Whenever the coefficients of one row are much larger (or much smaller) than those of the other rows, it is useful to *scale* each row by dividing by the element with the largest magnitude. This normalization technique is called *equilibration*. There is a useful variation of equilibration that avoids the round-off error introduced by the division operation itself. In this method the normalization factors (the largest elements in each row) are first computed. When the pivot is determined, the *choice* of pivot is made by dividing the pivot candidates by the normalization factors. However, no division is applied to the equations themselves. Instead, the elimination step is performed on the original equations, but with the pivot chosen as if the equations had been normalized. Since the equations themselves are not scaled, this method is referred to as *implicit equilibration*.

To facilitate comparison with other techniques, it is useful to examine the computational effort required by the Gauss-Jordan elimination method. Consider the number of floating point multiplications and divisions (FLOPs) required to implement Alg. 2.2.1. A similar exercise can be carried out for the number of floating point additions and subtractions (see Problem 2.5). In computing the number of arithmetic operations, the following integer summation formula is useful.

$$\boxed{\sum_{i=1}^{n} i = \frac{n(n+1)}{2}} \tag{2.2.7}$$

All of the arithmetic operations occur in step (2) of Alg. 2.2.1. Pivot selection does not require any multiplications or divisions. The normalization in step (2d) requires $n + 1 - j$ divisions because $C_{ij} = 0$ for $i < j$ and $C_{jj} = 1$. Thus using (2.2.7) with a change of variables, $i = n + 1 - j$, the total number of division operations associated with normalization for $1 \le j \le n$ is

$$c_{norm}(n) = \frac{n(n+1)}{2} \tag{2.2.8}$$

The elimination process in step (2e) requires $(n - 1)(n + 1 - j)$ multiplications in view of the fact that $C_{jk} = 0$ for $k < j$ and $C_{ij} = 0$ for $i \ne j$. Thus, the total number of multiplication operations associated with the elimination process for $1 \le j \le n$ is

$$c_{elim}(n) = (n-1)c_{norm}(n) \tag{2.2.9}$$

Combining the computational effort for normalization and elimination, and examining what happens for large values of n, we get the following number of FLOPs for Alg. 2.2.1.

$$\boxed{c_{GJ}(n) \approx \frac{n^3}{2} \quad \text{FLOPs}, \quad n \gg 1} \tag{2.2.10}$$

Thus, for large values of n, the computational effort required by Gauss-Jordan elimination grows as a *cubic* polynomial of n. In this case, we say the method is an n^3 computational process with a coefficient of proportionality of 0.5.

The Gauss-Jordan elimination method can also be used to find the inverse of a matrix. If the right-hand side vector b is selected to be i^k, the kth column of the identity matrix I, the resulting solution x is the kth column of the inverse of A. Consequently, A^{-1} can be computed column-by-column by solving the following sequence of linear algebraic systems:

$$Ax = i^k \quad , \quad 1 \le k \le n \tag{2.2.11}$$

The problem with this "brute force" method is that finding each column of A^{-1} requires approximately $n^3/2$ multiplications and divisions. Consequently, computing A^{-1} using (2.2.11) requires $n^4/2$ FLOPs, a number that grows very rapidly as n increases.

A more effective alternative is to modify Alg. 2.2.1 slightly so as to solve the n separate linear algebraic systems in (2.2.11) simultaneously. This is achieved by forming an augmented $n \times 2n$ matrix $C = [A, I]$ and then applying elementary row operations until the first n columns of C correspond to the identity matrix I as follows.

$$[A, I] \rightarrow [I, A^{-1}] \tag{2.2.12}$$

This is equivalent to solving a matrix equation of the form $AX = I$ where X is an unknown $n \times n$ matrix. The solution is then $X = A^{-1}$. The following algorithm is an implementation of this approach.

Alg. 2.2.2 **Matrix Inverse**

1. Set $E = 1$ and form the augmented $n \times 2n$ matrix $C = [A, I]$.
2. For $j = 1$ to n do

   {

   (a) Compute the pivot index $j \le p \le n$ such that:

   $$|C_{pj}| = \max_{i=j}^{n}\{|C_{ij}|\}$$

   (b) If $C_{pj} = 0$, set $E = 0$ and exit.

   (c) If $p > j$, interchange rows p and j.

   (d) Divide row j by the pivot C_{jj}.

   (e) For each $i \ne j$, subtract C_{ij} times row j from row i.

   }
3. Partition C as $C = [I, A^{-1}]$.

If Alg. 2.2.2 returns with $E = 1$, then the inverse, A^{-1}, has been successfully computed, and it is located in the last n columns of the augmented matrix C. Otherwise, A is singular. The following example illustrates the application of Alg. 2.2.2 to find the inverse.

EXAMPLE 2.2.5 **Matrix Inverse**

Consider the following 3×3 matrix.

$$A = \begin{bmatrix} 1 & -1 & 0 \\ 2 & 0 & 4 \\ 0 & 2 & -1 \end{bmatrix}$$

Applying Alg 2.2.2 yields

$$C = [A, I] = \left[\begin{array}{ccc|ccc} 1 & -1 & 0 & 1 & 0 & 0 \\ 2 & 0 & 4 & 0 & 1 & 0 \\ 0 & 2 & -1 & 0 & 0 & 1 \end{array}\right] \rightarrow \begin{bmatrix} 1 & 0 & 2 & 0 & 0.5 & 0 \\ 1 & -1 & 0 & 1 & 0 & 0 \\ 0 & 2 & -1 & 0 & 0 & 1 \end{bmatrix}$$

$$\rightarrow \begin{bmatrix} 1 & 0 & 2 & 0 & 0.5 & 0 \\ 0 & -1 & -2 & 1 & -0.5 & 0 \\ 0 & 2 & -1 & 0 & 0 & 1 \end{bmatrix} \rightarrow \begin{bmatrix} 1 & 0 & 2 & 0 & 0.5 & 0 \\ 0 & 1 & -0.5 & 0 & 0 & 0.5 \\ 0 & -1 & -2 & 1 & -0.5 & 0 \end{bmatrix}$$

$$\rightarrow \begin{bmatrix} 1 & 0 & 2 & 0 & 0.5 & 0 \\ 0 & 1 & -0.5 & 0 & 0 & 0.5 \\ 0 & 0 & -2.5 & 1 & -0.5 & 0.5 \end{bmatrix} \rightarrow \begin{bmatrix} 1 & 0 & 2 & 0 & 0.5 & 0 \\ 0 & 1 & -0.5 & 0 & 0 & 0.5 \\ 0 & 0 & 1 & -0.4 & 0.2 & -0.2 \end{bmatrix}$$

$$\rightarrow \left[\begin{array}{ccc|ccc} 1 & 0 & 0 & 0.8 & 0.1 & 0.4 \\ 0 & 1 & 0 & -0.2 & 0.1 & 0.4 \\ 0 & 0 & 1 & -0.4 & 0.2 & -0.2 \end{array}\right] = [I, A^{-1}]$$

Thus, the inverse of A is

$$A^{-1} = \begin{bmatrix} 0.8 & 0.1 & 0.4 \\ -0.2 & 0.1 & 0.4 \\ -0.4 & 0.2 & -0.2 \end{bmatrix}$$

The computational effort required to obtain the inverse is considerably more than that for solving the system with a single right-hand side vector. Again, all of the arithmetic operations occur in step (2). The normalization in step (2d) requires $2n - j$ divisions because $C_{ij} = 0$ for $i < j$ and $C_{jj} = 1$. Thus, using Equation (2.2.7) with the change of variables, $i = n - j$, the total number of division operations associated with normalization for $1 \leq j \leq n$ is

$$\begin{aligned} c_{norm}(n) &= \sum_{i=0}^{n-1} (n + i) \\ &= n^2 + \frac{(n-1)n}{2} \\ &= \frac{n(3n-1)}{2} \end{aligned} \tag{2.2.13}$$

The elimination process in step (2e) requires $(n - 1)(2n - j)$ multiplications, given that $C_{jk} = 0$ for $k < j$ and $C_{ij} = 0$ for $i \neq j$. Consequently, the total number of multiplication operations associated with the elimination process for $1 \leq j \leq n$ is

$$c_{elim}(n) = (n - 1)c_{norm}(n) \tag{2.2.14}$$

Combining the computational effort for normalization and elimination, and examining what happens for large values of n, yields the following number of FLOPs for Alg. 2.2.2.

$$\boxed{c_{inv}(n) \approx \frac{3n^3}{2} \text{ FLOPs}, \quad n \gg 1} \tag{2.2.15}$$

It follows that computing an inverse is also an n^3 computational process, but in this case the coefficient of proportionality is 1.5. Consequently, for large values of n, finding an inverse using Alg. 2.2.2 is about three times more expensive computationally than solving a linear algebraic system using Alg. 2.2.1. Gauss-Jordan elimination is a dependable and conceptually simple way to find an inverse. As we shall see, there is a somewhat more efficient way to find the inverse, called the *LU* decomposition method, that allows us to reduce the constant of proportionality from 3/2 to 4/3.

2.3 GAUSSIAN ELIMINATION

Next we examine a modification of the Gauss-Jordan method that reduces the number of operations. Observe how in step (2e) of Alg. 2.2.1, all of the entries in column j except for the pivot are eliminated. This is referred to as *complete* elimination. Complete elimination has the effect of reducing the first n columns of the augmented matrix C to a *diagonal* matrix. It is also possible to solve the system $Ax = b$ using only *partial* elimination, thereby reducing the first n columns of C to an *upper-triangular* matrix D, a matrix with zeros below the main diagonal as follows.

$$[A, b] \rightarrow [D, e] \tag{2.3.1}$$

This still leaves some work to be done in order to solve the simplified equivalent system $Dx = e$ for x, but the additional computational effort is not significant. The following algorithm is an implementation of the partial elimination approach called the *Gaussian elimination* method. Like Alg. 2.2.1, it returns an error code E that is zero if the coefficient matrix A is singular.

Alg. 2.3.1 Gaussian Elimination

1. Set $E = 1$ and form the $n \times (n + 1)$ augmented matrix $C = [A, b]$.
2. For $j = 1$ to n do

   {

   (a) Compute the pivot index $j \le p \le n$ such that:

   $$|C_{pj}| = \max_{i=j}^{n}\{|C_{ij}|\}$$

   (b) If $C_{pj} = 0$, set $E = 0$ and exit.

   (c) If $p > j$ interchange rows p and j.

   (d) For each $i > j$, subtract C_{ij}/C_{jj} times row j from row i.

   }
3. Partition C as $C = [D, e]$ where D is $n \times n$ and e is $n \times 1$.
4. For $j = n$ down to 1 compute

   {

   $$x_j = \frac{1}{D_{jj}}\left(e_j - \sum_{i=j+1}^{n} D_{ji}x_i\right)$$

   }

Alg. 2.3.1 is a two-pass algorithm. The first pass is the *forward elimination* process in steps (1) through (3) where the first n columns of C are row-reduced to an upper-triangular matrix D. The process is called forward elimination because the columns are processed in left-to-right order. Note that the diagonal elements of D are *not* normalized to one as in Alg. 2.2.1. At the completion of the first pass, the original system

$Ax = b$ is reduced to the simpler row-equivalent system $Dx = e$. Pass two consists of step (4). Since D is upper-triangular, the last equation is simply $D_{nn}x_n = e_n$ which has the solution, $x_n = e_n/D_{nn}$. Once x_n is known, it can be substituted into equation $n - 1$, which involves only x_{n-1} and x_n. Equation $n - 1$ can then be solved for x_{n-1}, etc. This process, which takes place in step (4), is called *back substitution* because the components of x are computed in reverse order using successive substitution. Note that when $j = n$, the summation in step (4) contains no terms. The following example illustrates the Gaussian elimination procedure.

EXAMPLE 2.3.1 Gaussian Elimination

Consider the following linear algebraic system.

$$\begin{bmatrix} 1 & 0 & 2 \\ 2 & -1 & 3 \\ 4 & 1 & 8 \end{bmatrix} x = \begin{bmatrix} 1 \\ -1 \\ 2 \end{bmatrix}$$

Applying step (2) of Alg. 2.3.1 yields

$$C = [A, b] = \left[\begin{array}{ccc|c} 1 & 0 & 2 & 1 \\ 2 & -1 & 3 & -1 \\ 4 & 1 & 8 & 2 \end{array}\right] \rightarrow \begin{bmatrix} 4 & 1 & 8 & 2 \\ 2 & -1 & 3 & -1 \\ 1 & 0 & 2 & 1 \end{bmatrix}$$

$$\rightarrow \begin{bmatrix} 4 & 1 & 8 & 2 \\ 0 & -1.5 & -1 & -2 \\ 0 & -0.25 & 0 & 0.5 \end{bmatrix} \rightarrow \left[\begin{array}{ccc|c} 4 & 1 & 8 & 2 \\ 0 & -1.5 & -1 & -2 \\ 0 & 0 & 0.1667 & 0.8333 \end{array}\right]$$

$$= [D, e]$$

This completes the forward elimination process. Applying step (4) of Alg. 2.3.1, then yields

$$x_3 = 0.8333/0.1667 = 5$$

$$x_2 = (-2 + 5)/(-1.5) = -2$$

$$x_1 = (2 - 8(5) + 2)/4 = -9$$

Thus the back substitution process yields $x = [-9, -2, 5]^T$.

Replacement of complete elimination in Alg. 2.2.1 with partial elimination plus back substitution in Alg. 2.3.1 reduces the number of computations for large values of n. To see this, we analyze the Gaussian elimination method to determine the number of floating point multiplications and divisions. To that end, the following integer summation formula is useful.

$$\sum_{i=1}^{n} i^2 = \frac{n(n + 1)(2n + 1)}{6} \tag{2.3.2}$$

All arithmetic operations occur in steps (2) and (4). The elimination process in step (2d) requires $n - j$ divisions plus $(n - j)(n + 1 - j)$ multiplications because

$C_{ik} = 0$ for $i > j$ and $k \le j$. Thus, using Equations (2.2.7) and (2.3.2) with the change of variables, $i = n - j$, the total number of operations associated with the elimination process for $1 \le j \le n$ is

$$c_{elim}(n) = \sum_{i=0}^{n-1} i(i+2) = \frac{(n-1)n(2n-1)}{6} + (n-1)n \tag{2.3.3}$$

Next, consider the back substitution process in step (4), which requires $(n - j)$ multiplications plus one division. Thus, using (2.2.7) and the change of variable $i = n + 1 - j$, the total number of operations associated with back substitution for $1 \le j \le n$ is

$$c_{back}(n) = \frac{n(n+1)}{2} \tag{2.3.4}$$

Combining the computational effort for the forward elimination with that for back substitution, and examining large values of n, yields the following number of FLOPs for Alg. 2.3.1.

$$\boxed{c_{GE}(n) \approx \frac{n^3}{3} \quad \text{FLOPs}, \quad n \gg 1} \tag{2.3.5}$$

Comparing (2.3.5) with (2.2.10) we see that for large values of n, the use of partial elimination reduces the computational cost by approximately 33%. Gauss-Jordan elimination and Gaussian elimination are both n^3 computational processes, but for Alg. 2.3.1 the coefficient of proportionality is reduced from 1/2 to 1/3.

One of the most important theoretical attributes of an $n \times n$ matrix is the *determinant*. The determinant, denoted $\det(A)$, has the property that $\det(A) = 0$ if and only if A is singular. For example, the determinant for the case $n = 2$ is simply

$$\boxed{\det(A) = A_{11}A_{22} - A_{21}A_{12}} \tag{2.3.6}$$

Determinants are very useful for developing theoretical properties of matrices. However, they are used less often in numerical work because their computation is expensive and is susceptible to accumulated round-off error. For example, if the determinant of an $n \times n$ matrix is computed by using the method of expansion by minors, the number of multiplications required is proportional to $n!$. Consequently, the computational effort grows very rapidly with the size of the matrix. Determinants have a number of useful properties, some of which are summarized in Table 2.3.1. The first two properties apply to general $n \times n$ matrices, the third to an upper-triangular matrix, and the remaining properties to the elementary matrices defined in Table 2.3.1.

By exploiting the properties in Table 2.3.1, the Gaussian elimination procedure can be modified so as to produce a more economical calculation of the determinant. Note from property three that the determinant of an upper-triangular matrix U is just the product of the diagonal elements. The Gaussian elimination method transforms a

TABLE 2.3.1 Properties of Determinant

Property	*Description*
1	$\det(A^T) = \det(A)$
2	$\det(AB) = \det(A)\det(B)$
3	$\det(U) = U_{11}U_{22}\cdots U_{nn}$
4	$\det[E_1(k,j)] = -1$
5	$\det[E_2(k,\alpha)] = \alpha$
6	$\det[E_3(k,j,\alpha)] = 1$

matrix into upper-triangular form using elementary row operations, which is equivalent to premultiplication by elementary row matrices.

$$U = E_q \cdots E_b E_a A \tag{2.3.7}$$

Using property two, we find that $\det(U)$ is equal to the product of the determinants of the matrices on the right-hand side of (2.3.7). Note from Alg. 2.3.1 that only two types of elementary row operations are used in the Gaussian elimination procedure: Multiplication by $E_1(k,j)$ is used to interchange rows for pivoting, and multiplication by $E_3(\alpha,k,j)$ is used to eliminate elements below the diagonal. From Table 2.3.1 the first of these operations introduces a minus sign, while the second has no effect on the determinant. Consequently, if r denotes the number of row interchanges to reduce A to upper-triangular form U, then from (2.3.7) and the properties in Table 2.3.1, the determinant of A can be expressed

$$\boxed{\det(A) = (-1)^r U_{11}U_{22}\cdots U_{nn}} \tag{2.3.8}$$

The following algorithm is a modification of the basic Gaussian elimination procedure to compute the determinant of an $n \times n$ matrix A using the formulation in (2.3.8).

Alg. 2.3.2 Matrix Determinant

1. Set $r = 0$.

2. For $j = 1$ to n do

{

(a) Compute the pivot index $j \le p \le n$ such that:

$$|A_{pj}| = \max_{i=j}^{n}\{|A_{ij}|\}$$

(b) If $A_{pj} = 0$, set $\Delta = 0$ and exit.

(c) If $p > j$ interchange rows p and j and set $r = r + 1$.

(d) For each $i > j$, subtract A_{ij}/A_{jj} times row j from row i.

}

3. Set $\Delta = (-1)^r A_{11}A_{22}\cdots A_{nn}$.

The value returned by Alg. 2.3.2 is $\Delta = \det(A)$. If A is singular, then Alg. 2.3.2 terminates prematurely with $\Delta = 0$. The following is an illustration of the use of Alg. 2.3.2.

EXAMPLE 2.3.2 Matrix Determinant

Consider the following 3×3 matrix.

$$A = \begin{bmatrix} 1 & 4 & 0 \\ 0 & 2 & 6 \\ -1 & 0 & 1 \end{bmatrix}$$

Applying Alg. 2.3.2 yields

$$A = \begin{bmatrix} 1 & 4 & 0 \\ 0 & 2 & 6 \\ -1 & 0 & 1 \end{bmatrix}_{r=0} \rightarrow \begin{bmatrix} 1 & 4 & 0 \\ 0 & 2 & 6 \\ 0 & 4 & 1 \end{bmatrix}_{r=0}$$

$$\rightarrow \begin{bmatrix} 1 & 4 & 0 \\ 0 & 4 & 1 \\ 0 & 2 & 6 \end{bmatrix}_{r=1} \rightarrow \begin{bmatrix} 1 & 4 & 0 \\ 0 & 4 & 1 \\ 0 & 0 & 5.5 \end{bmatrix}_{r=1}$$

Step (3) then yields

$$\det(A) = (-1)^1(1)(4)(5.5) = -22$$

The evaluation of the computational effort for calculating the determinant proceeds in a manner similar to that of Alg. 2.3.1. The elimination process in step (2d) requires $n - j$ divisions plus $(n - j)(n - j)$ multiplications because $A_{ik} = 0$ for $i > j$ and $k \leq j$. In addition, n multiplications are needed for step (3). Thus, using Equations (2.2.7) and (2.3.2) with the change of variables $i = n - j$, the total number of operations for computing the determinant is

$$\begin{aligned} c_{det}(n) &= \sum_{i=0}^{n-1} (i^2 + i + n) \\ &= \frac{(n-1)n(2n-1)}{6} + \frac{(n-1)n}{2} + n^2 \end{aligned} \tag{2.3.9}$$

Thus, the total number of FLOPs associated with Alg. 2.3.2 for large values of n is

$$c_{det}(n) \approx \frac{n^3}{3} \text{ FLOPs}, \quad n \gg 1 \tag{2.3.10}$$

2.4 *LU* DECOMPOSITION

When Gaussian elimination is used to solve a linear algebraic system, the elementary row operations that reduce the coefficient matrix to upper-triangular form are applied simultaneously to the right-hand side vector. Consequently, if the system is to be solved again, but with a different right-hand side vector, the row reduction process must be repeated.

2.4.1 *LU* Factorization

There is a useful variation of the Gaussian elimination method which eliminates the need to row-reduce the system each time a new right-hand side vector is used. The basic idea is to apply Gaussian elimination to the A matrix itself while recording the operations required to reduce A to upper-triangular form. In particular, suppose the matrix A is reduced to upper-triangular form U by eliminating each of the elements below the diagonal using a sequence of elementary row operations. If no pivoting is required, then the sequence of row operations can be represented by premultiplication by a lower-triangular matrix T as follows.

$$TA = U \tag{2.4.1}$$

Let $L = T^{-1}$ denote the inverse of the lower-triangular matrix. Then L is also lower-triangular, and multiplying (2.4.1) on the left by L yields the following representation called an *LU factorization* of A.

$$A = LU \tag{2.4.2}$$

To develop an efficient way to compute the lower-triangular factor L, and the upper-triangular factor U, we first examine the general form of L and U.

$$L = \begin{bmatrix} L_{11} & 0 & 0 & \cdots & 0 \\ L_{21} & L_{22} & 0 & \cdots & 0 \\ \vdots & \vdots & \ddots & \vdots & \vdots \\ L_{n1} & L_{n2} & L_{n3} & \cdots & L_{nn} \end{bmatrix}, \quad U = \begin{bmatrix} U_{11} & U_{12} & U_{13} & \cdots & U_{1n} \\ 0 & U_{22} & U_{23} & \cdots & U_{2n} \\ \vdots & \vdots & \ddots & \vdots & \vdots \\ 0 & 0 & 0 & \cdots & U_{nn} \end{bmatrix} \tag{2.4.3}$$

Using (2.2.7), the number of unknowns in L is $n(n+1)/2$, and similarly for U. Thus (2.4.2) consists of a total of n^2 equations in $n^2 + n$ unknowns. Since the number of unknowns exceeds the number of constraints by n, we can assign arbitrary values to n of the unknowns. In particular, suppose the diagonal elements of the upper-triangular factor U are set to unity.

$$U_{kk} = 1 \quad , \quad 1 \le k \le n \tag{2.4.4}$$

To compute the elements of L and U, we write out the product LU term by term. For illustrative purposes, it is sufficient to consider the case, $n = 3$. Using (2.4.3) and (2.4.4), and multiplying L times U yields

$$\begin{bmatrix} L_{11} & L_{11}U_{12} & L_{11}U_{13} \\ L_{21} & L_{21}U_{12} + L_{22} & L_{21}U_{13} + L_{22}U_{23} \\ L_{31} & L_{31}U_{12} + L_{32} & L_{31}U_{13} + L_{32}U_{23} + L_{33} \end{bmatrix} = \begin{bmatrix} A_{11} & A_{12} & A_{13} \\ A_{21} & A_{22} & A_{23} \\ A_{31} & A_{32} & A_{33} \end{bmatrix} \tag{2.4.5}$$

The system in (2.4.5) can be solved for L and U in a number of ways. Note that the first column of LU is identical to the first column of L. Equating the first columns of LU and A therefore yields

$$L_{11} = A_{11} \tag{2.4.6a}$$

$$L_{21} = A_{21} \tag{2.4.6b}$$

$$L_{31} = A_{31} \tag{2.4.6c}$$

Next we compute the first row of U. Recalling (2.4.4) and comparing the first rows of LU and A, if $L_{11} \neq 0$, then

$$U_{11} = 1 \tag{2.4.7a}$$
$$U_{12} = A_{12}/L_{11} \tag{2.4.7b}$$
$$U_{13} = A_{13}/L_{11} \tag{2.4.7c}$$

This completes the first column of L and the first row of U. Next equating the second column of LU with the second column of A and solving for the elements of L yields

$$L_{12} = 0 \tag{2.4.8a}$$
$$L_{22} = A_{22} - L_{21}U_{12} \tag{2.4.8b}$$
$$L_{32} = A_{32} - L_{31}U_{12} \tag{2.4.8c}$$

We then compute the second row of U. Using (2.4.4) and comparing the second rows of LU and A, if $L_{22} \neq 0$, then

$$U_{21} = 0 \tag{2.4.9a}$$
$$U_{22} = 1 \tag{2.4.9b}$$
$$U_{23} = (A_{23} - L_{21}U_{13})/L_{22} \tag{2.4.9c}$$

This completes the second column of L and the second row of U. Finally, equating the third column of LU with the third column of A and solving for the elements of L yields

$$L_{13} = 0 \tag{2.4.10a}$$
$$L_{23} = 0 \tag{2.4.10b}$$
$$L_{33} = A_{33} - L_{31}U_{13} - L_{32}U_{23} \tag{2.4.10c}$$

From (2.4.4), the elements in the third row of U are

$$U_{31} = 0 \tag{2.4.11a}$$
$$U_{32} = 0 \tag{2.4.11b}$$
$$U_{33} = 1 \tag{2.4.11c}$$

Thus the elements of L and U can be identified quite easily as long as they are computed in the proper order. This procedure, which can be readily extended to an $n \times n$ matrix, is called *Crout's method*. If we had instead elected to set the diagonal components of L equal to unity, the resulting formulation would be *Dolittle's method*. The general equations for Crout's method are as follows where it is understood that the summations have no terms when the upper limit is less that the lower limit.

$$L_{kj} = A_{kj} - \sum_{i=1}^{j-1} L_{ki}U_{ij} \;, \quad j \le k \le n \tag{2.4.12a}$$

$$U_{jk} = \frac{1}{L_{jj}}\left(A_{jk} - \sum_{i=1}^{j-1} L_{ji}U_{ik}\right) \;, \quad j < k \le n \tag{2.4.12b}$$

Since there are a total of n^2 variable elements in L and U, these two triangular matrices can be stored in a single compact $n \times n$ *storage matrix* Q as follows.

$$Q = [L \backslash U] = \begin{bmatrix} L_{11} & U_{12} & U_{13} & \cdots & U_{1,n-1} & U_{1n} \\ L_{21} & L_{22} & U_{23} & \cdots & U_{2,n-1} & U_{2n} \\ \vdots & \vdots & \vdots & \ddots & \vdots & \vdots \\ L_{n-1,1} & L_{n-1,2} & L_{n-1,3} & \cdots & L_{n-1,n-1} & U_{n-1,n} \\ L_{n1} & L_{n2} & L_{n3} & \cdots & L_{n-1,n} & L_{nn} \end{bmatrix} \tag{2.4.13}$$

Here the lower-triangular part of Q is L, while the elements of Q above the diagonal contain U. Recall from (2.4.4) that the diagonal elements of U do not have to be stored because they are known to be one.

Observe from (2.4.12) that when Q_{kj} is computed, only A_{kj} and previously-computed elements of Q are used. This means that a separate array Q does not have to be set aside for storage. Instead, the elements of Q can be stored in the corresponding positions in A, in which case we say that it is an *in place* computation. The resulting savings of n^2 floating point memory locations can be an important consideration when the value of n is very large.

It is evident from inspection of (2.4.12b) that the computations of the rows of U are valid only if the corresponding diagonal elements of L are nonzero. However, the diagonal elements of L are not guaranteed to be nonzero, even when A is nonsingular. To guard against division by zero, we must use pivoting. Recall that in Gaussian elimination the row interchanges associated with pivoting are performed simultaneously on the coefficient matrix and the right-hand side vector. In this case there is no right-hand vector, but it is still important to keep a record of the row interchanges so they can be performed later. This can be easily accomplished by introducing a *row permutation* matrix P which is first initialized to $P = I$. Whenever a row interchange is performed, the corresponding rows of P are also interchanged thereby recording the operation. Pivot selection is performed just before the pivot element is needed (Gerald and Wheatley, 1989). That is, after the kth column of Q is computed the element in column k at or below the diagonal with the largest magnitude is selected for the pivot. This row of Q is then interchanged with the kth row. A similar row interchange is performed on P.

The following algorithm is an implementation of Crout's method with partial pivoting. It starts with the matrix A and generates a scalar Δ, which is the determinant of A, and the matrices Q and P.

Alg. 2.4.1 ***LU* Factorization**

1. Set $\Delta = 1$ and form the $n \times 2n$ augmented matrix $D = [A, I]$.
2. For $j = 1$ to n do

{

(a) For $k = j$ to n compute

$$D_{kj} = D_{kj} - \sum_{i=1}^{j-1} D_{ki} D_{ij}$$

(b) Compute the pivot index $j \le q \le n$ such that:

$$|D_{qj}| = \max_{i=j}^{n}\{|D_{ij}|\}$$

(c) If $D_{qj} = 0$, set $\Delta = 0$ and exit.
(d) If $q > j$, interchange rows q and j of D and set $\Delta = -\Delta$.
(e) For $k = j + 1$ to n compute

$$D_{jk} = \frac{1}{D_{jj}}\left(D_{jk} - \sum_{i=1}^{j-1} D_{ji}D_{ik}\right)$$

(f) Set $\Delta = D_{jj}\Delta$

}

3. Partition D as $D = [Q, P]$ where Q and P are $n \times n$.

Again, it is understood that sums in steps (2a) and (2e) contain no terms when the upper limit is less than the lower limit. Partial pivoting based on the lower-triangular factor L is implemented in steps (2b) through (2d). If the matrix A is singular, then Alg. 2.4.1 returns with $\Delta = 0$. Otherwise, it returns with $Q = [L\backslash U]$ and P, in which case

$$\boxed{LU = PA} \tag{2.4.14}$$

Note from steps (2d) and (2f) that $\Delta = \det(A)$ is computed as the product of the diagonal elements of L times $(-1)^r$ where r is the number of row interchanges used to form P. The following example illustrates the use of Alg. 2.4.1 to obtain an LU factorization of a matrix.

EXAMPLE 2.4.1 *LU* Factorization

Consider the following 3×3 coefficient matrix.

$$A = \begin{bmatrix} 0 & 1 & -1 \\ 2 & -2 & 1 \\ 1 & 2 & 0 \end{bmatrix}$$

After step (1) of Alg. 2.4.1 we have, $\Delta = 1$ and

$$D = [A, I] = \left[\begin{array}{ccc|ccc} 0 & 1 & -1 & 1 & 0 & 0 \\ 2 & -2 & 1 & 0 & 1 & 0 \\ 1 & 2 & 0 & 0 & 0 & 1 \end{array}\right]$$

Applying step (2) with $j = 1$, the pivot index is $q = 2$. At the end of step (2), we have $\Delta = -2$ and

$$D = \left[\begin{array}{ccc|ccc} 2 & -1 & 0.5 & 0 & 1 & 0 \\ 0 & 1 & -1 & 1 & 0 & 0 \\ 1 & 2 & 0 & 0 & 0 & 1 \end{array}\right]$$

Next, applying step (2) with $j = 2$, the pivot index is $q = 3$. At the end of step (2), we have $\Delta = 6$ and

$$D = \left[\begin{array}{ccc|ccc} 2 & -1 & 0.5 & 0 & 1 & 0 \\ 1 & 3 & -0.1667 & 0 & 0 & 1 \\ 0 & 1 & -1 & 1 & 0 & 0 \end{array}\right]$$

Applying step (2) with $j = 3$, there is no need to perform pivoting because $j = n$. In this case, $\Delta = -5$ and

$$D = \left[\begin{array}{ccc|ccc} 2 & -1 & 0.5 & 0 & 1 & 0 \\ 1 & 3 & -0.1667 & 0 & 0 & 1 \\ 0 & 1 & -0.8333 & 1 & 0 & 0 \end{array}\right]$$

Finally, partitioning D as in step (3), the storage matrix Q and the row permutation matrix P are

$$Q = [L\backslash U] = \begin{bmatrix} 2 & -1 & 0.5 \\ 1 & 3 & -0.1667 \\ 0 & 1 & -0.8333 \end{bmatrix}, \quad P = \begin{bmatrix} 0 & 1 & 0 \\ 0 & 0 & 1 \\ 1 & 0 & 0 \end{bmatrix}$$

It is left as an exercise to verify that $LU = PA$ in this case.

2.4.2 Forward and Back Substitution

Once the coefficient matrix is factored into lower- and upper-triangular parts, a linear algebraic system can be solved very efficiently for a variety of right-hand side vectors. First, note that the system $Ax = b$ can be rewritten as $PAx = Pb$ where P is the row permutation matrix. Next, from Equation (2.4.14), the system $PAx = Pb$ can be rewritten as $LUx = Pb$. The system $LUx = Pb$ then can be solved in stages, using d and y as intermediate variables as follows.

$$d = Pb \tag{2.4.15a}$$

$$Ly = d \tag{2.4.15b}$$

$$Ux = y \tag{2.4.15c}$$

At first glance, this may not appear to be an improvement because we have replaced one system of n equations by a sequence of two systems of n equations. However, it is important to note that (2.4.15b) and (2.4.15c) have a very special structure, the left-hand side of (2.4.15b) is lower-triangular and the left-hand side of (2.4.15c) is upper-triangular. As a result, these equations can be easily solved using *forward substitution* for (2.4.15b) and then *back substitution* for (2.4.15c). That is, the elements of y can be computed starting with the first and ending with the last. Then y can be used on the right-hand side of (2.4.15c) and the elements of x can be computed starting with the last and ending with the first.

The procedure for solving $Ax = b$ using the *LU decomposition* technique is summarized in the following algorithm. This is a higher-level algorithm in the sense that it uses Alg. 2.4.1 as one of its steps.

Alg. 2.4.2 ***LU* Decomposition**

1. Apply Alg. 2.4.1 to compute Δ, Q, and P. If $\Delta = 0$, exit.
2. Compute $d = Pb$.
3. For $k = 1$ to n compute

{

$$y_k = \frac{1}{Q_{kk}}\left(d_k - \sum_{i=1}^{k-1} Q_{ki} y_i\right)$$

}

4. For $k = n$ down to 1 compute

{

$$x_k = y_k - \sum_{i=k+1}^{n} Q_{ki} x_i$$

}

Here it is understood that the summations in steps (3) and (4) have no terms when the upper limit is smaller than the lower limit. It is important to note that if $Ax = b$ is to be solved with more than one right-hand side vector b, then step (1) of Alg. 2.4.2 needs to be executed only *once*. Subsequent calls to Alg. 2.4.2 using the same coefficient matrix A can use step (2) as the entry point. This is a key qualitative feature of the *LU* decomposition method that sets it apart from the previous elimination methods. The following example illustrates the use of Alg. 2.4.2 to solve a linear algebraic system.

EXAMPLE 2.4.2 ***LU* Decomposition**

Consider the following system of linear equations.

$$\begin{bmatrix} 0 & 1 & -1 \\ 2 & -2 & 1 \\ 1 & 2 & 0 \end{bmatrix} x = \begin{bmatrix} -4 \\ 9 \\ 0 \end{bmatrix}$$

The coefficient matrix A has already been factored into lower- and upper-triangular parts in Example 2.4.1, where it was found that

$$P = \begin{bmatrix} 0 & 1 & 0 \\ 0 & 0 & 1 \\ 1 & 0 & 0 \end{bmatrix}, \quad Q = \begin{bmatrix} 2 & -1 & 0.5 \\ 1 & 3 & -0.1667 \\ 0 & 1 & -0.8333 \end{bmatrix}$$

Applying step (2), we reorder the rows of b by premultiplying by the row permutation matrix P.

$$d = Pb = \begin{bmatrix} 0 & 1 & 0 \\ 0 & 0 & 1 \\ 1 & 0 & 0 \end{bmatrix} \begin{bmatrix} -4 \\ 9 \\ 0 \end{bmatrix} = \begin{bmatrix} 9 \\ 0 \\ -4 \end{bmatrix}$$

Next, the forward substitution in step (3) yields the following vector of intermediate variables y.

$$y_1 = d_1/Q_{11} = 4.5$$

$$y_2 = (d_2 - Q_{21}y_1)/Q_{22} = -1.5$$
$$y_3 = (d_3 - Q_{31}y_1 - Q_{32}y_2)/Q_{33} = 3$$

The back substitution in step (4) then yields the components of the solution vector x in reverse order.

$$x_3 = y_3 = 3$$
$$x_2 = y_2 - Q_{23}x_3 = -1$$
$$x_1 = y_1 - Q_{12}x_2 - Q_{13}x_3 = 2$$

Thus the solution is $x = [2, -1, 3]^T$. It is left as an exercise for the reader to verify that $Ax = b$ in this case.

The computational effort required by the *LU* decomposition method can be divided into two parts. First we examine the number of floating point multiplications and divisions required to call Alg. 2.4.1 which factors A. All of the arithmetic operations occur in step (2). The computation of column j of D in step (2a) requires $(j-1)(n-j+1)$ multiplications. Next the computation of row j of D in step (2e) requires $(j-1)(n-j)$ multiplications plus $(n-j)$ divisions. Finally, the computation of Δ in step (2f) requires one multiplication. If we sum these numbers from 1 to n, multiply out all the terms, and apply Equations (2.2.7) and (2.3.2), then after simplification, the total number of floating point multiplications and divisions for factorization is

$$c_{fac}(n) = n^2(n+1) + n - \frac{n(n+1)(2n+1)}{3} \tag{2.4.16}$$

Next, we examine the number of multiplications and divisions required to implement the remaining steps of the *LU* decomposition method. The matrix multiplication by the row permutation matrix P in step (2) of Alg. 2.4.2 requires n^2 multiplications. However, due to the special structure of P, an actual matrix multiplication does not have to be performed. Instead the components of b have to be reordered based upon the locations of the ones in the rows of P. For the forward substitution in step (3), the computation of y_k requires one division plus $k-1$ multiplications. Similarly, for the back substitution in step (4), the computation of x_k requires $n-k$ multiplications. It follows that for each k, the total number of operations is n. Thus, the total number of floating point multiplications and divisions required for the forward and back substitution for $1 \le k \le n$ is

$$c_{sub}(n) = n^2 \tag{2.4.17}$$

For large values of n, the factorization of A dominates the computational effort, and the total number of FLOPs to implement Alg. 2.4.2 is

$$\boxed{c_{LU}(n) \approx \frac{n^3}{3} \quad \text{FLOPs}, \quad n \gg 1} \tag{2.4.18}$$

Consequently, for large values of n, the *LU* decomposition method requires the same computational effort as Gaussian elimination. If multiple right-hand side vectors are used, then LU decomposition is the method of choice because the LU factorization must be performed only once. One example of this is the computation of the inverse of A. Recall from Equation (2.2.11) that the kth column of A^{-1} can be obtained by solving $Ax = b$ with b replaced by the kth column of the identity matrix.

$$Ax = i^k \quad , \quad 1 \leq k \leq n \tag{2.4.19}$$

Suppose LU decomposition is used to solve the n systems in (2.4.19). From (2.4.18), we conclude that for large values of n approximately $n^3/3$ FLOPs are required to factor A. Once this is done, only forward and back substitution are needed to solve the n systems in (2.4.19). From (2.4.17), this requires n^3 FLOPs. Thus, for large values of n, the total number of FLOPs needed to compute A^{-1} using LU decomposition is

$$\boxed{c_{inv}(n) \approx \frac{4n^3}{3} \quad \text{FLOPs}, \quad n \gg 1} \tag{2.4.20}$$

The Gauss-Jordan method for finding the inverse in Alg. 2.2.2 required approximately $3n^3/2$ FLOPs for large n. Thus, the LU method is about 11% more efficient.

Recall that a matrix is symmetric if $A^T = A$ and positive-definite if $x^T A x > 0$ for $x \neq 0$. When A is a symmetric positive-definite matrix, the LU decomposition can be simplified in the sense that the upper-triangular factor can be selected to be the transpose of the lower-triangular factor.

$$A = LL^T \tag{2.4.21}$$

This factorization of a symmetric positive-definite A is referred to as *Cholesky's method*. Using a procedure similar to that presented for *Crout's method,* we find that the nonzero components of L can be computed as follows for $1 \leq j \leq n$.

$$L_{jj} = A_{jj} - \sum_{i=1}^{j-1} L_{ji}^2 \tag{2.4.22a}$$

$$L_{kj} = \frac{1}{L_{jj}}\left(A_{jk} - \sum_{i=1}^{j-1} L_{ji}L_{ki}\right) \quad , \quad j+1 \leq k \leq n \tag{2.4.22b}$$

Here it is understood that the summations contain no terms when the lower limit exceeds the upper limit. It can be shown that the Cholesky factorization requires only about $n^3/6$ FLOPs for large values of n. Thus, it is twice as fast as the general LU factorization.

2.4.3 Tridiagonal Systems

In engineering applications, there are many instances where the coefficient matrix A is *sparse* in the sense that it has a large number of zero elements arranged in some pattern. There are numerous ways a matrix can be sparse (see, e.g., Press et al., 1992). Diagonal matrices, lower-triangular matrices, and upper-triangular matrices are familiar examples of sparse matrices. Another common case is for a matrix to have a *banded* structure in which only elements within a certain radius of the main diagonal are nonzero. In particular, a *tridiagonal* matrix is a banded matrix in which nonzero elements appear only on the main diagonal, just above it on the first superdiagonal, and just below it on the first subdiagonal.

$$A = \begin{bmatrix} b_1 & a_1 & 0 & \cdots & 0 & 0 & 0 \\ c_1 & b_2 & a_2 & \cdots & 0 & 0 & 0 \\ 0 & c_2 & b_3 & \cdots & 0 & 0 & 0 \\ \vdots & \vdots & \vdots & \ddots & \vdots & \vdots & \vdots \\ 0 & 0 & 0 & \cdots & b_{n-2} & a_{n-2} & 0 \\ 0 & 0 & 0 & \cdots & c_{n-2} & b_{n-1} & a_{n-1} \\ 0 & 0 & 0 & \cdots & 0 & c_{n-1} & b_n \end{bmatrix} \tag{2.4.23}$$

Recall that a practical illustration of a tridiagonal coefficient matrix arose in connection with the digital-to-analog converter circuit in (2.1.3). The *LU* decomposition method simplifies considerably when the coefficient matrix A is tridiagonal. In this case the lower and upper-triangular factors are also tridiagonal. In particular, one can verify through direct multiplication that if A is tridiagonal and if the diagonal elements of U are set to unity as in Equation (2.4.4), then L and U must have the following structure:

$$L = \begin{bmatrix} L_{11} & 0 & \cdots & 0 & 0 \\ c_1 & L_{22} & \cdots & 0 & 0 \\ \vdots & \vdots & \ddots & \vdots & \vdots \\ 0 & 0 & \cdots & L_{n-1,n-1} & 0 \\ 0 & 0 & \cdots & c_{n-1} & L_{nn} \end{bmatrix}, U = \begin{bmatrix} 1 & U_{12} & \cdots & 0 & 0 \\ 0 & 1 & \cdots & 0 & 0 \\ \vdots & \vdots & \ddots & \vdots & \vdots \\ 0 & 0 & \cdots & 1 & U_{n-1,n} \\ 0 & 0 & \cdots & 0 & 1 \end{bmatrix} \tag{2.4.24}$$

Note that the "ones" on the main diagonal of U dictate that the first subdiagonal of L is the same as the first subdiagonal of A. Consequently, for an $n \times n$ tridiagonal matrix there are only $2n - 1$ unknown elements of L and U to be determined. Multiplying L times U and equating the product with A reveals that the following equations can be used to solve for the elements of L and U. First we initialize using $L_{11} = b_1$. The remaining unknowns are then computed successively for $1 \le k < n$ assuming $L_{kk} \ne 0$.

$$U_{k,k+1} = \frac{a_k}{L_{kk}} \tag{2.4.25a}$$

$$L_{k+1,k+1} = b_{k+1} - c_k U_{k,k+1} \tag{2.4.25b}$$

The computation of L and U for a tridiagonal matrix is an n (linear) process, unlike the general case were it is an n^3 process. In particular, it can be shown (see Problem 2.18), that the number of FLOPs required to solve a tridiagonal system by *LU* decomposition is

$$c_{tri}(n) = 5n - 4 \quad \text{FLOPs} \tag{2.4.26}$$

This results in a dramatic reduction in computational effort when n is large. It is clear from (2.4.24) that the storage requirements for L and U can also be reduced significantly because there is no need to store the zero elements. As we shall see, tridiagonal systems arise repeatedly in the solution of linear partial differential equations by the finite-difference method.

2.5 ILL-CONDITIONED SYSTEMS

All the algorithms discussed thus far use pivoting, which tends to reduce the detrimental effects of accumulated round-off error. Some linear algebraic systems are more sensitive to round-off error than others. Indeed, for some systems a small change in one of the values of the coefficient matrix or the right-hand side vector can give rise to a large change in the solution vector. The following simple example illustrates this phenomenon.

EXAMPLE 2.5.1 Ill Conditioning

Consider the following system of two linear equations in two unknowns.

$$\begin{bmatrix} 100 & -200 \\ -200 & 401 \end{bmatrix} x = \begin{bmatrix} 100 \\ -100 \end{bmatrix}$$

This system can be solved using any of the methods covered thus far with the exact solution being $x = [201, 100]^T$. Suppose the elements of the coefficient matrix A are altered or perturbed slightly to yield the following generally similar system.

$$\begin{bmatrix} 101 & -200 \\ -200 & 400 \end{bmatrix} x = \begin{bmatrix} 100 \\ -100 \end{bmatrix}$$

Notice that A_{11} has been increased by 1.0%, and A_{22} has been decreased by less than 0.254%. With these rather modest changes in A, one might expect that there would be a correspondingly small change in the solution x. However, this is *not* the case. The exact solution of the perturbed system is $x = [50, 24.75]^T$, which means that changes on the order of 1% in the equations generate changes on the order of 75% in the solution! Clearly, the solution is very *sensitive* to the values of A in this case.

When the solution is highly sensitive to the values of the coefficient matrix A or the right-hand side vector b, the equations are said to be *ill-conditioned*. This can be a serious numerical problem for two reasons. Often the equations are generated from a mathematical model of an underlying physical system. In these instances only *estimates* of the values of A and b are known. If the system $Ax = b$ is ill-conditioned, then the solution is very sensitive to the accuracy of these estimates. Even when exact values of A and b are known, we can not *represent* these values exactly using a computer with finite precision. Therefore, ill-conditioned systems are more susceptible to the effects of accumulated round-off error, particularly when the dimension n is large.

2.5.1 Vector and Matrix Norms

The reliability of a numerical solution to an ill-conditioned system of equations must be regarded with skepticism. The problem is how do we know when a linear algebraic sys-

TABLE 2.5.1 Properties of Norms

Property	*Description*
1	$\|x\| > 0$ for $x \neq 0$
2	$\|x\| = 0$ for $x = 0$
3	$\|\alpha x\| = \|\alpha\| \cdot \|x\|$
4	$\|x + y\| \leq \|x\| + \|y\|$

tem is ill-conditioned? To develop a concise measure, it is useful to first consider the notion of vector and matrix norms. A norm of a vector is a generalization of the absolute value of a scalar. That is, a norm is a measure of the length or magnitude of the vector. There are several valid ways to define a norm of an $n \times 1$ vector x. For the purposes of evaluating the condition of a linear system, the following computationally simple formulation can be used.

$$\|x\| \triangleq \max_{k=1}^{n} \{|x_k|\} \tag{2.5.1}$$

Here, $\triangleq$ means equals by definition, and the notation *max*{} denotes the maximum element of the set. Notice how $\|x\|$ reduces to the absolute value of x when $n = 1$. The formulation in Equation (2.5.1) is sometimes called the *infinity norm* of x. Norms of vectors have the same basic properties as the absolute value of scalars as can be seen from Table 2.5.1. The last property, namely that the norm of the sum is bounded from above by the sum of the norms, is called the *triangle inequality*.

To illustrate the use of the vector norm, suppose we want to check the accuracy of a numerical solution to the linear algebraic system $Ax = b$. Let the *residual error* vector be defined as the difference between the right- and left-hand sides of the system.

$$r(x) \triangleq b - Ax \tag{2.5.2}$$

Clearly x is a solution of $Ax = b$ if and only if $r(x) = 0$. Of course, round-off error will typically cause $r(x)$ to be nonzero. As a measure of the accuracy of the solution, we can use $\|r(x)\|$.

In order to develop a measure of the sensitivity of a system, we also need to consider the notion of a matrix norm. A matrix norm can be defined in terms of a vector norm as follows:

$$\|A\| \triangleq \sup_{x \neq 0} \left\{ \frac{\|Ax\|}{\|x\|} \right\} \tag{2.5.3}$$

Here the notation *sup* $\{\cdot\}$ denotes the supremum or smallest upper bound. In practical terms, the effect of (2.5.3) is to ensure that the matrix norm has the following important property.

$$\|Ax\| \leq \|A\| \cdot \|x\| \tag{2.5.4}$$

Since the matrix norm is defined in terms of the vector norm, we refer to $\|A\|$ as an induced norm. Given the vector norm in (2.5.1), the induced matrix norm can be expressed directly in terms of the components of the matrix as follows.

$$\|A\| = \max_{k=1}^{n} \left\{ \sum_{j=1}^{n} |A_{kj}| \right\} \tag{2.5.5}$$

Note that the expression for $\|A\|$ involves summing the absolute values of elements in the rows of A. Consequently, it is referred to as the *row-sum* norm of A. The matrix norm satisfies all the same properties as the vector norm listed in Table 2.5.1. In addition, the vector norm and its induced matrix norm are related to one another through (2.5.4).

EXAMPLE 2.5.2 Norms

As an illustration of the use of norms, consider the following linear algebraic system.

$$\begin{bmatrix} 4 & -6 & 9 \\ 5 & 8 & -2 \\ 7 & -3 & 1 \end{bmatrix} x = \begin{bmatrix} -1 \\ 12 \\ 13 \end{bmatrix}$$

From (2.5.1), we have $\|b\| = 13$ and from (2.5.5), $\|A\| = \max\{19, 15, 11\} = 19$. The solution of this system is $x = [2, 0, -1]^T$, which means $\|x\| = 2$. It follows that $\|Ax\| \leq \|A\| \cdot \|x\|$ as required by (2.5.4).

2.5.2 Condition Number

To investigate the condition of a linear algebraic system, we perturb the equations and then examine the size of resulting change in the solution. First, consider the effect of changes in the right-hand side vector, say from b to $b + \Delta b$. Let x denote the solution to the original system, and let $x + \Delta x$ denote the solution to the perturbed system. Then $A(x + \Delta x) = b + \Delta b$ which reduces to $A\Delta x = \Delta b$ because $Ax = b$. It follows that $\Delta x = A^{-1}\Delta b$, and using Equation (2.5.4), this yields

$$\|\Delta x\| \leq \|A^{-1}\| \cdot \|\Delta b\| \tag{2.5.6}$$

This provides an upper bound on the variation in the solution x caused by the perturbation Δb. It is more meaningful to investigate the relative error caused by a relative change in b. Using the properties of norms, the relative change in the solution can be expressed in terms of the perturbation as follows.

$$\frac{\|\Delta x\|}{\|x\|} \leq K(A) \frac{\|\Delta b\|}{\|b\|} \tag{2.5.7}$$

Here the scalar $K(A)$ is called the *condition number* of the matrix A and is defined as follows, assuming A is nonsingular.

$$K(A) \triangleq \|A\| \cdot \|A^{-1}\| \tag{2.5.8}$$

Note that $K(A)$ is a measure of the relative sensitivity of the solution to changes in the right-hand side vector b. It is also possible to express changes in the solution caused by changes in the coefficient matrix A.

$$\frac{\|\Delta x\|}{\|x + \Delta x\|} \leq K(A)\frac{\|\Delta A\|}{\|A\|} \tag{2.5.9}$$

When the condition number $K(A)$ becomes large, the system is regarded as being *ill-conditioned*. The following example illustrates the computation of the condition number.

EXAMPLE 2.5.3 Condition Number

Consider the linear algebraic system from Example 2.5.1 where the coefficient matrix is

$$A = \begin{bmatrix} 100 & -200 \\ -200 & 401 \end{bmatrix}$$

For the simple case of a 2×2 matrix, a *short cut* method can be used to compute the inverse. All we have to do is interchange the diagonal elements, change the signs of the off-diagonal elements, and divide by the determinant.

$$\boxed{A^{-1} = \frac{1}{\det(A)}\begin{bmatrix} A_{22} & -A_{12} \\ -A_{21} & A_{11} \end{bmatrix}}$$

where

$$\det(A) = A_{11}A_{22} - A_{12}A_{21}$$

In this case, $\det(A) = 100$ and the inverse is

$$A^{-1} = \begin{bmatrix} 4.01 & 2 \\ 2 & 1 \end{bmatrix}$$

It remains to compute the norms of A and A^{-1}. Using (2.5.5) the row-sum norms are $\|A\| = \max\{300, 601\}$ and $\|A^{-1}\| = \max\{6.01, 3\}$. Thus the condition number of this ill-conditioned system is

$$K(A) = 601(6.01) \approx 3612$$

EXAMPLE 2.5.4 Equilibration

Large condition numbers can also arise from equations that are in need of scaling. Consider the following coefficient matrix which corresponds one "regular" equation and one "large" equation.

$$A = \begin{bmatrix} 1 & -1 \\ 1000 & 1000 \end{bmatrix}$$

In this case, $\det(A) = 2000$ and the inverse of A is

$$A^{-1} = \begin{bmatrix} 0.5 & 0.0005 \\ -0.5 & 0.0005 \end{bmatrix}$$

To determine $K(A)$, we use (2.5.5) to compute $\|A\| = \max\{2,2000\}$ and then $\|A^{-1}\| = \max\{0.5005,0.5005\}$. Thus, the condition number of this system is

$$K(A) = 2000(0.5005) = 1001$$

The equilibration method, in which the equations are normalized, can be used to reduce the condition number for a system that is poorly scaled. If each row of A is scaled by its largest element, then the new A and its inverse are as follows.

$$A = \begin{bmatrix} 1 & -1 \\ 1 & 1 \end{bmatrix}, \quad A^{-1} = \begin{bmatrix} 0.5 & 0.5 \\ -0.5 & 0.5 \end{bmatrix}$$

The norms of the scaled A and its inverse, are $\|A\| = \max\{2,2\}$ and $\|A^{-1}\| = \max\{1,1\}$. Thus the condition number of the scaled system is

$$K(A) = 2(1) = 2$$

Clearly there is a dramatic improvement in the condition number in this case because the original equations were so poorly scaled.

A question that has not yet been answered is how large does $K(A)$ have to be before a system is regarded as ill-conditioned? Although there is no clear threshold, we can make some useful observations. The most well-conditioned matrix is $A = I$ and $K(I) = 1$. At the other extreme, as A approaches a singular matrix we see from Equation (2.5.8) that $K(A) \to \infty$. Thus, the range of possible values for $K(A)$ is

$$\boxed{1 \le K(A) < \infty} \tag{2.5.10}$$

To assess the effects of ill-conditioning, a simple rule of thumb can be used (Hultquist, 1988). For a system with condition number $K(A)$, one can expect a loss of roughly $\log[K(A)]$ decimal places in the accuracy of the solution. Consequently, for the system in Example 2.5.1, we would expect to lose perhaps three to four decimal places in accuracy because $K(A) = 3.612 \times 10^3$.

2.5.3 Approximate Condition Number

There is one clear drawback to using $K(A)$ to measure the reliability of a numerical solution. To compute $K(A)$, we have to first compute the inverse of A, a process that requires on the order of $4n^3/3$ FLOPs. But this is more computational effort than finding the solution vector x itself! It is therefore worthwhile to consider ways in which the condition number of A can be *approximated* without going through all of the computations implied in Equation (2.5.8). To estimate $K(A)$, note that if $x = A^{-1}b$, then $\|x\| \le \|A^{-1}\| \cdot \|b\|$. Consequently, for a given right-hand side b, a lower bound on $\|A^{-1}\|$ can be obtained by using

$$\|A^{-1}\| \ge \frac{\|x\|}{\|b\|} \tag{2.5.11}$$

Since (2.5.11) must hold for *every* b, the estimate can be improved by considering several right-hand side vectors, solving the system for each, and then taking the largest ratio. In particular, suppose $\{b^1, b^2, \ldots, b^m\}$ is a set of right-hand side vectors and let $\{x^1, x^2, \ldots, x^m\}$ be the corresponding solutions of $Ax = b^k$ for $1 \le k \le m$. Then the following can be used to approximate the condition number.

$$K(A) \approx \|A\| \cdot \max_{k=1}^{n} \left\{ \frac{\|x^k\|}{\|b^k\|} \right\} \tag{2.5.12}$$

If an LU factorization of A is used to solve $Ax = b^k$, then relatively little additional effort is required for repeated solutions. The estimate of $K(A)$ in (2.5.12) is a conservative one in the sense that the exact condition number is at least as large as the estimate. Consequently, if the estimate is large then the system is in fact ill-conditioned. The test vectors $\{b^1, b^2, \ldots, b^m\}$ can be chosen to be random vectors. Alternatively, more sophisticated methods can be used to improve the estimate while still keeping m relatively small (Cline et al., 1979). The following example illustrates the use of (2.5.12) to obtain an estimate of the condition number.

EXAMPLE 2.5.5 Approximate Condition Number

Again consider the linear algebraic system from Example 2.5.1 where the coefficient matrix is

$$A = \begin{bmatrix} 100 & -200 \\ -200 & 401 \end{bmatrix}$$

Using (2.5.5), we have $\|A\| = 601$. Next suppose we solve the system with $b^1 = [1,0]^T$. In this case, the solution is $x^1 = [4.01, 2]^T$. Thus, the estimate of $K(A)$ based on $m = n/2$ samples is

$$K(A) \approx \frac{601(4.01)}{1} \approx 2410 \gg 1$$

This estimate of $K(A)$ is smaller than the actual value of 3612 found in Example 2.5.3, as expected. Still, it does provide all the information needed to conclude that the matrix A is indeed ill-conditioned.

2.5.4 Iterative Improvement

The solution techniques that we have covered thus far are all examples of *direct* methods—methods that take a fixed number of steps to produce a solution. Another class of solution techniques, called the *iterative* methods, instead use a *variable* number of steps to produce a sequence of increasingly accurate estimates of the solution. As a bridge between the direct methods and the iterative methods, we consider a hybrid method that can be applied as a final step to correct or improve the numerical solution generated by a direct method. This approach, called *iterative improvement*, is particularly useful for ill-conditioned systems where the effects of accumulated round-off error can be significant.

In general, there will be some error in a numerical solution to the system $Ax = b$. Suppose x^e denotes the *exact* solution and x denotes the *numerical* solution generated by a direct method. Then

$$x^e = x + \Delta x \tag{2.5.13}$$

where Δx is the *error* in the numerical solution. In this case neither x^e nor Δx are known, only x. However, it is clear that if Δx can be determined, then x^e can be recovered from x. To find Δx note that $Ax = Ax^e - A\Delta x$. But $Ax^e = b$, which means than we can solve for Δx by solving the following *auxiliary* system.

$$\boxed{A\Delta x = r(x)} \tag{2.5.14}$$

Recall from Equation (2.5.2) that $r(x) = b - Ax$ is the residual error vector, a quantity which is often quite small. For this reason the solution of (2.5.14) is often computed using *high precision* arithmetic. Once a solution to (2.5.14) is obtained, it can be used to correct or improve the original solution as follows:

$$\boxed{x^c = x + \Delta x} \tag{2.5.15}$$

If the exact solution to (2.5.14) is obtained, then from (2.5.13) we have $x^c = x^e$. In practice, x^c is merely an improved approximation to the solution. The application of iterative improvement is particularly suitable when *LU* decomposition of A is used. In this case, the extra computational cost of computing a solution to (2.5.14) is relatively cheap since only forward and back substitution, an n^2 process, is required.

Typically the iterative improvement step is performed only once to *tune* the solution. Of course, the procedure can be repeated by using x^c as the new starting point. If subsequent applications generate a correction Δx that has a sufficiently small norm, then the procedure has effectively converged. Usually, the original solution from a direct method is relatively close to the exact solution, in which case convergence requires, at most, a few iterations.

2.6 ITERATIVE METHODS

Iterative improvement is a hybrid technique for solving $Ax = b$ that starts with a direct method and then uses one or more iterative steps at the end to correct or improve the numerical solution. In essence the direct method is used to get an initial estimate of the solution, in most cases a very good estimate. It is also possible to start with an *arbitrary* initial estimate (an outright guess) and then proceed with as many iterative steps as needed. Iterative methods are attractive for solving very large systems where performing $n^3/3$ floating-point operations can be prohibitive. In many practical applications, large systems have coefficient matrices that are sparse with relatively few nonzero entries. Here iterative methods are particularly appealing because the sparse structure of the coefficient matrix is preserved (Kincaid, et al. 1982). Iterative methods are the methods of choice for solving certain types of linear partial differential equations using finite differences as will be seen in Chapter 9.

2.6.1 Jacobi Method

To apply any of the iterative techniques, the equations should first be *preprocessed* with a row permutation matrix P, if needed, to ensure that the diagonal elements of the coefficient matrix are nonzero. The rate of convergence can be improved if the row interchanges are selected so as to make the diagonal elements as large as possible. The coefficient matrix A is then decomposed into a sum of lower-triangular, diagonal, and upper-triangular terms as follows:

$$\boxed{A = L + D + U} \tag{2.6.1}$$

Here L is lower-triangular with zeros on the diagonal, D is diagonal with nonzero diagonal elements, and U is upper-triangular with zeros on the diagonal. Note that the L and U *terms* in Equation (2.6.1) are different from the lower- and upper-triangular *factors* used in the LU decomposition method. Indeed, the L, D, and U in (2.6.1) are obtained directly from inspection of A. Using (2.6.1) we have $Ax = Dx + (L + U)x$. But since $Ax = b$, this means that $Dx = b - (L + U)x$. This is the basis for the *Jacobi* method. In particular, if x^k denotes the kth estimate of the solution of $Ax = b$, then the $(k + 1)$th estimate is obtained as follows.

$$Dx^{k+1} = b - (L + U)x^k \quad , \quad k \geq 0 \tag{2.6.2}$$

Since D is diagonal, the Jacobi iteration in (2.6.2) can be written as a set of scalar equations expressed directly in terms of the components of A and b.

$$\boxed{x_i^{k+1} = \frac{1}{A_{ii}}\left(b_i - \sum_{j \neq i} A_{ij} x_j^k\right) \quad , \quad 1 \leq i \leq n} \tag{2.6.3}$$

To start the iterative scheme, an initial guess x^0 must be chosen. For example, if absolutely nothing is known about the solution, the initial guess $x^0 = 0$ is as good as any. Typically, the iterations are repeated until the norm of the residual error vector is sufficiently small, or until the norm of the difference between successive estimates is sufficiently small.

Although the iterative methods are easy to implement, one drawback is that they do not always converge. However, there is a simple sufficient condition that can be tested which guarantees that the estimates converge to the unique solution of $Ax = b$. Solving (2.6.2) for x^{k+1}, the norm of the coefficient of x^k must be less than unity. That is, if x^k is the kth estimate in (2.6.3) and x^0 is an *arbitrary* initial guess, then $x^k \to x$ as $k \to \infty$ if

$$\boxed{\|D^{-1}(L + U)\| < 1} \tag{2.6.4}$$

Recall from (2.5.5) that $\|D^{-1}(L + U)\|$ is the row-sum norm of the matrix $D^{-1}(L + U)$. Since D is diagonal, the condition $\|D^{-1}(L + U)\| < 1$ is equivalent to saying that for each row of A, the diagonal element must dominate the remaining elements in the following sense:

$$|A_{kk}| > \sum_{j \neq k} |A_{kj}| \quad , \quad 1 \leq k \leq n \tag{2.6.5}$$

In this case, we say that the coefficient matrix A is *strictly diagonally dominant* in a row sense. It is also possible to define diagonal dominance in a column sense. The degree to which the constraint in (2.6.4) is satisfied is a rough measure of how fast the estimates of x converge.

EXAMPLE 2.6.1 Jacobi's Iterative Method

Consider the following linear algebraic system of equations.

$$\begin{bmatrix} 5 & 0 & -2 \\ 3 & 5 & 1 \\ 0 & -3 & 4 \end{bmatrix} x = \begin{bmatrix} 7 \\ 2 \\ -4 \end{bmatrix}$$

It is clear from inspection that this system is strictly diagonally dominant. Using Equation (2.6.3), the iterative formulas in this case are

$$x_1^{k+1} = \frac{1}{5}(7 + 2x_3^k)$$

$$x_2^{k+1} = \frac{1}{5}(2 - 3x_1^k - x_3^k)$$

$$x_3^{k+1} = \frac{1}{4}(-4 + 3x_2^k)$$

Suppose we start the iterative sequence with the initial guess $x^0 = 0$. The residual error of the kth estimate is

$$r(x^k) = b - Ax^k$$

Example program *e261* on the distribution CD uses the Jacobi method to compute the first few estimates and the norms of their residual errors as summarized in the Table 2.6.1.

The convergence condition in (2.6.4) is satisfied in this case with $\|D^{-1}(L + U)\| = 0.8$. It is apparent that the estimates are converging to the exact solution, which is $x = [1, 0, -1]^T$.

TABLE 2.6.1 Jacobi Estimates

k	x_1	x_2	x_3	$\|r(x)\|$
0	0.0000000	0.0000000	0.0000000	7.0000000
1	1.4000000	0.4000000	−1.0000000	3.1999998
2	1.0000000	−0.2400000	−0.7000000	1.9200001
3	1.1200000	−0.0600000	−1.1799999	0.9600000
4	0.9280000	−0.0360000	−1.0450000	0.4410001
5	0.9820000	0.0522000	−1.0270000	0.2645998
6	0.9892000	0.0162000	−0.9608500	0.1322999
7	1.0156599	−0.0013500	−0.9878500	0.0539994
8	1.0048599	−0.0118259	−1.0010124	0.0455624
9	0.9995950	−0.0027135	−1.0088694	0.0273371

2.6.2 Gauss-Seidel Method

The convergence rate of the Jacobi method can be improved by a simple modification. The idea behind the modification is the observation that all of the components of the new estimate x^{k+1} are computed using only the current estimate x^k. This is true in spite of the fact that when x_i^{k+1} is computed in Equation (2.6.3), the updated estimates $\{x_1^{k+1}, x_2^{k+1}, \ldots, x_{i-1}^{k+1}\}$ are already available. In general, it is better to take advantage of the most recent information when updating an estimate of a solution as follows.

$$x_i^{k+1} = \frac{1}{A_{ii}}\left(b_i - \sum_{j=1}^{i-1} A_{ij}x_j^{k+1} - \sum_{j=i+1}^{n} A_{ij}x_j^k\right) \quad , \quad 1 \le i \le n \tag{2.6.6}$$

This modification of the Jacobi method is called the *Gauss-Seidel* method. The scalar equations in (2.6.6) can be written in vector form as $x^{k+1} = D^{-1}(b - Lx^{k+1} - Ux^k)$. Collecting the terms involving x^{k+1} on the left then yields

$$(D + L)x^{k+1} = b - Ux^k \quad , \quad k \ge 0 \tag{2.6.7}$$

Although (2.6.7) is a linear algebraic system that must be solved at each step, it can be solved easily with forward substitution as in (2.6.6) because the coefficient matrix $(D + L)$ is lower-triangular.

There is a simple sufficient condition which guarantees that the Gauss-Seidel estimates will converge to the unique solution of $Ax = b$. Solving for x^{k+1} in (2.6.7), the norm of the matrix coefficient of x^k must be less than unity. That is, if x^k is the kth estimate in (2.6.6) and x^0 is an arbitrary initial guess, then $x^k \to x$ as $k \to \infty$ if

$$\|(D + L)^{-1}U\| < 1 \tag{2.6.8}$$

The use of the most recently available information to update each component of x^k tends to increase the rate at which the sequence of estimates converges. The following example illustrates this point.

EXAMPLE 2.6.2 Gauss-Seidel Iterative Method

For comparison, we consider the same system of equations that was investigated in Example 2.6.1 where

$$A = \begin{bmatrix} 5 & 0 & -2 \\ 3 & 5 & 1 \\ 0 & -3 & 4 \end{bmatrix} \quad , \quad b = \begin{bmatrix} 7 \\ 2 \\ -4 \end{bmatrix}$$

From (2.6.6), the equations that must be solved at each iteration are

$$x_1^{k+1} = \frac{1}{5}(7 + x_3^k)$$

$$x_2^{k+1} = \frac{1}{5}(2 - 3x_1^{k+1} - x_3^k)$$

$$x_3^{k+1} = \frac{1}{4}(-4 + 3x_2^{k+1})$$

TABLE 2.6.2 Gauss-Seidel Estimates

k	x_1	x_2	x_3	$\|r(x)\|$
0	0.0000000	0.0000000	0.0000000	7.0000000
1	1.4000000	−0.4400000	−1.3299999	2.6599998
2	0.8680000	0.1452000	−0.8911000	0.8778000
3	1.0435599	−0.0479159	−1.0359370	0.2896738
4	0.9856252	0.0158123	−0.9881408	0.0955925
5	1.0047437	−0.0052180	−1.0039135	0.0315456
6	0.9984345	0.0017220	−0.9987085	0.0104103
7	1.0005165	−0.0005682	−1.0004262	0.0034351
8	0.9998295	0.0001875	−0.9998593	0.0011339
9	1.0000563	−0.0000619	−1.0000464	0.0003738

To facilitate comparison with Example 2.6.1, we again start the iterative sequence with the initial guess $x^0 = 0$. Example program *e262* on the distribution CD uses the Gauss-Seidel method to compute the first few estimates and the norms of their residual errors as shown in Table 2.6.2. A comparison with Table 2.6.1 reveals that the convergence to the exact solution, $x = [1, 0, -1]^T$, is clearly faster. To evaluate the convergence condition in Equation (2.6.8), we can use Gauss-Jordan elimination to compute $(D + L)^{-1}U$.

$$[D + L, U] = \left[\begin{array}{ccc|ccc} 5 & 0 & 0 & 0 & 0 & -2 \\ 3 & 5 & 0 & 0 & 0 & 1 \\ 0 & -3 & 4 & 0 & 0 & 0 \end{array}\right] \rightarrow \begin{bmatrix} 1 & 0 & 0 & 0 & 0 & -0.4 \\ 3 & 5 & 0 & 0 & 0 & 1 \\ 0 & -3 & 4 & 0 & 0 & 0 \end{bmatrix}$$

$$\rightarrow \begin{bmatrix} 1 & 0 & 0 & 0 & 0 & -0.4 \\ 0 & 5 & 0 & 0 & 0 & 2.2 \\ 0 & -3 & 4 & 0 & 0 & 0 \end{bmatrix} \rightarrow \begin{bmatrix} 1 & 0 & 0 & 0 & 0 & -0.4 \\ 0 & 1 & 0 & 0 & 0 & 0.44 \\ 0 & -3 & 4 & 0 & 0 & 0 \end{bmatrix}$$

$$\rightarrow \begin{bmatrix} 1 & 0 & 0 & 0 & 0 & -0.4 \\ 0 & 1 & 0 & 0 & 0 & 0.44 \\ 0 & 0 & 4 & 0 & 0 & 1.32 \end{bmatrix} \rightarrow \left[\begin{array}{ccc|ccc} 1 & 0 & 0 & 0 & 0 & -0.4 \\ 0 & 1 & 0 & 0 & 0 & 0.44 \\ 0 & 0 & 1 & 0 & 0 & 0.33 \end{array}\right]$$

$$= [I, (D + L)^{-1}U]$$

Thus, $\|(D + L)^{-1}U\| = 0.44$, which is indeed smaller than $\|D^{-1}(L + U)\| = 0.8$ from Example 2.6.1, hence the faster rate of convergence.

2.6.3 Relaxation Methods

The Gauss-Seidel method can be generalized by adding a new parameter (Allen, 1954). To see this, it is useful to rewrite Equation (2.6.6) by adding and subtracting the term x_i^k from the right-hand side.

$$x_i^{k+1} = x_i^k + \frac{1}{A_{ii}}\left(b_i - \sum_{j=1}^{i-1} A_{ij}x_j^{k+1} - \sum_{j=i}^{n} A_{ij}x_j^k\right), \quad 1 \le i \le n \tag{2.6.9}$$

Note how the last summation in (2.6.9) now starts at $j = i$. The term divided by A_{ii} represents the *correction* to x_i^k needed to produce x_i^{k+1}. The convergence rate can sometimes be increased by introducing a *relaxation factor* $\alpha > 0$, which modifies the size of the correction.

$$x_i^{k+1} = x_i^k + \frac{\alpha}{A_{ii}}\left(b_i - \sum_{j=1}^{i-1} A_{ij}x_j^{k+1} - \sum_{j=i}^{n} A_{ij}x_j^k\right) \quad , \quad 1 \le i \le n \qquad (2.6.10)$$

The formulation in (2.6.10) is called the *successive relaxation* or *SR* method. Note how the *SR* method reduces to the Gauss-Seidel method as a special case when $\alpha = 1$. If the relaxation factor is in the range $0 < \alpha < 1$, then the *SR* method is referred to as *under-relaxation*, whereas for $\alpha > 1$ it is called *over-relaxation*. The term in (2.6.10) corresponding to $j = i$ can be removed from the summation and combined with the x_i^k term.

$$x_i^{k+1} = (1-\alpha)x_i^k + \frac{\alpha}{A_{ii}}\left(b_i - \sum_{j=1}^{i-1} A_{ij}x_j^{k+1} - \sum_{j=i+1}^{n} A_{ij}x_j^k\right) \; , \; 1 \le i \le n \qquad (2.6.11)$$

The scalar equation in (2.6.11) can also be written in vector form as $x^{k+1} = (1-\alpha)x^k + \alpha D^{-1}(b - Lx^{k+1} - Ux^k)$. Collecting terms involving x^{k+1} on the left and x^k on the right, then results in the following vector formulation.

$$(D + \alpha L)x^{k+1} = \alpha b + [(1-\alpha)D - \alpha U]x^k \quad , \quad k \ge 0 \qquad (2.6.12)$$

Just as with the Jacobi and Gauss-Seidel methods, there is a simple sufficient condition that guarantees that the *SR* estimates converge to the unique solution of $Ax = b$. Solving for x^{k+1}, the norm of the matrix coefficient of x^k must be less than unity. That is, if x^k is the kth estimate in (2.6.12) and x^0 is an arbitrary initial guess, then $x^k \to x$ as $k \to \infty$ if

$$\|(D + \alpha L)^{-1}[(1-\alpha)D - \alpha U]\| < 1 \qquad (2.6.13)$$

Note that (2.6.13) reduces to the simpler Gauss-Seidel converge condition in (2.6.8) when $\alpha = 1$.

EXAMPLE 2.6.3 Successive Under-Relaxation

For comparison, we consider the same equations that were investigated in Examples 2.6.1 and 2.6.2 where

$$A = \begin{bmatrix} 5 & 0 & -2 \\ 3 & 5 & 1 \\ 0 & -3 & 4 \end{bmatrix} \quad , \quad b = \begin{bmatrix} 7 \\ 2 \\ -4 \end{bmatrix}$$

Using (2.6.11), the equations that must be solved at each iteration are

$$x_1^{k+1} = (1-\alpha)x_1^k + \frac{\alpha}{5}(7 - 2x_3^k)$$

$$x_2^{k+1} = (1-\alpha)x_2^k + \frac{\alpha}{5}(2 - 3x_1^{k+1} - x_3^k)$$

$$x_3^{k+1} = (1-\alpha)x_3^k + \frac{\alpha}{4}(-4 + 3x_2^{k+1})$$

TABLE 2.6.3 Successive Relaxation Residual Errors

k	$\alpha = 0.7$	$\alpha = 0.8$	$\alpha = 0.9$	$\alpha = 1.0$	$\alpha = 1.1$
0	7.0000000	7.0000000	7.0000000	7.0000000	7.0000000
1	1.3184400	0.9305599	1.5325403	2.6599998	3.8510609
2	0.3638902	0.2955074	0.0736614	0.8778000	2.2898731
3	0.1531762	0.0639529	0.0368419	0.2896738	1.4727535
4	0.0589825	0.0181828	0.0009806	0.0955925	0.9555669
5	0.0148927	0.0070615	0.0006390	0.0315456	0.6204176
6	0.0073280	0.0009792	0.0000191	0.0104103	0.4028330
7	0.0037150	0.0001864	0.0000119	0.0034351	0.2615576
8	0.0011320	0.0001135	0.0000005	0.0011339	0.1698284
9	0.0003142	0.0000210	0.0000000	0.0003738	0.1102681

To facilitate comparisons with Example 2.6.1 and Example 2.6.2, we again start the iterative sequence with the initial guess $x^0 = 0$. Example program *e263* on the distribution CD uses the successive relaxation method to compute the norms of the first few residual error vectors, for several values of α, as shown in Table 2.6.3. The case $\alpha = 1$ corresponds to the Gauss-Seidel method in Example 2.6.2. It is clear from inspection that in this instance the estimates converge fastest when $\alpha = 0.9$, which corresponds to under-relaxation.

For the SR method there is also a second very simple sufficient condition for convergence that does not involve norms. Recall that A is symmetric if $A^T = A$ and positive-definite if $x^T Ax > 0$ for $x \neq 0$. If the coefficient matrix A is a symmetric positive-definite matrix, then the SR method in Equation (2.6.11) converges to the unique solution of $Ax = b$ starting from an arbitrary initial guess x^0 for *all* relaxation factors in the following open interval (see Varga, 1962).

$$\boxed{0 < \alpha < 2} \tag{2.6.14}$$

EXAMPLE 2.6.4 Successive Over-Relaxation

In many instances, successive over-relaxation or SOR generates superior convergence. To illustrate this, consider the following linear algebraic system:

$$\begin{bmatrix} 1 & 2 & 3 \\ 2 & 9 & 6 \\ 3 & 6 & 27 \end{bmatrix} x = \begin{bmatrix} 2 \\ -4 \\ 24 \end{bmatrix}$$

Since this coefficient matrix is a symmetric positive-definite matrix, from Equation (2.6.14) the SR method converges, starting from an arbitrary initial guess x^0 for all relaxation parameters in the interval $(0, 2)$. Suppose $x^0 = 0$. Let $N(\alpha)$ be the number of iterations needed to converge using the following convergence criterion.

$$\|r(x^k)\| < \varepsilon$$

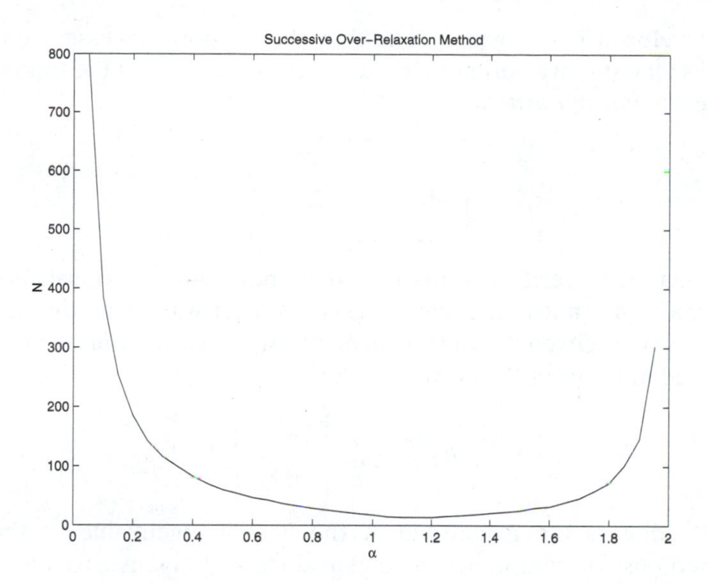

FIGURE 2.6.1 Optimal Relaxation Factor for Example 2.6.4

Here $r(x) = b - Ax$ is the residual error vector and $\varepsilon > 0$ is an error tolerance. Example program *e264* on the distribution CD uses $\varepsilon = 10^{-5}$, and computes $N(\alpha)$ for 40 values of the relaxation parameter uniformly distributed over (0,2). The resulting plot of $N(\alpha)$ is shown in Figure 2.6.1. In this case, the optimal relaxation factor is $\alpha^* = 1.15$, which corresponds to successive over-relaxation. Observe that any value of α in the range from 0.4 to 1.8 works quite well for this low-dimensional example.

2.6.4 Convergence

The three iterative methods that we have examined can all be formulated in the following general way where the $n \times 1$ vector x^0 is an initial guess.

$$x^{k+1} = Bx^k + c \quad , \quad k \geq 0 \tag{2.6.15}$$

The iteration matrix, B, and offset vector, c, corresponding to the Jacobi method, the Gauss-Seidel method, and the *SR* method are summarized in Table 2.6.4.

A simple sufficient condition for the iterative methods to converge starting from an arbitrary initial guess is that the $\|B\| < 1$ where $\|B\|$ is the row-sum norm defined in Equation (2.5.5). By using a more sophisticated formulation for the matrix norm, it is

TABLE 2.6.4 Explicit Formulation of Iterative Methods

Method	*B*	*c*
Jacobi	$-D^{-1}(L + U)$	$D^{-1}b$
Gauss-Seidel	$-(D + L)^{-1}U$	$(D + L)^{-1}b$
Successive Relaxation	$(D + \alpha L)^{-1}[(1 - \alpha)D - \alpha U]$	$\alpha(D + \alpha L)^{-1}b$

possible to develop a convergence condition that is both necessary and sufficient. Rather than use the infinity norm of a vector x, consider instead the following alternative called the *Euclidean norm* of x.

$$\|x\|_2 \triangleq \sqrt{\sum_{k=1}^{n} x_k^2} \tag{2.6.16}$$

It is a simple matter to verify that the Euclidean norm in (2.6.16) satisfies the fundamental properties of a norm in Table 2.5.1. A compact way of writing the Euclidean norm is $\|x\|_2 = \sqrt{x^T x}$. Given the Euclidean norm, the *spectral radius* of an $n \times n$ matrix B can be defined in terms of the norm as follows:

$$\rho(B) \triangleq \max_{x \neq 0} \left\{ \frac{\|Bx\|_2}{\|x\|_2} \right\} \tag{2.6.17}$$

The spectral radius is the magnitude of the largest eigenvalue of the matrix B. Numerical methods for computing the eigenvalues and eigenvectors of a matrix are covered in detail in Chapter 3. The importance of the spectral radius lies in the following fact. If A is nonsingular, then the general iterative method in (2.6.15) converges to the unique solution x, starting from an arbitrary initial guess x^0, if and only if

$$\rho(B) < 1 \tag{2.6.18}$$

Therefore, (2.6.18) provides a convergence condition that is both necessary and sufficient. It is a less conservative condition than the sufficient condition $\|B\| < 1$ because $\rho(B) \leq \|B\|$.

The successive relaxation or *SR* method is the most effective of the three iterative methods in Table 2.6.4. However, there remains the question of how to choose a suitable value for the relaxation parameter α. The smaller the spectral radius, $\rho(B)$, the faster the iterative method converges. Consequently, the optimal value, α_{opt}, is the value of α which minimizes $\rho(B)$. Let ρ_J denote the spectral radius associated with Jacobi's method.

$$\rho_J = \rho[D^{-1}(L + U)] \tag{2.6.19}$$

For the special case when A is a symmetric positive-definite tridiagonal matrix, the optimal value for the relaxation parameter can be expressed in terms of ρ_J as follows (Young, 1971):

$$\alpha_{opt} = \frac{2}{1 + \sqrt{1 - \rho_J^2}} \tag{2.6.20}$$

If ρ_J lies in the interval (0,1), then it is clear from (2.6.20) that the optimal relaxation parameter lies in the interval (1,2) which corresponds to over-relaxation. When $\alpha = \alpha_{opt}$, the spectral radius of the successive over-relaxation method is (Young, 1971)

$$\rho_{SR} = \left(\frac{\rho_J}{1 + \sqrt{1 - \rho_J^2}} \right)^2$$

The spectral radius of the Gauss-Seidel method can be expressed as $\rho_{GS} = \rho_J^2$, thus confirming that the Gauss-Seidel method is faster than the Jacobi method (Young, 1971). In the next chapter we examine several numerical techniques for computing the spectral radius, $\rho(B)$.

2.7 APPLICATIONS

The following examples illustrate applications of linear algebraic system techniques using both MATLAB and C. The relevant MATLAB functions in the NLIB toolbox are described in Section 1.3 of Appendix 1, while the corresponding C functions in the NLIB library are described in Section 2.6 of Appendix 2.

2.7.1 Chemical Absorption Process: MATLAB

Chemical separation processes typically consist of a sequence of stages in which materials are brought into contact with one another. Examples include distillation, extraction, and absorption processes. An absorption process with $n = 5$ stages is shown in Figure 2.7.1. Here a liquid is introduced at the top with a molar flow rate of L, and a gas is introduced at the bottom with a molar flow rate of G. For example a liquid absorbent might be used to remove sulfur dioxide (SO_2) from combustion gas. Here x_k and y_k denote the concentration of the absorbed component in the liquid and gas, respectively, at stage k where it is assumed that

$$y_k = ax_k + b \quad , \quad 1 \le k \le n \tag{2.7.1}$$

Let x_f denote the concentration of the absorbed component in the feed liquid, y_f denote the concentration of the absorbed component in the feed gas, and H denote the liquid holdup. Using simplifying assumptions and applying a component material

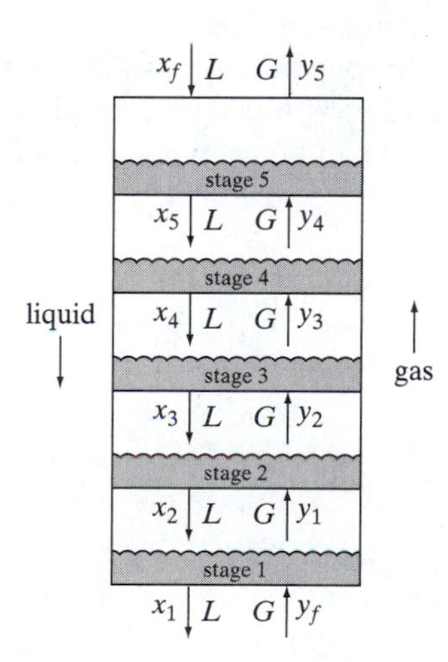

FIGURE 2.7.1 A Five-Stage Chemical Absorption Process

balance at each stage then yields the following equations for the absorption process (Seborg et al., 1989).

$$\tau \frac{dx_1}{dt} = K(y_f - b) - (1 + \delta)x_1 + x_2$$

$$\tau \frac{dx_2}{dt} = \delta x_1 - (1 + \delta)x_2 + x_3$$

$$\tau \frac{dx_3}{dt} = \delta x_2 - (1 + \delta)x_3 + x_4$$

$$\tau \frac{dx_4}{dt} = \delta x_3 - (1 + \delta)x_4 + x_5$$

$$\tau \frac{dx_5}{dt} = \delta x_4 - (1 + \delta)x_5 + x_f$$

Here $\tau = H/L$ is the stage liquid residence time, $\delta = aG/L$ is the stripping factor, and $K = G/L$ is the gas to liquid ratio. To analyze the steady-state operation of the absorption process (the constant solutions), we set $dx_k/dt = 0$ for $1 \leq k \leq n$. Let x be the 5×1 liquid concentration vector. Moving the terms that do not involve x_k to the right side of the equations, the system of equations can be expressed as a coefficient matrix: A times the liquid concentration vector x equals a right-hand-side vector c.

$$\begin{bmatrix} -(1+\delta) & 1 & 0 & 0 & 0 \\ \delta & -(1+\delta) & 1 & 0 & 0 \\ 0 & \delta & -(1+\delta) & 1 & 0 \\ 0 & 0 & \delta & -(1+\delta) & 1 \\ 0 & 0 & 0 & \delta & -(1+\delta) \end{bmatrix} x = \begin{bmatrix} K(b - y_f) \\ 0 \\ 0 \\ 0 \\ -x_f \end{bmatrix} \qquad (2.7.2)$$

Notice the special tridiagonal structure of the coefficient matrix A. It should be clear from inspection how to extend this model beyond five stages.

To make the chemical absorption example specific, suppose the process parameters are $a = 0.72, b = 0, K = 1.63, x_f = 0.01$, and $y_f = 0.06$ (Wong and Luus, 1980). The following MATLAB script on the distribution CD can be used to solve this problem.

```
%----------------------------------------------------------------------
% Example 2.7.1: Chemical Absorption Process
%----------------------------------------------------------------------

% Initialize

   clc                         % clear command window
   clear                       % clear variables
   n = 5;                      % number of stages
   a = 0.72;                   % slope
   b = 0.0;                    % intercept
```

```
   K = 1.63;               % gas to liquid ratio
   xf = 0.01;              % liquid feed concentration
   yf = 0.06;              % gas feed concentration
   delta = a*K;            % stripping factor
   A = zeros (n,n);
   c = zeros (n,1);

% Compute A and c

   fprintf ('Example 2.7.1: Chemical Absorption Process\n');
   for i = 1 : n
      A(i,i) = -(1 + delta);
      if i < n
         A(i,i+1) = 1;
         A(i+1,i) = delta;
      end
   end
   c(1) = K*(b - yf);
   c(n) = -xf;
   show ('A',A)
   show ('c',c)

% Find liquid concentrations

   show ('K(A)',condnum(A,0))
   x = gauss (A,c);
   show ('x',x)
   y = a*x + b;
   show ('y',y)
   show ('||r(x)||',residual(A,c,x))
%----------------------------------------------------------------------
```

When script *e271.m* is executed, it uses the Gaussian elimination to generate the following solutions for the liquid concentrations x and the gas concentrations y.

$$x = \begin{bmatrix} 0.07544 \\ 0.06618 \\ 0.05531 \\ 0.04255 \\ 0.02757 \end{bmatrix} \qquad y = \begin{bmatrix} 0.05432 \\ 0.04765 \\ 0.03982 \\ 0.03063 \\ 0.01985 \end{bmatrix} \tag{2.7.3}$$

It is evident from y that the concentration of the substance in the gas goes down with each stage as desired. Similarly, from x it is clear that as the liquid goes from stage five to stage one (see Figure 2.7.1), the concentration increases as the substance is absorbed. In this case, the condition number of the coefficient matrix A was found to be $K(A) = 17.6996$, and the norm of the residual error of the solution was $\|r(x)\| = 1.38778 \times 10^{-17}$. A solution to Example 2.7.1 using C can be found in file *e271.c* on the distribution CD.

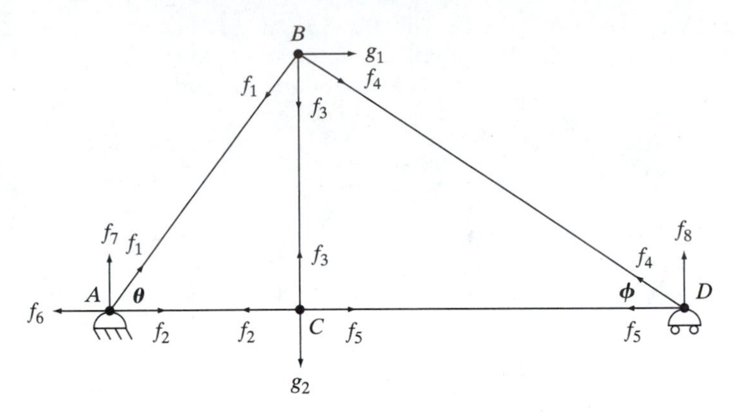

FIGURE 2.7.2 A Planar Truss with Five Members and Four Joints

2.7.2 Planar Truss: C

A *truss* is a mechanical structure that consists of rigid members interconnected to one another through joints. Trusses are common in the construction of bridges and other applications where mechanical support is required. A simple example of a planar truss is shown in Figure 2.7.2. This is a *statically determinate* truss with the number of members, $m = 5$, related to the number of joints, $j = 4$, by the relationship

$$m = 2j - 3 \tag{2.7.4}$$

The internal forces in the five members are $\{f_1, f_2, \ldots, f_5\}$. As drawn, they are all treated as compression forces because the arrows point away from the joints. If the solution is negative, this represents a tension force. The left end of the truss is supported by a fixed joint with external horizontal and vertical forces, f_6 and f_7, respectively, while the right end is supported by a roller joint with vertical external force f_8. Two external load forces are placed on the truss, a horizontal load g_1 at joint B, say from the wind, and a vertical load g_2 at joint C, say from an object supported by the truss.

Given values for the load forces, the problem is to find the five internal member forces and the three external support forces, assuming static equilibrium conditions hold at each joint. This is done by setting the sum of the horizontal forces to zero and the sum of the vertical forces to zero at each joint. Let θ be the angle between the members at joint A, and let ϕ be the angle between the members at joint D. If joint C is directly below joint B, summing the horizontal and vertical forces at each joint yields the following equations.

$$
\begin{aligned}
f_1 \cos\theta + f_2 - f_6 &= 0 \quad \}\text{joint A} \\
f_1 \sin\theta + f_7 &= 0 \\
-f_1 \cos\theta + f_4 \cos\phi + g_1 &= 0 \quad \}\text{joint B} \\
-f_1 \sin\theta - f_3 - f_4 \sin\phi &= 0 \\
-f_2 + f_5 &= 0 \quad \}\text{joint C} \\
f_3 - g_2 &= 0 \\
-f_4 \cos\phi - f_5 &= 0 \quad \}\text{joint D} \\
f_4 \sin\phi + f_8 &= 0
\end{aligned}
$$

Let f denote the 8×1 vector of unknown forces. By moving the load forces g_1 and g_2 to the right-hand side, this linear system can be rewritten as a coefficient matrix A times the force vector f equals a load vector b.

$$\begin{bmatrix} \cos\theta & 1 & 0 & 0 & 0 & -1 & 0 & 0 \\ \sin\theta & 0 & 0 & 0 & 0 & 0 & 1 & 0 \\ -\cos\theta & 0 & 0 & \cos\phi & 0 & 0 & 0 & 0 \\ -\sin\theta & 0 & -1 & -\sin\phi & 0 & 0 & 0 & 0 \\ 0 & -1 & 0 & 0 & 1 & 0 & 0 & 0 \\ 0 & 0 & 1 & 0 & 0 & 0 & 0 & 0 \\ 0 & 0 & 0 & -\cos\phi & -1 & 0 & 0 & 0 \\ 0 & 0 & 0 & \sin\phi & 0 & 0 & 0 & 1 \end{bmatrix} f = \begin{bmatrix} 0 \\ 0 \\ -g_1 \\ 0 \\ 0 \\ g_2 \\ 0 \\ 0 \\ 0 \end{bmatrix} \tag{2.7.5}$$

To make the truss example specific, suppose the angle at joint A is $\theta = \pi/3$ and the angle at joint D is $\phi = \pi/6$. Let the horizontal load force at joint B be $g_1 = 250$ N and the vertical load force at joint C be $g_2 = 1500$ N. The following C program on the distribution CD can be used to solve this problem.

```
/* ------------------------------------------------------------------ */
/* Example 2.7.2: Planar Truss                                        */
/* ------------------------------------------------------------------ */

#include "c:\nlib\util.h"

int main (void)
{
   int           n = 8;                          /* number of forces */
   float         theta = PI/3.0f,                /* joint A angle */
                 phi   = PI/6.0f;                /* joint D angle */
   vector        f = vec (n,""),
                 g = vec (n,"0 0 -250 0 0 1500 0 0 0");
   matrix        A = mat (n,n,"");

/* Initialize A */

   printf ("\nExample 2.7.2: Planar Truss\n");
   A[1][1] = (float) cos(theta);
   A[2][1] = (float) sin(theta);
   A[3][1] = (float) -cos(theta);
   A[4][1] = (float) -sin(theta);
   A[3][4] = (float) cos(phi);
   A[4][4] = (float) -sin(phi);
   A[8][4] = (float) sin(phi);
   A[7][4] = (float) -cos(phi);
   A[1][2] = A[2][7] = A[5][5] = A[6][3] = A[8][8] = 1;
   A[1][6] = A[4][3] = A[5][2] = A[7][5] = -1;
```

```
   showmat ("A",A,n,n,0);
   showvec ("g",g,n,0);

/* Find forces */

   shownum ("K(A)",condnum(A,n,0));
   ludec(A,g,n,f);
   showvec ("f",f,n,0);
   shownum ("||r(f)||",residual(A,g,f,n));
   return 0;
}
/* ------------------------------------------------------------- */
```

When program *e272.c* is executed, it uses the *LU* decomposition method to produce the following solution for the internal and external forces:

$$f = [-1174 \quad 837 \quad 1500 \quad -966.5 \quad 837 \mid 250 \quad 1017 \quad 483.3]^T \text{N} \tag{2.7.6}$$

The condition number of the coefficient matrix A was found to be $K(A) = 10$, and the norm of the residual error of the solution was $\|r(f)\| = 1.221 \times 10^{-4}$ N. Note that member forces f_2, f_3, and f_5 are positive and therefore compressions, while member forces f_1 and f_4 represent tensions. A solution to Example 2.7.2 using MATLAB can be found in file *e272.m* on the distribution CD.

2.7.3 DC Bridge Circuit: MATLAB

Linear circuits consist of an interconnection of resistors (R), capacitors (C), inductors (L), and voltage and current sources. If the sources used to power the circuit are DC sources that produce constant voltages and currents, the circuit will achieve a steady-state operation in which all voltages and currents are constant. To analyze the steady-state solution of a linear circuit, inductors are replaced by short circuits (no voltage drop) and capacitors are replaced by open circuits (no current flow). In this case the circuit reduces to one with resistors and constant sources. An example of a DC bridge circuit driven by a single constant voltage source is shown in Figure 2.7.3.

Suppose we are interested in solving for the voltages at the nodes labeled 1 through 3. The node labeled 0 is the *ground* or reference node and all voltages are expressed with respect to this node. Once the node voltages are known the current through any resistor can be easily computed using Ohm's law,

$$I = \frac{V}{R} \tag{2.7.7}$$

where V denotes the voltage drop across the resistor. Let v_k denote the voltage at node k for $1 \le k \le 3$. The relationships between the node voltages are obtained by applying Kirchhoff's current law, which says that the sum of the currents leaving each node must be zero. This yields the following system of three equations in three unknowns.

$$\frac{v_1 - E}{R_s} + \frac{v_1 - v_2}{R_1} + \frac{v_1 - v_3}{R_2} = 0\}\text{node 1}$$

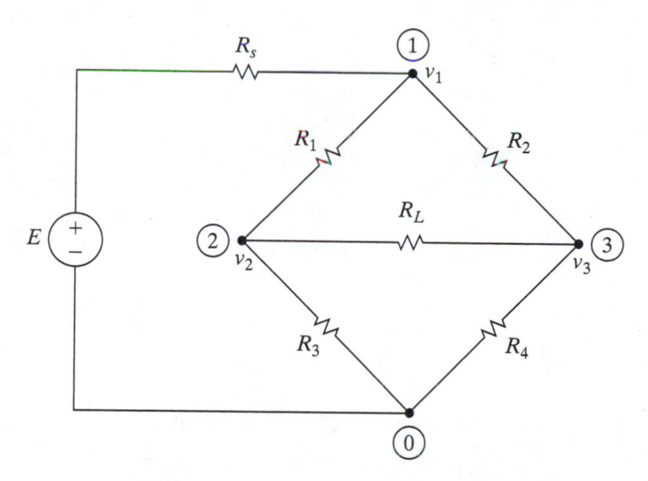

FIGURE 2.7.3 A DC Bridge Circuit

$$\frac{v_2 - v_1}{R_1} + \frac{v_2 - v_3}{R_L} + \frac{v_2}{R_3} = 0\} \text{ node 2}$$

$$\frac{v_3 - v_1}{R_2} + \frac{v_3 - v_2}{R_L} + \frac{v_3}{R_4} = 0\} \text{ node 3}$$

Let v denote the 3×1 node voltage vector. By moving the source voltage term E to the right-hand side and combining the coefficients of v_k, this linear system can be rewritten as a coefficient matrix A times the voltage vector v equals a source vector b. The final equations are simplified by replacing each resistance by its equivalent conductance: $G_s = 1/R_s$, $G_L = 1/R_L$, and $G_k = 1/R_k$ for $1 \le k \le 4$. This yields the following linear algebraic system.

$$\begin{bmatrix} G_s + G_1 + G_2 & -G_1 & -G_2 \\ -G_1 & G_1 + G_3 + G_L & -G_L \\ -G_1 & -G_L & G_2 + G_4 + G_L \end{bmatrix} v = \begin{bmatrix} G_s E \\ 0 \\ 0 \end{bmatrix} \qquad (2.7.8)$$

Unlike the two previous examples, the 3×3 coefficient matrix A in this case is not sparse because there is no pattern of zeros. To make the DC circuit example specific, suppose the source resistance is $R_s = 10\Omega$, the load resistance is $R_L = 1000\Omega$, the bridge resistances are $R_1 = 220\Omega$, $R_2 = 330\Omega$, $R_3 = 470\Omega$ and $R_4 = 560\Omega$, and the applied voltage is $E = 12$ V. The following MATLAB script on the distribution CD can be used to solve this problem.

```
%--------------------------------------------------------------------------
% Example 2.7.3: DC Bridge Circuit
%--------------------------------------------------------------------------

% Initialize

   clc                        % clear screen
   clear                      % clear variables
```

```
    n = 3;                      % number of nodes
    E  = 12;                    % applied voltage
    Gs = 1/10;                  % source conductance
    GL = 1/1000;                % load conductance
    G1 = 1/220;                 % bridge conductances
    G2 = 1/330;
    G3 = 1/470;
    G4 = 1/560;
    A = zeros (n,n);
    b = zeros (n,1);

% Initialize A and b

    fprintf ('Example 2.7.3: DC Bridge Circuit\n');
    A(1,1) = Gs + G1 + G1;
    A(1,2) = -G1;
    A(1,3) = -G2;
    A(2,1) = -G1;
    A(2,2) = G1 + G3 + GL;
    A(2,3) = -GL;
    A(3,1) = -G1;
    A(3,2) = -GL;
    A(3,3) = G2 + G4 + GL;
    b(1) = Gs*E;
    show ('A',A)
    show ('b',b)

% Find node voltages

    show ('K(A)',condnum(A,0))
    v = gauss (A,b);
    show ('v',v)
    show ('||r(v)||',residual(A,b,v))
%-------------------------------------------------------------
```

When script *e273.m* is executed, it uses Gaussian elimination to generate the following solution for the node voltages.

$$v = [11.6364 \quad 8.2636 \quad 10.5152]^T \text{ V} \tag{2.7.9}$$

The condition number of the coefficient matrix A was $K(A) = 25.4033$, and the norm of the residual error of the solution was $\|r(v)\| = 6.93889 \times 10^{-18}$ A. A solution to Example 2.7.3 using C can be found in file *e273.c* on the distribution CD.

2.7.4 Mass-Spring-Damper System: C

Physical systems that exhibit translational mechanical motion are modeled with masses interconnected by springs and viscous dampers. An example of a translational mechanical system with two masses is shown in Figure 2.7.4. Applying Newton's second law to

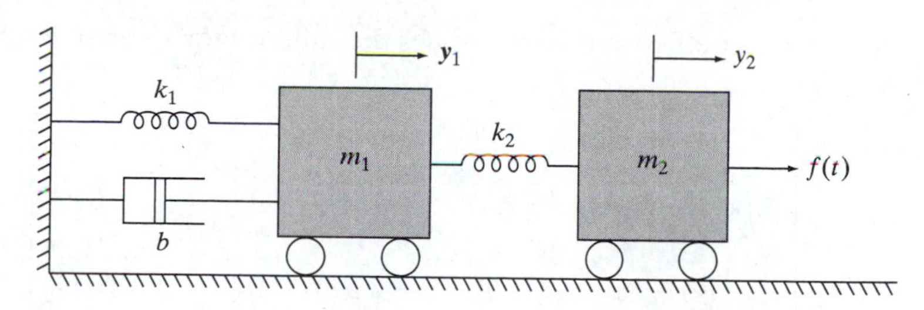

FIGURE 2.7.4 A Translational Mechanical System

each mass, we set mass times acceleration equal to the sum of the forces acting on that mass, which yields

$$m_1 \frac{dy_1^2}{dt^2} = k_2(y_2 - y_1) - k_1 y_1 - b\frac{dy_1}{dt} \text{ } \} \text{ mass 1} \tag{2.7.10a}$$

$$m_2 \frac{dy_2^2}{dt^2} = f(t) - k_2(y_2 - y_1) \text{ } \} \text{ mass 2} \tag{2.7.10b}$$

Here y_1 and y_2 denote the positions of masses m_1 and m_2, respectively; b is the viscous friction coefficient of the damper; k_1 and k_2 are the spring constants; and $f(t)$ is the applied force. This system of two second-order differential equations can be recast as a system of four first-order differential equations by introducing a new vector of variables called *state variables* consisting of the positions and velocities of the two masses (Ogata, 1997).

$$x = [y_1, dy_1/dt, y_2, dy_2/dt]^T \tag{2.7.11}$$

The technique for reformulating a higher-order differential equation as a system of first-order differential equations is explained in detail in Section 8.1. In this case, to derive the kth equation we differentiate the kth component of x with respect to time t. Substituting into the resulting expression using Equations (2.7.11) and (2.7.10), the right-hand side of each equation can be expressed entirely in terms of x and f as follows.

$$\frac{dx_1}{dt} = x_2$$

$$m_1\left(\frac{dx_2}{dt}\right) = -(k_1 + k_2)x_1 - bx_2 + k_2 x_3$$

$$\frac{dx_3}{dt} = x_4$$

$$m_2\left(\frac{dx_4}{dt}\right) = k_2 x_1 - k_2 x_3 + f$$

Suppose a constant force $f(t) = F$ is applied. The masses will undergo some transient motion after which time they will settle to constant steady-state positions. To analyze the steady-state solution of the mechanical translation system, we set $dx_k/dt = 0$ for $1 \le k \le 4$. Moving the force term to the right side of the equations, this linear system can be written as a coefficient matrix A times the state vector x equals a right-hand side vector b.

$$\begin{bmatrix} 0 & 1 & 0 & 0 \\ -(k_1+k_2)/m_1 & -b/m_1 & k_2/m_1 & 0 \\ 0 & 0 & 0 & 1 \\ k_2/m_2 & 0 & -k_2/m_2 & 0 \end{bmatrix} x = \begin{bmatrix} 0 \\ 0 \\ 0 \\ -F/m_2 \end{bmatrix} \tag{2.7.12}$$

Suppose we are interested in determining the steady-state distance between the masses, $\Delta y = y_2 - y_1$, when a constant force F is applied. To make the mechanical translation example specific, suppose the masses are $m_1 = 15$ kg and $m_2 = 30$ kg, the spring constants are $k_1 = 3.2$ N/m and $k_2 = 2.4$ N/m, the damping coefficient is $b = 0.5$ N-s/m, the applied force is $F = 1.2$ N, and the equilibrium distance between the masses when no force is applied is $d = 0.5$ m. The following C program on the distribution CD can be used to solve this problem.

```
/* ---------------------------------------------------------------- */
/* Example 2.7.4: Mass-Spring-Damper System                         */
/* ---------------------------------------------------------------- */

#include "c:\nlib\util.h"

int main (void)
{
   int          n = 4;                /* system order */
   float        m1 = 15.0f,           /* mass 1 */
                m2 = 30.0f,           /* mass 2 */
                k1 = 3.2f,            /* spring constant 1 */
                k2 = 2.4f,            /* spring constant 2 */
                b = 0.5f,             /* viscous friction coefficient */
                F = 1.2f;             /* applied force */
   vector       x = vec (n,""),
                y = vec (n,"");
   matrix       A = mat (n,n,"");

/* Initialize A and y */

   printf ("\nExample 2.7.4: Mass-Spring-Damper System\n");
   A[1][2] = A[3][4] = 1;
   A[2][1] = -(k1+k2)/m1;
   A[2][2] = -b/m1;
   A[2][3] = k2/m1;
   A[4][1] = k2/m2;
   A[4][3] = -k2/m2;
   y[4] = -F/m2;
   showmat ("A",A,n,n,0);
   showvec ("y",y,n,0);
```

```
/* Find positions and velocities */

   shownum ("K(A)",condnum(A,n,0));
   ludec(A,y,n,x);
   showvec ("x",x,n,0);
   shownum ("||r(x)||",residual(A,y,x,n));
   return 0;
}
/* ---------------------------------------------------------------- */
```

When program *e274.c* is executed, it uses the *LU* decomposition method to produce the following solution for the state vector.

$$x = [0.375 \quad 0 \quad 0.875 \quad 0]^T \tag{2.7.13}$$

Note that $x_2 = x_4 = 0$ as expected because these components represent velocities and the masses are at rest in the steady state. The steady-state separation between the two masses is $\Delta y = d + (x_3 - x_1) = 1$ m. In this case, the condition number of the coefficient matrix A was found to be $K(A) = 26.72$, and the norm of the residual error of the solution was $\|r(x)\| = 3.949 \times 10^{-9}$ N/kg. A solution to Example 2.7.4 using MATLAB can be found in file *e274.m* on the distribution CD.

2.8 SUMMARY

Numerical methods for solving the linear algebraic system, $Ax = b$, are summarized in Table 2.8.1. The last column indicates the approximate number of floating-point multiplications and divisions or FLOPs required, assuming the number of unknowns, n, is large.

The first three methods in Table 2.8.1 are direct methods based on elementary row operations. They all use partial pivoting to detect a singular matrix, prevent division by zero, and reduce accumulated round-off error. Round-off error is also reduced by prescaling each equation so the largest element in each row of A has a magnitude of one. This reduces the condition number of a poorly scaled system.

TABLE 2.8.1 Linear Algebraic System Techniques

Method	*Type*	*FLOPs*
Gauss-Jordan elimination	direct	$n^3/2$
Gaussian elimination	direct	$n^3/3$
LU decomposition	direct	$n^3/3$
Tridiagonal *LU* decomposition	direct	$5n - 4$
Jacobi	iterative	—
Gauss-Seidel	iterative	—
Successive relaxation	iterative	—

The Gauss-Jordan elimination method uses complete elimination to row-reduce the augmented $n \times (n+1)$ matrix as follows.

$$[A, b] \rightarrow [I, x] \tag{2.8.1}$$

Complete elimination requires approximately $n^3/2$ FLOPs to find x. If b is replaced by the identity matrix I, then Gauss-Jordan elimination can be used to find the inverse of A using approximately $3n^3/2$ FLOPs.

The Gaussian elimination technique uses partial elimination to row-reduce the augmented matrix to the form $[D, e]$ where D is upper-triangular.

$$[A, b] \rightarrow [D, e] \tag{2.8.2}$$

The solution to $Dx = e$ is obtained using back substitution. The use of partial elimination results in a reduction in the number of FLOPs from $n^3/2$ to $n^3/3$. The determinant of A can also be computed by Gaussian elimination using $n^3/3$ FLOPs.

The LU decomposition method is a variation of the Gaussian elimination method that factors a row-permuted version of A into the product LU where L is lower-triangular and U is upper-triangular.

$$PA = LU \tag{2.8.3}$$

The solution is then obtained using forward and back substitution. The number of FLOPs required for large values of n is again $n^3/3$. The advantage of the LU decomposition method is that, once A is factored, the system can be solved many times with different right-hand side vectors. In this case only n^2 additional FLOPs are required for each solution. Using this technique, a tridiagonal system can be solved using only $5n - 4$ FLOPs, and the inverse of A can be computed using approximately $4n^3/3$ FLOPs.

The next three solution methods in Table 2.8.1 are indirect or iterative methods. For each of these methods, row permutations should first be performed on $[A, b]$ to ensure that the diagonal elements of A are all nonzero. The iterative methods decompose A into a sum of lower-triangular, diagonal, and upper-triangular terms.

$$A = L + D + U \tag{2.8.4}$$

Jacobi's method is the simplest iterative method, but it also has the slowest rate of convergence. The Gauss-Seidel method requires solving a lower-triangular system at each iteration using forward substitution. It converges in fewer iterations than Jacobi's method because it makes use of more recent information.

The successive relaxation or SR method is a generalization of the Gauss-Seidel method, which includes a relaxation parameter α. When $\alpha = 1$, the SR method reduces to the Gauss-Seidel method. Faster converge rates can often be obtained for values of $\alpha > 1$ which is successive over-relaxation, and sometimes for values of $\alpha < 1$ which is successive under-relaxation. When the coefficient matrix A is a symmetric positive-definite matrix, the SR method is guaranteed to converge for any relaxation parameter in the open interval

$$0 < \alpha < 2 \tag{2.8.5}$$

Iterative methods are useful for solving very large systems where performing $n^3/3$ FLOPs may be prohibitively expensive. They are particularly attractive for large sparse systems because, unlike the direct methods, they preserve the sparse structure of the coefficient matrix. An important class of applications for iterative methods is the solution of linear partial differential equations using the finite-difference method discussed in Chapter 9.

PROBLEMS

The problems are divided into Analysis problems, which can be solved by hand, and Computation problems, which require the use of MATLAB or C. Solutions to selected problems can be found in Appendix 4. Students are encouraged to use these selected problems, which are identified with an (S), as a check on their understanding of the material. Problems marked with a (P) are programming problems that require the student to implement one or more of the algorithms discussed in the chapter. The remaining Computation problems require the student to write a main program that uses one or more of the NLIB functions discussed in Appendix 1 or Appendix 2.

Analysis

2.1 Consider the following system of linear algebraic equations. Write these equations in matrix form. Specify the coefficient matrix A and the right-hand-side vector b.

$$\begin{aligned} x_1 - 2x_2 + 4x_5 - 3 &= 0 \\ 3x_2 + 2x_3 - x_4 + 1 &= 0 \\ 2x_1 + x_2 - 4x_3 + 3x_4 - 6 &= 0 \\ x_3 + x_4 + x_5 &= 0 \\ x_1 - x_2 + x_3 - 5x_4 + x_5 + 4 &= 0 \end{aligned}$$

2.2 Specify the coefficient matrix A and the right-hand-side vector b for the following linear algebraic systems.

(a) The robotic arm in Equation (2.1.1) when $x = [F_x, F_y]^T$

(b) The converter circuit in Equation (2.1.3) when $x = [i_1, i_2, \ldots, i_8]^T$

(c) The DC motor in Equation (2.1.6) when $x = [I_a, \omega]^T$

2.3 Consider the following linear algebraic system. Find the solution using the Gauss-Jordan method. For convenience, you can use any nonzero pivot.

$$\begin{bmatrix} 2 & -1 & 1 \\ 1 & 0 & 3 \\ 0 & 4 & -2 \end{bmatrix} x = \begin{bmatrix} 6 \\ -2 \\ 4 \end{bmatrix}$$

2.4 Find the inverse of the following matrix, using the Gauss-Jordan method. For convenience, you can use any nonzero pivot.

$$A = \begin{bmatrix} 1 & 0 & 2 \\ 2 & -1 & 3 \\ 4 & 1 & 8 \end{bmatrix}$$

2.5 Most computers have a floating-point coprocessor chip that makes the computational effort for multiplication and division about the same as for addition and subtraction. Determine the number of floating-point additions and subtractions needed for the Gauss-Jordan method in Alg. 2.2.1. What does it approach when $n \gg 1$?

2.6 Consider the following linear algebraic system. Find the solution using the Gaussian elimination method. For convenience, you can use any nonzero pivot.

$$\begin{bmatrix} -1 & 2 & 0 \\ 4 & 1 & 2 \\ 0 & -1 & 3 \end{bmatrix} x = \begin{bmatrix} -3 \\ 7 \\ 7 \end{bmatrix}$$

2.7 Find the determinant of the matrix in problem 2.6 using Gaussian elimination. For convenience, you can use any nonzero pivot.

2.8 Determine the number of floating-point additions and subtractions needed for the Gaussian elimination method in Alg. 2.3.1. What does it approach when $n \gg 1$?

2.9 Find the infinity norm of the following vector.

$$x = [2, 0, -3, 8, 5, -4, 2, -1]^T$$

2.10 Find the row-sum norm of the following matrix.

$$A = \begin{bmatrix} 2 & 1 & -3 & 4 & 2 \\ -3 & 2 & 5 & -3 & 1 \\ 8 & 1 & -3 & 2 & 4 \\ -4 & 2 & 3 & -1 & 5 \\ 6 & 2 & 1 & -5 & 9 \end{bmatrix}$$

(S)2.11 Consider the following coefficient matrix.

$$A = \begin{bmatrix} 10 & -1 \\ 0.1 & 0.01 \end{bmatrix}$$

(a) Find the condition number $K(A)$.

(b) Find the condition number $K(A)$ after equilibration.

2.12 Consider the following matrix. Compute the LU factorization of A. That is, find Q, and P.

$$A = \begin{bmatrix} 2 & -1 & 0 \\ 0 & 4 & 3 \\ 1 & 0 & -1 \end{bmatrix}$$

2.13 Use the LU decomposition method to solve the system $Ax = b$ if A is as given in problem 2.12 and

$$b = [3, 2, -1]^T$$

(S)2.14 Find the inverse of the following matrix A by reducing $C = [A, I]$ to upper-triangular form with elementary row operations and then solving for each column of A^{-1} using back substitution.

$$A = \begin{bmatrix} 2 & -1 & 0 \\ 0 & 4 & 3 \\ 4 & 0 & -1 \end{bmatrix}$$

2.15 Determine the number of floating-point additions and subtractions needed for the LU factorization in Alg. 2.4.1. What does it approach when $n \gg 1$?

2.16 Consider the following linear algebraic system. Suppose $x = [2.8, -0.1, 2.2]^T$ is an initial estimate of the solution. Use the iterative correction method to find an improved estimate x^c. Find $\|r(x)\|$ and $\|r(x^c)\|$ where $r(x) = b - Ax$ is the residual error vector.

$$\begin{bmatrix} 4 & 0 & -2 \\ 1 & -1 & 0 \\ 0 & 5 & 3 \end{bmatrix} x = \begin{bmatrix} 8 \\ 3 \\ 6 \end{bmatrix}$$

2.17 Suppose a linear algebraic system is solved by Gaussian elimination using approximately 243,000 floating-point multiplications and divisions.

(a) What is the approximate number of unknowns n ?

(b) How many floating-point multiplications and divisions would be needed if Gauss-Jordan elimination were used instead?

(c) In general, what computational advantage does LU decomposition offer over Gaussian elimination?

2.18 Determine the number of floating-point multiplications and divisions needed to solve a tridiagonal system of dimension n using LU decomposition. What does it approach when $n \gg 1$?

2.19 Suppose the iteration matrix B of the iterative solution method in Equation (2.6.15) has a spectral radius of $\rho(B) < 1$. Then the Euclidean error in the kth estimate is related to the error in the previous estimate as follows where x denotes the exact solution:

$$\|x^k - x\|_2 \leq \rho(B)\|x^{k-1} - x\|_2$$

Let $\varepsilon > 0$ be an error tolerance. Find number of iterations N needed to ensure that $\|x^N - x\|_2 < \varepsilon$. Express your answer in terms of x^0, B, and ε.

2.20 Suppose U is an $n \times n$ upper-triangular matrix. Write an algorithm that finds the inverse of U or determines that U is singular.

2.21 Estimate the number of floating-point multiplications and divisions needed for the algorithm in problem 2.20 when n is large.

Computation

(P)2.22 Write a function called *gauss_el* which implements the Gaussian elimination method to solve $Ax = b$ as described in Alg. 2.3.1. The pseudo-prototype for the function is

```
[vector x,float Delta] = gauss_el (matrix A,vector b,int n)
```

On entry to *gauss_el:* A is an $n \times n$ coefficient matrix, b is an $n \times 1$ right-hand-side vector, and $n \geq 1$ is the dimension of the system. On exit, *Delta* is the determinant of A. If *Delta* $\neq 0$, then the $n \times 1$ vector x is the solution. Test *gauss_el* by writing a program that uses it to compute the solution to the following linear algebraic system. Print det(A), x and that $\|r(x)\|$ where $r(x)$ is the residual error vector.

$$\begin{bmatrix} 1 & -1 & 4 & 0 & 2 \\ 0 & 5 & -2 & 7 & 8 \\ 1 & 0 & 5 & 7 & 3 \\ 6 & -1 & 2 & 3 & 0 \\ -4 & 2 & 0 & 5 & -5 \end{bmatrix} x = \begin{bmatrix} 2 \\ -1 \\ 4 \\ 0 \\ -5 \end{bmatrix}$$

(S)2.23 Write a program that uses the NLIB functions to compute and print the inverse, determinant, and condition number of the coefficient matrix A in problem 2.22.

2.24 Write a program that uses the NLIB functions to solve the system in problem 2.22. Print the solution x and $\|r(x)\|$.

2.25 For every $n \times n$ matrix A, there is an associated polynomial of degree n called the *characteristic polynomial* of A.

$$\Delta(s) \triangleq \det(sI - A) = \sum_{k=0}^{n} a_k s^k$$

One way to find the coefficients of the characteristic polynomial is to define an $(n + 1) \times 1$ coefficient vector $x \triangleq [a_0, a_1, \ldots, a_n]^T$ and solve the following linear algebraic system.

$$\sum_{k=1}^{n+1} (s_j^{k-1}) x_k = \Delta(s_j) \quad , \quad 1 \le j \le (n + 1)$$

Here, the values of s_j must be distinct; for example, $s_j = j$ for $1 \le j \le (n + 1)$. Write a program which uses the NLIB functions to find the coefficients of the characteristic polynomial of the coefficient matrix in problem 2.10.

2.26 Write a program that uses the NLIB functions *lufac* and *lusub* to find the inverse of the coefficient matrix A in problem 2.22. Print L, U, P and A^{-1}.

2.27 Write a program that uses the NLIB function *tridec* to solve the following tridiagonal system. Print the solution x and $\|r(x)\|$.

$$\begin{bmatrix} 3 & 1 & 0 & 0 & 0 & 0 & 0 & 0 \\ 1 & 3 & 1 & 0 & 0 & 0 & 0 & 0 \\ 0 & 1 & 3 & 1 & 0 & 0 & 0 & 0 \\ 0 & 0 & 1 & 3 & 1 & 0 & 0 & 0 \\ 0 & 0 & 0 & 1 & 3 & 1 & 0 & 0 \\ 0 & 0 & 0 & 0 & 1 & 3 & 1 & 0 \\ 0 & 0 & 0 & 0 & 0 & 1 & 3 & 1 \\ 0 & 0 & 0 & 0 & 0 & 0 & 1 & 3 \end{bmatrix} x = \begin{bmatrix} 1 \\ -1 \\ 2 \\ 0 \\ 3 \\ -4 \\ -2 \\ 5 \end{bmatrix}$$

2.28 Write a program that uses the NLIB function *condnum* to estimate the condition number $K(A)$ of the following matrix for $1 \le m \le n - 1$. Use *randinit* to initialize the random number generator with a seed of 1000 prior to each call of *condnum*. Plot the estimate of $K(A)$ versus m and compare the results with the exact condition number.

$$A = \begin{bmatrix} 1 & -1 & 4 & 0 & 2 & 9 \\ 0 & 5 & -2 & 7 & 8 & 4 \\ 1 & 0 & 5 & 7 & 3 & -2 \\ 6 & -1 & 2 & 3 & 0 & 8 \\ -4 & 2 & 0 & 5 & -5 & 3 \\ 0 & 7 & -1 & 5 & 4 & -2 \end{bmatrix}$$

(S)2.29 Write a function called *jac* that implements the Jacobi iterative method to solve the system $Ax = b$. The pseudo-prototype for the function is

```
[vector x,int k] = jac (vector x,matrix A,vector b,float tol,int
n,int m)
```

On entry to *jac*: the $n \times 1$ vector x is an initial guess, A is an $n \times n$ coefficient matrix, b is an $n \times 1$ right-hand side vector, $tol \geq 0$ is an error tolerance used to terminate the search, $n \geq 1$ is the number of unknowns, and $m \geq 1$ is an upper bound on the number of iterations. On exit from *jac*: k is the number of iterations performed, and the $n \times 1$ vector x contains the updated estimate of the solution. If $0 < k < m$, then the following error criterion must be satisfied where $r = b - Ax$ is the residual error vector:

$$\|r\| < tol$$

If $k = 0$, one of the diagonal elements of A was found to be zero. Test the function *jac* by writing a program that uses it to solve the following system. Compute and display the norms of the residual error vectors for the first six solution estimates $\{x^1, x^2, \ldots, x^6\}$ and display the final estimate, x^6, using $x^0 = 0$.

$$\begin{bmatrix} 10 & 1 & -1 & 2 & 0 \\ 0 & -12 & 2 & -3 & 1 \\ 2 & -2 & -11 & 0 & 1 \\ 4 & 0 & 1 & 8 & -1 \\ -3 & 1 & 2 & 0 & 9 \end{bmatrix} x = \begin{bmatrix} 19 \\ 2 \\ 13 \\ -7 \\ -9 \end{bmatrix}$$

2.30 Write a program that uses the NLIB function *sr* to solve the system in problem 2.29 using the Gauss-Seidel method. Compute the first twenty estimates $\{x^1, x^2, \ldots, x^{20}\}$ and print the norms of their residual error vectors using $x^0 = 0$.

2.31 Write a program that uses NLIB to compute and display the determinant, condition number, and inverse of the matrix in problem 2.10.

CHAPTER 3

Eigenvalues and Eigenvectors

Many dynamic physical systems can be modeled by systems of linear differential equations or systems of linear difference equations. The dynamic nature of the system typically arises from energy storage elements within the system. Examples include rotating masses, compressed springs, charged capacitors, and pressurized containers. In the absence of an external input or excitation, the distribution of energy within the system will decay to a minimum energy state if the system is stable, or it may oscillate between states if the system is unstable. The rate of decay of the natural modes of a stable system, and the frequency of oscillation or rate of growth of the natural modes of an unstable system, are determined by the eigenvalues of the coefficient matrix of the system. For every $n \times n$ real matrix A, there is a set of n complex numbers $\{\lambda_1, \lambda_2, \ldots, \lambda_n\}$ called the *eigenvalues* of A. Corresponding to each eigenvalue λ_k, there is a nonzero $n \times 1$ complex vector x^k called an *eigenvector.* When an eigenvector x^k is operated on by the matrix A, the resulting product Ax^k is simply a scaled version of the eigenvector.

$$\boxed{Ax^k = \lambda_k x^k \quad , \quad 1 \leq k \leq n}$$

Thus, the eigenvectors of A are special vectors whose *directions* are invariant when they are multiplied by A. Instead, the eigenvectors are merely scaled, with the scale factor being the eigenvalue. The eigenvalues of A are the roots of a special polynomial of degree n called the characteristic polynomial, which can be defined as follows.

$$\boxed{\Delta(\lambda) = \det(\lambda I - A)}$$

In this chapter, we examine numerical techniques for computing eigenvalues and eigenvectors. To motivate the discussion, we begin by investigating the notion of stability for both continuous-time and discrete-time dynamic systems. We present practical examples that illustrate how the essential qualitative behavior of a system can be determined from its eigenvalues. Next, we consider the characteristic polynomial and present a numerical technique called Leverrier's method for finding its coefficients. We then examine the power methods, which are basic iterative techniques for approximating the largest eigenvalue and its eigenvector and the smallest eigenvalue and its eigenvector. The magnitude of the largest eigenvalue is the spectral radius, $\rho(A)$. By using the spectral radius, we can determine the rate of convergence of linear iterative methods. We then examine an iterative technique for finding all of the eigenvalues and eigenvectors of a symmetric matrix—Jacobi's method. Jacobi's method is based on the notion of similarity transformations, which are used to simplify the structure of a matrix while preserving its eigenvalues. The rate of convergence of the iterative methods is increased by first preprocessing the matrix with the Householder transformation, which converts the matrix to a special form called the upper-Hessenberg form.

We then present a very effective iterative technique for finding the eigenvalues of a general matrix, the orthogonal-triangular QR method. Next, we investigate Danilevsky's method for finding the eigenvalues and eigenvectors of a general matrix. Danilevsky's method is based on the idea of transforming a matrix to a special companion form from which the eigenvalues and eigenvectors can be readily computed. The chapter concludes with a summary of eigenvalue and eigenvector methods and a presentation of engineering applications. The selected examples include transient analysis of a four-stage chemical absorption process, population growth and age distribution in plants and animals, position control of a telescope, and oscillations in rotating masses interconnected by torsional springs.

3.1 MOTIVATION AND OBJECTIVES

Eigenvalues and eigenvectors arise in numerous engineering applications. Given the eigenvalues of a physical system, engineers can often draw conclusions about the qualitative behavior of the system without actually solving the equations. The following example is a case in point.

3.1.1 Seismograph

A linear n-dimensional continuous-time dynamic system can be modeled with a system of n first-order ordinary differential equations of the following general form.

$$\frac{dx(t)}{dt} = Ax(t) + Bu(t) \quad , \quad t \geq 0 \tag{3.1.1}$$

Here the scalar t denotes time, and the $n \times 1$ vector $x(t)$ denotes the state of the system at time t. The scalar $u(t)$ represents the excitation or input to the system. The system in Equation (3.1.1) is said to be *stable* if and only if for every initial state $x(0)$, the solution corresponding to $u(t) = 0$ converges to the zero vector.

$$x(t) \rightarrow 0 \quad \text{as} \quad t \rightarrow \infty \tag{3.1.2}$$

It is an easy matter to determine the stability of Equation (3.1.1) once the eigenvalues of A are known. A necessary and sufficient condition for the linear dynamic system in (3.1.1) to be stable is that the eigenvalues of A all lie in the open left half of the complex plane (Chen 1984). Thus, the system is stable if and only if

$$\boxed{\text{Re}(\lambda_k) < 0 \quad , \quad 1 \leq k \leq n} \tag{3.1.3}$$

As an example of engineering analysis based on eigenvalues, consider the seismograph system shown in Figure 3.1.1 (Ogata, 1997). Here a mass m is supported by a spring with spring constant k and a damper with damping constant f. The vertical position of the case in inertial space is y_g and the vertical position of the mass relative to the case is y, with $y = 0$ corresponding to the equilibrium position. If an earthquake occurs, the case will be displaced in inertial space which causes the mass to start to oscillate with the amplitude of the oscillation being a measure of the strength of the quake.

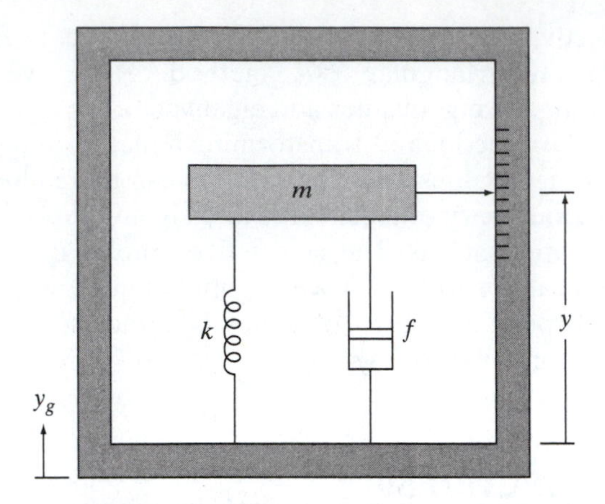

FIGURE 3.1.1
Seismograph

To develop a dynamic model of the seismograph, we apply Newton's second law and set mass times acceleration equal to the force acting on the mass. One component of the force is the restoring force due to deformation of the spring, while another is a viscous friction force associated with the damper which resists motion. The excitation term is the force generated by the acceleration of the case in inertial space.

$$m\frac{d^2y(t)}{dt^2} + f\frac{dy(t)}{dt} + ky(t) = -m\frac{d^2y_g(t)}{dt^2} \tag{3.1.4}$$

This second-order differential equation can be recast in the form of the first-order system in Equation (3.1.1) by introducing the vector of state variables $x = [y,\ dy/dt]^T$ (see Section 8.1). This results in the following linear continuous-time system.

$$\frac{dx(t)}{dt} = \underbrace{\begin{bmatrix} 0 & 1 \\ -k/m & -f/m \end{bmatrix}}_{A} x(t) + \underbrace{\begin{bmatrix} 0 \\ -1 \end{bmatrix}}_{B} \frac{d^2y_g(t)}{dt^2} \tag{3.1.5}$$

The characteristic polynomial of the seismograph is

$$\begin{aligned}
\Delta(\lambda) &= \det(\lambda I - A) \\
&= \det\left\{\begin{bmatrix} \lambda & -1 \\ k/m & \lambda + f/m \end{bmatrix}\right\} \\
&= \lambda(\lambda + f/m) + k/m \\
&= \lambda^2 + (f/m)\lambda + k/m \\
&= (\lambda - \lambda_1)(\lambda - \lambda_2)
\end{aligned} \tag{3.1.6}$$

Applying the quadratic formula, the two eigenvalues of the seismograph occur at

$$\lambda_{1,2} = \frac{-f \pm \sqrt{f^2 - 4mk}}{2m} \tag{3.1.7}$$

Thus, the seismograph has two real eigenvalues when the spring constant k and damping constant f satisfy $4mk \le f^2$. Otherwise, the eigenvalues form a complex conjugate pair. In any event, the system is stable as long as $f > 0$, which means the mass will eventually return to its equilibrium state $x = 0$. It is of interest to examine the case $f = 0$, which corresponds to no friction. In this case, the system is not stable because the eigenvalues are

$$\lambda_{1,2} = \pm j\sqrt{\frac{k}{m}} \tag{3.1.8}$$

When $f = 0$, the mass oscillates up and down with a period of

$$T = 2\pi\sqrt{\frac{m}{k}} \tag{3.1.9}$$

Thus, the imaginary part of the eigenvalue determines the frequency of oscillation. More generally, the imaginary part determines the mechanical frequency of vibration while the real part determines how fast the vibration dies out.

3.1.2 Convergence of Iterative Methods

Iterative computational techniques are often used to generate sequences of increasingly accurate approximations to solutions of numerical problems. In many cases the *errors* in the approximations can be modeled as follows where e^k denotes the error vector associated with the kth iteration, and A is an $n \times n$ matrix that depends on the details of the iterative technique used.

$$e^{k+1} = Ae^k \quad , \quad k \geq 0 \tag{3.1.10}$$

The system of equations in (3.1.10) can be thought of as a linear discrete-time dynamic system with the iteration index k denoting discrete time. A linear discrete-time system is *stable* if and only if for every initial condition e^0 the solution e^k converges to the zero vector as time approaches infinity.

$$e^k \rightarrow 0 \quad \text{as} \quad k \rightarrow \infty \tag{3.1.11}$$

Again, one can easily determine whether or not the system in Equation (3.1.10) is stable once the eigenvalues of A are known. By repeatedly applying (3.1.10), observe that the error associated with the kth iteration is

$$e^k = A^k e^0 \tag{3.1.12}$$

For the scalar case, $A^k = \alpha^k$, it is clear that the system is stable if and only if $|\alpha| < 1$. This generalizes to the case $n \geq 1$ in the following way. A necessary and sufficient condition for the system in Equation (3.1.10) to be stable is that the eigenvalues of A all have magnitudes less than unity (Franklin et al., 1990). Thus, the system is stable if and only if

$$\boxed{|\lambda_i| < 1 \quad , \quad 1 \leq i \leq n} \tag{3.1.13}$$

In geometric terms, the linear discrete-time dynamic system in Equation (3.1.10) is stable if and only if the eigenvalues of A all lie inside the unit circle of the complex plane. Note that to check (3.1.13) it is sufficient to compute only the dominant eigenvalue. The magnitude of the largest eigenvalue of A is the *spectral radius* of A introduced previously in Chapter 2. That is,

$$\boxed{\rho(A) \triangleq \max_{k=1}^{n}\{|\lambda_k|\}} \tag{3.1.14}$$

Thus a discrete-time linear system is stable if and only if the spectral radius of the coefficient matrix is less than unity. More generally, it can be shown that the rate of convergence increases as $\rho(A) \to 0$.

As an illustration of the use of the stability criterion in (3.1.13), consider the problem of solving a linear algebraic system of the form $Cx = b$. Recall from Chapter 2 that Jacobi's iterative method for finding x is as follows where x^0 is an initial guess.

$$Dx^{k+1} = b - (L + U)x^k \quad , \quad k \geq 0 \tag{3.1.15}$$

Here D, L, and U are the diagonal, lower-triangular, and upper-triangular parts of C, respectively. That is, $C = L + D + U$. Next, suppose x is the solution of $Cx = b$. Then

$$\begin{aligned} b &= Cx \\ &= (L + D + U)x \\ &= Dx + (L + U)x \end{aligned} \tag{3.1.16}$$

It follows that $Dx = b - (L + U)x$. Let e^k denote the error in the kth estimate of x.

$$e^k \triangleq x^k - x \tag{3.1.17}$$

Subtracting $Dx = b - (L + U)x$ from $Dx^{k+1} = b - (L + U)x^k$, and multiplying both sides by D^{-1}, we get the following equation for the error:

$$e^{k+1} = D^{-1}(L + U)e^k \tag{3.1.18}$$

Comparing this with Equation (3.1.10), we can conclude that Jacobi's iterative method converges if and only if the eigenvalues of the following matrix lie inside the unit circle.

$$A = D^{-1}(L + U) \tag{3.1.19}$$

This convergence criterion involving the spectral radius of A is more complete (both necessary and sufficient) than the simple sufficient condition involving $\|A\|$ cited in Chapter 2. The same convergence analysis can also be applied to the Gauss-Seidel method and the successive relaxation (SR) methods discussed in Chapter 2. The results are summarized in Table 3.1.1. The convergence criterion for the SR method reduces to that of the Gauss-Seidel method when the relaxation parameter is set to $\alpha = 1$.

TABLE 3.1.1 Convergence Criteria for Linear Iterative Methods

Method	*Convergence Criterion*
Jacobi	$\rho\{D^{-1}(L + U)\} < 1$
Gauss-Seidel	$\rho\{(D + L)^{-1}U\} < 1$
Successive relaxation	$\rho\{(D + \alpha L)^{-1}[(1 - \alpha)D - \alpha U]\} < 1$

3.1.3 Chapter Objectives

When you finish this chapter, you will be able to efficiently compute the eigenvalues and the eigenvectors of a real $n \times n$ matrix. You will understand how eigenvalues can be used to determine whether or not linear dynamic systems are stable, and whether or not linear iterative methods converge. You will be able to find both the dominant eigenvalue and its eigenvector, and the least dominant eigenvalue and its eigenvector, using the power method and inverse power method, respectively. You will know how to find the real eigenvalues of a symmetric matrix using the Jacobi method, and the eigenvalues of a general matrix using the popular orthogonal-triangular QR method. You will also be able to find the eigenvectors of a general matrix using the Danilevsky method. Finally, you will be able to find the roots of a general nth degree polynomial by converting it to an equivalent eigenvalue problem. These overall goals will be achieved by mastering the following chapter objectives.

Objectives for Chapter 3

- Know how to define and compute the eigenvalues of a square matrix.
- Know how to determine the stability of a linear dynamic system by examining its eigenvalues.
- Know how to determine the convergence rate of a linear iterative method by examining its eigenvalues.
- Know how to define the characteristic polynomial of a matrix and compute its coefficients using Leverrier's method.
- Know how to define and compute the eigenvectors of a matrix.
- Be able to estimate the dominant eigenvalue and its eigenvector using the direct power method.
- Be able to estimate the least dominant eigenvalue and its eigenvector using the inverse power method.
- Know when the direct and inverse power methods will converge.
- Be able to identify a symmetric matrix, and understand what this implies in terms of its eigenvalues and eigenvectors.
- Understand which types of matrices are diagonalizable and how to use eigenvectors and similarity transformations to diagonalize them.
- Be able to find the eigenvalues and eigenvectors of a symmetric matrix using Jacobi's iterative method.
- Know how to define the upper Hessenberg form and convert a general matrix to that form using the Householder transformation.
- Know how to find the eigenvalues of a general square matrix using the iterative QR method.
- Know how to find the eigenvectors of a general square matrix using Danilevsky's method.
- Be able to find the roots of a polynomial by converting it to an eigenvalue problem using the companion matrix.

- Understand how eigenvalue and eigenvector techniques can be used to solve practical engineering problems.
- Understand the relative strengths and weaknesses of each computational method and know which are most applicable for a given problem.

3.2 THE CHARACTERISTIC POLYNOMIAL

If x is an eigenvector of an $n \times n$ matrix A associated with the eigenvalue λ, then $Ax = \lambda x$. To compute the eigenvalues, it is useful to reformulate $Ax = \lambda x$ as a linear algebraic system in the unknown components of x. Moving both terms to one side of the equation and factoring out the vector x yields

$$(\lambda I - A)x = 0 \tag{3.2.1}$$

Here I is the $n \times n$ identity matrix. Note that if the matrix $\lambda I - A$ is nonsingular, then Equation (3.2.1) has the unique solution $x = 0$. Since eigenvectors must be nonzero, this means that eigenvalues are scalars λ for which the matrix $\lambda I - A$ is singular. Recall that a matrix is singular if and only if its determinant is zero. That is, eigenvalues of A are solutions of the equation $\Delta(\lambda) = 0$ where

$$\boxed{\Delta(\lambda) \triangleq \det(\lambda I - A)} \tag{3.2.2}$$

The equation $\Delta(\lambda) = 0$ is called the *characteristic equation* of A. By expanding the determinant in (3.2.2), one can show that $\Delta(\lambda)$ is a polynomial of degree n of the form

$$\Delta(\lambda) = \lambda^n + a_n\lambda^{n-1} + \cdots + a_2\lambda + a_1 \tag{3.2.3}$$

The polynomial $\Delta(\lambda)$ is called the *characteristic polynomial* of the matrix A. It follows that the eigenvalues of A are the *roots* of the characteristic polynomial. From the fundamental theorem of algebra, every polynomial of degree n has exactly n roots in the complex plane. Consequently, $\Delta(\lambda)$ can be written in factored form as

$$\boxed{\Delta(\lambda) = (\lambda - \lambda_1)(\lambda - \lambda_2)\cdots(\lambda - \lambda_n)} \tag{3.2.4}$$

It is evident from (3.2.4) that the $n \times n$ matrix A has n eigenvalues $\{\lambda_1, \lambda_2, \ldots, \lambda_n\}$. However, the n eigenvalues need not be distinct from one another. If an eigenvalue occurs m times as a root of the characteristic polynomial, then it is referred to as an eigenvalue of *multiplicity* m. Eigenvalues of multiplicity one are called *simple* eigenvalues. Note that eigenvalues can be complex even though the matrix A is real. However, when A is real, complex eigenvalues always occur in *complex-conjugate* pairs $\lambda = \sigma \pm j\omega$ where $j = \sqrt{-1}$.

Determination of the characteristic polynomial in Equation (3.2.2) requires the evaluation of a determinant of order n, a computationally expensive task. The problem is made all the more difficult by the fact that the matrix is not purely numerical. Instead, a *symbolic* calculation is required due to the presence of the parameter λ. Fortunately, there are a number of alternative ways to determine the characteristic polynomial and its roots. Consider the following function called the *trace* of an $n \times n$ matrix A.

$$\text{trace}(A) \triangleq A_{11} + A_{22} + \cdots + A_{nn} \tag{3.2.5}$$

Note that the trace of a matrix is just the sum of the diagonal elements. If a matrix is triangular, then the eigenvalues are simply the diagonal elements (see Problem 3.4). Consequently, for triangular matrices the trace is the *sum* of the eigenvalues and the determinant is the *product* of the eigenvalues. As it turns out, these two properties hold in general for *any* $n \times n$ matrix.

$$\boxed{\text{trace}(A) = \lambda_1 + \lambda_2 + \cdots + \lambda_n} \tag{3.2.6}$$

$$\boxed{\det(A) = \lambda_1\lambda_2 \cdots \lambda_n} \tag{3.2.7}$$

The relationships between the trace, determinant, and eigenvalues of a matrix can be used as a partial check on the validity of numerical estimates of eigenvalues. In particular, the trace check in Equation (3.2.6) is inexpensive.

EXAMPLE 3.2.1 Characteristic Polynomial

As an example of a characteristic polynomial and its roots, consider the following 2×2 matrix.

$$A = \begin{bmatrix} 1 & 1 \\ 6 & 2 \end{bmatrix}$$

From Equation (3.2.2), the characteristic polynomial of this matrix is

$$\begin{aligned} \Delta(\lambda) &= \det(\lambda I - A) \\ &= \det\left\{\begin{bmatrix} \lambda - 1 & -1 \\ -6 & \lambda - 2 \end{bmatrix}\right\} \\ &= (\lambda - 1)(\lambda - 2) - 6 \\ &= \lambda^2 - 3\lambda - 4 \\ &= (\lambda - \lambda_1)(\lambda - \lambda_2) \end{aligned}$$

Factoring the characteristic polynomial using the quadratic formula yields

$$\lambda_{1,2} = \frac{3 \pm \sqrt{9 + 16}}{2} = \{4, -1\}$$

Thus, the matrix A has two real eigenvalues, each of multiplicity one. As a check, we can apply conditions (3.2.6) and (3.2.7). In this case,

$$\begin{aligned} \text{trace}(A) &= 3, \\ \det(A) &= -4 \end{aligned}$$

Using matrix multiplications and the trace function, it is possible to determine the coefficients of the characteristic polynomial of an $n \times n$ matrix A using the following procedure.

Alg. 3.2.1 Leverrier's Method

1. Set $B_n = A$ and $a_n = -\text{trace}(B_n)$.
2. For $k = n - 1$ down to 1 compute

{

$$B_k = A(B_{k+1} + a_{k+1}I)$$

$$a_k = -\frac{\text{trace}(B_k)}{n - k + 1}$$

}

The procedure in Alg. 3.2.1 is called *Leverrier's method.* It generates the coefficients $\{a_1, a_2, \ldots, a_n\}$ of the characteristic polynomial $\Delta(\lambda)$ in Equation (3.2.3). Note that each matrix multiplication in step (2) requires n^3 floating-point multiplications. Consequently, the number of FLOPs for the entire algorithm for large values of n is approximately n^4.

EXAMPLE 3.2.2 Leverrier's Method

As an example of an application of Leverrier's method for finding the characteristic polynomial, consider the following 3×3 matrix.

$$A = \begin{bmatrix} 1 & -1 & 0 \\ 0 & 2 & -1 \\ -1 & 0 & 1 \end{bmatrix}$$

Applying step (1) of Alg. 3.2.1 with $n = 3$ yields $B_3 = A$ and

$$a_3 = -(1 + 2 + 1) = -4$$

Next, applying step (2) with $k = 2$, we have

$$B_2 = \begin{bmatrix} 1 & -1 & 0 \\ 0 & 2 & -1 \\ -1 & 0 & 1 \end{bmatrix} \begin{bmatrix} -3 & -1 & 0 \\ 0 & -2 & 0 \\ -1 & 0 & -3 \end{bmatrix} = \begin{bmatrix} -3 & 1 & 1 \\ 1 & -4 & 1 \\ 2 & 1 & -3 \end{bmatrix}$$

$$a_2 = -\frac{-3 - 4 - 3}{3 - 2 + 1} = 5$$

Finally, applying step (2) with $k = 1$ yields

$$B_1 = \begin{bmatrix} 1 & -1 & 0 \\ 0 & 2 & -1 \\ -1 & 0 & 1 \end{bmatrix} \begin{bmatrix} 2 & 1 & 1 \\ 1 & 1 & 1 \\ 2 & 1 & 2 \end{bmatrix} = \begin{bmatrix} 1 & 0 & 0 \\ 0 & 1 & 0 \\ 0 & 0 & 1 \end{bmatrix}$$

$$a_1 = -\frac{1 + 1 + 1}{3 - 1 + 1} = -1$$

It follows that the characteristic polynomial of the 3×3 matrix A is

$$\Delta(\lambda) = \lambda^3 - 4\lambda^2 + 5\lambda - 1$$

Although the Leverrier algorithm is a conceptually simple technique for identifying the coefficients of the characteristic polynomial, it requires a substantial computational effort. Approximately n^4 FLOPs are required when n is large. As a consequence, Leverrier's method is *not* a useful technique for large matrices (say, $n > 20$) due to the accumulated round-off error associated with the large number of floating-point operations. A somewhat more efficient technique called *Danilevsky's method* is examined later that requires approximately n^3 FLOPs.

It is tempting to think that once the characteristic polynomial is determined, the problem of computing eigenvalues is effectively solved because all one has to do is find its roots. Unfortunately, the roots of a higher-degree polynomial can be extremely sensitive to the coefficient values. Consequently, a significant loss of accuracy can arise due to accumulated round-off error, particularly when the degree of the polynomial is large. As a result, the direct approach of finding eigenvalues by factoring $\Delta(\lambda)$ is effective only for relatively small values of n. In the remainder of this chapter, we examine alternative techniques, based on similarity transformations, that are less susceptible to the effects of accumulated round-off error.

3.3 POWER METHODS

Before we examine general techniques for finding all the eigenvalues and eigenvectors of a matrix, we first consider the simpler problem of finding a single eigenvalue and its eigenvector. Suppose A is an $n \times n$ matrix whose eigenvalues can be ordered from largest to smallest as follows.

$$|\lambda_1| > |\lambda_2| \geq \cdots \geq |\lambda_{n-1}| > |\lambda_n| \tag{3.3.1}$$

The eigenvalue whose magnitude is largest is called the *dominant* eigenvalue. If there is a single eigenvalue that dominates the others as in Equation (3.3.1), then it must be a simple real eigenvalue with a real eigenvector. Let x^k denote an eigenvector associated with the eigenvalue λ_k. That is,

$$Ax^k = \lambda_k x^k \quad , \quad 1 \leq k \leq n \tag{3.3.2}$$

Note that a superscript is used to denote the kth eigenvector, rather than a subscript, because subscripts are reserved to denote the elements of a vector.

3.3.1 Direct Power Method

Suppose that the n eigenvectors $\{x^1, x^2, \ldots, x^n\}$ in (3.3.2) form a linearly independent set (see Appendix 3). This assumption is satisfied, for example, when A has n distinct eigenvalues and also when A is symmetric. However, it does not hold in general. If A has n linearly independent eigenvectors, then they can be used as a basis for the space of $n \times 1$ column vectors, R^n. That is, for an arbitrary $n \times 1$ vector y^0 there exists an $n \times 1$ coordinate vector c such that

$$y^0 = \sum_{k=1}^{n} c_k x^k \tag{3.3.3}$$

If y^0 is selected such that $(y^0)^T x^1 \neq 0$, then $c_1 \neq 0$. Suppose we generate a sequence of vectors y^k by successive multiplication by the matrix A. That is,

$$y^{k+1} = Ay^k \quad , \quad k \geq 0 \tag{3.3.4}$$

Using the expression for y^0 in Equation (3.3.3), and the fundamental property of eigenvectors in (3.3.2), the vector y^1 can be expressed as follows.

$$\begin{aligned} y^1 &= Ay^0 \\ &= A\left(\sum_{k=1}^{n} c_k x^k\right) \\ &= \sum_{k=1}^{n} c_k A x^k \\ &= \sum_{k=1}^{n} c_k \lambda_k x^k \end{aligned} \tag{3.3.5}$$

Applying a similar analysis starting with y^1 instead of y^0, it follows that subsequent vectors in the sequence also can be represented in terms of the eigenvectors of A. In particular,

$$y^i = \sum_{k=1}^{n} c_k \lambda_k^i x^k \quad , \quad i \geq 0 \tag{3.3.6}$$

Recall from Equation (3.3.1) that λ_1 is the dominant eigenvalue. Factoring out λ_1^i from each term in (3.3.6) yields

$$y^i = \lambda_1^i \left[c_1 x^1 + \sum_{k=2}^{n} c_k \left(\frac{\lambda_k}{\lambda_1} \right)^i x^k \right] \quad , \quad i \geq 0 \tag{3.3.7}$$

But from (3.3.1), $|\lambda_k/\lambda_1| < 1$ for $2 \leq k \leq n$. It follows that $(\lambda_k/\lambda_1)^i \to 0$ as $i \to \infty$, which means that

$$y^i \approx \lambda_1^i c_1 x^1 \quad , \quad i \gg 1 \tag{3.3.8}$$

But if x^1 is an eigenvector of A associated with the eigenvalue λ_1, then so is αx^1 for any $\alpha \neq 0$. Therefore, when i is sufficiently large, y^i approximates an eigenvector of A associated with the dominant eigenvalue λ_1. Of course, the problem with the formulation in (3.3.8) is that y^i can become very large if $|\lambda_1| > 1$ or very small if $|\lambda_1| < 1$. This practical difficulty can be alleviated by normalizing the length of y^k at each step by modifying (3.3.4) as follows.

$$y^{k+1} = \frac{Ay^k}{\|Ay^k\|} \tag{3.3.9}$$

Once an eigenvector is known, its corresponding eigenvalue can be easily computed. Suppose x is an eigenvector of A associated with the eigenvalue λ; that is, $Ax = \lambda x$. Multiplying both sides on the left by x^T then yields the scalar equation

$x^T Ax = \lambda x^T x$. Solving for λ, we arrive at the following expression for the eigenvalue called the *Rayleigh quotient.*

$$\boxed{\lambda = \frac{x^T Ax}{x^T x}} \tag{3.3.10}$$

Since y^k approximates the dominant eigenvector when k is large, we can substitute y^k for x in Equation (3.3.10) to get an approximation for the dominant eigenvalue.

$$\mu_k = \frac{(y^k)^T A y^k}{(y^k)^T y^k}, \quad k \geq 0 \tag{3.3.11}$$

This iterative technique for estimating the dominant eigenvalue and its eigenvector is called the *power method.* To implement the power method we must establish a termination criterion for the iterative procedure. This can be achieved by making use of the *residual error* vector, r^k, at the kth iteration, which is defined as

$$r^k \triangleq (\mu_k I - A)y^k \tag{3.3.12}$$

It is clear that $r^k = 0$ if μ_k is an eigenvalue of A and y^k is its eigenvector. Thus, when $\|r^k\|$ is sufficiently small, the process can be terminated. Of course, if the error tolerance is made too small, then it is possible that the convergence criterion might never be met due to accumulated round-off error. Therefore, one should always include a fixed upper limit on the number of iterations. The following algorithm is an implementation of the power method.

Alg. 3.3.1 Power Method

1. Pick $\varepsilon > 0, m > 0$ and $y \neq 0$. Set $k = 0$ and compute $x = Ay$.
2. Do

   {

   $$y = \frac{x}{\|x\|}$$
   $$x = Ay$$
   $$\mu = \frac{y^T x}{y^T y}$$
   $$r = \mu y - x$$
   $$k = k + 1$$

   }
3. While ($\|r\| > \varepsilon$) and ($k < m$)

Note that Alg. 3.3.1 is implemented in such a way that it requires only one matrix multiplication per iteration. If Alg. 3.3.1 terminates in fewer than m iterations, then convergence was achieved, in which case μ is an estimate of the dominant eigenvalue λ_1 and y is an estimate of it eigenvector x^1.

Although the power method allows us to find only one eigenvalue and its eigenvector, the eigenvalue found is an important one. The magnitude of the dominant eigenvalue is the *spectral radius* of A and is denoted $\rho(A)$.

$$\rho(A) = |\lambda_1| = \max_{k=1}^{n} |\lambda_k| \tag{3.3.13}$$

Consequently, Alg. 3.3.1 provides a means of estimating the spectral radius of a matrix. Recall from Section 3.1.2 that the spectral radius is important in determining whether or not a linear iterative sequence converges. In particular, consider the following sequence of vectors starting from an arbitrary initial guess y^0.

$$y^{k+1} = By^k + c \tag{3.3.14}$$

This sequence converges to the solution of $(I - B)y = c$ if and only if the spectral radius of B satisfies $\rho(B) < 1$. The Jacobi, Gauss-Seidel, and successive relaxation methods for solving $Ax = b$ are all special cases of the iterative system in (3.3.14).

EXAMPLE 3.3.1 Power Method

As an example of an application of the power method for computing the dominant eigenvalue and its eigenvector, consider the following 4×4 matrix.

$$A = \begin{bmatrix} -3.9 & -0.6 & -3.9 & 0.4 \\ 1 & 0 & 0 & 0 \\ 0 & 1 & 0 & 0 \\ 0 & 0 & 1 & 0 \end{bmatrix}$$

By construction, this matrix has eigenvalues at $\lambda_1 = -4, \lambda_{2,3} = \pm j, \lambda_4 = 0.1$. Example program *e331* on the distribution CD uses a random initial vector y^0, and an error tolerance of $\varepsilon = 10^{-6}$

continued

TABLE 3.3.1 Power Method Estimates

k	μ_k	$(y^k)^T$				$\lVert r^k \rVert$
1	−3.6180851	−1.0000000	−0.2692910	−0.7325776	0.6021505	3.5414021
2	−3.7240922	1.0000000	−0.1396748	−0.0376132	−0.1023226	0.4798381
3	−3.8731112	−1.0000000	0.2695103	−0.0376438	−0.0101371	0.1237117
4	−4.0103736	1.0000000	−0.2576622	0.0694426	−0.0096994	0.0333218
5	−4.0077777	−1.0000000	0.2487495	−0.0640934	0.0172738	0.0081224
6	−3.9993191	1.0000000	−0.2495244	0.0620691	−0.0159929	0.0020722
7	−3.9995131	−1.0000000	0.2500780	−0.0624006	0.0155221	0.0005061
8	−4.0000424	1.0000000	−0.2500297	0.0625269	−0.0156020	0.0001296
9	−4.0000305	−1.0000000	0.2499951	−0.0625062	0.0156314	0.0000317
10	−3.9999979	1.0000000	−0.2499981	0.0624983	−0.0156264	0.0000080
11	−3.9999981	−1.0000000	0.2500003	−0.0624996	0.0156246	0.0000020
12	−4.0000000	1.0000000	−0.2500001	0.0625001	−0.0156249	0.0000005

to produce the sequence in Table 3.3.1, which includes the iteration number, the eigenvalue estimate, the eigenvector estimate, and the norm of the residual error vector.

From Table 3.3.1, we see that the method converged in 12 iterations. The estimate for the dominant eigenvalue is $\lambda_1 = -4.0000000$, and its estimated eigenvector is

$$x^1 = [1.0000000, -0.2500001, 0.0625001, -0.0156249]^T$$

The rate of convergence of the power method depends on the spread between the dominant eigenvalue and the next largest eigenvalue. In this case, $|\lambda_1|/|\lambda_2| = 4$ is relatively large, so the convergence is reasonably rapid. From Equation (3.3.13) we see that the spectral radius of A is $\rho(A) = 4$.

3.3.2 Inverse Power Method

It is also possible to find the eigenvalue whose magnitude is smallest. This *least-dominant* eigenvalue can be computed with a simple modification of the power method. If the matrix A is singular, then from Equation (3.2.7) the least-dominant eigenvalue in (3.3.1) is $\lambda_n = 0$. Consequently, suppose A is nonsingular and suppose there is a single least-dominant eigenvalue λ_n as in (3.3.1). In this case, λ_n and its eigenvector x^n must be real. If we premultiply both sides of (3.3.2) by A^{-1} and then divide both sides by λ_k, it is apparent that λ_k^{-1} is an eigenvalue of A^{-1} associated with the eigenvector x^k.

$$A^{-1}x^k = \lambda_k^{-1}x^k \quad , \quad 1 \le k \le n \tag{3.3.15}$$

That is, the eigenvalues of A^{-1} are the reciprocals of the eigenvalues of A. Consequently, the least-dominant eigenvalue of A is the dominant eigenvalue of A^{-1}. Using this fact, we can replace A in (3.3.9) and (3.3.11) with A^{-1}. The resulting formulation is called the *inverse power method.*

$$y^k = \frac{A^{-1}y^{k-1}}{\|A^{-1}y^{k-1}\|} \tag{3.3.16a}$$

$$\mu_k = \frac{(y^k)^T A^{-1} y^k}{(y^k)^T y^k} \tag{3.3.16b}$$

Of course, to implement (3.3.16) one does not have to explicitly calculate A^{-1}. Instead, it is more efficient to solve the linear algebraic system $Ax = y^{k-1}$ for x and then set $y^k = x/\|x\|$. If the initial guess y^0 is a nonzero vector that is not orthogonal to x^n, then the sequence of scalars μ_k converges to the reciprocal of the least-dominant eigenvalue λ_n, and the sequence of vectors y^k converges to its eigenvector x^n. The following algorithm is an implementation of the inverse power method.

Alg. 3.3.2 Inverse Power Method

1. Pick $\varepsilon > 0, m > 0$ and $y \neq 0$. Set $k = 0$,
2. Solve $Ax = y$ by factoring A into L and U using Alg. 2.4.2.
3. Do

   {

   (a) Compute $y = x/\|x\|$.

(b) Solve $Ax = y$ using L and U.

(c) Compute

$$\mu = \frac{y^T x}{y^T y}$$

$$r = \mu y - x$$

$$k = k + 1$$

}

4. While ($\|r\| > \varepsilon$) and ($k < m$)

Again, Alg. 3.3.2 is implemented in such a way that it requires solving only one linear algebraic system per iteration. For each iteration, it is the *same* linear algebraic system that must be solved, but with a different right-hand side vector. Consequently, the *LU* decomposition method is used. If Alg. 3.3.2 terminates in fewer than m iterations, then convergence was achieved, in which case $1/\mu$ is an estimate of the least-dominant eigenvalue λ_n and y is an estimate of it eigenvector x^n.

EXAMPLE 3.3.2 Inverse Power Method

As an example of an application of the inverse power method for computing the least-dominant eigenvalue and its eigenvector, consider the 4×4 matrix A examined previously in Example 3.3.1. Example program *e332* on the distribution CD uses a random initial vector y^0, and an error tolerance of $\varepsilon = 10^{-6}$, to produce the sequence in Table 3.3.2, which includes the iteration number, the reciprocal of the eigenvalue estimate, the eigenvector estimate, and the norm of the residual error vector.

From Table 3.3.2, we see that the method converged in 7 iterations. The estimate for the least-dominant eigenvalue is $\lambda_n = 1/\mu = 0.1000000$ and its estimated eigenvector is

$$x^n = [-0.0009999, -0.0100001, -0.1000001, -1.0000000]^T$$

The rate of convergence of the inverse power method depends on the spread between the least-dominant eigenvalue and the next smallest eigenvalue. In this case, the remaining eigenvalues occur at $\lambda_1 = -4$ and $\lambda_{2,3} = \pm j$. Since the ratio $|\lambda_3|/|\lambda_4| = 10$ is quite large, the convergence is faster in this case than it was for Example 3.3.1.

TABLE 3.3.2 Inverse Power Method Estimates

k	μ_k	$(y^k)^T$				$\|r^k\|$
1	9.3500261	− 0.0707181	0.0581276	− 0.0341231	− 1.0000000	0.7193441
2	9.9233665	0.0061764	− 0.0036258	− 0.1062560	− 1.0000000	0.0702760
3	10.0054798	− 0.0003652	− 0.0107013	− 0.1007121	− 1.0000000	0.0076727
4	10.0007572	− 0.0010694	− 0.0100648	− 0.0999369	− 1.0000000	0.0007193
5	9.9999447	− 0.0010064	− 0.0099930	− 0.0999929	− 1.0000000	0.0000766
6	9.9999924	− 0.0009993	− 0.0099994	− 0.1000006	− 1.0000000	0.0000072
7	10.0000010	− 0.0009999	− 0.0100001	− 0.1000001	− 1.0000000	0.0000010

3.4 JACOBI'S METHOD

Next, we examine the problem of finding all the eigenvalues and eigenvectors of an $n \times n$ matrix. To simplify the discussion, we initially restrict our attention to *symmetric* matrices, matrices for which $A^T = A$. Symmetric matrices have the property that all of the eigenvalues are *real.* In addition, every symmetric $n \times n$ matrix has a set of n linearly independent eigenvectors. By exploiting these properties, we can convert a symmetric matrix to a *diagonal* form, in which case the eigenvalues are the diagonal elements.

The key to converting A to a diagonal form is to perform operations that simplify the structure of A, yet preserve its eigenvalues. Suppose P is a nonsingular $n \times n$ matrix. Consider the following matrix B, which is obtained from A through a *similarity transformation.*

$$\boxed{B = P^{-1}AP} \tag{3.4.1}$$

Similarity transformations preserve the eigenvalues (but not the eigenvectors) of a matrix. To see this, let $\Delta_B(\lambda)$ denote the characteristic polynomial of B, and let $\Delta_A(\lambda)$ denote the characteristic polynomial of A. Then, using the fact that the determinant of the product of matrices equals the product of the determinants, we have

$$\begin{aligned}\Delta_B(\lambda) &= \det(\lambda I - P^{-1}AP)\\ &= \det(P^{-1}(\lambda I - A)P)\\ &= \det(P^{-1})\det(\lambda I - A)\det(P)\\ &= \det(P^{-1}P)\det(\lambda I - A)\\ &= \Delta_A(\lambda)\end{aligned} \tag{3.4.2}$$

Thus, the eigenvalues of B are identical to the eigenvalues of A. To determine the eigenvectors of A in terms of the eigenvectors of B, suppose y^k is an eigenvector of B associated with the eigenvalue λ_k. That is, $By^k = \lambda_k y^k$. Premultiplying (multiplying on the left) both sides of Equation (3.4.1) by P yields $AP = PB$. It then follows that

$$\begin{aligned}A(Py^k) &= P(By^k)\\ &= \lambda_k(Py^k)\end{aligned} \tag{3.4.3}$$

Consequently, if y^k is an eigenvector of B associated with the eigenvalue λ_k, then $x^k = Py^k$ is an eigenvector of A associated with the eigenvalue λ_k. Thus the eigenvectors of A are obtained from the eigenvectors of B by premultiplication by the matrix P. When B is diagonal, the eigenvectors of A are the columns of P.

By performing a sequence of orthogonal similarity transformations ($P^{-1} = P^T$), the symmetric matrix A can be transformed into a diagonal matrix. The eigenvalues of A are then the diagonal elements. The tool that we use to transform A to diagonal form is a *rotation matrix* denoted $R(\phi)$. The rotation matrix $R(\phi)$, associated with row p and column q, is the $n \times n$ identity matrix I except for the elements in row p and column q, which satisfy

$$R_{pp} = R_{qq} = \cos\phi \tag{3.4.4a}$$

$$R_{pq} = -R_{qp} = -\sin\phi \tag{3.4.4b}$$

Here ϕ is the angle of rotation. Rotation matrices have a number of useful properties. One is that $R^T(\phi)R(\phi) = I$, in which case we say that $R(\phi)$ is an *orthogonal matrix*. Since $R(\phi)$ is orthogonal, computing its inverse is simple and numerically accurate, namely, $R^{-1}(\phi) = R^T(\phi)$. Given the symmetry properties of the sine and cosine functions, this can be further simplified because $R^T(\phi) = R(-\phi)$. If the similarity transformation in Equation (3.4.1) is applied using a rotation matrix, the result is

$$B = R^T(\phi)AR(\phi) \tag{3.4.5}$$

Through a judicious choice of the angle ϕ, element B_{pq} in row p and column q can be nulled or made zero. Since A is assumed to be symmetric, B will also be symmetric, which means $B_{qp} = 0$ as well. Using (3.4.4) and (3.4.5), one finds that

$$B_{pq} = (A_{qq} - A_{pp})\sin\phi\cos\phi + A_{pq}(\cos^2\phi - \sin^2\phi) \tag{3.4.6}$$

Setting $B_{pq} = 0$, we can then solve for ϕ. The equation can be simplified by first using the trigonometric identities, $\sin(2\phi) = 2\sin\phi\cos\phi$ and $\cos(2\phi) = \cos^2\phi - \sin^2\phi$. Substituting these into (3.4.6) and recalling that $\tan(2\phi) = \sin(2\phi)/\cos(2\phi)$ then yields

$$\phi = \frac{1}{2}\arctan\left\{\frac{2A_{pq}}{A_{pp} - A_{qq}}\right\} \tag{3.4.7}$$

The basic idea is to zero all of the off-diagonal elements of A by performing a sequence of similarity transformations based on rotation matrices. The following algorithm is an implementation of this technique applied to a symmetric $n \times n$ matrix A.

Alg. 3.4.1 **Jacobi's Method**

1. Pick $\varepsilon > 0$ and set $P = I$. Find the element A_{pq} above the diagonal whose magnitude is largest.
2. While $|A_{pq}| > \varepsilon$ do

   {

   (a) Compute

   $$\phi = \frac{1}{2}\arctan\left\{\frac{2A_{pq}}{A_{pp} - A_{qq}}\right\}$$

   (b) Set $R = I$ and then set $R_{pp} = R_{qq} = \cos\phi$ and $R_{pq} = -R_{qp} = -\sin\phi$.

   (c) Compute $P = PR$ and $A = R^TAR$.

   (d) Find the element A_{pq} above the diagonal whose magnitude is largest.

   }
3. Set $\lambda_k = A_{kk}$ for $1 \le k \le n$ and partition P as $P = [x^1, x^2, \ldots, x^n]$.

Alg. 3.4.1 is called *Jacobi's method* for computing the eigenvalues and eigenvectors of a real symmetric matrix. Note that it is an open-ended iterative method rather than a direct method requiring a fixed number of steps. This is because the off-diagonal elements of A that have been set to zero on one step do not necessarily remain zero on subsequent steps although they typically remain small. In some cases, the Jacobi method requires many steps before all of the off-diagonal elements have magnitudes below the user-selected threshold ε. If σ_k denotes the sum of the squares of the off-diagonal elements after the kth iteration, then

$$\sigma_{k+1} = \sigma_k - 2A_{pq}^2 \quad , \quad k \geq 0 \tag{3.4.8}$$

Since σ_k is bounded from below by zero, it follows that σ_k decreases monotonically to zero as $k \to \infty$. Therefore Jacobi's method is guaranteed to converge.

It should be pointed out that implementation of Alg. 3.4.1 does not require the full $3n^3$ multiplications suggested by the notation in step (2c). Since R is identical to I except for columns p and q, only columns p and q of PR and AR have to be computed. Similarly only rows p and q of $R^T(AR)$ have to be computed. This reduction to $6n$ multiplications results in a significant improvement in execution speed. There are also techniques available for computing $\cos\phi$ and $\sin\phi$ more efficiently (see, e.g., Lastman and Sinha, 1989).

EXAMPLE 3.4.1 Jacobi's Method

As an illustration of Jacobi's method for computing the eigenvalues and eigenvectors of a symmetric matrix, consider the following 2×2 matrix.

$$A = \begin{bmatrix} -1 & 2 \\ 2 & 2 \end{bmatrix}$$

When $n = 2$, Jacobi's method converges in a single step because there is only one element above the diagonal. This is in contrast to the case $n > 2$ which can take many steps. From step (2a) of Alg. 3.4.1 the angle is

$$\phi = \frac{1}{2}\arctan\left\{\frac{2(2)}{-1-2}\right\} = 1.107$$

The rotation matrix from step (2b) is then

$$R = \begin{bmatrix} 0.4472 & -0.8944 \\ 0.8944 & 0.4472 \end{bmatrix}$$

Applying the similarity transformation $A = R^T AR$ in step (2c) then zeros A_{12} and A_{21} to yield

$$A = \begin{bmatrix} 3.0000 & 0.0000 \\ 0.0000 & -2.0000 \end{bmatrix}$$

Since A is now diagonal, it follows that the eigenvalues are $\lambda_1 = 3$ and $\lambda_2 = -2$. From step (2c), we also have $P = R$. Therefore, the eigenvectors are $x^1 = [0.4472, 0.8944]^T$ and $x^2 = [-0.8944, 0.4472]^T$.

3.5 HOUSEHOLDER TRANSFORMATION

Jacobi's method for finding the eigenvalues and eigenvectors of a symmetric matrix suffers from the drawback that it can take many iterations to converge. The convergence rate can be improved if the matrix A is first preprocessed. A preliminary step that is helpful is to first transform the matrix A to *tridiagonal form* where nonzero elements appear only on the diagonal, just above it on the first superdiagonal, and just below it on the first subdiagonal. If Jacobi's method is applied to a tridiagonal matrix, then the elements that are already zero are not disturbed by the process. Consequently, the only nonzero elements that have to be eliminated by Jacobi's method are the $n - 1$ elements along the first superdiagonal, a substantial reduction in computational effort. The technique for reducing a symmetric matrix to tridiagonal form can also be applied to nonsymmetric matrices. In this case, the transformed matrix is upper-triangular except for nonzero elements along the first subdiagonal as in (3.5.1). A matrix with this structure is called an *upper-Hessenberg matrix*.

$$B = \begin{bmatrix} B_{11} & B_{12} & B_{13} & \cdots & B_{1,n-1} & B_{1n} \\ B_{21} & B_{22} & B_{23} & \cdots & B_{2,n-1} & B_{2n} \\ 0 & B_{32} & B_{33} & \cdots & B_{3,n-1} & B_{3n} \\ 0 & 0 & B_{43} & \cdots & B_{4,n-1} & B_{4n} \\ \vdots & \vdots & \vdots & \ddots & \vdots & \vdots \\ 0 & 0 & 0 & \cdots & B_{n,n-1} & B_{nn} \end{bmatrix} \tag{3.5.1}$$

The technique for transforming a general $n \times n$ real matrix to upper-Hessenberg form is called the *Householder transformation* method. Like the Jacobi method, which transforms a matrix to diagonal form, it is based on similarity transformations so as to preserve the eigenvalues. However, while the Jacobi method employs a rotation matrix to zero individual elements, the Householder method uses a *reflection matrix* to simultaneously zero a column of elements. Consider, in particular, the following matrix where u is an arbitrary $n \times 1$ nonzero vector.

$$P(u) = I - \frac{2uu^T}{u^T u} \tag{3.5.2}$$

Here I is the $n \times n$ identity matrix, uu^T is the $n \times n$ *outer* product of the vector u with itself, and $u^T u$ is the 1×1 *inner* product of u with itself. The matrix $P(u)$ has a number of useful properties. For example, direct calculation reveals that $P^2(u) = I$ which means that $P^{-1}(u) = P(u)$. Furthermore, applying the matrix transpose one finds that $P^T(u) = P(u)$. It follows that $P^T(u) = P^{-1}(u)$ which makes $P(u)$ an orthogonal matrix.

The key to reducing a matrix A to upper-Hessenberg form is to use a sequence of similarity transformations of the form $B = P(u^k)AP(u^k)$ where the vector u^k is chosen so as to eliminate the elements in column k below the first subdiagonal. This can be achieved if the vector u^k is chosen as follows:

$$\alpha = \operatorname{sgn}(A_{k+1,k})\sqrt{\sum_{j=k+1}^{n} A_{jk}^2} \tag{3.5.3a}$$

$$u^k = [0, \ldots, 0, A_{k+1,k} + \alpha, A_{k+2,k}, \ldots, A_{n,k}]^T \tag{3.5.3b}$$

Here $sgn(a)$ denotes the sign of a with $\operatorname{sgn}(a) = 1$ when $a \geq 0$ and $\operatorname{sgn}(a) = -1$ when $a < 0$. The vector u^k is constructed such that if $B = P(u^k)AP(u^k)$, then $B_{jk} = 0$ for $k < j \leq n$. That is, the nonzero elements in column k below the first subdiagonal are eliminated. If a sequence of Householder transformations is performed, starting with column 1 and ending with column $n - 2$, then the matrix A is transformed into upper-Hessenberg form. This technique is implemented in the following algorithm.

Alg. 3.5.1 **Householder's Transformation**

1. Set $Q = I$ where I is the $n \times n$ identity matrix.
2. For $k = 1$ to $n - 2$ compute

{

$$\alpha = \text{sgn}(A_{k+1,k})\sqrt{\sum_{j=k+1}^{n} A_{jk}^2}$$

$$u = [0, \ldots, 0, A_{k+1,k} + \alpha, A_{k+2,k}, \ldots, A_{n,k}]^T$$

$$P = I - \frac{2uu^T}{u^T u}$$

$$Q = QP$$

$$A = PAP$$

}

3. Set $B = A$.

Note that the Householder method in Alg. 3.5.1 converts the matrix A into upper-Hessenberg form B in a *fixed* number of steps. Therefore, the Householder transformation method is a direct method as opposed to an iterative method, which requires an undetermined number of steps. Of course, the Householder method does not find the eigenvalues of A. Instead, it transforms A to a simpler form from which the eigenvalues of A can be efficiently obtained. If the matrix A is symmetric, then Alg. 3.5.1 converts A to tridiagonal form. The preferred way to find the eigenvalues of a symmetric matrix A is to first preprocess A with the Householder transformation method, and then apply Jacobi's method to the resulting tridiagonal matrix B.

The matrix Q generated by Alg. 3.5.1 can be used to recover the eigenvectors. The upper-Hessenberg matrix B is related to the original matrix A through the composite similarity transformation

$$B = Q^T A Q \tag{3.5.4}$$

Suppose x^k is an eigenvector of B associated with the eigenvalue λ_k. That is, $Bx^k = \lambda_k x^k$. Premultiplying both sides by Q and noting from (3.5.4) that $QB = AQ$, we then have

$$\begin{aligned} A(Qx^k) &= Q(Bx^k) \\ &= \lambda_k(Qx^k) \end{aligned} \tag{3.5.5}$$

It follows from (3.5.5) that if x^k is an eigenvector of B associated with the eigenvalue λ_k, then $y^k = Qx^k$ is an eigenvector of A associated with λ_k. Thus, the eigenvectors of A can be obtained from the eigenvectors of B by multiplication by Q.

EXAMPLE 3.5.1 Householder's Transformation

As an illustration of Householder's method to convert a matrix to upper-Hessenberg form, consider the following symmetric 3×3 matrix.

$$A = \begin{bmatrix} 1 & -2 & 3 \\ -2 & 4 & 1 \\ 3 & 1 & 2 \end{bmatrix}$$

When $n = 3$, the Householder transformation method requires only a single iteration. Applying step (2) with $k = 1$ yields

$$\alpha = \text{sgn}(-2)(4 + 9)^{1/2} = -3.606$$

Next, the vector u is

$$u = [0, -5.606, 3]^T$$

The orthogonal matrix $P(u)$ is then

$$P = I - \frac{2uu^T}{u^T u} = \begin{bmatrix} 1 & 0 & 0 \\ 0 & -0.5547 & 0.8321 \\ 0 & 0.8321 & 0.5547 \end{bmatrix}$$

Finally, we have $Q = P$ and applying the similarity transformation yields

$$A = PAP = \begin{bmatrix} 1 & 3.606 & 0.0000 \\ 3.606 & 1.692 & -0.5385 \\ 0.0000 & -0.5385 & 4.308 \end{bmatrix}$$

Note that because the original matrix A was symmetric, the upper-Hessenberg form in this case is a tridiagonal matrix.

3.6 *QR* METHOD

We now examine an effective method for finding the eigenvalues of a general $n \times n$ real matrix A. The first step is to transform A to upper-Hessenberg form using the Householder transformation method in Alg. 3.5.1. The structure of A is then further simplified with an iterative procedure which eliminates some of the nonzero elements along the first subdiagonal. This results in a matrix B with an *upper block triangular* structure.

$$B = \begin{bmatrix} B_1 & X_1 & X_2 & \cdots & X_{q-2} & X_{q-1} \\ 0 & B_2 & Y_1 & \cdots & Y_{q-3} & Y_{q-2} \\ \vdots & \vdots & \vdots & \ddots & \vdots & \vdots \\ 0 & 0 & 0 & \cdots & B_{q-1} & Z_1 \\ 0 & 0 & 0 & \cdots & 0 & B_q \end{bmatrix} \tag{3.6.1}$$

Each of the components of B is a submatrix. The kth diagonal block, B_k, is either a 1×1 submatrix or a 2×2 submatrix. The submatrices X_k, Y_k, and Z_k contain elements which can be nonzero. The characteristic polynomial of a block triangular matrix is just the product of the characteristic polynomials of the diagonal blocks.

$$\Delta_B(\lambda) = \det(\lambda I - B_1)\det(\lambda I - B_2)\cdots\det(\lambda I - B_q) \tag{3.6.2}$$

As a result, the eigenvalues of B can be obtained directly from the diagonal blocks. If B_k is a 1×1 block, then its associated eigenvalue is simply $\lambda_k = B_k$. The other possibility is that B_k is a 2×2 diagonal block. In this case, there are two eigenvalues associated with B_k, and they can be obtained by factoring its characteristic polynomial using the quadratic formula. For the 2×2 case, this yields,

$$\lambda_{k,k+1} = \frac{\text{trace}(B_k) \pm \sqrt{\text{trace}^2(B_k) - 4\det(B_k)}}{2} \tag{3.6.3}$$

The technique for transforming A from upper-Hessenberg form to upper block triangular form is to factor A into a product of an orthogonal matrix Q times an upper-triangular matrix R.

$$A = QR \tag{3.6.4}$$

Since Q is an orthogonal matrix, $Q^T = Q^{-1}$. Therefore, premultiplying both sides of Equation (3.6.4) by Q^T yields

$$Q^T A = R \tag{3.6.5}$$

From the formulation in (3.6.5), it is clear that to factor A it is sufficient to find an orthogonal Q^T that reduces A to upper-triangular form R. An updated matrix B is then obtained by postmultiplying R by Q.

$$\begin{aligned} B &= RQ \\ &= Q^T A Q \end{aligned} \tag{3.6.6}$$

Thus, B and A are related through a similarity transformation, which means they have the same eigenvalues. The process in Equations (3.6.4) through (3.6.6) is repeated until B approaches upper block triangular form.

Since the matrix A is assumed to start out in upper-Hessenberg form, a sequence of $n - 1$ rotations can be used to transform A to an upper-triangular matrix R. The product of these rotation matrices is the orthogonal matrix Q^T. The following algorithm is an implementation of this technique.

Alg. 3.6.1 ***QR* Method**

1. Apply Alg. 3.5.1 to transform A to upper-Hessenberg form, B.
2. Pick $\varepsilon > 0$ and $m > 0$. Set $i = 0$.
3. Do

   {

   (a) Set $Q^T = I$.

   (b) For $k = 1$ to $n - 1$ do

   {

(i) Compute

$$c = \frac{B_{kk}}{\sqrt{B_{kk}^2 + B_{k+1,k}^2}}$$

$$s = \frac{B_{k+1,k}}{\sqrt{B_{kk}^2 + B_{k+1,k}^2}}$$

(ii) Set $P = I$, then set $P_{kk} = P_{k+1,k+1} = c$ and $P_{k+1,k} = -P_{k,k+1} = -s$.

(iii) Compute

$$B = PB$$

$$Q^T = PQ^T$$

}

(c) Compute $B = BQ$.

(d) Set $i = i + 1$.

}

4. While (B is not upper block triangular) and ($i < m$).

5. Compute λ_k for $k = 1$ to n from the diagonal blocks of B.

Alg. 3.6.1 is called the *QR method* for finding the eigenvalues of an $n \times n$ matrix. The iterations in step (3) continue until either B becomes upper block triangular or the upper bound m on the number of iterations is reached. To determine if B is upper block triangular, the elements along the first subdiagonal are evaluated to make sure that there are no adjacent elements with magnitudes greater than ε. As with Jacobi's method, the QR method does not require full matrix multiplication. For example, in step (3), only rows k and $k + 1$ of B and Q^T have to be updated.

3.6.1 Deflation

The basic QR method in Alg. 3.6.1 can be modified in a number of ways to improve its rate of convergence. The first modification is based on the observation that, with successive iterations, the subdiagonal elements of B approach zero, starting with row n and ending with row 2. If $|B_{n,n-1}| < \varepsilon$ after several iterations, then this indicates that $\lambda_n \approx B_{nn}$. At this point, the dimension of the problem can be reduced by one. That is, the remaining $n - 1$ eigenvalues can be obtained by using the $(n - 1) \times (n - 1)$ submatrix in the upper-left corner of B. This process of reducing the size of B as eigenvalues are identified and extracted is called *deflation*.

If some of the eigenvalues of B are complex, then $|B_{n-1,n}|$ may never become small. Therefore, $B_{n-1,n-2}$ should be checked as well because if $|B_{n-1,n-2}| < \varepsilon$ after several iterations, then λ_{n-1} and λ_n can be obtained by finding the eigenvalues of the 2×2 submatrix in the lower-right corner of B. This is easily done by factoring the characteristic polynomial using the quadratic formula as in Equation (3.6.3). The matrix B is then deflated by removing the last two rows and columns. That is, the remaining $n - 2$ eigenvalues are obtained by using the $(n - 2) \times (n - 2)$ submatrix in the upper-left corner of B.

The process of deflating B by extracting either a real eigenvalue or a complex-conjugate pair of eigenvalues continues until the remaining submatrix in the upper left corner of B is either 1×1 or 2×2. The deflation process does not necessarily reduce the total number of QR iterations, but does substantially reduce the average number of FLOPs per iteration. Consequently, there is a marked increase in speed.

3.6.2 Shifting

Another modification to the basic algorithm which improves the convergence rate involves *shifting* the eigenvalues of A. Rather than factor the matrix A as in Equation (3.6.4), one can instead factor the matrix $A - \alpha I$ for some scalar α.

$$A - \alpha I = QR \tag{3.6.7}$$

If λ_k is an eigenvalue of A, then $\lambda_k - \alpha$ is an eigenvalue of $A - \alpha I$. The original eigenvalues are restored by shifting them back at the end of the iteration, using

$$B = QR + \alpha I \tag{3.6.8}$$

Suppose the eigenvalues of A are ordered according to decreasing magnitude: $|\lambda_1| \geq |\lambda_2| \geq \cdots \geq |\lambda_n|$. In the absence of shifting, the rate of converge to eigenvalue λ_k depends on the ratio $|\lambda_{k+1}|/|\lambda_k|$ with smaller values generating faster rates of convergence. When shifting is used, the convergence to the eigenvalue $\lambda_k - \alpha$ depends on the ratio $|(\lambda_{k+1} - x)/(\lambda_k - x)|$. Consequently, choosing α close to λ_{k+1} can accelerate the convergence. An effective technique is to select α based on the eigenvalues of the 2×2 submatrix, C, appearing in the lower-right corner of the current B (Wilkinson, 1965). If the eigenvalues of this 2×2 submatrix are real, then α is set to the eigenvalue closest to C_{22}. For complex conjugate eigenvalues, α is set to one of the eigenvalues for the first iteration and then the other for the second iteration. This double QR step can be implemented using only real arithmetic (see, e.g., Press et al., 1992).

EXAMPLE 3.6.1 *QR* Method

As an illustration of the QR method, consider the problem of finding the eigenvalues of the 4×4 matrix A investigated in Example 3.3.1 and Example 3.3.2. By applying the power methods to this example, it was determined that the dominant eigenvalue is $\lambda_1 = -4$ and the least dominant eigenvalue is $\lambda_4 = 0.1$. Example program *e361* on the distribution CD finds the remaining eigenvalues by applying the QR method in Alg. 3.6.1 with $\varepsilon = 10^{-6}$. In this case, the QR method converges in 10 iterations to the following upper block triangular matrix B.

$$B = \left[\begin{array}{c|cc|c} -4.0000005 & -3.0400488 & -1.9111767 & -1.4310256 \\ \hline 0.0000003 & 0.1993588 & 0.8285232 & -0.1273743 \\ -0.0000000 & -1.2549365 & -0.1993586 & 0.7941328 \\ \hline 0.0000000 & -0.0000000 & 0.0000000 & 0.1000000 \end{array}\right]$$

Thus, there are two 1×1 diagonal blocks with real eigenvalues and one 2×2 diagonal block with a complex conjugate pair of eigenvalues. Computing the eigenvalues of these blocks, we get the results listed in Table 3.6.1.

TABLE 3.6.1 *QR* Estimates of Eigenvalues

k	λ_k
1	– 4.0000005 +0.0000000j
2	0.0000001 –1.0000000j
3	0.0000001 +1.0000000j
4	0.1000000 +0.0000000j

3.7 DANILEVSKY'S METHOD

Next, we examine the problem of finding the eigenvectors of a general $n \times n$ matrix A by using a two-step process. First, we investigate the properties of a matrix of a special form. Then we consider the problem of transforming a general matrix to this form. Given an $n \times 1$ vector a, consider the following matrix constructed from a.

$$B = \begin{bmatrix} -a_n & -a_{n-1} & \cdots & -a_2 & -a_1 \\ 1 & 0 & \cdots & 0 & 0 \\ 0 & 1 & \cdots & 0 & 0 \\ \vdots & \vdots & \ddots & \vdots & \vdots \\ 0 & 0 & \cdots & 1 & 0 \end{bmatrix} \tag{3.7.1}$$

A matrix with the special structure in (3.7.1) is called a *companion matrix*. The companion matrix in (3.7.1) has the interesting property that its characteristic polynomial is simply

$$\boxed{\Delta_B(\lambda) = \lambda^n + a_n\lambda^{n-1} + \ldots + a_1} \tag{3.7.2}$$

That is, the coefficients of the characteristic polynomial of a companion matrix can be obtained directly from inspection of the first row of the matrix. Of more interest for the present discussion is the simple relationship between the eigenvalues and eigenvectors of a companion matrix. Recall from the Raleigh quotient in (3.3.10) that eigenvalues can be easily computed once the eigenvectors are known. In the case of a companion matrix, the opposite is also true. Suppose λ_k is the kth eigenvalue of the companion matrix B. Then the kth eigenvector, y^k, can be computed as follows (see, e.g., Lastman and Sinha, 1989):

$$\boxed{y^k = [\lambda_k^{n-1}, \lambda_k^{n-2}, \ldots, \lambda_k, 1]^T \quad , \quad 1 \leq k \leq n} \tag{3.7.3}$$

Consequently, (3.7.3) provides a simple means of computing the eigenvectors of B once the eigenvalues are known.

To exploit the special properties of the companion matrix, we next consider the problem of converting a general $n \times n$ matrix A to companion form B. If this is done with a similarity transformation, then the eigenvalues of B will be identical to the eigenvalues of A. Furthermore, the eigenvectors of A can be reconstructed from the eigenvectors of B by reversing the transformation.

To convert A to the companion form B, we start with the last row of A and work backwards one row at a time. Initially, suppose $A_{n,n-1} \neq 0$. Element $A_{n,n-1}$ can then be normalized to unity by dividing column $n - 1$ of A by

$$\alpha = A_{n,n-1} \tag{3.7.4}$$

Next the kth element of the last row of A can be made zero by subtracting A_{nk} times column $n - 1$ from column k for $k \neq n - 1$. These elementary column operations can be combined into a single operation represented by postmultiplication of A (multiplication on the right) by the following matrix.

$$P_n = \begin{bmatrix} 1 & 0 & \cdots & 0 & 0 & 0 \\ 0 & 1 & \cdots & 0 & 0 & 0 \\ \vdots & \vdots & \ddots & \vdots & \vdots & \vdots \\ 0 & 0 & \cdots & 1 & 0 & 0 \\ -A_{n1}/\alpha & -A_{n2}/\alpha & \cdots & -A_{n,n-2}/\alpha & 1/\alpha & -A_{nn}/\alpha \\ 0 & 0 & \cdots & 0 & 0 & 1 \end{bmatrix} \tag{3.7.5}$$

A direct calculation verifies that the last row of AP_n is identical to the last row of B in (3.7.1). To complete the similarity transformation (and thus preserve the eigenvalues), we must also premultiply A by the inverse of P_n. The following expression for the inverse of P_n can be verified through direct multiplication.

$$P_n^{-1} = \begin{bmatrix} 1 & 0 & \cdots & 0 & 0 & 0 \\ 0 & 1 & \cdots & 0 & 0 & 0 \\ \vdots & \vdots & \ddots & \vdots & \vdots & \vdots \\ 0 & 0 & \cdots & 1 & 0 & 0 \\ A_{n1} & A_{n2} & \cdots & A_{n,n-2} & \alpha & A_{nn} \\ 0 & 0 & \cdots & 0 & 0 & 1 \end{bmatrix} \tag{3.7.6}$$

When AP_n is premultiplied by P_n^{-1}, the last row of AP_n is not affected. Therefore, the last row of $B_{n-1} \triangleq P_n^{-1}AP_n$ is identical to the last row of the companion matrix B. This process can now be repeated by working on row $n - 1$ of B_{n-1} and so on until rows 2 through n of B_1 have the required companion form structure. The final result after $n - 1$ similarity transformations is

$$B = P^{-1}AP \tag{3.7.7}$$

Here P is the composite similarity transformation matrix $P = P_n P_{n-1} \cdots P_2$. The technique for converting A to companion matrix form using a sequence of $n - 1$ similarity transformations is called *Danilevsky's method.* Note that it is analogous to performing Gaussian elimination except that elementary column operations (postmultiplication) are used in addition to elementary row operations (premultiplication). As with Gaussian elimination, the process can be made more robust by using pivoting. In the case of constructing P_k, the pivot element is in row k and column $k - 1$ of B_k. To avoid possible division by zero, we search row k to the left of column k for the element with the largest magnitude. If this pivot candidate occurs in column j, it is brought into position by interchanging columns j and $k - 1$ of B_k. This is followed by an interchange of rows j and $k - 1$ of B_k in order to preserve the eigenvalues.

Pivoting with the largest element tends to reduce the effects of accumulated round-off error. In most instances, it also eliminates the possibility of division by zero. However, there are *degenerate* cases when all the elements in row k to the left of column k are zero. When this occurs, it indicates that the matrix B_k has a special upper block triangular structure as follows.

$$B_k = \begin{bmatrix} F & G \\ 0 & H \end{bmatrix} \tag{3.7.8}$$

Here F is a $k \times k$ diagonal block and H is an $(n - k) \times (n - k)$ diagonal block. Recall from Equation (3.6.2) that the characteristic polynomial of a block triangular matrix is simply the product of the characteristic polynomials of the diagonal blocks.

$$\Delta(\lambda) = \Delta_F(\lambda)\Delta_H(\lambda) \tag{3.7.9}$$

Moreover, the diagonal block H is already in companion form, so its characteristic polynomial, $\Delta_H(\lambda)$, can be determined from inspection of the first row of H. To find the characteristic polynomial of F, one can apply Danilevsky's method to the $k \times k$ block F. The degenerate case has the interesting property that the overall characteristic polynomial is already partially factored.

The eigenvectors of the companion matrix B are easily computed using Equation (3.7.3) once the eigenvalues are known. Recall from (3.4.3) that the eigenvectors of A can then be obtained from the eigenvectors of B by premultiplication by the similarity transformation matrix P. The following algorithm is an implementation of Danilevsky's method for finding the eigenvectors of a general $n \times n$ matrix A. The eigenvalues are computed using the QR method. The algorithm returns an error code E that is nonzero if the transformation to companion form is successful.

Alg. 3.7.1 **Danilevsky's Method**

1. Set $E = 1$, and form the augmented $2n \times n$ matrix $Q = \begin{bmatrix} A \\ I \end{bmatrix}$.
2. For $k = n$ down to 2 do

{

(a) Compute the pivot index $1 \le p < k$ such that

$$|Q_{kp}| = \max_{j=1}^{k-1} \{|Q_{kj}|\}$$

(b) If $|Q_{kp}| = 0$, set $E = 0$ and exit.

(c) Interchange columns p and $k - 1$ of Q.

(d) Interchange rows p and $k - 1$ of Q.

(e) Save row k as $r_j = Q_{kj}$ for $j = 1$ to n.

(f) Divide column $k - 1$ of Q by $Q_{k,k-1}$.

(g) For $j \ne k - 1$ subtract Q_{kj} times column $k - 1$ from column j.

(h) For $j = 1$ to n do

{

(1) Save column j as $c_i = Q_{ij}$ for $i = 1$ to n.
(2) Compute $Q_{k-1,j} = r^T c$.
}

}

3. Partition Q as $Q = \begin{bmatrix} B \\ P \end{bmatrix}$ where B and P are $n \times n$.
4. Apply the QR method in Alg. 3.6.1 to compute the eigenvalues $\{\lambda_1, \lambda_2, \ldots, \lambda_n\}$ of A.
5. For $k = 1$ to n compute

{

$$y^k = [\lambda_k^{n-1}, \lambda_k^{n-2}, \ldots, \lambda_k, 1]^T$$
$$x^k = P y^k$$

}

Steps (2a) through (2d) of Alg. 3.7.1 implement the pivoting operation. Postmultiplication by P_k is performed in steps (2f) and (2g), and premultiplication by P_k^{-1} is performed in step (2h). If Alg. 3.7.1 terminates with $E \neq 0$, then the eigenvalues of A are $\{\lambda_1, \lambda_2, \ldots, \lambda_n\}$ and the corresponding eigenvectors are $\{x^1, x^2, \ldots, x^n\}$. A return code of $E = 0$ indicates that the degenerate case has been detected.

EXAMPLE 3.7.1 Danilevsky's Method

As an illustration of Danilevsky's method, consider the problem of finding the eigenvalues and eigenvectors of the following 4×4 matrix.

$$A = \begin{bmatrix} 2 & -1 & 3 & 4 \\ 0 & 7 & 1 & -6 \\ 5 & 8 & 0 & -3 \\ 1 & 4 & -3 & 9 \end{bmatrix}$$

Example program *e371* on the distribution CD applies Alg. 3.7.1 to A with $\varepsilon = 10^{-6}$. The resulting companion form B and transformation matrix P are as follows.

$$B = \begin{bmatrix} 18.0000 & -83.0000 & -199.0000 & 1656.0001 \\ 1.0000 & 0.0000 & 0.0000 & 0.0000 \\ 0.0000 & 1.0000 & 0.0000 & 0.0000 \\ 0.0000 & -0.0000 & 1.0000 & 0.0000 \end{bmatrix}$$

$$P = \begin{bmatrix} 0.0300 & -0.3723 & 0.6513 & 6.9935 \\ 0.0260 & -0.2409 & 0.0227 & 5.2376 \\ 0.0446 & -0.4453 & -0.0860 & 12.3147 \\ 0.0000 & -0.0000 & 0.0000 & 1.0000 \end{bmatrix}$$

Thus, the characteristic polynomial of A is

$$\Delta(\lambda) = \lambda^4 - 18\lambda^3 + 83\lambda^2 + 199\lambda + 1656$$

The QR method converged in 11 iterations to the eigenvalues shown in Table 3.7.1. Also shown are the associated eigenvectors. We can use the trace condition in Equation (3.2.6) as a partial check of the numerical accuracy of the estimated eigenvalues.

TABLE 3.7.1 Danilevsky Estimates of Eigenvalues and Eigenvectors

λ	x^1	x^2	x^3	x^4
7.5812 + 2.8019j	1.1751 − 0.1531j	1.1751 + 0.1531j	3.6868 + 0.0000j	−2.5616 + 0.0000j
7.5812 − 2.8019j	0.1311 + 1.7976j	0.1311 − 1.7976j	2.3687 + 0.0000j	0.2127 + 0.0000j
6.6499 + 0.0000j	1.0394 + 1.4119j	1.0394 − 1.4119j	5.1706 + 0.0000j	3.7004 + 0.0000j
−3.8122 + 0.0000j	1.0000 + 0.0000j	1.0000 + 0.0000j	1.0000 + 0.0000j	1.0000 + 0.0000j

$$\left|\text{trace}(A) - \sum_{k=1}^{4} \lambda_k\right| = 4.768 \times 10^{-7}$$

Although Danilevsky's method provides us with a simple technique for computing the eigenvectors of an $n \times n$ matrix, the results can be sensitive to the effects of accumulated round-off error. If the accuracy of the eigenvector estimates is an important consideration, then it may be useful to carry out the computations in Alg. 3.7.1 using high-precision arithmetic.

3.8 POLYNOMIAL ROOTS

The highly effectively techniques for computing eigenvalues can be put to work to solve an important class of problems that occur repeatedly in engineering applications. Let $f(x)$ denote the following polynomial of degree n.

$$f(x) = a_1 + a_2 x + \cdots + a_{n+1} x^n \tag{3.8.1}$$

The problem of interest is to find all of the solutions of the equation $f(x) = 0$. Each solution is called a *root* of the polynomial. From the fundamental theorem of algebra, a polynomial of degree n has exactly n roots in the complex plane. Furthermore, if the coefficients are real, then complex roots occur in conjugate paris. The polynomial $f(x)$ can be written in factored form as follows.

$$f(x) = a_{n+1}(x - r_1)(x - r_2)\cdots(x - r_n) \tag{3.8.2}$$

The objective is to find the $n \times 1$ vector of roots r. Knowing the locations of the roots in the complex plane provides valuable *qualitative* information about certain engineering problems. For example, the poles of a transfer function of a linear dynamical system tell us whether or not that system is stable (Ogata, 1997).

For convenience, we assume that the coefficient of the highest power of x in Equation (3.8.1) has been normalized to $a_{n+1} = 1$. If this is not the case, one can always divide both sides of (3.8.1) by a_{n+1}. This process simplifies the polynomial without changing the roots. An nth degree polynomial for which $a_{n+1} = 1$ is called a *monic* polynomial. A familiar example of a monic polynomial is the characteristic polynomial of the matrix A.

$$\Delta(\lambda) = \det(\lambda I - A) \tag{3.8.3}$$

Just as every matrix has associated with it a characteristic polynomial, every polynomial has associated with it a corresponding matrix. Consider, in particular, the following matrix associated with the monic polynomial $f(x)$.

$$A = \begin{bmatrix} -a_n & -a_{n-1} & \cdots & -a_2 & -a_1 \\ 1 & 0 & \cdots & 0 & 0 \\ 0 & 1 & \cdots & 0 & 0 \\ \vdots & \vdots & \ddots & \vdots & \vdots \\ 0 & 0 & \cdots & 1 & 0 \end{bmatrix} \tag{3.8.4}$$

Recall from Section 3.7 that a matrix with this special structure is called a companion matrix. It is possible to show, using induction, that the characteristic polynomial of the companion matrix is

$$\Delta(\lambda) = \lambda^n + a_n\lambda^{n-1} + \cdots + a_2\lambda + a_1 \tag{3.8.5}$$

Comparing (3.8.5) with (3.8.1) assuming $a_{n+1} = 1$, we see that the characteristic polynomial of the companion matrix is *identical* to the polynomial used to generate that matrix. As a result, finding the eigenvalues of the matrix A in (3.8.4) is equivalent to finding the roots of the polynomial $f(x)$. The eigenvalues of A can be found efficiently using the iterative QR method. Note that the companion matrix is already in upper-Hessenberg form as all of the elements below the first subdiagonal are zero. Consequently, the first step in Alg. 3.6.1 can be skipped in this case.

The following algorithm uses the QR method to find the roots of a monic polynomial $f(x)$ of degree n. It terminates once QR convergence has been obtained or a maximum number of QR iterations m has been performed.

Alg. 3.8.1 **Polynomial Roots: *QR***

1. Pick $\varepsilon > 0, m > 0$.
2. Form the $n \times n$ companion matrix.

$$A = \begin{bmatrix} -a_n & -a_{n-1} & \cdots & -a_2 & -a_1 \\ 1 & 0 & \cdots & 0 & 0 \\ 0 & 1 & \cdots & 0 & 0 \\ \vdots & \vdots & \ddots & \vdots & \vdots \\ 0 & 0 & \cdots & 1 & 0 \end{bmatrix}$$

3. Apply Alg. 3.6.1 starting at step (2) to find the $n \times 1$ vector of eigenvalues λ.
4. Set $r = \lambda$.

The following example illustrates the use of Alg. 3.8.1. Examples of more general root-finding techniques can be found in Chapter 5.

EXAMPLE 3.8.1 Polynomial Roots: *QR*

Consider the following polynomial of degree $n = 5$.

$$f(x) = x^5 - x^4 - 4x^3 + 4x^2 - 5x - 75$$

From (3.8.4), the companion matrix of $f(x)$ is

$$A = \begin{bmatrix} -1 & -4 & 4 & -5 & -75 \\ 1 & 0 & 0 & 0 & 0 \\ 0 & 1 & 0 & 0 & 0 \\ 0 & 0 & 1 & 0 & 0 \\ 0 & 0 & 0 & 1 & 0 \end{bmatrix}$$

Example program *e381* on the distribution CD applies the *QR* method to the matrix A with the results summarized in Table 3.8.1 which shows the root number, the estimated root, and the magnitude of f at the estimated root. In this case 42 iterations of the *QR* method were required for convergence. The resulting roots are exact when rounded to six digits.

TABLE 3.8.1 Polynomial Roots: *QR* Method

k	r_k	$\lvert f(r_k)\rvert$
1	3.0000019 + 0.0000000j	0.0003967
2	1.0000002 + 2.0000000j	0.0000389
3	1.0000002 − 2.0000000j	0.0000389
4	−2.0000000 + 0.9999998j	0.0000295
5	−2.0000000 − 0.9999998j	0.0000295

3.9 APPLICATIONS

The following examples illustrate applications of eigenvalue and eigenvector techniques using both MATLAB and C. The relevant MATLAB functions in the NLIB toolbox are described in Section 1.4 of Appendix 1, while the corresponding C functions in the NLIB library are described in Section 2.7 of Appendix 2.

3.9.1 Transient Analysis of an Absorption Process: C

In Chapter 2, the steady-state operation of a multi-stage chemical absorption process was analyzed. A diagram of a four-stage version of the absorption process is shown in Figure 3.9.1. Recall that when the gas and liquid come into contact at stage k, a component of the gas (e.g. sulfur dioxide in combustion gas) is absorbed into the liquid where x_k denotes the concentration of the absorbed component in the liquid and y_k denotes the concentration of the absorbed component in the gas.

The analysis in Section 2.8.1 showed that the rate of change of the concentration of the absorbed component in the liquid satisfied the following system of first-order differential equations.

$$\tau \frac{dx_1}{dt} = K(y_f - b) - (1 + \delta)x_1 + x_2 \qquad (3.9.1a)$$

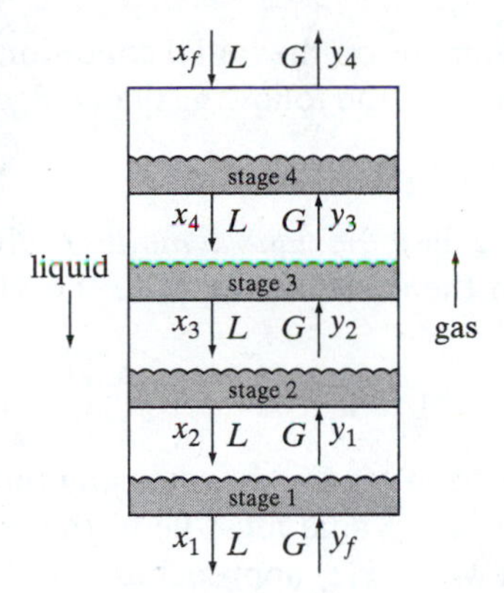

FIGURE 3.9.1 A Four-Stage Chemical Absorption Process

$$\tau \frac{dx_2}{dt} = \delta x_1 - (1 + \delta)x_2 + x_3 \tag{3.9.1b}$$

$$\tau \frac{dx_3}{dt} = \delta x_2 - (1 + \delta)x_3 + x_4 \tag{3.9.1c}$$

$$\tau \frac{dx_4}{dt} = \delta x_3 - (1 + \delta)x_4 + x_f \tag{3.9.1d}$$

This system of four differential equations can be recast in terms of vectors and matrices as follows.

$$\frac{dx}{dt} = \frac{1}{\tau}\begin{bmatrix} -(1+\delta) & 1 & 0 & 0 \\ \delta & -(1+\delta) & 1 & 0 \\ 0 & \delta & -(1+\delta) & 1 \\ 0 & 0 & \delta & -(1+\delta) \end{bmatrix} x + \frac{1}{\tau}\begin{bmatrix} K(y_f - b) \\ 0 \\ 0 \\ x_f \end{bmatrix} \tag{3.9.2}$$

Here the gas and liquid feed concentrations, y_f and x_f, are regarded as inputs to the system. In Section 2.8.1, these inputs were assumed constant. The derivative vector dx/dt was then set to zero, and the resulting linear algebraic system was solved for x to find the steady-state concentrations. Unfortunately, this *static* analysis does not provide any information about the transient performance of the process, specifically: how *fast* is the steady-state solution reached? To answer this question, we must look at the eigenvalues of the differential equation coefficient matrix.

$$A = \frac{1}{\tau}\begin{bmatrix} -(1+\delta) & 1 & 0 & 0 \\ \delta & -(1+\delta) & 1 & 0 \\ 0 & \delta & -(1+\delta) & 1 \\ 0 & 0 & \delta & -(1+\delta) \end{bmatrix} \tag{3.9.3}$$

Let $\lambda = [\lambda_1, \lambda_2, \lambda_3, \lambda_4]^T$ and $X = [x^1, x^2, x^3, x^4]$ denote the eigenvalues and eigenvectors of A, and let x^s denote the steady-state solution. If the eigenvalues are distinct and the inputs are constant, the complete solution $x(t)$ can be expressed as follows.

$$x(t) = \sum_{k=1}^{4} c_k \exp(\lambda_k t)x^k + x^s \tag{3.9.4}$$

Here the constant vector $c = [c_1, c_2, c_3, c_4]^T$ depends on the initial condition $x(0)$. In particular, setting $t = 0$, we find that c is the solution of the following linear algebraic system.

$$Xc = x(0) - x^s \tag{3.9.5}$$

The four terms in the summation of $x(t)$ are called the *natural modes* of the system and together they make up the transient part of the solution. Suppose the kth eigenvalue has a real part a_k and imaginary part b_k.

$$\lambda_k = a_k + jb_k \quad , \quad 1 \le k \le 4 \tag{3.9.6}$$

For the transient part of the solution to approach zero with increasing time, it is necessary that $a_k < 0$ for all k. That is, all of the eigenvalue must lie in the left half of the complex plane. Furthermore, the speed with which $x(t)$ approaches x^s is determined by how deep the eigenvalues are located into the left-half plane. Suppose we are interested in determining how fast it takes for the transient part of the solution to decay to, say, p percent of its initial value. The kth natural mode term decays to zero at the rate $\exp(a_k t)$. Setting this to $p/100$ and solving for t yields the following expression for the decay time of the kth natural mode.

$$T_k = \frac{\ln(0.01p)}{a_k} \quad , \quad 1 \le k \le 4 \tag{3.9.7}$$

The speed with which the overall solution $x(t)$ approaches its steady-state value x^s is then determined by the largest T_k, which is the decay time associated with the *dominant* or slowest mode.

To make the transient analysis example specific, suppose the stage liquid residence time is $\tau = 1.86$ minutes and the remaining parameters are as in Section 2.8.1. Let the transient decay percentage be $p = 1$ percent. The following C program on the distribution CD can be used to solve this problem.

```
/* ---------------------------------------------------------------- */
/* Example 3.9.1: Chemical Absorption Transients */
/* ---------------------------------------------------------------- */

#include "c:\nlib\util.h"

int main (void)
{
   int        n = 4,               /* number of stages */
              m = 500,             /* maximum iterations */
              i;
   float      a = 0.72f,           /* slope */
              K = 1.63f,           /* gas to liquid ratio */
              tau = 1.86f,         /* stage liquid residence time */
              delta = a*K,         /* stripping factor */
              eps = 1.e-5f,        /* convergence threshold */
              p = 1.0f,            /* decay percentage */
              t = 0.0f;
```

```
   vector     T = vec (n,"");             /* decay times */
   cvector    lambda = cvec (n,"");       /* eigenvalues */
   matrix     A = mat (n,n,"");

/* Initialize A */

   printf ("\nExample 3.9.1: Chemical Absorption Transients\n");
   for (i = 1; i <= n; i++)
   {
      A[i][i] = -(1 + delta);
      if (i < n)
      {
        A[i][i+1] = 1;
        A[i+1][i] = delta;
      }
   }
   showmat ("A",A,n,n,0);

/* Find eigenvalues and decay times */

   shownum ("QR iterations",(float) qr(A,n,m,eps,lambda));
   showcvec ("Eigenvalues",lambda,n,0);
   for (i = 1; i <= n; i++)
      t += A[i][i] - lambda[i].x;
   shownum ("Trace check",(float) fabs(t));
   for (i = 1; i <= n; i++)
      T[i] = ((float) log(p/100))/lambda[i].x;
   showvec ("Decay times",T,n,0);
   return 0;
}
/* ------------------------------------------------------------ */
```

When program *e391.c* is executed, it uses the *QR* method to generate the following set of eigenvalues and decay times for *A*.

$$\lambda = \begin{bmatrix} -3.926 \\ -2.843 \\ -1.504 \\ -0.4207 \end{bmatrix}, \quad T = \begin{bmatrix} 1.173 \\ 1.62 \\ 3.062 \\ 10.95 \end{bmatrix} \tag{3.9.8}$$

In this case, all of the eigenvalues happen to be real and all are negative. The eigenvalue with the largest real part is $\lambda_4 = -0.4207$. Thus, the last mode dominates the transient response, which decays to within $p = 1$ percent of its initial value after $T_4 = 10.95$ minutes. As a partial check on the accuracy of λ, the sum of the eigenvalues was subtracted from the trace of A to yield

$$\left| \text{trace}(A) - \sum_{k=1}^{4} \lambda_k \right| = 7.451 \times 10^{-7} \approx 0 \tag{3.9.9}$$

Note that the trace check is about as close to zero as can be expected using single-precision arithmetic. A solution to Example 3.9.1 using MATLAB can be found in the file *e391.m* on the distribution CD.

3.9.2 Population Growth Model: MATLAB

Some physical phenomena are naturally modeled as discrete-time systems with difference equations rather than as continuous-time systems with differential equations. As an example of an ecological system, consider the problem of modeling the population growth of a plant or animal species. Let $x_i(k)$ denote the number of individuals of age i that are alive at the end of year k for $1 \leq i \leq n$. Thus, $x(k)$ is an $n \times 1$ vector that represents the *age distribution* of the population at the end of year k.

The population level of each age group changes through births, which add to the population, and deaths due to predators or severe weather, which subtract from the population. Let $\alpha_i x_i(k)$ denote the number of individuals who are born from age group i during year k. These individuals are regarded as one year old at the end of year k and therefore

$$x_1(k) = \alpha_1 x_1(k-1) + \alpha_2 x_2(k-1) + \cdots + \alpha_n x_n(k-1) \tag{3.9.10}$$

The birth rate parameters α_k will vary with age. For example, the very young and the very old may be less likely to reproduce than those in their optimal reproductive years. As the population ages, only a certain fraction will survive from one year to the next. Let $0 < \beta_i \leq 1$ denote the fraction of individuals of age i that live to be age $i + 1$. This can be expressed

$$x_{i+1}(k) = \beta_i x_i(k-1) \quad , \quad 1 \leq i \leq n-1 \tag{3.9.11}$$

A block diagram of the population growth model is shown in Figure 3.9.2. Here z^{-1} denotes a *delay* operator which delays the discrete-time signal by one year. Thus, the output of the first block is $x_2(k) = \beta_1 x_1(k-1)$.

The n difference equations can be combined into a single vector equation in which the population distribution at the end of year k equals a coefficient matrix A times the population distribution at the end of year $k - 1$.

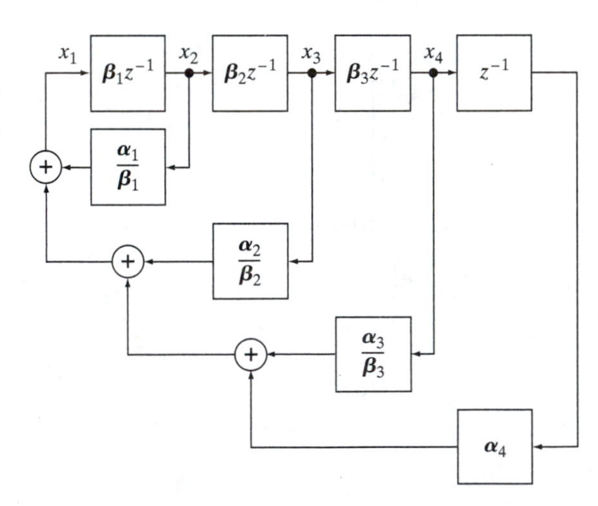

FIGURE 3.9.2 Block Diagram of Population Growth Model

$$x(k) = \underbrace{\begin{bmatrix} \alpha_1 & \alpha_2 & \cdots & \alpha_{n-1} & \alpha_n \\ \beta_1 & 0 & \cdots & 0 & 0 \\ 0 & \beta_2 & \cdots & 0 & 0 \\ \vdots & \vdots & \ddots & \vdots & 0 \\ 0 & 0 & \cdots & \beta_n & 0 \end{bmatrix}}_{A} x(k-1) \quad , \quad k \geq 1 \tag{3.9.12}$$

The $n \times 1$ vector $x(0)$ denotes the initial population distribution introduced into the environment. In order to see how the population grows, note that $x(1) = Ax(0)$, which means that $x(2) = Ax(1) = A[Ax(0)] = A^2x(0)$, and so on. If this procedure is repeated k times, the population distribution at the end of year k is

$$x(k) = A^k x(0) \quad , \quad k \geq 0 \tag{3.9.13}$$

To determine whether the species flourishes or dies out, it is helpful to examine the eigenvalues $\lambda = [\lambda_1, \lambda_2, \ldots, \lambda_n]^T$ and the eigenvectors $X = [x^1, x^2, \ldots, x^n]$ of the coefficient matrix A. If the n eigenvalues are distinct, the solution of the difference equation can be expressed as follows.

$$x(k) = \sum_{i=1}^{n} c_i \lambda_i^k x^i \quad , \quad k \geq 0 \tag{3.9.14}$$

Here the constant vector $c = [c_1, c_2, \ldots, c_n]^T$ depends on the initial condition $x(0)$. In particular, setting $k = 0$ in (3.9.14) makes it evident that c is the solution of the following linear algebraic system.

$$Xc = x(0) \tag{3.9.15}$$

Each term in the solution $x(k)$ is a natural mode of system. Since there is no constant input to this ecological system, the steady-state part of the solution is zero. Note that if $|\lambda_i| < 1$ for $1 \leq i \leq n$, then all of the natural mode terms will decay to zero. Consequently, if the spectral radius $\rho(A)$ is less than one, the species will not survive. However, if $\rho(A) > 1$ the species will continue to grow and flourish. Suppose the eigenvalue λ_j is larger in magnitude than all the other eigenvalues and $|\lambda_j| > 1$. Then, as time increases, the solution will approach an age distribution $x(k)$ that is proportional to the eigenvector x^j associated with the dominant eigenvalue λ_j.

To make the ecological system specific, suppose $n = 5$, the birth rate parameter vector is $\alpha = [1.1, 1.5, 2.2, 2.7, 1.3]^T$ and the death rate parameter vector is $\beta = [0.4, 0.1, 0.1, 0.5]^T$. Then the growth model coefficient matrix is

$$A = \begin{bmatrix} 1.1 & 1.5 & 2.2 & 2.7 & 1.3 \\ 0.4 & 0 & 0 & 0 & 0 \\ 0 & 0.1 & 0 & 0 & 0 \\ 0 & 0 & 0.1 & 0 & 0 \\ 0 & 0 & 0 & 0.5 & 0 \end{bmatrix} \tag{3.9.16}$$

The following MATLAB script on the distribution CD can be used to solve this problem.

```
%----------------------------------------------------------------------
% Example 3.9.2: Population Growth Model
%----------------------------------------------------------------------

% Initialize

   clc                                   % clear screen
   clear                                 % clear variables
   n = 5;                                % number of stages
   alpha = [1.1 1.5 2.2 2.7 1.3]';       % birth vector
   beta = [0.4 0.1 0.1 0.5]';            % death vector
   A = zeros (n,n);                      % growth coefficient matrix
   tol = 1.e-6;                          % error tolerance

% Compute A

   fprintf ('Example 3.9.2: Population Growth Model\n');
   for i = 1: n
      A(1,i) = alpha(i);
      if i < n
         A(i+1,i) = beta(i);
      end
   end
   show ('A',A)

% Find dominant eigenvalue and eigenvector

   [c,x,k] = powereig (A,tol,50,1);
   show ('Power method iterations',k)
   show ('Dominant eigenvalue',c)
   show ('Dominant eigenvector',x)
%----------------------------------------------------------------------
```

When script *e392.m* is executed, it uses the iterative power method to produce the following dominant eigenvalue for the coefficient matrix A.

$$\lambda_1 = 1.532 \tag{3.9.17}$$

In this case, the spectral radius is $\rho(A) = 1.532 > 1$. Thus, the species flourishes. The stable age distribution that is approached with increasing time is given by the eigenvector x^1 associated with eigenvalue λ_1. This was also generated by the power method and was found to be

$$x^1 = [1 \quad 0.261 \quad 0.01703 \quad 0.001111 \quad 0.0003627]^T \tag{3.9.18}$$

It is clear that the age distribution in this case is skewed toward the younger members of the species. A solution to Example 3.9.2 using C can be found in the file *e392.c* on the distribution CD.

3.9.3 Telescope Position Control: C

Engineers are sometimes called upon to design feedback systems for precise position control. For example, suppose a large telescope must be pointed at precise coordinates in the sky. Systems of this nature are modeled as an electrical motor turning a rotating mechanical load. A schematic diagram of a position control feedback system is shown in Figure 3.9.3.

For the DC motor, R_a is the armature resistance, L_a is the armature inductance, I_a is the armature current, and τ is the torque developed at the motor shaft. The motor shaft is coupled to a mechanical load, possibly through a gear train, which is modeled by a moment of inertia J and a viscous friction coefficient b. The variable to be controlled is the load shaft angle θ. The system in Figure 3.9.3 is a feedback system because the shaft angle θ is fed back and compared with the desired shaft angle r which is the control system input. The difference between the desired and measured angles is the *error* signal $e = r - \theta$. The error is amplified by an amplifier of gain k_c to produce the armature voltage V_a applied to the motor.

$$V_a = k_c(r - \theta) \tag{3.9.19}$$

When the error becomes nonzero, the motor activates and turns the shaft in a direction that reduces the error. Since the control voltage V_a is proportional to the error signal e, this is called a *proportional* feedback control system.

The lone design parameter that the engineer has control over is the amplifier gain k_c. As k_c increases, the overall steady-state tracking performance improves. However, if the controller gain k_c is made too large, the feedback system will become unstable and not work at all. To determine whether or not a given k_c is too large, we must examine the eigenvalues of this system. Applying Kirchhoff's voltage law to the armature winding yields the following electrical equation of motion.

$$V_a = R_a I_a + L_a \frac{dI_a}{dt} + k_b \frac{d\theta}{dt} \tag{3.9.20}$$

Here the constant k_b is called the back emf constant of the motor. Next, applying Newton's second law to the rotating mass yields the following mechanical equation of motion.

$$J \frac{d^2\theta}{dt^2} = \tau - b \frac{d\theta}{dt} \tag{3.9.21}$$

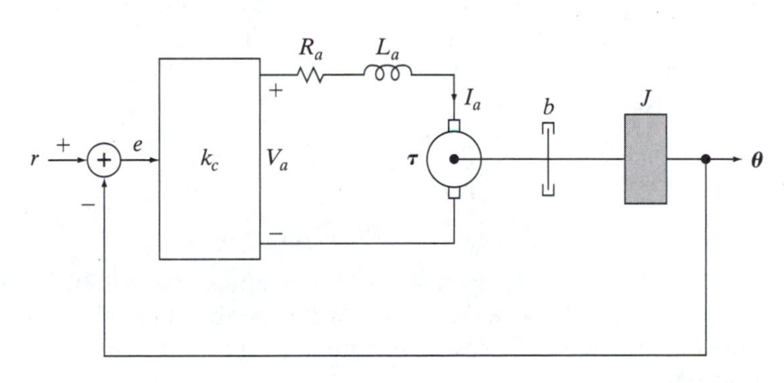

FIGURE 3.9.3 Position Control Feedback System

The torque developed at the motor shaft is proportional to the armature current I_a. Thus, the algebraic equation that couples the mechanical system to the electrical system is the following.

$$\tau = k_t I_a \tag{3.9.22}$$

The constant of proportionality, k_t, is called the torque constant of the motor. The four equations which model the feedback control system can be combined and recast as a first order system of differential equations (see Section 8.1). Let x denote the following 3×1 vector of state variables (Ogata, 1997).

$$x = [\theta,\ d\theta/\,dt,\ I_a]^T \tag{3.9.23}$$

Thus, x specifies the position and velocity of the rotating mass and the armature winding current. Differentiating the components of x and substituting using the model equations, we arrive at the following mathematic model.

$$\frac{dx_1}{dt} = x_2 \tag{3.9.24a}$$

$$\frac{dx_2}{dt} = \frac{k_t x_3 - b x_2}{J} \tag{3.9.24b}$$

$$\frac{dx_3}{dt} = \frac{k_c(r - x_1) - R_a x_3 - k_b x_2}{L_a} \tag{3.9.24c}$$

This system of three equations can be rewritten using vectors as matrices as

$$\frac{dx}{dt} = \begin{bmatrix} 0 & 1 & 0 \\ 0 & -b/J & k_t/J \\ -k_c/L_a & -k_b/L_a & -R_a/L_a \end{bmatrix} x + \begin{bmatrix} 0 \\ 0 \\ k_c/L_a \end{bmatrix} r \tag{3.9.25}$$

Before we examine the eigenvalues, it is useful to obtain the steady-state solution when the input r is constant. Setting $dx/dt = 0$ and solving the first equation yields $x_2 = 0$. Substituting this into the second equation then yields $x_3 = 0$. Finally, substituting into the third equation yields $x_1 = r$. Thus, the steady-state solution of the feedback system is

$$x^s = [r, 0, 0]^T \tag{3.9.26}$$

Recalling that $x_1 = \theta$, it follows that when r is constant the steady-state angle of the load shaft is $\theta = r$. However, this analysis of the steady-state assumes that the transient part of the solution goes to zero, that is, it assumes that the system is stable. To determine whether or not the feedback system is stable, we must look at the eigenvalues of the coefficient matrix.

$$A = \begin{bmatrix} 0 & 1 & 0 \\ 0 & -b/J & k_t/J \\ -k_c/L_a & -k_b/L_a & -R_a/L_a \end{bmatrix} \tag{3.9.27}$$

For some values of the controller gain k_c, the eigenvalues of A will be in the left half of the complex plane, and therefore the system will be stable. For other values of k_c, one or more of the eigenvalues will cross over into the right half plane, thereby rendering the control system unstable.

To make the problem specific, suppose the armature resistance is $R_a = 10\Omega$, the armature inductance is $L_a = 100$ mH, the back emf constant is $k_b = 0.1$ A-s, the torque constant is $k_t = 0.32$ N-m/A, the viscous friction coefficient is $b = 0.1$ N-m-s, and the moment of inertia of the load is $J = 2$ N-m-s^2. Three values of controller gain are considered: $k_c = \{100, 200, 400\}$ V-s. The following C program on the distribution CD can be used to solve this problem.

```
/* ----------------------------------------------------------- */
/* Example 3.9.3: Telescope Position Control */
/* ----------------------------------------------------------- */

#include "c:\nlib\util.h"

int main (void)
{
   int       n = 3,                   /* order of system */
             m = 500,                 /* maximum iterations */
             i,j;
   float     Ra = 10.0f,              /* armature resistance */
             La = 0.100f,             /* armature inductance */
             kb = 0.1f,               /* back emf constant */
             kt = 0.32f,              /* torque constant */
             b = 0.1f,                /* friction coefficient */
             J = 2.0f,                /* moment of inertia */
             eps = 1.e-5f,            /* convergence threshold */
             t = 0.0f;
   vector    kc = vec (3,"100 200 400");    /* controller gains */
   cvector   lambda = cvec (n,"");          /* eigenvalues */
   matrix    A = mat (n,n,"");

/* Initialize A */

   printf ("\nExample 3.9.3: Telescope Position Control\n");
   A[1][2] = 1;
   A[2][2] = -b/J;
   A[2][3] = kt/J;
   A[3][2] = -kb/La;
   A[3][3] = -Ra/La;
   showmat ("A",A,n,n,0);

/* Find eigenvalues and oscillation frequencies */

   for (i = 1; i <= 3; i++)
   {
      A[3][1] = -kc[i]/La;
      shownum ("QR iterations",(float) qr(A,n,m,eps,lambda));
      showcvec ("Eigenvalues",lambda,n,0);
      t = 0;
      for (j = 1; j <= n; j++)
         t += A[j][j] - lambda[j].x;
```

```
        t += A[j][j] - lambda[j].x;
      shownum ("Trace check",(float) fabs(t));
    }
    return 0;
}
/* ------------------------------------------------------------ */
```

When program *e393.c* is executed, it uses the *QR* method to produce the following set of eigenvalues of *A* for the three cases.

$$\lambda = \left[\begin{array}{cc|cc|cc} -100 & +0j & -100 & +0j & -100.1 & +0j \\ -0.0178 & -1.265j & -0.009797 & -1.789j & 0.006197 & -2.529j \\ -0.0178 & +1.265j & -0.009797 & +1.789j & 0.006197 & +2.529j \end{array}\right] \tag{3.9.28}$$

Notice that the first eigenvalue $\lambda_1 \approx -100$ is deep inside the left-half plane and barely moves as k_c is increased from 100 to 400. This very fast natural mode is due to the electrical part of the system. The remaining pair of complex conjugate eigenvalues gravitate to the right as the controller gain is increased. For $k_c = 100$ and $k_c = 200$, the real parts are negative, and therefore, the system is stable. However, for $k_c = 400$, the eigenvalues have crossed over into the right-hand plane. Consequently, there is a critical gain somewhere in the interval $[200, 400]$ where the feedback system goes unstable. Due to the small size of the *A* matrix, the *QR* method converged in, at most, three iterations in each case. A solution to Example 3.9.3 using MATLAB can be found in the file *e393.m* on the distribution CD.

3.9.4 Rotating Masses and Torsional Springs: MATLAB

Rotational mechanical motion is common in engineering applications. Examples range from high-speed flywheels used to store energy, to inertial guidance systems with gyroscopes. An example of two rotating masses interconnected by torsional springs is shown in Figure 3.9.4. To obtain the equations of motion, we apply Newton's second law to

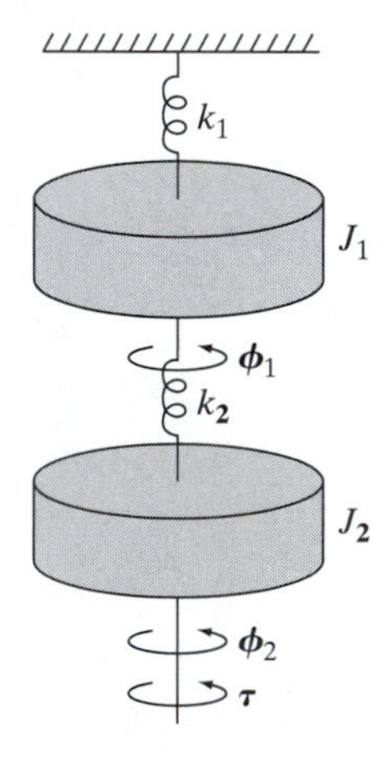

FIGURE 3.9.4 A Rotational Mechanical System

each mass and set the moment of inertia times angular acceleration equal to the sum of the torques.

$$J_1 \frac{d\phi_1^2}{dt^2} = k_2(\phi_2 - \phi_1) - k_1\phi_1 \quad \}\text{mass 1} \tag{3.9.29a}$$

$$J_2 \frac{d\phi_2^2}{dt^2} = \tau(t) - k_2(\phi_2 - \phi_1) \quad \}\text{ mass 2} \tag{3.9.29b}$$

Here ϕ_1 and ϕ_2 denote the angular positions of the masses whose moments of inertia are J_1 and J_2, respectively. The masses are interconnected with torsional springs with spring constants, k_1 and k_2, respectively. Finally, an external torque τ is applied to the system. This system of two second-order differential equations can be recast as a system of four first-order differential equations by introducing the following vector of state variables consisting of the angular positions and velocities of the two rotating masses (Ogata, 1997).

$$x = [\phi_1,\ d\phi_1/dt,\ \phi_2,\ d\phi_2/dt]^T \tag{3.9.30}$$

In this case, x is a 4×1 vector that specifies the angular positions and velocities of the two masses. If the expression for x is differentiated with respect to time t, after substitutions the right-hand side can be expressed entirely in terms of x and τ as follows.

$$\frac{dx_1}{dt} = x_2 \tag{3.9.31a}$$

$$J_1\left(\frac{dx_2}{dt}\right) = -(k_1 + k_2)x_1 + k_2x_3 \tag{3.9.31b}$$

$$\frac{dx_3}{dt} = x_4 \tag{3.9.31c}$$

$$J_2\left(\frac{dx_4}{dt}\right) = k_2x_1 - k_2x_3 + \tau \tag{3.9.31d}$$

This system of four differential equations can be recast in terms of vectors and matrices as

$$\frac{dx}{dt} = \begin{bmatrix} 0 & 1 & 0 & 0 \\ -(k_1 + k_2)/J_1 & 0 & k_2/J_1 & 0 \\ 0 & 0 & 0 & 1 \\ k_2/J_2 & 0 & -k_2/J_2 & 0 \end{bmatrix} x + \begin{bmatrix} 0 \\ 0 \\ 0 \\ \tau/J_2 \end{bmatrix} \tag{3.9.32}$$

Suppose the two masses start out at rest at the equilibrium position $x = 0$. If a constant torque is applied, then the two rotating masses will begin to oscillate back and forth. To verify that they will oscillate and determine the frequency of oscillation, it is necessary to examine the eigenvalues of the coefficient matrix.

$$A = \begin{bmatrix} 0 & 1 & 0 & 0 \\ -(k_1 + k_2)/J_1 & 0 & k_2/J_1 & 0 \\ 0 & 0 & 0 & 1 \\ k_2/J_2 & 0 & -k_2/J_2 & 0 \end{bmatrix} \tag{3.9.33}$$

Let $\lambda = [\lambda_1, \lambda_2, \lambda_3, \lambda_4]^T$ be the eigenvalues of A. Suppose the kth eigenvalue is expressed in terms of its real part a_k and its imaginary part b_k.

$$\lambda_k = a_k + jb_k \quad , \quad 1 \le k \le 4 \tag{3.9.34}$$

The rotational mechanical system will exhibit sustained oscillations if the eigenvalues of A are purely imaginary. In this case, the period of the oscillation is determined by the imaginary part b_k. In particular, the kth natural mode oscillates with period

$$T_k = \frac{2\pi}{|b_{2k-1}|} \quad , \quad 1 \le k \le 2 \tag{3.9.35}$$

Note that complex eigenvalues always occur in conjugate pairs for a real A. Consequently, there are two periods of oscillation rather than four.

To make the mass-spring example specific, suppose the moments of inertia are $J_1 = 1.5$ N-m-s^2 and $J_2 = 3.8$ N-m-s^2, and the spring constants are $k_1 = 5.1$ N-m and $k_2 = 8.3$ N-m. The following MATLAB script on the distribution CD can be used to solve this problem.

```
%----------------------------------------------------------------------
% Example 3.9.4: Rotating Masses and Torsional Springs
%----------------------------------------------------------------------

% Initialize

   clc                          % clear screen
   clear                        % clear variables
   n = 4;                       % order of system
   J1 = 1.5;                    % moment of inertia 1
   J2 = 3.8;                    % moment of inertia 2
   k1 = 5.1;                    % spring constant 1
   k2 = 8.3;                    % spring constant 2
   A = zeros (n,n);

% Compute A

   fprintf ('Example 3.9.4: Rotating Masses and Torsional Springs\n');
   A(1,2) = 1;
   A(2,1) = -(k1 + k2)/J1;
   A(2,3) = k2/J1;
   A(3,4) = 1;
   A(4,1) = k2/J2;
   A(4,3) = -k2/J2;
   show ('A',A)

% Find eigenvalues and oscillation periods

   lambda = eig(A);
   show ('lambda',lambda)
   t = abs(trace(A) - sum(lambda));
   show ('Trace check',t)
```

```
    show ('T1 (sec)',abs(2*pi/imag(lambda(1))))
    show ('T2 (sec)',abs(2*pi/imag(lambda(3))))
%---------------------------------------------------------------
```

When script *e394.m* is executed, it uses the QR method to generate the following set of eigenvalues for A:

$$\lambda = \begin{bmatrix} 0 + 3.225\text{j} \\ 0 - 3.225\text{j} \\ 0 + 0.8449\text{j} \\ 0 - 0.8449\text{j} \end{bmatrix} \tag{3.9.36}$$

In this case, the eigenvalues are all imaginary. Thus, from (3.9.35) the motion of the masses can be expressed as the sum of two periodic terms with the periods $T = [T_1, T_2]^T$ being

$$T = \begin{bmatrix} 1.94798 \\ 7.43683 \end{bmatrix} \tag{3.9.37}$$

Thus, there is a relatively fast oscillation with period $T_1 = 1.94798$ seconds and a slower oscillation with period $T_2 = 7.43683$ seconds. Finally, as a partial check on the accuracy of λ, the sum of the eigenvalues was subtracted from the trace of A to yield

$$\left| \text{trace}(A) - \sum_{k=1}^{4} \lambda_k \right| = 0 \tag{3.9.38}$$

One might be tempted to conclude, based on the trace check, that the computed eigenvalues are accurate to within the machine precision (2.22×10^{-16} for MATLAB). This is *not* necessarily the case. Indeed, the trace check is only a partial check because it only checks the accuracy of the *sum* of the eigenvalues, not the eigenvalues themselves.

There is a simple physical interpretation to explain why the system in Figure 3.9.4 should exhibit sustained oscillation when set into motion. As it oscillates, the energy of the system transfers back and forth between the kinetic energy contained in the rotating masses and the potential energy contained in the twisted springs. A more complete model of a rotating mechanical system would include viscous friction terms in the form of dampers. In this case, the masses would still oscillate, but the amplitudes of the oscillations would gradually decrease as energy is converted to heat. Eventually, the masses would come to rest at a constant steady-state solution x^s that depends on the applied torque. A solution to Example 3.9.3 using C can be found in the file *e394.c* on the distribution CD.

3.10 SUMMARY

Numerical methods for finding the eigenvalues and eigenvectors of an $n \times n$ matrix A are summarized in Table 3.10.1.

Leverrier's method is a simple but computationally expensive method for computing the coefficients of the characteristic polynomial, $\Delta(\lambda)$. There are no restrictions on the matrix A. In theory, the eigenvalues of A can be obtained by factoring $\Delta(\lambda)$ using polynomial root-finding techniques (see Chapter 5). However, this approach is practical only for relatively small values of n (say, $n \leq 20$) because polynomial roots can be

TABLE 3.10.1 Eigenvalue and Eigenvector Techniques

Method	*Result*	*A*
Leverrier	$\Delta(\lambda)$	general
Power	(λ_1, x^1)	diagonalizable
Inverse power	(λ_n, x^n)	diagonalizable
Jacobi	$\{(\lambda_k, x^k) \mid 1 \le k \le n\}$	symmetric
QR	$\{\lambda_k \mid 1 \le k \le n\}$	general
Danilevsky	$\Delta(\lambda), \{x^k \mid 1 \le k \le n\}$	not deficient
Polynomial roots	$\{r_k \mid 1 \le k \le n\}$	companion

extremely sensitive to the coefficient values, hence the technique is plagued by the effects of accumulated round-off error.

The power method is used to estimate the dominant eigenvalue, λ_1, and its eigenvector, x^1. Thus, the power method can be used to compute the spectral radius, $\rho(A)$. The power method is guaranteed to converge if the matrix A has n linearly independent eigenvectors and a single dominant eigenvalue, and the initial guess y^0 satisfies $(y^0)^T x^1 \neq 0$. An $n \times n$ matrix with n linearly independent eigenvectors is called a *diagonalizable* matrix. The name arises from the observation that if $X = [x^1, x^2, \ldots, x^n]$ where x^k is the kth eigenvector, then the following similarity transformation converts A to a diagonal matrix B whose diagonal elements are the eigenvalues of A.

$$B = X^{-1} A X \tag{3.10.1}$$

Matrices that are not diagonalizable are sometimes referred to as *defective* matrices. The set of diagonalizable matrices includes matrices with n distinct eigenvalues and symmetric matrices as shown in Figure 3.10.1.

The inverse power method is used to estimate the least-dominant eigenvalue, λ_n, and its eigenvector, x^n. It does so by estimating the dominant eigenvalue of A^{-1}. Consequently, for the inverse power method to be applicable, the matrix A must be nonsingular. The inverse power method is guaranteed to converge if the matrix A is diagonalizable with a single least-dominant nonzero eigenvalue, and the initial guess y^0 satisfies $(y^0)^T x^n \neq 0$. The inverse power method requires more computational effort than the power method because a linear algebraic system must be solved at each itera-

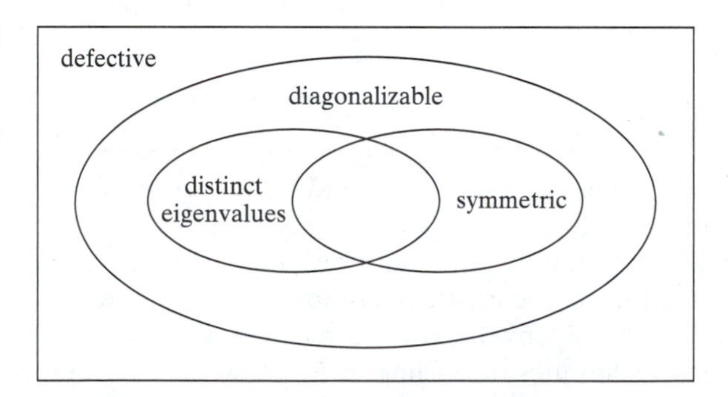

FIGURE 3.10.1 Diagonalizable Matrices

tion. The LU decomposition method (Chapter 2) can be used to efficiently accomplish this task.

The Jacobi method is an iterative method that uses orthogonal similarity transformations to find the eigenvalues and eigenvectors of a symmetric matrix A. For symmetric matrices, the eigenvalues and eigenvectors are real. The rate of convergence of Jacobi's method is increased by first preprocessing the matrix A by the Householder transformation. This uses an orthogonal similarity transformation to convert A to a tridiagonal matrix.

The QR method is the most effective general method for finding the eigenvalues of an $n \times n$ matrix A. It does so by first using the Householder transformation to convert A to an upper-Hessenberg matrix with zeros below the first subdiagonal. This matrix is then transformed to an upper block triangular matrix with 1×1 and 2×2 blocks along the diagonal, the 1×1 blocks corresponding to real eigenvalues, and the 2×2 blocks corresponding to complex conjugate pairs of eigenvalues. Transformation to upper block triangular form is achieved by factoring A into a product of an orthogonal matrix Q times an upper-triangular matrix R. The new A is then computed as $A = Q^T A Q$ and the process is repeated. Deflation and eigenvalue shifting techniques are used to increase the rate of convergence of the QR method.

Danilevsky's method is a direct method that uses a similarity transformation to convert the $n \times n$ matrix A to companion form. The characteristic polynomial of A can then be obtained directly from inspection of the first row of the companion matrix. This method of finding $\Delta(\lambda)$ is more efficient than Leverrier's method. If the eigenvalues of A are computed using the QR method, the eigenvectors of A can be obtained by reversing the transformation used to find the companion matrix. For some matrices, the transformation to companion form fails due to the absence of a nonzero pivot. These matrices are referred to as *deficient* matrices. The eigenvectors obtained by Danilevsky's method are sensitive to the effects of accumulated round-off error.

The polynomial root finding problem can be converted into an eigenvalue problem. Given a monic polynomial of degree n, an associated $n \times n$ companion matrix is constructed. The eigenvalues of the companion matrix are the roots of the polynomial. The companion matrix is already in upper-Hessenberg form. Consequently, the QR method can be applied starting at step (2) of Alg. 3.6.1. This is a highly effective way to find the roots of a polynomial. More general root-finding techniques are introduced in Chapter 5.

PROBLEMS

The problems are divided into Analysis problems, which can be solved by hand, and Computation problems, which require the use of MATLAB or C. Solutions to selected problems can be found in Appendix 4. Students are encouraged to use these problems, which are identified with an (S), as a check on their understanding of the material. Problems marked with a (P) are programming problems that require the student to implement one or more of the algorithms discussed in the chapter. The remaining Computation problems require the student to write a main program that uses one or more of the NLIB functions discussed in Appendix 1 or Appendix 2.

Analysis

3.1 Show that a matrix A is singular if and only if it has an eigenvalue at $\lambda = 0$.

3.2 Suppose A is an $n \times n$ matrix with real components. Show that if n is odd, at least one eigenvalue of A is real.

3.3 Consider the following matrix.

$$A = \begin{bmatrix} 1 & 2 \\ 3 & 5 \end{bmatrix}$$

(a) Find the characteristic polynomial.

(b) Find the eigenvalues.

(c) Find the spectral radius.

(S)3.4 Suppose A is a triangular $n \times n$ matrix (either upper-triangular or lower-triangular). Show that the eigenvalues of A are the diagonal elements.

3.5 Apply the Leverrier algorithm to find the characteristic polynomial of the following matrix.

$$A = \begin{bmatrix} 1 & 0 & -1 \\ -1 & 1 & 0 \\ 0 & -1 & 1 \end{bmatrix}$$

3.6 Consider the following matrix. Find an approximation to the dominant eigenvalue and its eigenvector using the first two iterations of the power method and an initial guess of $y = [1, 0]^T$. What is the exact value of the dominant eigenvalue in this case?

$$A = \begin{bmatrix} 3 & 1 \\ 5 & 2 \end{bmatrix}$$

3.7 Consider the matrix in problem 3.6. Find an approximation to the least-dominant eigenvalue and its eigenvector using the first two iterations of the inverse power method and an initial guess of $y = [1, 0]^T$. What is the exact value of the least-dominant eigenvalue in this case?

3.8 Consider the following symmetric matrix. Apply Jacobi's method in Alg. 3.4.1 to find the eigenvalues and eigenvectors of this matrix.

$$A = \begin{bmatrix} 1 & 2 \\ 2 & 1 \end{bmatrix}$$

3.9 Consider the following matrix. Apply the Householder method in Alg. 3.5.1 to find the upper-Hessenberg form of this matrix.

$$A = \begin{bmatrix} 3 & -2 & 0 \\ 2 & 1 & 4 \\ 1 & -1 & 2 \end{bmatrix}$$

(S)3.10 Consider the following linear discrete-time system. Determine whether or not this system is stable. Explain.

$$x^{k+1} = \begin{bmatrix} 0 & -0.25 \\ 1 & -1 \end{bmatrix} x^k \quad , \quad k \geq 0$$

3.11 Consider the following linear continuous-time dynamic system. Determine whether or not this system is stable. Explain.

$$\frac{dx(t)}{dt} = \begin{bmatrix} 2 & 3 \\ 1 & 2 \end{bmatrix} x(t) \quad , \quad t \ge 0$$

Computation

3.12 Write a program that uses the NLIB function *poly.m* or *leverrier.c* to find the characteristic polynomial of the following matrix. Print the coefficient vector.

$$A = \begin{bmatrix} 1 & -2 & 3 & 0 & 4 \\ 0 & 4 & -6 & 1 & 2 \\ 1 & -1 & -3 & 5 & 4 \\ 2 & 0 & 5 & 9 & -2 \\ 4 & -3 & 1 & 7 & 5 \end{bmatrix}$$

3.13 Write a program that uses the NLIB function *powereig* to find the dominant eigenvalue λ_1 and eigenvector x^1 of the following matrix. Print λ_1, x^1, the spectral radius, and the norm of the residual error vector $r^1 = (\lambda_1 I - A)x^1$.

$$A = \begin{bmatrix} 5 & 4 & 3 & 2 & 1 \\ 4 & 3 & 2 & 1 & 0 \\ 3 & 2 & 1 & 0 & -1 \\ 2 & 1 & 0 & -1 & -2 \\ 1 & 0 & -1 & -2 & -3 \end{bmatrix}$$

3.14 Write a program that uses the NLIB function *powereig* to find the least-dominant eigenvalue λ_n and eigenvector x^n of the matrix in problem 3.13. Print λ_n, x^n, and the norm of the residual error vector $r^n = (\lambda_n I - A)x^n$.

3.15 Write a program that uses the NLIB function *eig.m* or *jacobi.c* to find the eigenvalues and eigenvectors of the following symmetric matrix. Use an error tolerance of $\varepsilon = 10^{-6}$ with *jacobi.c.* Print the vector of eigenvalues, the matrix of eigenvectors, and the magnitude of the difference between trace(A) and the sum of the eigenvalues.

$$A = \begin{bmatrix} 2 & -1 & 4 & 9 & 3 & 7 \\ -1 & 6 & 8 & -7 & 4 & 2 \\ 4 & 8 & 0 & 9 & 2 & -6 \\ 9 & -7 & 9 & 5 & 4 & -2 \\ 3 & 4 & 2 & 4 & 7 & 1 \\ 7 & 2 & -6 & -2 & 1 & 8 \end{bmatrix}$$

3.16 Write a program that uses the NLIB functions *eig.m* or *qr.c* to find the eigenvalues of the following matrix. Use an error tolerance of $\varepsilon = 10^{-6}$ with *qr.c.* Print the vector of eigenvalues, and the magnitude of the difference between trace(A) and the sum of the eigenvalues.

$$A = \begin{bmatrix} 1 & -2 & 5 & 7 & 4 & 3 \\ 0 & 9 & 4 & -4 & 2 & 1 \\ 9 & -3 & 3 & 2 & 4 & 7 \\ -2 & 7 & 8 & 4 & 1 & -1 \\ 8 & -6 & 5 & 9 & 4 & 3 \\ 2 & 4 & -7 & 8 & 1 & 0 \end{bmatrix}$$

3.17 Write a program that uses the NLIB function *eig.m* or *danilevsky.c* to find the eigenvalues and eigenvectors of the matrix in problem 3.16. Use an error tolerance of $\varepsilon = 10^{-6}$ with *danilevsky*. Print the vector of eigenvalues, the matrix of eigenvectors, and the magnitude of the difference between trace(A) and the sum of the eigenvalues.

3.18 Consider the following polynomial. Write a program which uses the NLIB function *roots* to find the roots of this polynomial. Print the roots, and the polynomial evaluated at each of the roots.

$$f(x) = 4x^5 - 2x^4 + x^3 + 5x^2 - 7$$

(S)3.19 Write a program that uses NLIB to compute and display the characteristic polynomial and the spectral radius of the following matrix.

$$A = \begin{bmatrix} 4 & -1 & 3 & 9 & -5 \\ -1 & 2 & 7 & 12 & 7 \\ 3 & 7 & 0 & -11 & 8 \\ 2 & -4 & 14 & 1 & -5 \\ 11 & -8 & 6 & -2 & 13 \end{bmatrix}$$

3.20 Write a program that uses NLIB to compute and display the eigenvalues and eigenvectors of the matrix in problem 3.19.

(P)3.21 Let A be an $m \times n$ matrix with $m \leq n$ and let $B = AA^T$. The eigenvalues $\{\mu_1, \mu_2, \ldots, \mu_m\}$ of the $m \times m$ matrix B are called the *singular values* of A. The matrix B is symmetric and positive semidefinite, which means that the singular values are real and $\mu_k \geq 0$. The number of nonzero singular values is called the *rank* of A. Write a function called *singval* that uses the NLIB function *eig.m* or *qr.c* to find the singular values and rank of the matrix A. The pseudo-prototype for the function *singval* is

```
[vector c,int r,int k] = singval (matrix A,int m,int n,int p,float
tol)
```

On entry to *singval*: A is an $m \times n$ matrix, m is the number of rows of A, $n \geq m$ is the number of columns of A, p is the maximum number of QR iterations, and $tol \geq 0$ is an error tolerance used to terminate the search. On exit, the $m \times 1$ vector c contains the singular values of A, r is the rank of A, and k is the number of QR iterations used. If $k < p$, then the termination criterion specified in QR must be satisfied. For the MATLAB version, the scalar arguments m, n, p, tol, and k are not needed. Test *singval* by writing a program which uses it to find the singular values and the rank of the following matrix. Use a tolerance of $100\varepsilon_M$ to detect nonzero singular values where ε_M is the machine epsilon.

$$A = \begin{bmatrix} 4 & -1 & 3 & 9 & -2 \\ -1 & 2 & 7 & 12 & -15 \\ 3 & 7 & 0 & -11 & 3 \\ -5 & 9 & -6 & -29 & 18 \end{bmatrix}$$

CHAPTER 4

Curve Fitting

Engineering problems often involve the collection of experimental data. To make full use of the data, the development of a mathematical model is often helpful. Curve fitting provides a means of obtaining a mathematical model from discrete data. Suppose (x_k, y_k) for $1 \leq k \leq n$ consists of a set of n measurements. Here x_k denotes the kth value of the independent variable and y_k denotes the corresponding value of the dependent variable. Typically, we order the independent variables so that $x_1 < x_2 < \cdots < x_n$. If the number of points n is small, the simplest way to fit a curve to the data D is to find a function f which passes through each of the data samples.

$$\boxed{f(x_k) = y_k \quad , \quad 1 \leq k \leq n}$$

Once f is determined, we can use it to predict the value of y at points other than the samples. When the point of evaluation x lies inside $[x_1, x_n]$, the process is called *interpolation* otherwise it is called *extrapolation*. If the number of samples is large or if the dependent variables contain measurement noise, it is often better to find a function f that approximates the data by minimizing an error criterion such as

$$\boxed{E = \sum_{k=1}^{n} [f(x_k) - y_k]^2}$$

A function f that minimizes the value of E is called a *least-squares* curve fit. This approach makes sense when one wants to represent only the underlying trend of the data. Once we fit a curve to the data, various types of analysis can be performed such as finding minimum values, maximum values, slopes, and areas. We can also use curve fitting to approximate complicated functions with simpler functions. In this case, the data points in D are generated analytically rather than from physical measurements. This can be useful for developing numerical approximation formulas.

In this chapter, we examine numerical curve fitting techniques. We begin by considering some simple examples of curve fitting problems from science and engineering. Next, we investigate basic interpolation techniques, including piecewise-linear interpolation and polynomial interpolation. We then discuss the Newton difference formula, which represents interpolating polynomials in terms of numerical differences. We present a highly effective interpolation technique called cubic spline interpolation. Next, we examine least-squares curve fitting using polynomials to model the underlying trend of data. We follow with a discussion of two-dimensional curve fitting where surfaces are fitted to data having two independent variables. The chapter concludes with a summary of curve fitting methods, and a presentation of engineering applications. The selected examples include pressure-temperature curves, water resource management, a voltage regulator circuit, and a nonlinear friction model.

4.1 MOTIVATION AND OBJECTIVES

There are many instances in engineering where a relationship between two variables can be characterized experimentally, but no underlying theory is available to accurately represent it mathematically. Curve fitting is a natural way to obtain a working mathematical description in these cases.

4.1.1 Gravitational Acceleration

The law of universal gravitation says that the force of attraction between two masses m_1 and m_2 is proportional to the product of the masses divided by the square of the distance r between them.

$$F = \frac{Gm_1m_2}{r^2} \tag{4.1.1}$$

Here, $G = 6.673 \times 10^{-11}\text{nt-m}^2/\text{kg}^2$ is the constant of gravitation. Suppose m_1 represents the mass of the earth and m_2 represents the mass of an object on the surface of the earth. In this case, the force on the object caused by the gravitational attraction of the earth reduces to

$$F = m_2 g \tag{4.1.2}$$

The new parameter $g = Gm_1/r^2$ is the acceleration due to gravity measured at the surface of the earth. Since the earth is not perfectly spherical, the acceleration due to gravity is not a constant, but instead varies with latitude. Measurements of the mean value of g at sea level at various latitudes are listed in Table 4.1.1 (Halliday and Resnick, 1965).

Suppose a scientific experiment that requires a highly accurate value for acceleration due to gravity is performed on ship. In this case, it is appropriate to fit a curve f to the data in Table 4.1.1 and then use $g = f(x)$ where x is the latitude. An example of a least-squares curve fit to the data in Table 4.1.1 using a polynomial of degree four is shown in Fig. 4.1.1. Least-squares curve fitting is discussed in Section 4.5.

TABLE 4.1.1 Variation in g with Latitude at Sea Level

Latitude (deg)	g (m/s^2)
0	9.78039
10	9.78195
20	9.78641
30	9.79329
40	9.80171
50	9.81071
60	9.81918
70	9.82608
80	9.83059
90	9.83217

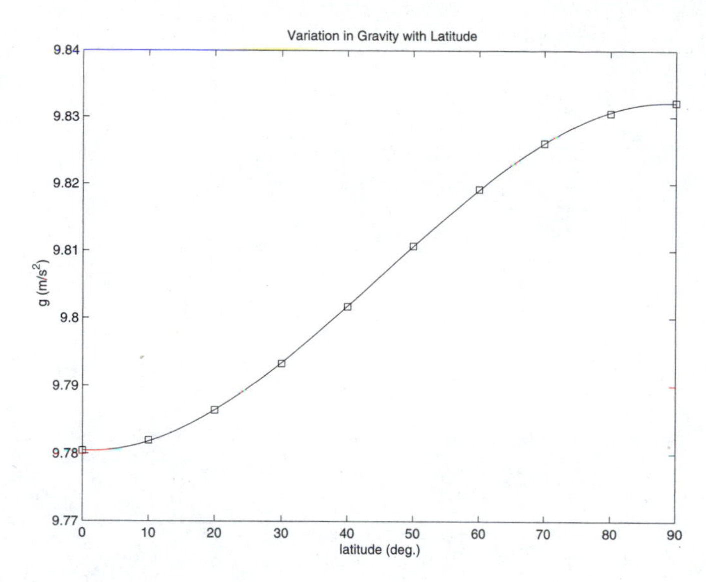

FIGURE 4.1.1 Variation in g with Latitude at Sea Level

4.1.2 Circadian Rhythms

Circadian rhythms are daily cycles in living organisms. In humans, they include variations in temperature, pulse rate, alcohol metabolism, pain tolerance, and sensitivity to drugs (Palmer, 1976). These rhythms, which are driven by an internal biological clock, are synchronized with daily cycles in light and temperature. Measurements of a circadian rhythm, the average daily variation in body temperature of 70 English seamen, are listed in Table 4.1.2 (Colquhoun, 1971).

A graphical display of the daily body temperature rhythm is shown in Figure 4.1.2. Note how the temperature begins to rise just before the subjects awake at 6:30 A.M. and is in a steep decline when they go to sleep at 11:30 P.M. In order to mathematically model the temperature rhythm, a curve must be fitted to the discrete data samples.

TABLE 4.1.2 Daily Variation in Body Temperature

Time (hr)	*Temperature (°F)*	*Time (hr)*	*Temperature (°F)*
1	97.46	14	98.05
3	97.22	15	98.11
5	97.17	16	98.14
7	97.26	17	98.20
8	97.48	18	98.28
9	97.69	19	98.32
10	97.88	20	98.36
11	97.94	21	98.35
12	97.98	22	98.19
13	98.02	23	97.87

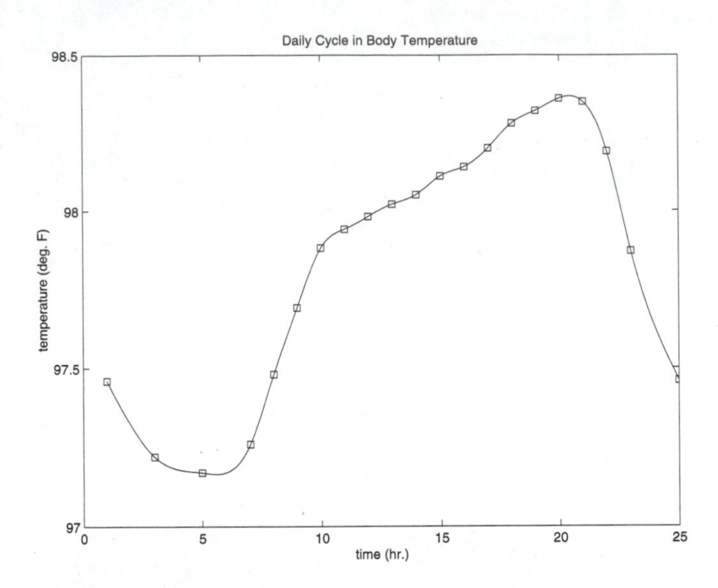

FIGURE 4.1.2 Daily Variation in Body Temperature

The curve shown in Figure 4.1.2 is a cubic spline fit, a technique that is discussed in Section 4.4.

4.1.3 Chapter Objectives

When you finish this chapter, you will be able to construct empirical mathematical models by fitting curves and surfaces to experimental data. You will understand what an interpolating polynomial is and how to efficiently evaluate it. You will know how to express an interpolating polynomial directly in terms of the data points using Newton's difference formula. You will understand why higher-order polynomials are not good candidates for curve fitting and how this problem can be corrected by using cubic splines. You will also know how to model the underlying trend in noisy data using least-squares techniques. Finally, you will learn how to fit a surface to two-dimensional data using bilinear interpolation. These overall goals will be achieved by mastering the following chapter objectives.

Objectives for Chapter 4

- Understand the difference between interpolation and extrapolation.
- Know how to express piecewise-linear interpolation parametrically.
- Know why interpolation with high-degree polynomials is undesirable.
- Know why finding the interpolating polynomial by solving a linear algebraic system can be inaccurate.
- Know how to construct and use Lagrange interpolating polynomials to perform interpolation.
- Be able to efficiently evaluate polynomials using Horner's recursive method.
- Know how to construct the interpolating polynomial using Newton's forward difference formula.

- Know how to fit low-order interpolating polynomials together using cubic spline interpolation.
- Be able to fit curves to the underlying trend of noisy data using least-squares techniques.
- Know how to construct and use orthogonal polynomials to find a least-squares curve fit.
- Be able to fit a surface to two-dimensional data using bilinear interpolation.
- Understand how curve fitting techniques can be used to solve practical engineering problems.
- Understand the relative strengths and weaknesses of each computational method and know which are most applicable for a given problem.

4.2 INTERPOLATION

The curve fitting problem is posed by starting with a set of measurements. Each experiment has an *independent* variable or input, x, whose value we can adjust, and a *dependent* variable or output, y, whose value we measure as shown in Figure 4.2.1.

The details of the relationship between x and y depend on the underlying physical system that generates the data. For convenience, we refer to this system as f where it is understood that the function f is *unknown*, but it can be estimated by measuring it at discrete points. Let x_k denote the value of the independent variable for the kth measurement, and let $y_k = f(x_k)$ denote the corresponding value observed for the dependent variable. The experimental data can then be represented as a set of n input/output pairs.

$$D \triangleq \{(x_k, y_k) \mid 1 \le k \le n\} \tag{4.2.1}$$

That is, D is defined to be the set of points (x_k, y_k) with k ranging from 1 to n. The $n \times 1$ vector x is the vector of independent variables, and the $n \times 1$ vector y is the vector of dependent variables. We assume that the independent variables have been *ordered* such that $x_1 < x_2 < \cdots < x_n$. The most straightforward way to fit a curve to the data D is to find a function f thatpasses through each of the n data points or samples. That is, find a function f such that

$$f(x_k) = y_k \quad , \quad 1 \le k \le n \tag{4.2.2}$$

Once a function is determined that satisfies (4.2.2), the value of y at points other than the samples can be estimated. For example, if x lies in the interval $[x_k, x_{k+1}]$, then $y = f(x)$ is an estimate. In this case, the process of estimating y is referred to as *interpolation* between x_k and x_{k+1}. In general, interpolation is a process of approximating the dependent variable between two known data points. There are also instances where

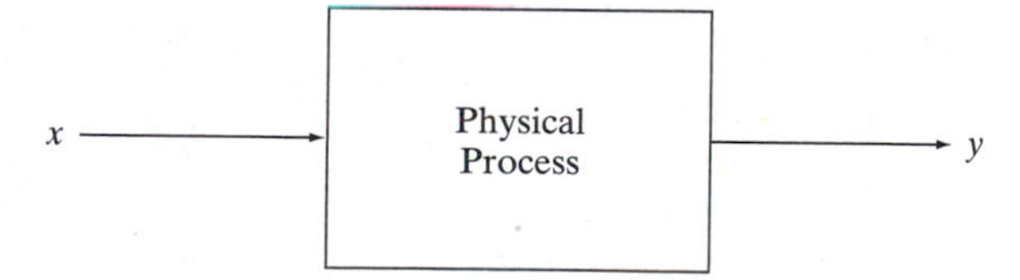

FIGURE 4.2.1 Input/Output Measurements of a Physical Process

one is interested in estimating the value of the dependent variable at points which lie outside the range of measurements. This process, which is typically less reliable, is referred to as *extrapolation*. Thus, interpolation fills in between measurements, while extrapolation attempts to extend the estimates beyond the range of measurements.

4.2.1 Piecewise-Linear Interpolation

The simplest form of interpolation over a set of data points is called *piecewise-linear* interpolation. In this case, a straight line segment is used between each adjacent pair of data points as shown in Figure 4.2.2. The virtue of piecewise-linear interpolation lies in the fact that the function f can be expressed in closed form. To see this, suppose (x, y) represents a point on the line segment connecting point (x_k, y_k) to point (x_{k+1}, y_{k+1}) as shown in Figure 4.2.3. Using similar triangles and setting the ratios of the two sides equal, we get

$$\frac{y - y_k}{x - x_k} = \frac{y_{k+1} - y_k}{x_{k+1} - x_k}, \quad 1 \le k < n \tag{4.2.3}$$

If $f_k(x)$ denotes the value for y between points k and $k + 1$, then solving Equation (4.2.3) for y, we get the following interpolating function for data segment k.

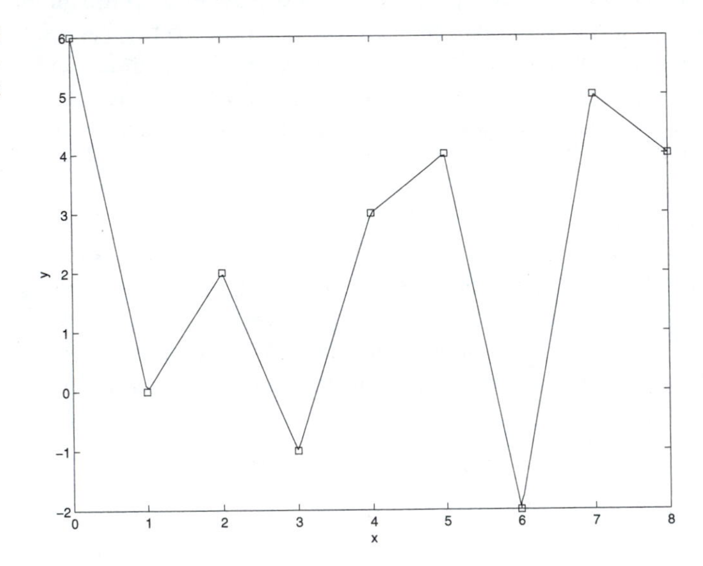

FIGURE 4.2.2 Piecewise-Linear Fit

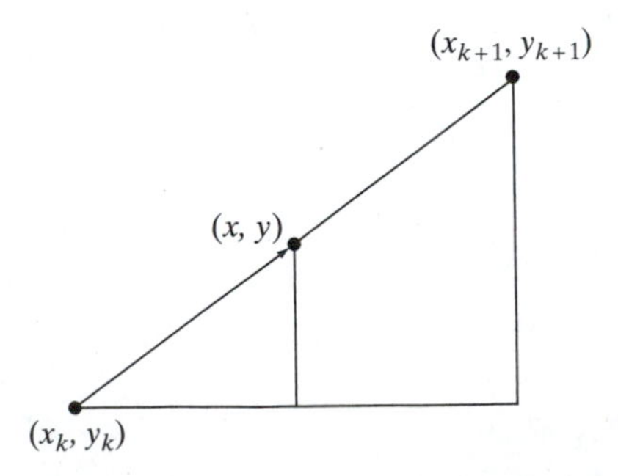

FIGURE 4.2.3 Straight-Line Interpolation Between Points k and $k + 1$

$$f_k(x) = y_k + \frac{x - x_k}{x_{k+1} - x_k}(y_{k+1} - y_k) \quad , \quad 1 \le k < n \tag{4.2.4}$$

Note that the coefficient $\sigma_k(x) \triangleq (x - x_k)/(x_{k+1} - x_k)$ is a fraction that specifies how far x has traversed as it moves from x_k to x_{k+1}. It can be thought of as a normalized step length. Since $\sigma_k(x_k) = 0$, it follows that $f_k(x_k) = y_k$. Similarly, from $\sigma_k(x_{k+1}) = 1$, it follows that $f_k(x_{k+1}) = y_{k+1}$. To extrapolate beyond the range of data points, we use the nearest segment. That is, $f_1(x)$ is used when $x < x_1$, while $f_{n-1}(x)$ is used when $x > x_n$.

EXAMPLE 4.2.1 Piecewise-Linear Fit

As an illustration of a piecewise-linear fit, let D denote the data used to generate the points in Figure 4.2.2.

$$D = \{(0,6),(1,0),(2,2),(3,-1),(4,3),(5,4),(6,-2),(7,5),(8,4)\}$$

In this case, $x_k = k - 1$, and the normalized step length function is

$$\begin{aligned}\sigma_k(x) &= \frac{x - x_k}{x_{k+1} - x_k} \\ &= x - k + 1\end{aligned}$$

The piecewise-linear interpolation, say, between the points $(x_6, y_6) = (5,4)$ and $(x_7, y_7) = (6,-2)$, is

$$\begin{aligned}f_6(x) &= 4 + (x - 5)(-2 - 4) \\ &= 34 - 6x\end{aligned}$$

Direct substitution verifies that $f_6(5) = 4$ and $f_6(6) = -2$ as required.

4.2.2 Polynomial Interpolation

Although piecewise-linear interpolation is easily implemented, it is a relatively crude way to fit a curve to experimental data. The resulting curve is clearly not smooth; instead it is characterized by sharp corners at the data points where the slope changes abruptly as can be seen in Figure 4.2.2. Piecewise-linear interpolation employs a first-degree polynomial, $f(x) = a_1 + a_2 x$, to interpolate over each segment. To obtain a smoother curve that goes through the n data points, one can use a higher-degree polynomial. Since there are n constraints to satisfy in (4.2.2), we use a polynomial of degree $n - 1$ that has n coefficients or degrees of freedom.

$$f(x) = \sum_{k=1}^{n} a_k x^{k-1} \tag{4.2.5}$$

The n interpolation constraints in (4.2.2) can now be formulated as a linear algebraic system of n equations in the n unknown coefficients $a = [a_1, a_2, \ldots, a_n]^T$.

$$\begin{bmatrix} 1 & x_1 & x_1^2 & \cdots & x_1^{n-1} \\ 1 & x_2 & x_2^2 & \cdots & x_2^{n-1} \\ \vdots & \vdots & \vdots & \ddots & \vdots \\ 1 & x_n & x_n^2 & \cdots & x_n^{n-1} \end{bmatrix} a = \begin{bmatrix} y_1 \\ y_2 \\ \vdots \\ y_n \end{bmatrix} \tag{4.2.6}$$

The system in Equation (4.2.6) can be solved for the unknown vector a. If the independent variables are distinct, then the $n \times n$ coefficient matrix will be nonsingular. Consequently, a unique solution exists when $x_1 < x_2 < \cdots < x_n$. The coefficient matrix in (4.2.6) is called the *Vandermonde* matrix. As we shall see, the Vandermonde matrix becomes very ill-conditioned as n increases. Consequently, (4.2.6) should be used to compute the coefficients of the interpolating polynomial only when the value of n is small.

EXAMPLE 4.2.2 Polynomial Fit

As a simple illustration of a low-order interpolating polynomial, suppose $n = 3$ measurements are performed to generate the following set of input/output pairs.

$$D = \{(0,6), (1,0), (2,2)\}$$

In this case, a quadratic polynomial is sufficient to interpolate between these points. Applying (4.2.6) the equations for the unknown coefficients of this polynomial are

$$\begin{bmatrix} 1 & 0 & 0 \\ 1 & 1 & 1 \\ 1 & 2 & 4 \end{bmatrix} a = \begin{bmatrix} 6 \\ 0 \\ 2 \end{bmatrix}$$

Solving this system yields $a = [6, -10, 4]^T$. Consequently, the interpolating polynomial for the data set D is

$$f(x) = 4x^2 - 10x + 6$$

It is a simple matter to verify that $f(0) = 6, f(1) = 0$, and $f(2) = 2$ as required. A graph of $f(x)$ and the data samples is shown in Figure 4.2.4.

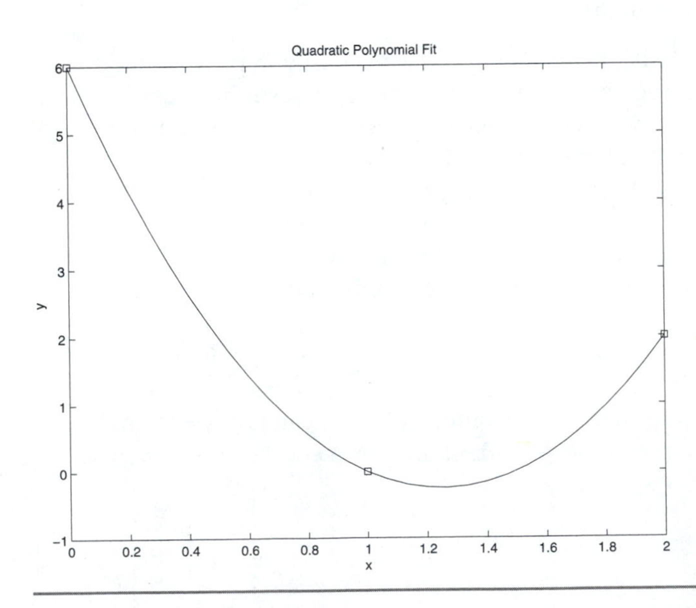

FIGURE 4.2.4 Quadratic Polynomial Fit

The problem with finding an interpolating polynomial by solving Equation (4.2.6) is that the coefficient matrix is very ill-conditioned for moderate to large values of n. The following example illustrates this point for the case of uniformly-spaced measurements.

EXAMPLE 4.2.3 Condition Number

Consider the problem of interpolating with a polynomial using equally-spaced measurements, $x_k = k - 1$ for $1 \leq k \leq n$. From (4.2.6), the Vandermonde matrix of the associated linear algebraic system is

$$A_n \triangleq \begin{bmatrix} 1 & 0 & 0 & \cdots & 0 \\ 1 & 1 & 1 & \cdots & 1 \\ 1 & 2 & 4 & \cdots & 2^{n-1} \\ \vdots & \vdots & \vdots & \cdots & \vdots \\ 1 & n-1 & (n-1)^2 & \cdots & (n-1)^{n-1} \end{bmatrix}$$

Example program *e423* on the distribution CD computes the condition number $K(A_n) = \|A_n\| \cdot \|A_n^{-1}\|$ for several values of n. The results, shown in Table 4.2.1, reveal that $K(A_n)$ grows extremely rapidly with n; hence the system quickly becomes very ill-conditioned.

TABLE 4.2.1 Condition Number of Vandermonde Matrix

n	$K(A_n)$
1	1.0000e+00
2	4.0000e+00
3	2.8000e+01
4	2.6667e+02
5	4.5467e+03
6	1.0416e+05
7	2.8566e+06
8	9.2066e+07
9	3.4282e+09

4.2.3 Lagrange Interpolating Polynomials

One way to circumvent the need to solve the ill-conditioned system in (4.2.6) is to reformulate the problem in another coordinate system. Rather than solve for the coefficients of terms of the form $\{1, x, x^2, \ldots, x^{n-1}\}$, we can use alternative polynomials of x. Given a set of n data points D, the kth *Lagrange interpolating* polynomial is defined as follows for $1 \leq k \leq n$.

$$Q_k(x) \triangleq \frac{(x - x_1)\cdots(x - x_{k-1})(x - x_{k+1})\cdots(x - x_n)}{(x_k - x_1)\cdots(x_k - x_{k-1})(x_k - x_{k+1})\cdots(x_k - x_n)} \tag{4.2.7}$$

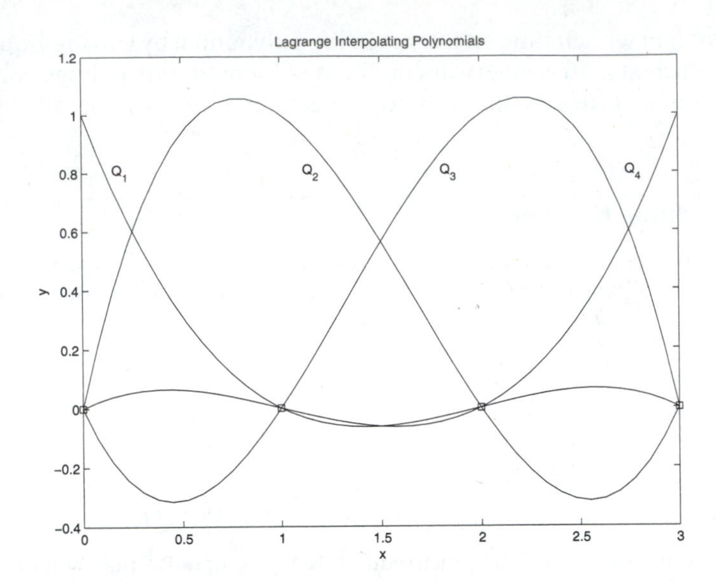

FIGURE 4.2.5 Lagrange Interpolating Polynomials

The Lagrange interpolating polynomials have a number of interesting properties. Since the denominator of $Q_k(x)$ is a constant and the numerator is the product of $n-1$ linear terms, it follows that $Q_k(x)$ is a polynomial of degree $n-1$. By construction, the Lagrange polynomials have the following *orthogonal* property with respect to the data set D.

$$Q_k(x_j) = I_{kj} \quad , \quad 1 \leq k, j \leq n \tag{4.2.8}$$

Recall that I_{kj} is the element in row k and column j of the identity matrix. Thus $Q_k(x_j)$ is equal to one when $j = k$ and zero otherwise. However, $Q_k(x)$ is nonzero *between* the data points. This is apparent from Figure 4.2.5, which is a graph of the first four Lagrange interpolating polynomials using the samples $x_k = k - 1$.

The orthogonality property in Equation (4.2.8) makes it very simple to write down the interpolating polynomial $f(x)$ in terms of the original data D. In particular, since $Q_k(x_k) = 1$ and $Q_k(x_j) = 0$ for $j \neq k$, it follows that the polynomial that passes through the points in the data set D is

$$\boxed{f(x) = \sum_{k=1}^{n} y_k Q_k(x)} \tag{4.2.9}$$

Consequently, if the set $\{Q_1(x), Q_2(x), \ldots, Q_n(x)\}$ is used for interpolation, the unknown coefficients needed to construct $f(x)$ are simply the dependent variables $y = [y_1, y_2, \ldots, y_n]^T$.

EXAMPLE 4.2.4 Lagrange Polynomials

As an illustration of the use of Lagrange interpolating polynomials, consider the interpolation problem posed in Example 4.2.2. Given $D = \{(0,6), (1,0), (2,2)\}$, the three Lagrange interpolation polynomials are

$$Q_1(x) = \frac{(x-1)(x-2)}{(0-1)(0-2)} = \frac{x^2-3x+2}{2}$$

$$Q_2(x) = \frac{(x-0)(x-2)}{(1-0)(1-2)} = \frac{x^2-2x}{-1}$$

$$Q_3(x) = \frac{(x-0)(x-1)}{(2-0)(2-1)} = \frac{x^2-x}{2}$$

The interpolating polynomial is then

$$\begin{aligned} f(x) &= 6Q_1(x) + 0Q_2(x) + 2Q_3(x) \\ &= 3(x^2-3x+2) + (x^2-x) \\ &= 4x^2 - 10x + 6 \end{aligned}$$

This is identical to the result obtained in Example 4.2.2, which is to be expected because there is exactly one polynomial of degree $n-1$ that passes through n data points.

The basic interpolation problem in Equation (4.2.2) can be generalized to include specifying the derivatives of the function f at the n data points as well. For example, if the first $p-1$ derivatives are specified, the problem becomes one of finding a function f such that

$$f^{(j)}(x_k) = y_{kj} \quad , \quad 1 \le k \le n, 0 \le j < p \qquad (4.2.10)$$

Here $f^{(j)}(x)$ denotes the jth derivative of f evaluated at x. The approach taken is essentially the same as with (4.2.2), but now the dimension of the problem is increased from n to pn. Polynomials that are fitted to a function and its derivatives are referred to as *Hermite interpolating* polynomials. For a discussion of Hermite interpolating polynomials, see Nakamura (1993) or Lastman and Sinha (1989).

4.2.4 Polynomials

Polynomials are used quite often for interpolation and extrapolation because they are simple functions that have a number of useful properties. For example, polynomials are *smooth* functions in the sense that they can be differentiated an arbitrary number of times. Furthermore, a polynomial of degree $n-1$ can be represented exactly with a set of n numbers. From the point of view of curve fitting, perhaps the most important property of polynomials is summarized in the *Weierstrass approximation* theorem, which says that if $F(x)$ is any continuous function defined over a finite interval $[a, b]$, then for each $\varepsilon > 0$, there exists a polynomial $f(x)$ such that

$$|f(x) - F(x)| < \varepsilon \quad , \quad a \le x \le b \qquad (4.2.11)$$

Since ε can be chosen arbitrarily small, this means that polynomials can be used to approximate continuous functions to an arbitrary degree of accuracy over any finite interval. In view of the fact that polynomials are used extensively, it is worthwhile to examine the computational effort required to evaluate a polynomial. Consider, in particular, the following polynomial of degree n.

$$f(x) = \sum_{k=1}^{n+1} a_k x^{k-1} \qquad (4.2.12)$$

On the surface, it would appear that the evaluation of $f(x)$ requires first computing the powers $\{x^2, x^3, \ldots, x^n\}$ with $n-1$ multiplications, then performing n multiplications, and finally performing n additions. Hence, a total of $2n-1$ multiplications and n additions are required to evaluate $f(x)$. However, a more careful examination of the calculation reveals that the number of floating-point operations can be reduced. In particular, a significant reduction in the number of multiplications can be achieved if the calculation is restructured in the following nested fashion, which is referred to as *Horner's rule*.

Alg. 4.2.1 **Horner's Rule**

1. Set $f = a_{n+1}$
2. For $k = n$ down to 1 do

{

$$f = fx + a_k$$

}

From Alg. 4.2.1, we see that the number of floating-point multiplications has been reduced from $2n-1$ to n, while n floating-point additions are still required. Thus, Horner's procedure cuts the number of FLOPs approximately in half.

EXAMPLE 4.2.5 **Horner's Rule**

As an illustration of Horner's Rule, consider the following polynomial of degree four.

$$f(x) = 8x^4 + 3x^3 - 5x^2 + 2x - 7$$

Applying Alg. 4.2.1, we get the following sequence of values for $f(x)$.

$$\begin{aligned}
k = 4, \quad & f = 8x + 3 \\
k = 3, \quad & f = (8x + 3)x - 5 \\
k = 2, \quad & f = ((8x + 3)x - 5)x + 2 \\
k = 1, \quad & f = (((8x + 3)x - 5)x + 2)x - 7
\end{aligned}$$

The final nested form, $k = 1$, requires only four multiplications as compared to seven multiplications needed by the standard brute force method.

4.3 NEWTON'S DIFFERENCE FORMULA

Functions which have sufficiently many derivatives can be approximated by using a Taylor series expansion. Recall from Chapter 1 that if $f(x)$ has n derivatives at $x = x_1$, the value of $f(x)$ can be approximated by using the first n terms of a Taylor series expansion of $f(x)$ about the point $x = x_1$ as follows.

$$f(x) \approx \sum_{k=0}^{n-1} \frac{f^{(k)}(x_1)(x - x_1)^k}{k!} \tag{4.3.1}$$

Here $f^{(k)}(x_1)$ denotes the kth derivative of $f(x)$ evaluated at $x = x_1$ and $0! \triangleq 1$. Since the only information available to us concerning the function f is the set of discrete input/output pairs D, wecan not use the Taylor series directly to construct $f(x)$. However, there is an analogous formulation that is based on discrete data. Rather than work with derivatives of f, one can instead use differences. More specifically, the first *forward difference* of f evaluated at x is defined as follows.

$$\boxed{\Delta f(x) \triangleq f(x+h) - f(x)} \tag{4.3.2}$$

Here the constant $h > 0$ denotes the *step size*. Suppose the samples in D are uniformly spaced with step size h.

$$x_k = x_1 + (k-1)h \quad , \quad 1 \le k \le n \tag{4.3.3}$$

Then the following $n - 1$ forward differences are available from the data set D.

$$\Delta f(x_k) = y_{k+1} - y_k \quad , \quad 1 \le k < n \tag{4.3.4}$$

The difference in Equation (4.3.2) is called a *first-order* forward difference. It is also possible to compute higher-order forward differences by evaluating differences of differences. For example, the *second-order* forward difference of f evaluated at x_k is as follows.

$$\begin{aligned} \Delta^2 f(x_k) &\triangleq \Delta[\Delta f(x_k)] \\ &= (y_{k+2} - y_{k+1}) - (y_{k+1} - y_k) \\ &= y_{k+2} - 2y_{k+1} + y_k \end{aligned} \tag{4.3.5}$$

The process can be repeated for third- and higher-order forward differences. It can be shown that the jth-order forward difference of f evaluated at x_k can be expressed

$$\boxed{\Delta^j f(x_k) = \sum_{i=0}^{j} (-1)^i \binom{j}{i} y_{k+j-i}} \tag{4.3.6}$$

Here the *binomial coefficient* is defined:

$$\binom{j}{i} \triangleq \frac{j!}{i!(j-i)!} \quad , \quad 0 \le i \le j \tag{4.3.7}$$

The forward differences of f evaluated at $x = x_1$ can be used to obtain a polynomial approximation of f that is useful for interpolation. In particular, the following formulation, called *Newton's forward difference* formula, can be used.

$$\boxed{f(x_1 + \alpha h) \approx f(x_1) + \alpha \Delta f(x_1) + \cdots + \frac{\alpha(\alpha-1)\cdots(\alpha-(n-2))}{(n-1)!} \Delta^{n-1} f(x_1)} \tag{4.3.8}$$

As the parameter α ranges over the interval $[0, n-1]$, the argument $x_1 + \alpha h$ spans the interval $[x_1, x_n]$. The formulation in (4.3.8) only approximates f because the series is truncated after n terms. The error due to truncation can be expressed as follows.

$$E_n(\alpha) = \frac{\alpha(\alpha-1)\cdots(\alpha-(n-1))f^{(n)}(\xi)h^n}{n!} \tag{4.3.9}$$

In this case, the parameter ξ lies somewhere in the interval $[x_1, x_n]$. If the nth derivative of $f(x)$ is bounded in the interval $[x_1, x_n]$, then an upper bound on the size of truncation error can be obtained.

Note that when $\alpha = k$, the argument of f in (4.3.8) is $x_1 + kh = x_{k+1}$. But from (4.3.9), we see that $E_n(k) = 0$ for $0 \le k < n$. Consequently, it follows that the Newton difference formula in (4.3.8) is yet another formulation of the interpolating polynomial that passes through the n samples in the data set D.

EXAMPLE 4.3.1 Newton's Difference Formula

As an illustration of the use of Newton's difference formula, consider the interpolation problem posed in Example 4.2.2. Given $D = \{(0,6),(1,0),(2,2)\}$, the forward differences at $x_1 = 0$ using (4.3.4) and (4.3.5) are

$$\Delta f(x_1) = y_2 - y_1 = -6$$
$$\Delta^2 f(x_1) = y_3 - 2y_2 + y_1 = 8$$

Now $x_1 = 0$ and $h = 1$. Hence, from Equation (4.3.8), the Newton difference formula generated by D is

$$\begin{aligned} f(\alpha) &= 6 + \alpha(-6) + \frac{\alpha(\alpha-1)8}{2} \\ &= 4\alpha^2 - 10\alpha + 6 \end{aligned}$$

Again, this is identical to the result in Example 4.2.2 because the interpolating polynomial is unique.

4.4 CUBIC SPLINES

At this point, we have a number of ways of finding the polynomial of degree $n-1$, which passes through or interpolates the n data points in the set D. Although this solves the interpolation problem as posed, it does not offer a satisfactory solution to the more fundamental problem of approximating the function f *between* the samples. To see why, it is only necessary to consider the following simple rational function.

$$g(x) = \frac{1}{1+x^2} \tag{4.4.1}$$

Suppose the data points are generated by sampling this well-behaved function at nine points uniformly distributed over the interval $[-4, 4]$. These points can be interpolated with a polynomial of degree eight as shown in Figure 4.4.1. Although the polynomial $f(x)$ does indeed pass through the data points as required, it is apparent that it does so

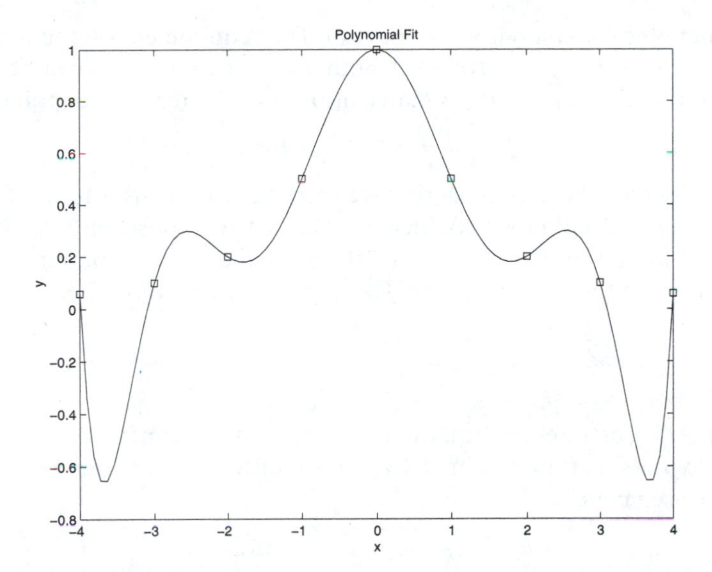

FIGURE 4.4.1 Interpolation with an Eighth Degree Polynomial

by oscillating wildly between the samples, particularly near the ends of the data range. This is a characteristic feature of *all* high-order polynomials, a feature which makes them unattractive for interpolation applications.

In order to reduce the large swings between samples, it is useful to reexamine the piecewise-linear approach used in Figure 4.2.2. Recall that the reason for going to high-order polynomials was to make $f(x)$ smooth. Rather than go all the way to a single polynomial of degree $n-1$, another approach is to piece together low-order polynomials such as quadratics or cubics. This approach has the advantage that the extra degrees of freedom can be used to impose additional constraints at the data points. In particular, the constraints can be chosen so as to ensure smoothness. For example, using cubic polynomials one can require continuity of the zeroth, first, and second derivatives. Polynomial functions that satisfy these smoothness conditions are called *splines*. To construct a set of cubic splines, let $r_k(x)$ denote the cubic polynomial associated with the kth segment $[x_k, x_{k+1}]$.

$$r_k(x) = a_k(x - x_k)^3 + b_k(x - x_k)^2 + c_k(x - x_k) + y_k \quad , \quad 1 \le k < n \qquad (4.4.2)$$

By construction, r_k satisfies the basic *interpolation* constraint, $r_k(x_k) = y_k$ for $1 \le k < n$. Since there are $n-1$ segments and three coefficients per segment, the system in (4.4.2) introduces a total of $3n-3$ unknown coefficients to be determined. To find these coefficients, we impose an equal number of constraints. The basic *continuity* constraint requires that at the end of each of the $n-1$ segments the function pass through the next data point.

$$r_k(x_{k+1}) = y_{k+1} \quad , \quad 1 \le k < n \qquad (4.4.3)$$

Next, we generate additional constraints by requiring that the *first derivatives* match at the $n-2$ interior data points where the segments join.

$$r'_{k-1}(x_k) = r'_k(x_k) \quad , \quad 1 < k < n \qquad (4.4.4)$$

Here $r'(x)$ denotes the first derivative of $r(x)$. The requirement on the first derivatives yields an additional $n - 2$ constraints. Another $n - 2$ constraints can be obtained by requiring that the *second derivatives* match at the $n - 2$ interior data points.

$$r''_{k-1}(x_k) = r''_k(x_k) \quad , \quad 1 < k < n \tag{4.4.5}$$

Again, $r''(x)$ denotes the second derivative of $r(x)$. This yields a total of $3n - 5$ constraints on the $3n - 3$ unknown coefficients. The last two constraints can be generated in a number of ways. The simplest is to set the second derivative, or curvature, equal to zero at both ends of the data sequence. This is called a *natural* spline.

$$r''_1(x_1) = 0 \tag{4.4.6a}$$

$$r''_{n-1}(x_n) = 0 \tag{4.4.6b}$$

The $3n - 3$ constraints in Equations (4.4.3) to (4.4.6) can be made specific by substituting the expression for $r_k(x)$ in (4.4.2). To simplify the equations, let h_k be the sample spacing for segment k.

$$h_k \triangleq x_{k+1} - x_k \quad , \quad 1 \le k < n \tag{4.4.7}$$

Differentiating $r_k(x)$ as needed and substituting into Equations (4.4.3) through (4.4.6), then yields the following equations, which must be solved simultaneously.

$$a_k h_k^3 + b_k h_k^2 + c_k h_k = y_{k+1} - y_k \quad , \quad 1 \le k < n \tag{4.4.8a}$$

$$3a_{k-1}h_{k-1}^2 + 2b_{k-1}h_{k-1} + c_{k-1} - c_k = 0 \quad , \quad 1 < k < n \tag{4.4.8b}$$

$$6a_{k-1}h_{k-1} + 2b_{k-1} - 2b_k = 0 \quad , \quad 1 < k < n \tag{4.4.8c}$$

$$2b_0 = 0 \tag{4.4.8d}$$

$$6a_{n-1}h_{n-1} + 2b_{n-1} = 0 \tag{4.4.8e}$$

The linear algebraic system in (4.4.8) consists of $3n - 3$ equations in the vector of $3n - 3$ unknowns $[a_1, b_1, c_1, a_2, \ldots, c_{n-1}]^T$. The coefficient matrix of this system is sparse, as can be seen from the following example, which corresponds to the special case $n = 3$.

$$\begin{bmatrix} h_1^3 & h_1^2 & h_1 & 0 & 0 & 0 \\ 0 & 0 & 0 & h_2^3 & h_2^2 & h_2 \\ 3h_1^2 & 2h_1 & 1 & 0 & 0 & -1 \\ 6h_1 & 2 & 0 & 0 & -2 & 0 \\ 0 & 2 & 0 & 0 & 0 & 0 \\ 0 & 0 & 0 & 6h_2 & 2 & 0 \end{bmatrix} \begin{bmatrix} a_1 \\ b_1 \\ c_1 \\ a_2 \\ b_2 \\ c_2 \end{bmatrix} = \begin{bmatrix} y_2 - y_1 \\ y_3 - y_2 \\ 0 \\ 0 \\ 0 \\ 0 \end{bmatrix} \tag{4.4.9}$$

If the linear algebraic system in (4.4.9) is used to find the coefficients of the cubic spline interpolating polynomials, a total of $3n - 3$ equations must be solved. However, the number of equations can be reduced if the problem is reformulated in a different manner. Consider, in particular, the following alternative formulation for the kth spline function for $1 \le k < n$ (Hultquist, 1988).

$$\boxed{s_k(x) = a_k(x - x_k) + b_k(x_{k+1} - x) + \frac{(x - x_k)^3 c_{k+1} + (x_{k+1} - x)^3 c_k}{6h_k}} \tag{4.4.10}$$

To interpret the unknowns $\{a_k, b_k, c_k\}$, we differentiate $s_k(x)$.

$$s'_k(x) = a_k - b_k + \frac{(x - x_k)^2 c_{k+1} - (x_{k+1} - x)^2 c_k}{2h_k} \quad , \quad 1 \le k < n \qquad (4.4.11a)$$

$$s''_k(x) = \frac{(x - x_k)c_{k+1} + (x_{k+1} - x)c_k}{h_k} \quad , \quad 1 \le k < n \qquad (4.4.11b)$$

It is clear from Equations (4.4.11b) and (4.4.7) that $s''_k(x_k) = c_k$. Consequently, c_k represents the second derivative of the spline function at data point k. Note that, by construction, the second derivative is continuous at the interior data points because $s''_{k-1}(x_k) = c_k$ for $1 < k < n$. The equations in (4.4.10) contain $3n - 2$ unknowns $\{a_1, \ldots, a_{n-1}, b_1, \ldots, b_{n-1}, c_1, \ldots, c_n\}$. We can eliminate $n - 1$ of these unknowns by applying the basic interpolation constraint. Setting $s_k(x_k) = y_k$ in (4.4.10) yields $b_k h_k + h_k^2 c_k/6 = y_k$ or

$$b_k = \frac{6y_k - h_k^2 c_k}{6h_k} \quad , \quad 1 \le k < n \qquad (4.4.12)$$

Another $n - 1$ of the unknowns can be eliminated by applying the basic continuity constraint. Setting $s_k(x_{k+1}) = y_{k+1}$ in (4.4.10) yields $a_k h_k + h_k^2 c_{k+1}/6 = y_{k+1}$ or

$$a_k = \frac{6y_{k+1} - h_k^2 c_{k+1}}{6h_k} \quad , \quad 1 \le k < n \qquad (4.4.13)$$

It remains to determine the second derivatives at the data points, the vector $c = [c_1, c_2, \ldots, c_n]^T$. This is done by applying the derivative constraint. Setting $s'_{k-1}(x_k) = s'_k(x_k)$ using (4.4.11a) yields

$$a_{k-1} - b_{k-1} + \frac{h_{k-1}c_k}{2} = a_k - b_k - \frac{h_k c_k}{2} \quad , \quad 1 < k < n \qquad (4.4.14)$$

Recall from Equations (4.4.12) and (4.4.13) that a_k and b_k depend on c_k. Substituting the expressions for a_k and b_k, collecting the coefficients of c_{k-1}, c_k, and c_{k+1}, and simplifying, we get the following system of equations for $1 < k < n$.

$$h_{k-1}c_{k-1} + 2(h_{k-1} + h_k)c_k + h_k c_{k+1} = 6\left(\frac{y_{k+1} - y_k}{h_k} - \frac{y_k - y_{k-1}}{h_{k-1}}\right) \qquad (4.4.15)$$

The system in (4.4.15) consists of $n - 2$ equations in the vector of n unknowns c. The two extra constraints needed to solve the system uniquely come from the boundary conditions in (4.4.6). Since $c_k = s''_k(x_k)$, it follows that the boundary conditions for a natural spline are simply

$$c_1 = 0 \qquad (4.4.16a)$$

$$c_n = 0 \qquad (4.4.16b)$$

The linear algebraic system in Equations (4.4.15) and (4.4.16) can be recast in matrix form. To simplify the final result, let

$$w_k \triangleq \frac{y_{k+1} - y_k}{h_k} \quad , \quad 1 \le k < n \qquad (4.4.17)$$

In this case, the coefficient matrix is *tridiagonal*, as can be seen from the following example, which corresponds to the special case $n = 5$.

$$\begin{bmatrix} 1 & 0 & 0 & 0 & 0 \\ h_1 & 2(h_1 + h_2) & h_2 & 0 & 0 \\ 0 & h_2 & 2(h_2 + h_3) & h_3 & 0 \\ 0 & 0 & h_3 & 2(h_3 + h_4) & h_4 \\ 0 & 0 & 0 & 0 & 1 \end{bmatrix} c = \begin{bmatrix} 0 \\ 6(w_2 - w_1) \\ 6(w_3 - w_2) \\ 6(w_4 - w_3) \\ 0 \end{bmatrix} \tag{4.4.18}$$

The coefficients of the cubic spline segments are obtained by solving (4.4.18) for the vector c and then using (4.4.13) and (4.4.12) to obtain the vectors a and b, respectively. The segments in (4.4.10) can then be pieced together to form the spline function $s(x)$. This procedure is summarized in the following algorithm, which performs an n-point cubic spline interpolation at $x = \alpha$.

Alg. 4.4.1 **Cubic Spline Fit**

1. For $k = 1$ to $n - 1$ compute

{

$$h_k = x_{k+1} - x_k$$

$$w_k = \frac{y_{k+1} - y_k}{h_k}$$

}

2. Use LU decomposition to solve the tridiagonal system $Ac = d$ where

$$A = \begin{bmatrix} 1 & 0 & 0 & \cdots & 0 & 0 & 0 \\ h_1 & 2(h_1 + h_2) & h_2 & \cdots & 0 & 0 & 0 \\ 0 & h_2 & 2(h_2 + h_3) & \cdots & 0 & 0 & 0 \\ \vdots & \vdots & \vdots & \ddots & \vdots & \vdots & \vdots \\ 0 & 0 & 0 & \cdots & 2(h_{n-3} + h_{n-2}) & h_{n-2} & 0 \\ 0 & 0 & 0 & \cdots & h_{n-2} & 2(h_{n-2} + h_{n-1}) & h_{n-1} \\ 0 & 0 & 0 & \cdots & 0 & 0 & 1 \end{bmatrix}$$

$$d = [0, 6(w_2 - w_1), 6(w_3 - w_2), \ldots, 6(w_{n-1} - w_{n-2}), 0]^T$$

3. For $k = 1$ to $n - 1$ compute

{

$$a_k = \frac{6y_{k+1} - h_k^2 c_{k+1}}{6h_k}$$

$$b_k = \frac{6y_k - h_k^2 c_k}{6h_k}$$

}

4. Compute $1 \le k < n$ such that $x_k \le \alpha \le x_{k+1}$, and

$$\beta = a_k(\alpha - x_k) + b_k(x_{k+1} - \alpha) + \frac{(\alpha - x_k)^3 c_{k+1} + (x_{k+1} - \alpha)^3 c_k}{6h_k}$$

The value returned by Alg. 4.4.1 is $\beta = f(\alpha)$. Typically, Alg. 4.4.1 is called once to compute the coefficients $\{a, b, c\}$ and β. Subsequent calls, for different values of α, then use step (4) as the entry point.

EXAMPLE 4.4.1 Spline Functions

As an illustration of Alg. 4.4.1, suppose data set D consists of the following input/output pairs:

$$D = \{(0,2), (1,3), (2,1)\}$$

In this case, the step size is constant with $h_1 = h_2 = 1$. From step (1), the right-hand side variables are $w_1 = 1$ and $w_2 = -2$. Thus, from step (2), the tridiagonal linear algebraic system that must be solved is

$$\begin{bmatrix} 1 & 0 & 0 \\ 1 & 4 & 1 \\ 0 & 0 & 1 \end{bmatrix} c = \begin{bmatrix} 0 \\ -18 \\ 0 \end{bmatrix}$$

Since $c_1 = 0$ and $c_3 = 0$, it follows that $4c_2 = -18$ or $c_2 = -4.5$. Applying step (3), we obtain the remaining unknowns as follows.

$$a_1 = \frac{6(3) - (-4.5)}{6} = 3.75$$

$$a_2 = \frac{6(1) - 0}{6} = 1$$

$$b_1 = \frac{6(2) - 0}{6} = 2$$

$$b_2 = \frac{6(3) - (-4.5)}{6} = 3.75$$

Consequently, from (4.4.10), the two spline functions are

$$s_1(x) = 3.75x + 2(1 - x) - (4.5/6)x^3$$

$$s_2(x) = (x - 1) + 3.75(2 - x) - (4.5/6)(2 - x)^3$$

It is a simple matter to verify that $s_1(0) = 2$, $s_1(1) = s_2(1) = 3$, and $s_2(2) = 1$. Similarly, $s_1'(x) = -2.25x^2 + 1.75$ and $s_2'(x) = 2.25x^2 - 9x + 6.25$. Thus, $s_1'(1) = s_2'(1) = -0.5$. Finally, $s_1''(x) = -4.5x$ and $s_2''(x) = 4.5x - 9$. Thus, $s_1''(1) = s_2''(1) = -4.5$ and the natural spline boundary conditions are $s_1''(0) = s_2''(2) = 0$.

The formulation in Equation (4.4.18) is valid for natural splines, spline functions that satisfy the boundary conditions $s''(x_1) = s''(x_n) = 0$. An alternative way to produce the boundary constraints is to pass a cubic interpolating polynomial $f_1(x)$ through the first four data points and then set $s''(x_1) = f_1''(x_1)$. Similarly, a second cubic interpolating polynomial $f_n(x)$ can be fitted to the last four data points and we can then set $s''(x_n) = f_n''(x_n)$. The virtue of using natural splines is that the resulting function tends to minimize the curvature. More specifically, it can be shown that, among functions $s(x)$ having a continuous second derivative, the natural cubic spline minimizes the following integral.

$$J = \int_{x_1}^{x_n} [s''(x)]^2 \, dx \tag{4.4.19}$$

Since J is minimized, this tends to reduce the size of the oscillations between the data points. Therefore, cubic splines are more attractive for interpolation purposes than higher-order polynomials. The following example illustrates this point.

EXAMPLE 4.4.2 Cubic Spline

As an illustration of the use of cubic splines, consider the problem of fitting a curve to the discrete data points generated by the rational function $g(x)$ in Equation (4.4.1). Example program *e442* on the distribution CD uses $n = 9$ points uniformly distributed over the interval $[-4, 4]$. The resulting cubic spline fit is shown in Figure 4.4.2. Comparing this fit with the eighth degree polynomial interpolation in Figure 4.4.1, it is evident that there is a dramatic improvement in the behavior of the function between the samples.

FIGURE 4.4.2 Interpolation with Cubic Splines

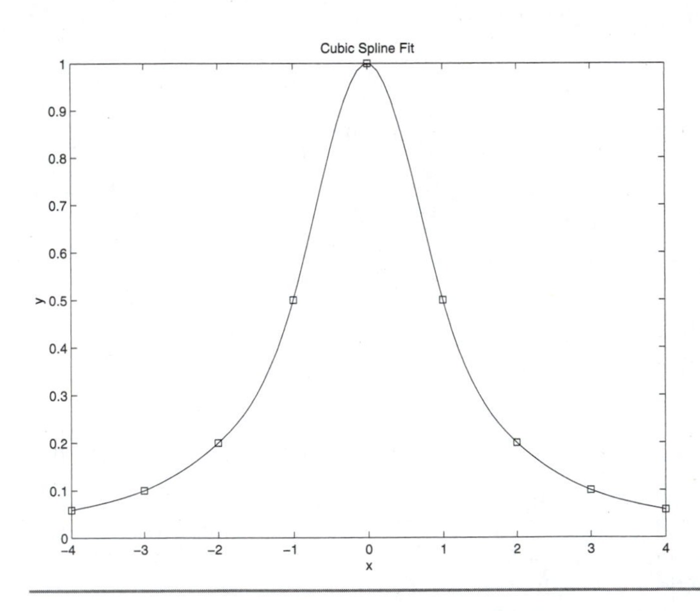

Cubic splines provide one means of improving the poor fit near the ends of the data interval that is characteristic of higher-order interpolating polynomials. Another approach is to control the sample spacing. Although this is not always possible, there are applications, such as the approximation of a complex function with a simpler function, where one is at liberty to vary the sample spacing. Suppose the independent variable x ranges over the interval $[a, b]$. Rather than use uniform spacing, it makes more sense to put additional points out near the ends of the interval where the poor fit tends to occur. A particularly effective way to do this is to use the roots of the *Chebyshev* polynomials. The first two Chebyshev polynomials are $T_0(x) = 1$ and $T_1(x) = x$, and subsequent Chebyshev polynomials are generated by the recurrence relation

$$T_n(x) = 2xT_{n-1}(x) - T_{n-2}(x) \quad , \quad n \geq 2 \tag{4.4.20}$$

The Chebyshev polynomials have many interesting properties, one of which is that the n roots of $T_n(x)$ all lie in the interval $[-1, 1]$, the jth root being

$$r_k = \cos\left[\frac{(2k-1)\pi}{2n}\right] \quad , \quad 1 \leq k \leq n \tag{4.4.21}$$

These roots can be mapped from the interval $[-1, 1]$ into the interval $[a, b]$ by using scaling and offset as follows.

$$x_k = \frac{(b-a)r_k + a + b}{2} \quad , \quad 1 \leq k \leq n \tag{4.4.22}$$

If the points of interpolation are selected as in (4.4.22), then the amplitudes of the oscillations between the data points are minimized. This is a consequence of the fact that, among all monic polynomials of degree n, the magnitude of the polynomial $g(x) = T_n(x)/2^{n-1}$ over the interval $[-1, 1]$ is smallest. This is called the *minimax property* of Chebyshev polynomials. Additional discussion of Chebyshev interpolation can be found in Lastman and Sinha (1989).

4.5 LEAST SQUARES

Although cubic splines provide an effective means of passing a smooth curve through a set of data points, there are instances where a different approach is needed. If the data set D consists of experimental measurements, then the data may contain a random component that represents *measurement noise* generated by a sensor as shown in Figure 4.5.1. In these instances, it is more appropriate to find a curve that describes the underlying *trend* of the data without necessarily passing through each data point.

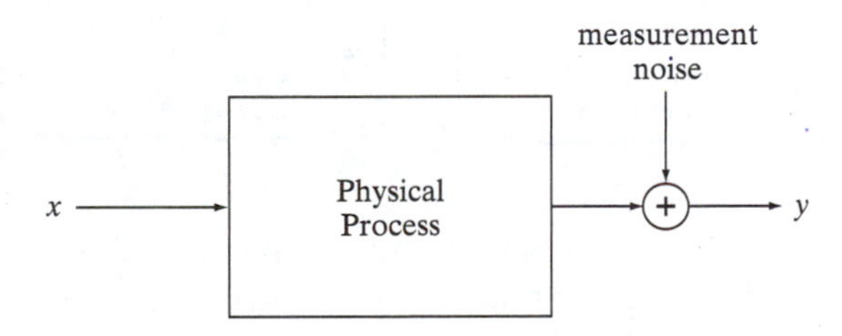

FIGURE 4.5.1 Input/Output Measurements with Noise

4.5.1 Straight Line Fit

Noise-corrupted experimental measurements are common in engineering. The type of curve used to describe the trend will depend on the nature of the underlying physical system that generates the data. In the absence of specialized knowledge about that system, the simplest trend curve is a linear polynomial.

$$f(x) = a_1 + a_2 x \tag{4.5.1}$$

To illustrate how this curve might be fitted to a "cloud" of noisy data, consider the following measure of the error associated with the n points in the data set D.

$$E(a) \triangleq \sum_{k=1}^{n} w_k [f(x_k) - y_k]^2 \tag{4.5.2}$$

Here $w_k > 0$ denotes a *weight* associated with the kth data sample, with the most common case being *uniform* weighting:

$$w_k = 1 \quad , \quad 1 \le k \le n \tag{4.5.3}$$

There may be instances where some of the measurements are known to be more accurate or more important. In these cases, larger values can be used for those weights. The objective is to find a 2×1 coefficient vector a that minimizes the error.

$$\begin{aligned} \text{minimize:} \quad & E(a) \\ \text{subject to:} \quad & a \in R^2 \end{aligned}$$

The polynomial f whose coefficients minimize $E(a)$ is referred to as the weighted least-squares fit or simply the *least-squares* fit. Although alternative error criteria can be used, the least-squares objective in Equation (4.5.2) is preferred because a solution can be easily found. Indeed, if we differentiate $E(a)$ with respect to the unknown components of the coefficient vector a, this yields

$$\frac{\partial E}{\partial a_1} = 2 \sum_{k=0}^{n} w_k [f(x_k) - y_k] \tag{4.5.4a}$$

$$\frac{\partial E}{\partial a_2} = 2 \sum_{k=0}^{n} w_k [f(x_k) - y_k] x_k \tag{4.5.4b}$$

Setting $\partial E / \partial a = 0$ then results in two equations in the unknowns a_1 and a_2. After rearrangement, these equations can be expressed as a linear algebraic system of the form $Ca = b$.

$$\underbrace{\begin{bmatrix} \sum_{k=1}^{n} w_k & \sum_{k=1}^{n} w_k x_k \\ \sum_{k=1}^{n} w_k x_k & \sum_{k=1}^{n} w_k x_k^2 \end{bmatrix}}_{C} \begin{bmatrix} a_1 \\ a_2 \end{bmatrix} = \underbrace{\begin{bmatrix} \sum_{k=1}^{n} w_k y_k \\ \sum_{k=1}^{n} w_k x_k y_k \end{bmatrix}}_{b} \tag{4.5.5}$$

Note that the coefficient matrix C is symmetric, and it depends only on the weighting vector w and the vector of independent variables x. This is in contrast to the right-hand-side vector b, which depends on w, x, and the vector of dependent variables y.

EXAMPLE 4.5.1 Straight Line Fit

As an illustration of a straight line fit, consider the following set of input/output pairs, which contain noise in the dependent variable.

$$D = \{(0, 2.10), (1, 2.85), (2, 1.10), (3, 3.20), (4, 3.90)\}$$

Suppose uniform weighting ($w_k = 1$) is used. Using Equation (4.5.5), the coefficient matrix C and the right-hand-side vector b are

$$C_{11} = \sum_{k=1}^{5} 1 = 5$$

$$C_{12} = \sum_{k=1}^{5} x_k = 1 + 2 + 3 + 4 = 10$$

$$C_{22} = \sum_{k=1}^{5} x_k^2 = 1 + 4 + 9 + 16 = 30$$

$$b_1 = \sum_{k=1}^{5} y_k = 2.10 + 2.85 + 1.10 + 3.20 + 3.90 = 13.15$$

$$b_2 = \sum_{k=1}^{5} x_k y_k = 2.85 + 2(1.10) + 3(3.20) + 4(3.90) = 30.25$$

From symmetry, $C_{21} = C_{12} = 10$. Thus, the linear algebraic system that must be solved to find the coefficient vector a is

$$\begin{bmatrix} 5 & 10 \\ 10 & 30 \end{bmatrix} a = \begin{bmatrix} 13.15 \\ 30.25 \end{bmatrix}$$

In this case, the solution is $a = [1.84, 0.395]^T$. Thus, the linear polynomial that best fits the data, in a uniform least-squares sense, is

$$f(x) = 1.84 + 0.395x$$

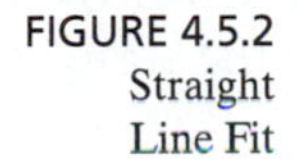

FIGURE 4.5.2 Straight Line Fit

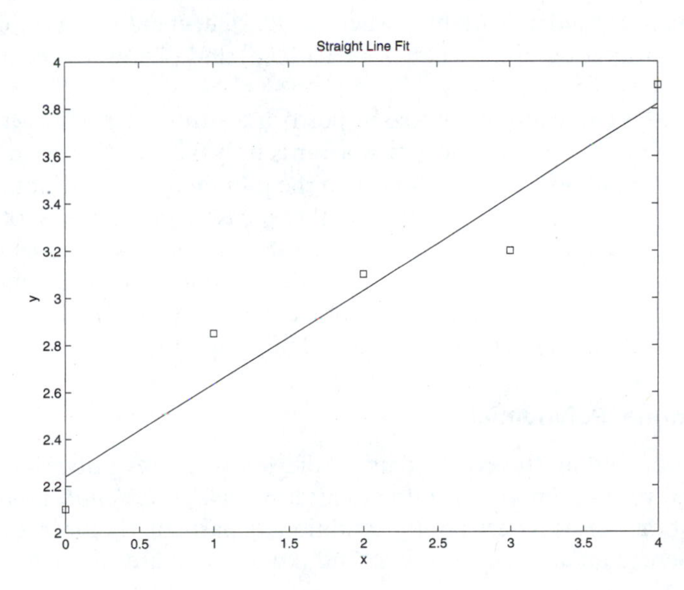

Example program *e451* on the distribution CD creates the graph shown in Figure 4.5.2, which shows the linear polynomial $f(x)$ and the data points in D.

4.5.2 Polynomial Fit

In many cases, the underlying trend of the data cannot be modeled effectively with a straight line. Since a straight line fit corresponds to using a polynomial of degree one, a natural generalization is to use a polynomial of degree $m - 1$.

$$f(x) = \sum_{k=1}^{m} a_k x^{k-1} \tag{4.5.6}$$

Given n data points in D, we select $m \le n$. Note that when $m = n$, the least-squares fit becomes the interpolating polynomial with $E(a) = 0$. As with a straight line fit, the optimal value for a is obtained by computing $\partial E/\partial a$. In this case, the derivative of the error E with respect to the kth coefficient of f is

$$\frac{\partial E}{\partial a_j} = 2 \sum_{k=1}^{n} w_k[f(x_k) - y_k]x_k^{j-1} \quad , \quad 1 \le j \le m \tag{4.5.7}$$

Setting $\partial E/\partial a = 0$ results in m equations in the m unknown components of the coefficient vector a. Upon rearrangement, these equations can be cast as a linear algebraic system $Ca = b$ where

$$C_{kj} = \sum_{i=1}^{n} w_i x_i^{k+j-2} \quad , \quad 1 \le k, j \le m \tag{4.5.8}$$

$$b_k = \sum_{i=1}^{n} w_i x_i^{k-1} y_i \quad , \quad 1 \le k \le m \tag{4.5.9}$$

Again it is clear that the coefficient matrix is symmetric and depends only on the weights and the independent variables, whereas the right-hand-side vector depends on the dependent variables as well. When $m = 2$, (4.5.8) and (4.5.9) reduce to the straight line fit equations in (4.5.5).

Least-squares curve fitting can also be performed using non-polynomial functions (e.g., exponentials) when the application warrants it. If the function used for curve fitting is linear in the unknown parameters, then the parameters can be obtained by solving a suitable linear algebraic system. In these cases, the process of finding the least-squares fit is called *linear regression*. When the function used for curve fitting is a nonlinear function of the unknown parameters, the parameters can be obtained by solving an optimization problem where the objective is to minimize the error $E(a)$. Optimization problems are considered in detail in Chapter 6.

4.5.3 Orthogonal Polynomials

The technique of finding the coefficients of the least-squares polynomial by solving $Ca = b$ suffers from an important numerical drawback, the Vandermonde type matrix C in Equation (4.5.8) is very poorly conditioned, particularly for large values of m. Fortunately, there is an alternative way to find the least-squares fit that avoids a linear algebraic system altogether. The key is to use polynomials other than $\{1, x, x^2, \ldots, x^{m-1}\}$ for the terms of $f(x)$.

Let f and g be two polynomials. Suppose w is the $n \times 1$ weight vector, and D is the set of n data points. Then the *discrete inner product* of f and g associated with D is defined as follows.

$$\langle f, g \rangle \triangleq \sum_{k=1}^{n} w_k f(x_k) g(x_k) \tag{4.5.10}$$

The inner product has a number of useful properties. For example, the inner product is a linear function of its first argument, which means that for all scalars α and β,

$$\langle \alpha f_1 + \beta f_2, g \rangle = \alpha \langle f_1, g \rangle + \beta \langle f_2, g \rangle \tag{4.5.11}$$

The inner product is also a linear function of its second argument. We say that the polynomials f and g are *orthogonal* if and only if $\langle f, g \rangle = 0$. Our objective is to construct a set of m orthogonal polynomials $\{\phi_1(x), \phi_2(x), \ldots, \phi_m(x)\}$. One important property of orthogonal polynomials is that they can be generated by a three-term recurrence relation.

$$\phi_{k+1}(x) = (x - b_k)\phi_k(x) - c_k \phi_{k-1}(x) \quad , \quad k \geq 2 \tag{4.5.12}$$

Thus, to compute the sequence, it is necessary to know the first two polynomials $\{\phi_1, \phi_2\}$ and the coefficients $\{b_k, c_k\}$ for $k \geq 2$. The first two polynomials are

$$\phi_1(x) = 1 \tag{4.5.13a}$$

$$\phi_2(x) = x - b_1 \tag{4.5.13b}$$

Since ϕ_1 and ϕ_2 are orthogonal, setting $\langle \phi_1, \phi_2 \rangle = 0$ and using (4.5.11) to solve for b_1 yields

$$b_1 = \frac{\langle x\phi_1, \phi_1 \rangle}{\langle \phi_1, \phi_1 \rangle} \tag{4.5.14}$$

The coefficients needed to generate the remaining polynomials now can be determined by induction. Suppose $\{\phi_1(x), \phi_2(x), \ldots, \phi_k(x)\}$ is an orthogonal set. To find (b_k, c_k) we make sure that the expression for $\phi_{k+1}(x)$ in Equation (4.5.12) is orthogonal to both $\phi_k(x)$ and $\phi_{k-1}(x)$. Setting $\langle \phi_{k+1}, \phi_k \rangle = 0$ and solving for b_k yields

$$b_k = \frac{\langle x\phi_k, \phi_k \rangle}{\langle \phi_k, \phi_k \rangle} \quad , \quad k \geq 2 \tag{4.5.15}$$

Note that to obtain b_k we have used (4.5.11) and have exploited the fact that $\langle \phi_k, \phi_{k-1} \rangle = 0$. Similarly, setting $\langle \phi_{k+1}, \phi_{k-1} \rangle = 0$ and solving for c_k yields

$$c_k = \frac{\langle x\phi_k, \phi_{k-1} \rangle}{\langle \phi_{k-1}, \phi_{k-1} \rangle} \quad , \quad k \geq 2 \tag{4.5.16}$$

EXAMPLE 4.5.2 Orthogonal Polynomials

As a simple illustration of the construction of orthogonal polynomials, consider the following set of input/output pairs.

$$D = \{(0,-3),(1,0),(2,5)\}$$

Suppose uniform weighting ($w_k = 1$) is used. The first orthogonal polynomial is simply

$$\phi_1(x) = 1$$

Thus, $\langle \phi_1, \phi_1 \rangle = 3$ and $\langle x\phi_1, \phi_1 \rangle = \Sigma_{k=1}^3 x_k = 3$. Hence, from Equation (4.5.14), we have $b_1 = 3/3$, and the second orthogonal polynomial is

$$\phi_2(x) = x - 1$$

Next, $\langle \phi_2, \phi_2 \rangle = \Sigma_{k=1}^3 (x_k - 1)^2 = 2$, and using (4.5.15) and (4.5.10), we have

$$b_2 = \frac{\langle x\phi_2, \phi_2 \rangle}{\langle \phi_2, \phi_2 \rangle} = \frac{\Sigma_{k=1}^3 x_k(x_k - 1)^2}{2} = 1$$

Similarly, using (4.5.16) and (4.5.10),

$$c_2 = \frac{\langle x\phi_2, \phi_1 \rangle}{\langle \phi_1, \phi_1 \rangle} = \frac{\Sigma_{k=1}^3 x_k(x_k - 1)}{3} = \frac{2}{3}$$

Hence, from (4.5.12), the third orthogonal polynomial is

$$\begin{aligned} \phi_3(x) &= (x - b_2)\phi_2(x) - c_2\phi_1(x) \\ &= (x - 1)^2 - 2/3 \\ &= x^2 - 2x + 1/3 \end{aligned}$$

As a check, the inner products are

$$\begin{aligned} \langle \phi_1, \phi_2 \rangle &= \sum_{k=1}^{3} (x_k - 1) = -1 + 1 = 0 \\ \langle \phi_1, \phi_3 \rangle &= \sum_{k=1}^{3} (x_k^2 - 2x_k + 1/3) = 1/3 - 2/3 + 1/3 = 0 \\ \langle \phi_2, \phi_3 \rangle &= \sum_{k=1}^{3} (x_k - 1)(x_k^2 - 2x_k + 1/3) = -1/3 + 1/3 = 0 \end{aligned}$$

The motivation for finding an orthogonal set of polynomials lies in the fact that the computation of a least-squares fit to the data D then becomes extremely simple, and it does not suffer from ill conditioning. Indeed, consider the following alternative formulation of the least-squares polynomial.

$$\boxed{f(x) = \sum_{k=1}^{m} a_k \phi_k(x)} \tag{4.5.17}$$

To find the optimal value for the coefficient vector a, we again compute $\partial E/\partial a$. Using Equation (4.5.2), the derivative of the error E with respect to the kth coefficient of f is

$$\frac{\partial E}{\partial a_j} = 2\sum_{k=1}^{n} w_k[f(x_k) - y_k]\phi_j(x_k) \quad , \quad 1 \le j \le m \tag{4.5.18}$$

If we now substitute the expression for $f(x)$ from (4.5.17) and recall the definition of the inner product from (4.5.10), we find that the expression for the derivative of E with respect to a_j reduces to

$$\frac{\partial E}{\partial a_j} = 2\sum_{i=1}^{m} a_i\langle\phi_i, \phi_j\rangle - 2\sum_{k=1}^{n} y_k\phi_j(x_k) \quad , \quad 1 \le j \le m \tag{4.5.19}$$

But the $\{\phi_k\}$ are orthogonal, which means that $\langle\phi_i, \phi_j\rangle = 0$ for $i \ne j$. Setting $\partial E/\partial a_j = 0$, we can then solve for a_j directly as

$$\boxed{a_j = \frac{\sum_{k=1}^{n} y_k\phi_j(x_k)}{\langle\phi_j, \phi_j\rangle} \quad , \quad 1 \le j \le m} \tag{4.5.20}$$

This effectively eliminates the need to solve an ill-conditioned linear algebraic system. By using orthogonal polynomials, the coefficient matrix C becomes diagonal and each component of a is uncoupled from the others; hence it can be solved for independently as in Equation (4.5.20). This procedure is summarized in the following algorithm; which performs a least-squares fit using a polynomial of degree $m - 1$.

Alg. 4.5.1 **Least-Squares Fit**

1. Set $\phi_1(x) = 1$ and compute

{

$$d_1 = \sum_{i=1}^{n} w_i$$

$$b_1 = \frac{1}{d_1}\sum_{i=1}^{n} w_i x_i$$

$$\phi_2(x) = x - b_1$$

}

2. For $k = 2$ to m compute

{

$$d_k = \sum_{i=1}^{n} w_i\phi_k^2(x_i)$$

$$b_k = \frac{1}{d_k}\sum_{i=1}^{n} w_i x_i\phi_k^2(x_i)$$

$$c_k = \frac{1}{d_{k-1}} \sum_{i=1}^{n} w_i x_i \phi_k(x_i)\phi_{k-1}(x_i)$$

$$\phi_{k+1}(x) = (x - b_k)\phi_k(x) - c_k\phi_{k-1}(x)$$

}

3 Compute

$$\beta = \sum_{k=1}^{m} \left(\frac{y_k}{d_k}\right)\phi_k(\alpha)$$

The value returned by Alg. 4.5.1 is $\beta = f(\alpha)$. Typically, Alg. 4.5.1 is called once to compute the coefficients $\{d, b, c\}$ and β. Subsequent calls, for different values of α, then use step (3) as the entry point.

EXAMPLE 4.5.3 Polynomial Fit

To illustrate the use of orthogonal polynomials to find a least-squares fit, consider the following set of input/output pairs.

$$D = \{(0, -3), (1, -1), (2, 5)\}$$

This data set was used in Example 4.5.2 where it was found that, with uniform weighting, the independent variables generated the following set of orthogonal polynomials.

$$\begin{aligned} \phi_1(x) &= 1 \\ \phi_2(x) &= x - 1 \\ \phi_3(x) &= x^2 - 2x + 1/3 \end{aligned}$$

Using (4.5.20), the coefficients of the least-squares polynomial are

$$\begin{aligned} a_1 &= \frac{\sum_{k=1}^{3} y_k \phi_1(x_k)}{\langle \phi_1, \phi_1 \rangle} = \frac{-3 - 1 + 5}{3} = \frac{1}{3} \\ a_2 &= \frac{\sum_{k=1}^{3} y_k \phi_2(x_k)}{\langle \phi_2, \phi_2 \rangle} = \frac{3 + 5}{2} = 4 \\ a_3 &= \frac{\sum_{k=1}^{3} y_k \phi_3(x_k)}{\langle \phi_3, \phi_3 \rangle} = \frac{-3(1/3) - (-2/3) + 5(1/3)}{2/3} = 2 \end{aligned}$$

Thus, the polynomial of degree zero that best fits the data, in a uniform least-squares sense, is

$$f_1(x) = a_1\phi_1(x) = 1/3$$

Next, the polynomial of degree one that best fits the data is

$$\begin{aligned} f_2(x) &= a_1\phi_1(x) + a_2\phi_2(x) \\ &= 1/3 + 4(x - 1) \\ &= 4x - 11/3 \end{aligned}$$

Finally, the polynomial of degree two that best fits the data is

$$\begin{aligned} f_3(x) &= a_1\phi_1(x) + a_2\phi_2(x) + a_3\phi_3(x) \\ &= 1/3 + 4(x-1) + 2(x^2 - 2x + 1/3) \\ &= 2x^2 - 3 \end{aligned}$$

Note that since $n = 3$, the polynomial $f_3(x)$ is in fact the interpolating polynomial of the data set D. Example program *e453* on the distribution CD creates the graph shown in Figure 4.5.3, which shows the constant, linear, and quadratic fits along with the data points in D.

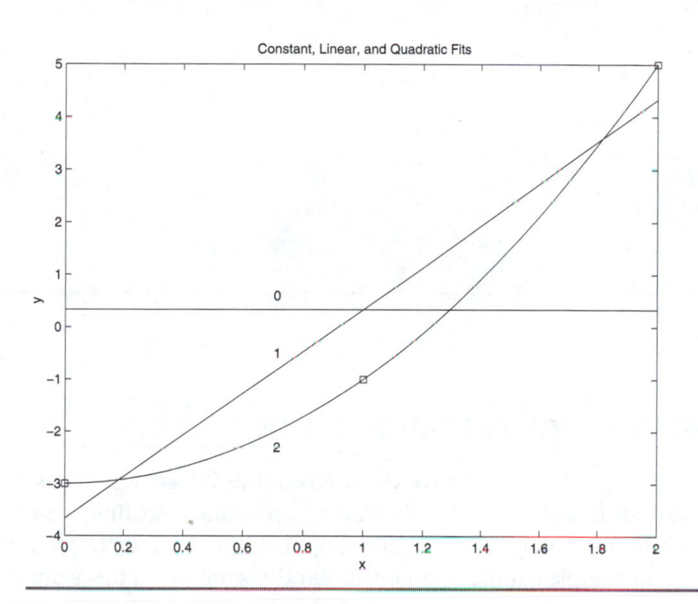

FIGURE 4.5.3 Constant, Linear, and Quadratic Fits

The use of orthogonal polynomials provides a numerically well-conditioned way to find both the least-squares polynomial (when $m < n$) and the interpolating polynomial (when $m = n$). Of course, higher-order interpolating polynomials are rarely used because cubic splines provide a smoother fit between samples, particularly near the ends of the data interval.

EXAMPLE 4.5.4 Least-Squares Fit

As an illustration of a least-squared fit to noise-corrupted data, consider a set of data points generated by sampling the following function.

$$f(x) = x^3 - 4x^2 + 3x + v(x)$$

Here $v(x)$ represents random noise uniformly distributed over the interval $[-0.5, 0.5]$. Example program *e454* on the distribution CD samples the function at $n = 100$ values of x uniformly distributed over the interval $[0, 3.2]$. The resulting least-squares fit using a polynomial with $m = 4$ coefficients is shown in Figure 4.5.4. It is apparent that the third-order least-squares fit follows the underlying trend of the data in this case.

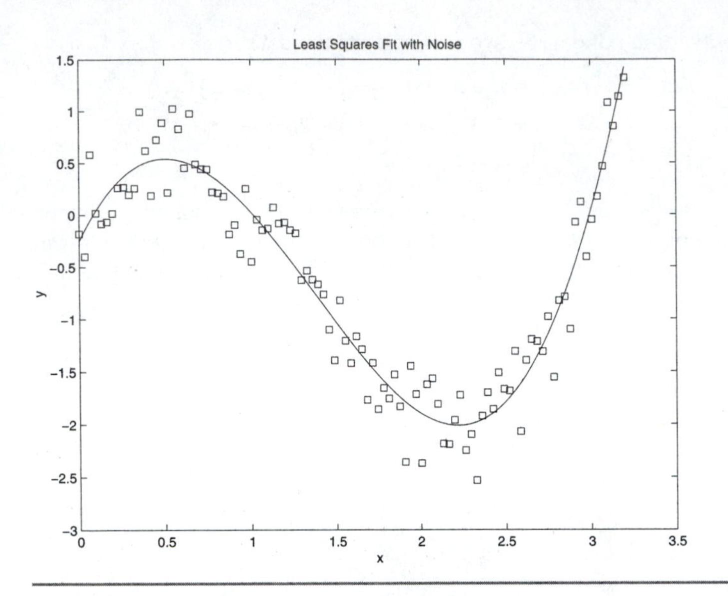

FIGURE 4.5.4 Least-Squares Fit of Noise-Corrupted Data

4.6 TWO-DIMENSIONAL INTERPOLATION

All of the curve fitting methods we have considered thus far are applicable to functions of a single variable. In this section, we briefly examine how these techniques can be extended to functions with two independent variables. Suppose the data set D consists of a total of mn samples with m samples in the x dimension and n samples in the y dimension.

$$D = \{(x_k, y_j, z_{kj}) \mid 1 \le k \le m,\ 1 \le j \le n\} \tag{4.6.1}$$

The independent variables (x_k, y_j) form a grid and the value of the dependent variable, z_{kj}, can be thought of as the height of a surface above this grid. The basic interpolation problem in two dimensions is to find a function $f(x, y)$ that passes through the mn data points.

$$f(x_k, y_j) = z_{kj} \quad , \quad 1 \le k \le m,\ 1 \le j \le n \tag{4.6.2}$$

We assume the independent variables have been ordered such that $x_1 < x_2 < \cdots < x_m$ and $y_1 < y_2 < \cdots < y_n$. Let (x, y) be an arbitrary point within the data range $[x_1, x_m] \times [y_1, y_n]$. To estimate a value for $f(x, y)$, we begin by determining the grid element that contains the point (x, y). For example, suppose

$$x_k \le x \le x_{k+1} \tag{4.6.3a}$$

$$y_j \le y \le y_{j+1} \tag{4.6.3b}$$

Thus, the point (x, y) lies in the grid element whose lower-left corner is (x_k, y_j) as shown in Figure 4.6.1. Next, let (u, v) represent the normalized coordinates of the location of point (x, y) within the grid element.

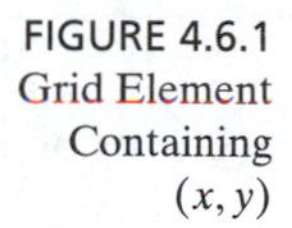
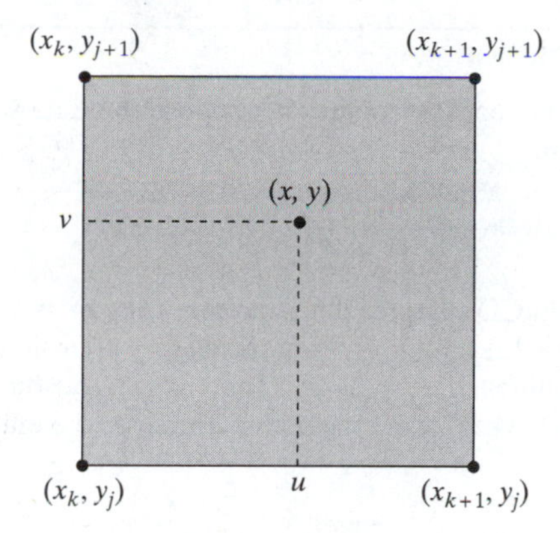

FIGURE 4.6.1 Grid Element Containing (x, y)

$$u(x) \triangleq \frac{x - x_k}{x_{k+1} - x_k} \tag{4.6.4a}$$

$$v(y) \triangleq \frac{y - y_j}{y_{j+1} - y_j} \tag{4.6.4b}$$

Note that $u(x)$ ranges from 0 to 1 as x moves from x_k to x_{k+1}. A similar remark holds for $v(y)$ as y moves from y_j to y_{j+1}.

The simplest form of two-dimensional interpolation is a generalization of piecewise-linear interpolation called *bilinear interpolation*. The value of $f(x, y)$ is expressed in terms of u, v, and the dependent variables at the four corners of the grid element.

$$\boxed{f(x, y) = (1 - u)(1 - v)z_{k,j} + (1 - u)vz_{k,j+1} + u(1 - v)z_{k+1,j} + uvz_{k+1,j+1}} \tag{4.6.5}$$

Using the fact that $u(x_k) = v(y_j) = 0$ and $u(x_{k+1}) = v(y_{j+1}) = 1$, it is a simple matter to verify that the surface patch in Equation (4.6.5) passes through the four corner points in Figure 4.6.1. The formulation in (4.6.5) is called bilinear because when y fixed $f(x, y)$ is a linear polynomial in x, and when x fixed $f(x, y)$ is a linear polynomial in y. The bilinear interpolation scheme is very simple to implement, and the surface that it produces is continuous. However, the surface gradient, $\nabla f(x, y) = [\partial f/\partial x, \partial f/\partial y]^T$, is not continuous at the boundaries between grid elements.

Improved accuracy and smoother surfaces can be obtained by using higher-order interpolation methods. For example, to perform an interpolation of order $p - 1$ in each dimension, we begin by locating a p element by p element block that contains the evaluation point (x, y). We then perform p one-dimensional interpolations, one for each value of y_j. These p interpolating functions are then evaluated at x to generate a new set of points. A one-dimensional interpolation is then performed on this new set of points and the result is evaluated at y to produce the estimate of $f(x, y)$. A discussion of higher-order two-dimensional interpolation methods can be found in (Press et al., 1992).

EXAMPLE 4.6.1 Two-Dimensional Interpolation

As an illustration of a two-dimensional bilinear interpolation, suppose the data set consists of samples of the following rational function.

$$f(x, y) = \frac{0.1}{x^2 + y^2 + 0.1}$$

Example program *e461* on the distribution CD samples this function using $m = 9$ points in the x coordinate uniformly distributed over $[-1, 1]$ and $n = 9$ points in the y coordinate uniformly distributed over $[-1, 1]$. The resulting bilinear interpolation function, based on 81 points, is shown in Figure 4.6.2. Careful inspection reveals some of the individual grid elements that are

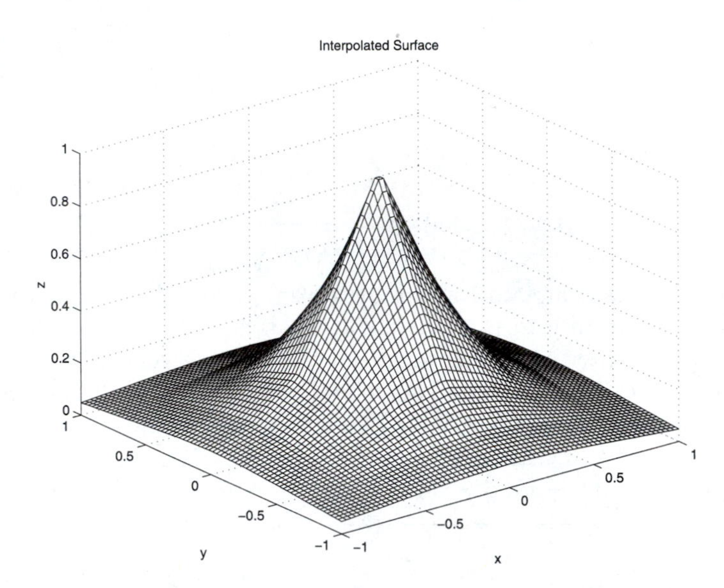

FIGURE 4.6.2 Bilinear Interpolation with 81 Points

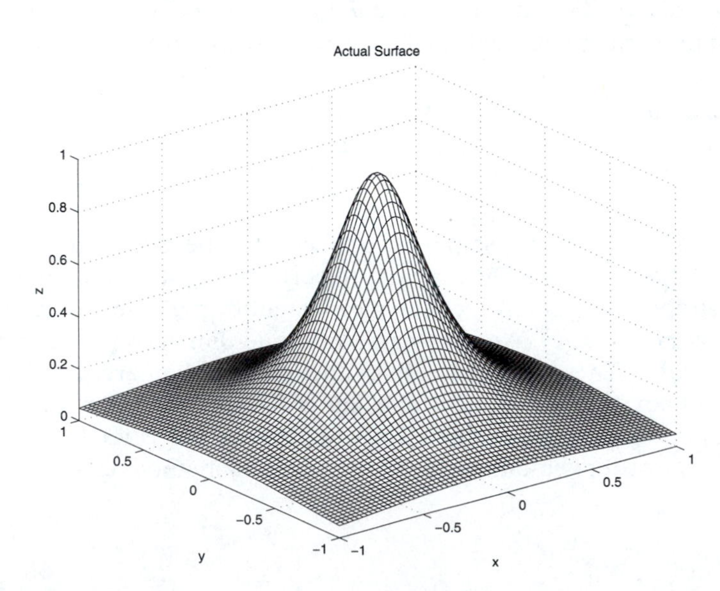

FIGURE 4.6.3 The Surface $f(x, y)$ Using 5184 Points

of dimension 0.25×0.25. For comparison, the actual surface, plotted to a resolution of 90 points by 90 points, is shown in Figure 4.6.3. Although this is clearly smoother, the bilinear representation in Figure 4.6.2 is a reasonable approximation, particularly in light of the fact that it is based on only one percent of the samples plotted in Figure 4.6.3.

4.7 APPLICATIONS

The following examples illustrate applications of curve fitting techniques using both MATLAB and C. The relevant MATLAB functions in the NLIB toolbox are described in Section 1.5 of Appendix 1, while the corresponding C functions in the NLIB library are described in Section 2.8 of Appendix 2.

4.7.1 Pressure-Temperature Curves: MATLAB

Chemical engineers use tabulated pressure-volume-temperature (pvT) data to analyze the characteristics of various liquids such as ethyl alcohol, refrigerant, and water, which change phases between solid, liquid, and gas. One such problem is to determine the change in enthalpy of saturated water as it is vaporized to steam. Suppose the system consists of one kg of saturated water. An appropriate model to use in this instance is the Clapeyron equation (Moran and Shapiro, 1995).

$$\Delta h = T(v_g - v_f)\frac{dp}{dT} \tag{4.7.1}$$

Here Δh is the change in enthalpy between the liquid and gas phase as the water vaporizes. The parameter T denotes the temperature in degrees Kelvin, and v_g and v_f are the specific volumes of the gas and liquid, respectively. The derivative, dp/dT denotes the rate of change of pressure with respect to temperature. Numerical pressure-temperature points are available from standard steam tables and are summarized in Table 4.7.1.

Values for the constants T, v_g, and v_f also can be obtained directly from the steam table data. In this case, 100° C corresponds to $T = 373.15°$ K, while $v_g = 1.63$ m^3/kg, and $v_f = 0.00104$ m^3/kg. To estimate the value of dp/dT, a number of approaches

TABLE 4.7.1 Pressure-Temperature Data for Saturated Water

T (°C)	p (bars)
50	0.1235
60	0.1994
70	0.3119
80	0.4739
90	0.7014
100	1.014
110	1.433
120	1.958
130	2.701
140	3.613
150	4.758

might be used. One possibility is to compute a numerical derivative directly from the data in Table 4.7.1. Numerical differentiation is a topic that is covered in Chapter 7. Another approach is to fit a curve to the data and then differentiate the curve. This is the approach taken in the following MATLAB script on the distribution CD.

```
%------------------------------------------------------------------
% Example 4.7.1: Pressure-Temperature Curves
%------------------------------------------------------------------

% Initialize

   clc                        % clear screen
   clear                      % clear variables
   n = 11;                    % number of data points
   q = 10*n;                  % number of plot points
   m = 4;                     % number of coefficients
   f = 1.e5;                  % conversion factor
   vg = 1.673;                % specific volume (gas)
   vf = 0.00104;              % specific volume (liquid)
   b = zeros (m,1);
   d = zeros (m-1,1);
   C = zeros (m,m);
   X = zeros (q,2);
   Y = zeros (q,2);
   T = [50 60 70 80 90 100 110 120 130 140 150]';
   p = [.1235 .1994 .3119 .4739 0.7014 1.014 ...
       1.433 1.958 2.701 3.613 4.758]';

% Find least-squares polynomial using equations (4.5.9)-(4.5.10) */

   fprintf ('Example 4.7.1: Pressure-Temperature Curves\n');
   for k = 1 : m
      for i = 1 : n
         b(k) = b(k) +  p(i)*T(i)^(k-1);
      end
      for j = 1 : m
         for i = 1 : n
            C(k,j) = C(k,j) + T(i)^(k+j-2);
         end
      end
   end
   a = gauss (C,b);

% Find slope at T = 100

   show ('Polynomial coefficients',a)
   for i = 1 : m-1
      d(i) = i*a(i+1);
   end
```

```
   show ('Derivative coefficients',d)
   s = polynom (d,100);
   show ('Slope (bars/K)',s);

% Find change in enthalpy

   dh = 373.15*(vg - vf)*f*s;
   show ('Change in enthalpy (kJ/kg)',dh/1000);

% Plot curve and data

   for i = 1 : q
      j = min ([i,n]);
      X(i,1) = T(1) + (T(n) - T(1))*(i-1)/(q-1);
      Y(i,1) = polynom (a,X(i,1));
      X(i,2) = T(j);
      Y(i,2) = p(j);
   end
   graphxy (X,Y,'Pressure-Temperature Curves','T (^oC)','p
(bars)','s')
%----------------------------------------------------------------------
```

When script *e471.m* is executed, it fits the following least-squares cubic polynomial to the data in Table 4.7.1.

$$p(T) = -1.061 + 4.73(T/100) - 6.869(T/100)^2 + 4.199(T/100)^3 \qquad (4.7.2)$$

A plot of the least-squares cubic polynomial $p(T)$ along with the tabulated data points from Table 4.7.1 is shown in Figure 4.7.1. Differentiating the expression for $p(T)$,

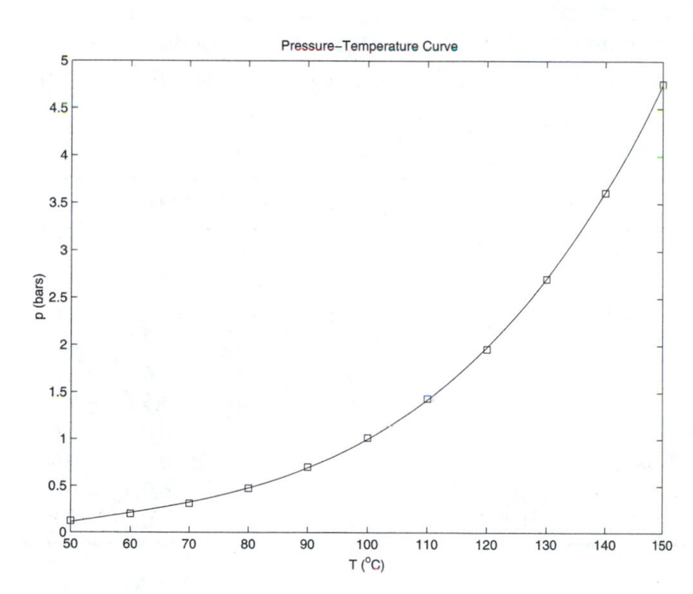

FIGURE 4.7.1 Least-Squares Cubic Polynomial Fit to Discrete Pressure-Temperature Data

and evaluating the result at $T = 100°$ C, then yields the following numerical estimate of the slope.

$$\frac{dp}{dT} = 0.0358839 \text{ bars}/° \text{C} \quad \text{at} \quad T = 100° \text{C} \tag{4.7.3}$$

Finally, to get the change in enthalpy associated with the vaporization of water we return to Clapeyron's equation, (4.7.1). To express the answer in units of J/kg, we use the conversion factor 10^5 N/m^2 per bar. This results in the following estimate of the change in enthalpy based on the curve fit.

$$\Delta h = 2238.77 \text{ kJ/kg} \tag{4.7.4}$$

In Chapter 7, this example is revisited where an alternative numerical technique is used to compute the change in enthalpy. A solution to Example 4.7.1 using C can be found in the file *e471.c* on the distribution CD.

4.7.2 Water Resource Management: C

Civil and environmental engineers are often concerned with the planning and construction of dams and the management of water resources. Suppose it is desired to dam a river in order to prevent floods and generate hydro electric power. When the dam is placed across the river, it will create a large pond behind it as the water backs up. In order to estimate the size and shape of the pond, surveyors perform a number of topographical measurements to determine the elevation of the land along a grid of points behind the planned location for the dam.

To make the problem specific, consider the depth measurements summarized in Table 4.7.2. Here a 10 × 10 grid of locations was measured with each elevation point specified in meters below the planned nominal height for the water.

Once the discrete set of pond depth measurements is available, a number of calculations might be made. For example, the total volume of water held in the pond could be estimated. Another important calculation for planning purposes is the total force of

TABLE 4.7.2 Pond Depth Measurements

	0.0	20.0	40.0	60.0	80.0	100.0	120.0	140.0	160.0	180.0
0.0	0.0	0.0	0.0	0.0	0.0	0.0	0.0	0.0	0.0	0.0
20.0	0.0	0.0	0.0	0.0	0.0	0.0	0.0	0.0	0.0	0.0
40.0	−2.7	0.0	−2.1	−3.4	−2.6	0.0	0.0	0.0	0.0	0.0
60.0	−6.1	−4.5	−6.6	−7.3	−6.5	−8.3	−1.4	0.0	0.0	0.0
80.0	−5.6	−8.3	−11.4	−9.6	−7.1	−9.3	−5.4	−2.2	0.0	0.0
100.0	−10.7	−8.4	−5.3	−7.3	−8.1	−6.4	−3.5	−1.3	0.0	0.0
120.0	−7.2	−6.1	−9.8	−4.3	−6.4	−1.7	0.0	0.0	0.0	0.0
140.0	−2.3	−1.3	0.0	−2.4	−3.5	0.0	0.0	0.0	0.0	0.0
160.0	0.0	0.0	0.0	0.0	0.0	0.0	0.0	0.0	0.0	0.0
180.0	0.0	0.0	0.0	0.0	0.0	0.0	0.0	0.0	0.0	0.0

the water that the dam must hold back. In order to make calculations of this type, it is helpful to develop a mathematical model that shows how the pond depth changes with location behind the dam.

$$z = f(x, y) \tag{4.7.5}$$

The following C program on the distribution CD can be used to solve this problem.

```
/* ---------------------------------------------------------------- */
/* Example 4.7.2: Water Resource Management */
/* ---------------------------------------------------------------- */

#include "c:\nlib\util.h"

int main (void)
{
   int          m,                      /* number of x data points */
                n,                      /* number of y data points */
                q = 40,                 /* number of x plot points */
                r = 80,                 /* number of y plot points */
                i,j;
   vector       x1 = vec (q,""),
                y1 = vec (r,""),
                x,y;
   matrix       Z1 = mat (q,r,""),
                Z;
   FILE         *f = fopen ("data472.m","r");

/* Get depth data from data file */

   printf ("\nExample 4.7.2: Water Resource Management\n");
   x = getvec (f,&m);
   y = getvec (f,&n);
   Z = getmat (f,&m,&n);
   showvec ("x",x,m,0);
   showvec ("y",y,n,0);
   showmat ("Z",Z,m,n,0);

/* Evaluate surface using bilinear interpolation */

   for (i = 1; i <= q; i++)
   {
      x1[i] = x[1] + (x[m] - x[1])*(i-1)/(q - 1);
      for (j = 1; j <= r; j++)
      {
         y1[j] = y[1] + (y[n] - y[1])*(j-1)/(r - 1);
         Z1[i][j] = bilin (x1[i],y1[j],x,y,Z,m,n);
      }
   }
```

```
/* Plot surface */

   plotxyz (x1,y1,Z1,q,r,"","x (m)","y (m)","z (m)","fig472.m");
   return 0;
}
/* ------------------------------------------------------------ */
```

When program *e472.c* is executed, it first reads the file *data472.m*, which contains the depth measurements shown in Table 4.7.2. It then computes the pond depth at arbitrary points by fitting a surface $f(x, y)$ to the data using bilinear interpolation. A plot of the resulting surface that shows the size and shape of the pond and the profile of water immediately behind the dam (at $y = 0$) is shown in Figure 4.7.2.

The problems of finding the total volume V of water stored in the pond and the total force F on the dam due to the water both involve computing numerical integrals. The topic of numerical integration is covered in Chapter 8. A solution to Example 4.7.2 using MATLAB can be found in the file *e472.m* on the distribution CD.

4.7.3 Voltage Regulator Circuit: MATLAB

Electrical instruments that operate from standard 120 volt 60 Hz AC power outlets typically contain a transformer-rectifier-filter circuit that converts high voltage AC power to low voltage DC power. Typically, the output u of this DC power supply circuit is a constant voltage plus a small sinusoidal ripple voltage. To refine the DC power signal u, it is sent through a voltage regulator circuit that steps the DC voltage down to the final level needed and also filters out the residual AC ripple voltage. An example of a voltage regulator circuit using a zener diode is shown in Figure 4.7.3

Here the resistor R and capacitor C form a low pass filter that reduces the small sinusoidal ripple that may be present in u. The resistor R_L represents the load that the

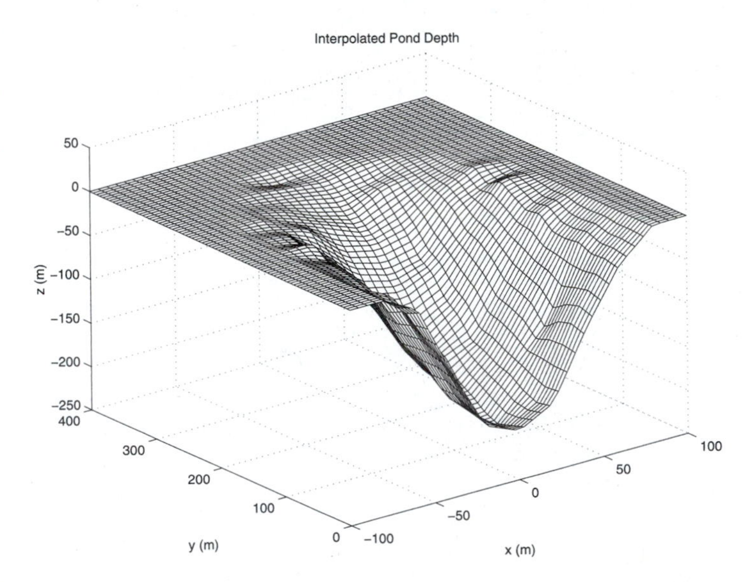

FIGURE 4.7.2 Bilinear Interpolation of Pond Depth Measurements

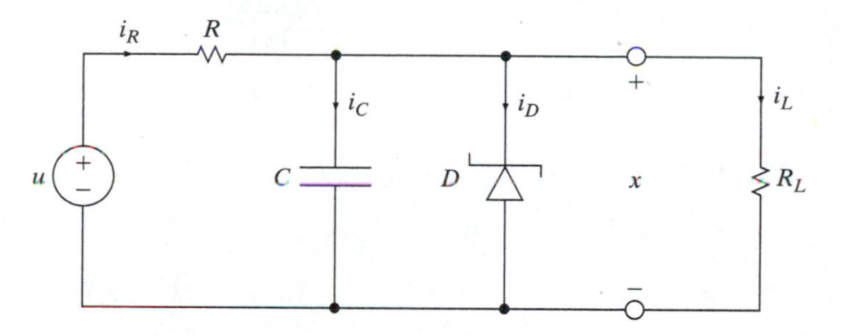

FIGURE 4.7.3 A DC Voltage Regulator Circuit Using a Zener Diode

circuit must drive. The diode D is a zener diode whose current i_D is a nonlinear function of the voltage drop x across it.

$$i_D = f(x) \tag{4.7.6}$$

The equation of motion that describes the operation of the voltage regulator circuit is obtained by observing that the capacitance C times the rate of change of the voltage across the capacitor, dx/dt, equals the current flowing through the capacitor, i_C. The capacitor current is obtained by applying Kirchhoff's current law to the RC node, which yields $i_R = i_C + i_D + i_L$. Applying Ohm's law to compute $i_R = (v - x)/R$ and $i_L = x/R_L$, one arrives at the following mathematical model for the operation of the voltage regulator circuit.

$$C \frac{dx}{dt} = \frac{u - x}{R} - f(x) - \frac{x}{R_L} \tag{4.7.7}$$

In Chapter 8, we examine techniques for solving this equation for an arbitrary input $u(t)$. At present, suppose the input u is constant. The steady-state voltage drop across the load resistor can then be obtained by setting $dx/dt = 0$ and solving for x. Setting $dx/dt = 0$ results in the following nonlinear algebraic equation.

$$f(x) + \frac{x}{R} + \frac{x}{R_L} - \frac{u}{R} = 0 \tag{4.7.8}$$

The voltage-current characteristic $f(x)$ for a zener diode is not amenable to representation with a simple analytical expression. However, measurements are available for a variety of zener diodes in data books. Consequently, a curve can be fitted to the voltage-current data and a mathematical model for the zener diode can be obtained in this manner. To make the problem specific, suppose the measurements associated with a 5.1-volt zener diode are as summarized in Table 4.7.3.

The following MATLAB script on the distribution CD can be used to fit a curve to the zener diode data.

```
%----------------------------------------------------------------
% Example 4.7.3: Voltage Regulator Circuit
%----------------------------------------------------------------

% Initialize
```

```
    clc                          % clear screen
    clear                        % clear variables
    q = 200;                     % number of plot points
    X = zeros (q,2);
    Y = zeros (q,2);

% Get data for diode I-V characteristic

    fprintf ('Example 4.7.3: Voltage Regulator Circuit\n');
    data473                      % get data for zener diode
    show ('x',x);
    show ('i',i);
    n = length (x);

% Compute spline and graph curve with data

    for k = 1 : q
       j = min ([k,n]);
       X(k,1) = x(1) + (x(n) - x(1))*(k-1)/(q-1);
       X(k,2) = x(j);
       Y(k,2) = i(j);
    end
    Y(:,1) = spline (x,i,X(:,1));
    graphxy (X,Y,'Current-Voltage Characteristic','x (V)','i
(mA)','s')

%-----------------------------------------------------------------------
```

TABLE 4.7.3 Voltage-Current Data for Zener Diode

$x\,(V)$	$i\,(mA)$	$x\,(V)$	$i\,(mA)$
−1.0000	−14.5773	1.2750	0.0000
−0.8750	−9.7656	2.5500	0.0000
−0.7500	−6.1498	3.8250	0.0000
−0.6250	−3.5589	4.9215	0.8791
−0.5000	−1.8222	4.9470	1.3960
−0.3750	−0.7687	4.9725	2.4123
−0.2500	−0.2278	4.9980	4.7116
−0.1250	−0.0285	5.0235	11.1683
0.0000	0.0000	5.0490	37.6926

When script *e473.m* is executed, it uses cubic spline interpolation to fit a curve to the data in Table 4.7.3. A plot of the resulting curve and the tabulated data points is shown in Figure 4.7.4. Note that this is referred to as a 5.1 volt zener diode because the slope of $f(x)$ is close to vertical near $x = 5.1$, which represents the desired voltage for driving the load. Zener diodes are available for many operating voltages.

The cubic spline curve $f(x)$ in Figure 4.7.4 can be inserted into the nonlinear algebraic steady-state equation in (4.7.8) to solve for the steady-state voltage x for differ-

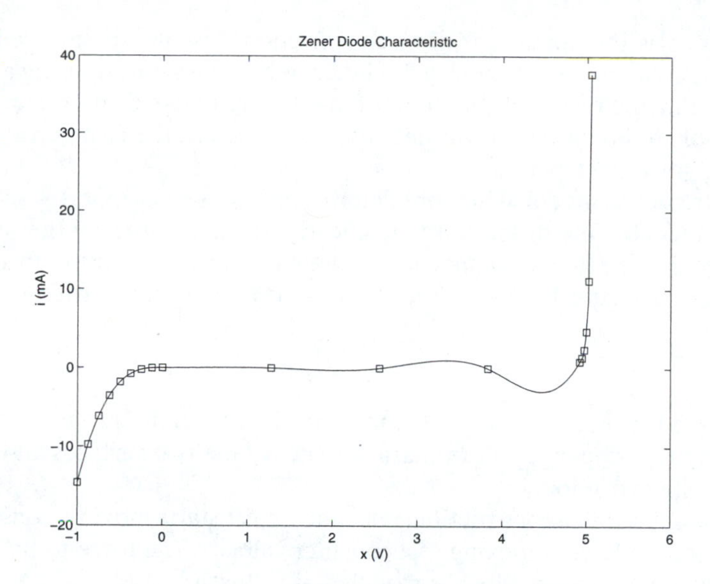

FIGURE 4.7.4 Cubic Spline Interpolation of Zener Diode Measurements

ent values of u. Solving a nonlinear algebraic equation is referred to as root-finding. The topic of root finding is covered in detail in Chapter 5, where a number of effective numerical techniques are presented. A solution to Example 4.7.3 using C can be found in the file *e473.c* on the distribution CD.

4.7.4 Nonlinear Friction Model: C

Mechanical engineers use mathematical models to describe the motion of objects under the influence of forces. A simple translational mechanical system containing a single mass m acted on by a force f is shown in Figure 4.7.5. For the model to be realistic, it should include the effects of friction. Applying Newton's second law, we set mass times acceleration equal to the sum of the forces acting on the mass.

$$m \frac{dy^2}{dt^2} = f - F \tag{4.7.9}$$

Here F denotes the frictional force that opposes motion. As a matter of convenience, friction is often modeled with a *viscous* friction term that is proportional to the *velocity* $v = dy/dt$.

$$F_v = \mu_v \frac{dy}{dt} \tag{4.7.10}$$

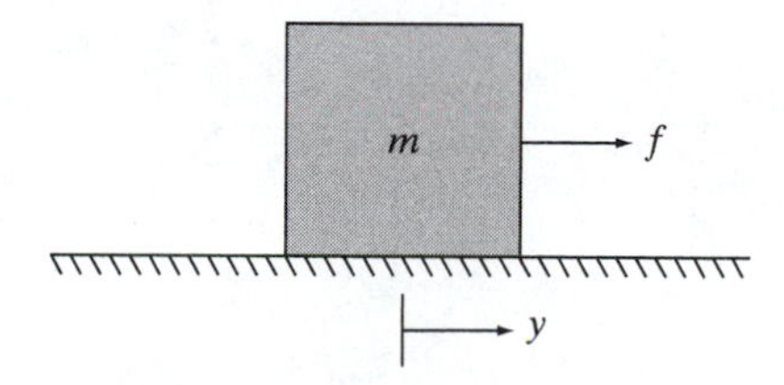

FIGURE 4.7.5 A Translational Mechanical System with Friction

The constant μ_v is the coefficient of viscous friction. Viscous friction can be used to model the effects of moving through a fluid such as air. The virtue of using a viscous friction model is that the resulting equations of motion are linear. Consequently, analytical tools applicable to linear time-invariant equations, such as the Laplace transform, can be applied to solve the equations.

When friction is modeled in more detail, it becomes clear that it is actually a nonlinear phenomenon. One of the most significant effects ignored by the linear viscous friction model is the effect of *static* friction, which corresponds to the threshold force, F_s, that is needed to get a body that is at rest to start moving. Static friction can be modeled as follows.

$$F_s = \mu_s N \tag{4.7.11}$$

Here N is the normal force between the moving object and the surface that it is sliding on. The constant μ_s depends on the characteristics of the two surfaces and is called the coefficient of static friction.

In the absence of viscous friction, once the object starts moving, considerably less force is required to keep it moving at a constant velocity. The force to be overcome in this case is *dynamic* friction, which is modeled as follows.

$$F_d = \mu_d N \tag{4.7.12}$$

Again, N is the normal force between the moving object and the sliding surface, and the constant μ_d depends on the characteristics of the two surfaces and is called the coefficient of dynamic friction. In many cases, $\mu_d \ll \mu_s$. When all three frictional terms are present, the function which represents the frictional force is awkward to express analytically. Consequently, this is an instance where it is useful to fit a curve through numerical data.

To make the problem specific, suppose a set of velocity-force measurements are performed, resulting in the experimental data points listed in Table 4.7.4.

It is assumed that the points corresponding to negative velocities are the mirror image of those in Table 4.7.4 because friction always opposes motion. That is, the friction curve $F(v)$ has odd symmetry.

TABLE 4.7.4 Force-Velocity Data for Friction Model

v (m/s)	F (N)
0	32
0.1	4.0
0.2	5.6
0.3	7.1
0.4	8.4
0.5	10.1
0.6	11.5
0.7	13.1
0.8	14.4
0.9	16.1
1.0	17.5

$$F(-v) = -F(v) \tag{4.7.13}$$

The following C program on the distribution CD can be used to solve this problem.

```
/* ------------------------------------------------------------------- */
/* Example 4.7.4: Nonlinear Friction Model */
/* ------------------------------------------------------------------- */

#include "c:\nlib\util.h"

int main (void)
{
   int          n = 11,                        /* number of data points */
                q = 10*n,                      /* number of plot points */
                i,j;
   vector       v = vec (n,"0 .1 .2 .3 .4 .5 .6 .7 .8 .9 1."),
                F = vec (n,"32 4 6.5 9 11.5 14 16.5 19 21.5 24 26.5"),
                c = vec (n,"");
   matrix       X = mat (2,q,""),
                Y = mat (2,q,"");

/* Find cubic spline fit */

   printf ("\nExample 4.7.4: Nonlinear Friction Model\n");
   spline (0,v,F,c,n,1);
   showvec ("v",v,n,0);
   showvec ("F",F,n,0);
   showvec ("Spline coefficients",c,n,0);

/* Plot curve and data */

   for (i = 1; i <= q; i++)
   {
      j = MIN (i,n);
      X[1][i] = v[1] + (v[n] - v[1])*(i-1)/(q-1);
      Y[1][i] = spline (X[1][i],v,F,c,n,0);
      X[2][i] = v[j];
      Y[2][i] = F[j];
   }
   graphmm (X,Y,2,q,"","v (m/s)","p (N)","fig474.m",1);
   return 0;
}
*/ ------------------------------------------------------------------- */
```

When program *e474.c* is executed, it uses cubic spline interpolation to fit a curve to the data in Table 4.7.4. A plot of the resulting curve together with the tabulated data points from Table 4.7.4 is shown in Figure 4.7.6. The cubic spline fit undershoots the data somewhat in the range $0.1 \le v \le 0.2$, and it is continuously differentiable in the region $v > 0$. However, there is a jump discontinuity at $v = 0$ due to the presence of static friction.

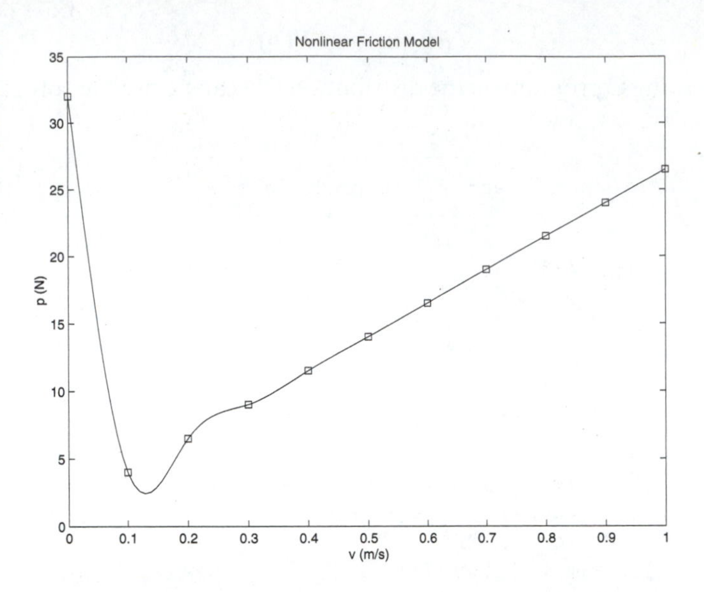

FIGURE 4.7.6 Cubic Spline Interpolation of Friction Measurements

A still more realistic friction model can be obtained by taking into account the fact that the surfaces may not be uniform. For example, there may be a rough spot for a certain value of y or a shaft may be out of round and exhibit more friction at certain angular positions. In these instances, the friction term should be generalized from a curve $F(v)$ to a surface $F(y, v)$. A solution to Example 4.7.4 using MATLAB can be found in the file *e474.m* on the distribution CD.

4.8 SUMMARY

Numerical methods for one- and two-dimensional curve fitting are summarized in Table 4.8.1. The Dimension column indicates the number of independent variables, and the Smooth column specifies whether or not the curve or surface has at least one continuous derivative.

Given a set of n data points (x_k, y_k) for $1 \le k \le n$, a function $f(x)$ that passes through all n points is said to perform interpolation on the n points:

$$f(x_k) = y_k \quad , \quad 1 \le k \le n \tag{4.8.1}$$

TABLE 4.8.1 Curve Fitting Techniques

Method	*Dimension*	*Smooth*
Piecewise-linear	1	no
Interpolating polynomial	1	yes
Cubic spline	1	yes
Least squares	1	yes
Bilinear	2	no

Piecewise-linear interpolation is a simple explicit method that fits a curve to the data by connecting the points with straight line segments. The curve $f(x)$ is not smooth because the slope changes abruptly at the sample points.

Polynomial interpolation produces a smooth curve that can be differentiated as many times as necessary. However, if the number of samples n is not small, the interpolating polynomial tends to oscillate wildly between the samples, particularly near the ends of the data range. The amplitude of the oscillations can be minimized by selecting the sample points to be distributed in a nonuniform way based on the roots of the Chebyshev polynomials. Although the coefficients of the interpolating polynomial can be obtained by solving the Vandermonde system, this is not the recommended method because the coefficient matrix becomes very ill-conditioned as the number of samples increases. Instead, the Lagrange interpolating polynomials, Newton's difference formula, or orthogonal polynomials can be used to compute the coefficients explicitly.

Cubic spline interpolation is an effective curve fitting technique based on splicing cubic polynomials together in such a way that the zeroth, first, and second derivatives are continuous. For an n point interpolation, a tridiagonal linear algebraic system of dimension n must be solved to find the cubic spline coefficients. Cubic splines are better behaved between the data points than higher-order interpolating polynomials.

A second way to fit a curve to a set of data points is to find a function $f(x)$ that minimizes the following least-squares error criterion:

$$E = \sum_{k=1}^{n} [f(x_k) - y_k]^2 \qquad (4.8.2)$$

Least-squares curve fitting is a useful technique when there are a large number of samples and one is only interested in representing the underlying trend of the data. This is appropriate, for example, when the dependent variable contains a random component due to measurement noise. Typically, polynomials are used to find a least-squares fit, but the technique can be extended to other types of functions as well. When the degree of the polynomial is one less than the number of samples, the least-squares fit reduces to the interpolating polynomial. Although least-squares coefficients can be found by solving a linear algebraic system, this is not recommended because the coefficient matrix becomes very ill-conditioned as the degree increases. Instead, the coefficients can be computed explicitly using orthogonal polynomials.

The bilinear method is a two-dimensional interpolation method that is analogous to the one-dimensional piecewise-linear method. It is a simple explicit method that produces a continuous interpolation surface. However, the surface is not smooth because the gradient vector of partial derivatives is not continuous at the grid element boundaries.

PROBLEMS

The problems are divided into Analysis problems, which can be solved by hand, and Computation problems, which require the use of MATLAB or C. Solutions to selected problems can be found in Appendix 4. Students are encouraged to use these problems, which are identified with an (S), as a check on their understanding of the material. Problems marked with a (P) are programming problems that require the student to implement one or more of the algorithms discussed in the chapter. The remaining

Computation problems require the student to write a main program that uses one or more of the NLIB functions discussed in Appendix 1 or Appendix 2.

Analysis

4.1 Consider the two-point interpolation problem of finding a linear polynomial $f_k(x) = a_1 + a_2 x$ such that $f_k(x_k) = y_k$ and $f_k(x_{k+1}) = y_{k+1}$. In this case the coefficient vector $a = [a_1, a_2]^T$ must satisfy the linear algebraic system

$$\begin{bmatrix} 1 & x_k \\ 1 & x_{k+1} \end{bmatrix} a = \begin{bmatrix} y_k \\ y_{k+1} \end{bmatrix}$$

Solve this system, and show that the resulting polynomial $f_k(x)$ is identical to the parametric piecewise-linear fit in Equation (4.2.4).

4.2 Use Equation (4.2.6) to find the interpolating polynomial for the following data set.

$$D = \{(1,-10),(2,-6),(3,0)\}$$

(S)4.3 Find the Lagrange interpolating polynomials $\{Q_1(x), Q_2(x), Q_3(x), Q_4(x)\}$ for the following data set. Use the Lagrange interpolating polynomials to find the interpolating polynomial $f(x)$. Simplify your final answers as much as possible.

$$D = \{(0,-18),(1,-10),(2,0),(3,18)\}$$

4.4 Find the forward differences $\{\Delta f(x_1), \Delta^2 f(x_1), \Delta^3 f(x_1)\}$ for the data set D in problem 4.3. Find the Newton difference formula $f(x_1 + \alpha h)$ for this data.

4.5 Recall from (2.4.26) that the number of FLOPs required to solve a tridiagonal linear system of dimension n using LU decomposition is $5n - 4$. What is the percent savings in computational effort obtained by computing a cubic spline function using the approach in (4.4.18), rather than the method in (4.4.9) when n is large?

4.6 Find the natural cubic spline function for the following data set D.

$$D = \{(0,-18),(1,-10),(2,0)\}$$

4.7 Find the linear polynomial $f(x) = a_1 + a_2 x$ that best fits the following data in a least-squares sense, assuming uniform weighting.

$$D = \{(1,-2.95),(2,-1.15),(3,0.90),(4,3.15),(5,4.95),(6,7.05)\}$$

(S)4.8 Some noisy experimental data (e.g., radioactive decay) can be effectively modeled by using an exponential fit of the following form where u and v are unknown parameters.

$$z = u \exp(-vx)$$

This problem can be converted to an equivalent problem of finding a straight-line fit by applying the natural logarithm to both sides of the equation. Let $y = \ln(z)$ and $a = [\ln(u), -v]^T$. Then the above equation can be written as

$$y = a_1 + a_2 x$$

This is identical to Equation (4.5.1). Suppose the data samples have uniform weighting as in (4.5.3). Then, from (4.5.5), the optimal least-squares coefficient vector a is the solution of

$$\underbrace{\begin{bmatrix} n & \sum_{k=1}^n x_k \\ \sum_{k=1}^n x_k & \sum_{k=1}^n x_k^2 \end{bmatrix}}_{C} \begin{bmatrix} a_1 \\ a_2 \end{bmatrix} = \underbrace{\begin{bmatrix} \sum_{k=1}^n y_k \\ \sum_{k=1}^n x_k y_k \end{bmatrix}}_{bX}$$

Use this result to find the exponential function $f(x) = u \exp(-vx)$ that best fits the following data in a least-squares sense.

$$D = \{(1,1.84),(2,0.91),(3,0.45),(4,0.26),(5,0.13)\}$$

4.9 Find the orthogonal polynomials $\{\phi_1(x), \phi_2(x), \phi_3(x)\}$ for the following data set D, assuming uniform weighting. Use these polynomials to find the least-squares fits of degree zero, one, and two. Verify that the second-degree least-squares fit is the interpolating polynomial.

$$D = \{(1,-2),(2,1),(3,7)\}$$

Computation

(P)4.10 Write a function called *intpoly* that computes the coefficients of the interpolating polynomial by solving Equation (4.2.6). The pseudo-prototype for the function is

```
[vector a,float c] = intpoly (matrix D,int n)
```

On entry to *intpoly:* D is a $2 \times n$ data matrix containing the strictly increasing independent variables in row one and the dependent variables in row two, and $n \geq 1$ is the number of samples. On exit from *intpoly:* a is the $n \times 1$ coefficient vector of the interpolating polynomial, and c is the condition number of the coefficient matrix used to find a. Test the function with the following data set D.

$$D = \{(0,5),(1,-2),(2,3),(3,4),(4,-1),(5,7),(6,5),(7,2)\}$$

Print a and the condition number. Plot the interpolating polynomial over $[x_1, x_n]$ and the data points in D on the same graph.

(P)4.11 Write a function called *lagrange* that returns the value of an interpolating polynomial $f(x)$ by using the Lagrange interpolating polynomials as in Equation (4.2.9). The pseudo-prototype for the function is

```
[float y] = lagrange (float x,matrix D,int n)
```

On entry to *lagrange:* x is the evaluation point, D is a $2 \times n$ data matrix containing the strictly increasing independent variables in row one and the dependent variables in row two, and $n \geq 2$ is the number of samples. On exit from *lagrange,* y is the value $f(x)$. Test the function using the data set in problem 4.10. Plot $f(x)$ over $[x_1, x_n]$ and the data points in D on the same graph.

4.12 Write a program that uses the NLIB function *spline* to find the natural cubic spline fit for the data set D in problem 4.10. Plot the spline function $s(x)$ over $[x_1, x_n]$ and the data points in D on the same graph.

(SP)4.13 Write a function called *expfit* that computes the coefficients of the least-squares exponential fit $f(x)$ using the results of problem 4.8. The pseudo-prototype for the function is

```
[float u,float v] = expfit (matrix D,int n)
```

On entry to *expfit:* D is a $2 \times n$ data matrix containing the strictly increasing independent variables in row one and the dependent variables in row two, and $n \geq 2$ is the number of samples. On exit from *expfit:* u is the amplitude and v is the exponent of the least-squares fit $f(x) = u \exp(-vx)$. Test the function with the data set D in problem 4.8. Plot $f(x)$ over $[x_1, x_n]$ and the data points in D on the same graph, and print u and v.

4.14 Write a program that uses the NLIB function *polyfit* to compute a least-squares fit to the noise-corrupted data obtained by sampling the following function.

$$f(x) = \frac{\cos(\pi x)}{1 + x^2 + x^4} + v(x)$$

Here $v(x)$ represents random numbers uniformly distributed over the interval $[-.1, .1]$. Suppose the data set D is constructed by sampling $f(x)$ at $n = 16$ points uniformly distributed over the interval $[0, 4]$. Plot the data points in D and the least-squares polynomials of degree two, four, and eight on the same graph.

(S)4.15 Consider the problem of constructing an interpolating polynomial for the following rational function.

$$f(x) = \frac{1}{1 + x^2}$$

Suppose $n = 9$ samples in the interval $[-4, 4]$ are used. Write a program that uses the roots of the Chebyshev polynomials with appropriate offset and scaling as in Equation (4.4.22) to generate the n samples. Use *polyfit* to compute an interpolating polynomial based on these points. Plot the interpolating polynomial and the data points on the same graph. Compare your results with the case of uniformly distributed samples in Figure 4.4.1.

4.16 Consider the problem of constructing a continuous interpolation surface based on samples from the following function.

$$f(x, y) = \frac{\sin(\pi x)\cos(\pi y)}{1 + x^2 + y^2}$$

Suppose this surface is sampled at $m = 11$ points uniformly distributed over $-1 \le x \le 1$ and $n = 11$ points uniformly distributed over $-1 \le y \le 1$. Write a program which uses the NLIB function *bilin* to construct a bilinear fit. Plot the bilinear function over the range $-1 \le x \le 1$ and $-1 \le y \le 1$ using a precision of 50 points in each dimension.

4.17 Write a program that uses NLIB to compute and display the interpolating polynomial for the data in problem 4.3.

4.18 Write a program that uses NLIB to compute and display a cubic spline fit to the variation in gravity data in Table 4.1.1.

4.19 Write a program that uses NLIB to compute and display the least-squares straight line fit for the data in problem 4.7.

4.20 Write a program that uses NLIB to compute and display the bilinear surface generated by the following data.

$$x = [0 \quad 1 \quad 2 \quad 3 \quad 4]^T$$

$$y = [0 \quad 1 \quad 2 \quad 3 \quad 4]^T$$

$$Z = \begin{bmatrix} 0.0 & 0.7 & 1.0 & 0.7 & 0.0 \\ 0.1 & 0.9 & 1.2 & 0.9 & 0.1 \\ 0.0 & 0.7 & 1.0 & 0.7 & 0.0 \\ 0.1 & 0.9 & 1.2 & 0.9 & 0.1 \\ 0.0 & 0.7 & 1.0 & 0.7 & 0.0 \end{bmatrix}$$

Plot the surface over the region $0 \le x \le 4$ and $0 \le y \le 4$ using a precision of 40 points in each dimension.

CHAPTER 5

Root Finding

Many of the mathematical models used in engineering problems are linear. For linear equations, a rich theoretical framework has been developed over many years with entire books devoted to the topic. Engineers commonly constrain their models to be linear so the numerous results from linear theory can be used in analysis and design. Linear models are typically obtained by simply ignoring the effects of nonlinear terms that would otherwise appear in a more complete mathematical description of the physical system. Linear models are useful when limited accuracy is required and when the domain of application is restricted. However, some physical systems are inherently nonlinear, and linearization would destroy the essential characteristics of the system. This chapter is concerned with these types of applications. More specifically, we examine the problem of finding solutions to the following n-dimensional *nonlinear algebraic system.*

$$\boxed{f(x) = 0}$$

There is no loss of generality in the convention of setting the right-hand side to zero, because any terms appearing on the right-hand side can always be brought over to the left-hand side and absorbed into the definition of f. We investigated the special case of linear algebraic systems, $f(x) = Ax - b$, in Chapter 2. In this chapter, we examine functions f which contain terms that are neither constant nor proportional to x. Because this is a more difficult problem, we focus primarily on the special case, $n = 1$. That is, we initially investigate the case when $f(x) = 0$ represents a single nonlinear equation rather than a system of equations. The solutions of a nonlinear algebraic equation are often called the *roots* of the equation and therefore solving $f(x) = 0$ is called *root finding.* Since there are few constraints on the form of the function f, there are relatively few general results that are applicable to root finding. Indeed, depending on the details of f, the equation can have no roots, a single root, several roots, or an infinite number of roots. Furthermore, the roots can be real or complex. In most instances, we will be content to estimate a single real root using an iterative procedure. The convergence of the iterative process can be greatly facilitated if a good initial guess for the root is available as a starting point. If the equation comes from a mathematical model of a physical system, our understanding of that system may allow us to generate a reasonable initial estimate of the solution. The one case where a more complete theory of root finding is available is when f is a polynomial. From the fundamental theorem of algebra, it is known that every polynomial of degree n has exactly n roots in the complex plane. Furthermore, if the coefficients of the polynomial are real, then complex roots appear in conjugate pairs.

We begin this chapter by considering a number of applications where root finding problems arise. We examine two simple bracketing methods for root finding—the bisection method and the false position method. These methods are reliable, but relatively slow. Next, we investigate a general theoretical method called the contraction mapping method, along with a technique for improving its rate of convergence called Aitken extrapolation. We then investigate a family of three relatively fast root finding methods: the secant method, Muller's method, and Newton's method. Next, we consider the important special case of finding roots of polynomials. We introduce a higher-order technique called Laguerre's method for finding individual roots. Using synthetic division and deflation, we extend this technique to finding all of the roots of a polynomial. We then generalize Newton's method to the problem of solving a system of n nonlinear equations. The chapter concludes with a summary of root finding methods, and a presentation of engineering applications. The selected examples include a propane cylinder, a bacterial chemostat, an industrial high-temperature oven, and a cable suspension system.

5.1 MOTIVATION AND OBJECTIVES

Nonlinear equations occur naturally in many applications in engineering. Before we examine numerical techniques for finding roots of nonlinear equations, it is instructive to consider some examples where nonlinear equations arise.

5.1.1 Tunnel Diode Circuit

Practical electrical circuits typically include a combination of linear and nonlinear circuit elements. One of the most fundamental nonlinear circuit elements is the *diode* which is essentially a switch which allows electrical current to flow in one direction, but not the other. The voltage-current relationship of a tunnel diode is shown in Figure 5.1.1. Since the curve is not a straight line through the origin, it is clear from inspection that this is a *nonlinear* device.

The nonmonotonic relationship between voltage x and current y of a tunnel diode can be modeled mathematically by fitting a curve $g(x)$ to the measured data supplied by the manufacturer using the curve fitting techniques from Chapter 4.

$$y = g(x) \tag{5.1.1}$$

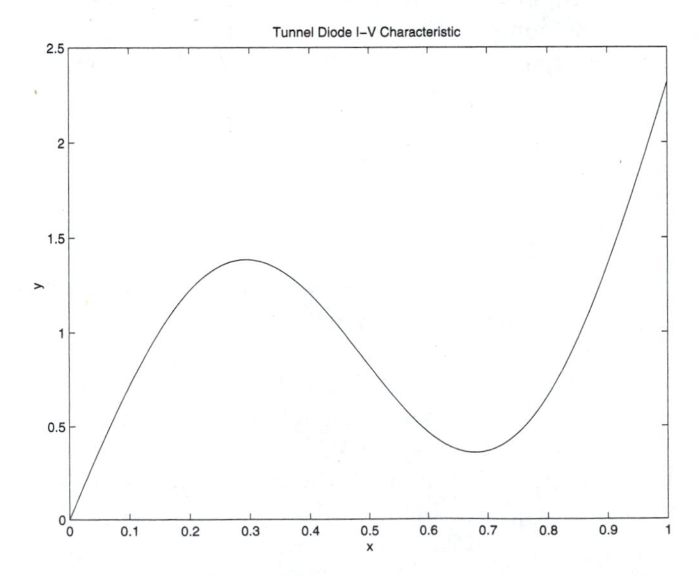

FIGURE 5.1.1 Tunnel Diode Voltage-Current Characteristic g

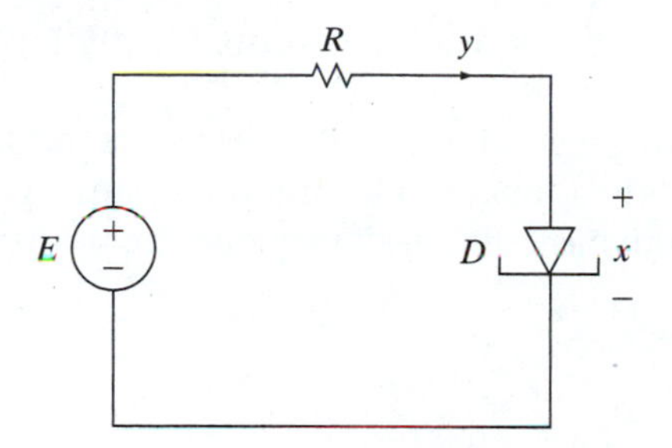

FIGURE 5.1.2 A Tunnel Diode Circuit

Consider the circuit in Figure 5.1.2, which employs the diode D in series with a resistor with resistance R and an applied voltage source E. The current in this circuit y can be easily determined once the voltage drop across the diode, x, is known. Applying Kirchhoff's voltage law, and using the voltage-current relationship of the diode, the diode voltage x must satisfy the following nonlinear equation:

$$Rg(x) + x - E = 0 \tag{5.1.2}$$

It is clear that Equation (5.1.2) is a special case of $f(x) = 0$. It is of interest to note that a good approximation to the diode voltage can be obtained graphically in this case. First note that (5.1.2) can be rewritten as

$$g(x) = \frac{E - x}{R} \tag{5.1.3}$$

If the two terms, $g(x)$ and $(E - x)/R$, are plotted on the same graph as shown in Figure 5.1.3, then intersections of the two curves are roots. This graphical procedure can be used to get a good initial estimate of the solution. Iterative techniques then can be used to refine this estimate.

Observe from Figure 5.1.3 that when $E = 1.5$ and $R = 1$, there are three roots. If the resistance R is sufficiently large, there is a range of values $E_0 < E < E_1$ over which

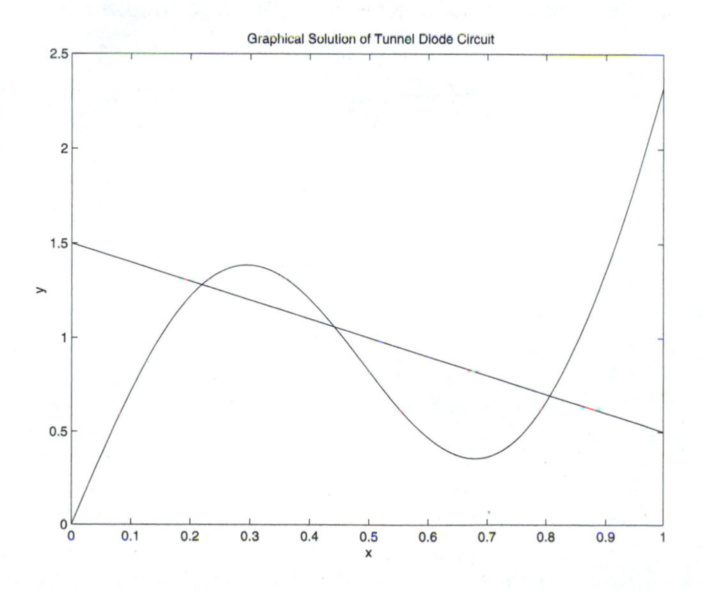

FIGURE 5.1.3 Graphical Solution of Tunnel Diode Circuit

three roots exist. At the limits of this range there are two roots. Finally, for $0 \le E < E_0$ or $E > E_1$, there is a single root.

Another approach to constructing a good initial guess is to use our knowledge of the physical system. For example, for the tunnel diode circuit the initial guess $x \approx E/2$ might be used because it is known from basic circuit theory that the solution must lie in the interval $0 \le x \le E$.

5.1.2 Leaky Tank

As a second illustration of a nonlinear system, consider the liquid level control system shown in Figure 5.1.4. Liquid is added to the top of the tank through a valve and simultaneously flows out of the bottom of the tank through a hole. If x denotes the height of liquid in the tank, then the rate of change of the height can be modeled using a conservation of energy argument which describes how fast the liquid leaves the tank. The general form of the resulting equation is as follows (McClamroch, 1980).

$$\frac{dx}{dt} = -\alpha\sqrt{x} + \beta u \qquad (5.1.4)$$

The variable u represents the rate at which liquid enters the tank through the valve, while α and β are positive constants that depend on the tank geometry, friction, and gravity. Suppose the valve setting is controlled with feedback by measuring the liquid level and opening or closing the valve as follows.

$$u = \gamma\left(1 - \frac{x}{h}\right) \qquad (5.1.5)$$

Here γ is a positive constant which represents the maximum flow rate and h is the height of the tank. Note that the feedback in Equation (5.1.5) provides for a maximum input when the tank is empty ($x = 0$), but turns the input flow off when the tank is full ($x = h$).

An important design question when choosing the maximum flow rate γ is, how high will the liquid be when the tank reaches a steady-state condition? To determine the

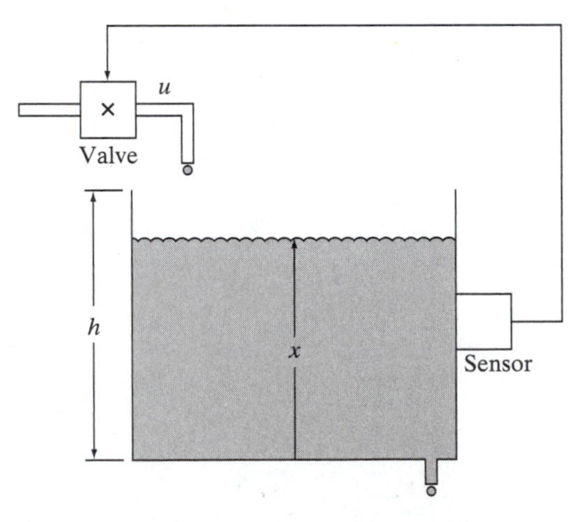

FIGURE 5.1.4 Liquid Level Control System

steady-state height of the liquid, we substitute Equation (5.1.5) into (5.1.4) and set the rate of change, dx/dt, equal to zero. This results in the following nonlinear equation.

$$\beta\gamma\left(1 - \frac{x}{h}\right) - \alpha\sqrt{x} = 0 \qquad (5.1.6)$$

Again, it is clear that Equation (5.1.6) is a special case of $f(x) = 0$. Just as with the tunnel diode circuit in (5.1.2), this nonlinear equation can be solved by plotting the two terms $\beta\gamma(1 - x/h)$ and $\alpha\sqrt{x}$ on the same graph and finding intersections of the two curves. The graphical solution can then be used as an initial estimate for a more accurate iterative method. Alternatively, the initial guess $x \approx h/2$ can be used as a starting point because it is known from physical considerations that the solution must lie in the interval $0 \le x \le h$.

For this particular nonlinear equation, a "trick" can be used to find a closed-form solution. If the change of coordinates $z = \sqrt{x}$ is used, then the new equation in the variable z is a second degree polynomial that can be factored using the quadratic formula. Once z is determined, x can be recovered using $x = z^2$ (see problem 5.1).

5.1.3 Bacterial Chemostat

As a third illustration of a nonlinear system, consider the chemostat shown in Figure 5.1.5. A *chemostat* is a device that is used for growing and harvesting bacteria. Nutrient of a given concentration C_0 is pumped into the bacterial culture chamber from a reservoir at an input flow rate of F. An output valve is used to regulate an outflow at the same rate so as to maintain a constant culture volume V within the growth chamber.

Let x_1 and x_2 denote the growth chamber *bacterial density* and *nutrient concentration,* respectively. The following is a mathematical model of the rate at which bacteria grow as they consume the nutrient (Edelstein-Keshet, 1988).

$$\frac{dx_1}{dt} = \frac{\gamma x_1 x_2}{1 + x_2} - x_1 \qquad (5.1.7a)$$

$$\frac{dx_2}{dt} = \frac{-x_1 x_2}{1 + x_2} - x_2 + \delta \qquad (5.1.7b)$$

Here γ and δ are two positive constants that depend on such things as the reservoir concentration C_0, the flow rate F, the volume V, and the type of bacteria and nutrient. For

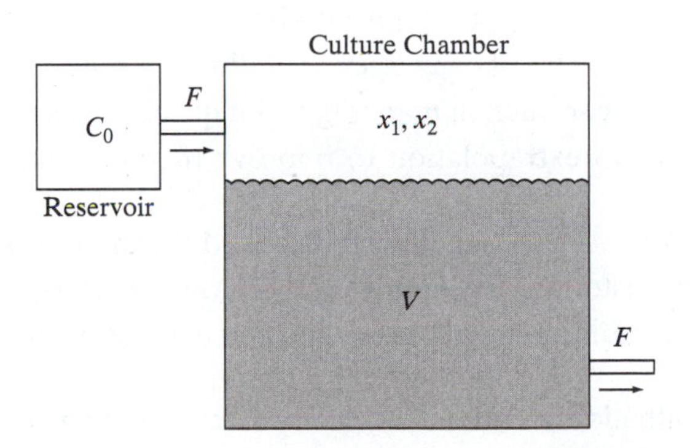

FIGURE 5.1.5 Bacterial Growth in a Chemostat

a biologically meaningful steady-state solution to occur it is necessary that parameters γ and δ satisfy the following constraints.

$$\gamma > 1 \tag{5.1.8a}$$

$$\delta > \frac{1}{\gamma - 1} \tag{5.1.8b}$$

The steady-state levels of bacteria density x_1 and nutrient concentration x_2 are obtained by setting the derivatives in Equation (5.1.7) to zero and solving the resulting equations for x_1 and x_2. After simplification, this leads to the following system of two nonlinear equations.

$$\gamma x_1 x_2 - x_1(1 + x_2) = 0 \tag{5.1.9a}$$

$$-x_1 x_2 + (\delta - x_2)(1 + x_2) = 0 \tag{5.1.9b}$$

In this case, we see that Equation (5.1.9) takes the form $f(x) = 0$ where x and $f(x)$ are 2×1 vectors. Since this is a two-dimensional system, simple graphical techniques no longer suffice to approximate the solution. Instead, iterative numerical techniques must be used.

5.1.4 Chapter Objectives

When you finish this chapter, you will be able to find roots of nonlinear algebraic equations. You will know how to bracket a root and then narrow down the interval of uncertainty using the bisection and false position methods. You will understand how to determine the rates of convergence of different root finding methods. You will learn how to use Aitken extrapolation to improve the convergence rate of linear methods including the contraction mapping method. You will understand how different curve fitting techniques are used to develop superlinear root-finding methods including the secant, Muller, and Newton methods. You will know how to deflate a polynomial as its roots are found and find those roots with Laguerre's method. Finally, you will know how to generalize Newton's method to solve systems of nonlinear algebraic equations. These overall goals will be achieved by mastering the following chapter objectives.

Objectives for Chapter 5

- Know how to identify a pair of points that bracket a root.
- Be able to apply the bisection and false position methods to reduce the interval of uncertainty within which a root is known to lie.
- Know how to specify the rates of convergence of different root finding methods.
- Know how to apply the contraction mapping technique to find a root.
- Be able to apply Aitken extrapolation to improve the performance of linear methods.
- Know how to find roots using the superlinear secant, Muller, and Newton methods.
- Understand why the faster nonbracketing methods do not always converge.
- Be able to deflate polynomials by removing linear and quadratic roots using synthetic division.
- Understand how synthetic division can be used to efficiently evaluate polynomials.

- Be able to find polynomial roots using Laguerre's third order method.
- Know how to define and numerically estimate the Jacobian matrix of partial derivatives of a function.
- Know how to solve nonlinear algebraic systems of equations iteratively using Newton's method.
- Understand how root finding techniques can be used to solve practical engineering problems.
- Understand the relative strengths and weaknesses of each computational method and know which are most applicable for a given problem.

5.2 BRACKETING METHODS

Perhaps the simplest numerical method for finding a root is to start with an *interval* within which a root is known to lie and then gradually reduce the size of the interval. This general approach is called a *bracketing* method. We assume that the function f is *continuous* and a root of f is known to lie in the interval

$$x_0 \le x \le x_1 \tag{5.2.1}$$

We can often come up with a relatively small starting interval that brackets the root by exploiting our understanding of the underlying physical system. Alternatively, we can systematically search f for a starting interval. A simple way to verify that a continuous function has at least one root in an interval is to examine the sign of the function at each end of the interval. If the function changes sign, then at least one root is bracketed by the interval. Consequently, the interval end-points x_0 and x_1 should be chosen such that

$$\boxed{f(x_0)f(x_1) < 0} \tag{5.2.2}$$

If Equation (5.2.2) holds, the function f crosses the x axis with a negative slope if $f(x_0) > 0$ or a positive slope if $f(x_0) < 0$. Of course, there may be multiple roots inside $[x_0, x_1]$ because (5.2.2) only guarantees that there is at least one. If $f(x_0)f(x_1) = 0$, then at least one of the endpoints is a root.

5.2.1 Bisection Method

The simplest bracketing method is called the *bisection method.* The basic idea behind the bisection method is to cut the interval of uncertainty in half with each iteration (hence the term *bisection*). This is easily accomplished by evaluating the *sign* of f at the midpoint of the interval.

$$x_2 = \frac{x_0 + x_1}{2} \tag{5.2.3}$$

If $f(x_2) = 0$, then x_2 is root, so consider the case when $f(x_2) \neq 0$. If $f(x_0)f(x_2) < 0$, then a root must lie in the subinterval $[x_0, x_2]$; otherwise one must lie in the subinterval $[x_2, x_1]$. Thus, we either set $x_1 = x_2$ or $x_0 = x_2$, as appropriate, and repeat the process.

One of the attractive features of the bisection method is that size of the interval of uncertainty is known after each iteration. In particular, if n iterations are performed, then a root is known to lie within an interval of length $e_n = (x_1 - x_0)/2^n$. Given an error tolerance, $\varepsilon > 0$, we can set $e_n \le \varepsilon$ and solve for the number of iterations required to achieve that error bound.

$$n \ge \frac{\ln (x_1 - x_0) - \ln \varepsilon}{\ln 2} \tag{5.2.4}$$

If the number of iterations n satisfies Equation (5.2.4), then the final midpoint is guaranteed to lie within a radius of $\varepsilon/2$ of a root. However, this does not guarantee that the value of $|f(x)|$ at the estimated root is small, because $|f(x)|$ could have a very large slope at the root. Conversely, if $|f(x)|$ is small, this does not guarantee that x is close to the root because $|f(x)|$ could have a very small slope at the root. The iterative search might also be terminated when the difference between successive estimates becomes sufficiently small. In the algorithms presented in this chapter, we will use the following *termination criterion:*

$$(|f(x)| < \varepsilon) \quad \text{or} \quad (k > m) \tag{5.2.5}$$

Here k is the iteration number, m is the maximum number of iterations to perform, and $\varepsilon > 0$ is an *error tolerance* on $|f(x)|$. If the procedure terminates in fewer than m iterations, then the convergence criterion $|f(x)| \le \varepsilon$ was achieved. It is important to place an upper bound m on the number of iterations to ensure that the procedure does not loop endlessly. For example, suppose x_k and x_{k+1} are successive estimates, and $|x_{k+1} - x_k| < \varepsilon_M |x_k|$ where ε_M denotes the machine precision. In this case, x_k and x_{k+1} are indistinguishable, as are $f(x_k)$ and $f(x_{k+1})$, which means that the procedure will *not* terminate unless an upper limit is placed on the number of iterations. The following algorithm is an implementation of the bisection method using the termination criterion in Equation (5.2.5).

Alg. 5.2.1 **Bisection**

1. Pick $\varepsilon > 0$, $m > 0$, and $x_1 > x_0$ such that $f(x_0)f(x_1) < 0$. Set $k = 1$ and compute $f_0 = f(x_0)$.
2. Do

   {

   (a) Compute $x_2 = (x_0 + x_1)/2$ and $f_2 = f(x_2)$.

   (b) If $f_0 f_2 < 0$ set $x_1 = x_2$. Otherwise, set $x_0 = x_2$ and $f_0 = f_2$.

   (c) Set $k = k + 1$.

   }
3. While ($|f_2| > \varepsilon$) and ($k \le m$)
4. Set $x = x_2$.

Note that Alg. 5.2.1 requires one function evaluation per iteration. The number of function evaluations required for convergence is a measure of the amount of computational effort required to find a root.

EXAMPLE 5.2.1 Bisection Method

As an illustration of the bisection method, consider the tunnel diode circuit in Figure 5.1.2. The nonlinear characteristic of the tunnel diode element, shown in Figure 5.1.1, can be modeled mathematically as follows.

$$g(x) = \sin(2\pi x) + \exp(1.2x) - 1$$

Suppose the circuit parameters are $E = 1.5$ and $R = 1$. To find the voltage drop across the tunnel diode, we find a root of $f(x) = 0$ where, from Equation (5.1.2),

$$\begin{aligned} f(x) &= Rg(x) + x - E \\ &= \sin(2\pi x) + \exp(1.2x) + x - 2.5 \end{aligned}$$

To start the bisection method, we need an interval which contains a root. From physical considerations, the tunnel diode voltage x must lie in the interval $[x_0, x_1] = [0, E]$. This can be verified mathematically by using Equation (5.2.2), which yields

$$f(0)f(1.5) = -0.5 \exp(1.8) < 0$$

Example program *e521* on the distribution CD applies the bisection method in Alg. 5.2.1 with $x_0 = 0, x_1 = 1.5$ and $\varepsilon = 10^{-5}$. The results are summarized in Table 5.2.1, which shows the iteration number k, the midpoint x_2, the interval length $x_1 - x_0$, and the magnitude of f at the midpoint. In this case, 19 iterations are required for convergence. From the midpoint column of Table 5.2.1 we see that there is a root of f at $x = 0.8063421$. From inspection of Figure 5.1.3 it is apparent that this is the largest of three roots.

TABLE 5.2.1 Bisection Method Estimates

k	x_2	$x_1 - x_0$	$\lvert f(x_2)\rvert$
1	0.7500000	0.7500000	0.2903968
2	1.1250000	0.3750000	3.1895328
3	0.9375000	0.1875000	1.1350337
4	0.8437500	0.0937500	0.2647542
5	0.7968750	0.0468750	0.0581442
6	0.8203125	0.0234375	0.0924621
7	0.8085938	0.0117188	0.0143700
8	0.8027344	0.0058594	0.0225926
9	0.8056641	0.0029297	0.0042867
10	0.8071289	0.0014648	0.0049979
11	0.8063965	0.0007324	0.0003447
12	0.8060303	0.0003662	0.0019738
13	0.8062134	0.0001831	0.0008152
14	0.8063049	0.0000916	0.0002354
15	0.8063507	0.0000458	0.0000546
16	0.8063278	0.0000229	0.0000904
17	0.8063393	0.0000114	0.0000179
18	0.8063450	0.0000057	0.0000183
19	0.8063421	0.0000029	0.0000002

In order to compare one root finding algorithm with another, it is useful to develop a measure of how fast the algorithm converges to a root. Let $e_k = x_1 - x_0$ denote the length of the interval of uncertainty after iteration k. We say that an algorithm converges with *order p*, with respect to this error interval, if and only if there exists a constant α such that

$$e_{k+1} \approx \alpha e_k^p \quad , \quad k \gg 1 \tag{5.2.6}$$

Since the length of the interval of uncertainty gets cut in half with each iteration of the bisection method, we have $\alpha = 0.5$ and $p = 1$. In this case, we say that the bisection method converges *linearly*. There are faster algorithms for which $1 < p < 2$, and they are said to exhibit *superlinear* convergence. Still faster are algorithms for which $p = 2$ and $p = 3$, which exhibit *quadratic* and *cubic* convergence, respectively. Of course, for two algorithms which converge with the same order p, the one with the smaller asymptotic error constant, α, is faster. To make a meaningful comparison between algorithms, it is also important to examine the computational effort (number of function evaluations) per iteration.

5.2.2 False Position Method

The bisection method is very reliable, but quite slow because the only information it uses is the *sign* of f at each end of the interval. If the magnitude of f at one end of the interval is very small, then it is likely that the root is closer to this end. The *false position* method is a modification of the bisection method that exploits this concept to improve performance. The basic idea is to interpolate a straight line through the two end points as shown in Figure 5.2.1 where

$$f_k \triangleq f(x_k) \quad k \geq 0 \tag{5.2.7}$$

The intersection of the straight line in Figure 5.2.1 with the x axis provides the new end point. The other end point is chosen from among the two original end points using Equation (5.2.2) to ensure that a root remains bracketed. The process is then repeated. The false position method is sometimes called the *regula falsi* method, which is the Latin formulation of the name.

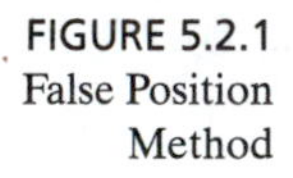

FIGURE 5.2.1 False Position Method

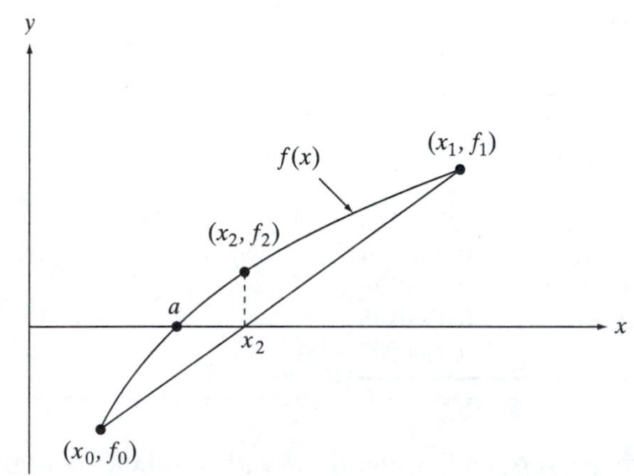

To obtain an expression for the new end point x_2, first note that a straight line through the two original end points (x_0, f_0) and (x_1, f_1) can be expressed

$$y = f_1 + \left(\frac{f_1 - f_0}{x_1 - x_0}\right)(x - x_1) \tag{5.2.8}$$

The new end point x_2 is determined by setting $y = 0$ in Equation (5.2.8) and solving for x, which yields

$$\boxed{x_2 = x_1 - \left(\frac{x_1 - x_0}{f_1 - f_0}\right)f_1} \tag{5.2.9}$$

The computation in (5.2.9) involves somewhat more work than the computation of the midpoint x_2, but in many cases the number of iterations required for convergence is reduced. The following algorithm is an implementation of the false position method. It terminates when an error tolerance has been met by $|f(x)|$ or a maximum number of iterations m has been performed.

Alg. 5.2.2 False Position Method

1. Pick $\varepsilon > 0$, $m > 0$, and $x_1 > x_0$ such that $f(x_0)f(x_1) < 0$. Set $k = 1$ and compute $f_0 = f(x_0)$ and $f_1 = f(x_1)$.
2. Do

   {

   (a) Compute $x_2 = x_1 - (x_1 - x_0)f_1/(f_1 - f_0)$ and $f_2 = f(x_2)$.
   (b) If $f_0 f_2 < 0$ set $x_1 = x_2$ and $f_1 = f_2$. Otherwise, set $x_0 = x_2$ and $f_0 = f_2$.
   (c) Set $k = k + 1$.

   }
3. While $(|f_2| \geq \varepsilon)$ and $(k \leq m)$
4. Set $x = x_2$.

Note that the false position method, as implemented in Alg. 5.2.2, requires only one function evaluation per iteration. The false position method often exhibits *super-linear* convergence, but this is not always the case.

EXAMPLE 5.2.2 False Position Method

To compare the false position method with the bisection method, again consider the tunnel diode circuit in Figure 5.2.2 with $E = 1.5$ and $R = 1$. From Example 5.2.1, the nonlinear function in this case is

$$f(x) = \sin(2\pi x) + \exp(1.2x) + x - 2.5$$

Example program *e522* on the distribution CD applies the false position method in Alg. 5.2.2 with $x_0 = 0$, $x_1 = 1.5$ and $\varepsilon = 10^{-5}$. The results are summarized in Table 5.2.2, which shows the

TABLE 5.2.2 False Position Method Estimates

k	x_2	$x_1 - x_0$	$\|f(x_2)\|$
1	1.5000000	1.5000000	5.0496478
2	0.3435299	0.3435299	0.1859576
3	0.3056393	0.3056393	0.1882137
4	0.2715645	0.2715645	0.1476448
5	0.2472297	0.2472297	0.0924571
6	0.2328757	0.2328757	0.0494963
7	0.2254368	0.2254368	0.0242021
8	0.2218572	0.2218572	0.0112976
9	0.2201987	0.2201987	0.0051580
10	0.2194441	0.2194441	0.0023309
11	0.2191037	0.2191037	0.0010484
12	0.2189506	0.2189506	0.0004706
13	0.2188819	0.2188819	0.0002110
14	0.2188512	0.2188512	0.0000946
15	0.2188374	0.2188374	0.0000424
16	0.2188312	0.2188312	0.0000190
17	0.2188284	0.2188284	0.0000085

iteration number k, the new end point x_2, the interval length $x_1 - x_0$, and the magnitude of f at the new end point. From Table 5.2.2, we see that 17 iterations are required for convergence to $x = 0.2188284$. This is somewhat faster than the bisection method in Example 5.2.1, which required 19 iterations. Interestingly enough, the false position method converged to a *different* root than the bisection method. From Figure 5.1.3, we see that the false position method converged to the smallest of the three roots rather than to the largest. Note, also, that the length of the interval of uncertainty does not decrease very much in this case because the root is approached from one side.

The false position method uses more information than the bisection method and, as a result, tends to converge faster in most cases. However, it is not difficult to construct a counter example which shows that this is not always the case. Consider, in particular, the following function.

$$f(x) = x^9 - 1 \tag{5.2.10}$$

This function is relatively flat over a large range and then increases abruptly as can be seen by its graph in Figure 5.2.2. This function has a single real root at $x = 1$. If the bisection and false position methods are both applied with $[x_0, x_1] = [0, 1.5]$ and $\varepsilon = 10^{-5}$, then the bisection method converges in 19 iterations, while the false position method requires 129 iterations! This is because the false position method steps slowly along the flat portion of f.

When the false position method approaches a root from one side, the speed can be increased by using the *modified* false position method, which replaces the stationary end point $f(x)$ by $f(x)/2$. This is done repeatedly until the computed end point x_2 even-

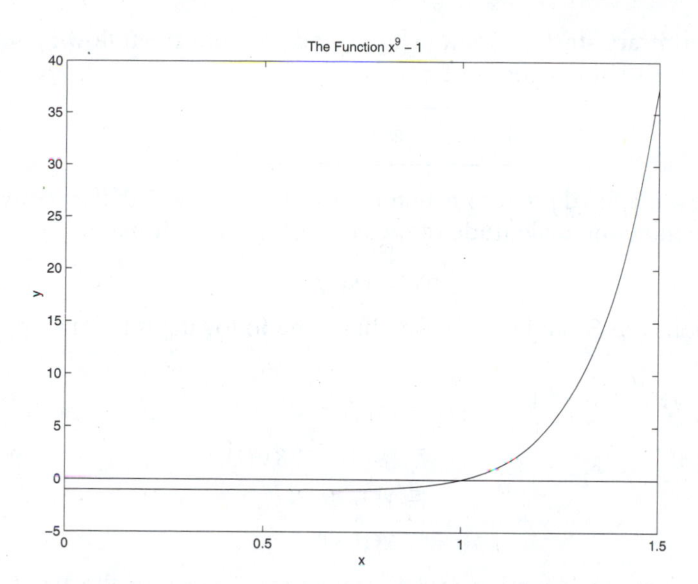

FIGURE 5.2.2 The Function $f(x) = x^9 - 1$

tually crosses over to the other side of the root. For a further discussion of the modified false position method, see Gerald and Wheatley (1989).

5.3 CONTRACTION MAPPING METHOD

In this section, we examine a very general method for root finding called the contraction mapping or successive substitution method. The contraction mapping formulation provides us with a theoretical framework within which the convergence properties of subsequent methods can be evaluated. Before attacking the main problem, we first digress to consider a related problem, namely, that of finding a solution x to the following equation.

$$g(x) = x \tag{5.3.1}$$

Solutions of Equation (5.3.1) are called *fixed points* of the function g because when they are "operated on" by g they remain stationary; they do not move. There is a simple sufficient condition which guarantees that g has a unique fixed point. Suppose the function g maps the interval $[\alpha, \beta]$ into itself. Let $g'(x)$ denote the derivative of g and suppose the magnitude of the derivative is bounded in the following manner.

$$|g'(x)| \leq \gamma \quad , \quad \alpha \leq x \leq \beta \tag{5.3.2}$$

If $\gamma < 1$, then the function g is called a *contraction mapping* on $[\alpha, \beta]$ or simply a *contraction*. Note that for any x and y in $[\alpha, \beta]$, we have

$$|g(x) - g(y)| \leq \gamma |x - y| \tag{5.3.3}$$

Since the constant $\gamma < 1$, it follows that the distance between the points x and y "contracts" when the points are operated on by the function g, hence the term *contraction.*

Let x_0 be an arbitrary starting point in $[\alpha, \beta]$, and consider the following sequence generated by successive applications of g.

$$\boxed{x_{k+1} = g(x_k) \quad , \quad k \geq 0} \tag{5.3.4}$$

Next, suppose a is a fixed point of g, that is, $g(a) = a$. To show that x_k converges to a as $k \to \infty$, let e_k denote the magnitude of the *error* of the kth estimate.

$$e_k \triangleq |x_k - a| \tag{5.3.5}$$

Using Equations (5.3.3) and (5.3.4), we obtain the following relationship between successive errors.

$$\begin{aligned} e_{k+1} &= |x_{k+1} - a| \\ &= |g(x_k) - g(a)| \\ &\leq \gamma |x_k - a| \\ &= \gamma e_k \end{aligned} \tag{5.3.6}$$

Thus, the sequence in Equation (5.3.4) converges *linearly* to the fixed point $x = a$. Applying (5.3.6) repeatedly, one can show that the error in the kth estimate is bounded from above as follows.

$$e_k \leq \frac{\gamma^k |x_1 - x_0|}{1 - \gamma} \quad , \quad k \geq 1 \tag{5.3.7}$$

As was the case with the bisection method, a bound can be found on the number of iterations n required to achieve a specified error tolerance ε. Setting $e_n = \varepsilon$ in (5.3.7) and solving for n yields

$$\boxed{n \geq \frac{\ln[(1 - \gamma)\varepsilon] - \ln|x_1 - x_0|}{\ln \gamma}} \tag{5.3.8}$$

EXAMPLE 5.3.1 Contraction Mapping

As an illustration of a contraction mapping, consider the following polynomial function.

$$g(x) = x + \frac{1 - x^2}{4}$$

To determine the fixed points of g, we find the roots of the polynomial $f(x) = g(x) - x$. From inspection, the roots are $x = \pm 1$. To see if g is a contraction in the neighborhood of either of these points, we examine the slope.

$$g'(x) = 1 - \frac{x}{2}$$

At the first fixed point $g'(1) = 1/2$, which means g, is a contraction near $x = 1$ because $|g(1)| < 1$. However, $g'(-1) = 3/2$, which indicates that g is not a contraction near $x = -1$. Suppose that x is restricted to the following interval about the fixed point, $a = 1$.

$$[\alpha, \beta] = [0.5, 2]$$

It then follows that $0 \le g'(x) \le 0.75$ for $\alpha \le x \le \beta$. Furthermore, $g(0.5) = 0.61875$ and $g(2) = 1.25$. Since g is nondecreasing in $[\alpha, \beta]$, this means that g maps $[\alpha, \beta]$ into itself. Hence, g is a contraction mapping on $[\alpha, \beta]$. Consequently, for every initial guess $\alpha \le x_0 \le \beta$, the following iterative sequence converges to the fixed point $x = 1$.

$$x_{k+1} = x_k - \frac{1 - x_k^2}{4} \quad , \quad k \ge 0$$

5.3.1 Root Finding

To put the contraction mapping method to work in solving the root finding problem, we must find an interval $[\alpha, \beta]$ and a function g whose fixed point is a root of f. To that end, consider the following candidate for g.

$$\boxed{g(x) = x - bf(x)} \tag{5.3.9}$$

Here, b is a nonzero constant whose value remains to be determined. It is clear from Equation (5.3.9) that if $g(x) = x$, then $f(x) = 0$. That is, fixed points of g are roots of f. Next we determine conditions under which g is a contraction. The magnitude of $g'(x)$ must be less than one. Thus, from (5.3.9), we have

$$-1 < 1 - bf'(x) < 1 \quad , \quad \alpha \le x \le \beta \tag{5.3.10}$$

We can subtract 1 from the expressions in Equation (5.3.10) and then divide by -1, which changes the direction of the inequalities. This results in the following bounds on b.

$$\boxed{0 < bf'(x) < 2 \quad , \quad \alpha \le x \le \beta} \tag{5.3.11}$$

Consequently, if $f'(x)$ is positive in an interval $[\alpha, \beta]$ containing a root, the constant b should lie in the interval $0 < b < 2/f'(x)$. If $f'(x)$ is negative in $[\alpha, \beta]$, b should be in the interval $2/f'(x) < b < 0$.

The following algorithm is an implementation of the contraction mapping method for root finding. It terminates when a prescribed error tolerance on $|f(x)|$ has been achieved or a maximum number of iterations m has been performed.

Alg. 5.3.1 **Contraction Mapping**

1. Determine $[\alpha, \beta]$ and b using (5.3.11). Pick $\varepsilon > 0$, $m > 0$, $\alpha \le x_0 \le \beta$. Compute $f_0 = f(x_0)$.
2. Do

   {

   $$x_1 = x_0 - bf_0$$
   $$x_0 = x_1$$
   $$f_0 = f(x_0)$$
   $$k = k + 1$$

   }

3. While ($|f_0| \ge \varepsilon$) and ($k \le m$)
4. Set $x = x_1$.

The key to a meaningful application of Alg. 5.3.1 is to pick b to suit the problem at hand. The following example illustrates how this might be done.

EXAMPLE 5.3.2 Contraction Mapping

As an illustration of the contraction mapping, consider the following nonlinear function whose roots can not be determined analytically.

$$f(x) = \exp(x) + x - 2$$

Note that $f(0) = -1$ and $f(1) = e - 1$. Since $f(0)f(1) < 0, f(x)$ has a root in the interval $[0, 1]$. Applying Equation (5.3.9), we get the following contraction mapping function associated with f.

$$g(x) = x - b(\exp(x) + x - 2)$$

Consider the interval $[\alpha, \beta] = [0, 1]$ which brackets a root. The slope of $f(x)$ is

$$f'(x) = 1 + \exp(x)$$

Thus, $2 \le f'(x) \le 1 + e$ for $\alpha \le x \le \beta$. Since the maximum slope is $1 + e$, suppose we take the following value for b to satisfy (5.3.11).

$$b = \frac{1}{1 + e}$$

Finally, to verify that g maps $[\alpha, \beta]$ into itself, note that $g(x)$ is strictly increasing, $g(0) = 1/(1 + e)$ and $g(1) = 1 - (e - 1)/(e + 1)$. Example program *e532* on the distribution CD applies Alg. 5.3.1 with an initial guess of $x^0 = 0.5$ and $\varepsilon = 10^{-6}$. The results are summarized in Table 5.3.1, which shows the iteration number k, the current estimate x, and the magnitude of f at the current estimate. In this case, 11 iterations are required for convergence to the root $x = 0.4428546$.

TABLE 5.3.1 Contraction Mapping Method Estimates

k	x	$\|f(x)\|$
0	0.5000000	0.1487213
1	0.4600027	0.0440809
2	0.4481475	0.0135571
3	0.4445014	0.0042138
4	0.4433682	0.0013140
5	0.4430148	0.0004101
6	0.4429045	0.0001280
7	0.4428701	0.0000400
8	0.4428593	0.0000125
9	0.4428559	0.0000039
10	0.4428549	0.0000012
11	0.4428546	0.0000004

The relatively slow convergence of the contraction mapping method exhibited in Example 5.3.2 is to be expected because from Equation (5.3.6) we see that this process exhibits only linear convergence ($p = 1$). The utility of the contraction method lies not in its convergence rate but in the fact that it can be easily generalized to higher dimensional systems. For example, the Jacobi, Gauss-Seidel, and SR methods for solving $Ax = b$ are all contraction mapping methods.

5.3.2 Aitken Extrapolation

For any sequence that converges linearly, an *extrapolation* technique can be used to accelerate the rate of convergence. Let $x = a$ be the limit of a linearly convergent sequence $\{x_k\}$. For large values of k, successive errors will be approximately proportional to one another.

$$\frac{x_{k+2} - a}{x_{k+1} - a} \approx \frac{x_{k+1} - a}{x_k - a}, \quad k \gg 1 \tag{5.3.12}$$

Here, for convenience, we have assumed that the sign of the error does not change. Clearing the denominators of (5.3.12) by cross multiplying and then simplifying, we get

$$x_k x_{k+2} - (x_k + x_{k+2})a \approx x_{k+1}^2 - 2x_{k+1}a \tag{5.3.13}$$

Solving (5.3.13) for the limit a then yields the following *extrapolated estimate* x_e:

$$\boxed{x_e = \frac{x_k x_{k+2} - x_{k+1}^2}{x_k - 2x_{k+1} + x_{k+2}}} \tag{5.3.14}$$

Thus, once three points in the sequence of estimates are known, a fourth point can be computed using Equation (5.3.14). The extrapolation technique in (5.3.14) is known as *Aitken's* method. It can be formulated in a more compact form by using the forward difference operator. Recall from the discussion of Newton's difference formula (Sec. 4.3) that the first and second forward differences of x evaluated at $x = x_k$ are

$$\Delta x_k \triangleq x_{k+1} - x_k \tag{5.3.15a}$$

$$\Delta^2 x_k \triangleq x_{k+2} - 2x_{k+1} + x_k \tag{5.3.15b}$$

By adding and subtracting x_k from the expression for x_e in Equation (5.3.14), one can reformulate it as a *correction* to x_k using forward differences as follows.

$$\boxed{x_e = x_k - \frac{(\Delta x_k)^2}{\Delta^2 x_k}} \tag{5.3.16}$$

The extrapolation technique in (5.3.16) can be used to transform a linearly convergent sequence into a sequence with superlinear convergence.

EXAMPLE 5.3.3 Aitken Extrapolation

For comparison purposes, consider the nonlinear function that was investigated in Example 5.3.2.

$$f(x) = \exp(x) + x - 2$$

Example program *e533* on the distribution CD applies the Aitken extrapolation method, starting with the third iteration, using the same initial condition ($x^0 = 0.5$) and error tolerance ($\varepsilon = 10^{-6}$). The results are summarized in Table 5.3.2, which shows the iteration number k, the extrapolated estimate x_e, and the magnitude of f at the extrapolated estimate.

TABLE 5.3.2 Aitken Extrapolation Estimates

k	x_e	$\lvert f(x_e) \rvert$
0	0.5000000	0.1487213
1	0.4600027	0.0440809
2	0.4481475	0.0135571
3	0.4428820	0.0000707
4	0.4428571	0.0000069
5	0.4428546	0.0000006

In this case, five iterations are required for convergence to $x = 0.4428546$ in comparison with 11 iterations for the contraction mapping method without Aitken extrapolation.

5.4 SECANT METHOD

Recall that the false position method approximates the function f with a straight line through the two end points. The intersection of this straight line with the x axis is then chosen as a new end point. The second end point is selected from among the two original ones so as to obtain a sign change in $f(x)$, thereby ensuring that the root remains bracketed. As an alternative, we can relax the bracketing constraint and instead simply choose the last two computed points as the end points. This is referred to as the *secant* method and is illustrated in Figure 5.4.1 where $f_k = f(x_k)$.

It is of interest to note that the secant method no longer involves only interpolation because, in some cases, the intersection with the x axis is extrapolated outside the interval. The secant method starts with two distinct points, x_0 and x_1, that are close to

FIGURE 5.4.1 Secant Method

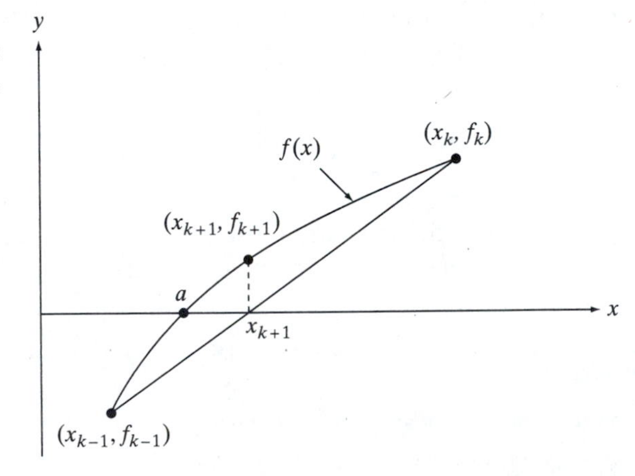

the root of interest, but do not necessarily bracket it. Subsequent points are then generated iteratively as follows.

$$x_{k+1} = x_k - \left(\frac{x_k - x_{k-1}}{f_k - f_{k-1}}\right) f_k \quad , \quad k \geq 1 \tag{5.4.1}$$

Comparing Equations (5.4.1) and (5.2.9), we see that the secant method looks very similar to the false position method. There are two differences. The first is that with the secant method the two starting points can be picked arbitrarily as long as $x_1 \neq x_0$. There is no need to have them bracket a root. The second difference is that after x_{k+1} is computed, the oldest point x_{k-1} is automatically dropped from subsequent computations when k is incremented. For the false position method, either x_k or x_{k-1} is retained so as to ensure that the retained point and x_{k+1} bracket a root. Since root bracketing is no longer preserved, the secant method is not guaranteed to converge to a root. However, the secant method is simple to implement and is often very effective. When it does converge, it is faster than the false position method. In particular, let $x = a$ be the root of f to which the secant method converges, and let e_k denote the magnitude of the *error* of the kth estimate.

$$e_k = |x_k - a| \quad , \quad k \geq 1 \tag{5.4.2}$$

If the function f is sufficiently smooth in the neighborhood of the root $x = a$, then it can be shown that the secant method is a *superlinear* method with order of convergence $p = (1 + \sqrt{5})/2$. That is, there exists a constant α such that

$$e_{k+1} \approx \alpha e_k^{1.618} \quad , \quad k \gg 1 \tag{5.4.3}$$

The following algorithm is an implementation of the secant method. It terminates when the prescribed error tolerance on $|f(x)|$ has been achieved or a maximum number of iterations m has been performed.

Alg. 5.4.1 **Secant Method**

1. Pick $\varepsilon > 0, m > 0$, and $x_1 \neq x_0$. Set $k = 1$ and compute $f_0 = f(x_0)$ and $f_1 = f(x_1)$.
2. Do

{

$$x_2 = x_1 - \left(\frac{x_1 - x_0}{f_1 - f_0}\right) f_1$$

$$x_0 = x_1$$

$$f_0 = f_1$$

$$x_1 = x_2$$

$$f_1 = f(x_2)$$

$$k = k + 1$$

}

3. While ($|f_1| \geq \varepsilon$) and ($k \leq m$)
4. Set $x = x_2$.

Alg. 5.4.1 is implemented in such a way that it requires only one function evaluation per iteration. This is accomplished at the expense of storing a few intermediate variables, which is a good bargain, particularly when f is a complicated function.

EXAMPLE 5.4.1 Secant Method

To illustrate the secant method, consider the following nonlinear function.

$$f(x) = x^2 \exp(-x/2) - 1$$

Note that $f(0) = -1$ and $f(2) = 4/e - 1 = 0.472$. Since $f(0)f(2) < 0$, it follows that there is a root in the interval $[0, 2]$. Furthermore, $f(10) = 100/e^5 - 1 = -0.326$. Thus, $f(2)f(10) < 0$, and there is also a root in the interval $[2, 10]$. To start the secant method, any two distinct initial guesses can be used. Example program *e541* on the distribution CD applies Alg. 5.4.1 using $x_0 = 0$, $x_1 = 2$, and an error tolerance of $\varepsilon = 10^{-6}$. The results are summarized in Table 5.4.1, which shows the iteration number k, the current estimate x, and the magnitude of f at the current estimate. In this case, four iterations are required for convergence to the root at $x = 1.4296118$.

TABLE 5.4.1 Secant Method Estimates

k	x	$\lvert f(x)\rvert$
0	0.0000000	−1.0000000
1	2.0000000	0.4715178
2	1.3591409	0.0637426
3	1.4354589	0.0052535
4	1.4296479	0.0000325
5	1.4296118	0.0000000

5.5 MULLER'S METHOD

The secant method uses a linear approximation to the function f by interpolating a straight line through the last two computed points. A more accurate approximation to f in the neighborhood of the root can be obtained by interpolating a quadratic polynomial through the last three computed points and then determining where this curve crosses the x axis. This is the basis for *Muller's method,* which is illustrated in Figure 5.5.1.

Let $\{x_0, x_1, x_2\}$ denote three distinct points. Consider the following quadratic polynomial, which can be fitted to f.

$$y = c_1 + c_2(x - x_2) + c_3(x - x_2)^2 \tag{5.5.1}$$

Let $f_k = f(x_k)$ for $1 \leq k \leq 3$. Evaluating Equation (5.5.1) at $x = x_2$ yields $c_1 = f_2$. To find the remaining two unknown coefficients c_2 and c_3, let $h_k = x_k - x_2$ for $0 \leq k \leq 1$. Evaluating (5.5.1) at $x = x_0$ and $x = x_1$, we get the following linear algebraic system.

FIGURE 5.5.1 Muller's Method

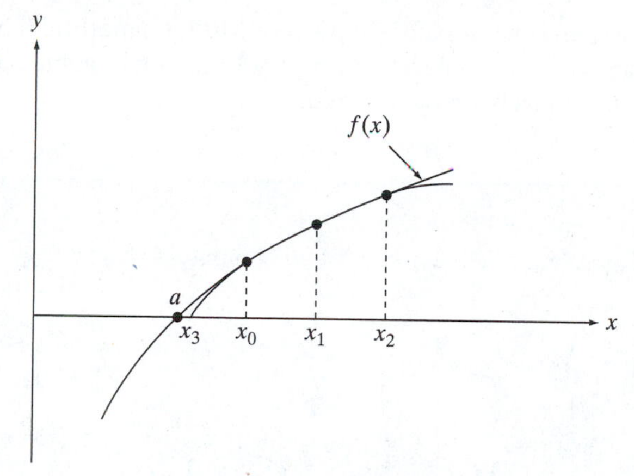

$$\begin{bmatrix} h_0 & h_0^2 \\ h_1 & h_1^2 \end{bmatrix} \begin{bmatrix} c_2 \\ c_3 \end{bmatrix} = \begin{bmatrix} f_0 - f_2 \\ f_1 - f_2 \end{bmatrix} \tag{5.5.2}$$

Since this system is only two-dimensional, it can be solved in closed form, which yields

$$c_2 = \frac{h_1^2(f_0 - f_2) - h_0^2(f_1 - f_2)}{h_0 h_1 (h_1 - h_0)} \tag{5.5.3a}$$

$$c_3 = \frac{-h_1(f_0 - f_2) + h_0(f_1 - f_2)}{h_0 h_1 (h_1 - h_0)} \tag{5.5.3b}$$

Once the coefficients of the quadratic interpolating polynomial are determined, we must then find its roots in order to generate a new estimate of the root of f. Setting $y = 0$ in Equation (5.5.1) and solving for $x - x_2$ using the quadratic formula yields

$$x - x_2 = \frac{-c_2 \pm \sqrt{c_2^2 - 4c_1c_3}}{2c_3} \tag{5.5.4}$$

A potential problem with the quadratic formula is that it can produce inaccurate numerical results under certain conditions. If $c_2^2 \gg c_1c_3$, then one of the two roots in (5.5.4) is obtained by subtracting two nearly identical numbers. Recall from Chapter 1 that this can cause *catastrophic cancellation* and can result in a significant loss of numerical accuracy. This situation can be avoided by using an alternative formulation of the quadratic formula. In particular, suppose we multiply the top and bottom of the right-hand side of Equation (5.5.4) by $-c_2 \mp \sqrt{c_2^2 - 4c_1c_3}$. After simplification, this results in the following equivalent formulation for the updated estimate.

$$x_3 = x_2 - \frac{2c_1}{c_2 \pm \sqrt{c_2^2 - 4c_1c_3}} \tag{5.5.5}$$

It remains to determine which of the two roots to select. The root closest to $x = x_2$ can be obtained by maximizing the denominator of Equation (5.5.5). Once x_3 is found, the oldest point x_0 is discarded, the new points are relabeled $\{x_0, x_1, x_2\}$, and the process is repeated.

The following algorithm is an implementation of Muller's method. This algorithm terminates when a prescribed error tolerance on $|f(x)|$ has been achieved or a maximum number of iterations m has been performed.

Alg. 5.5.1 Muller's Method

1. Pick $\varepsilon > 0$, $m > 0$, $\{x_0, x_1, x_2\}$. Set $k = 1$ and compute $f_0 = f(x_0)$, $f_1 = f(x_1)$, and $f_2 = f(x_2)$.
2. Do

   {

   (a) Compute

   {

$$h_0 = x_0 - x_2$$
$$h_1 = x_1 - x_2$$
$$h_3 = h_0 h_1 (h_1 - h_0)$$
$$c_1 = f_2$$
$$c_2 = [h_1^2(f_0 - f_2) - h_0^2(f_1 - f_2)]/h_3$$
$$c_3 = [-h_1(f_0 - f_2) + h_0(f_1 - f_2)]/h_3$$
$$d_1 = c_2 + \sqrt{c_2^2 - 4c_1c_3}$$
$$d_2 = c_2 - \sqrt{c_2^2 - 4c_1c_3}$$

   }

   (b) If $|d_1| \geq |d_2|$ then $x_3 = x_2 - 2c_1/d_1$ else $x_3 = x_2 - 2c_1/d_2$.

   (c) Compute

   {

$$x_0 = x_1$$
$$f_0 = f_1$$
$$x_1 = x_2$$
$$f_1 = f_2$$
$$x_2 = x_3$$
$$f_2 = f(x_3)$$
$$k = k + 1$$

   }

   }
3. While $(|f_2| \geq \varepsilon)$ and $(k \leq m)$
4. Set $x = x_3$.

Alg. 5.5.1 is formulated in such a way that it requires only one function evaluation per iteration. The convergence rate for Muller's method is *superlinear* but close to quadratic. In particular, Atkinson (1978) has shown that it has an order of convergence of $p \approx 1.85$. Another distinctive feature of Muller's method is that it can be used to find *complex* roots, even when the initial guesses are all real. This can be seen in Equation (5.5.5), which returns a complex estimate x_3 whenever $c_2^2 < 4c_1c_3$. The following example illustrates Muller's method.

EXAMPLE 5.5.1 Muller's Method

To facilitate comparison of Muller's method with the secant method, we again consider the nonlinear function

$$f(x) = x^2 \exp(-x/2) - 1$$

From Example 5.4.1, this function is known to have at least two real roots, one in the interval [0, 2] and another in the interval [2, 10]. Three distinct points are required to start Muller's method. Example program *e551* on the distribution CD applies Alg. 5.5.1 using $x_0 = 2$, $x_1 = 6$, $x_2 = 10$, and an error tolerance of $\varepsilon = 10^{-6}$. The results are summarized in Table 5.5.1, which shows the iteration number k, the current estimate x, and the magnitude of f at the current estimate. In this case, four iterations are required for convergence to the second root $x = 8.6131697$. A plot of the function $f(x)$ showing the two roots is shown in Figure 5.5.2.

TABLE 5.5.1 Muller's Method Estimates

k	x	$\lvert f(x) \rvert$
0	2.0000000 + 0.0000000j	0.4715178
1	6.0000000 + 0.0000000j	0.7923344
2	10.0000000 + 0.0000000j	0.3262053
3	9.2325182 + 0.0000000j	0.1570059
4	8.5947618 + 0.0000000j	0.0049372
5	8.6131382 + 0.0000000j	0.0000083
6	8.6131697 + 0.0000000j	0.0000001

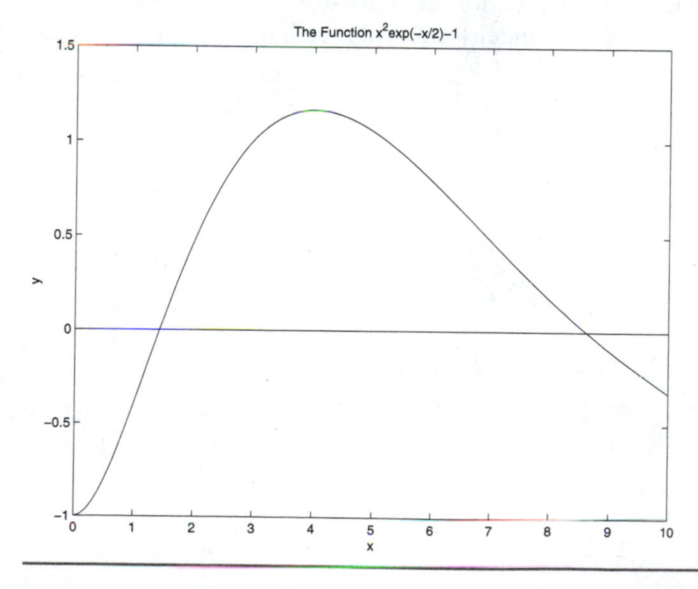

FIGURE 5.5.2 The Function $f(x) = x^2 \exp(-x/2) - 1$

5.6 NEWTON'S METHOD

Another way to approximate a curve in the vicinity of a root is to use the tangent to the curve at the current estimate. This is the basis of a popular method called the *Newton-Raphson* method or, simply, *Newton's* method. Let $f'(x_k)$ denote the *derivative* of $f(x)$ evaluated at $x = x_k$. Then the function f can be approximated by the following straight line generated by the tangent to f at x_k.

$$y = f(x_k) + f'(x_k)(x - x_k) \tag{5.6.1}$$

This approximation to $f(x)$ in the neighborhood of $x = x_k$ is equivalent to performing a Taylor series expansion of $f(x)$ about the point $x = x_k$ and dropping all but the constant and linear terms. The Newton estimate of the root is obtained by setting $y = 0$ in Equation (5.6.1) and solving for x. This yields the following iterative formula.

$$\boxed{x_{k+1} = x_k - \frac{f(x_k)}{f'(x_k)}} \tag{5.6.2}$$

A geometric illustration of Newton's method is shown in Figure 5.6.1. Note that the secant method in Equation (5.4.1) can be recast in the form $x_{k+1} = x_k - f(x_k)/\alpha$ where

$$\alpha = \frac{f(x_k) - f(x_{k-1})}{x_k - x_{k-1}} \tag{5.6.3}$$

In this case, α is a backward difference approximation to $f'(x)$ evaluated at $x = x_k$. Consequently, the secant method is a finite difference approximation of Newton's method. Finite difference approximations to derivatives are investigated in Chapter 7.

Whereas the secant method is a superlinear method with order of convergence $p \approx 1.618$, Newton's method is a *quadratic* method with order of convergence $p = 2$. To see this, suppose f has at least two derivatives and suppose $x = a$ is a simple root of $f(x)$. We expand $f(x)$ into a Taylor series about $x = x_k$ and evaluate the result at $x = a$. Using a second-order remainder term, and recalling that $f(a) = 0$, this yields

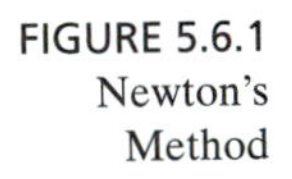

FIGURE 5.6.1 Newton's Method

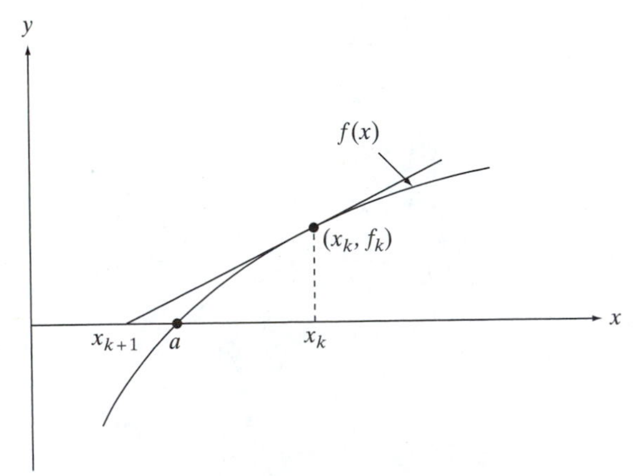

$$0 = f(x_k) + f'(x_k)(a - x_k) + \frac{f''(\xi)(a - x_k)}{2} \tag{5.6.4}$$

Here $f''(\xi)$ denotes the second derivative of $f(x)$ evaluated at $x = \xi$ where ξ lies somewhere in the interval from x_k to a. Solving Equation (5.6.4) for $f(x_k)$, substituting the result into Equation (5.6.2), and simplifying then yields

$$x_{k+1} - a = \frac{f''(\xi)}{2f'(x_k)}(x_k - a)^2 \tag{5.6.5}$$

Next, let $e_k = |x_k - a|$ denote the magnitude of the error of the kth estimate as in Equation (5.3.5). Taking the absolute value of both sides of (5.6.5) and reformulating in terms of the error, we get

$$\boxed{e_{k+1} = \left|\frac{f''(\xi)}{2f'(x_k)}\right| e_k^2 \quad , \quad k \geq 0} \tag{5.6.6}$$

If $f''(x)$ is bounded and $f'(a) \neq 0$, then we see from Equation (5.6.6) that Newton's method exhibits *quadratic* convergence. Note that if $f(x)$ has a double root at $x = a$, then $f'(a) = 0$. In this case the convergence is only linear. Refer to Ralston and Rabinowitz (1978) for a discussion of modifications applicable to multiple roots. When Newton's method converges, it does so very rapidly. However, like the other non-bracketing methods, it is not guaranteed to converge in all cases.

The following algorithm is an implementation of Newton's method. This algorithm terminates when a prescribed error tolerance on $|f(x)|$ has been achieved or a maximum number of iterations m has been performed.

Alg. 5.6.1 Newton's Method

1. Pick $\varepsilon > 0, m > 0, x_0$. Set $k = 1$ and compute $f_0 = f(x_0)$.
2. Do

   {

   $$q = f'(x_0)$$
   $$x_1 = x_0 - f_0/q$$
   $$x_0 = x_1$$
   $$f_0 = f(x_0)$$
   $$k = k + 1$$

   }
3. While $(|f_0| \geq \varepsilon)$ and $(k \leq m)$
4. Set $x = x_1$.

Unlike the previous algorithms, Alg. 5.6.1 requires two function evaluations per iteration, one for the function f itself, and the other for the derivative f'.

EXAMPLE 5.6.1 Newton's Method

To facilitate comparison with the secant method (Example 5.4.1) and Muller's method (Example 5.5.1), we again consider the nonlinear function

$$f(x) = x^2 \exp(-x/2) - 1$$

Example program *e561* on the distribution CD applies Alg. 5.6.1 using an initial guess of $x_0 = 1$ and an error tolerance of $\varepsilon = 10^{-6}$. The results are summarized in Table 5.6.1, which shows the iteration number k, the current estimate x, and the magnitude of f at the current estimate. In this case only two iterations are required for convergence to $x = 1.4296111$ in comparison with four iterations for the secant and Muller methods. The speed of Newton's method in Table 5.6.1 is with respect to the number of iterations, rather than the number of function evaluations. If the latter is used as a measure of the rate of convergence, then Newton's method, the secant method, and Muller's method are roughly comparable. Newton's method can be regarded as having an order of convergence of $p = \sqrt{2}$ with respect to the number of function evaluations.

TABLE 5.6.1 Newton's Method Estimates

k	x	$\lvert f(x)\rvert$
0	1.0000000	0.3934693
1	1.4324808	0.0025785
2	1.4296111	0.0000007

The number of function evaluations for Newton's method can be reduced by evaluating $f'(x)$ only once at $x = x_0$. This *quasi-Newton* method tends to take more iterations, but with fewer function evaluations per iteration.

The conditions under which Newton's method converges can be developed by analyzing Newton's method within the framework provided by the contraction mapping method. From Equation (5.6.2), Newton's method can be viewed as the following contraction mapping.

$$g(x) = x - \frac{f(x)}{f'(x)} \tag{5.6.7}$$

Taking the first derivative of g and simplifying yields

$$g'(x) = \frac{f(x)f''(x)}{[f'(x)]^2} \tag{5.6.8}$$

The function g is a contraction in the vicinity of the root $x = a$ if the magnitude of the slope is less than one. Thus for Newton's method to converge, a sufficient condition is that x_0 lies in a small interval $[\alpha, \beta]$ containing the root $x = a$ and that in this interval

$$\boxed{|f(x)f''(x)| < |f'(x)|^2} \tag{5.6.9}$$

There are a number of additional methods that have been proposed for root finding. Included among these are Ridder's method and the Van Wijngaarden-Dekker-Brent method (Brent, 1973). Ridder's method is a variation of the false position method that achieves quadratic convergence, while the Van Wijngaarden-Dekker-Brent method uses a combination of bisection and inverse quadratic interpolation (x as a function of y). Implementations of both of these methods can be found in Press et al. (1992).

5.7 POLYNOMIAL ROOTS

In engineering applications, there is a class of root finding problems that merits special attention. It corresponds to the case when the function f is a *polynomial* of degree n.

$$f(x) = a_1 + a_2 x + \cdots + a_{n+1} x^n \tag{5.7.1}$$

For the purpose of root finding, it is often convenient to *normalize* the polynomial by dividing its coefficients by $a_{n+1} \neq 0$. This process does not change the roots, yet it simplifies the polynomial by making the highest-power coefficient equal to one. An nth degree polynomial for which $a_{n+1} = 1$ is called a *monic* polynomial. Recall from Chapter 3 that the characteristic polynomial of a matrix is a monic polynomial. The roots of this polynomial are the eigenvalues of the matrix.

For a general function f, there is relatively little that can be said concerning the existence and nature of the roots. However, for polynomials, certain theoretical results are available that characterize the roots. The main result is the fundamental theorem of algebra that states that every polynomial of degree n has exactly n roots in the complex plane. Therefore, the polynomial in (5.7.1) can be written in *factored* form as follows where $x = x_k$ is the kth root.

$$f(x) = a_{n+1}(x - x_1)(x - x_2)\cdots(x - x_n) \tag{5.7.2}$$

The n roots of the polynomial f need not be distinct from one another. If a root occurs m times, it is referred to as a root of *multiplicity* m. Roots of multiplicity one are called *simple* roots. If the coefficients of a polynomial are real, then complex roots must occur in conjugate pairs. It follows that an odd-degree polynomial with real coefficients must have at least one real root. We assume that the coefficients of f are real.

5.7.1 Quadratic Formula

For lower-order polynomials, closed-form expressions are available for the roots. The simplest special case is the *linear* polynomial, $n = 1$. When $a_2 \neq 1$, the single real root occurs at $x_1 = -a_1/a_2$. It is also possible to find expressions for the roots of a *quadratic* polynomial, $n = 2$. When $a_3 \neq 1$, the two roots are

$$x_{1,2} = \frac{-a_2 \pm \sqrt{a_2^2 - 4a_1 a_3}}{2a_3} \tag{5.7.3}$$

The roots are real and distinct if the *discriminant*, $\delta \triangleq a_2^2 - 4a_1 a_3$, is positive. If $\delta = 0$, there is a double real root. Otherwise, the roots form a complex conjugate pair.

Recall that the *quadratic formula* in Equation (5.7.3) can sometimes give rise to a significant loss of numerical accuracy for one of the two roots. In particular, if $a_2^2 \gg 4a_1a_3$, then one of the roots is obtained by taking the difference of two nearly identical numbers. This process can cause catastrophic cancellation and can lead to significant round-off error. There are a number of ways to avoid this problem. One is to multiply the top and bottom of the right-hand side of Equation (5.7.3) by $-a_2 \mp \sqrt{a_2^2 - 4a_1a_3}$. This leads to the alternative formulation used in Muller's method and summarized in (5.5.5). Another approach is to start with the factored form of $f(x)$ and specify the coefficients in terms of the roots.

$$a_3(x - x_1)(x - x_2) = a_3[x^2 - (x_1 + x_2)x + x_1x_2] \tag{5.7.4}$$

Comparing Equation (5.7.4) with (5.7.1), we see that the constant coefficient is a_3 times the product of the roots, while the linear coefficient is a_3 times the sum of the roots.

$$x_1x_2 = \frac{a_1}{a_3} \tag{5.7.5a}$$

$$x_1 + x_2 = \frac{a_2}{a_3} \tag{5.7.5b}$$

To reduce the effects of round-off error when $a_2^2 \gg 4a_1a_3$, we can first use (5.7.3) to compute the larger of the two roots. The larger root arises when the sign of the discriminant term corresponds to the sign of $-a_2$.

$$\boxed{x_1 = \frac{-a_2 - \text{sgn}(a_2)\sqrt{a_2^2 - 4a_1a_3}}{2a_3}} \tag{5.7.6}$$

Recall that $\text{sgn}(x) = 1$ for $x \geq 0$ and $\text{sgn}(x) = -1$ for $x < 0$. If $x_1 \neq 0$, then the other root is computed using Equation (5.7.5a).

$$\boxed{x_2 = \frac{a_1}{a_3x_1}} \tag{5.7.7}$$

Note that it does not make sense to use (5.7.5b) because we would again be subtracting two nearly identical numbers. If $x_1 = 0$, then $x_2 = 0$.

EXAMPLE 5.7.1 Quadratic Roots

To illustrate the problem caused by catastrophic cancellation, consider the following quadratic polynomial.

$$f(x) = x^2 - (10^4 + 10^{-4})x + 1$$

This problem was first posed at the start of Chapter 1. The *exact* roots in this case are $x_1 = 10^4$ and $x_2 = 10^{-4}$. The discriminant is

$$\begin{aligned}\delta &= a_2^2 - 4a_1a_2 \\ &= (10^4 + 10^{-4})^2 - 4\end{aligned}$$

$$= (10^8 + 2 + 10^{-8}) - 4$$
$$= 100000002.00000001 - 4$$
$$\approx 1.000000 \times 10^8$$

The approximate value of δ is obtained by retaining seven significant digits, which is the number of digits supported by a four-byte single precision floating point data type. Using seven significant digits the value of a_2 in this case is

$$\begin{aligned} a_2 &= -(10^4 + 10^{-4}) \\ &= -10000.0001 \\ &\approx -1.000000 \times 10^4 \end{aligned}$$

It follows that using 7-digit precision in quadratic formula (5.7.3) yields $x_1 = 10^4$, which is exact, and $x_2 = 0$, which has a relative error of 100 percent! However, if Equation (5.7.7) is used to compute the second root instead, the exact answer of $x_2 = 10^{-4}$ is obtained.

It is possible to obtain closed-form expressions for the roots of a *cubic* polynomial using *Cardan's* method (see, e.g., Lastman and Sinha, 1989) and also for polynomials of degree four. However, these methods are somewhat tedious, and consequently, polynomials of degree three or more will be treated as *higher-order* polynomials whose roots are found by iterative methods.

5.7.2 Synthetic Division

The roots of higher-order polynomials ($n \geq 3$) can be determined by using iterative methods to find one root at a time. In order to prevent the iterative method from converging to the same root twice, we remove the roots as they are found using a process known as *synthetic division.* To illustrate the concept, let $f(x)$ be a polynomial of degree n and consider division of $f(x)$ by a factor $x - c$.

$$\boxed{f(x) = (x - c)g(x) + d} \tag{5.7.8}$$

The polynomial $g(x)$ is a polynomial of degree $n - 1$ called the *quotient* polynomial, and the constant d is the *remainder* term. Interestingly enough, synthetic division provides an efficient way to *evaluate* a polynomial. Note that if Equation (5.7.8) is evaluated at $x = c$, the result is

$$f(c) = d \tag{5.7.9}$$

Consequently, the remainder after division by $(x - c)$ is the original polynomial f evaluated at $x = c$. As we shall see, this technique for evaluating $f(x)$ is just as efficient as (and is equivalent to) the nested multiplications of Horner's rule in Alg. 4.2.1. Of course, if x is a root of $f(x)$, then the remainder is $d = 0$.

The technique in (5.7.8) can be extended to the evaluation of the derivative of a polynomial as well. Differentiating (5.7.8), we have

$$f'(x) = (x - c)g'(x) + g(x) \tag{5.7.10}$$

Evaluating (5.7.10) at $x = c$, we see that the derivative of f evaluated at $x = c$ is the quotient polynomial g evaluated at $x = c$.

$$f'(c) = g(c) \tag{5.7.11}$$

To evaluate $g(x)$ at $x = c$, we can perform a second synthetic division. That is, we can divide $g(x)$ by $(x - c)$ and find the remainder.

When $x = c$ is a root of f, the process of dividing a polynomial by the factor $(x - c)$ to produce a polynomial of lesser degree is called *deflation*. Before we put this technique to work to find all the roots of a polynomial, we must develop an algorithm for synthetic division, which starts with $f(x)$ and c and computes $g(x)$ and d. The basic idea is to equate the coefficients of like powers of x on both sides of (5.7.8). To that end, suppose $f(x)$ is as in (5.7.1) and $g(x)$ is as follows.

$$g(x) = b_1 + b_2 x + \cdots + b_n x^{n-1} \tag{5.7.12}$$

Substituting (5.7.12) into (5.7.8) and collecting the coefficients of each power of x, we get the following formulation of $f(x)$.

$$f(x) = (d - cb_1) + (b_1 - cb_2)x + \cdots + (b_{n-1} - cb_n)x^{n-1} + b_n x^n \tag{5.7.13}$$

Since Equation (5.7.13) must hold for *all* values of x, it follows that we can equate coefficients of like powers of x in (5.7.1) and (5.7.13). This results in the following linear synthetic division algorithm, which takes $f(x)$ and c as inputs and generates $g(x)$ and d as outputs.

Alg. 5.7.1 Linear Synthetic Division

1. Set $b_n = a_{n+1}$.
2. For $k = 1$ to $n - 1$ compute

   {

   $$b_{n-k} = a_{n-k+1} + cb_{n-k+1}$$

   }
3. Set $d = a_1 + cb_1$.

Alg. 5.7.1 is referred to as linear synthetic division because $f(x)$ is being divided by a linear factor, $x - c$.

EXAMPLE 5.7.2 Linear Synthetic Division

As an illustration of linear synthetic division, consider the following cubic polynomial.

$$f(x) = x^3 + 6x - 20$$

This polynomial has a root at $x = 2$. To see this, we can perform synthetic division by $(x - 2)$ and verify that the remainder is $d = 0$. Applying Alg. 5.7.1 with $n = 3$ and $c = 2$, we have

$$b_3 = 1$$
$$b_2 = 0 + 2(1) = 2$$
$$b_1 = 6 + 2(2) = 10$$
$$d = -20 + 2(10) = 0$$

Thus, the remainder is $d = 0$, which confirms that $x_1 = 2$ is a root of f. In this case, the deflated quotient polynomial is

$$g(x) = x^2 + 2x + 10$$

Since $g(x)$ is of degree two, we can apply the quadratic formula to determine the remaining two roots, which are as follows where $j = \sqrt{-1}$.

$$x_{2,3} = -1 \pm j\sqrt{3}$$

Synthetic division is used to deflate the polynomial $f(x)$ each time a new root is found by an iterative method. Since the deflated polynomial is of one degree less than the original polynomial, repeating this process eventually yields a lower-order polynomial whose roots can be determined in closed form. When a complex root $x = c + jd$ is found, we know that its conjugate, $x = c - jd$, is also a root because the coefficients of f are assumed to be real. Consequently, for complex roots, it makes sense to deflate the polynomial f by a quadratic factor that takes both roots into account. This has the added benefit of avoiding complex arithmetic. The quadratic factor is as follows.

$$\begin{aligned}(x - c - jd)(x - c + jd) &= (x - c)^2 + d^2 \\ &= x^2 - 2cx + c^2 + d^2 \end{aligned} \tag{5.7.14}$$

For convenience, suppose we consider the synthetic division of $f(x)$ by the notationally simpler quadratic factor $(x^2 - px - q)$. In this case, the quotient $g(x)$ is a polynomial of degree $n - 2$ and the remainder is a linear polynomial.

$$f(x) = (x^2 - px - q)g(x) + rx + s \tag{5.7.15}$$

The quadratic polynomial, $x^2 - px - q$, is a factor of $f(x)$ if and only if $r = 0$ and $s = 0$. As was done earlier for the linear case, we can determine r, s, and the coefficients of g by equating the coefficients of like powers of x in (5.7.15). In this case, $g(x)$ is as in Equation (5.7.12), but with $b_n = 0$. Substituting (5.7.12) with $b_n = 0$ into (5.7.15) and collecting the coefficients of each power of x, we get the following formulation of $f(x)$.

$$\begin{aligned} f(x) = (s - qb_1) + (r - pb_1 - qb_2)x + (b_1 - pb_2 - qb_3)x^2 + \cdots \\ + (b_{n-3} - pb_{n-2} - qb_{n-1})x^{n-2} + (b_{n-2} - pb_{n-1})x^{n-1} + b_{n-1}x^n \end{aligned} \tag{5.7.16}$$

Equating the coefficients of like powers of x in Equations (5.7.1) and (5.7.16) results in the following quadratic synthetic division algorithm, which takes $f(x)$, p, and q as inputs and generates $g(x)$, r, and s as outputs.

Alg. 5.7.2 Quadratic Synthetic Division

1. Set $b_{n-1} = a_{n+1}$ and $b_{n-2} = a_n + pb_{n-1}$.
2. For $k = 3$ to $n - 1$ compute

{

$$b_{n-k} = a_{n-k+2} + pb_{n-k+1} + qb_{n-k+2}$$

}

3. Set $r = a_2 + pb_1 + qb_2$ and $s = a_1 + qb_1$.

Alg. 5.7.2 is referred to as quadratic synthetic division because $f(x)$ is being divided by a quadratic factor, $x^2 - px - q$.

EXAMPLE 5.7.3 Quadratic Synthetic Division

As an illustration of quadratic synthetic division, consider the cubic polynomial examined in Example 5.7.2.

$$f(x) = x^3 + 6x - 20$$

Recall that this polynomial has roots at $x_1 = 2$, and $x_{2,3} = -1 \pm j\sqrt{3}$. To verify the complex roots, we divide by the corresponding quadratic factor and show that the remainder polynomial is zero. In this case, the quadratic factor is

$$\begin{aligned}(x + 1 + j\sqrt{3})(x + 1 - j\sqrt{3}) &= (x + 1)^2 + 3^2 \\ &= x^2 + 2x + 10\end{aligned}$$

Thus, we apply Alg. 5.7.2 to $f(x)$ using $n = 3$, $p = -2$, and $q = -10$. This yields

$$\begin{aligned}b_2 &= 1 \\ b_1 &= 0 + (-2)(1) = -2 \\ r &= 6 + (-2)(-2) + (-10)(1) = 0 \\ s &= -20 + (-10)(-2) = 0\end{aligned}$$

Thus, the remainder is $rx + s = 0$, which confirms that $x_{2,3} = -1 \pm j\sqrt{3}$ are roots of f. In this case, the deflated quotient polynomial is

$$g(x) = x - 2$$

5.7.3 Laguerre's Method

There are a number of iterative methods that can be used to estimate the roots of polynomials. For example, if complex arithmetic is used, Newton's method can be employed. Another possibility is Muller's method, which can converge to a complex root even when the initial guess is real. Of course, Newton's method and Muller's method are general root finding techniques applicable to nonpolynomial functions. In this section, we discuss a very fast specialized technique called *Laguerre's* method, which is highly effective for finding roots of polynomials. Laguerre's iterative method is based on the following two equations (see, e.g., Ralston and Rabinowitz, 1978).

$$g(x) \triangleq [f'(x)]^2 - \frac{nf(x)f''(x)}{n-1} \tag{5.7.17}$$

$$x_{k+1} = x_k - \frac{nf(x_k)}{f'(x_k) \pm (n-1)\sqrt{g(x_k)}} \tag{5.7.18}$$

Here the sign of the square root term in Equation (5.7.18) is chosen so as to maximize the magnitude of the denominator. Like Muller's method, Laguerre's method requires complex arithmetic because $g(x_k)$ in (5.7.18) can be negative.

Recall that Newton's method uses the first derivative of f and exhibits quadratic convergence to simple roots. From Equation (5.7.17), we see that Laguerre's method uses both the first and second derivatives of f. This results in a very fast algorithm that exhibits *cubic* convergence to simple roots that may be real or complex. Requiring the use of both the first and second derivatives of a general function f can be problematic, but in the case of polynomials, simple closed-form expressions for these derivatives are easily obtained. For polynomials with only real roots, Laguerre's method is guaranteed to converge to a root from *any* starting point x_0.

The following algorithm is an implementation of Laguerre's method. It terminates when the prescribed error tolerance on $|f(x)|$ has been achieved or a maximum number of iterations m has been performed.

Alg. 5.7.3 Laguerre's Method

1. Pick $\varepsilon > 0, m > 0, x_0$. Set $k = 1$. Compute $f_0 = f(x_0)$.
2. If $n = 1$, set $x_1 = -a_1/a_2$ and return.
3. Do

{

(a) Compute

{

$$f_1 = f'(x_0)$$
$$f_2 = f''(x_0)$$
$$g = f_1^2 - nf_0f_2/(n-1)$$
$$d_1 = f_1 + (n-1)\sqrt{g}$$
$$d_2 = f_1 - (n-1)\sqrt{g}$$

}

(b) If $|d_1| \geq |d_2|$ then $x_1 = x_0 - nf_0/d_1$ else $x_1 = x_0 - nf_0/d_2$.

(c) Compute

{

$$x_0 = x_1$$
$$f_0 = f(x_0)$$

$k = k + 1$

}

}

4. While ($|f_0| \geq \varepsilon$) and ($k \leq m$)

5. Set $x = x_1$.

Although Alg. 5.7.3 exhibits cubic convergence to simple roots, it does require three function evaluations per iteration, one for f, another for f', and the third for f''. These polynomials can be evaluated efficiently using either Horner's rule or linear synthetic division.

EXAMPLE 5.7.4 Laguerre's Method

As an illustration of Laguerre's method, consider the following polynomial of degree $n = 5$.

$$f(x) = x^5 - x^4 - 4x^3 + 4x^2 - 5x - 75$$

By construction, this polynomial has a real root at $x_1 = 3$ and complex roots at $x_{2,3} = -2 \pm j$ and $x_{4,5} = 1 \pm j2$. Example program *e574* on the distribution CD applies Alg. 5.7.3 using $\varepsilon = 10^{-6}$ and $x_0 = 0$. The results are summarized in Table 5.7.1, which shows the iteration number k, the current estimate x, and the magnitude of f at the current estimate. In this case, five iterations are required for convergence to $x = -2.0000000 + j1.0000000$. Since the coefficients of f are real, we know that $x = -2 - j$ is also a root. Note that rapid convergence to a complex root occurred even though the initial guess was real.

TABLE 5.7.1 Laguerre's Method Estimates

k	x	$\lvert f(x)\rvert$
0	0.0000000 + 0.0000000j	75.0000000
1	−3.2228875 + 0.0000000j	339.0397034
2	−2.1010365 + 1.5534867j	113.2877731
3	−2.0142701 + 0.9999696j	1.9708816
4	−1.9999999 + 1.0000004j	0.0000454
5	−2.0000000 + 1.0000000j	0.0000000

Our ultimate objective is to find *all* of the roots of the polynomial f. This can be achieved with a higher-level algorithm that uses Laguerre's method as one step. The basic approach is to find one root at a time, deflating the polynomial f after each root is found. Subsequent roots are found using the deflated polynomial. If a real root is found, the polynomial is deflated by a linear factor using linear synthetic division. Otherwise, it is deflated by a quadratic factor using quadratic synthetic division. Ultimately, the deflated polynomial is of degree one or two, at which point closed-form expressions for the roots can be used. The deflation process works best when the roots with smaller magnitude are removed first. Consequently, $x_0 = 0$ is used as an initial guess when Laguerre's method is applied.

In order to increase the numerical accuracy of the estimated roots, a two-step process is used. First, a root is found by applying Laguerre's method to the deflated polynomial. This root is then used as an initial guess and Laguerre's method is applied a second time, this time to the original polynomial f. This latter step, in which the estimate is refined or corrected, is sometimes called root *polishing*. It helps improve the numerical accuracy of the roots because cumulative errors are introduced as a result of the deflation process.

The following algorithm finds all n roots of the polynomial f. For each root, it terminates when the error tolerance on $|f(x)|$ is achieved or the maximum number of iterations m has been performed.

Alg. 5.7.4 **Polynomial Roots**

1. Pick $\varepsilon > 0, m > 0$. Set $h(x) = f(x)$ and $k = 1$.
2. Do

{

(a) Using $x_0 = 0$, apply Alg. 5.7.3 to $h(x)$ to find a root y.

(b) Using $x_0 = y$, apply Alg. 5.7.3 to $f(x)$ to find a root x_k.

(c) If x_k is real then

{

(1) Apply Alg. 5.7.1 to divide $h(x)$ by $(x - x_k)$

(2) Set $h(x) = g(x)$ where $g(x)$ is the quotient polynomial.

(3) Set $k = k + 1$.

}

else

{

(1) Set $x_{k+1} = x_k^*$.

(2) Apply Alg. 5.7.2 to divide $h(x)$ by $(x - x_k)(x - x_{k+1})$.

(3) Set $h(x) = g(x)$ where $g(x)$ is the quotient polynomial.

(4) Set $k = k + 2$.

}

}

3. While ($k \leq n$)

In Alg. 5.7.4, $h(x)$ is the deflated polynomial, root polishing occurs in step (2b), deflation occurs in step (2c), and x_k^* denotes the *complex conjugate* of x_k.

EXAMPLE 5.7.5 Polynomial Roots: Laguerre

As an illustration of the repetitive application of Laguerre's method with deflation, consider the polynomial of degree $n = 5$ previously examined in Example 5.7.4.

$$f(x) = x^5 - x^4 - 4x^3 + 4x^2 - 5x - 75$$

Example program *e575* on the distribution CD applies Alg. 5.7.4 with $\varepsilon = 10^{-6}$. The results are summarized in Table 5.7.2 which shows the root number, the estimated root, and the magnitude of f at the estimated root. In this case, a maximum of five Laguerre iterations are required for convergence to each of the roots.

TABLE 5.7.2 Polynomial Roots

Root	x	$\lvert f(x) \rvert$
1	3.0000000 + 0.0000000j	0.0000000
2	−2.0000000 − 1.0000000j	0.0000000
3	1.0000000 − 2.0000000j	0.0000000
4	1.0000000 + 2.0000000j	0.0000000
5	−2.0000000 + 1.0000000j	0.0000000

There are a number of other methods that have been developed for finding roots of polynomials. Included among these are *Bairstow's* method (Gerald and Wheatley, 1989), the *Jenkins-Traub* method (Ralston and Rabinowitz, 1978), and the *Lehmer-Schur* method (Acton, 1970). Barstow's method is based on the idea that if a polynomial $f(x)$ is divided by a quadratic factor $(x^2 - px - q)$, the remainder $rx + s$ will be zero. If $r = r(p, q)$ and $s = s(p, q)$ are not zero, then values for p and q are updated so as to reduce r and s. The Jenkins-Traub method is a complicated method that is applicable to real polynomials. It attempts to compute roots of increasing magnitude using a special sequence of polynomials. The Lehmer-Schur method is a generalization of the bracketing methods to the complex plane which is based on circles rather than intervals. Finally, an excellent method for finding the roots of a polynomial is to find the eigenvalues of the corresponding companion matrix using the QR method as described in Alg. 3.8.1.

5.8 NONLINEAR SYSTEMS OF EQUATIONS

All of the root finding techniques covered thus far are applicable only to a *single* nonlinear equation. The more general root finding problem consists of solving an n-dimensional nonlinear system of equations.

$$f(x) = 0 \tag{5.8.1}$$

Here it is understood that both x and $f(x)$ are $n \times 1$ vectors. All the previous methods examined the special case $n = 1$. Solving n nonlinear equations is considerably more difficult than solving either one nonlinear equation or n linear equations. Indeed, as we shall see, solving n linear equations constitutes just one step of an iterative technique for solving Equation (5.8.1). There are several approaches available for solving a nonlinear algebraic system. One technique is to convert (5.8.1) to an equivalent *optimization* problem:

$$\begin{aligned} \text{minimize:} \quad & F(x) \triangleq f^T(x)f(x) \\ \text{subject to:} \quad & x \in R^n \end{aligned}$$

This is an example of an unconstrained minimization problem because there are no constraints placed on the value of the $n \times 1$ vector x. Since the objective function $F(x) \geq 0$ and $F(x) = 0$ if and only if $f(x) = 0$, it follows that solutions to the optimization problem are roots of f. Optimization techniques are treated in detail in Chapter 6.

A more direct approach to solving Equation (5.8.1) is to generalize one of the one-dimensional root finding methods. The contraction mapping, secant, and Newton methods are all candidates for generalization. We focus our attention on Newton's method, which has the highest rate of convergence. Recall that Newton's method achieves quadratic convergence by making use of the derivative of the function f. When x and $f(x)$ are $n \times 1$ vectors, the appropriate generalization of the derivative of f is the $n \times n$ *Jacobian matrix* of partial derivatives defined as follows.

$$\boxed{J_{kj}(x) \triangleq \frac{\partial f_k(x)}{\partial x_j} \quad , \quad 1 \leq k,j \leq n} \tag{5.8.2}$$

Thus, the element of $J(x)$ appearing in the kth row and jth column is the partial derivative of $f_k(x)$ with respect to x_j. When $n = 1$, the Jacobian matrix reduces to $J(x) = f'(x)$. Given this generalization of $f'(x)$, we extend Newton's iterative formula in Equation (5.6.2) to n dimensions as follows.

$$x^{k+1} = x^k - J^{-1}(x^k)f(x^k) \quad , \quad k \geq 0 \tag{5.8.3}$$

Note that a *superscript* notation x^k is used for the kth estimate rather than x_k because subscripts are reserved to denote elements of the vector x. Although the formulation in (5.8.3) is a literal generalization of Equation (5.6.2), there is no need to find the inverse of $J(x^k)$ to obtain x^{k+1}. Instead, one can reformulate (5.8.3) by subtracting x^k from both sides and then multiplying both sides by $J(x^k)$. This leads to the following more efficient formulation, which uses a *correction* term, $\Delta x^k = x^{k+1} - x^k$.

$$\boxed{J(x^k)\Delta x^k = -f(x^k)} \tag{5.8.4}$$

$$\boxed{x^{k+1} = x^k + \Delta x^k} \tag{5.8.5}$$

Recall from Chapter 2 that for large values of n, finding $J^{-1}(x^k)$ requires about $4n^3/3$ FLOPs. Solving the linear algebraic system in Equation (5.8.4) requires only $n^3/3$ FLOPs. Thus, (5.8.4) and (5.8.5) are about four times faster than (5.8.3) for large values of n.

The following algorithm is an implementation of Newton's method for solving a nonlinear algebraic system. This algorithm terminates when a prescribed error tolerance on $\|f(x)\|$ has been achieved or a maximum number of iterations m has been performed.

Alg. 5.8.1 **Newton's Method: Vector**

1. Pick $\varepsilon > 0, m > 0, x^0$. Set $k = 1$ and compute $f^0 = f(x^0)$.
2. Do

   {

   (a) Compute the Jacobian matrix $J(x^0)$.

   (b) Solve the linear algebraic system $J(x^0)\Delta x = -f^0$.

   (c) Set

$$x^0 = x^0 + \Delta x$$

$$f^0 = f(x^0)$$

$$k = k + 1$$

   }
3. While ($\|f^0\| \geq \varepsilon$) and ($k \leq m$)
4. Set $x = x^0$.

In comparison with the scalar case of Alg. 5.6.1, the execution of Newton's method in Alg. 5.8.1 is quite expensive. There are a total of $n^2 + n$ function evaluations per iteration. In addition, step (2b) requires approximately $n^3/3$ FLOPs per iteration.

One of the drawbacks of Newton's method is the need to obtain the Jacobian matrix of partial derivatives of f. Not only is this computationally expensive for large values of n, but in some instances closed-form expressions for the partial derivatives may simply not be available. If only the values of $f(x)$ are available, then the partial derivatives can be approximated *numerically* using two function evaluations. Let i^j denote the jth column of the $n \times n$ identity matrix I and let $\varepsilon > 0$ denote a small perturbation. Then the derivative of $f_k(x)$ with respect to x_j can be approximated using central differences as follows.

$$J_{kj}(x) \approx \frac{f_k(x + \varepsilon i^j) - f_k(x - \varepsilon i^j)}{2\varepsilon} \tag{5.8.6}$$

By definition, the approximation in Equation (5.8.6) approaches the exact value $\partial f_k(x)/\partial x_j$ as $\varepsilon \to 0$. However, if ε is made too small in (5.8.6), then the numerator involves taking a difference of two nearly identical numbers, a process that is susceptible to round-off error.

The numerical approximation to $J(x)$ in (5.8.6) eliminates the need for a closed-form expression for the partial derivatives, but it increases the number of function evaluations per iteration. To reduce the number of function evaluations, a number of modifications to the basic algorithm have been proposed. The simplest approach is to refrain from updating the Jacobian matrix at each iteration. Instead, the initial value $J(x^0)$ is used, say, for the first n iterations, then $J(x^n)$ is used for the next n iterations, and so on. This effectively reduces the average number of function evaluations per iteration from $n^2 + n$ to $2n$.

EXAMPLE 5.8.1 Newton's Method: Vector

As an illustration of Newton's vector method, consider the following nonlinear algebraic system of dimension $n = 3$.

$$f(x) = \begin{bmatrix} \exp(2x_1) - x_2 + 4 \\ x_2 - x_3^2 - 1 \\ x_3 - \sin x_1 \end{bmatrix}$$

Example program *e581* on the distribution CD applies Newton's vector method in Alg. 5.8.1 using an initial guess of $x^0 = 0$, an error tolerance of $\varepsilon = 10^{-6}$, and a numerical approximation to the Jacobian matrix as in Equation (5.8.6). The results are summarized in Table 5.8.1, which shows the iteration number k, the current estimate x, and the norm of f at the current estimate.

TABLE 5.8.1 Newton's Vector Method Estimates

k		x		$\|f(x)\|$
0	0.0000000	0.0000000	0.0000000	3.0000000
1	2.0001705	1.0000205	2.0001705	49.6167488
2	1.5494552	1.3870203	1.0968041	16.7867565
3	1.1841072	1.9730949	0.9919639	4.7052126
4	0.9499587	1.6782521	0.8378814	1.0070899
5	0.8673295	1.5804416	0.7653263	0.0865539
6	0.8590704	1.5733788	0.7572618	0.0007761
7	0.8589966	1.5733330	0.7571875	0.0000000

Here, seven iterations are required to converge to the root

$$x = \begin{bmatrix} 0.8589966 \\ 1.5733330 \\ 0.7571875 \end{bmatrix}$$

Another approach to reducing the computational effort is to construct a sequence of estimates of the Jacobian matrix iteratively using *Broyden's* method (Broyden, 1965). Let $\Delta f^k = f(x^k) - f(x^{k-1})$ and let $J_0 = J(x^0)$. Then the estimate of $J(x^k)$ is computed iteratively as follows.

$$J_k = J_{k-1} + \frac{(\Delta f^k - J_{k-1}\Delta x^k)(\Delta x^k)^T}{(\Delta x^k)^T \Delta x^k} \tag{5.8.7}$$

This is an example of a *quasi-Newton* method. The computational efficiency of this approach can be improved by using the *Sherman-Morrison* matrix inversion formula, which eliminates the need to solve the linear algebraic system $J_k \Delta x^k = f(x^k)$ in step (2) of Alg. 5.8.1. See, for example, Lastman and Sinha (1989) for more details.

5.9 APPLICATIONS

The following examples illustrate applications of root finding techniques using both MATLAB and C. The relevant MATLAB functions in the NLIB toolbox are described in Section 1.6 of Appendix 1, while the corresponding C functions in the NLIB library are described in Section 2.9 of Appendix 2.

5.9.1 Propane Cylinder: C

Chemical engineers often work with hydrocarbons such as butane (C_4H_{10}), methane (CH_4), and propane (C_3H_8). To analyze the behavior of these compressible gases under a variety of operating conditions, it is useful to determine the molar specific volume v given the pressure p and temperature T. The mathematical relationship between p, v, and T is referred to as the equation of state. The simplest and best known equation of state is the ideal gas law,

$$pv = RT \tag{5.9.1}$$

where R is the universal gas constant. When the pressure p is expressed in units of bars, the appropriate value for the universal gas constant is $R = 0.08314$ bar-m^3/kmol-K. In 1873, van der Waals proposed the following more realistic model for the equation of state of a compressible gas (Morgan and Shapiro, 1995).

$$p = \frac{RT}{v - b} - \frac{a}{v^2} \tag{5.9.2}$$

The constant a accounts for the attractive forces between molecules, which reduce the overall pressure on the container wall. Similarly, the constant b accounts for the fact that the molecules themselves occupy some small, but not infinitesimal, volume. Note that when $a = b = 0$, the van der Waals model in Equation (5.9.2) reduces to the ideal gas law in (5.9.1).

The van der Waals model was refined by Redlich and Kwong in 1949 when they introduced the following empirical two-parameter model, which is generally considered to be more accurate.

$$p = \frac{RT}{v - b} - \frac{a}{v(v + b)\sqrt{T}} \tag{5.9.3}$$

For both the van der Waals model and the Redlich-Kwong model, the constants a and b can be determined in terms of the critical pressure and the critical temperature of the substance. The parameters of the Redlich-Kwong model, and the molecular mass M, for three hydrocarbon substances are summarized in Table 5.9.1 (Nelson and Obert, 1954).

Suppose we are interested in determining the molar specific volume, v, given the temperature, T, and the pressure, p, using the Redlich-Kwong model. To make the problem specific, consider a gas cylinder whose exterior dimensions are a height of $h = 1.35$ m and a diameter of $d = 0.25$ m as shown in Figure 5.9.1. Suppose the cylinder contains

TABLE 5.9.1 Redlich-Kwong Parameters for Hydrocarbon Gases

Substance	*Formula*	*a (bar-m⁶-K.⁵/kmol²)*	*b (m³/kmol)*	*M (kg/kmol)*
Butane	C_4H_{10}	289.55	0.08060	58.12
Methane	CH_4	32.11	0.02965	16.04
Propane	C_3H_8	182.23	0.06242	44.09

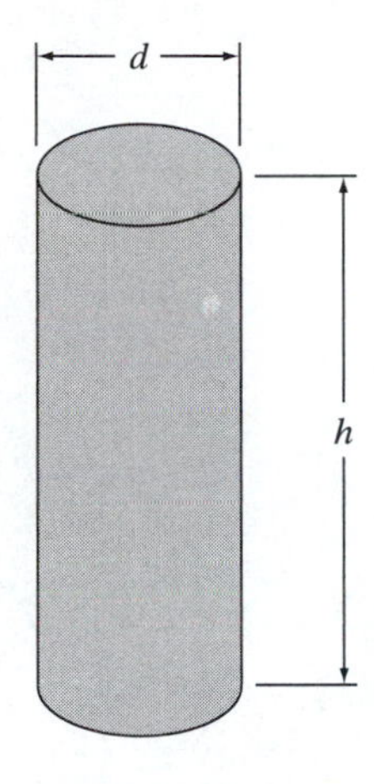

FIGURE 5.9.1 A Gas Cylinder Containing 2.9 kg of Propane at 20° C

$m = 2.9$ kg of propane at $T = 20°$C. The objective is to determine the interior volume of the cylinder and from that the wall thickness. If the wall thickness is zero, then from Figure 5.9.1, the volume of propane would be

$$V_0 = \frac{\pi d^2 h}{4} = 0.06627 \text{ m}^3 \tag{5.9.4}$$

The molar specific volume is then obtained by dividing by the mass and converting to units of m^3/kmol using the molecular mass of propane $M = 44.09$ kg/kmol, which yields

$$v_0 = \frac{MV_0}{m} = 0.9130 \text{ m}^3/\text{kmol} \tag{5.9.5}$$

Suppose that readings from a pressure gauge attached to the cylinder indicate that the pressure of the propane is $p = 30$ bars. We then use the Redlich-Kwong model in Equation (5.9.3) to solve for v given $T = 293.15°$K and $p = 30$ bars. In this case, the molar specific volume v_0 corresponding to infinitesimally thin cylinder walls can be used as an initial guess. The following C program on the distribution CD can be used to solve this problem.

```
/*----------------------------------------------------------------*/
/* Example 5.9.1: Propane Cylinder                                */
/*----------------------------------------------------------------*/

#include "c:\nlib\util.h"
```

```
float f (float v)     /* bisection, secant */
{
/* Redlich-Kwong equation of state */

   float  R = 0.08314f,  /* universal gas constant (bar-m^3/kmol-K) */
          T = 393.15f,   /* temperature (K) */
          a = 182.23f,   /* bar-m^6K^.5/kmol^2 */
          b = 0.06242f,  /* m^3/kmol */
          p = 30.0f,
          q;

   q = R*T/(v - b) - a/(v*(v + b)*((float) sqrt(T)));
   return p - q;
}

void g(vector x,int n,vector y)       /* Newton */
{
   y[1] = f(x[1]);
}

int main (void)
{
   int        r = 100,          /* maximum iterations */
              n = 1,            /* number of equations */
              i;                /* number of iterations */
   float      v0 = 0.1f,        /* lower bound (must be > 0) */
              v1 = 0.9130f,     /* upper bound, initial guess */
              eps = 1.e-6f,     /* error tolerance */
              M = 44.09f,       /* kg/kmol for propane */
              h = 1.35f,        /* height of cylinder (m) */
              d = 0.25f,        /* diameter of cylinder (m) */
              m = 2.9f,         /* mass of propane (kg) */
              V_0,              /* exterior volume of cylinder */
              V_1,              /* interior volume of cylinder */
              A,                /* exterior surface area of cylinder */
              t,                /* wall thickness (m) */
              v;
   vector     x = vec (n,""); /* used by Newton */

/* Compute and display molar specific volume v */

   printf ("\nExample 5.9.1: Propane Cylinder\n");
   printf ("\n-----------------------------------------------");
   i = bisect (v0,v1,eps,r,&v,f);
   printf ("\nBisection: v = %.5f m^3/kmol, %2i iterations",v,i);
   i = secant (v0,v1,eps,r,&v,f);
   printf ("\nSecant:    v = %.5f m^3/kmol, %2i iterations",v,i);
   x[1] = v0;
   i = newton (x,eps,n,r,g,NULL);
   printf ("\nNewton:    v = %.5f m^3/kmol, %2i iterations",x[1],i);
   printf ("\n-----------------------------------------------");
```

```
/* Compute cylinder volume and wall thickness */

   V_0 = PI*d*d*h/4;
   V_1 = m*v/M;
   shownum ("Exterior volume (m^3)",V_0);
   shownum ("Interior volume (m^3)",V_1);
   A = PI*d*h + 2*(PI*d*d/4);
   shownum ("Exterior surface area (m^2)",A);
   t = (V_0 - V_1)/A;
   shownum ("Wall thickness (cm)",100*t);
   return 0;
}
/*-------------------------------------------------------------------*/
```

When program *e591.c* is executed, it uses three numerical root finding methods to solve the Redlich-Kwong model for v, starting with the initial guess $v_0 = 0.9130$ m^3/kmol. A summary of the results is provided in Table 5.9.2, which shows the method used, the estimate of v, and the number of iterations required for convergence using an error tolerance of $\varepsilon = 10^{-6}$.

In this instance all three methods arrive at the same numerical estimate of the molar specific volume, $v = 0.83665$ m^3/kmol, although the secant and Newton methods do so with fewer iterations. Knowing the molar specific volume, the actual volume inside the cylinder can be found by solving Equation (5.9.5) for V_0 with $v_0 = v$, which yields

$$V = \frac{mv}{M} = 0.05503 \text{ m}^3 \tag{5.9.6}$$

The wall thickness can then be obtained by dividing the difference between the exterior and interior volumes by the surface area. If the walls are sufficiently thin, the exterior surface area can be used, namely,

$$A_0 = \pi dh + \frac{\pi d^2}{2} = 1.158 \text{ m}^2 \tag{5.9.7}$$

Finally, the estimate of the wall thickness of the cylinder based on the computed specific volume is

$$t = \frac{V_0 - V}{A_0} = 0.9701 \text{ cm} \tag{5.9.8}$$

This analysis infers the wall thickness of the cylinder from the pressure reading, the cylinder dimensions, and the properties of propane. It does not attempt to answer

TABLE 5.9.2 Specific Volume of Propane

Method	*v (m^3/kmol)*	*Iterations*
Bisection	0.83665	24
Secant	0.83665	5
Newton	0.83665	9

the question of whether or not the wall is sufficiently thick to withstand the pressure. This will depend on the material from which the cylinder is constructed. A solution to Example 5.9.1 using MATLAB can be found in the file *e591.m* on the distribution CD.

5.9.2 Bacterial Chemostat: MATLAB

One of the techniques that environmental engineers employ for the remediation of toxic wastes is to use special genetically engineered bacteria that convert the waste from a toxic state to an environmentally benign state. Consider the chemostat system used for growing and harvesting bacteria introduced in Section 5.1 and shown in Figure 5.1.5. Recall that the bacterial density x_1 and nutrient concentration x_2 in the growth chamber are obtained by solving the following system of two equations.

$$\frac{dx_1}{dt} = \frac{\gamma x_1 x_2}{1 + x_2} - x_1 \qquad (5.9.9a)$$

$$\frac{dx_2}{dt} = \frac{-x_1 x_2}{1 + x_2} - x_2 + \delta \qquad (5.9.9b)$$

The parameters γ and δ are positive constants that depend on the reservoir concentration C_0, the flow rate F, the volume V, and the type of bacteria and nutrient. Suppose both Equations (5.9.9a) and (5.9.9b) are multiplied by $1 + x_2$ to clear the denominators. The steady-state levels of bacterial density and nutrient concentration can be obtained by setting the rates of change dx_1/dt and dx_2/dt to zero. This results in the following nonlinear algebraic system of two equations that describe the steady-state operation of the chemostat.

$$\gamma x_1 x_2 - x_1(1 + x_2) = 0 \qquad (5.9.10a)$$

$$-x_1 x_2 + (\delta - x_2)(1 + x_2) = 0 \qquad (5.9.10b)$$

Recall from Section 5.1.3 that to obtain a physically meaningful solution, the parameters γ and δ must satisfy the constraints in (5.1.8). To make the problem specific, suppose $\gamma = 5$ and $\delta = 1$, which yields the following two-dimensional nonlinear function.

$$f(x) = \begin{bmatrix} 4x_1 x_2 - x_1 \\ 1 - x_1 x_2 - x_2^2 \end{bmatrix} \qquad (5.9.11)$$

To find a root of $f(x)$, Newton's vector method can be used. From (5.8.2), the Jacobian matrix of partial derivatives of $f(x)$ with respect to the components of x is

$$J(x) = \begin{bmatrix} 4x_2 - 1 & 4x_1 \\ -x_2 & -x_1 - 2x_2 \end{bmatrix} \qquad (5.9.12)$$

Suppose the initial guess is

$$x^0 = \begin{bmatrix} 1.0 \\ 0.1 \end{bmatrix}$$

The following MATLAB script on the distribution CD can be used to solve this problem.

```
%----------------------------------------------------------------------
% Example 5.9.2: Bacterial Chemostat
%----------------------------------------------------------------------

% Initialize

   clc                          % clear screen
   clear                        % clear variables
   m   = 50;                    % maximum iterations
   tol = 1.e-6;                 % error tolerance
   x0  = [1 0.1]';              % initial guess

   g = inline ('norm(funf592([a;b]),2)^2','a','b');    % error surface

% Compute and display steady-state solution

   fprintf ('Example 5.9.2: Bacterial Chemostat\n');
   [x,i] = newton (x0,tol,m,'funf592');
   show ('Number of iterations',i)
   show ('Steady-state solution',x)

% Plot error surface e(x) = ||f(x)|| */

   plotfun (0,3,-1,1.5,'Error Surface','x_1','x_2','g(x)',g)


function y = funf592 (x)
%----------------------------------------------------------------------
% Bacterial chemostat
%----------------------------------------------------------------------
   gamma = 5;
   delta = 1;
   y = zeros(2,1);         % column vector
   y(1) = gamma*x(1)*x(2) - x(1)*(1 + x(2));
   y(2) = -x(1)*x(2) + (delta - x(2))*(1 - x(2));
%----------------------------------------------------------------------
```

When script *e592.m* is executed, it uses Newton's vector method with an error tolerance of $\varepsilon = 10^{-6}$. In this case, Newton's method converges in five iterations to the following steady-state solution.

$$x = \begin{bmatrix} 2.25 \\ 0.25 \end{bmatrix} \tag{5.9.14}$$

A plot of $g(x) = f^T(x)f(x)$ is shown in Figure 5.9.2. Note that constant solutions for the chemostat are points on the surface where $g(x) = 0$. It is of interest to note that if a different initial guess is used, a different root of $f(x)$ can be reached. For example, if $x^0 = [0.5, 2]^T$, then the output converges in 11 iterations to $x = [0, 1]$. This steady-state

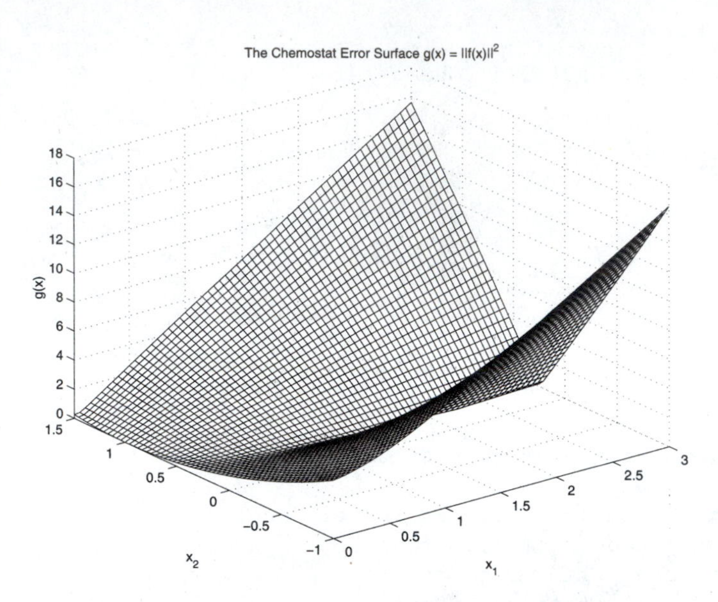

FIGURE 5.9.2 The Chemostat Surface $g(x) = f^T(x)f(x)$

solution is not of practical interest because the population of bacteria has died out and the nutrient concentration is the same as that of the reservoir.

A solution to Example 5.9.2 using C can be found in the file *e592.c* on the distribution CD.

5.9.3 Industrial High-Temperature Oven: C

Electrical engineers are involved with the manufacture of semiconductor devices such as solar cells and integrated circuits. Part of the manufacturing process involves treating the materials in an industrial high-temperature oven as shown in Figure 5.9.3. Heat is added to the oven through the heating element and is simultaneously lost to the environment due to the temperature drop across the oven walls. The variable x denotes the temperature inside the oven, which is assumed to be uniform, while T denotes the ambi-

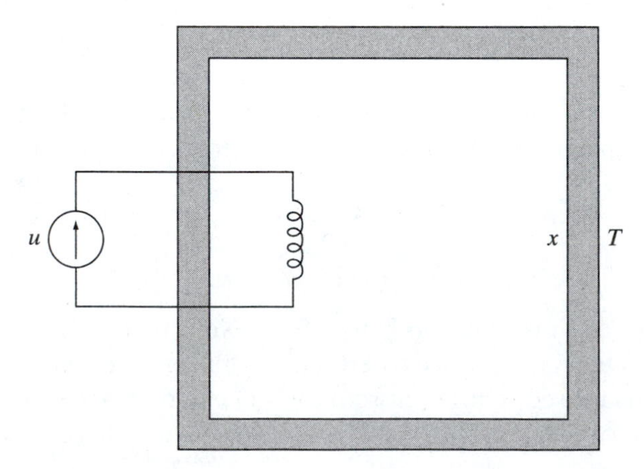

FIGURE 5.9.3 Industrial High-Temperature Oven

ent temperature of the air outside the oven. Both are expressed in degrees K. The rate of change of the oven temperature x is proportional to the rate of heat entering the oven from the heating element minus the heat leaving the oven due to convection and radiation. Based on experimental observations, this process can be modeled as a nonlinear system of the following general form (McClamroch, 1980).

$$\frac{dx}{dt} = \alpha(T - x) + \beta(T^4 - x^4) + \gamma u^2 \tag{5.9.15}$$

Here α, β, and γ are positive constants that depend on the thermal capacity of the oven. The term $\alpha(T - x)$ represents heat loss due to convection, while the term $\beta(T^4 - x^4)$ represents heat loss due to radiation. The input variable u denotes the current in the heating element. For many manufacturing processes, such as annealing, the temperature must be carefully controlled. Suppose the heating element current is held constant. To obtain the steady-state temperature x, we set $dx/dt = 0$ in Equation (5.9.15), which yields the following nonlinear equation.

$$\alpha(T - x) + \beta(T^4 - x^4) + \gamma u^2 = 0 \tag{5.9.16}$$

To make the problem specific, suppose the heating element current is $u = 21$ A, the ambient temperature is $T = 20°\text{C} = 293.15°\text{K}$, and the oven parameters are $\alpha = 50 \text{ min}^{-1}$, $\beta = 2 \times 10^{-7} \text{ min}^{-1} - °\text{K}^{-3}$, and $\gamma = 800\,°\text{ K/min-A}^2$. In this case, solving (5.9.16) for x is equivalent to finding the roots of

$$f(x) = 3.689 \times 10^5 - 50x - (2 \times 10^{-7})x^4 \tag{5.9.17}$$

A plot of the oven function $f(x)$ is shown in Figure 5.9.4. Note that temperatures where $f(x) = 0$ represent constant steady-state solutions of (5.9.15). From the point of

FIGURE 5.9.4 The Oven Function $f(x)$

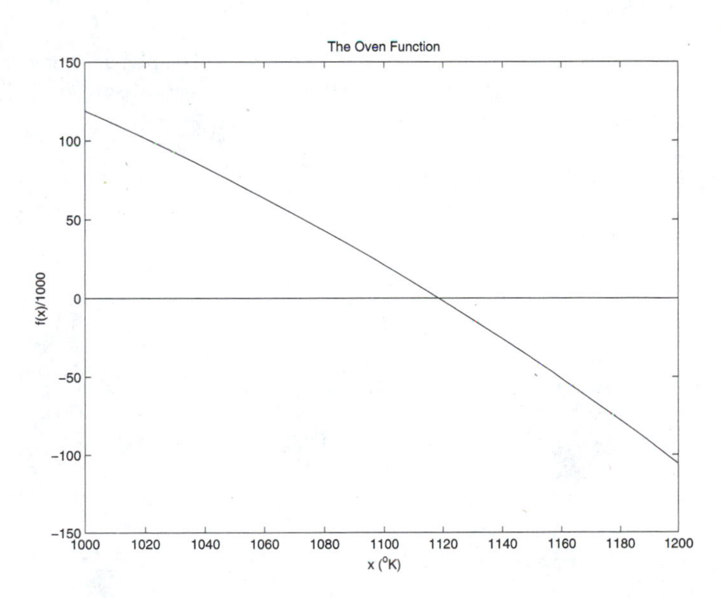

view of root finding, the fourth-degree polynomial $f(x)$ is quite challenging in that the magnitudes of its coefficients differ by twelve orders of magnitude!

The following C program on the distribution CD can be used to solve this problem.

```
/*----------------------------------------------------------------------*/
/* Example 5.9.3: Industrial High-Temperature Oven                       */
/*----------------------------------------------------------------------*/

#include "c:\nlib\util.h"

int main (void)
{
   int          m = 100,                   /* maximum iterations */
                n = 4,                     /* polynomial degree */
                p = 200,                   /* number of plot points */
                i;
   float        eps = 1.e-6f,              /* error tolerance */
                T = 293.15f,               /* ambient temperature (K) */
                u = 21.0f,                 /* heater current */
                alpha = 50.0f,
                beta = 2.0e-7f,
                gamma = 800.0f,
                xmin = 1000.0f,
                xmax = 1200.0f;
   vector       c = vec (n+1,""),          /* polynomial coefficients */
                x = vec (p,"");
   cvector      z = cvec (n,"");           /* polynomial roots */
   matrix       Y = mat (2,p,"");

/* Initialize coefficients */

   printf ("\nExample 5.9.3: Industrial High-Temperature Oven\n");
   c[1] = (float) (alpha*T + beta*pow(T,4) + gamma*pow(u,2));
   c[2] = -alpha;
   c[5] = -beta;
   showvec ("Coefficients",c,n+1,0);

/* Display function */

   for (i = 1; i <= p; i++)
   {
      x[i]      = xmin + (i-1)*(xmax - xmin)/(p-1);
      Y[1][i] = polynom (x[i],c,n)/1000;
   }
   graphxm (x,Y,2,p,"","x (^oK)","f(x)/1000","fig594");

/* Compute and display roots */

   i = polyroot (c,eps,n,m,z);
```

```
    shownum ("Number of iterations",(float) i);
    showcvec ("Roots",z,n,0);
    return 0;
}
/*-------------------------------------------------------------------*/
```

When program *e593.c* is executed, it uses the polynomial root finding method in Alg. 5.7.4 to produce the following four roots of $f(x)$.

$$x = \begin{bmatrix} -1211 & & \\ 46.02 & - & 1166j \\ 46.02 & + & 1166j \\ 1118 & & \end{bmatrix} \tag{5.9.18}$$

Clearly, the first three roots are not physically meaningful because x represents absolute temperature. Thus, the steady-state temperature in the oven associated with a heating element current of $u = 21$ A is

$$x_4 = 1118°\text{K} \tag{5.9.19}$$

A solution to Example 5.9.3 using MATLAB can be found in the file *e593.m* on the distribution CD.

5.9.4 Suspension Cable: MATLAB

One of the problems that arises in the field of engineering mechanics is the question of determining the tension in overhead suspension cables when they are used to support a weight such as the roadway for a bridge. As an example, consider the cable suspension system shown in Figure 5.9.5. Here a denotes the span of the suspension section, x is the sag of the cable, and m is the mass being supported by the cable.

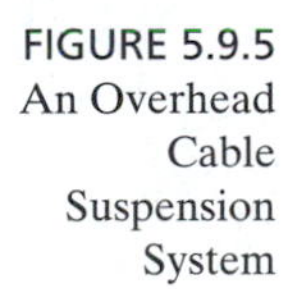
FIGURE 5.9.5 An Overhead Cable Suspension System

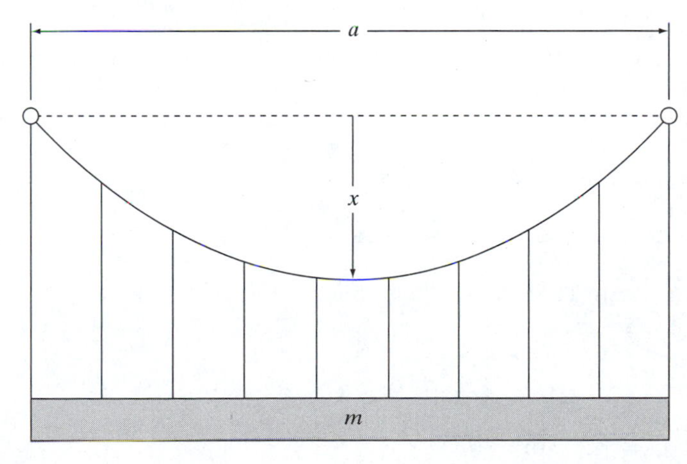

Suppose the weight being supported by the cable is uniformly distributed over the span of length a. If $g = 9.78$ m/s^2 denotes acceleration due to gravity, then the load associated with the mass m is

$$w = \frac{mg}{a} \tag{5.9.20}$$

If we neglect the effects of changes in temperature, the tension T at the end points where the cable is supported can be computed as follows (McLean and Nelson, 1978).

$$T = \frac{wa}{2}\sqrt{1 + \left(\frac{a}{x}\right)^2} \tag{5.9.21}$$

Thus, to design a structure which supports the cable at each end, we must first know the sag x. Let $L > a$ denote the length of the cable. The cable sag is determined by solving the following nonlinear algebraic equation.

$$L = a\left[1 + \frac{8}{3}\left(\frac{x}{a}\right)^2 - \frac{32}{5}\left(\frac{x}{a}\right)^4 + \frac{256}{7}\left(\frac{x}{a}\right)^6 - \cdots\right] \tag{5.9.22}$$

As more terms are added to Equation (5.9.22), the solution becomes increasingly more precise. To make the problem specific, suppose $a = 100$ m, $L = 125$ m, and $m = 1200$ kg. Then the load on the cable is $w = 117.4$ N/m.

The following MATLAB script on the distribution CD can be used to solve this problem.

```
%-------------------------------------------------------------------
% Example 5.9.4: Suspension Cable
%-------------------------------------------------------------------

% Initialize

   clc                              % clear screen
   clear                            % clear variables
   global L a                       % used by funf594.m
   a   = 100;                       % cable span (m)
   x0  = a/2;                       % initial guess at cable sag
   m   = 1200;                      % supported mass (kg)
   g   = 9.78;                      % acceleration due to gravity
   tol = 1.e-4;                     % relative error tolerance
   h   = fopen ('e594.dat','w');

% Compute cable sag x, and tension T/

   fprintf ('Example 5.9.4: Suspension Cable\n');
   w = m*g/a;
   show ('Load (N/m)',w);
   for j = 1 : 10
      L = a + 5*j;
      [x,i] = newton (x0,tol,100,'funf594');
```

```
        if j == 5
            k = i;
        end
        T = (w*a/2)*sqrt(1 + (a/(4*x))^2);
        fprintf (h,'\n %.0f & %.2f & %.1f \\\\',L,x,T);
    end
    show ('Iterations',k);
    fclose (h);
```

```
function y = funf594 (x)
%---------------------------------------------------------------------
% Sag in suspension cable
%---------------------------------------------------------------------
    global L a;
    q = x/a;
    y = a*(1 + (8/3)*q^2 - (32/5)*q^4 + (256/7)*q^6) - L;
%---------------------------------------------------------------------
```

When script *e594.m* on the distribution CD is executed, it uses Newton's method with starting point $x_0 = a/2$. Using an error tolerance of $\varepsilon = 10^{-5}$, it converges in six iterations to the following solution for the cable sag.

$$x = 32.30 \text{ m} \tag{5.9.23}$$

Once the sag is known, the tension at the supports can be computed using (5.9.21), which yields

$$T = 7420.8 \text{ N} \tag{5.9.24}$$

Script *e594.m* also shows that as the length of the cable L increases, the sag increases, but the tension needed to support the cable decreases at each end. The numerical relationship between L, x, and T produced by the script is shown in Table 5.9.3.

TABLE 5.9.3 Sag x and Tension T in Suspension Cables of Different Length L

L (m)	x (m)	T (N)
105	13.99	12018.0
110	20.13	9355.8
115	24.95	8307.6
120	28.94	7753.9
125	32.29	7420.8
130	35.11	7203.5
135	37.49	7052.8
140	39.53	6943.0
145	41.30	6859.4
150	42.84	6794.0

A solution to Example 5.9.4 using C can be found in the file *e594.c* on the distribution CD.

5.10 SUMMARY

Numerical iterative methods for root finding are summarized in Table 5.10.1. The higher-order methods tend to converge faster, but they can involve more computational effort per iteration.

The bisection method is a simple, reliable method that exhibits linear convergence ($p = 1$). It requires two initial points x_0 and x_1 that must bracket a root of $f(x)$. This is easily checked using the condition

$$f(x_0)f(x_1) < 0 \tag{5.10.1}$$

The bisection method reduces the interval of uncertainty by a factor of two with each iteration, so its progress is predictable, though slow. If f is continuous, the bisection method is guaranteed to find a root.

The false position or regula falsi method is similar to the bisection method in that it requires two initial points, which must bracket a root of $f(x)$ as in Equation (5.10.1). It uses a straight-line fit through the two end points to generate a new point. The false position method can exhibit superlinear convergence ($p > 1$), but there are examples where its performance is inferior to the bisection method. Like the bisection method, the false position brackets a root and is guaranteed to converge if f is continuous.

The contraction mapping method is based on finding a fixed point of the following function.

$$g(x) = x - bf(x) \tag{5.10.2}$$

The gain parameter b must be selected such that $0 < bf'(x) < 2$ for all x in an interval $[\alpha, \beta]$, which includes the root of interest. The contraction mapping method converges linearly if g maps $[\alpha, \beta]$ into itself, and $\alpha \le x_0 \le \beta$. By applying Aitken extrapolation starting with the third estimate, the convergence becomes superlinear ($p > 1$). The Aitken extrapolation technique can be applied to any linear iterative method. The contraction mapping method generalizes easily to systems of higher dimension, and it pro-

TABLE 5.10.1 Root Finding Techniques

Method	*Order*	*Initial Points*	*Converges*	*Roots*
Bisection	$p = 1$	2	always	real
False position	$p > 1$	2	always	real
Aitken	$p > 1$	1	sometimes	real
Secant	$p \approx 1.618$	2	sometimes	real
Muller	$p \approx 1.85$	3	sometimes	complex
Newton	$p = 2$	1	sometimes	real
Laguerre	$p = 3$	1	sometimes	complex

vides a theoretical framework for analyzing the convergence properties of other methods such as Newton's method and the linear iterative methods from Chapter 2. Practical difficulties in identifying b and the interval $[\alpha, \beta]$ tend to limit the applications of the contraction mapping method.

The secant method is an effective root finding technique that is based on a linear polynomial approximation of the function $f(x)$ in the neighborhood of the root. The method requires two distinct starting points, but they do not have to bracket a root as in Equation (5.10.1). Although the secant method does not always converge, when it does, it converges rapidly with an order of convergence of $p = (1 + \sqrt{5})/2$. The secant method can be regarded as a finite difference approximation to Newton's method. As such, its convergence rate is not as fast as Newton's method, but it requires less computational effort per iteration.

Muller's method is an effective root finding technique that is based on a quadratic polynomial approximation of the function $f(x)$ in the neighborhood of the root. Muller's method requires three distinct starting points. Muller's method does not always converge, but when it does, it converges rapidly with an order of convergence of $p \approx 1.85$. Unlike the previous methods, Muller's method can find complex roots, even when the starting points are real. Complex arithmetic is required to implement Muller's method.

Newton's method is a very fast root finding method based on approximating $f(x)$ locally with a two-term Taylor series. A single starting point is required. Like the other non-bracketing methods, Newton's method is not guaranteed to converge, but when it does, it exhibits quadratic convergence $(p = 2)$ to simple roots. Although Newton's method is a quadratic method, it requires two function evaluations per iteration, one for $f(x)$ and the other for the derivative, $f'(x)$. Consequently, with regard to the number of function evaluations, the order of convergence is $p = \sqrt{2}$. Newton's method can be easily generalized to find roots of an n dimensional nonlinear algebraic system. It then requires the computation of the $n \times n$ Jacobian matrix

$$J(x) = \frac{\partial f(x)}{\partial x} \tag{5.10.3}$$

but this matrix can be approximated numerically using central differences. A linear algebraic system of dimension n must be solved at each iteration using techniques from Chapter 2.

Laguerre's method is a highly effective root finding technique that is applicable to polynomials. Although Laguerre's method does not always converge to a root, it is guaranteed to converge from an arbitrary initial guess when all the roots of the polynomial are real. When Laguerre's method does converge, it exhibits an extremely rapid cubic convergence $(p = 3)$ to simple roots. Moreover, the roots can be real or complex. All n roots of a polynomial of degree n can be computed by successively removing the roots as they are found using synthetic division, and then applying Laguerre's method to the resulting lower-order polynomial. This process is called deflation. Alternatively, polynomial roots can be computed by finding the eigenvalues of the corresponding companion matrix as in Alg. 3.8.1. The eigenvalue technique is particularly attractive for high order polynomials because it is less sensitive to the effects of accumulated round-off error.

PROBLEMS

The problems are divided into Analysis problems, which can be solved by hand and Computation problems, which require the use of MATLAB or C. Solutions to selected problems can be found in Appendix 4. Students are encouraged to use these problems, which are identified with an (S), as a check on their understanding of the material. Problems marked with a (P) are programming problems that require the student to implement one or more of the algorithms discussed in the chapter. The remaining Computation problems require the student to write a main program that uses one or more of the NLIB functions discussed in Appendix 1 or Appendix 2.

Analysis

5.1 Verify that the solution to the leaky tank problem in Equation (5.1.6) can be obtained in closed form with the substitution $z = \sqrt{x}$.

(S)5.2 Consider the following function.

$$f(x) = \frac{\exp(-x) - x^3}{5}$$

(a) Show that this function has at least one root in the interval $[-1, 1]$.

(b) Show that f maps the interval $[-1, 1]$ into itself by plotting $f(x)$ for $-1 \le x \le 1$.

(c) Find bounds γ and δ on the derivative $f'(x)$ such that the following inequality holds.

$$\gamma \le f'(x) \le \delta < 0 \quad , \quad -1 \le x \le 1$$

(d) For what values of b is the following function a contraction mapping?

$$g(x) = x - bf(x) \quad , \quad -1 \le x \le 1$$

5.3 Suppose $f(x)$ has a *double* root at $x = a$. Show that $f'(a) = 0$ where $f'(x) = df(x)/\,dx$.

5.4 Consider the following polynomial which has a root at $x = 1$.

$$f(x) = x^2 - 2x + 1$$

(a) Why are the bisection and false position methods not applicable to this function?

(b) Find a simplified expression for x_{k+1} if Newton's method is applied to this function.

(c) What is the order of convergence of Newton's method in this case?

5.5 Consider the following quadratic polynomial. Sketch the locus of roots in the complex plane for $0 \le \alpha \le 1$. For what values of α are the roots real, complex, distinct?

$$f(x) = x^2 + x + \alpha$$

5.6 Derive the alternative form of the quadratic formula by multiplying Equation (5.7.3) by $-a_2 \mp \sqrt{a_2^2 - 4a_1a_3}$ and simplifying. Verify the result by computing the roots of the following quadratic using both (5.7.3) and the alternative form.

$$f(x) = x^2 + 4x + 13$$

(S)5.7 Consider the following cubic polynomial. Apply linear synthetic division to verify that $(x + 4)$ is a factor. Find the roots of $f(x)$.

$$f(x) = x^3 + 4x^2 + 9x + 36$$

5.8 Consider the following fourth-degree polynomial. Apply quadratic synthetic division to verify that $(x^2 + 2x + 5)$ is a factor. Find the roots of $f(x)$.

$$f(x) = x^4 + 4x^3 + 6x^2 + 4x - 15$$

5.9 Suppose linear synthetic division in Alg. 5.7.1 is used to evaluate an nth degree polynomial $f(x)$ at $x = c$ as in Equations (5.7.8) and (5.7.9). How many floating-point multiplications are required? How many floating-point additions? Compare the number of FLOPs with Horner's rule in Alg. 4.2.1.

5.10 Consider the following system of two equations in two unknowns. Find the Jacobian matrix $J(x)$.

$$x_1 \cos(\pi x_2) + x_2^2 = 1$$

$$x_1^3 - x_2 \sin(\pi x_1) = -1$$

Computation

(S)5.11 Write a program that uses the NLIB function *bisect* to find a root of the following function f using $[x_0, x_1] = [0, 10]$ and an error tolerance of $tol = 10^{-6}$. Print the x, $|f(x)|$, and the number of iterations needed for convergence. Verify the result by plotting $f(x)$ over $[0, 10]$ using the NLIB function *graphfun*.

$$f(x) = \ln(2 + x) - \sqrt{x}$$

(P)5.12 Write a function called *falsepos* that implements the false position method as described in Alg 5.2.2. The pseudo-prototype for *falsepos* is

```
[float x,int i]=falsepos (float x0,float x1,float tol,int m,
                          function f);
```

On entry to *falsepos:* $x0$ and $x1$ are initial guesses that bracket the root of interest as in Equation (5.10.1), $tol \geq 0$ is an error tolerance used to terminate the search, $m \geq 1$ is the maximum number of iterations, and f is the name of a user-supplied function which specifies the function to be searched. The pseudo-prototype of f is as follows.

```
[float y]=f (float x);
```

When f is called, it must return the value $y = f(x)$. On exit from *falsepos:* x is an estimate of a root of f, and i is the number of iterations performed. If $i < m$, the following convergence criterion should be satisfied:

$$|f(x)| < tol$$

Test *falsepos* by using it to solve problem 5.11.

(P)5.13 Write a program that uses the contraction mapping method to find a root of the function $f(x)$ in problem 5.2 using $x_0 = 0$, an error tolerance of $tol = 10^{-6}$, and a suitable value for b of your choosing. Print the x, $|f(x)|$, and the number of iterations needed for convergence. Solve the problem with and without Aitken extrapolation. What is the percentage improvement in speed (measured in iterations) when Aitken extrapolation is used?

5.14 Write a program that uses the NLIB function *secant* to find a root of the following function using $x_0 = 0, x_1 = 4$, and an error tolerance of $tol = 10^{-6}$. Print the $x, |f(x)|$, and the number of iterations needed for convergence. Verify the result by plotting $f(x)$ over $[0,4]$ using the NLIB function *graphfun*. How many roots are there?

$$f(x) = 10x \exp(-x) - 1$$

5.15 Repeat problem 5.14, but using the NLIB function *muller* in place of *secant*. Use $(x_0, x_1, x_2) = (0, 2, 4)$.

5.16 Repeat problem 5.14, but using the NLIB function *newton* in place of *secant* and $x_0 = 0$. How many function evaluations are required?

5.17 If a second-order underdamped linear dynamic system with zero initial conditions is driven with a constant input $u(t) = 1$ for $t \geq 0$, the resulting output, called the *step response*, is of the following general form.

$$y(t) = \alpha\{1 - \exp(-\beta t)[\cos(\gamma t) + \delta \sin(\gamma t)]\}$$

One measure of the *speed* of the system is the time t_0 it takes the output to rise from zero to $\alpha/2$ where α is the steady-state value of $y(t)$ in the limit as $t \to \infty$. Write a program that finds the smallest t such that $y(t) = \alpha/2$. You can assume $\alpha = 1, \beta = 1, \gamma = 3\pi$ and $\delta = 0.5$

5.18 Write a program that uses the NLIB functions *secant*, *muller*, and *newton* to find a root of the following function using $(x_0, x_1, x_2) = (-1, 1, 3)$ for initial guesses, as needed, and an error tolerance of $\varepsilon = 10^{-6}$. Print the x, $|f(x)|$, and the number of iterations needed for convergence in each case. Verify the result by plotting $f(x)$ over $[-1,3]$ using the NLIB function *graphfun*. How many roots are there? Which method is fastest in terms of iterations? Which is fastest in terms of function evaluations?

$$f(x) = 2^x - x^2$$

5.19 Write a program that uses the NLIB function *muller.m* or *laguerre.c* to find a root of the following polynomial using $x_0 = 0$ and an error tolerance of $\varepsilon = 10^{-6}$. Print x, $|f(x)|$, and the number of iterations needed for convergence. Find an $x_0 \neq 0$ that converges to a different root of f. Print x, $|f(x)|$, and the number of iterations needed for convergence.

$$f(x) = x^9 - x^8 + x^7 - x^6 + x^5 - x^4 + x^3 - x^2 + x - 1$$

5.20 Write a program that uses the NLIB function *roots.m* or *polyroot.c* to find all the roots of the following polynomial using an error tolerance of $\varepsilon = 10^{-5}$ for *polyroot*. Print the roots and the values of $f(x)$ at the roots.

$$f(x) = x^7 - 2x^6 + 3x^5 - 4x^4 + 5x^3 - 6x^2 + 7x - 8$$

(S)5.21 Write a program that uses the NLIB function *newton* to find a root of the nonlinear algebraic system in problem 5.10 using the initial condition $x^0 = [1, -1]^T$ and an error tolerance of $\varepsilon = 10^{-6}$. Print $x, \|f(x)\|$, and the number of iterations needed for convergence.

5.22 Write a program that uses NLIB to compute and display the roots of the following polynomial. The program should also display the values of $f(x)$ at the roots as a vector.

$$(x) = 8x^7 - 7x^6 + 6x^5 - 5x^4 + 4x^3 - 3x^2 + 2x - 1$$

5.23 Write a program that uses NLIB to compute and display a root of the following nonlinear function. Use an error tolerance of $\varepsilon = 10^{-6}$ and an initial guess of $x_0 = 0$.

$$f(x) = \ln(1 + x^2) - \exp(-x) + x$$

5.24 Write a program that uses NLIB to compute and display the solution to the following nonlinear algebraic system. Plot the two x_2 vs. x_1 curves on the same graph. How many solutions are there?

$$x_1^2 - x_2 = 1$$

$$10x_1 \exp(-x_1) - x_2 = 0$$

5.25 Write a program that uses NLIB to compute and display the solution to the following nonlinear algebraic system. Plot the two x_2 vs. x_1 curves on the same graph. How many solutions are there?

$$x_1^2 \exp(-x_1) - x_2/5 = 0$$

$$\ln (1 + x_1) - x_2 = -1$$

CHAPTER 6

Optimization

Engineers are sometimes required to select, from among a set of feasible designs, the design that is *best* in some sense, such as the least expensive design or the design with the highest performance. This process is referred to as optimal design or simply *optimization*. Optimization problems can be formulated mathematically in the following way.

$$\begin{aligned} \text{minimize:} \quad & f(x) \\ \text{subject to:} \quad & x \in S \end{aligned}$$

Here x is an $n \times 1$ vector of design parameters whose values are to be determined. The real-valued function to be minimized, $f(x)$, is called the *objective function*. For a solution to be *feasible*, it must belong to the constraint set S, which is a subset of the space of $n \times 1$ column vectors, R^n. Often $S = R^n$, in which case we say that the problem is an unconstrained optimization. More generally, the constraint set is characterized implicitly by a collection of r equality constraints and s inequality constraints on x.

$$\boxed{S = \{x \mid p(x) = 0,\ q(x) \geq 0\}}$$

Note that there is no loss of generality in considering only the minimization of $f(x)$ because maximizing a function is equivalent to minimizing the negative of the function. Optimization is a complex topic that is the subject of entire books (see, e.g., Luenberger, 1973; Zangwill, 1969). Relatively few optimization techniques are available for finding the overall or global minimum of a function. Instead, it is more typical to search for a local minimum, an x^* which minimizes $f(x)$ in some neighborhood of x^*. Of course, for some objective functions there is only one local minimum; finding it is equivalent to finding the overall global minimum.

The problem of finding the unconstrained minimum of $f(x)$ is closely related to the root finding problem discussed in Chapter 5. To see this, suppose $f(x)$ is differentiable. Let the $n \times 1$ vector $\nabla f = \partial f(x)/\partial x$ denote the *gradient vector* of partial derivatives of $f(x)$ with respect to the components of x. If x is an unconstrained minimum of $f(x)$, then it is a root or solution of the following n-dimensional nonlinear algebraic system.

$$\boxed{\nabla f(x) = 0}$$

The problem with finding a root of $\nabla f(x) = 0$ to minimize $f(x)$ is that we are just as likely to find an x that maximizes $f(x)$ instead. Rather than use this approach, we develop more robust iterative methods that generate a sequence of estimates x^k that produce successively smaller values of the objective function.

We begin this chapter by posing a number of optimization problems that arise in engineering. We then examine the distinction between local and global minima, along with some necessary and some sufficient conditions. We follow with a discussion of line search techniques, which are one-dimensional minimizations to determine the optimal step length along a given direction vector in R^n. The line search techniques include the golden section, bisection, and inverse parabolic interpolation searches. Next, the simplest unconstrained minimization technique, the steepest descent method, is presented. We then investigate the more powerful Fletcher-Reeves conjugate-gradient method. We follow with a discussion of the popular Davidon-Fletcher-Powell quasi-Newton method with self-scaling. Both of these methods converge to the minimum of a quadratic objective in, at most, n iterations. Next, we present a technique applicable to problems with equality and inequality constraints—the penalty function method. The penalty function method converts a constrained optimization into an equivalent unconstrained optimization by adding terms to the objective function; the terms become large when the constraints are violated. In this way, the minimization process itself tends to seek out feasible solutions because unfeasible solutions that violate the constraints are penalized. The last optimization technique we investigate a statistical technique called simulated annealing that is used to find a global minimum for a problem that has lower and upper bounds on x and $f(x)$. The chapter concludes with a summary of optimization methods and a presentation of engineering applications. The selected examples include a heat exchanger, transportation planning, maximum power extraction, and optimal container design.

6.1 MOTIVATION AND OBJECTIVES

Many engineering design problems can be formulated as optimization problems. In this section, we examine a number of representative examples.

6.1.1 Nonlinear Regression

Recall that in Chapter 4 we considered the problem of finding a least-square fit of a polynomial to a set of data points. To determine the polynomial coefficients that best fit the data, a linear algebraic system must be solved. Depending on the nature of the experimental data, a function other than a polynomial may provide for a more effective fit of the data. For example, suppose D represents the following set of m measurements.

$$D = \{(t_k, y_k) \mid 1 \le k \le m\} \tag{6.1.1}$$

Here t_k is the independent variable, and y_k is the dependent variable. Suppose the data represent radioactive decay or some other exponentially damped physical process. In this case, the following nonlinear function can be used to model the data.

$$y_k \approx x_1 \exp(-x_2 t_k) + x_3 \quad , \quad 1 \le k \le m \tag{6.1.2}$$

To find the best design parameters, x, we use an objective function $f(x)$, which consists of a sum of squares of the errors between the value predicted by the model and the actual data.

$$f(x) = \frac{1}{m} \sum_{k=1}^{m} [x_1 \exp(-x_2 t_k) + x_3 - y_k]^2 \tag{6.1.3}$$

Finding an optimal x tkhat minimizes $f(x)$ is then equivalent to finding a *least-squares fit* of the data. Note that $f(x)$ represents the average of the squared error. In

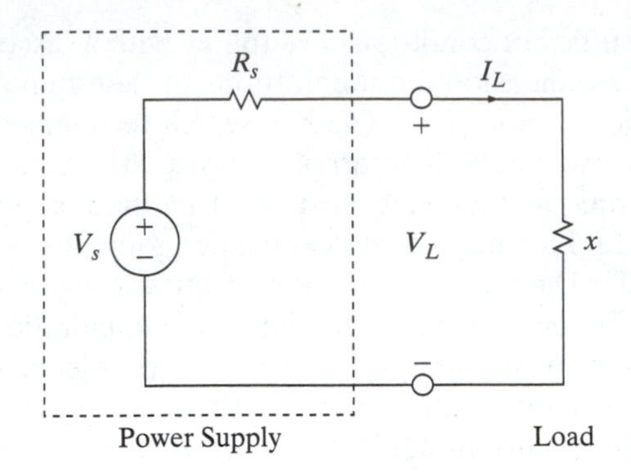

FIGURE 6.1.1 Power Supply Circuit

this case, there are no constraints on the components of the 3×1 vector x. Therefore, this is an example of an unconstrained optimization problem with $S = R^3$.

6.1.2 Electrical Load Design

Consider the electrical power supply circuit shown in Figure 6.1.1. The part enclosed in the dashed lines represents a DC power supply that is modeled as a constant voltage source V_s in series with a small source resistance R_s. The power supply drives a load resistor with resistance x. By Ohm's law, the current drawn by the load resistor is

$$I_L(x) = \frac{V_s}{R_s + x} \tag{6.1.4}$$

The power dissipated by the load resistor is the resistance times the square of the resistor current. Thus,

$$P_L(x) = \frac{xV_s^2}{(R_s + x)^2} \tag{6.1.5}$$

Suppose our objective is to design a load that is optimal in the sense that it maximizes the amount of power extracted from the power supply. In this case, there is a single design parameter, so $n = 1$. The objective function and constraint functions are as follows.

$$f(x) = -P_L(x) \tag{6.1.6a}$$

$$q(x) = x \tag{6.1.6b}$$

Notice that minimizing $f(x)$ is equivalent to maximizing $P_L(x)$. The inequality constraint $q(x) \geq 0$ ensures that the load resistance is not negative. This problem is sufficiently simple that it can be solved analytically rather than numerically. If x^* represents the optimal value, then it must satisfy the *necessary condition* $\nabla f(x^*) = 0$ where $\nabla f(x) = df(x)/dx$. Taking the derivative of both sides of Equation (6.1.5), setting the result to zero, and solving for x, we find that the optimal load resistance is

$$x^* = R_s \tag{6.1.7}$$

In electrical engineering, this is an illustration of the principle that for maximum power transfer, the load impedance must match the source impedance.

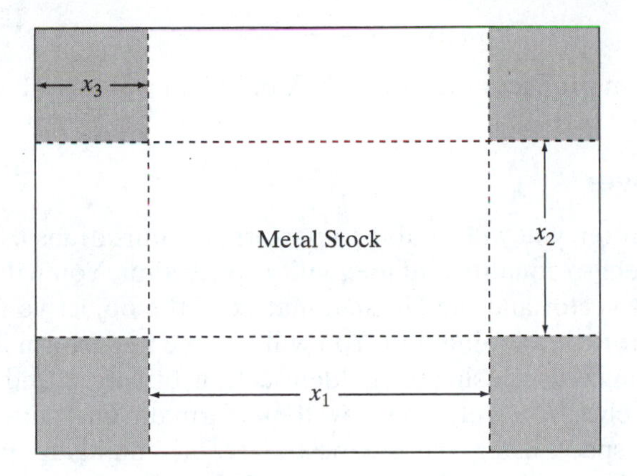

FIGURE 6.1.2 Container Design

6.1.3 Container Design

Suppose a company is interested in constructing a container by starting with a flat sheet of metal stock as shown in Figure 6.1.2. The basic approach is to remove a square of dimension x_3 from each corner of the rectangular sheet and then bend the four sides up and seal them to form a box. When the sides are bent up along the dashed lines in Figure 6.1.2, this adds x_3 units of depth. Thus, the volume of the resulting box is

$$V(x) = x_1 x_2 x_3 \tag{6.1.8}$$

The material not used in the four corners is melted down and recycled to form new sheets. Thus, the area of the metal stock used to construct the box is

$$A(x) = (x_1 + 2x_3)(x_2 + 2x_3) - 4x_3^2 \tag{6.1.9}$$

The cost of the material is proportional to the area, while the value of the final product is proportional to the volume that it holds. Suppose we try to maximize the volume, subject to the constraint that the area equal α. Furthermore, suppose the height of the box must be, at most, β. The dimension of this optimization problem is $n = 3$, and the objective function is as follows.

$$f(x) = -V(x) \tag{6.1.10}$$

By making the objective equal to the negative of the volume, minimizing $f(x)$ is equivalent to maximizing $V(x)$. Next, the constraint on the area can be written as $p(x) = 0$ where

$$p(x) = A(x) - \alpha \tag{6.1.11}$$

Finally, there are additional inequality constraints. Clearly, none of the box dimensions can be negative. In addition, there is an upper bound of β on the height, x_3. These bounds on the components of x can be achieved with the inequality constraint, $q(x) \geq 0$, where

$$q_1(x) = x_1 \tag{6.1.12a}$$

$$q_2(x) = x_2 \tag{6.1.12b}$$

$$q_3(x) = x_3 \tag{6.1.12c}$$

$$q_4(x) = \beta - x_3 \tag{6.1.12d}$$

A solution to this problem produces the largest box of height β or less that has a fixed materials cost.

6.1.4 Chapter Objectives

When you finish this chapter, you will be able to perform optimizations by minimizing objective functions subject to equality and inequality constraints. You will understand how to find the gradient vector and the Hessian matrix of the objective function and how to use them to test for a local minimum. You will be able to perform line searches or one-dimensional minimizations using the golden section, bisection, and inverse parabolic interpolation searches. You will know how to perform unconstrained minimizations in n-dimensional space using the steepest descent, conjugate-gradient, and quasi-Newton methods. You will also be able to apply penalty functions to convert constrained minimization problems to unconstrained minimizations. Finally, you will know how to perform global optimizations using the statistical simulated annealing method. These overall goals will be achieved by mastering the chapter objectives summarized in the following list.

Objectives for Chapter 6

- Know how to define and compute the gradient vector and the Hessian matrix of a function.
- Understand the difference between local and global minima.
- Know how to convert a maximization problem into an equivalent minimization problem.
- Understand how a search in n-dimensional space can be reformulated as a sequence of one-dimensional line searches.
- Know how to construct a three-point pattern containing a local minimum.
- Know how to perform the golden section line search and why it is optimal.
- Know how to perform a bisection line search using the derivative.
- Know how to perform an inverse parabolic interpolation line search.
- Be able to apply the steepest descent unconstrained optimization method.
- Be able to apply the conjugate gradient unconstrained optimization method.
- Be able to apply the DFP unconstrained optimization method.
- Know when the steepest descent method will take a long time to converge.
- Understand the difference between constrained and unconstrained optimization.
- Be able to construct penalty functions and use them to perform constrained optimization.
- Know how to perform global optimization using the simulated annealing technique.
- Understand how optimization techniques can be used to solve practical engineering design problems.
- Understand the relative strengths and weaknesses of each computational method and know which methods are applicable for a given problem.

6.2 LOCAL AND GLOBAL MINIMA

Before we explore numerical methods for minimizing functions, it is useful to clarify some terms that will be used in the remainder of the chapter. Suppose the optimization problem is formulated in the following way where S is a subset of R^n.

$$\begin{aligned} \text{minimize:} \quad & f(x) \\ \text{subject to:} \quad & x \in S \end{aligned}$$

A point x^* in the feasible region S is called a *local minimum* of $f(x)$ if and only if there exists a neighborhood Ω containing x^* such that

$$f(x^*) \leq f(x) \quad , \quad x \in \Omega \tag{6.2.1}$$

Thus, a local minimum is a minimum point over some, perhaps small, neighborhood Ω. If the inequality in Equation (6.2.1) is strengthened to read $f(x^*) < f(x)$ for $x \in \Omega$ with $x \neq x^*$, then x^* is referred to as a *strict local minimum* of $f(x)$. It is not difficult to construct functions that have several or even an infinite number of local minima or strict local minima. For example, $f(x) = \exp(-x) \sin(x)$ has an infinite number of strict local minima over the feasible region $S = [0, \infty)$.

The ultimate goal of optimization is to find the best local minimum, a point referred to as a global minimum. More specifically, x^* is a *global minimum* of $f(x)$ over the feasible region S if x^* satisfies (6.2.1) when Ω is replaced by S. Again, if the inequality in (6.2.1) is strengthened to $f(x^*) < f(x)$ for $x \in S$ with $x \neq x^*$, then x^* is referred to as a *strict global minimum*. Although there can be an infinite number of global minima, there is at most one strict global minimum.

EXAMPLE 6.2.1 Strict Global Minimum

Suppose $n = 1$ and the feasible region is the entire real line, $S = R$. The simplest polynomial that can have a strict global minimum is a quadratic polynomial.

$$f(x) = a + bx + \frac{c}{2}x^2$$

To find a minimum or a maximum, we set the slope of $f(x)$ to zero and solve for x. Taking the first derivative yields $f'(x) = b + cx$. Thus,

$$x^* = -\frac{b}{c}$$

To determine if x^* is a minimum, we examine the sign of the second derivative, $f''(x) = c$. If $c > 0$, which corresponds to positive curvature, then the point x^* is a strict global minimum of $f(x)$.

Example 6.2.1 is instructive because we can expand the same ideas to the general n-dimensional case. Suppose $f(x)$ has at least two derivatives. Then $f(x)$ can be approximated by the first three terms of its Taylor series expansion about the point $x = x^*$ as follows.

$$f(x) \approx f(x^*) + g(x^*)(x - x^*) + \frac{1}{2}(x - x^*)^T H(x^*)(x - x^*) \tag{6.2.2}$$

Here $g(x^*) = \nabla f(x^*)$ is the $n \times 1$ *gradient vector* of partial derivatives of $f(x)$ with respect to the components of x. That is,

$$g_k(x) \triangleq \frac{\partial f(x)}{\partial x_k} \quad , \quad 1 \le k \le n \tag{6.2.3}$$

Similarly, $H(x^*)$ is the $n \times n$ *Hessian matrix* of second partial derivatives of $f(x)$ with respect to the components of x. The entries in the rows and columns of $H(x)$ are computed as follows.

$$H_{kj}(x) \triangleq \frac{\partial^2 f(x)}{\partial x_k \, \partial x_j} \quad , \quad 1 \le k, j \le n \tag{6.2.4}$$

Note that, because changing the order of taking partial derivatives does not change the result, $H(x)$ is a symmetric matrix. The gradient vector $g(x)$, which is also denoted $\nabla f(x)$, plays the role of the slope of a function, while the Hessian matrix $H(x)$ plays the role of the second derivative or curvature. Consequently, if x^* is a minimum point of $f(x)$, then it is *necessary* that

$$\boxed{\nabla f(x^*) = 0} \tag{6.2.5}$$

That is, if x^* is a minimum point of $f(x)$, then the slope of $f(x)$ in each dimension must be zero at x^*.

To distinguish between a minimum or a maximum, we must look at the second derivative of $f(x)$. Recall that a matrix A is *positive-definite* if and only if $x^T A x > 0$ for $x \ne 0$. This is the condition that must be applied to the Hessian matrix $H(x^*)$. In particular, for x^* to be a strict local minimum of $f(x)$, it is *sufficient* that x^* satisfy Equation (6.2.5) and that the Hessian matrix, $H(x)$, be positive-definite at x^*.

$$\boxed{x^T H(x^*) x > 0 \quad , \quad x \ne 0} \tag{6.2.6}$$

There is a simple, but important, special case when the existence of a strict global minimum of $f(x)$ is guaranteed. Suppose the function $f(x)$, can be represented exactly by the first three terms of its Taylor series. That is, suppose $f(x)$ is a *quadratic* objective function.

$$f(x) = d + g^T x + \frac{1}{2} x^T H x \tag{6.2.7}$$

In this case, the Hessian matrix is the constant H. If H is positive-definite as in Equation (6.2.6), then $f(x)$ has a strict global minimum x^*. To find x^*, we compute the gradient of $f(x)$, which in this case is $\nabla f(x) = g + Hx$. Setting this to zero and solving for x then yields the following strict global minimum of the quadratic objective $f(x)$.

$$x^* = -H^{-1} g \tag{6.2.8}$$

Note how Equation (6.2.8) is the n-dimensional equivalent of the scalar result obtained in Example 6.2.1.

6.3 LINE SEARCHES

The majority of optimization algorithms make use of the following basic idea. Given an initial guess x^0, find an $n \times 1$ *direction vector* d^0 along which the value of $f(x)$ decreases. Next, search for the minimum value of $f(x)$ along this direction starting at x^0. This is equivalent to solving the following *one-dimensional* minimization problem.

$$\text{minimize:} \qquad F(\alpha) \triangleq f(x^0 + \alpha d^0) \tag{6.3.1}$$

$$\text{subject to:} \qquad \alpha \geq 0 \tag{6.3.2}$$

The one-dimensional optimization in Equation (6.3.1) is called a *line search*, and the value α_0 that minimizes $F(\alpha)$ is called the optimal *step length*. Once α_0 is found, the process is then repeated, starting from the new point $x^1 = x^0 + \alpha_0 d^0$. The sequence of solution estimates generated in this manner is

$$x^{k+1} = x^k + \alpha_k d^k \quad , \quad k \geq 1 \tag{6.3.3}$$

The virtue of this basic approach is that it replaces the original n-dimensional optimization by a sequence of one-dimensional optimizations, which are much simpler. The various optimization techniques that have been developed differ mainly in the way they construct the direction vectors d^k. Note that if a point x is reached where, in every direction, the value of $F(\alpha)$ does not decrease, then a local minimum has been found.

6.3.1 Golden Section

A reliable way to find an α that minimizes $F(\alpha)$ is to construct an interval $[a, c]$ that contains the solution and then to successively reduce the length of this interval until it is sufficiently small that the optimal α is known within a specified tolerance. This general approach is called *bracketing*.

First, we examine the problem of finding an initial interval $[a, c]$ that contains the solution. The basic idea is to find three points $a < b < c$ such that $F(b) < F(a)$ and $F(b) \leq F(c)$. That is, we find three points where the value of F at the middle point is smaller than the value of F at the first point and no larger than the value of F at the last point. A triplet of points satisfying this property will be referred to as a *three-point pattern*. Recall that the direction vector d^0 in Equation (6.3.1) is selected such that $F'(0) < 0$ where

$$F'(\alpha) \triangleq \frac{dF(\alpha)}{d\alpha} \tag{6.3.4}$$

Consequently, we can start with $a = 0$ and search for b and c. To ensure that b and c can be found, we assume that the function $F(\alpha)$ is *unimodal*. That is, we assume that there exists a unique point, $\alpha^* > 0$, such that $F(\alpha)$ decreases for $0 \leq \alpha < \alpha^*$ and $F(\alpha)$ increases for $\alpha > \alpha^*$. Under these conditions, the following algorithm constructs an interval $[a, c]$ that contains the optimal step length α^*.

Alg. 6.3.1 **Bracket**

1. Pick $h_{max} > h_{min} > 0$, and $h_{min} < h < h_{max}$. Set $a = 0, b = h, c = 0$.

2. While ($F(b) \geq F(a)$) and ($b > h_{min}$) do

{

$$c = b$$

$$b = b/2$$

}

3. If $c > 0$ stop, else set $c = 2b$.
4. While ($F(c) < F(b)$) and ($c < h_{max}$) do

{

$$a = b$$

$$b = c$$

$$c = 2c$$

}

If the function f is not unimodal, then it is possible that a bracket containing x^* might not be found. The lower limit on b in step (2) is included to guard against the possibility of Alg. 6.3.1 being called with $F'(0) \geq 0$, while the upper limit on c in step (4) is included in case $F'(\alpha) < 0$ for all $\alpha \geq 0$.

Once an interval $[a, c]$ is found that contains the solution, we can address the problem of decreasing the size of this *interval of uncertainty*. If we attempt to evaluate $F(\alpha)$ at the center of $[a, c]$, this does not provide us with sufficient information to determine if α^* lies in the left half or the right half. Instead, we must evaluate the function at *two* interior points, b_1 and b_2, as shown in Figure 6.3.1. If $F(b_1) \leq F(b_2)$, then α^* must lie at or to the left of b_2. Hence, point c can be discarded and the new interval is $[a, b_2]$. Alternatively, if $F(b_1) > F(b_2)$, then α^* must lie to the right of b_1, in which case point a can be discarded and the new interval is $[b_1, c]$. The process is then repeated using the new interval.

At this point, a natural question arises. Can we pick b_1 and b_2 in such a way that the computed values of $F(b_k)$ can be reused? This would have the effect of reducing the

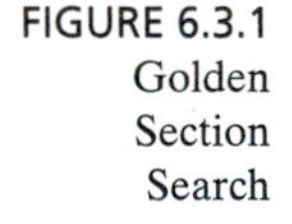

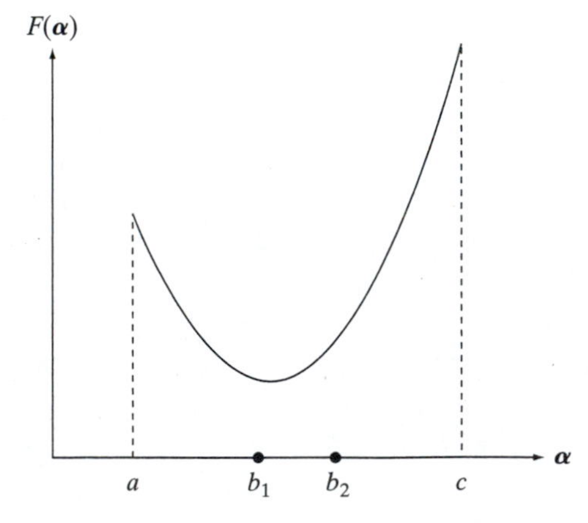

FIGURE 6.3.1 Golden Section Search

number of function evaluations per iteration. Suppose b_1 is selected to be a fraction $0 < \gamma < 0.5$ of the distance from a to c, and b_2 is selected to be the complementary fraction $1 - \gamma$. Thus,

$$b_1 = a + \gamma(c - a) \tag{6.3.5a}$$

$$b_2 = a + (1 - \gamma)(c - a) \tag{6.3.5b}$$

Because of the *symmetry* of the points b_1 and b_2 about the midpoint, we see that in both cases the length of the new interval, $[a, b_2]$ or $[b_1, c]$, is $1 - \gamma$ times the length of the previous interval. That is, if L_k denotes the length of the solution interval after iteration k, then

$$\boxed{L_{k+1} = (1 - \gamma)L_k \quad , \quad k \geq 0} \tag{6.3.6}$$

It follows that the process of successively reducing the solution interval by the factor $1 - \gamma$ using pairs of symmetrically distributed test points converges *linearly*. That is, it is a first-order method ($p = 1$).

Next, consider the selection of the fraction γ. If the new interval is $[a, b_2]$, then the previously computed point b_1 lies inside. To reuse the value of $F(b_1)$, we make b_1 at iteration k correspond to b_2 at iteration $k + 1$. Since the new interval is of length $(1 - \gamma)(c - a)$, this is equivalent to choosing γ such that

$$\gamma(c - a) = (1 - \gamma)[(1 - \gamma)(c - a)] \tag{6.3.7}$$

The left-hand side of Equation (6.3.7) corresponds to b_1 at iteration k, while the right-hand side corresponds to b_2 at iteration $k + 1$. The factor $(c - a)$ cancels from both sides of (6.3.7), and, we are left with the quadratic equation

$$\gamma^2 - 3\gamma + 1 = 0 \tag{6.3.8}$$

This quadratic polynomial has two real roots. The root that is less than 0.5 is

$$\boxed{\gamma = \frac{3 - \sqrt{5}}{2} \approx 0.382} \tag{6.3.9}$$

Thus, if γ is selected as in (6.3.9), there will be only *one* function evaluation per iteration, rather than two, when the new interval is $[a, b_2]$.

Next, we examine the case when the new interval is $[b_1, c]$. Here the previously computed point b_2 lies inside. To reuse the value of $F(b_2)$, we make b_2 at iteration k correspond to b_1 at iteration $k + 1$. This is equivalent to choosing γ such that

$$(1 - 2\gamma)(c - a) = \gamma[(1 - \gamma)(c - a)] \tag{6.3.10}$$

The left-hand side of (6.3.10) represents the distance from b_1 (which is the new a) to b_2 at iteration k, while the right-hand side corresponds to b_1 for iteration $k + 1$. Canceling the factor $(c - a)$ from both sides of Equation (6.3.10), we again arrive at the quadratic equation (6.3.8). It follows that the γ in (6.3.9) can be used in both cases. The process of successively narrowing the interval by the factor $1 - \gamma \approx 0.618$ using one function evaluation per iteration is called the *golden section search*. The name arises from the fact that the ancient Greeks used the number $1/(1 - \gamma)$ in architectural designs and called it the *golden ratio*.

The following algorithm implements the golden section search. It terminates when the relative length of the interval of uncertainty is less than a user specified *error tolerance* ε.

Alg. 6.3.2 Golden Section Search

1. Pick $\varepsilon > 0$. Apply Alg. 6.3.1 to find a three-point pattern $\{a, b, c\}$ bracketing the solution.
2. Compute

   {

   $$b_1 = a + \gamma(c - a)$$
   $$b_2 = a + (1 - \gamma)(c - a)$$
   $$F_1 = F(b_1)$$
   $$F_2 = F(b_2)$$

   }
3. While $(c - a) > \varepsilon(a + c)$ do

   {

   (a) If $(F_1 \leq F_2)$ set

   {

   $$c = b_2$$
   $$b_2 = b_1$$
   $$F_2 = F_1$$
   $$b_1 = a + \gamma(b_2 - a)$$
   $$F_1 = F(b_1)$$

   }

   (b) else set

   {

   $$a = b_1$$
   $$b_1 = b_2$$
   $$F_1 = F_2$$
   $$b_2 = a + (1 - \gamma)(c - b_1)$$
   $$F_2 = F(b_2)$$

   }

   }
4. Set $\alpha = (a + c)/2$

It is evident from step (3) of Alg. 6.3.2 that there is only one function evaluation per iteration once the process is initialized. The final estimate is the midpoint of the interval of uncertainty and is therefore within about $\pm\varepsilon\alpha^*$ of the optimal step size. It might be tempting to think that the error tolerance ε could be set to something on the order of the machine epsilon ε_M. This is actually too small and will result in unnecessary function evaluations. Indeed, since $F'(\alpha^*) = 0$, it follows from a Taylor series expansion of $F(\alpha)$ about $\alpha = \alpha^*$ that, in the neighborhood of the solution,

$$F(\alpha) - F(\alpha^*) \approx \beta(\alpha - \alpha^*)^2 \tag{6.3.11}$$

If we set $|F(\alpha) - F(\alpha^*)| = \varepsilon_M |F(\alpha^*)|$ and use Equation (6.3.11) to solve for $(\alpha - \alpha^*)$, we find that it is more appropriate to set the error tolerance ε proportional to the square root of ε_M. Consequently, as a rule of thumb, one might use

$$\boxed{\varepsilon \approx \sqrt{\varepsilon_M}} \tag{6.3.12}$$

EXAMPLE 6.3.1 Golden Section Search

To illustrate the use of the golden section search to solve a one-dimensional minimization problem, consider the following objective function.

$$F(\alpha) = -10\alpha^3 \exp(-2\alpha)$$

Example program *e631* on the distribution CD applies Alg. 6.3.1 with an initial step size of $h = 1$. The resulting output was the three-point pattern $\{1, 2, 4\}$, which required four function evaluations. Next, Alg. 6.3.2 was called with the following error tolerance.

$$\varepsilon = \sqrt{\varepsilon_M} = 3.453 \times 10^{-4}$$

This resulted in convergence in 19 iterations to $\alpha^* = 1.4997735$, which corresponds to an objective value of $F(\alpha^*) = -1.6803135$. These values can be confirmed from inspection of the graph of $F(\alpha)$ shown in Figure 6.3.2. A total of 23 function evaluations were required to find the minimum.

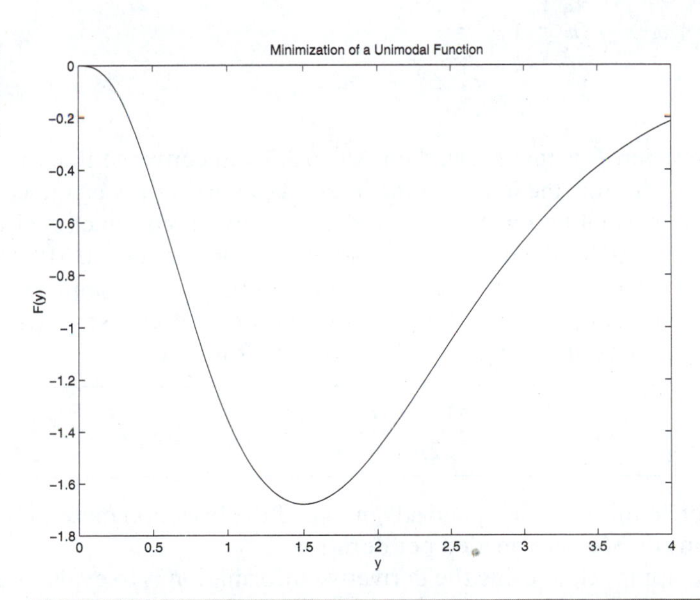

FIGURE 6.3.2 Minimization of a Unimodal Function

6.3.2 Derivative Bisection

In many instances an expression for the *derivative* of the objective function, $F'(\alpha)$, is available to aid the search for a minimum. One way to use the derivative information is to solve the following root finding problem.

$$F'(\alpha) = 0 \tag{6.3.13}$$

The drawback of converting a minimization problem into an equivalent root finding problem is that the condition in Equation (6.3.13) is only a *necessary* condition for the minimum, not a sufficient condition. Roots of (6.3.13) can just as well be local maxima or points of inflection of F.

A less direct approach is to use the *sign* of the derivative. Suppose $\{a, b, c\}$ is a three-point pattern that brackets the minimum. By examining the sign of $F'(b)$, we can determine if the minimum is to the left of b or the right of b. That is, if $F'(b) > 0$, then the solution lies in the subinterval $[a, b]$; otherwise, it is in the subinterval $[b, c]$. To reduce the size of successive intervals as fast as possible, we bisect the new interval and then repeat the processes. The following simple algorithm is an implementation of the *derivative bisection method*. It terminates when the relative length of the interval of uncertainty is less than a user-prescribed error tolerance, ε.

Alg. 6.3.3 Derivative Bisection

1. Pick $\varepsilon > 0$. Apply Alg. 6.3.1 to find a three-point pattern $\{a, b, c\}$ bracketing the solution. Set $b = (a + c)/2$.
2. While $(c - a) > 2\varepsilon b$ do

   {

   (a) If $F'(b) > 0$, set $c = b$

   (b) else set $a = b$.

   (c) Set $b = (a + c)/2$

   }

If the function F is unimodal, then Alg. 6.3.3 will converge linearly to the minimum. For each iteration, the length of the interval of uncertainty is reduced by a factor of 0.5. This is somewhat better than the golden section search, which is also first-order, but reduces the length of the interval by a factor of approximately 0.618. Thus, when an explicit expression for $F'(\alpha)$ is available, and it is comparable in complexity to $F(\alpha)$, the bisection method requires fewer function evaluations. Of course, $F'(\alpha)$ can also be approximated numerically using *central differences* as follows:

$$\boxed{F'(\alpha) \approx \frac{F(\alpha + h) - F(\alpha - h)}{2h}, \qquad 0 < h \ll 1} \tag{6.3.14}$$

However, in this case, the speed advantage of the bisection method is lost because two evaluations of $F(\alpha)$ are needed per iteration.

Another approach to using the derivative information is to evaluate $F'(\alpha)$ at two points and then apply the secant method from Chapter 5 to $F'(\alpha)$ to extrapolate to the point where the derivative is zero. This method of finding the roots of Equation (6.3.13) is called *Brent's derivative method.*

EXAMPLE 6.3.2 Derivative Bisection Search

To illustrate the use of the derivative bisection method to solve a one-dimensional minimization problem, consider the following objective function.

$$F(\alpha) = \frac{\cos(\pi\alpha)}{1 + \alpha^2}$$

Example program *e632* on the distribution CD applies Alg. 6.3.1 with an initial step size of $h = 1$. The resulting three-point pattern was $\{0, 1, 2\}$ and required three function evaluations. Next, Alg. 6.3.3 was called with the error tolerance as in Equation (6.3.12). It required 12 iterations to converge to $\alpha^* = 0.9025879$, which corresponds to an objective value of $F(\alpha^*) = -0.5254620$. These values can be confirmed from inspection of the graph of $F(\alpha)$ shown in Figure 6.3.3. A total of 15 function evaluations were required to find the minimum. It is of interest to note that the function F in Figure 6.3.3 is not unimodal, but instead features a number of local minima and maxima. In this instance, the bisection method succeeded in finding the global minimum. However, if Alg. 6.3.1 is called with a larger initial step, $h = 1.5$, a different three-point pattern is produced: $\{1.5, 3, 6\}$. This, in turn, results in Alg 6.3.3 returning the local minimum, $\alpha = 2.93866$, which corresponds to an objective value of $F(\alpha) = -0.101859$.

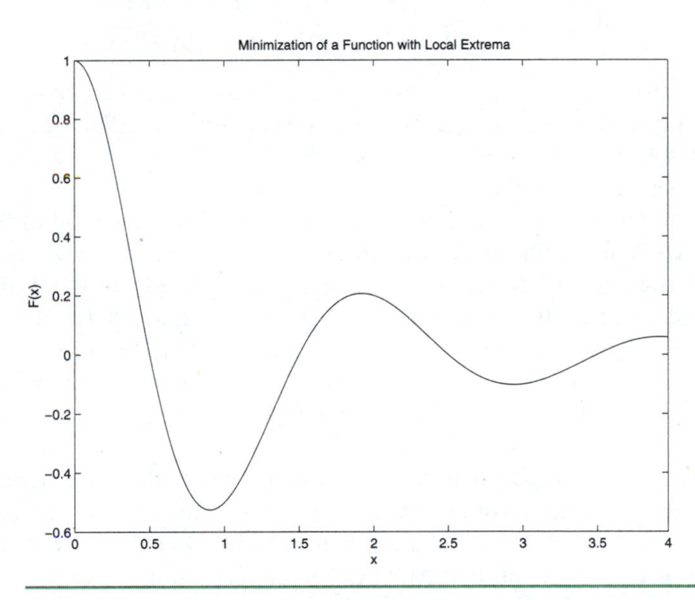

FIGURE 6.3.3 Minimization of a Function with Local Extrema

6.3.3 Inverse Parabolic Interpolation

The golden section and bisection methods are bracketing methods that are relatively slow but reliable. The rate of convergence for a line search can be increased by using a simple interpolation scheme. Suppose a three-point pattern $\{a, b, c\}$ is found, which

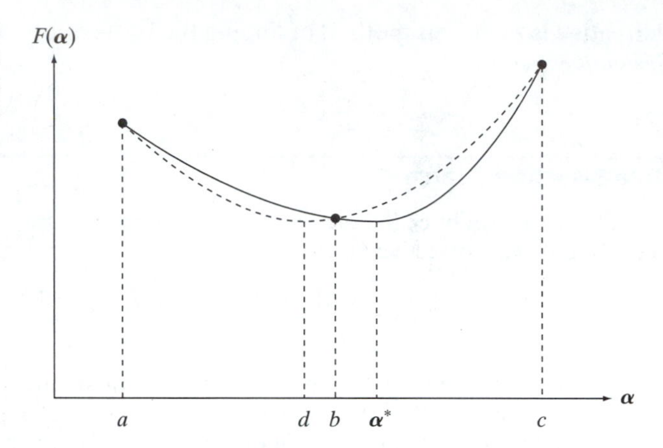

FIGURE 6.3.4 Inverse Parabolic Interpolation

brackets the minimum of $F(\alpha)$ as in Alg. 6.3.1. Rather than successively shorten the interval of uncertainty with pairs of interior points, suppose we instead pass a quadratic polynomial, $G(\alpha)$, through the three points. Since $F(b) < F(a)$ and $F(b) \le F(c)$, the polynomial $G(\alpha)$ will have a minimum in the interval $[a, c]$. This minimum can be found by differentiating $G(\alpha)$, setting the result to zero, and solving for α. The resulting value that minimizes the quadratic interpolating polynomial is (see Problems 6.2 and 6.3):

$$d = b + 0.5\left(\frac{(c-b)^2[F(a)-F(b)] - (a-b)^2[F(c)-F(a)]}{(c-b)[F(a)-F(b)] - (a-b)[F(c)-F(b)]}\right) \tag{6.3.15}$$

The minimization technique in Equation (6.3.14) is called *inverse parabolic interpolation*. An illustration of the method is shown in Figure 6.3.4. The only time the denominator in (6.3.15) goes to zero is when the three points are in a straight line that corresponds to a minimum at infinity.

If the function $F(\alpha)$ is sufficiently smooth in the neighborhood of the minimum, then the interpolated estimate d will be close to the minimum. At this point, the inverse quadratic interpolation can be repeated by replacing b with d discarding one of the original end points to form a new triplet. For smooth functions, successive inverse parabolic interpolation can be quite effective and rapidly converge to the minimum. However, for functions that are not well behaved, problems can arise, such as a d located outside the interval $[a, c]$ or a d approximating a local maximum rather than a local minimum.

Brent (1973) developed a hybrid technique that makes the inverse parabolic interpolation method more robust by combining it with the golden section search method. The basic idea of *Brent's method* is first to bracket the solution with a three-point pattern $\{a, b, c\}$. Next, an inverse parabolic interpolation step is attempted, as in Equation (6.3.15). For d to be accepted as a new point, it must lie inside the current interval of uncertainty, lead to a decrease in the objective function, and generate a change in estimates that is less than half of what it was at the previous iteration. The last condition ensures that the sequence of interpolated estimates converges to a point. If any of these three conditions is violated, then a golden section search step is performed. Part of the challenge in implementing Brent's method is to keep track of past values so that extra function evaluations are not performed when switching back and forth between the two methods.

6.4 STEEPEST DESCENT METHOD

At this point, we turn our attention to the n-dimensional unconstrained optimization problem. That is, suppose x is an $n \times 1$ vector of design parameters. Our goal is to find an x^* that minimizes the objectives function $f(x)$.

$$\begin{array}{ll} \text{minimize:} & f(x) \\ \text{subject to:} & x \in R^n \end{array}$$

Starting with an initial guess, x^0, we select a search direction, d^0, and perform a line search along that direction as in Equation (6.3.1). The result of the line search is taken as an updated estimate, and the process is then repeated. In order to determine a suitable direction to search, suppose the objective function f is differentiable. Let $\nabla f(x)$ denote the *gradient vector* of partial derivatives of f with respect to the components of x.

$$\boxed{\nabla f_k(x) \triangleq \frac{\partial f(x)}{\partial x_k} \quad , \quad 1 \le k \le n} \tag{6.4.1}$$

The gradient vector has a simple geometric interpretation when $n = 2$. For the two-dimensional case, the objective function $f(x)$ can be thought of as a surface with hills and valleys. At each point x, the vector $\nabla f(x)$ points uphill in the direction of steepest ascent starting from x. A direction d is downhill or a *descent* direction if

$$d^T \nabla f(x) < 0 \tag{6.4.2}$$

Since we are interested in finding a minimum of $f(x)$, and $\nabla f(x)$ is the direction of steepest ascent, a logical direction to search in is $d^0 = -\nabla f(x)$. This is the basis of the method of *steepest descent*, a technique first proposed by Cauchy in 1845. If α_k denotes the optimal step length resulting from searching along the direction $d^k = -\nabla f(x^k)$ starting from the point x^k, then the updated estimate of the solution is computed as follows.

$$\boxed{x^{k+1} = x^k - \alpha_k \nabla f(x_k) \quad , \quad k \ge 0} \tag{6.4.3}$$

The steepest descent sequence in Equation (6.4.3) can be terminated in a number of ways. One possibility is to stop when the relative step size falls below the machine epsilon, ε_M. Another approach is based on the observation that if x^* is local minimum of $f(x)$, then $\nabla f(x^*) = 0$. Consequently, we can continue to generate new estimates until the norm of $\nabla f(x^k)$ falls below a user-specified *error tolerance*, ε. The following error criterion combines these two approaches.

$$(\|\nabla f(x^k)\| < \varepsilon) \quad \text{or} \quad (\|x^k - x^{k-1}\| < \varepsilon_M \|x^k\|) \tag{6.4.4}$$

The following algorithm is an implementation of the steepest descent method that terminates when the error criterion in Equation (6.4.4) is satisfied or when a maximum number of iterations, m, is performed.

Alg. 6.4.1 **Steepest Descent**

1. Pick $\varepsilon > 0, m > 0$, and $x^0 \in R^n$. Set $k = 0$.
2. Do

   {

   (a) Compute $d^k = -\nabla f(x^k)$

   (b) Find $\alpha_k > 0$ to minimize $F(\alpha) = f(x^k + \alpha d^k)$

   (c) Compute $x^{k+1} = x^k + \alpha_k d^k$

   (d) Set $k = k + 1$.

   }
3. While $(k \le m)$ and $(\|d^k\| \ge \varepsilon)$ and $(\|x^k - x^{k-1}\| \ge \varepsilon_M \|x^k\|)$

The one-dimensional minimization in step (2b) of Alg. 6.4.1 can be performed using any line search technique such as the golden section search in Alg. 6.3.2 or the bisection search in Alg. 6.3.3. If a closed-form expression for the gradient is available, then the derivative of $F(\alpha)$ needed for the bisection method can be computed using the chain rule as follows.

$$\boxed{F'(\alpha) = (d^k)^T \nabla f(x^k + \alpha d^k)} \tag{6.4.5}$$

EXAMPLE 6.4.1 Steepest Descent

To illustrate the use of the steepest descent method to solve an unconstrained optimization problem, consider the following objective function of dimension $n = 2$.

$$f(x) = \lambda_1 (x_1 - 3)^2 + \lambda_2 (x_2 - 2)^2$$

Suppose $\lambda_1 = 1000$ and $\lambda_2 = 1$. It is clear from inspection that the optimal value is $x^* = [3, 2]^T$ and that the value of the objective function at this point is $f(x^*) = 0$. Example program *e641* on the distribution CD applies Alg. 6.4.1 with an initial guess of $x^0 = 0$ and an error tolerance of $\varepsilon = 10^{-4}$. In this case, the steepest descent method converges to

$$x = \begin{bmatrix} 3.0000000 \\ 1.9999994 \end{bmatrix}$$

This required 302 iterations and 10,273 scalar function evaluations. A plot of the objective surface is shown in Figure 6.4.1.

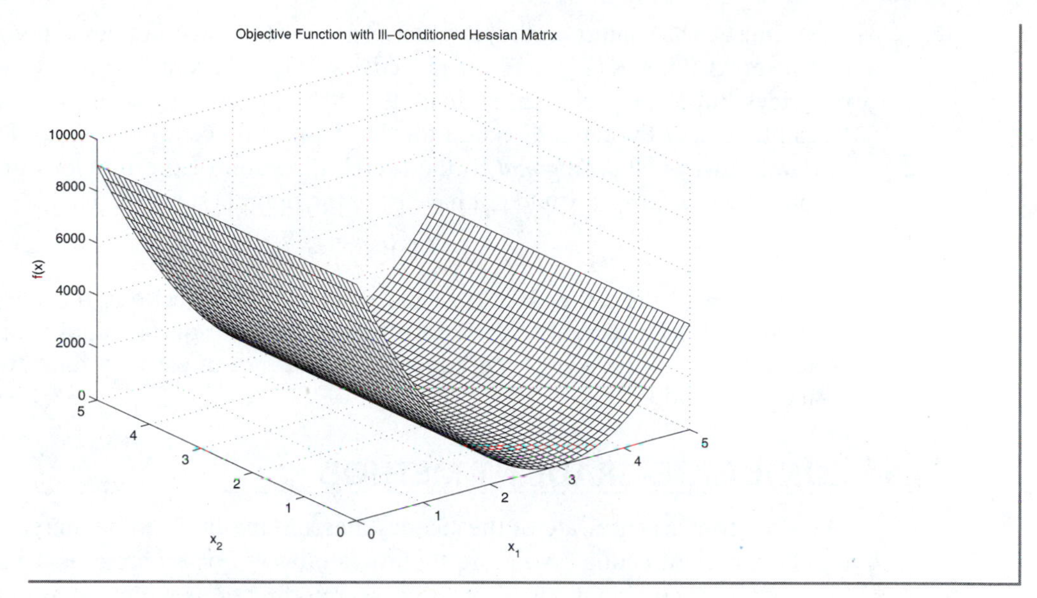

FIGURE 6.4.1 Objective with Ill-Conditioned Hessian Matrix

The appeal of the steepest descent method is that it is conceptually simple, and it is easy to program. However, for many problems it can take a large number of iterations to converge. This was the case even for the simple uncoupled quadratic objective function in Example 6.4.1, which took 302 iterations. If the first coefficient of the objective function in Example 6.4.1 is changed from $\lambda_1 = 1000$ to $\lambda_1 = 1$, the steepest descent method converges to the exact answer (seven digits) in just two iterations. To see why this makes such a difference, consider the $n \times n$ Hessian matrix of f.

$$H_{kj}(x) = \frac{\partial^2 f(x)}{\partial x_k \, \partial x_j} \quad , \quad 1 \le k, j \le n \tag{6.4.6}$$

For the quadratic objective function in Example 6.4.1, the Hessian matrix is a constant 2×2 diagonal matrix with diagonal elements λ_1 and λ_2. Recall from Equation (2.5.8) that the condition number of the Hessian matrix is

$$\begin{aligned} K(H) &= \|H\| \cdot \|H^{-1}\| \\ &= \max\{|\lambda_1|, |\lambda_2|\} \max\{|\lambda_1^{-1}|, |\lambda_2^{-1}|\} \\ &= \left(\frac{|\lambda_1|}{|\lambda_2|}\right) \end{aligned} \tag{6.4.7}$$

Thus, when $\lambda_1 = 1000$ and $\lambda_2 = 1$, we have $K(H) = 1000$, which makes the Hessian matrix ill-conditioned. Conversely, when $\lambda_1 = 1$ and $\lambda_2 = 1$, the Hessian matrix is as well conditioned as possible with $K(H) = 1$. In general, when the Hessian matrix is ill-conditioned, the steepest descent method converges very slowly. Observe from Equation (6.4.7) that, in this case, the condition number is the ratio of the magnitude of the largest eigenvalue to the magnitude of the smallest eigenvalue. This ratio is a commonly used alternative way to characterize the condition number of a general $n \times n$ matrix.

One can also interpret the slow convergence of the steepest descent method in geometric terms. When $K(H) \gg 1$, the objective surface will contain a valley that is steep on two sides but slopes very gently along the other dimension, as shown in Figure 6.4.1. Recall that the kth search direction for the steepest descent method is $d^k = -\nabla f(x^k)$. This will always be *orthogonal* to the search direction of the previous iteration, d^{k-1}, because otherwise α_{k-1} would not have been the optimal step size. That is,

$$(d^k)^T d^{k-1} = 0 \tag{6.4.8}$$

In view of the orthogonality of successive search directions, the steepest descent method tends to traverse back and forth across the floor of the valley much like a sailboat tacking into the wind. Consequently, it does not move very far down the valley with each iteration.

6.5 CONJUGATE-GRADIENT METHOD

The slow convergence rate of the steepest descent method can be increased by choosing the search directions in a more sophisticated way. For example, as the solution progresses, linear combinations of previous search directions might be used to form new search directions. An effective approach of this form is the *conjugate-gradient method*. The first step in the conjugate-gradient method is a steepest descent step. That is, starting with an initial guess x^0, set $d^0 = -g^0$ where $g^k = \nabla f(x^k)$. Next, for $k = 0$ compute

$$x^{k+1} = x^k + \alpha_k d^k \tag{6.5.1}$$

Here α_k is the optimal step length obtained by performing a line search along direction, d^k, starting at point x^k. To generate the kth search direction, d^k, for $1 \le k < n$, we form the following linear combination of g^k and d^{k-1}.

$$\beta_k = \frac{(g^k)^T g^k}{(g^{k-1})^T g^{k-1}} \tag{6.5.2a}$$

$$d^k = -g^k + \beta_k d^{k-1} \tag{6.5.2b}$$

Note that computing d^k for the conjugate-gradient method involves only slightly more effort than computing d^k for the steepest descent method. Suppose the function $f(x)$ is quadratic with Hessian matrix H as in Equation (6.2.7). Then the first n search directions of the conjugate-gradient method can be shown to be H-orthogonal or *conjugate* in the following sense.

$$(d^k)^T H d^j = 0 \quad , \quad k \neq j \tag{6.5.3}$$

When the objective function f is quadratic with a positive-definite Hessian matrix, the conjugate-gradient method converges to the minimum x^* in, at most, n iterations (see Luenberger, 1973). Of course, in many cases, $f(x)$ is not quadratic. For the more general case, the following adaptation of the conjugate-gradient method developed by Fletcher and Reeves can be used. Like the steepest descent method, it terminates when the error criterion in Equation (6.4.4) is satisfied or a maximum number of iterations, m, is performed.,

Alg. 6.5.1 Fletcher-Reeves Method

1. Pick $\varepsilon > 0, m > 0$, and $x^0 \in R^n$. Set $j = 0$ and $k = 0$.
2. Do

   {

   (a) Compute $g^k = \nabla f(x^k)$ and $\gamma = \|g^k\|$

   (b) If $(k < 1)$ set $d^k = -g^k$, else compute

   {

   $$\beta_k = \frac{(g^k)^T g^k}{(g^{k-1})^T g^{k-1}}$$

   $$d^k = -g^k + \beta_k d^{k-1}$$

   }

   (c) Find $\alpha_k > 0$ to minimize $F(\alpha) = f(x^k + \alpha d^k)$

   (d) Compute $x^{k+1} = x^k + \alpha_k d^k$

   (e) Set $k = k + 1$ and $j = j + 1$

   (f) If $(k = n)$, set $k = 0$ and $x^0 = x^n$

   }
3. While $(j \le m)$ and $(\gamma \ge \varepsilon)$ and $(\|x^k - x^{k-1}\| \ge \varepsilon_M \|x^k\|)$

The one-dimensional minimization in step (2c) of Alg. 6.5.1 can be performed using any line search technique. Notice that after n iterations, the iterative procedure is reset in step (2f), which ensures that every nth step is a steepest descent step. This is important for nonquadratic objective functions. Steady progress toward a local minimum of $f(x)$ is assured because the value of the objective function is guaranteed to decrease every nth step and will not increase during the intervening $n - 1$ steps. The steepest descent step is sometimes called a *spacer* step (Luenberger, 1973). A detailed discussion of conjugate direction methods in general can be found in Polak (1971).

EXAMPLE 6.5.1 Fletcher-Reeves Method

To illustrate the use of the Fletcher-Reeves conjugate-gradient method to solve an unconstrained optimization problem, consider the following objective function of dimension $n = 2$.

$$f(x) = (x_1 - x_2)^4 + (x_1^2 + x_2 - 2)^2 + (x_1 x_2 - 1)^2$$

This is a nonquadratic (fourth-order) objective function that has a minimum of $f(x^*) = 0$ at $x^* = [1, 1]^T$. Example program *e651* on the distribution CD applies Alg. 6.5.1 with an initial guess of $x^0 = 0$ and an error tolerance of $\varepsilon = 10^{-5}$. In this case, the conjugate-gradient algorithm converged in nine iterations and used 279 scalar function evaluations to compute

$$x = \begin{bmatrix} 0.9999682 \\ 1.0000514 \end{bmatrix}$$

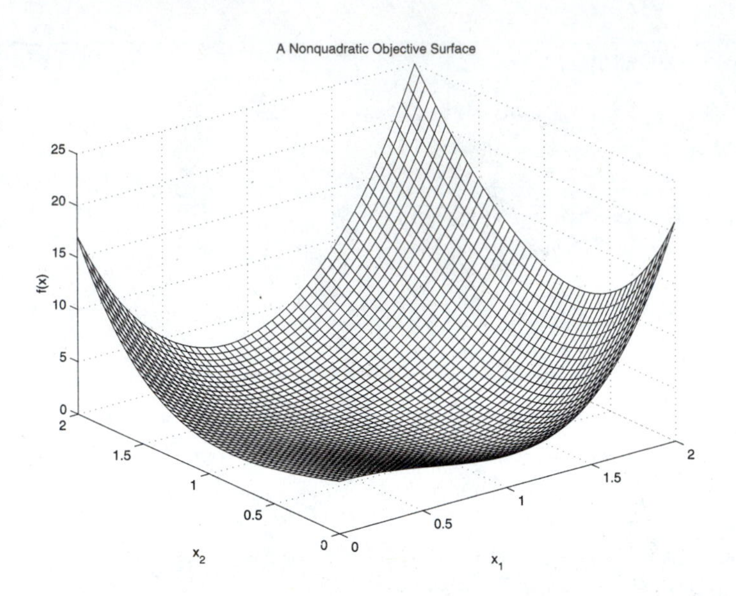

FIGURE 6.5.1 Nonquadratic Objective Surface

This corresponds to an objective function value of $f(x) = 0.0000000$. A plot of the objective surface $f(x)$ is shown in Figure 6.5.1.

If the conjugate-gradient method is compared with the steepest descent method by applying both techniques to the objective function in Example 6.4.1, we find that the conjugate-gradient method finds the optimal x in only 9 iterations, compared to 302 iterations of the steepest descent method. Consequently, an increase in speed of more than an order of magnitude was observed in this case. Of course, for problems that do not have an ill-conditioned Hessian matrix, the difference is less dramatic.

6.6 QUASI-NEWTON METHODS

Recall that if x^* is a local minimum of the objective function $f(x)$, then a necessary condition that x^* must satisfy is $\nabla f(x^*) = 0$ where f is the gradient vector. If $f(x)$ is twice differentiable, we can approximate the gradient at x^{k+1} by the first two terms of its Taylor series expansion about the point x^k.

$$\nabla f(x^{k+1}) \approx \nabla f(x^k) + H(x^k)(x^{k+1} - x^k) \qquad (6.6.1)$$

Here the $n \times n$ matrix $H(x)$ is the Hessian matrix of second partial derivatives of $f(x)$ with respect to the components of x as defined in Equation (6.2.4). If the objective function $f(x)$ is quadratic, then the approximation in Equation (6.6.1) is exact. Setting $\nabla f(x^{k+1}) = 0$ in (6.6.1) and solving for x^{k+1} yields

$$\boxed{x^{k+1} = x^k - H^{-1}(x^k)\nabla f(x^k) \quad , \quad k \geq 0} \qquad (6.6.2)$$

By comparing with Equation (5.8.3), we recognize (6.6.2) as *Newton's method* for solving the n-dimensional nonlinear algebraic system $\nabla f(x) = 0$. If the initial guess x^0 is sufficiently close to a local minimum x^*, then the sequence of estimates in (6.6.2) will exhibit *quadratic* convergence to x^*. Indeed, if $f(x)$ is a quadratic function with a positive-definite Hessian matrix, then Newton's method converges to x^* in a single step! Unfortunately, there are some practical drawbacks to Newton's method as formulated in (6.6.2). To begin with, a substantial amount of computational effort is required at each iteration to compute and invert the $n \times n$ Hessian matrix $H(x^k)$. It is a simple matter to eliminate the computation of the inverse. If $\Delta x^k = x^{k+1} - x^k$ denotes the *change* in x at step k, then (6.6.2) can be rewritten as

$$H(x^k)\Delta x^k = -\nabla f(x^k) \quad , \quad k \geq 0 \tag{6.6.3}$$

Consequently, it is only necessary to solve a linear algebraic system at each iteration rather than compute an explicit inverse. Recall that this reduces the number of FLOPs for large n from about $4n^3/3$ to about $n^3/3$. However, there remains the need to evaluate the n^2 components of $H(x^k)$, and this can be a computationally intensive task depending on the complexity of $f(x)$.

A more serious practical drawback of Newton's method is that it is just as likely to converge to a local maximum of $f(x)$ as a local minimum if x^0 does not start out sufficiently close to a local minimum. Recall from Equation (6.4.2) that a search direction, d^k, is a decent direction leading to a decrease in $f(x)$ if $(d^k)^T \nabla f(x^k) < 0$. The search direction for Newton's method is $d^k = \Delta x^k$. Multiplying both sides of (6.6.3) on the left by $(\Delta x^k)^T$, we find that the kth Newton step generates a decrease in $f(x)$ if

$$(\Delta x^k)^T H(x^k)\Delta x^k > 0 \tag{6.6.4}$$

That is, Newton's method converges to a local minimum if the Hessian matrix $H(x^k)$ is *positive-definite*. Although $H(x^k)$ will be positive-definite near a strict local minimum, there is no guarantee that it will remain so far from the local minimum.

Both the computational drawback and the convergence question can be addressed by modifying Newton's method. The basic idea, first developed by Davidon, is to construct a sequence of positive-definite approximations, Q_k, to the inverse of the Hessian matrix.

$$Q_k \approx H^{-1}(x^k) \quad , \quad k \geq 0 \tag{6.6.5}$$

The inverse of the Hessian matrix is then replaced by the increasingly accurate approximation Q_k. Since Q_k only approximates $H^{-1}(x^k)$, we also replace the unit step length in Equation (6.6.2) by a step of length $\alpha_k > 0$ obtained from a line search along $d = -Q_k \nabla f(x^k)$. To simplify the subsequent notation, let $g^k = \nabla f(x^k)$. This leads to the following modification of Newton's method, called a *quasi-Newton method*.

$$x^{k+1} = x^k - \alpha_k Q_k g^k \quad , \quad k \geq 0 \tag{6.6.6}$$

It remains to determine the $n \times n$ approximations Q_k to the inverse of the Hessian matrix. First, note that if the symmetric matrix Q_k is positive-definite, then $d^k = -Q_k g^k$ is a descent direction because $-(g^k)^T Q_k g^k < 0$. Furthermore, if Q_k is to approximate the inverse of the Hessian matrix, then it should satisfy Equation (6.6.1).

Let $\Delta g^k = g^{k+1} - g^k$ denote the *change* in the gradient vector at step k. Then, for Q_k to satisfy (6.6.1), we have

$$Q_k \Delta x^k = \Delta g^k \tag{6.6.7}$$

Davidon, Fletcher, and Powell developed an iterative formulation of Q_k that satisfies Equation (6.6.7) for $1 \le k \le n$. They start with a positive-definite initial guess Q_0. For example, $Q_0 = I$. Subsequent approximations are then formulated as follows where $r^k = Q_k \Delta g^k$.

$$Q_{k+1} = Q_k - \frac{r^k (r^k)^T}{(r^k)^T \Delta g^k} + \frac{\Delta x^k (\Delta x^k)^T}{(\Delta x^k)^T \Delta g^k} \quad , \quad 0 \le k < n \tag{6.6.8}$$

If Q_k is a symmetric positive-definite matrix, then so is Q_{k+1} (see Luenberger, 1973). Consequently, the search directions $d^k = -Q_k g^k$ are descent directions as long as Q_0 is a symmetric positive-definite matrix. For the special case when the objective function $f(x)$ is quadratic with Hessian matrix H, the approximations in Equation (6.6.8) can be shown to converge to the exact inverse of H in n steps.

$$Q_n = H^{-1} \tag{6.6.9}$$

When $f(x)$ is quadratic, it can also be shown that the direction vectors $d^k = -Q_k g^k$ are H-orthogonal or conjugate (Luenberger, 1973). Indeed, when $Q_0 = I$ and $f(x)$ is quadratic, the Davidon-Fletcher-Powell method is equivalent to the conjugate-gradient method. Consequently, when $f(x)$ is quadratic with a positive-definite Hessian matrix, the sequence in Equation (6.6.6) converges to the minimum x^* in, at most, n steps.

The following algorithm is an implementation of the Davidon-Fletcher-Powell method. It terminates when the criterion in Equation (6.4.4) is satisfied or a maximum number of iterations, m, is performed.

Alg. 6.6.1 **Davidon-Fletcher-Powell Method**

1. Pick $\varepsilon > 0, m > 0$, and $x^0 \in R^n$. Set $Q_0 = I, j = 0$ and $k = 0$. Compute $g^0 = \nabla f(x^0)$.
2. Do

   {

   (a) Find $\alpha_k > 0$ to minimize $F(\alpha) = f(x^k - \alpha Q_k g^k)$
   (b) Compute $x^{k+1} = x^k - \alpha_k Q_k g^k$ and $\Delta x^k = x^{k+1} - x^k$
   (c) Compute $g^{k+1} = \nabla f(x^{k+1})$, $\gamma = \|g^{k+1}\|$ and $\Delta g^k = g^{k+1} - g^k$
   (d) Compute $r^k = Q_k \Delta g^k$ and

   $$Q_{k+1} = Q_k - \frac{r^k (r^k)^T}{(r^k)^T \Delta g^k} + \frac{\Delta x^k (\Delta x^k)^T}{(\Delta x^k)^T \Delta g^k}$$

   (e) Set $k = k + 1$ and $j = j + 1$
   (f) If $(k = n)$, set $k = 0$, $x^0 = x^n$, and $g^0 = g^n$

   }
3. While $(j \le m)$ and $(\gamma \ge \varepsilon)$ and $(\|x^k - x^{k-1}\| \ge \varepsilon_M \|x^k\|)$

The one-dimensional minimization in step (2a) of Alg. 6.6.1 can be performed using any line search technique. Every nth iteration, the iterative procedure is reset in step (2f). This ensures that every nth step is a steepest descent step, a consideration that is important when $f(x)$ is nonquadratic. Continuous progress toward a local minimum of $f(x)$ is assured because the value of the objective function is guaranteed to decrease every nth step and will not increase during the intervening $n - 1$ steps.

The Davidon-Fletcher-Powell, or *DFP,* method in Alg. 6.6.1 is an effective method for solving unconstrained optimization problems. However, experience shows that the performance of the method is sensitive to the accuracy of the line search in step (2a). This sensitivity can be reduced and the algorithm made more robust by scaling the kth estimate of the inverse of the Hessian in step (2d) by β_k where

$$\beta_k = \frac{(\Delta x^k)^T \Delta g^k}{(r^k)^T \Delta g^k} \tag{6.6.10}$$

When $f(x)$ is quadratic with Hessian matrix H, the effect of the scaling is to relocate the eigenvalues of $Q_k H$ such that they occur both below and above unity (Luenberger, 1973). Replacing Q_k by $\beta_k Q_k$ in the update formula generates the following *self-scaled* version of Q_{k+1}.

$$Q_{k+1} = \beta_k \left(Q_k - \frac{r^k (r^k)^T}{(r^k)^T \Delta g^k} \right) + \frac{\Delta x^k (\Delta x^k)^T}{(\Delta x^k)^T \Delta g^k} \tag{6.6.11}$$

EXAMPLE 6.6.1 Davidon-Fletcher-Powell Method

To illustrate the use of the Davidon-Fletcher-Powell method to solve an unconstrained optimization problem, consider the following objective function of dimension $n = 2$.

$$f(x) = \frac{(x_1 - 3)^8}{1 + (x_1 - 3)^8} + \frac{(x_2 - 3)^4}{1 + (x_2 - 3)^4}$$

This eighth-order rational polynomial has a minimum of $f(x^*) = 0$ at $x^* = [3,3]^T$. Example program *e661* on the distribution CD applies Alg. 6.6.1 with an initial guess of $x^0 = 0$ and an error tolerance of $\varepsilon = 2 \times 10^{-7}$. In this case, the Davidon-Fletcher-Powell algorithm converged in four iterations and used 151 scalar function evaluations to compute

$$x = \begin{bmatrix} 3.0196705 \\ 3.0004380 \end{bmatrix}$$

This produced a value of $f(x) = 0.0000000$. This is a relatively challenging objective function in that it is almost flat over several regions of the $x_1 x_2$ plane, including the neighborhood of x^*, as can be seen by the surface plot in Figure 6.6.1. Although the algorithm converged in just a few iterations, in order to get a reasonably accurate value for x, it was necessary to make the error tolerance quite small. Decreasing the error tolerance below $\varepsilon = 2 \times 10^{-7}$ does not improve the estimate of x^* because $f(x)$ is already zero using single-precision arithmetic. In order to get further improvements in the accuracy of x, higher precision arithmetic is required.

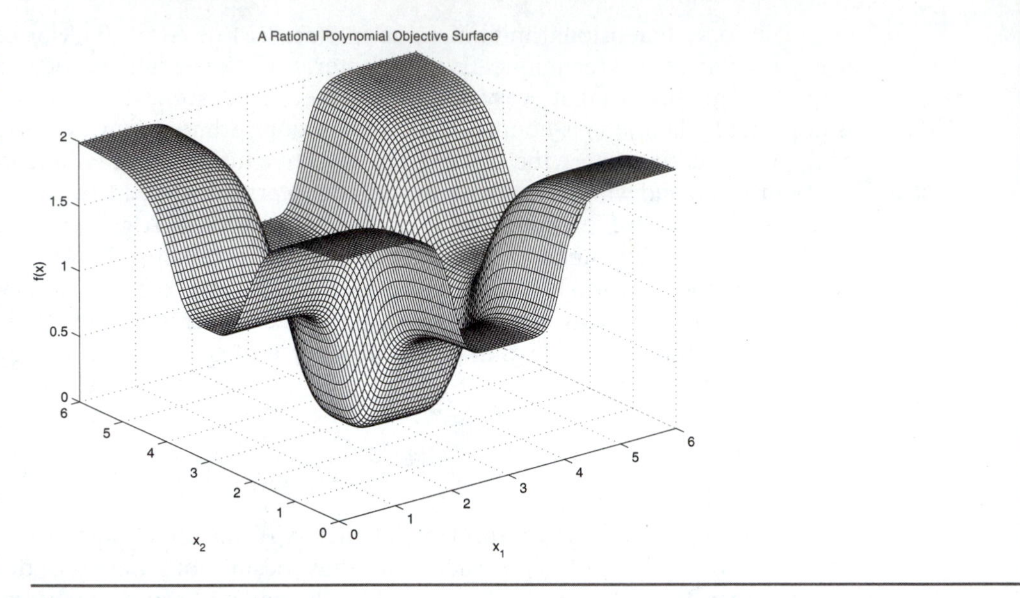

FIGURE 6.6.1 A Rational Polynomial Objective Surface

All of the objective functions considered for unconstrained optimizations thus far have been of dimension $n = 2$. There is nothing in the steepest descent, conjugate-gradient, or quasi-Newton methods that restricts n to 2. Instead, two-dimensional objective functions were used merely to facilitate *visualization* of the objective as a surface. Instances of higher-dimensional optimizations are examined later and are also included in the problems at the end of the chapter.

6.7 PENALTY FUNCTIONS

At this point, we turn our attention to the more general problem of *constrained* optimization. That is, suppose x is an $n \times 1$ vector of design parameters. The goal is to find an x^* that minimizes the objective function $f(x)$ while satisfying a set of equality and inequality constraints.

$$\begin{aligned} \text{minimize:} \quad & f(x) \\ \text{subject to:} \quad & p(x) = 0 \\ & q(x) \geq 0 \end{aligned}$$

Here, $p(x)$ is an $r \times 1$ vector that specifies r equality constraints, and $q(x)$ is an $s \times 1$ vector that specifies s inequality constraints. Recall that an x which satisfies the constraints is said to be a *feasible* vector. Thus, the problem is to find a feasible x that minimizes the objective function $f(x)$. In some instances, the equality constraints are sufficiently simple that $p(x) = 0$ can be solved for one or more components of x in terms of the remaining components. When this is possible, the isolated components of x can be substituted into $f(x)$ and $q(x)$. This not only reduces the number of constraints, but also lowers the dimension of the problem.

The constrained optimization problem is considerably more difficult to solve than the unconstrained problem. Although several general approaches are available to attack the problem (see, e.g., Luenberger, 1973; Zangwill, 1969), we restrict our investigation to a single technique, the method of *penalty functions*. The virtue of penalty functions is that they allow us to build directly on our previous results by converting constrained optimization problems into equivalent unconstrained problems. The basic idea is to make the constraints *implicit* by adding terms to the objective function. These terms are zero when the constraints are satisfied but become large when the constraints are violated. Consequently, the minimization process itself tends to seek out feasible solutions because solutions that violate the constraints are *penalized* in the sense that they incur a large cost in the generalized objective function.

To construct suitable penalty terms, we begin with the equality constraints. Suppose $P(x)$ denotes a penalty function associated with the constraint $p(x) = 0$. The basic requirement on $P(x)$ is that $P(x) = 0$ when $p(x) = 0$ and $P(x) > 0$ when $p(x) \neq 0$. This way, feasible solutions are not penalized, whereas unfeasible solutions that violate $p(x) = 0$ are penalized. A simple $P(x)$ that satisfies these properties is $P(x) = p^T(x)p(x)$, which can be written explicitly as

$$P(x) = \sum_{k=1}^{r} p_k^2(x) \tag{6.7.1}$$

Notice that not only is $P(x)$ positive when $p(x) \neq 0$, but the value of $P(x)$ grows as the violation of the constraint becomes more severe. Of course, there are many other functions that also qualify as penalty functions. For example, the terms in Equation (6.7.1) can be replaced by $|p_k(x)|$. This tends to penalize minor violations of the constraint more severely, but the gradient of $P(x)$ is not continuous.

Penalty functions for the inequality constraints can be constructed in a similar manner. If $Q(x)$ is a penalty function associated with the constraint, $q(x) \geq 0$, then it is required that $Q(x) = 0$ when $q(x) \geq 0$ and $Q(x) > 0$ otherwise. One such penalty function which satisfies these properties is

$$Q(x) = \sum_{k=1}^{s} \min{}^2\{0, q_k(x)\} \tag{6.7.2}$$

Again, notice that not only is $Q(x)$ positive when the constraint is violated, but the value of $Q(x)$ increases as the violation becomes more severe. Once penalty functions are constructed for the equality and inequality constraints, they can then be incorporated into a *generalized objective function*, $F(x)$, as follows.

$$F(x) \triangleq f(x) + \mu[P(x) + Q(x)] \tag{6.7.3}$$

Here the *penalty parameter* $\mu > 0$ controls the relative cost of violating the constraints. The original constrained optimization problem is then replaced by the following parameterized unconstrained optimization problem.

$$\begin{aligned} \text{minimize:} \quad & f(x) + \mu[P(x) + Q(x)] \\ \text{subject to:} \quad & x \in R^n \end{aligned}$$

As the penalty parameter $\mu > 0$ increases, solutions to the unconstrained problem become increasingly good approximations to the solution to the original constrained optimization problem. However, large values for μ can cause the generalized objective function and its gradient to become very large, particularly if the initial guess, x^0, violates the constraints.

EXAMPLE 6.7.1 Penalty Functions

To illustrate the formulation of a penalty function, consider the following two-dimensional constrained optimization problem.

$$\begin{aligned} \text{minimize:} \quad & 3x_1^2 + 2x_2^2 \\ \text{subject to:} \quad & x_1 + x_2 - 4 = 0 \\ & -3 \le x_1 - x_2 \le 2 \end{aligned}$$

The first thing to notice is that the equality constraint is sufficiently simple that one variable can be solved for in terms of the other. For example, $x_2 = 4 - x_1$. Substituting this into the objective function and the remaining constraints, we get the following simpler one-dimensional problem:

$$\begin{aligned} \text{minimize:} \quad & 3x_1^2 + 2(4 - x_1)^2 \\ \text{subject to:} \quad & 0.5 \le x_1 \le 3 \end{aligned}$$

Next, the double-sided inequality can be converted into the standard form of $q(x) \ge 0$ by writing it as two separate constraints.

$$q_1(x) = x_1 - 0.5$$
$$q_2(x) = 3 - x_1$$

If we use Equation (6.7.2) as a penalty function, then this yields

$$Q(x) = \min{}^2\{0, x_1 - 0.5\} + \min{}^2\{0, 3 - x_1\}$$

A graph of the penalty term, $\mu Q(x)$, for various values of μ is shown in Figure 6.7.1. It is clear that there is no penalty when x_1 is feasible, and the penalty becomes more severe both as μ is increased and as x_1 strays outside the feasible interval $[0.5, 3]$.

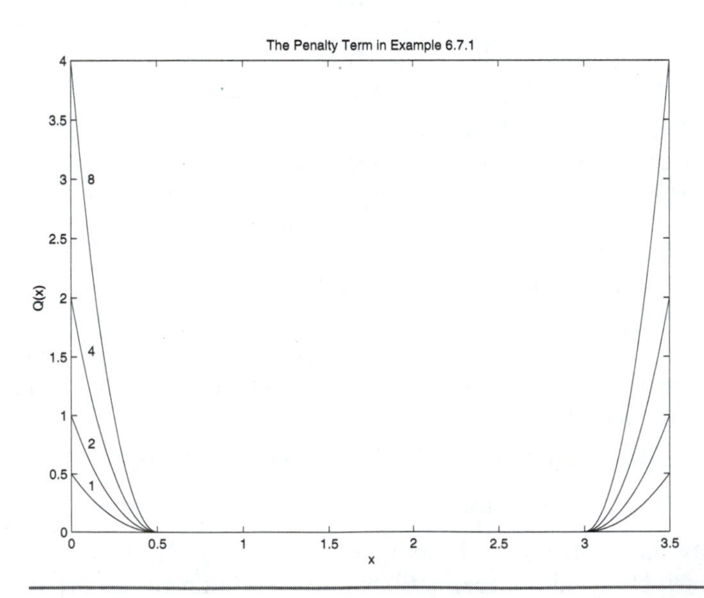

FIGURE 6.7.1 The Penalty Term in Example 6.7.1 for Four Values of μ

Closely related to the notion of a penalty function is the concept of a barrier function. A *barrier function* is a function whose value is near zero in the interior of the feasible region, but goes to infinity (forms a barrier) at the boundary of the feasible region. Barrier functions have the virtue that solutions produced with them are guaranteed to be feasible. However, they suffer from two drawbacks. The first drawback is that the initial guess, x^0, must itself be feasible so that it lies inside the barrier. Choosing a feasible x^0 can be a nontrivial task if the constraints are complex. The second drawback is that a barrier function is not defined when the interior of the feasible region associated with a constraint is empty.

The solution computed by minimizing $F(x)$ in Equation (6.7.3) often violates one or more of the constraints, but by a small amount. Increasing μ decreases the degree of the violation, but it does not completely remove it. If it is crucial that an inequality constraint be satisfied exactly, then a common approach is to *artificially* tighten the original constraint slightly. When the artificial constraints are violated by a small amount, the original constraint still may be satisfied. For example, the penalty function $Q(x)$ in (6.7.2) can be modified as follows for some small $\varepsilon > 0$.

$$Q_\varepsilon(x) = \sum_{k=1}^{s} \min^2\{0, q_k(x) - \varepsilon\} \tag{6.7.4}$$

This will tend to produce a *suboptimal* solution of the original problem, but this may be acceptable when complete satisfaction of $q(x) \geq 0$ is essential.

EXAMPLE 6.7.2 Constrained Optimization

As an illustration of the use of penalty functions to perform a constrained optimization, consider the following objective function of dimension $n = 3$.

$$f(x) = (x_1 - 1)^2 + (x_2 - 2)^2 + (x_3 - 3)^2$$

Clearly, the *unconstrained* optimum is $x^* = [1,2,3]^T$. Suppose the following equality and inequality constraints are imposed.

$$p(x) = x_1 + x_2 - 2 = 0$$

$$q(x) = x_2 - x_3 - 3 \geq 0$$

This problem can be solved using the penalty functions in Equations (6.7.1) and (6.7.2). Example program *e672* on the distribution CD applies the Davidon-Fletcher-Powell (DFP) method to minimize the generalized objective function using an error tolerance of $\varepsilon = 10^{-4}$ and an initial guess of $x^0 = 0$. A penalty parameter of $\mu = 500$ was used, and four separate cases were solved as summarized in Table 6.7.1. The first column in Table 6.7.1 is the type of

TABLE 6.7.1 Constrained Optimization

Constraints	*k*	x^T			$f(x)$	$p(x)$	$q(x)$
None	2	1.000000	2.000000	3.000000	0.000000	1.000000	−4.000000
Equality	7	0.500330	1.500677	3.001046	0.498995	0.001006	−4.500370
Inequality	3	1.000578	3.999435	1.003402	7.984148	3.000013	−0.003966
Both	16	−0.526733	2.529580	−0.463494	14.607163	0.002846	−0.006926

constraint, the second is the number of iterations performed k, the next three columns are the components of x, the sixth column is the value of the objective, $f(x)$, the seventh is the value of the equality constraint $p(x)$, and the last column is the value of the inequality constraint $q(x)$. Note how $f(x)$ increases as constraints on x are added. Although neither constraint is satisfied exactly, they are violated by only a small amount. Increasing μ reduces the degree of the violation. Alternatively, the modified penalty function in Equation (6.7.4) might be used.

6.8 SIMULATED ANNEALING

All of the optimization methods discussed thus far are aggressive in the sense that they attempt to move rapidly toward a minimum using a local approximation of the objective function. This has the advantage that it seeks out a minimum quickly without requiring added computational effort. Unfortunately, the minimum found by these methods is often a *local* minimum. One strategy for attempting to find a *global* minimum is to solve the optimization problem repeatedly, starting from different initial guesses. The best local minimum found can then be taken as an approximation to the global minimum. However, the problem of determining a suitable sequence of initial guesses is a daunting task. For example, there is no guarantee that two different initial guesses will converge to two different local minima. Moreover, an exhaustive search of initial guesses is computationally prohibitive, even for small values of n.

There is an interesting alternative approach that is based on an analogy from the field of thermodynamics. When metal is raised to a temperature above its melting point, its atoms generate highly-excited random motion. As the temperature is gradually lowered, the motion of the atoms becomes less violent as they lose thermal mobility and decay to lower and lower energy states. Eventually, the atoms settle into a global energy state minimum, a single highly-structured crystal. This process is known as annealing, and optimization techniques based on it are referred to as *simulated annealing* methods. The distribution of energy states in annealed metal can be described by the *Boltzmann distribution.*

$$p(e) = \alpha \exp\left(\frac{-e}{kT}\right) \tag{6.8.1}$$

Here, $e \geq 0$ denotes the energy level, k is the Boltzmann constant, and T is temperature in degrees Kelvin. When $\alpha = 1/(kT)$, the area under $p(e)$ is unity, and $p(e)$ can be regarded as a probability density function. A graph showing the Boltzmann distribution for several values of kT is shown in Figure 6.8.1. Notice that at high temperatures, $p(e)$ is relatively flat over a large range, which means that high energy states are almost as likely as low energy states. However, at low temperatures $p(e)$ falls off rapidly, which means that it is highly unlikely that the system will exist in a high energy state when T is small.

Annealing can be regarded as a natural physical optimization process (Metropolis et al., 1953). Suppose we associate the $n \times 1$ vector x with the configuration or state of the atoms, and the objective function $f(x)$ with the energy level. Then the effect of annealing is to perform the following unconstrained optimization.

$$\begin{aligned} \text{minimize:} \quad & f(x) \\ \text{subject to:} \quad & x \in R^n \end{aligned}$$

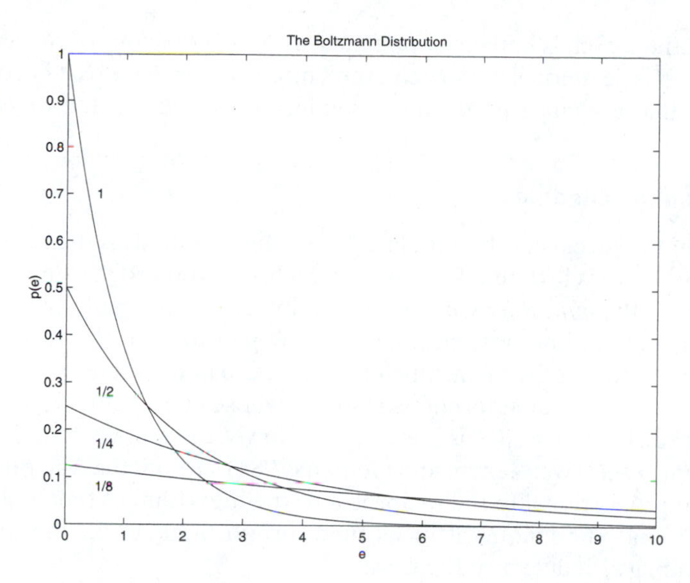

FIGURE 6.8.1 The Boltzmann Distribution for Four Values of kT

Let T be an artificial temperature and k be a constant analogous to Boltzmann's constant. We start out with a large value for T and an initial guess, x^0. Let Δx denote a random change in x. For example, the components of Δx might be distributed about a mean of zero with the following Gaussian distribution.

$$p(\Delta x_j) = \frac{1}{T\sqrt{2\pi}} \exp\left(\frac{-\Delta x_j^2}{2T^2}\right) \quad , \quad 1 \le j \le n \qquad (6.8.2)$$

Since the standard deviation of $p(\Delta x_j)$ is T, the random changes will initially be large but will gradually decrease in size as the artificial temperature T is lowered. Next, let Δf denote the change in the value of the objective function produced by Δx. That is,

$$\Delta f \triangleq f(x^0 + \Delta x) - f(x^0) \qquad (6.8.3)$$

If $\Delta f < 0$, this represents an improvement in the value of the objective, so Δx is accepted and used to produce a new point $x^1 = x^0 + \Delta x$. However, if $\Delta f \ge 0$, this corresponds to a lack of improvement or a deterioration in the value of the objective, but we *may* want to use Δx anyway because it could provide us with a means of escaping from a local minimum. We compute the probability of using Δx with the Boltzmann distribution.

$$p(\Delta f) = \frac{1}{kT} \exp\left(\frac{-\Delta f}{kT}\right) \qquad (6.8.4)$$

To determine whether Δx is to be used, select a random number z uniformly distributed over the interval $[0, 1/(kT)]$. If $p(\Delta f) > z$, set $x^1 = x^0 + \Delta x$. Otherwise, generate a new random step Δx and repeat the process until an acceptable Δx is found. The entire process is then repeated starting from x^1. As the iterations continue, the value of the artificial temperature T is gradually lowered. Initially, for large values of T, there is a relatively high probability that a change producing a $\Delta f > 0$ will be accepted.

Consequently, the search is initially *uninhibited* with many regions of R^n being visited, if only briefly. As T is decreased, the search grows more conservative as it becomes increasingly unlikely that a change producing a significant increase in the objective will be accepted.

6.8.1 Annealing Schedules

A key factor in the success of the simulated annealing methods is the rate at which the artificial temperature is lowered. The rate at which T is lowered starting from an initial value, T_0, is called the *annealing schedule*. If T is lowered too rapidly, then a global minimum will not be found because of inadequate exploration of the local minima. The physical analogy to this is the quenching process where liquid metal is rapidly cooled to produce a polycrystalline or amorphous material whose energy state is higher than that of a single crystal. Of course, if T is lowered too slowly, the already long computational time will become excessive. Geman and Geman (1984) showed that the rate of decrease in T should be proportional to the reciprocal of the logarithm of time if a global minimum is to be found. If t denotes artificial time (proportional to the iteration number), then the annealing schedule is of the form

$$T(t) = \frac{T_0}{1 + \ln(1 + t)} \tag{6.8.5}$$

This represents a very gradual reduction in T and therefore can result in substantial computational effort before a global minimum is found. Of course, this should come as no surprise because locating a global minimum is a much more difficult task than finding the nearest local minimum.

The speed of simulated annealing can be improved by modifying the basic method. Szu and Hartley (1987) have proposed replacing the Boltzmann distribution in Equation (6.8.4) with the following *Cauchy distribution*.

$$p(\Delta f) = \frac{2T}{\pi(T^2 + \Delta f^2)} \tag{6.8.6}$$

This distribution falls off much more gradually with Δf than the exponential Boltzmann distribution. As result, larger values of Δf are more likely to be accepted for a given T. This in turn means that the rate of annealing can be increased. In particular, the annealing schedule can be made proportional to the reciprocal of time, rather than the reciprocal of the logarithm of time.

$$T(t) = \frac{T_0}{1 + t} \tag{6.8.7}$$

Although $T(t)$ in Equation (6.8.7) decreases quite slowly, it is still much faster than the Boltzmann annealing schedule in (6.8.5). When the Cauchy distribution is used, $p(\Delta f)$ is computed using (6.8.6), the random number z is uniformly distributed over $[0, 2/(\pi T)]$, and T is updated using (6.8.7).

The speed of convergence of the basic simulated annealing technique can also be accelerated by making use of the notion of specific heat. In physics, the specific heat of a material is defined as $\sigma = (1/m)\ dQ/dt$ where m is the mass, T is the temperature, and

Q is the heat energy needed to bring about a change in temperature. Physical materials often go through phase changes at certain critical temperatures. Phase changes are characterized by abrupt changes in the value of the specific heat. For example, when water is cooled, it undergoes a phase change as it turns to ice, at which point the specific heat abruptly drops from unity to about 0.55. Phase changes represent changes in the energy level. Since the objective function is analogous to the system energy, we can define *artificial specific heat* as

$$\sigma(T) \triangleq \frac{df(x)}{dt} \tag{6.8.8}$$

When T is far from a critical temperature, the artificial specific heat changes slowly with T, in which case T can be reduced relatively quickly. A rapid change in $\sigma(T)$ indicates that a critical phase transition temperature is being approached, which means that T must be changed very slowly in order to successfully negotiate the phase transition to a lower energy state.

We can interpret critical temperatures in terms of optimization in the following way. Suppose there are two adjacent local minima a and b for a one-dimensional objective function, as shown in Figure 6.8.2. The change in energy required to go from a to b is $\Delta f_1 = f(c) - f(a)$ where c is a local maximum between a and b. Similarly, the change in energy required to go from b back to a is $\Delta f_2 = f(c) - f(b)$. If $f(b) < f(a)$, then there will be a small range of temperatures T over which the smaller transition from a to b is likely, whereas the larger change from b back to a is unlikely. It is at this point that the process is likely to achieve a transition from $f(a)$ to $f(b)$, thereby realizing a net decrease in energy of $f(a) - f(b)$.

6.8.2 Constrained Optimization

Another way to speed up the convergence of simulated annealing is to exploit information that may be available concerning lower and upper bounds on x and $f(x)$. Suppose the solution x is known to lie in the following rectangular region.

$$S = \{x \in R^n \mid a \leq x \leq b\} \tag{6.8.9}$$

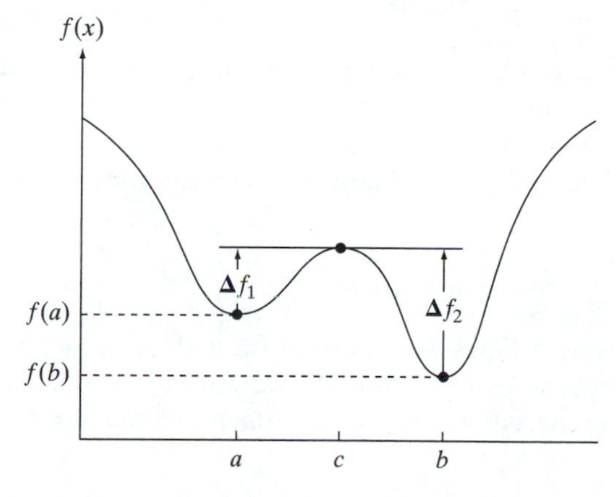

FIGURE 6.8.2 Transitions Between Local Minima

Thus, there are lower and upper bounds on each component of x. Next, suppose that similar lower and upper bounds are available for the objective function $f(x)$. That is,

$$f_0 \le f(x) \le f_1 \quad , \quad x \in S \tag{6.8.10}$$

Assuming that $f(x)$ is a *continuous* function, finite values for f_0 and f_1 always exist because S is a closed bounded subset of R^n. The bounds in Equation (6.8.10) should be made as tight as possible. It may be possible, for example, to set $f_0 = f(x^*)$ where x^* is the global minimum. That is, in some cases, the optimal *value* of the objective may be known even though the optimal x^* at which it occurs is unknown.

Given the lower and upper bounds on $f(x)$, we can use the value of $f(x)$ itself to play a role analogous to artificial temperature. Suppose Δf is the change in $f(x)$ associated with the random step Δx as in (6.8.3). If $\Delta f < 0$ this represents a descent, so Δx is used to produce a new point $x^1 = x^0 + \Delta x$. When $\Delta f \ge 0$, we use the following Boltzmann type distribution where $\alpha > 0$.

$$\Delta F \triangleq f(x^0) - f_0 \tag{6.8.11a}$$

$$p(\Delta f) = \exp\left(\frac{-\alpha \Delta f}{\Delta F}\right) \tag{6.8.11b}$$

Notice that $\Delta F \ge 0$ is a rough measure of degree to which x^0 approximates the global minimum x^*. In particular, if $f_0 = f(x^*)$, then $\Delta F \to 0$ as $x \to x^*$. To determine if Δx is to be used, we generate a random number z uniformly distributed over the interval $[0, 1]$. If $p(\Delta f) > z$, then Δx is used to produce a new point $x^1 = x_0 + \Delta x$. Otherwise, a new random Δx is generated, and the process is repeated until an acceptable Δx is found. Observe that as $f(x^0)$ approaches the lower bound f_0, the exponent in Equation (6.8.11b) becomes increasingly negative, thereby reducing the probability that a Δx producing a $\Delta f > 0$ will be used.

To select the parameter α in (6.8.11b), suppose it is desired that there be an 0.5 probability of using a Δx that generates a $\Delta f \ge \gamma \Delta F$ where $0 < \gamma < 1$ is a constant whose value will be determined empirically. Then, setting $p(\gamma \Delta F) = 0.5$ and solving for α yields

$$\alpha = \frac{\ln(2)}{\gamma} \tag{6.8.12}$$

The scalar γ can be thought of as a *localization parameter* because as γ is decreased the scope of the search becomes more restricted.

We can also take advantage of the lower and upper bounds on x to devise a strategy for generating random steps Δx. Let d denote the maximum size of the feasible region S. That is,

$$d \triangleq \max_{k=1}^{n} \{b_k - a_k\} \tag{6.8.13}$$

Suppose we generate Δx using a Gaussian distribution with zero mean and standard deviation σ. As $f(x)$ approaches the lower limit f_0, we want σ to approach zero so the search can become localized, whereas when $f(x)$ approaches the upper limit f_1, we want

σ to approach γd so a more exhaustive search can be made. This can be achieved by generating a random Δx using the following Gaussian probability density function.

$$\sigma = \frac{\gamma d \Delta F}{f_1 - f_0} \tag{6.8.14a}$$

$$p(\Delta x) = \frac{1}{\sigma\sqrt{2\pi}} \exp\left(\frac{-\Delta x^2}{2\sigma^2}\right) \tag{6.8.14b}$$

Of course, some of the random steps generated in this manner produce an unfeasible $x^1 = x^0 + \Delta x$ that violates the constraints. When this happens, the offending components of Δx are clipped so as to satisfy their respective constraints.

The simulated annealing method can be terminated when $f(x^k) - f_0$ falls below a user-specified error tolerance $\varepsilon > 0$ or a maximum number of iterations m is performed. In general, the error tolerance must satisfy $\varepsilon > f(x^*) - f_0$. When $f_0 = f(x^*)$, this reduces to $\varepsilon > 0$, but for very small values of ε, the method can take a large number of iterations. Instead, it is more efficient to choose ε sufficiently small to get somewhere *near* the global minimum but sufficiently large to limit the number of iterations required. Then an efficient local search technique such as the DFP method, with penalty functions, can be employed to finish the search. The penalty function in Equation (6.7.2) can be used where the $2n$ inequality constraints, $q(x) \geq 0$, are as follows for $1 \leq k \leq n$:

$$q_k(x) = x_k - a_k \tag{6.8.15a}$$

$$q_{n+k}(x) = b_k - x_k \tag{6.8.15b}$$

The overall procedure then consists of random global search based on simulated annealing followed by an efficient local search. The local search can be terminated when the norm of the gradient of $f(x)$ falls below an error tolerance η or when a maximum number of iterations has been performed. The following algorithm is an implementation of this two-stage global optimization procedure.

Alg. 6.8.1 Simulated Annealing

1. Pick $m > 0, \varepsilon > 0, \eta > 0, 0 < \gamma < 1, \mu > 0$, and $x^0 \in R^n$.
2. Set $j = 0, \bar{x} = x^0$ and $\Delta F = f(x^0) - f_0$.
3. Do

   {

   (a) Compute $\sigma = \gamma d \Delta F/(f_1 - f_0)$.

   (b) For $i = 1$ to n do

   {

   (i) Generate a random Δx_i using a Gaussian distribution with zero mean and standard deviation σ.

   (ii) Clip Δx_i if needed using $\Delta x_i = \max[a_i - x_i^j, \min(b_i - x_i^j, \Delta x_i)]$

   }

   (c) Compute $\Delta f = f(x^0 + \Delta x) - f(x^0)$.

(d) If ($\Delta f < 0$) do
{
(i) Set $x^{j+1} = x^j + \Delta x$ and $j = j + 1$.
(ii) If $f(x^j) < f(\overline{x})$, set $\overline{x} = x^j$
}
(e) else do
{
(i) Compute $p(\Delta f) = \exp(-\alpha \Delta f / \Delta F)$ where $\alpha = \ln(2)/\gamma$.
(ii) Generate a random z uniformly distributed over $[0, 1]$.
(iii) If $p(\Delta f) > z$, set $x^{j+1} = x^j + \Delta x$ and $j = j + 1$.
}
(f) Compute $\Delta F = f(x^j) - f_0$ and $\delta = \Delta F/(f_1 - f_0)$
}

4. While ($\delta \geq \varepsilon$) and ($j \leq m$)
5. Form the generalized objective function $F(x) = f(x) + \mu Q(x)$ using (6.7.2) for $Q(x)$ and (6.8.15) for $q(x)$.
6. Apply Alg. 6.6.1 to minimize $F(x)$ starting from $\overline{x}$, using m, and $\varepsilon = \eta$.

The global search using simulated annealing is performed in steps (1) through (4), whereas the local search using the DFP method is performed in steps (5) and (6). Notice that the starting point for the local search is $\overline{x}$ rather than x^j where $\overline{x}$ is the value of x that produces the lowest $f(x)$ encountered thus far. This is done to guard against the possibility that the estimate of f_0 is too conservative, which means that ΔF remains relatively large. When this happens, the search does not become localized, so it could jump away from the global minimum just before terminating with $j = m$. Many other variations of simulated annealing are possible. (See, for example, Press, et al. (1992) or Wasserman (1989) for other approaches.) The latter reference uses simulated annealing techniques to train artificial neural networks to find the optimal weights for the inputs to the neurons.

EXAMPLE 6.8.1 Simulated Annealing

To illustrate the simulated annealing optimization technique, consider the following objective function.

$$f(x) = \frac{-\cos[\pi(x_1 - 3)] \cos[2\pi(x_2 - 2)]}{1 + (x_1 - 3)^2 + (x_2 - 2)^2}$$

This function has an infinite number of strict local minima in the $x_1 x_2$ plane. However, it is not difficult to obtain tight bounds on the value of $f(x)$. Each factor in the numerator is the interval $[-1, 1]$, so the numerator itself lies in this range. Since the denominator is in the range $[1, \infty)$, it follows that $f(x)$ is bounded from below and above as follows:

$$-1 \leq f(x) \leq 1$$

FIGURE 6.8.3
The Objective Surface

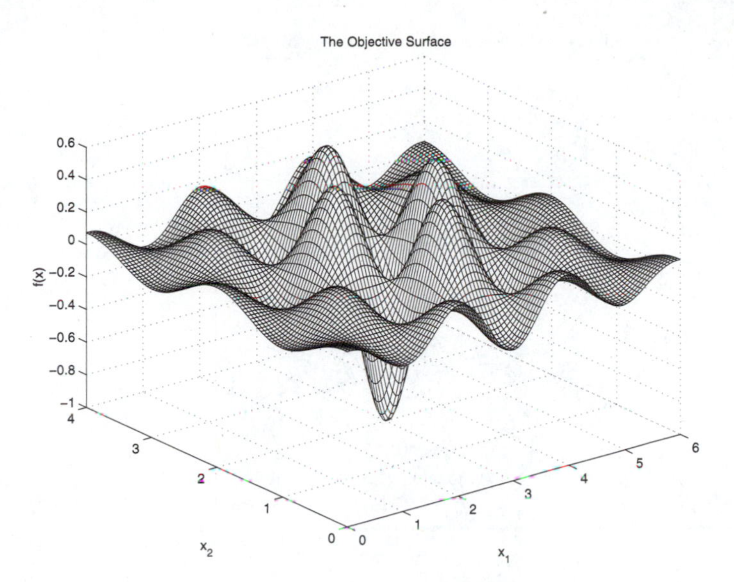

The numerator of $f(x)$ takes a value of -1 at the same point, $[3,2]^T$, that the denominator achieves its minimum value of unity. Thus, the global minimum for this problem occurs at $x^* = [3,2]^T$. A plot of the objective surface that includes the optimal point is shown in Figure 6.8.3.

In order to estimate x^* numerically, suppose we constrain each component of x to lie in an interval of length 20 centered at the origin. This restricts the scope of the search to 400 square units.

$$-10 \leq x_k \leq 10 \quad , \quad 1 \leq k \leq 2$$

Example program *e681* on the distribution CD applies Alg. 6.8.1 with an initial guess in the middle of the feasible region, $x^0 = 0$. The error tolerances used were $\varepsilon = 0.1$ and $\eta = 10^{-4}$, and the localization parameter was set to $\gamma = 0.2$. This resulted in the algorithm converging in 455 iterations and 533 scalar function evaluations. The simulated annealing part of the algorithm required 454 iterations to converge to

$$x = \begin{bmatrix} 2.8426495 \\ 1.9919055 \end{bmatrix}$$

$$f(x) = -0.8586850$$

The local search then took an additional two iterations to converge to the refined estimate

$$x = \begin{bmatrix} 3.0000777 \\ 1.9994377 \end{bmatrix}$$

$$f(x) = -0.9999981$$

The performance of the algorithm is somewhat sensitive to the value of the localization parameter, γ. As γ is made smaller, the procedure tends to spend more time exploring local minima, thus resulting in an increase in the number of iterations. For example, if a search is conducted over the region $0 \leq x_k \leq 5$, with $\gamma = 0.1$, the algorithm requires 245 iterations to

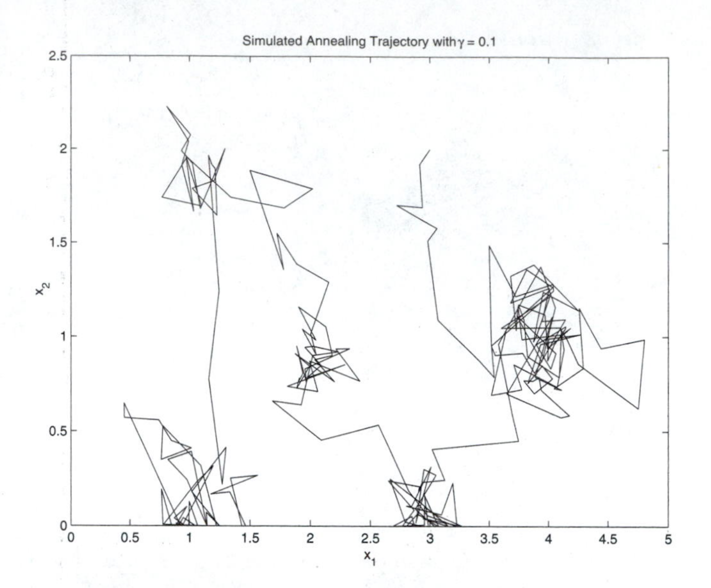

FIGURE 6.8.4 Simulated Annealing Trajectory with $\gamma = 0.1$

converge. An overhead view of the trajectory followed in this case is shown in Figure 6.8.4. Note how the search tends to pause at several points along the way before it moves on. This is in contrast to the case $\gamma = 0.2$, which is less localized and requires only 25 iterations to converge in this case.

6.9 APPLICATIONS

The following examples illustrate applications of optimization techniques using both MATLAB and C. The relevant MATLAB functions in the NLIB toolbox are described in Section 1.7 of Appendix 1, while the corresponding C functions in the NLIB library are described in Section 2.10 of Appendix 2.

6.9.1 Heat Exchanger: MATLAB

A popular use of optimization is to identify suitable values for the parameters of mathematical models based on measurements of a physical process. Suppose a heat exchanger is used to heat a solution of glycol with hot oil as shown in Figure 6.9.1 (Seborg et al., 1989). Let r denote the hot oil flow rate and let T denote the outlet temperature. Suppose that for a constant flow rate $r(t) = \alpha$, the corresponding constant steady-state outlet temperature is $T(t) = \beta$. Next, let u and y denote the *variations* in the flow rate and temperature, respectively, from their steady-state values. That is,

$$u(t) = r(t) - \alpha \qquad (6.9.1a)$$

$$y(t) = T(t) - \beta \qquad (6.9.1b)$$

Using the variational variables u and y, the heat exchanger can be modeled as a first order dynamic system with time delay as follows.

$$\frac{dy(t)}{dt} = \frac{1}{\tau}[au(t - T_d) - y(t)] \qquad (6.9.2)$$

FIGURE 6.9.1
A Glycol Hot Oil Heat Exchanger

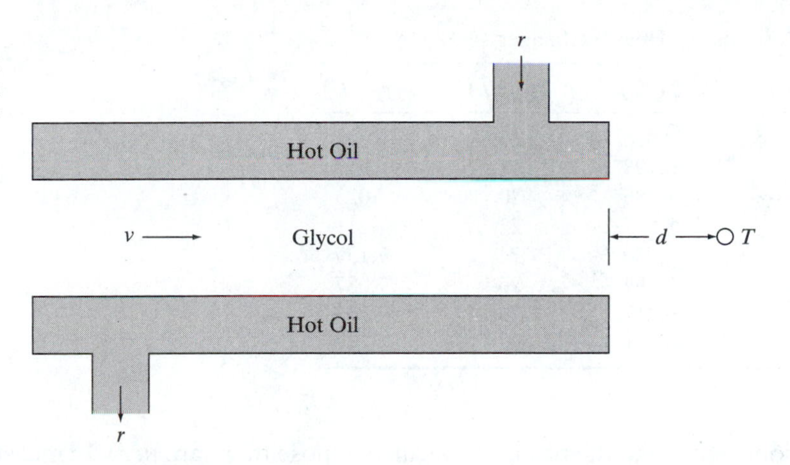

Here the model parameters are the gain a, the time constant τ, and the time delay T_d. The time delay arises because the thermocouple used to measure temperature is placed a distance d downstream from the outlet as shown in Figure 6.9.1. It is assumed that the dynamics of the thermocouple are fast in comparison with the dynamics of the heat exchanger. Let the parameters of the model be grouped into a 3×1 vector of unknowns.

$$x = [a \quad \tau \quad T_d]^T \tag{6.9.3}$$

Suppose there is an abrupt increase in the hot oil flow rate of one unit at time $t = 0$, which means $u(t) = 1$ for $t \geq 0$. Then the variation in the outlet temperature measured at the thermocouple predicted by the model will be as follows.

$$y(t, x) = \begin{cases} 0 & , \quad 0 \leq t \leq x_3 \\ x_1[1 - \exp(-(t - x_3)/x_2)] & , \quad x_3 < t < \infty \end{cases} \tag{6.9.4}$$

To test the effectiveness of the prediction, an experiment is performed in which the outlet temperature y_k is measured at various times t_k following the step change in the hot oil flow rate. If D denotes the set of m data points, then

$$D = \{(t_k, y_k) \mid 1 \leq k \leq m\} \tag{6.9.5}$$

The objective is to determine a suitable set of model parameters, x, such that the predicted temperature $y(t_k, x)$ matches the measured temperature y_k as closely as possible. A convenient measure of the difference between the predicted and measured values is

$$f(x) = \sum_{k=1}^{m} [y(t_k, x) - y_k]^2 \tag{6.9.6}$$

The problem of finding the best model then reduces to one of minimizing the objective function $f(x)$. This is essentially a curve fitting problem not unlike those discussed in Chapter 5. However, because the gradient $\nabla f(x)$ is not a linear function of x, this is an example of a *nonlinear* regression problem that must be solved using optimization

TABLE 6.9.1 Step Response of Heat Exchanger

t_k (min)	y_k (°C)	t_k (min)	y_k (°C)
0	−0.036	16	6.854
2	−0.098	18	6.948
4	1.386	20	7.025
6	3.577	22	7.211
8	4.950	24	7.159
10	5.794	26	7.087
12	6.317	28	7.203
14	6.671	30	7.242

techniques. To make the problem specific, suppose there are $m = 16$ measurements of the heat exchanger step response as summarized in Table 6.9.1.

The values for y include some random measurement noise. The following MATLAB script on the distribution CD can be used to solve this problem.

```
%-----------------------------------------------------------------------
% Example 6.9.1: Heat Exchanger
%-----------------------------------------------------------------------

% Initialize

   clc                         % clear screen
   clear                       % clear variables
   global  r t y               % used in funf691.m
   r    = 16;                  % number of data points
   n    = 3;                   % number of variables
   m    = 1500;                % maximum iterations
   p    = 10*r;                % number of plot points
   tol  = 1.0e-5;              % error tolerance
   T    = 30;                  % range of t (min)
   q    = [7.2 4.0 3.2]';      % nominal parameters
   x0   = [3 3 3]';            % initial guess
   X    = zeros (p,2);
   Y    = zeros (p,2);
   t    = zeros (r,1);
   y    = zeros (r,1);
   d    = fopen('e691.dat','w');

% Construct data

   fprintf ('Example 6.9.1: Heat Exchanger\n');
   randinit (100);
   for i = 1 : r
      t(i) = (i-1)*T/(r-1);
      y(i) = fung691 (t(i),q) + randu(1,1,-0.1,0.1);
```

```
   end
   for i = 1 : r/2
      fprintf (d,'\n %.0f & %.3f & %.0f & %.3f \\\\', ...
               t(i),y(i),t(r/2+i),y(r/2+i));
   end
   fclose (d);

% Find curve fit parameters

   disp ('Finding parameters ... ')
   [x,ev,j] = conjgrad (x0,tol,n,m,'funf691');
   show ('x',x);
   show ('f(x)',funf691(x));
   show ('function evaluations',ev)

% Display curves and data

   for i = 1 : p
      j = min ([i,r]);
      X(i,1) = (i-1)*T/(p-1);
      Y(i,1) = fung691 (X(i,1),x);
      X(i,2) = t(j);
      Y(i,2) = y(j);
   end
   graphxy (X,Y,'','t (min)','y (deg)',s)


function z = funf691 (x)
%----------------------------------------------------------------------
% Objective function: Example 6.9.1
%----------------------------------------------------------------------
   global  r t y
   q = 0;
   for i = 1 : r
      q = q + (fung691(t(i),x)-y(i))^2;
   end
   z = q/r;


function  y = fung691 (t,q)
%----------------------------------------------------------------------
% Heat Exchanger
%----------------------------------------------------------------------
   if t > q(3)
      a = -(t - q(3))/max([q(2),0.1]);
      y = q(1)*(1 - exp(a));
   else
      y = 0;
   end
%----------------------------------------------------------------------
```

When script *e691.m* is executed, it performs an unconstrained minimization of $f(x)$ starting from the initial guess

$$x^0 = [3 \quad 3 \quad 3]^T \tag{6.9.7}$$

In order to keep the exponent in $y(t,x)$ from getting too large during the search, the value of x_2 in Equation (6.9.4) is replaced by $\max(x_2, 0.1)$. Using an error tolerance of $\varepsilon = 10^{-5}$, the conjugate gradient method converges in seven iterations and uses 229 scalar function evaluations to produce the following set of optimal parameter values.

$$x^* = \begin{bmatrix} 7.179 \\ 3.879 \\ 3.198 \end{bmatrix} \tag{6.9.8}$$

For this set of parameters, the value of the objective function is

$$f(x^*) = 2.70022 \times 10^{-3} \tag{6.9.9}$$

Since $f(x^*) > 0$, the fit to the data points in Table 6.9.1 is not perfect. However, it is quite reasonable, as can be seen in Figure 6.9.2, which compares the measured step response data points from Table 6.9.1 (squares) with the predicted step response (solid line) using the optimal values for the model parameters.

Once a model is available, it can be used to analyze the behavior of the physical process. As a simple illustration, suppose the distance from the outlet to the thermocouple sensor in Figure 6.9.1 is $d = 2.0$ m. The time delay of $x_3 = 3.096$ min is caused by the glycol traveling the distance d with an average velocity of v. Thus, we can compute the velocity at which the glycol travels as

$$v = \frac{d}{x_3} = 0.6254 \text{ m/min}. \tag{6.9.10}$$

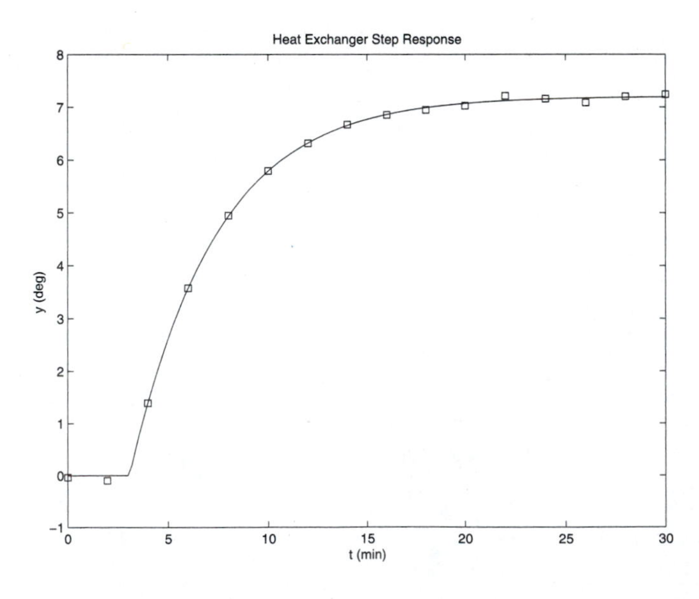

FIGURE 6.9.2 Nonlinear Regression of Heat Exchanger Step Response Data

A solution to Example 6.9.1 using C can be found in the file *e691.c* on the distribution CD.

6.9.2 Transportation Planning: C

Many optimization problems come down to a question of finding the "best" allocation of limited resources. As an example from the area of transportation planning, consider the problem of moving a given product from two source locations $\{s_1, s_2\}$ to three destination locations $\{d_1, d_2, d_3\}$ as shown in Figure 6.9.3. Let a_k denote the amount of the product that must be shipped from source location k, and let b_j denote the amount of product that must be received at destination location j. Since the product cannot accumulate on the transportation network, the total amount shipped must equal the total amount received.

$$a_1 + a_2 = b_1 + b_2 + b_3 \tag{6.9.11}$$

The planning problem is one of determining how much of the product should be shipped from a given source to a given destination. Let $x_{3(k-1)+j}$ denote the amount shipped from source k to destination j as shown in Figure 6.9.3, and let $c_{3(k-1)+j} > 0$ denote the cost of shipping one unit of the product from source k to destination j. Since there are two sources and three destinations, the vectors x and c are of dimension $n = 6$. The objective is to minimize the total shipping costs, which can be computed as follows.

$$f(x) = \sum_{i=1}^{6} c_i x_i = c^T x \tag{6.9.12}$$

The transportation problem is a constrained minimization in R^6. Indeed, the total amount shipped from source k must equal a_k, and the total amount received at destination j must equal b_j. Referring to Figure 6.9.3, this corresponds to the following five equality constraints.

$$x_1 + x_2 + x_3 = a_1 \tag{6.9.13a}$$

$$x_4 + x_5 + x_6 = a_2 \tag{6.9.13b}$$

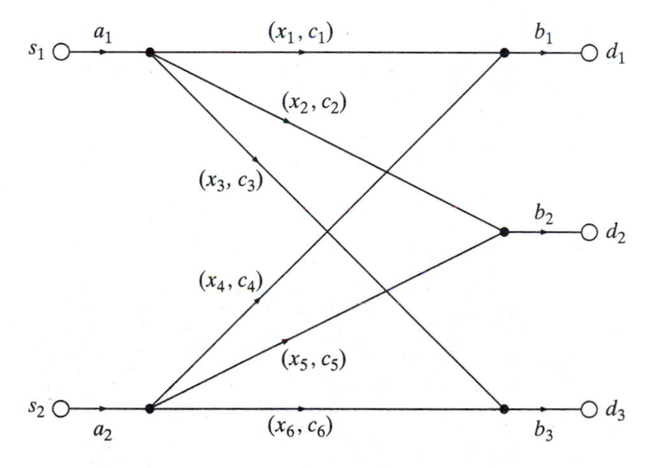

FIGURE 6.9.3 Transportation of a Product from Two Sources to Three Destinations

$$x_1 + x_4 = b_1 \tag{6.9.13c}$$

$$x_2 + x_5 = b_2 \tag{6.9.13d}$$

$$x_3 + x_6 = b_3 \tag{6.9.13e}$$

These equality constraints can be formulated as $p(x) = 0$ where p is the following function:

$$p(x) = \begin{bmatrix} x_1 + x_2 + x_3 - a_1 \\ x_4 + x_5 + x_6 - a_2 \\ x_1 + x_4 - b_1 \\ x_2 + x_5 - b_2 \\ x_3 + x_6 - b_3 \end{bmatrix} \tag{6.9.14}$$

In addition, the amounts shipped cannot be negative, $x \geq 0$. This corresponds to the standard from $q(x) \geq 0$ with

$$q(x) = x \tag{6.9.15}$$

To make the transportation planning problem specific, suppose the amount shipped, *a;* the amount received *b;* and the unit shipping costs *c;* are as follows.

$$a = [2000 \quad 1200]^T \text{ kg} \tag{6.9.16a}$$

$$b = [800 \quad 1400 \quad 1000]^T \text{ kg} \tag{6.9.16b}$$

$$c = [1.15 \quad 2.43 \quad 0.94 \quad 1.64 \quad 1.97 \quad 1.07]^T \text{ \$/kg} \tag{6.9.16c}$$

The following C program on the distribution CD can be used to solve this problem.

```
/*-----------------------------------------------------------------*/
/* Example 6.9.2: Transportation Planning                          */
/*-----------------------------------------------------------------*/

#include "c:\nlib\util.h"

float f (vector x,int n)
{
   float    c[7] = {0.0f,1.15f,2.43f,0.94f,
                    1.64f,1.97f,1.07f};         /* unit costs */
   return inner (c,x,n);
}

void  p (vector x,int n,int r,vector y)
{

   float    a[3] = {0,2000,1200},              /* amount shipped */
            b[4] = {0,800,1400,1000};          /* amount received */

/* Shipping */
   y[1] = x[1] + x[2] + x[3] - a[1];
```

```
   y[2] = x[4] + x[5] + x[6] - a[2];

/* Receiving */

   y[3] = x[1] + x[4] - b[1];
   y[4] = x[2] + x[5] - b[2];
   y[5] = x[3] + x[6] - b[3];
}

void  q (vector x,int n,int s,vector y)
{
   copyvec (x,s,y);
}

int main (void)
{
   int          n = 6,                  /* number of variables */
                r = 5,                  /* equality constraints */
                s = n,                  /* inequality constraints */
                m = 100,                /* maximum iterations */
                eval,                   /* function evaluations */
                trials = 5,             /* number of runs */
                i,j;                    /* iterations */
   float        eps = 1.0e-5f,          /* error tolerance */
                mu = 100.0f;            /* penalty parameter */
   vector       x = vec (n,""),         /* shipping schedule */
                y = vec (r,"");
   FILE         *g = fopen ("e692.dat","w");

/* Compute optimal solution for different penalty parameters */

   printf ("\nExample 6.9.2: Transportation Planning\n");
   for (j = 1; j <= trials; j++)
   {
      mu = (float) pow(10,j-2);
      zerovec (x,n);
      i = penalty (x,eps,mu,n,r,s,m,&eval,f,p,q);
      p(x,n,r,y);
      fprintf (g,"\n%10.1f & %10.2f & %10.4f
            \\\\",mu,f(x,n),vecnorm(y,r));
      fprintf (g,"\n & & \\\\");
      shownum ("Number of iterations",(float) i);
      shownum ("Number of scalar function evaluations",(float) eval);
      showvec ("Optimal x",x,n,0);
      shownum ("Optimal f(x)",f(x,n));
      shownum ("||p(x)||",vecnorm(y,r));
   }
   return 0;
}
/*------------------------------------------------------------*/
```

When program *e692.c* is executed, it uses the penalty function method to find an x that minimizes $f(x)$ subject to the constraints $p(x) = 0$ and $q(x) \geq 0$. An initial guess of $x^0 = 0$ is used. This satisfies $q(x^0) \geq 0$, but not $p(x^0) = 0$. However, as the iterative solution progresses, the costs associated with violating the constraints drive the solution toward an x that does satisfy the constraints. Using an error tolerance of $\varepsilon = 10^{-5}$ and a penalty parameter of $\mu = 100$, the method converges in four iterations and uses 187 scalar function evaluations to produce the following optimal shipping schedule:

$$x^* = \begin{bmatrix} 533.4 \\ 833.1 \\ 633.4 \\ 266.6 \\ 566.9 \\ 366.6 \end{bmatrix} \text{kg} \tag{6.9.17}$$

This corresponds to a total shipping cost of

$$f(x^*) = \$\,5179.50 \tag{6.9.18}$$

To check how well the solution satisfies the individual shipping and receiving constraints, we compute the norm of the equality constraint function at the solution.

$$\|p(x^*)\| = 0.0156 \tag{6.9.19}$$

Although this is not zero, it is very small in comparison with the components of the constraint vectors a and b. It is of interest to note that as the penalty parameter μ changes, the degree to which the constraints are satisfied also changes, as can be seen in Table 6.9.2.

The transportation planning problem is a special case of the general constrained optimization problem in the sense that both the objective function $f(x)$ and the constraint functions $p(x)$ and $q(x)$ are *linear* functions of the vector of unknowns x. For linear optimization problems, a very efficient optimization technique called the *simplex method* is available (Dantzig, 1951). For many engineering applications, the dimension of linear problems n can be very large. In these instances, it is best to use the more efficient simplex method. Detailed treatment of the simplex method can be found in Luenberger (1965).

A solution to Example 6.9.2 using MATLAB can be found in the file *e692.m* on the distribution CD.

TABLE 6.9.2 Optimal Shipping Schedules

μ	$f(x^*)$ (\$)	$\|p(x^*)\|$ (kg)
0.1	4686.32	7.2730
1.0	5177.77	0.6183
10.0	5179.43	0.0623
100.0	5179.50	0.0156
1000.0	5179.55	0.0044

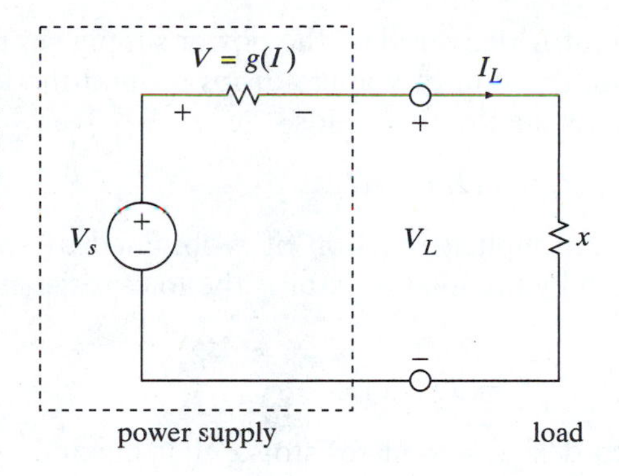

FIGURE 6.9.4 DC Power Supply with Nonlinear Source Resistance

6.9.3 Maximum Power Extraction: MATLAB

Virtually every piece of electrical and computer equipment requires some form of power supply circuit. Consider the DC power supply shown in Figure 6.9.4. This is similar to the circuit discussed in Section 6.1, but with one important change. The linear source resistance R_s has been replaced with a *nonlinear* resistor $V = g(I)$. Suppose the V versus I characteristic of the nonlinear source resistance is as follows.

$$g(I) = \alpha\left[\frac{1 - \exp(-\beta I)}{1 + \exp(-\beta I)}\right] \tag{6.9.20}$$

Here I denotes the current through the resistor, while $V = g(I)$ denotes the voltage drop across the resistor. The resistance characteristic g is a curve that goes through the origin with a slope of $\alpha\beta/2$. The curve "saturates" at $g(I) = \pm\alpha$ for large values of I. A plot of the nonlinear V versus I characteristic for the case $(\alpha, \beta) = (10, 1)$ is shown in Figure 6.9.5.

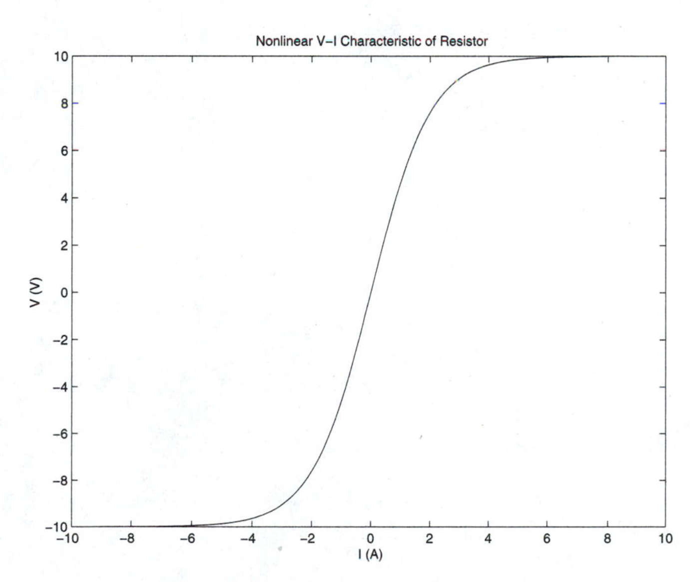

FIGURE 6.9.5 Nonlinear V versus I Characteristic g of Source Resistance

To determine the load current I_L delivered by the power supply circuit, we apply Kirchhoff's voltage law by setting the sum of voltage drops around the loop to zero. Assuming the load resistor has resistance x, this yields

$$V_s = g(I_L) + xI_L \tag{6.9.21}$$

Note that the load current I_L is an implicit function of x obtained by solving Equation (6.9.21). The power dissipated by the load resistor is the load resistance times the square of the load current.

$$P_L = xI_L^2(x) \tag{6.9.22}$$

Suppose the objective is to design a load resistor x that maximizes the power extracted from the power supply. This is equivalent to minimizing $f(x)$ where

$$f(x) = -P_L(x). \tag{6.9.23}$$

Since $I_L(x)$ is a solution of (6.9.21) for each x, the only constraint on x is that it be non-negative. That is, $q(x) \geq 0$ where

$$q(x) = x. \tag{6.9.24}$$

Thus, the problem of designing a load resistor that extracts the maximum amount of power from the nonlinear power supply reduces to one of minimizing $f(x)$ subject to $q(x) \geq 0$. This optimization problem is somewhat different from the others we have encountered thus far because to evaluate the objective function $f(x)$, we have to compute $I_L(x)$, which involves solving Equation (6.9.21). Consequently, each time $f(x)$ is evaluated, a root finding problem has to be solved using the techniques from Chapter 5.

To make the problem specific, suppose $V_s = 5$ V, $\alpha = 10$ V, and $\beta = 1$ A^{-1}. The following MATLAB script on the distribution CD can be used to solve this problem.

```
%----------------------------------------------------------------------
% Example 6.9.3: Maximum Power Extraction
%----------------------------------------------------------------------

% Initialize

   clc                        % clear screen
   clear                      % clear variables
   global R                   % used by funf693.m
   a = 0;                     % lower limit
   b = 10;                    % upper limit
   x0   = 2;                  % initial guess
   tol = 1.e-4;               % error tolerance
   m    = 500;                % maximum number of iterations

% Find optimal load resistance

   fprintf ('Example 6.9.3: Maximum Power Extraction\n');
   [x,ev,k] = dfp (x0,tol,1,m,'funf693');
```

```
   show ('Load Resistance R (ohm)',x);

% Compute power dissipated

   [i,k] = bisect (0,2*x,tol,m,'funp693');
   show ('Current I (A)',i)
   show ('Power Delivered P (watt)',-funf693(x))

% Plot objective function

   graphfun (a,b,'Objective Function','x','f(x)','funf693')


function y = funf693 (x)
%---------------------------------------------------------------------
% Objective function
%---------------------------------------------------------------------
   global R
   tol = 1.e-5;
   R = x;
   [i,k] = bisect (0,10,tol,100,'funp693');
   y = -R*i^2;

function v = funp693(i)
%---------------------------------------------------------------------
% Equality constraint: KVL
%---------------------------------------------------------------------
   global R
   Vs = 5;
   v = funr693(i) + R*i - Vs;
%---------------------------------------------------------------------
```

When script *e693.m* is executed, it produces the following optimal value for the load resistor.

$$x^* = 4.67324\ \Omega \tag{6.9.25}$$

Using the bisection root finding method from Chapter 5, the optimal load current is determined by solving Equation (6.9.21) with $x = x^*$, which yields

$$I_L(x^*) = 0.522882\ \text{A} \tag{6.9.26}$$

Finally, from (6.9.22) the maximum power that can be extracted from the nonlinear DC power supply is

$$P_L(x^*) = 1.2777\ \text{W} \tag{6.9.27}$$

A plot of the objective function for the case $(\alpha,\beta) = (10,1)$ is shown in Figure 6.9.6.

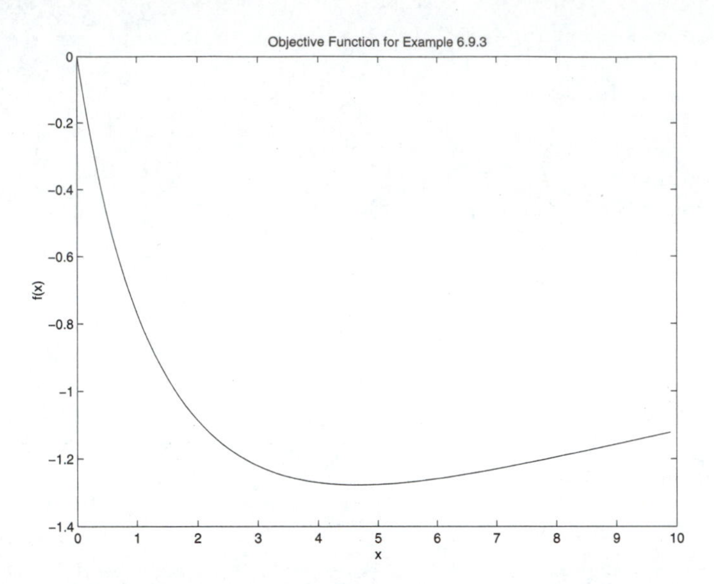

FIGURE 6.9.6 Objective Function for Example 6.9.3

It is of interest to note that if the nonlinear characteristic in (6.9.20) is replaced by a linear Ohm's law–type of characteristic such as $g(I) = R_s I$, the optimal load resistance found by script *e693* is $R_L = R_s$, as predicted by the analysis in Section 6.1.2.

A solution to Example 6.9.3 using C can be found in the file *e693.c* on the distribution CD.

6.9.4 Container Design: C

As part of a manufacturing process, a container must be designed to hold a given product. Recall that the container design problem was introduced in Section 6.1.3. The container is to be fabricated from a sheet of flat metal stock by folding the edges up as illustrated in Figure 6.1.1. The design goal is to maximize the container volume. Since the length, width, and height are x_1, x_2, and x_3, respectively, maximizing the volume is equivalent to minimizing the following objective function.

$$f(x) = -x_1 x_2 x_3 \tag{6.9.28}$$

From Figure 6.1.1, the metal sheet used to construct the box is of length $x_1 + 2x_3$ and of width $x_2 + 2x_3$. Suppose the area of the stock used to fabricate the container must be equal to α. This translates into the equality constraint function

$$p(x) = (x_1 + 2x_3)(x_2 + 2x_3) - 4x_3^2 - \alpha \tag{6.9.29}$$

Note that the x_3^2 term in Equation (6.9.29) represents the area of the four corner pieces, which are removed and recycled. The additional constraints are that all dimensions be non-negative and that the height of the box be no larger than β. This generates the following inequality constraint function.

$$q(x) = \begin{bmatrix} x_1 \\ x_2 \\ x_3 \\ \beta - x_3 \end{bmatrix} \tag{6.9.30}$$

Thus, the optimal container design problem reduces to minimize $f(x)$ subject to $p(x) = 0$ and $q(x) \geq 0$. To make the problem specific, suppose the area of stock used to construct the container is $\alpha = 80$ cm^2 and the maximum height is $\beta = 3$ cm. Example program *e694* on the distribution CD uses the penalty function method with a penalty parameter of $\mu = 100$ to convert this constrained problem into an unconstrained problem and employs the DFP method to solve the unconstrained minimization problem. It uses an error tolerance of $\varepsilon = 10^{-5}$ and an initial guess of

$$x^0 = [2 \quad 2 \quad 2]^T \tag{6.9.31}$$

Note that the initial guess satisfies the inequality constraint, $q(x) \geq 0$, but it violates the equality constraint, $p(x) = 0$. The following C program on the distribution CD can be used to solve this problem.

```
/*----------------------------------------------------------------------*/
/* Example 6.9.4: Container Design                                      */
/*----------------------------------------------------------------------*/

#include "c:\nlib\util.h"

float      alpha  = 80,                      /* area of sheet (cm^2) */
            beta  = 3.0;                     /* maximum height (cm)  */

float f (vector x,int n)                     /* objective function   */
{
   return -x[1]*x[2]*x[3];
}

void  p (vector x,int n,int r,vector y)      /* equality constraints */
{
   y[1] = (x[1] + 2*x[3])*(x[2] + 2*x[3])
           - 4*x[3]*x[3] - alpha;
}

void  q (vector x,int n,int s,vector y)    /* inequality constraints */
{
   copyvec (x,s,y);
   y[4] = beta - x[3];
}

int main (void)
{
   int      n = 3,               /* number of variables */
            r = 1,               /* number of equality constraints */
            s = 4,               /* number of inequality constraints */
            m = 250,             /* maximum iterations */
            i,                   /* number of iterations */
            ev;                  /* number of function evaluations */
   float    eps = 1.0e-5f,       /* error tolerance */
            mu = 40.0f;          /* penalty parameter */
   vector   x = vec(n,"2 2 2"),  /* initial guess */
```

```
            y = vec(r,"");

/* Find optimal box size */

   printf ("\nExample 6.9.4: Container Design\n");
   mu = prompt ("Enter penalty parameter mu: ",0.01f,1000.0f,100.0f);
   i = penalty (x,eps,mu,n,r,s,m,&ev,f,p,q);

/* Display results */

   shownum ("Iterations",(float) i);
   shownum ("Function evaluations",(float) ev);
   showvec ("Container dimensions x",x,n,0);
   shownum ("Container volume",-f(x,n));
   p(x,n,r,y);
   shownum ("||p(x)||",vecnorm(y,r));
   return 0;
}
/*-------------------------------------------------------------------*/
```

When program *e694.c* is executed, it converges in seven iterations and uses 363 scalar function evaluations to produce the following optimal dimensions for the design.

$$x^* = \begin{bmatrix} 4.72 \\ 4.72 \\ 3.057 \end{bmatrix} \text{cm} \tag{6.9.32}$$

The volume of the resulting container is

$$V = -f(x^*) = 68.11 \text{ cm}^3 \tag{6.9.33}$$

To check the degree to which the equality constraint on the sheet metal area is satisfied, we compute the norm of $p(x^*)$.

$$\|p(x^*)\| = 9.905 \times 10^{-4} \text{ cm} \tag{6.9.34}$$

Although $\|p(x^*)\|$ is not zero, it is very small in comparison with $\|x^*\| = 4.72$. It should be pointed out that the designed container is only guaranteed to approximate a local minimum of $f(x)$. Other starting conditions and parameter values may produce other solutions.

A solution to Example 6.9.4 using MATLAB can be found in the file *e694.m* on the distribution CD.

6.10 SUMMARY

Numerical methods for solving optimization problems are summarized in Table 6.10.1. The dimension column refers to the number of independent variables, and the derivative column indicates whether or not the gradient of the objective function is required. The gradient $\nabla f(x)$ can always be approximated numerically using central differences, but then $2n$ scalar function evaluations are required as opposed to n scalar function evaluations when an analytic expression for $\nabla f(x)$ is available.

TABLE 6.10.1 Optimization Techniques

Method	*Dimension*	*Constraints*	*Minimum*	*Derivative*
Golden section	1	inequality	local	no
Bisection	1	inequality	local	yes
Inverse parabolic interpolation	1	none	local	no
Steepest descent	n	none	local	yes
Conjugate-gradient	n	none	local	yes
Quasi-Newton	n	none	local	yes
Penalty function	n	equality, inequality	local	yes
Simulated annealing	n	inequality	global	yes

The first three methods are line search techniques that are used to find the optimal step length along a direction vector d. These methods begin by bracketing the minimum with a three-point pattern, $a < b < c$, where $f(b) < f(a)$ and $f(b) \le f(c)$. The golden section search narrows down the interval of uncertainty by the factor of $\gamma \approx 0.618$ using one function evaluation per iteration. When the gradient of $f(x)$ is available, the interval of uncertainty can be reduced by a factor of 0.5 per iteration using the bisection method. The inverse parabolic interpolation method fits a quadratic polynomial to the three-point pattern and then locates the minimum of the polynomial. A new three-point pattern is then generated, and the process is repeated. Inverse parabolic interpolation can be quite effective for smooth functions, but it is not guaranteed to converge in all cases. Brent's methods are hybrid methods that combine fast interpolation-based methods with slower, but more reliable, bracketing methods.

The next group of three methods in Table 6.10.1 solves n-dimensional unconstrained optimization problems. These methods can be compared in terms of how well they minimize the *quadratic* objective function:

$$f(x) = d + g^T x + \frac{1}{2} x^T H x \tag{6.10.1}$$

Nonquadratic objective functions can often be approximated by a quadratic in the neighborhood of a local minimum. The steepest descent method uses the search direction $d^k = -\nabla f(x^k)$ and is therefore very simple to implement. However, the method converges very slowly when the $n \times n$ Hessian matrix H is ill-conditioned. The Fletcher-Reeves conjugate-gradient method has superior convergence properties. For the objective function in Equation (6.10.1), the conjugate-gradient method converges in n iterations. Convergence in a single iteration can be achieved using Newton's method, which finds the roots of the nonlinear system, $\nabla f(x) = 0$, using the inverse of the Hessian matrix H. The Davidon-Fletcher-Powell (DFP) method is a quasi-Newton method that builds up an approximation to the inverse Hessian in stages. For a quadratic objective, the DFP method also converges in n iterations. The DFP method requires more storage than the conjugate-gradient method, an issue which can be important when n is large. A self-scaled version of the DFP method is available, which improves performance by making the method less sensitive to the accuracy of the line search.

The penalty function method is used to solve optimization problems with both equality and inequality constraints. This is achieved by constructing a generalized objective function $F(x)$ that includes additional penalty terms, which become large when the constraints are violated. In this way the minimization process itself tends to seek out feasible solutions because unfeasible solutions that violate the constraints are penalized. Minimization of the generalized objective function $F(x)$ can be performed with an unconstrained method such as the DFP or the conjugate-gradient method.

The simulated annealing method is different from the other methods in a number of respects. It is a statistical technique, based on an analogy with the annealing of metal, that searches for a global minimum of the objective function. The algorithm described in this chapter assumes lower and upper bounds on the components of x and also uses lower and upper bounds on the objective function:

$$f_0 \leq f(x) \leq f_1 \qquad (6.10.2)$$

A random global search, based on simulated annealing, gradually becomes more and more localized as $f(x)$ approaches the lower bound f_0. When $f(x)$ is sufficiently close to f_0, an efficient local search is initiated using the DFP method or the conjugate-gradient method to complete the search.

PROBLEMS

The problems are divided into Analysis problems, which can be solved by hand, and Computation problems, which require the use of MATLAB or C. Solutions to selected problems can be found in Appendix 4. Students are encouraged to use these problems, which are identified with an (S), as a check on their understanding of the material. Problems marked with a (P) are programming problems that require the student to implement one or more of the algorithms discussed in the chapter. The remaining Computation problems require the student to write a main program that uses one or more of the NLIB functions discussed in Appendix 1 or Appendix 2.

Analysis

6.1 Suppose the following one-dimensional functions are defined over the feasible region $S = R$ where R denotes the real line. In each case, determine the number of local minima, strict local minima, global minima, and strict global minima.

(a) $f(x) = 1$ **(b)** $f(x) = \cos(x)$

(c) $f(x) = -\sin(x)/x$ **(d)** $f(x) = 1 - x^2$

6.2 Suppose $F(y)$ is approximated by an interpolating polynomial through the points $a < b < c$ as follows.

$$F(y) = \alpha + \beta(y - b) + \gamma(y - b)^2$$

Find the coefficients (α, β, γ) in terms of $F(a)$, $F(b)$ and $F(c)$.

6.3 Suppose $F(y)$ is the quadratic polynomial in problem 6.2.

(a) For what values of y does $F(y)$ have a strict global minimum y^*?

(b) Find the point y^* that minimizes $F(y)$.

(c) Use the results of part (b) and problem 6.2 to derive the inverse parabolic interpolation formula in Equation (6.3.15).

(S) 6.4 Suppose the bisection method in Alg. 6.3.3 is used to perform a line search with the derivative approximated numerically, using a central difference as in (6.3.14).

$$F'(y) \approx \frac{F(y+h) - F(y-h)}{2h}$$

Show that the length of the interval of uncertainly is reduced by a factor of approximately 0.707 per function evaluation. How much slower is this (as a percentage) than the golden section search?

6.5 Suppose an expression for the derivative, $F'(y)$, is available. Derive a formula to iteratively estimate the minimum y^* by finding a root of $F'(y) = 0$ using the secant method discussed in Chapter 5.

6.6 Consider the following quadratic objective function of dimension $n = 4$. Suppose $d = 10$, $g = [1, -3, 2, 8]^T$ and H is diagonal with diagonal elements $(3, 9, 27, 81)$.

$$f(x) = d + g^T x + \frac{1}{2} x^T H x$$

Minimize $f(x)$ by differentiating $f(x)$, setting the result to zero, and solving for $x = x^*$.

(S) 6.7 Consider the following constrained optimization problem.

$$\begin{aligned} \text{minimize:} \quad & f(y,z) \\ \text{subject to:} \quad & Ay + Bz = c \\ & q(y,z) \ge 0 \end{aligned}$$

Here, A is $r \times r$, B is $r \times (n - r)$, and c is $r \times 1$. Thus, there are r linear equality constraints where $r < n$. Suppose the square matrix A is nonsingular. Convert this n-dimensional optimization problem with r equality constraints and s inequality constraints into a simpler problem of dimension $n - r$ that has no equality constraints. That is, find F, and Q such that the equivalent problem is as follows. Show how to determine y once z is known.

$$\begin{aligned} \text{minimize:} \quad & F(z) \\ \text{subject to:} \quad & Q(z) \ge 0 \end{aligned}$$

6.8 Consider the following constrained optimization problem.

$$\begin{aligned} \text{minimize:} \quad & f(x) \\ \text{subject to:} \quad & p(x) = 0 \\ & q(x) \ge 0 \end{aligned}$$

(a) Construct a penalty function $P(x)$ for the equality constraint, $p(x) = 0$, that is different from the one in Equation (6.7.1). The difference should be more than just a scale factor.

(b) Construct a penalty function $Q(x)$ for the inequality constraint, $q(x) \ge 0$, that is different from the one in Equation (6.7.2). The difference should be more than just a scale factor.

6.9 Consider the problem of solving a linear algebraic system $Ax = b$ where the $n \times n$ matrix A is nonsingular. Find an objective function $f(x)$ such that the unique solution of $Ax = b$ is a solution of the following unconstrained optimization problem.

$$\begin{aligned} \text{minimize:} \quad & f(x) \\ \text{subject to:} \quad & x \in R^n \end{aligned}$$

6.10 Consider the problem of solving a nonlinear algebraic system $g(x) = 0$ where x and $g(x)$ are $n \times 1$ vectors. Find an objective function $f(x)$ such that a solution of $g(x) = 0$ is a solution of the following unconstrained optimization problem.

$$\begin{aligned} \text{minimize:} \quad & f(x) \\ \text{subject to:} \quad & x \in R^n \end{aligned}$$

Computation

(P) 6.11 Write a function called *getstep* that performs a line search using successive inverse parabolic interpolation as in Equation (6.3.15). The pseudo-prototype of *getstep* is

```
[float d,int k] = getstep (float a,float b,float c,int m,float tol,
                           function F);
```

On entry to *getstep*: a, b and c form a three-point pattern as in Alg. 6.3.1, $m \geq 1$ is an upper bound on the number of iterations, $tol \geq 0$ is an error tolerance used to terminate the search, and F is the name of the user-supplied function to be minimized. The pseudo-prototype of F is

```
[float z] = F (float y);
```

When F is called, it must return the value $z = F(y)$. On exit from *getstep*: d is the value that minimizes $F(y)$, and k is the number of iterations performed. If $k < m$, then the following termination criterion should be satisfied where $F'(y)$ is a central difference approximation to the derivative as in (6.3.14).

$$|F'(y)| < tol$$

Test *getstep* by writing a program that finds the minimum of the following function. Use $tol = 10^{-6}$ and $(a,b,c) = (-1,1,5)$. Print d, $F(d)$, and the number of iterations performed, and plot $F(y)$ over the interval $[a, c]$.

$$F(y) = -y^2 \exp(-y)$$

(S) 6.12 Consider the quadratic objective function in problem 6.6. Write a program that uses the NLIB function *conjgrad* to find the minimum x^* starting from an initial guess of $x^0 = 0$ and using an error tolerance of $tol = 10^{-3}$. Solve the system twice, once with $v = 1$ corresponding to the steepest descent method, and once with $v = n$ corresponding to the full conjugate gradient method. Print the order v, the optimal x, the optimal $f(x)$, the number of iterations, and the number of scalar function evaluations in each case.

6.13 Repeat problem 6.12, but use the NLIB function *dfp* in place of the function *conjgrad*.

(P) 6.14 Write a function called *newt* that performs an unconstrained minimization of $f(x)$ using Newton's method as described (6.6.3). The pseudo-prototype of *newt* is

```
[vector x,int ev,int k] = newt (vector x0,float tol,int n,int m,
                                function f,function g);
```

On entry to *newt*: the $n \times 1$ vector $x0$ is the initial guess, $tol \geq 0$ is an error tolerance used to terminate the search, $n \geq 1$ is the number of independent variables, $m \geq 1$ is an upper bound on the number of iterations, and f and g are the names of user-supplied functions that specify the objective and its gradient, respectively. The pseudo-prototypes of f and g are

```
[float y] = f (vector x,int n);
[vector dy] = g (vector x,int n);
```

When f is called with the $n \times 1$ vector x, it must return the objective value $y = f(x)$. When g is called, it must compute the gradient $dy_k = \partial f(x)/\partial x_k$ for $1 \le k \le n$. On exit from *newt*: the $n \times 1$ vector x contains an estimate of the optimal x, ev is the number of scalar function evaluations, and k is the number of iterations performed. If $k < m$, then the following termination criterion should be satisfied where ε_M is the machine epsilon.

$$\|\nabla f(x^k)\| < tol \quad \text{or} \quad \|x^k - x^{k-1}\| < \varepsilon_M \|x^k\|$$

Approximate the Hessian matrix numerically using central differences as in (6.3.14). Use the NLIB function *ludec* from Chapter 2 to solve the linear algebraic system in (6.6.3). Test *newt* by writing a program which finds the minimum of the following function. Use $tol = 10^{-5}$ and an initial guess of $x^0 = 0$. Print x, $f(x)$, the number of iterations, and the number of scalar function evaluations.

$$f(x) = \frac{(x_1 - 2)^2(x_2 + 3)^4}{1 + x_1^2 + x_2^2}$$

6.15 Write a program that uses the NLIB function *conjgrad* to minimize the following objective function. Use an error tolerance of $tol = 10^{-5}$ and an initial guess of $x^0 = 0$. Print the optimal x, the optimal $f(x)$, the number of iterations, and the number of function evaluations. Plot $f(x)$ over a region that includes x^0 and the optimal x.

$$f(x) = \frac{-10}{(x_1^2 + 4x_1 + 5)(x_2^2 - 2x_2 + 3)}$$

(S) 6.16 Write a program that uses the NLIB function *dfp* to minimize the following objective function. Use an error tolerance of $tol = 10^{-4}$ and an initial guess of $x^0 = 0$. Print the optimal x, the optimal $f(x)$, the number of iterations, and the number of function evaluations.

$$f(x) = (x_1 + x_2 - 2)^2 + (x_1 - x_2 + 3)^4$$

6.17 You have one week left to study for final exams in n courses. Your goal is to allocate your review time so as to maximize your overall grade point average. Possible grades range from 0.0 for an F to 4.0 for an A. Let x_k denote the fraction of time spent reviewing the material for course k. You estimate that, at this point, the grade earned for course k is related to the time spent in review as follows:

$$g_k(x_k) = a_k + (b_k - a_k)[1 - \exp(-x_k/c_k)]$$

Here $0 \le a_k \le 4.0$ is the grade that you would get if no time were spent reviewing the material, and $a_k < b_k \le 4.0$ represents the maximum possible grade based on the work done so far and the weight of the final exam. Finally, $c_k > 0$ is a measure of the difficulty of the material with larger values for c_k corresponding to more difficult courses. The objective is to find the optimal review strategy; that is, to minimize the negative of the grade point average (GPA):

$$f(x) = -\frac{1}{n}\sum_{k=1}^{n} g_k(x_k)$$

This optimization is subject to a number of constraints. Since x_k represents a fraction of the total time available, we have $p(x) = 0$ where

$$p(x) = x_1 + x_2 + \cdots + x_n - 1$$

In addition, the minimum time that can be spent on any course is zero. Thus, $q(x) \geq 0$ where

$$q_k(x) = x_k \quad , \quad 1 \leq k \leq n$$

Write a program that uses the NLIB function *penalty* to find the optimal allocation of review time. For the parameter values, use $n = 5$ courses and

$$a = [1.5, 2.0, 1.0, 2.5, 0.5]^T$$
$$b = [4.0, 4.0, 3.5, 4.0, 3.2]^T$$
$$c = [0.7, 0.5, 0.5, 1.0, 0.9]^T$$

Print the optimal x and the optimal GPA. How much higher (as a percentage) is this than the grade point average obtained by allocating equal review time to the n courses?

6.18 Consider the following two-dimensional objective function.

$$f(x) = \frac{\sin(\pi x_1)[1 - \cos(\pi x_2)]}{1 + x_1^4 + x_2^4}$$

Suppose the values of x are restricted to $-5 \leq x_k \leq 5$ for $1 \leq k \leq 2$.

(a) Construct lower and upper bounds on the value of $f(x)$.

(b) Write a program that uses the NLIB function *anneal* to find a global minimum of $f(x)$ over $a \leq x \leq b$. Print the optimal x, the optimal $f(x)$, the number of iterations, the number of scalar function evaluations, and the localization parameter γ.

(c) Plot the objective surface $f(x)$ for $a \leq x \leq b$.

6.19 Consider the problem of finding the global minimum of the following two-dimensional objective function.

$$f(x) = \frac{x_1^2 - \cos(\pi x_2)}{1 + x_2^2}$$

Suppose the search is restricted to the region $-3 \leq x_k \leq 3$ for $1 \leq k \leq 2$.

(a) Construct lower and upper bounds on the value of $f(x)$.

(b) Write a program that uses the NLIB function *anneal* to find a global minimum of $f(x)$ over $a \leq x \leq b$. Print the optimal x, the optimal $f(x)$, the number of iterations, the number of scalar function evaluations, and the localization parameter γ. Use an initial guess of $x^0 = [3, 3]^T$, a global error tolerance of $tol = 0.01$, a local error tolerance of $\eta = 10^{-5}$, and a localization parameter of $\gamma = 0.3$.

(c) Plot the objective surface $f(x)$ for $a \leq x \leq b$.

CHAPTER 7

Differentiation and Integration

Engineers are often called upon to draw conclusions about a physical system from an analysis of its mathematical model. Two interrelated techniques are often used to perform the analysis. One is to compute the rate of change of one variable with respect to another—to compute a derivative. The other is to compute the area enclosed by a curve—to compute an integral. This chapter examines numerical techniques for obtaining derivatives and integrals of functions. Suppose $f(x)$ is the function of interest, and a is a point at which the derivative is desired. We develop numerical techniques for estimating

$$D(a) = \left. \frac{df(x)}{dx} \right|_{x=a}$$

Unlike explicit differentiation based on formulas from calculus, numerical differentiation produces an estimate of the derivative using only the *value* of the function at selected points. Consequently, numerical differentiation is applicable to functions that are available only in *tabulated* form as well as functions that can be expressed analytically. Although numerical differentiation is more widely applicable than explicit differentiation, it is less accurate. In some cases, numerical derivatives can be evaluated faster than explicit derivatives, a consideration that is important for real-time applications. Typical engineering applications of numerical differentiation include computing velocity from position measurements and computing acceleration from velocity measurements. A practical drawback of numerical differentiation is that it is highly sensitive to the effects of noise.

Suppose the function $f(x)$ is defined over an interval $[a,b]$. We also examine numerical techniques for estimating the integral

$$I(a,b) = \int_a^b f(x)\, dx$$

Again, there is no need to have an analytic expression for the function being integrated; only values of the function at selected points are required. Therefor, tabulated functions can be integrated as well. Typical applications in engineering include such things as determining the area of an irregular piece of land, computing the electrical charge on a capacitor, and finding the work performed by moving a mass. The integration of a function can be regarded as a special case of solving a differential equation. Consequently, numerical integration provides a natural lead into the solution of differential equations, a topic that is addressed in detail in Chapters 8 and 9. Unlike numerical differentiation, numerical integration is relatively insensitive to the effects of noise.

We begin this chapter by posing a number of problems from engineering that involve numerical differentiation and integration. We then examine a family of finite-difference techniques for numerical differentiation. We use the Taylor series to develop multipoint difference formulas for estimating the first and second derivatives. We examine the question of selecting step sizes with the view of controlling the formula truncation error. We also investigate the sensitivity of numerical differentiation to noise. Next, we develop the Newton-Cotes integration formulas—including the midpoint rule, the trapezoid rule, and Simpson's rules—using Newton's forward difference interpolating polynomial. We follow with a discussion of the variable step size Romberg integration technique. Then we examine the Gauss-quadrature formulas based on the roots of the classical orthogonal polynomials. We investigate techniques applicable to improper integrals, and we evaluate multidimensional integrals by the statistical Monte Carlo method. To conclude the chapter, we summarize differentiation and integration methods and present engineering applications. The selected examples include estimating the change in enthalpy as water evaporates, designing a dam, solving an RC electrical network, and analyzing a link of a robotic arm.

7.1 MOTIVATION AND OBJECTIVES

The need to compute derivatives and integrals is commonplace in engineering applications. The following examples are representative.

7.1.1 Magnetic Levitation

Consider the magnetic levitation feedback control system shown in Figure 7.1.1. A ball of mass m is suspended below an electromagnet. The upward magnetic force, $h(y,u)$, depends on both the ball position y and the current u in the electromagnet. The magnetic force is opposed by the force, mg, due to gravity. Applying Newton's second law, the equation of motion of the magnetic levitation system is

$$m\frac{d^2y}{dt^2} = h(y,u) - mg \tag{7.1.1}$$

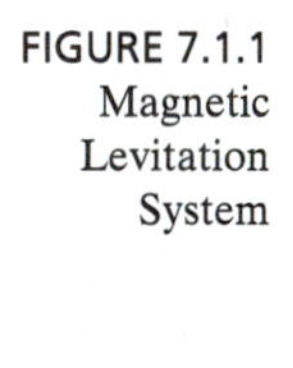

FIGURE 7.1.1 Magnetic Levitation System

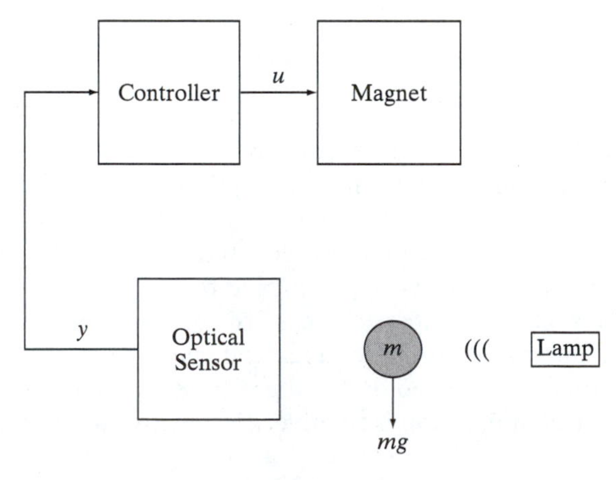

Suppose the current in the electromagnet is constant at $u(t) = u_0$. Then, setting the right-hand side of Equation (7.1.1) to zero and solving for y, we find that there is an *equilibrium position*, $y = y_0$, where the upward force from the electromagnet, $h(y_0, u_0)$, exactly cancels the downward force due to gravity. However, the ball will not remain levitated at $y = y_0$ because this equilibrium point is not stable. Indeed, if the ball drops slightly below y_0, then the electromagnetic force is weaker than gravity, and the ball falls. Similarly, if the ball moves slightly above $y = y_0$, then the electromagnetic force is stronger than gravity, and the ball accelerates upward until it sticks to the electromagnet.

To make the ball levitate, a control engineer must detect the ball position y and then use feedback to adjust the current u in the electromagnet. In particular, the following type of feedback control can be used to stabilize the system about the equilibrium point $y = y_0$.

$$u = u_0 - \alpha(y - y_0) - \beta\frac{dy}{dt} \tag{7.1.2}$$

Here $\alpha > 0$ and $\beta > 0$ are controller gains whose values must be properly adjusted. The ball position y can be measured with a light beam and a photocell. However, there is no simple reliable way to measure the ball velocity $y'(t) = dy(t)/dt$. Instead, the derivative, $y'(t)$, must be estimated numerically.

7.1.2 Mechanical Work

The mechanical work generated by an internal combustion engine is obtained through the expansion of gases within the cylinder as shown in Figure 7.1.2. As the gas expand, it causes the piston to move downward. The linear motion of the piston is then converted to rotary motion of the crank shaft through the connecting rod. The mechanical work associated with turning the crank shaft can be computed by measuring the pressure within the cylinder p as a function of the gas volume V. Measured values of p and V at different points in the expansion of the gas are listed in Table 7.1.1 (Moran and Shapiro, 1995).

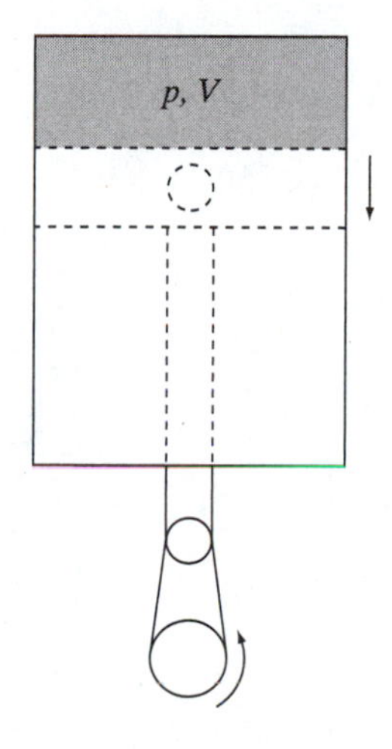

FIGURE 7.1.2 Mechanical Work due to Expanding Gas

TABLE 7.1.1 Pressure vs. Volume Measurements in Cylinder

p (bars)	V (cm^3)
20.0	454
16.1	540
12.2	668
9.9	780
6.0	1175
3.1	1980

Let W denote the work done as a result of the expansion of the gas. From thermodynamics, W is determined by evaluating the following integral. Since $p(V)$ is only available in tabulated form, this integral must be computed numerically.

$$W = \int_{V_0}^{V_1} p(V)dV \tag{7.1.3}$$

7.1.3 Water Management

Suppose an environmental engineer is asked to estimate the cost of constructing a dam on a river. One of the preliminary steps is to determine the cross-sectional area of the river at the point were the dam is to be constructed as shown in Figure 7.1.3.

In order to estimate the cross-sectional area A, depth measurements are performed with the results summarized in Table 7.1.2. The cross-sectional area is then

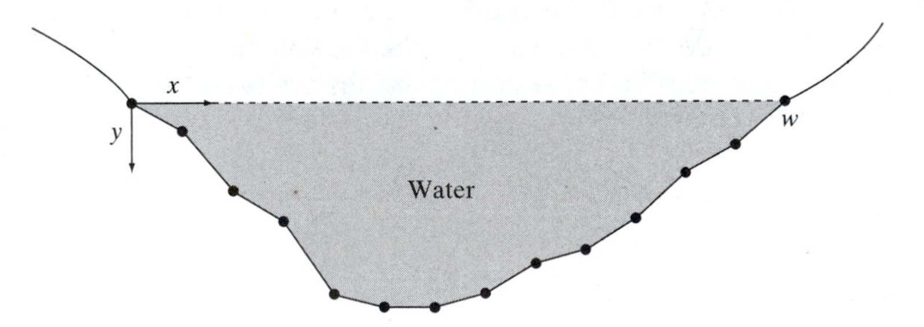

FIGURE 7.1.3 Cross Section of River

TABLE 7.1.2 Depth Measurements of River Cross Section

x (m)	y (m)	x (m)	y (m)
0	0.0	70	13.1
10	2.1	80	10.9
20	5.8	90	10.1
30	8.3	100	7.8
40	12.9	110	5.2
50	14.1	120	3.1
60	14.2	130	0.0

obtained by evaluating the following integral. Again, since the measurements are available only in tabulated form, this calculation must be performed numerically.

$$A = \int_0^w y(x)\, dx \tag{7.1.4}$$

7.1.4 Chapter Objectives

When you finish this chapter, you will be able compute numerical derivatives and integrals using discrete data points. You will know how to estimate the size of the truncation error for numerical differentiation and integration formulas. You will understand the detrimental effects of noise on numerical differentiation and be able to mitigate it with filtering. You will know how to perform numerical integration using both closed and open integration formulas, including the trapezoid, Simpson, and midpoint rules. You will use Richardson extrapolation to perform accurate numerical integration using Romberg's method. You will use orthogonal polynomials to find the optimal nonuniform spacing between points for the Gauss quadrature integration formulas. You will know how to integrate functions with singularities and how to use a change of variables to convert integrals with infinite limits to equivalent integrals with finite limits. Finally, you will be able to perform multidimensional integration using the statistical Monte Carlo method. These overall goals will be achieved by mastering the following chapter objectives.

Objectives for Chapter 7

- Understand the difference between forward, central, and backward difference formulas for numerical differentiation.
- Understand how to specify and derive formula truncation error.
- Be able to apply Richardson extrapolation to improve the accuracy of numerical derivatives.
- Understand why numerical differentiation is very sensitive to noise, while numerical integration is not.
- Understand the difference between open, closed, and semiopen numerical integration formulas.
- Be able to perform trapezoid, Simpson, and midpoint integration.
- Know how Richardson extrapolation is used to develop the Romberg integration technique.
- Understand how orthogonal polynomials are used to determine the evaluation points for the Gauss quadrature integration formulas.
- Know how to integrate functions containing singularities.
- Know how to convert integrals with infinite limits to equivalent integrals with finite limits using a change of variable.
- Be able to perform higher-dimensional numerical integration using the statistical Monte Carlo technique.
- Understand how numerical differentiation and integration techniques can be used to solve practical engineering problems.
- Understand the relative strengths and weaknesses of each computational method and know which are most applicable for a given problem.

7.2 NUMERICAL DIFFERENTIATION

We begin by examining techniques for numerical differentiation. Let $f(x)$ be a function mapping the real line R into R, and suppose f has n continuous derivatives in a neighborhood of the point $x = x_0$. Then $f(x)$ can be expanded into a Taylor series about $x = x_0$ as follows.

$$f(x) = \sum_{k=0}^{n-1} \frac{f^{(k)}(x_0)(x - x_0)^k}{k!} + E_n(x - x_0) \tag{7.2.1}$$

Here $f^{(k)}(x_0)$ denotes the kth derivative of $f(x)$ evaluated at $x = x_0$. The last term in Equation (7.2.1) is the formula *truncation error,* which takes the form of the first neglected term in the series. That is,

$$E_n(x - x_0) \triangleq \frac{f^{(n)}(\xi)(x - x_0)^n}{n!} \tag{7.2.2}$$

The constant ξ lies somewhere in the interval $[x_0, x]$ when $x > x_0$ or $[x, x_0]$ when $x < x_0$. From Equation (7.2.2), we see that the formula truncation error is of order $O(h^n)$ where $h = x - x_0$. That is, the error approaches zero at the same rate as h^n. Recall from Equation (1.5.7) that the notation $O(h^k)$ can be interpreted as follows where α is some constant.

$$\boxed{O(h^k) \approx \alpha h^k \quad , \quad |h| \ll 1} \tag{7.2.3}$$

Suppose the data samples of the function f are uniformly spaced. In particular, let h denote the step size between successive values of the independent variable x. The following notational shorthand can then be used to simplify subsequent formulations.

$$x_k \triangleq x_0 + kh \quad , \quad |k| = 0, 1, \ldots \tag{7.2.4a}$$

$$f_k \triangleq f(x_k) \quad , \quad |k| = 0, 1, \ldots \tag{7.2.4b}$$

7.2.1 First Derivative

If we expand $f(x)$ into a Taylor series about $x = x_0$ as in Equation (7.2.1), and evaluate the result at the two points adjacent to x_0, this yields the following pair of formulas, which can be used to estimate derivatives.

$$f_1 = f_0 + f_0^{(1)}h + \frac{f_0^{(2)}h^2}{2} + \frac{f_0^{(3)}h^3}{6} + \cdots \tag{7.2.5a}$$

$$f_{-1} = f_0 - f_0^{(1)}h + \frac{f_0^{(2)}h^2}{2} - \frac{f_0^{(3)}h^3}{6} + \cdots \tag{7.2.5b}$$

Here $f_0^{(k)}$ is understood to denote the kth derivative of $f(x)$ evaluated at $x = x_0$. The simplest formula for estimating $f_0^{(1)}$ is obtained directly from Equation (7.2.5a). Solving for $f_0^{(1)}$ and neglecting second- and higher-order terms yields

$$\boxed{f_0^{(1)} = \frac{f_1 - f_0}{h} - \frac{f^{(2)}(\xi)h}{2}} \tag{7.2.6}$$

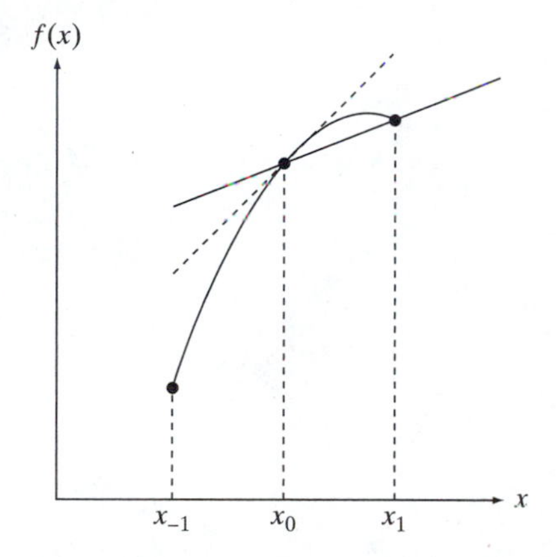

FIGURE 7.2.1 Two-Point Forward Difference Approximation

This is referred to as a *two-point forward difference* formula because it uses two values of $f(x)$, one at x_0 and the other in front at x_1. For the error term, ξ lies somewhere in the interval $[x_0, x_1]$. The two-point forward difference formula has a simple geometric interpretation as the slope of the forward secant line as shown in Figure 7.2.1.

Note that when both sides of (7.2.5a) are divided by h, the truncation error associated with the neglected terms becomes $-E_2(h)/h$. Consequently, the truncation error for the two-point forward difference approximation is of order $O(h)$. In the limit as h approaches zero, the two-point forward difference approximation reduces to the definition of the derivative.

$$\left.\frac{df(x)}{dx}\right|_{x=x_0} = \lim_{h\to 0} \frac{f(x_0+h) - f(x_0)}{h} \tag{7.2.7}$$

Next, suppose Equation (7.2.5b) is solved for $f_0^{(1)}$. Neglecting second- and higher-order terms, this yields the following alternative formulation for the derivative at x_0.

$$\boxed{f_0^{(1)} = \frac{f_0 - f_{-1}}{h} + \frac{f^{(2)}(\xi)h}{2}} \tag{7.2.8}$$

In this case, a value of x behind the point of interest is used, so this is referred to as a *two-point backward difference* approximation. The geometric interpretation of the two-point backward difference approximation, as the slope of the backward secant line, is shown in Figure 7.2.2.

The backward difference formula is useful in cases where the independent variable represents time. If x_0 denotes the present time, the backward difference formula uses only present and past samples; it does not rely on future data samples that may not yet be available in a real-time application.

If data samples on both sides of x_0 are available, a third formulation of the derivative can be obtained by subtracting Equation (7.2.5b) from (7.2.5a) and solving for $f_0^{(1)}$. Neglecting third- and higher-order terms, this yields

$$\boxed{f_0^{(1)} = \frac{f_1 - f_{-1}}{2h} - \frac{f^{(3)}(\xi)h^2}{6}} \tag{7.2.9}$$

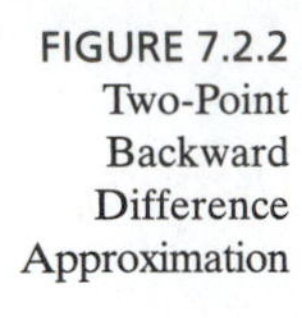

FIGURE 7.2.2 Two-Point Backward Difference Approximation

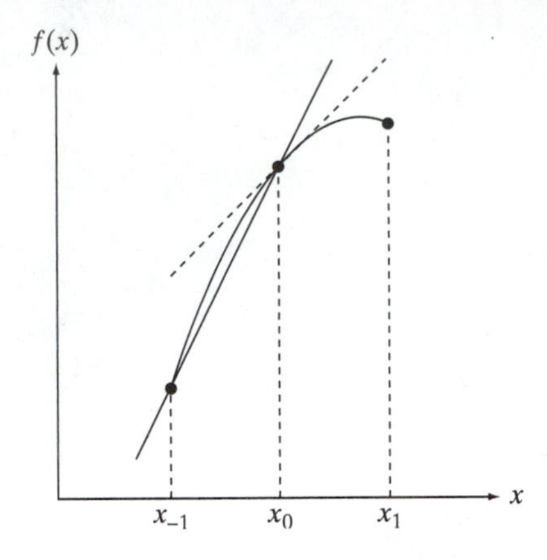

FIGURE 7.2.3 Two-Point Central Difference Approximation

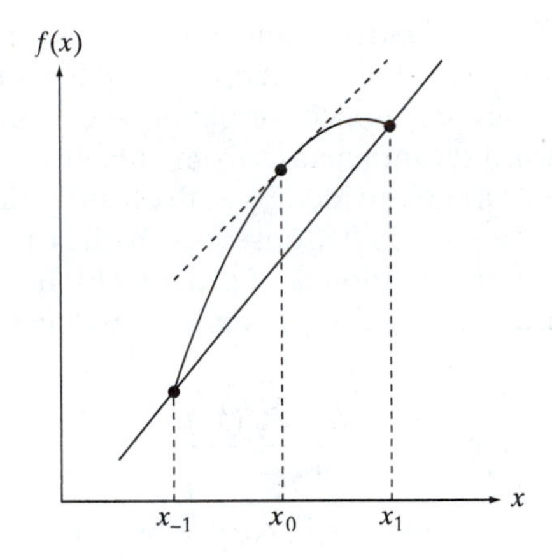

Since this formulation uses data points that are centered about the point of interest, it is referred to as a *three-point central difference* approximation. (The third point is x_0 even though it does not appear in the right-hand side.) The geometric interpretation of the central difference approximation is shown in Figure 7.2.3.

Observe that when (7.2.5b) is subtracted from (7.2.5a), the second-order terms cancel due to the symmetry of the points x_{-1} and x_1 about x_0. As a result, the truncation error representing the neglected terms is $-E_3(h)/h$, which means that the approximation is of order $O(h^2)$. Consequently, the central difference approximation is superior to both the forward difference and the backward difference in terms of accuracy. The central difference represents the *average* of the forward difference and the backward difference.

The accuracy of the numerical estimates of the derivative can be increased by using additional values of the function further away from the point of interest. To that end, consider the following two Taylor series expansions about x_0.

$$f_2 = f_0 + 2f_0^{(1)}h + 2f_0^{(2)}h^2 + \frac{4f_0^{(3)}h^3}{3} + \frac{2f_0^{(4)}h^4}{3} + \cdots \qquad (7.2.10a)$$

$$f_{-2} = f_0 - 2f_0^{(1)}h + 2f_0^{(2)}h^2 - \frac{4f_0^{(3)}h^3}{3} + \frac{2f_0^{(4)}h^4}{3} + \cdots \qquad (7.2.10b)$$

To develop a forward difference approximation, we use Equations (7.2.5a) and(7.2.10a). Notice that the $f_0^{(2)}$ term can be eliminated from (7.2.5a) by subtracting (7.2.10a) from four times (7.2.5a). The resulting equation can then be solved for $f_0^{(1)}$. Neglecting third- and higher-order terms, this yields

$$f^{(1)}(x_0) = \frac{-3f_0 + 4f_1 - f_2}{2h} + \frac{f^{(3)}(\xi)h^2}{3} \tag{7.2.11}$$

This is the *three-point forward difference* formula for estimating the derivative. Unlike the two-point forward difference formula, the truncation error in this case is of order $O(h^2)$. To develop the *three-point backward difference* formula, we eliminate the $f_0^{(2)}$ term in (7.2.5b) by subtracting four times (7.2.5b) from (7.2.10b). Solving the resulting equation for $f_0^{(1)}$ and neglecting third- and higher-order terms, this yields

$$f_0^{(1)} = \frac{f_{-2} - 4f_{-1} + 3f_0}{2h} + \frac{f^{(3)}(\xi)h^2}{3} \tag{7.2.12}$$

It is clear that this process can be extended by taking appropriate linear combinations of Taylor series expansions of $f(x_k)$ about $x = x_0$. For example, neglecting fifth- and higher-order terms, the five-point central difference formula for approximating the derivative is

$$f_0^{(1)} = \frac{f_{-2} - 8f_{-1} + 8f_1 - f_2}{12h} + \frac{f^{(5)}(\xi)h^4}{30} \tag{7.2.13}$$

In this case, the extra points on both sides of x_0 result in an estimate whose truncation error is of order $O(h^4)$. All of the estimates of the derivative are linear combinations of the values of $f(x)$ at points centered about $x = x_0$. They can be expressed in the general form

$$f_0^{(1)} = \frac{1}{h} \sum_{k=1-n}^{n-1} a_k f_k + O(h^{n-1}) \tag{7.2.14}$$

A compact way to summarize the different formulas is to use a *template* or stencil that contains the weighting coefficients, a_k, of the formula. Templates of the more popular approximations are summarized in Table 7.2.1. Note that in each case the coefficients in the templates sum to zero. This can be used as a partial check in deriving new formulas.

7.2.2 Second Derivative

It is also possible to estimate second- and higher-order derivatives numerically. For example, Equations (7.2.5a) and (7.2.5b) can be added, in which case the $f_0^{(1)}$ vanishes. The resulting equation can then be solved for $f_0^{(2)}$. Neglecting third- and higher-order terms, this yields the following three-point central difference approximation to the second derivative.

TABLE 7.2.1 Weighting Coefficients and Truncation Error: First Derivative

	Points	*Weighting Coefficients*							*Truncation*
Type	n	a_{-3}	a_{-2}	a_{-1}	a_0	a_1	a_2	a_3	*Error*
forward	2	0	0	0	−1	1	0	0	$-f^{(2)}(\xi)h/2$
	3	0	0	0	−3/2	2	−1/2	0	$f^{(3)}(\xi)h^2/3$
	4	0	0	0	−11/6	3	−3/2	1/3	$-f^{(4)}(\xi)h^3/4$
backward	2	0	0	−1	1	0	0	0	$f^{(2)}(\xi)h/2$
	3	0	1/2	−2	3/2	0	0	0	$f^{(3)}(\xi)h^2/3$
	4	−1/3	3/2	-3	11/6	0	0	0	$f^{(4)}(\xi)h^3/4$
central	3	0	0	−1/2	0	1/2	0	0	$-f^{(3)}(\xi)h^2/6$
	5	0	1/12	−2/3	0	2/3	−1/12	0	$f^{(5)}(\xi)h^4/30$

$$f_0^{(2)} = \frac{f_{-1} - 2f_0 + f_1}{h^2} - \frac{f^{(4)}(\xi)h^2}{12} \tag{7.2.15}$$

Due to cancellation of the third-order terms, the formula truncation error is of the form $E_4(h)/h^2$, which makes the approximation of order $O(h^2)$. The three-point forward difference approximation can be generated by eliminating $f_0^{(1)}$ from Equations (7.2.5a) and (7.2.10a), and similarly for the backward difference approximation, which uses (7.2.5b) and (7.2.10b). Again, each numerical estimate of the second derivative is a linear combination of values of the function at points centered about x_0. In this case, it takes the general form

$$f_0^{(2)} = \frac{1}{h^2} \sum_{k=1-n}^{n-1} b_k f_k + O(h^{n-2}) \tag{7.2.16}$$

Coefficient templates of the more commonly used numerical formulas for estimating the second derivative of a function are summarized in Table 7.2.2. Again, note that in each case the coefficients sum to zero.

TABLE 7.2.2 Weighting Coefficients and Truncation Error: Second Derivative

	Points	*Weighting Coefficients*							*Truncation*
Type	n	b_{-3}	b_{-2}	b_{-1}	b_0	b_1	b_2	b_3	*Error*
forward	3	0	0	0	1	−2	1	0	$-f^{(3)}(\xi)h$
	4	0	0	0	2	−5	4	−1	$11f^{(4)}(\xi)h^2/12$
backward	3	0	1	−2	1	0	0	0	$f^{(3)}(\xi)h$
	4	−1	4	−5	2	0	0	0	$11f^{(4)}(\xi)h^2/12$
central	3	0	0	1	−2	1	0	0	$-f^{(4)}(\xi)h^2/12$
	5	0	−1/12	4/3	−5/2	4/3	−1/12	0	$f^{(6)}(\xi)h^4/90$

The process used to obtain numerical formulas for first and second derivatives can be readily extended to third- and higher-order derivatives. A program that generates the weighting coefficients automatically can be found in Nakamura (1993). An alternative way to derive numerical differentiation formulas is to use interpolating polynomials rather than the Taylor series (Gerald and Wheatley, 1989).

7.2.3 Richardson Extrapolation

There are two parameters available to improve the accuracy of numerical estimates of derivatives. One is the number of points n, and the other is the step size h in those cases where it can be varied. For an n-point approximation of the kth derivative, the formula truncation error is of order $O(h^{n-k})$, which means that if $n > k$, the truncation error can be made arbitrarily small by decreasing the step size h. Unfortunately, this is only part of the picture. The total error also includes round-off error, and for numerical differentiation, the effects of round-off error are significant. To see why, let E_R denote the round-off error associated with forming the linear combination of values of $f(x)$ in Equation (7.2.14) or (7.2.16). If E_k denotes the total error associated with approximating the kth derivative with an n-point formula, then the two contributions to the error can be modeled as follows.

$$E_k(h) = \frac{E_R}{h^k} + O(h^{n-k}) \tag{7.2.17}$$

The round-off error, E_R, is substantial when small values of h are used because one is taking differences between nearly identical numbers. Recall from Chapter 1 that this can cause catastrophic cancellation, which results in a significant loss of accuracy. Furthermore, the effect is compounded by the fact that E_R is then divided by h^k when the kth derivative is estimated. In view of the relationship depicted in Equation (7.2.17), it is clear that for each instance of computing a numerical derivative, there will be an optimal step size h that minimizes the value of total error. For larger values of h, the truncation error will dominate, and for smaller values of h, the round-off error will dominate.

No general formula is available for determining the optimal sample spacing or step size because both E_R and the truncation error depend on the details of the function at the point of interest. However, Richardson's extrapolation method can be used to improve the estimate by varying h. For an n-point approximation, the truncation error is of the form $E(h) \approx \alpha h^{n-k}$ for $|h| \ll 1$ where α depends on $f^{(n)}(x)$. Using this approximation, the relationship between the truncation errors for two different step sizes h_1 and h_2 is

$$E(h_1) \approx \left(\frac{h_1}{h_2}\right)^{n-k} E(h_2) \tag{7.2.18}$$

Given the relationship in Equation (7.2.18), an estimate of the truncation error associated with the smaller step, h_2, can be found. Let $f_0^{(k)}(h)$ be the estimate of the kth derivative of $f(x)$ at $x = x_0$ using step size h. It then follows that

$$f_0^{(k)}(h_1) + E(h_1) = f_0^{(k)}(h_2) + E(h_2) \tag{7.2.19}$$

If the expression for $E(h_1)$ in (7.2.18) is now substituted into (7.2.19), the resulting equation can be solved for the error associated with step size h_2.

$$E(h_2) \approx \frac{f_0^{(k)}(h_1) - f_0^{(k)}(h_2)}{1 - (h_1/h_2)^{n-k}} \tag{7.2.20}$$

Thus, the truncation error can be estimated by combining two estimates of the derivative—one using step size h_1 and the other using step size h_2. An *extrapolated* estimate, $\bar{f}_0^{(k)}$, is then obtained by starting with $f_0^{(k)}(h_2)$ and adding the correction $E(h_2)$.

$$\bar{f}_0^{(k)} \triangleq f_0^{(k)}(h_2) + \frac{f_0^{(k)}(h_1) - f_0^{(k)}(h_2)}{1 - (h_1/h_2)^{n-k}} \tag{7.2.21}$$

For convenience, the relationship between successive step sizes is often taken to be $h_1 = 2h$, $h_2 = h$. In this case, the extrapolated estimate of the derivative in Equation (7.2.21) reduces to the following linear combination of the two original estimates.

$$\boxed{\bar{f}_0^{(k)} = \frac{2^{n-k} f_0^{(k)}(h) - f_0^{(k)}(2h)}{2^{n-k} - 1}} \tag{7.2.22}$$

It can be shown (Ralston and Rabinowitz, 1978) that if the errors $E(h_1)$ and $E(h_2)$ are of order $O(h^{n-k})$, then the truncation error of the extrapolated estimate in (7.2.22) is of order $O(h^{n+2-k})$. If two extrapolated estimates of order $O(h^{n+2-k})$ are constructed using different step sizes, then they can be combined to construct a third estimate whose truncation error is of order $O(h^{n+4-k})$. The process can be continued in this manner, but eventually the round-off error dominates and no further improvement is obtained (Gerald and Wheatley, 1989).

EXAMPLE 7.2.1 Step Size

To illustrate Richardson extrapolation and the effects of the step size, consider the problem of numerically estimating the first two derivatives of the following function at $x_0 = 0$.

$$f(x) = \frac{\exp(x)}{1 + x^2}$$

In this case, the exact values are $f^{(1)}(0) = 1$ and $f^{(2)}(0) = -1$. Example program *e721* on the distribution CD uses three-point central difference approximations with various values for the step size h. The computations are all performed using single precision arithmetic and the results are summarized in Table 7.2.3.

The first column of Table 7.2.3 is the step size h, the second is an estimate of the first derivative, the third is an extrapolated estimate of the first derivative (using $2h$ and h), the fourth is an estimate of the second derivative, and the fifth is an extrapolated estimate of the second derivative. Note that for the larger step sizes ($10^{-3} \le h \le 1$), the extrapolated estimates in columns three and five are superior to the unextrapolated estimates in columns two and four. For step sizes in the middle range ($10^{-6} \le h \le 10^{-4}$), the regular and extrapolated estimates are identical to seven decimal places. When the step size is smaller than the machine epsilon, $\varepsilon_M = 1.19 \times 10^{-7}$, the estimates for the second derivative (both regular and extrapolated) start to deteriorate due to round-off error. Based on Table 7.2.3, it would appear that a good choice for the step size in this instance is $h = 10^{-5}$.

TABLE 7.2.3 Three-Point Central Difference Estimates

h	$f_0^{(1)}$	$\overline{f}_0^{(1)}$	$f_0^{(2)}$	$\overline{f}_0^{(2)}$
1.0e + 00	0.5876006	0.6625721	−0.4569194	−0.5679657
1.0e − 01	0.9917500	0.9996794	−0.9892736	−0.9995885
1.0e − 02	0.9999167	0.9999999	−0.9998917	−1.0000000
1.0e − 03	0.9999992	1.0000000	−0.9999989	−0.9999999
1.0e − 04	1.0000000	1.0000000	−0.9999999	−0.9999999
1.0e − 05	1.0000000	1.0000000	−1.0000000	−1.0000000
1.0e − 06	1.0000000	1.0000000	−1.0000000	−1.0000000
1.0e − 07	1.0000000	1.0000000	−1.0000138	−1.0000184
1.0e − 08	1.0000000	1.0000000	−0.9985502	−0.9981888
1.0e − 09	1.0000000	1.0000000	−0.9757820	−0.9667470

7.3 NOISE-CORRUPTED DATA

The numerical differentiation discussed thus far has been somewhat idealized because it has ignored the effects of noise. If numerical differentiation is used to compute velocity from position measurements, for example, then the transducer used to generate the data samples will invariably add noise to the measurements. Thus, if x_k denotes the signal to be measured and y_k denotes the observed transducer output at time k, then

$$y_k = x_k + n_k \quad , \quad k \geq 0 \tag{7.3.1}$$

Here, n_k represents the measurement noise at time k. Clearly, the numerical derivative of the observed signal y_k is the numerical derivative of the underlying signal x_k plus the numerical derivative of the noise n_k. For a high-quality sensor, the average power of the measurement noise will be small in comparison to the average power of the underlying signal. However, the noise power is typically spread out over a range of frequencies, and therein lies the problem. Differentiation tends to amplify the high-frequency component of the noise. Indeed, if $n(t) = \alpha \sin(\omega t)$, then $dn(t)/dt = \omega\alpha \cos(\omega t)$. That is, the amplitude of the derivative of the noise is proportional to the frequency of the noise.

EXAMPLE 7.3.1 Noise-Corrupted Data

To illustrate the difficulties caused by noise-corrupted data, suppose the observed signal consists of a damped sinusoid plus a small random white noise signal (noise distributed over the entire frequency range).

$$y_k = \exp(-k/64)\sin(\pi k/32) + n_k$$

Here, n_k is a random number from a Gaussian distribution with zero mean and a standard deviation of 0.002. A graph of the noise-corrupted signal y_k is shown in Figure 7.3.1. Note that the noise is indeed small and difficult to detect except for the larger values of k.

Example program *e731* on the distribution CD computes the numerical derivative of the real-time signal y_k using a two-point backwards difference approximation. The sensitivity of the numerical derivative to noise becomes apparent as the signal amplitude decreases with increasing time as can be seen in Figure 7.3.2. When the signal becomes small, the differentiated high-frequency noise clearly manifests itself.

FIGURE 7.3.1
A Noise-Corrupted Signal

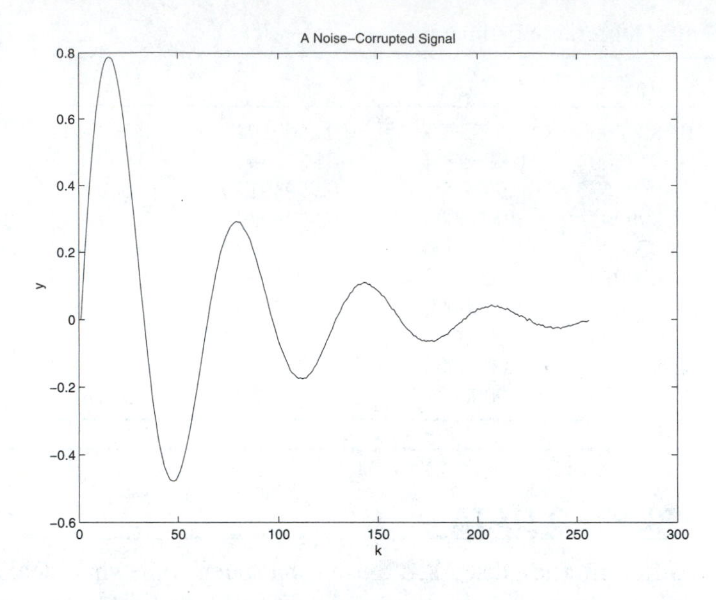

FIGURE 7.3.2
Derivative of a Noise-Corrupted Signal

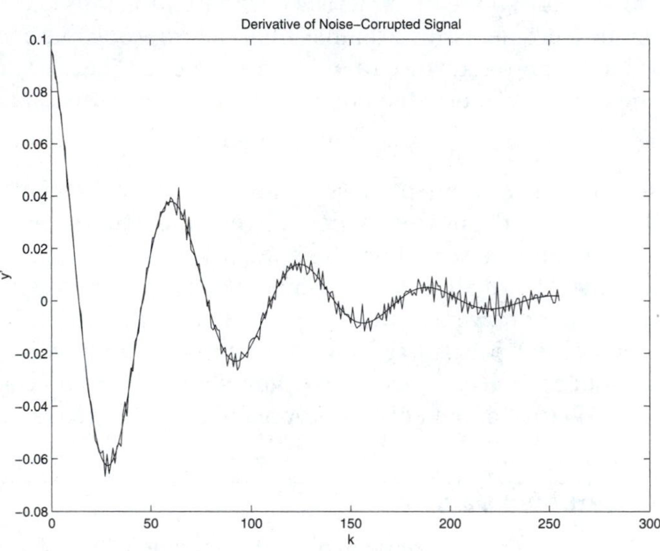

The problem posed by measurement noise can be addressed in two ways. One is to perform a least-squares curve fit to the data, as was done in Chapter 5, and then differentiate the curve. Care must be taken to capture only the underlying trend of the data. This way, the random effects of the noise are averaged out. If the fit is made too precise, the noise itself is being included in the fit.

The other approach is to pass the data through a low-pass filter to remove the most troublesome component of it, the high-frequency noise. Although this does not filter out all of the noise, the noise remaining is low-frequency noise, which is less troublesome. The filter output is then differentiated numerically. The following example illustrates this technique.

EXAMPLE 7.3.2 Filtering

Let y_k be the damped sinusoid examined in Example 7.3.1, where it is assumed, for convenience, that the sampling frequency is $f_s = 1$ sample/sec. Using a technique discussed in Chapter 10, a low-pass digital filter of order $p = 3$ with a gain of unity and a cutoff frequency of $f_c = f_s/4$ can be designed as follows.

$$z_k = -0.0085y_k - 0.001y_{k-1} + 0.2451y_{k-2} + 0.5y_{k-3} + 0.2451y_{k-4} - 0.001y_{k-5} - 0.0085y_{k-6}$$

This represents a compromise between making f_c small to get rid of as much noise as possible and making f_c large to preserve as much of the underlying signal x_k as possible. Example program *e732* on the distribution CD computes the filter output and its derivative. A plot of the derivative is shown in Figure 7.3.3. Comparison with Figure 7.3.2 shows that the effects of noise have been reduced but not removed. This has been obtained at the expense of generating a derivative signal that is delayed by $p = 3$ samples where p is the order of the low-pass filter. The derivative signal can be made smoother by increasing the filter order p, but this causes a larger delay. Thus, $p = 3$ represents a compromise.

FIGURE 7.3.3 Derivative of Filtered Signal

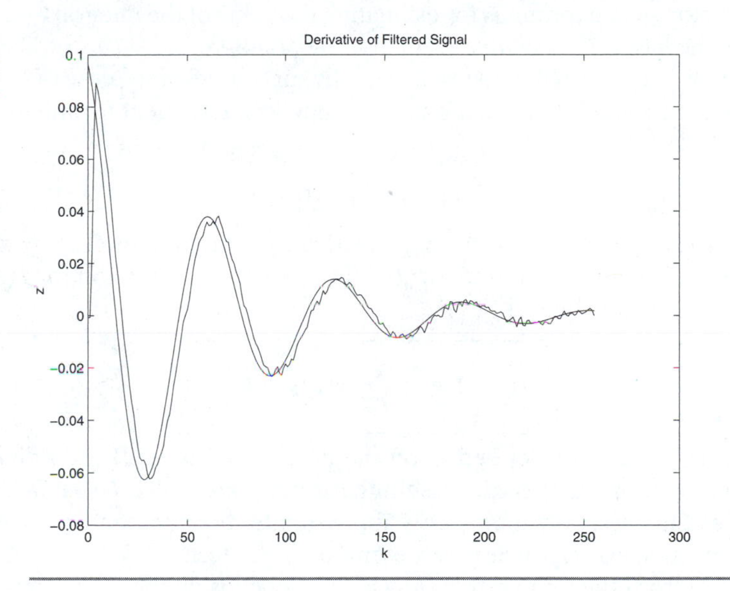

It is clear from the derivative of z_k in Figure 7.3.3 that it is difficult to obtain a clean accurate estimate of the derivative of a noise-corrupted signal. This is a consequence of the fact that numerical differentiation is an inherently ill-conditioned process. Engineers often try to avoid taking numerical derivatives, when possible, by using additional sensors. For example, a tachometer can be used to measure motor speed directly.

7.4 NEWTON-COTES INTEGRATION FORMULAS

Next, we examine the problem of evaluating the integral of a function $f(x)$ over an interval $[a, b]$. Initially, we assume that the limits of integration, a and b, are finite and that $f(x)$ does not have any singularities in the interval $[a, b]$. Thus, the objective is to develop numerical techniques that efficiently and accurately estimate the value of

$$I(a,b) = \int_a^b f(x)\,dx \tag{7.4.1}$$

Of course, many integrals can be evaluated in closed form by identifying the integrand as the derivative of another function and then applying the fundamental theorem of calculus. Handbooks of tables of integrals are available for this purpose (Beyer, 1991). There is also commercial software available for performing symbolic computations that include the evaluation of common derivatives and integrals (Moses, 1971). However, there are instances where closed-form expressions for an integral do not exist. For example, $f(x)$ may be generated from experimental measurements and is therefore available only in tabulated form. In this case, the integral must be evaluated numerically. Even when an analytical expression for $f(x)$ is available, there is no guarantee that a closed-form expression for its integral exists.

To develop simple formulas for estimating the value of the integral $I(a, b)$, we begin by sampling the interval $[a, b]$ with a set of discrete points $\{x_0, x_1, \ldots, x_n\}$ where $x_0 = a$ and $x_n = b$. We will assume that the points are equally spaced with a step size of $h = (b - a)/n$. The following notational shorthand is used to simplify subsequent formulations.

$$x_k \triangleq x_0 + kh \quad , \quad 0 \le k \le n \tag{7.4.2a}$$

$$f_k \triangleq f(x_k) \quad , \quad 0 \le k \le n \tag{7.4.2b}$$

The basic approach to estimating $I(a, b)$ is to approximate $f(x)$ by a polynomial $p_n(x)$ of degree n, and then integrate $p_n(x)$. The resulting expression for $I(a, b)$ consists of a weighted sum of values of $f(x)$.

$$I(a,b) = h \sum_{k=0}^{n} w_k f_k + O(h^j) \tag{7.4.3}$$

The weighting coefficients depend upon the polynomial used. If the polynomial is an interpolating polynomial, then the resulting formulas are called *Newton-Cotes* integration formulas. Since the polynomial only approximates the integrand $f(x)$, there is some formula truncation error, which is represented by the term $O(h^j)$. As h decreases, the error goes to zero at the same rate as h^j where j depends on the approximation used.

Recall from Equation (4.3.8) that an interpolating polynomial can be expressed in terms of forward differences using Newton's forward difference formula as follows.

$$f(x_0 + \alpha h) = f_0 + \alpha \Delta f_0 + \frac{\alpha(\alpha - 1)\Delta^2 f_0}{2!} + \cdots + E_n(\alpha) \tag{7.4.4a}$$

$$E_n(\alpha) = \frac{\alpha(\alpha - 1)\cdots[\alpha - (n - 1)]f^{(n)}(\xi)h^n}{n!} \tag{7.4.4b}$$

Here $E_n(\alpha)$ is the truncation error for an nth order forward difference formula, and $\Delta^k f_0$ denotes the kth forward difference of $f(x)$ evaluated at $x = x_0$. The first few forward differences are summarized in Table 7.4.1. Note that each new row in Table 7.4.1 can be generated from the previous row by applying the first forward difference operator Δ.

TABLE 7.4.1 Forward Differences of $f(x)$ at $x = x_0$

k	$\Delta^k f_0$
1	$f_1 - f_0$
2	$f_2 - 2f_1 + f_0$
3	$f_3 - 3f_2 + 3f_1 - f_0$
4	$f_4 - 4f_3 + 6f_2 - 4f_1 + f_0$
5	$f_5 - 5f_4 + 10f_3 - 10f_2 + 5f_1 - f_0$
6	$f_6 - 6f_5 + 15f_4 - 20f_3 + 15f_2 - 6f_1 + f_0$

7.4.1 Trapezoid Rule

The simplest of the Newton-Cotes integration formulas corresponds to the case $n = 1$, which is linear interpolation. Using the constant and linear terms in Equation (7.4.4a), and letting $x = x_0 + \alpha h$ and $dx = hd\alpha$ in (7.4.1), we get the following integral.

$$\begin{aligned} I(x_0, x_1) &= \int_0^1 [f_0 + \Delta f_0 \alpha + E_2(\alpha)] h d\alpha \\ &= \int_0^1 [f_0 + (f_1 - f_0)\alpha] h d\alpha + \int_0^1 \frac{f^{(2)}(\xi)\alpha(\alpha - 1)h^3}{2} d\alpha \\ &= \frac{h(f_0 + f_1)}{2} - \frac{f^{(2)}(\xi)h^3}{12} \end{aligned} \tag{7.4.5}$$

Thus, the integral of $f(x)$ from x_0 to x_1 is approximated by taking the area of a trapezoid of width h and average height $(f_0 + f_1)/2$ as shown in Figure 7.4.1. The derivative evaluation point, ξ, in the truncation error term lies somewhere in the interval $[x_0, x_1]$.

It is clear from Figure 7.4.1 that error in the approximation (the area between the two curves) can be substantial when large values of h are used. However, the approximation can be improved by partitioning the interval $[a, b]$ into n *panels* where $n > 1$. Linear interpolation can be used over each panel as in Equation (7.4.5), and the results can be summed to provide the following composite piecewise-linear interpolation formula known as the *trapezoid rule*.

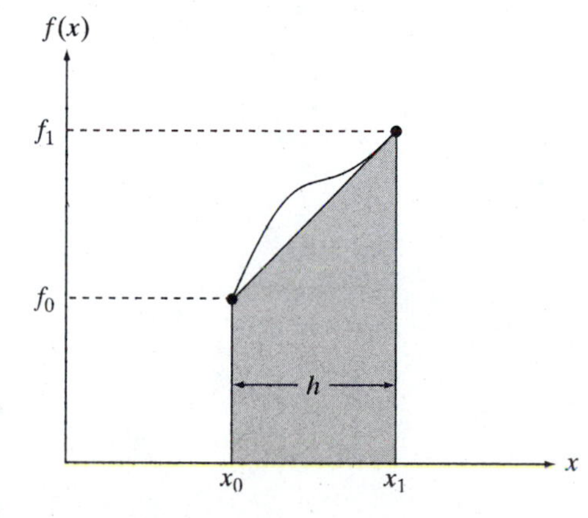

FIGURE 7.4.1 Trapezoid Rule Integration

$$I(x_0, x_n) = \frac{h}{2}(f_0 + 2f_1 + \cdots + 2f_{n-1} + f_n) - \frac{\overline{f}^{(2)}(b-a)h^2}{12} \tag{7.4.6}$$

Here $\overline{f}^{(2)}$, in the error term, represents the average of $f^{(2)}(\xi)$ over the n panels. Note that each of the interior points is counted twice and therefore has a coefficient of two. The global formula truncation error is of order $O(h^2)$, rather than $O(h^3)$, because the n local errors have been summed with $n = (b - a)/h$. Of course, it is also possible to express the truncation error in terms of n rather than h. Since $h = (b - a)/n$, it follows that the global truncation error is of order $O(n^{-2})$.

7.4.2 Simpson's Rules

The accuracy of the estimate of $I(a, b)$ can be improved by using a higher-degree interpolating polynomial. The case $n = 2$ corresponds to using a quadratic polynomial to approximate $f(x)$ over the interval $[x_0, x_2]$. Using the constant, linear, quadratic, and cubic terms in Equation (7.4.4a), and letting $x = x_0 + \alpha h$ and $dx = h d\alpha$ in (7.4.1), we get the following integral over two panels.

$$\begin{aligned} I(x_0, x_2) &= \int_0^2 \left[f_0 + \Delta f_0 \alpha + \frac{\Delta^2 f_0 \alpha(\alpha - 1)}{2} + \frac{\Delta^3 f_0 \alpha(\alpha-1)(\alpha-2)}{6} + E_4(\alpha) \right] h d\alpha \\ &= \int_0^2 \left[f_0 + (f_1 - f_0)\alpha + \frac{(f_2 - 2f_1 + f_0)\alpha(\alpha - 1)}{2} \right] h d\alpha + \\ &\quad \int_0^2 \left[\frac{f^{(4)}(\xi_0)\alpha(\alpha-1)(\alpha-2)(\alpha-3)h^5}{24} \right] d\alpha \\ &= \frac{h(f_0 + 4f_1 + f_2)}{3} - \frac{f^{(4)}(\xi)h^5}{90} \end{aligned} \tag{7.4.7}$$

In this case, the integral of $f(x)$ from x_0 to x_2 is approximated by taking the area under a quadratic polynomial as shown in Figure 7.4.2. Note that formula truncation error in Equation (7.4.7) is of order $O(h^5)$, instead of the expected $O(h^4)$. This rather nice result is a consequence of the fact that the integral of the first neglected term (the cubic term

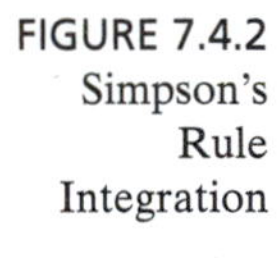

FIGURE 7.4.2 Simpson's Rule Integration

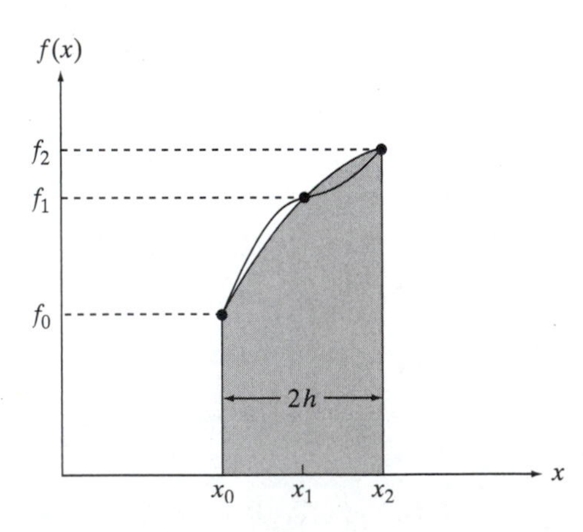

involving $\Delta^3 f_0$) turns out to be zero. The derivative evaluation point ξ lies somewhere in the interval $[x_0, x_2]$.

Just as with linear interpolation, the quadratic interpolation approximation can be improved by partitioning the interval into n panels where n is even and $n > 2$. Quadratic interpolation can be used over each pair of panels, and the results can be summed to provide the following composite piecewise-quadratic interpolation formula known as *Simpson's 1/3 rule* or simply Simpson's rule.

$$I(x_0, x_n) = \frac{h}{3}(f_0 + 4f_1 + 2f_2 + 4f_3 + \cdots + 4f_{n-1} + f_n) - \frac{\overline{f}^{(4)}(b-a)h^4}{180} \tag{7.4.8}$$

Here $\overline{f}^{(4)}$ represents the average of $f^{(4)}(\xi)$ over the $n/2$ pairs of panels. Again, note that the global truncation error is of order $O(h^4)$ because the $n/2$ local truncation errors have to be summed with $n = (b-a)/h$. Having a global truncation error of order $O(h^4)$ is still quite good because it means that reducing h by a factor of two reduces the error by a factor of 16.

The procedure for developing Newton-Cotes integration formulas based on an interpolating polynomial approximation of the integrand can be extended to higher-degree polynomials (see, e.g., Nakamura, 1993). Using a third-degree polynomial results in the following formulation for three panels.

$$I(x_0, x_3) = \frac{3h}{8}(f_0 + 3f_1 + 3f_2 + f_3) - \frac{3f^{(4)}(\xi)h^5}{80} \tag{7.4.9}$$

Notice that the local truncation error is still of order $O(h^5)$, just as with Simpson's 1/3 rule in Equation (7.4.7). The formulation in (7.4.5) is useful when there are three panels. More generally, if there are a multiple of three panels, then the following composite form known as *Simpson's 3/8* rule can be used.

$$I(x_0, x_n) = \frac{3h}{8}(f_0 + 3f_1 + 3f_2 + 2f_3 + 3f_4 + \cdots + 3f_{n-1} + f_n) - \frac{\overline{f}^{(4)}(b-a)h^4}{80} \tag{7.4.10}$$

Here the interior end-point terms, $\{f_{3i}\}$, have a coefficient of two because they are counted twice. If a fourth-degree polynomial approximation is used, it generates *Boole's rule,* which has a global truncation error of order $O(h^5)$. Although it might appear that it would be advantageous to use higher and higher degree approximations for $f(x)$, one quickly encounters a point of diminishing return. This is a consequence of the observation, first made in Chapter 4, that higher-order interpolating polynomials often do not provide good approximations to functions because they tend to oscillate wildly between the samples.

7.4.3 Midpoint Rule

The formulas for approximating an integral developed thus far, the trapezoid rule and Simpson's rules, are examples of *closed* integration formulas because they include evaluations of $f(x)$ at the two endpoints x_0 and x_n. It is also possible to develop analogous *open* formulas that only require values of $f(x)$ in the interior of the interval $[x_0, x_n]$. Open integration formulas are useful in those instances where $f(x)$ has a singularity at one or both of the end points. To develop open integration formulas, it is useful to perform a forward difference expansion of $f(x)$ about the point x_1 as follows.

$$f(x_1 + \alpha h) = f_1 + \alpha \Delta f_1 + \frac{\alpha(\alpha - 1)\Delta^2 f_1}{2!} + \cdots + E_n(\alpha) \tag{7.4.11}$$

Here $\Delta f_1 = f_2 - f_1$ denotes the first order forward difference of $f(x)$ evaluated at $x = x_1$, and $\Delta^2 f_1 = f_3 - 2f_2 + f_1$ denotes the second order forward difference of $f(x)$ at $x = x_1$.

The simplest case of an open integration formula is generated by approximating $f(x)$ over the interval $[x_0, x_2]$ using an interpolating polynomial of degree zero (a constant) passing through the midpoint. Using the constant and linear terms in Equation (7.4.11) and letting $x = x_1 + \alpha h$ and $dx = hd\alpha$ in (7.4.1), we get the following integral over two panels.

$$\begin{aligned} I(x_0, x_2) &= \int_{-1}^{1} [f_1 + \Delta f_1 \alpha + E_2(\alpha)]hd\alpha \\ &= \int_{-1}^{1} [f_1 + (f_2 - f_1)\alpha]hd\alpha + \int_{-1}^{1} \left[\frac{f^{(2)}(\xi)\alpha(\alpha - 1)h^3}{2}\right] d\alpha \\ &= 2hf_1 + \frac{f^{(2)}(\xi)h^3}{3} \end{aligned} \tag{7.4.12}$$

In this case, the integral of $f(x)$ from x_0 to x_2 is approximated by taking the area of a rectangle of width $2h$ and height f_1 as shown in Figure 7.4.3.

Interestingly enough, though the polynomial degree in this case is less than that for the trapezoid rule, the local truncation error is still $O(h^3)$. This is because the integral of the first neglected term (the linear term) is zero. The midpoint method is similar to Simpson's 1/3 rule in that regard. In general, if an *even* degree interpolating polynomial is used, the integral of the first neglected term will be zero, which means the order of the truncation error will be one higher than would otherwise be expected.

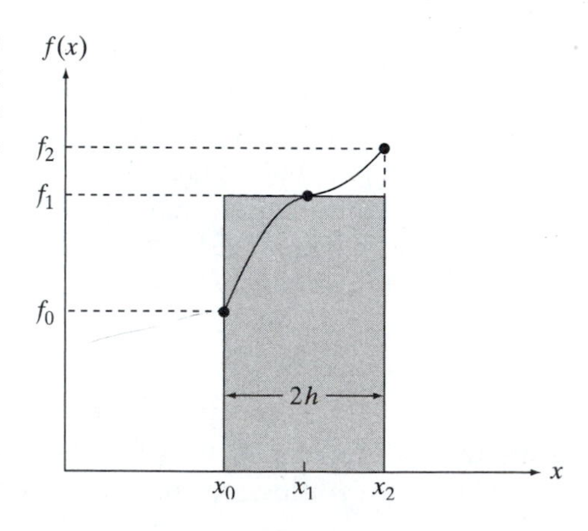

FIGURE 7.4.3 Midpoint Rule Integration

As with the closed formulas, the constant interpolation approximation can be improved by partitioning the interval into n panels where n is even and $n > 2$. Constant interpolation can be used over each pair of panels, and the results can be summed to provide the following composite piecewise-constant interpolation formula known as the *midpoint rule*.

$$I(x_0, x_n) = 2h(f_1 + f_3 + \cdots + f_{n-3} + f_{n-1}) + \frac{\overline{f}^{(2)}(b-a)h^2}{6} \tag{7.4.13}$$

Here $\overline{f}^{(2)}$ represents the average of $f^{(2)}(\xi)$ over the $n/2$ pairs of panels. The global truncation error is of order $O(h^2)$, rather than $O(h^3)$, because the $n/2$ local truncation errors have to be summed with $n = (b-a)/h$. Open integration formulas based on higher-degree interpolating polynomials can also be developed in an analogous manner (Hultquist, 1988). If an mth degree interpolating polynomial is used, the total number of panels must be a multiple of $m + 2$. Alternatively, one can develop hybrid integration formulas that combine different integration rules so as to treat the end panels differently than the interior panels. For example, if the number of panels is at least five, an open formula can be generated by using the midpoint rule at each end, and using the trapezoid rule for the remaining interior panels. If the number of panels is even, then Simpson's rule can be used on the interior panels, thereby increasing the order of the truncation error. By combining integration rules, it is also possible to develop *semi-open* formulas, formulas that are open at one end and closed at the other end.

EXAMPLE 7.4.1 Newton-Cotes Integration

To illustrate the relative convergence rates of the Newton-Cotes integration formulas, consider the following function, which has a simple closed-form integral.

$$f(x) = \alpha \exp(x)$$

If the scalar α is chosen to be $\alpha = 1/[\exp(b) - \exp(a)]$, then the exact value of the integral of $f(x)$ is $I(a,b) = 1$. Example program *e741* on the distribution CD applies the trapezoid rule, the midpoint rule, and Simpson's 1/3 rule using $a = 0$ and $b = 1$ with the results summarized in Table 7.4.2.

TABLE 7.4.2 Newton-Cotes Formulas

n	*Trapezoid*	*Midpoint*	*Simpson*
2	0.9595174	1.0207471	1.0003371
4	0.9896588	1.0052029	1.0000216
8	0.9974006	1.0013016	1.0000014
16	0.9993493	1.0003256	1.0000001
32	0.9998373	1.0000815	1.0000000
64	0.9999593	1.0000203	1.0000000
128	0.9999897	1.0000051	1.0000000
256	0.9999974	1.0000011	0.9999996
512	0.9999995	1.0000006	1.0000004
1024	0.9999995	1.0000005	1.0000004

The first column is the number of panels n. In this case, the panel width is $h = 1/n$. The remaining columns show the estimates of $I(0,1)$ using the three integration rules. It is evident that Simpson's 1/3 rule converges faster than the other two rules, as expected. Note that its results are exact to seven decimal places for $32 \le n \le 128$, but for larger values of n, the effects of round-off error begin to manifest themselves.

7.5 ROMBERG INTEGRATION

When an analytical representation of $f(x)$ is available, the integral of $f(x)$ can approximated using a variety of step sizes h. By comparing the estimates obtained using different step sizes, we can employ the Richardson extrapolation technique used in Section 7.2.3 to estimate what the integral would be in the limit as the step size approaches zero. This is similar to the extrapolation approach used in Equation (7.2.22) to improve the estimate of the derivative. Consider a sequence of m step sizes starting with the original interval $b - a$ and reducing the step size by a factor of two with each iteration.

$$h_k \triangleq \frac{b-a}{2^{k-1}} \quad , \quad 1 \le k \le m \tag{7.5.1}$$

Next, let R_{k1} denote the estimate of the integral using step size h_k, and let E_{k1} denote the associated truncation error. Since the exact value of the integral is $R_{k1} + E_{k1}$, it follows that

$$R_{k1} + E_{k1} = R_{k-1,1} + E_{k-1,1} \tag{7.5.2}$$

Suppose the trapezoid rule is used to compute the estimates of the integral. In this case, the global truncation error can be approximated as $E_{k1} \approx \alpha h_k^2$ when $|h_k| \ll 1$. It is clear from Equation (7.5.1) that $|h_k| \ll 1$ for sufficiently large values of k. Since $h_k = h_{k-1}/2$, it follows that successive estimates of the truncation error satisfy

$$E_{k1} \approx \frac{E_{k-1,1}}{4} \tag{7.5.3}$$

Substituting the expression for $E_{k-1,1}$ from (7.5.3) into (7.5.2), and solving for E_{k1} then yields

$$E_{k1} \approx \frac{R_{k1} - R_{k-1,1}}{3} \tag{7.5.4}$$

The estimate of the truncation error, E_{k1}, can now be added to the original estimate of the integral, R_{k1}, thereby producing a new *extrapolated* estimate R_{k2} as follows.

$$R_{k2} = \frac{4R_{k1} - R_{k-1,1}}{3} \tag{7.5.5}$$

Whereas the original estimates, R_{k1}, have truncation errors of order $O(h^2)$, it is possible to show (Faires and Burden, 1993) that the new extrapolated estimates, R_{k2}, have truncation errors of order $O(h^4)$. Moreover, there is nothing to stop us from extrapolating from the extrapolated estimates to produce higher-level extrapola-

FIGURE 7.5.1 Romberg Array

$$\begin{bmatrix} R_{11} & & & & \\ R_{21} & R_{22} & & & \\ R_{31} & R_{32} & R_{33} & & \\ \vdots & \vdots & \vdots & \ddots & \\ R_{m1} & R_{m2} & R_{m3} & \cdots & R_{mm} \end{bmatrix}$$

tions, and so on up to extrapolations of order m. In general, the truncation error E_{kj} for the jth order extrapolation will be of order $O(h^{2j})$. This technique is known as *Romberg integration*. The formula for computing estimate R_{kj} in terms of previous estimates is

$$R_{kj} = \frac{4^{j-1}R_{k,j-1} - R_{k-1,j-1}}{4^{j-1} - 1}, \quad 1 \le k \le m, \quad 2 \le j \le k \tag{7.5.6}$$

Note that as the row index k ranges from 1 to m, the column index j ranges from 2 to k. Thus the extrapolated estimates can be arranged into a lower-triangular matrix known as the *Romberg array* shown in Figure 7.5.1. The first column represents trapezoid rule estimates with successively smaller step sizes. Each of the remaining entries is obtained by forming a linear combination of two adjacent estimates, to the west and the northwest. A sequence of extrapolations of increasing order is then obtained from the diagonal elements $\{R_{11}, R_{22}, \ldots, R_{mm}\}$. The final estimate, R_{mm}, has a truncation error, E_{mm}, of order $O(h^{2m})$. Of course as the step size decreases, and the order of extrapolation increases, at some point the effects of round-off error begin to dominate and no further increase in accuracy is obtained.

It is clear from Equation (7.5.6) that the computation of the estimates in the Romberg array can be carried out one row at a time. Consequently, there is no need to know m ahead of time. Instead, the truncation error $E_{k,k-1}$ of estimate $R_{k,k-1}$ can be used as part of the termination criterion.

$$E_{k,k-1} = \frac{R_{k,k-1} - R_{k-1,k-1}}{4^{k-1} - 1}, \quad 2 \le k \le m \tag{7.5.7}$$

The process can be terminated when $|E_{k,k-1}|$ falls below an error threshold ε, or when a maximum extrapolation level m has been reached. The following algorithm uses this approach.

Alg. 7.5.1 **Romberg Integration**

1. Pick $\varepsilon > 0$ and $m \ge 2$.
2. Set $k = 1, n = 1, h = b - a$, and $R_{11} = h(f_0 + f_1)/2$.
3. Do

   {

   (a) Set $k = k + 1, n = 2n$, and $h = h/2$.

(b) Compute the trapezoid rule estimate

$$R_{k1} = \frac{h}{2}(f_0 + 2f_2 + \cdots + 2f_{n-1} + f_n)$$

(c) For $j = 2$ to k compute the extrapolated estimates
{

$$R_{kj} = \frac{4^{j-1}R_{k,j-1} - R_{k-1,j-1}}{4^{j-1} - 1}$$

}

(d) Compute the estimated truncation error

$$E_{k,k-1} = \frac{R_{k,k-1} - R_{k-1,k-1}}{4^{k-1} - 1}$$

}

4. While $(|E_{k,k-1}| \geq \varepsilon)$ and $(k < m)$.

When Alg. 7.5.1 terminates, the estimated value of the integral is R_{kk}. If $k < m$, the termination criterion $|E_{k,k-1}| < \varepsilon$ has been achieved. In steps (2) and (3) of Alg. 7.5.1, the shorthand notation $f_k = f(a + kh)$ is used. It is possible to restructure the calculation of R_{k1} in step (3) such that it uses $R_{k-1,1}$ plus terms involving only new values of $f(x)$ (Hultquist, 1988).

EXAMPLE 7.5.1 Romberg Integration

Romberg's extrapolation method converges very rapidly. To illustrate just how fast, consider the following function.

$$f(x) = \alpha \exp(x) \quad , \quad 0 \leq x \leq 1$$

This function was considered previously in Example 7.4.1 with the exact value of the integral being unity when $\alpha = 1/(e - 1)$. Example program *e751* on the distribution CD applies Romberg integration, with the results summarized in Table 7.5.1. The first column is the extrapolation level k, the next five are the entries in the Romberg table, and the last column is the actual error E. This table was generated by applying Alg. 7.5.1 with $\varepsilon = 0$ and $m = 5$. Larger values of m do not improve the total error due to the contributions of round-off error.

TABLE 7.5.1 Romberg Integration Table

k	R_{k1}	R_{k2}	R_{k3}	R_{k4}	R_{k5}	E
1	1.0819767					0.0819767
2	1.0207471	1.0003372				0.0003372
3	1.0052029	1.0000215	1.0000004			0.0000004
4	1.0013016	1.0000012	0.9999998	0.9999998		−0.0000002
5	1.0003256	1.0000002	1.0000001	1.0000001	1.0000001	0.0000001

7.6 GAUSS QUADRATURE

Numerical integration is often referred to by the term *quadrature*. This usage has a long history, which arose from an ancient problem in geometry that involved constructing a square with an area the same as that of a given circle (Kahaner et al., 1989). Recall that each of the Newton-Cotes quadrature formulas was based on using an interpolating polynomial to approximate the integrand. As a consequence, the Newton-Cotes formulas are exact when the integrand is a low-order polynomial. For example, the midpoint and trapezoid rules are exact for linear polynomials, while Simpson's rule is exact for cubic polynomials. Each of these integration formulas was based on the assumption that the points at which $f(x)$ is evaluated are uniformly spaced. If we now relax this assumption, it is possible to obtain still higher accuracy using the same number of points. The basic idea, called the *Gauss quadrature method*, is to choose the evaluation points x_k and coefficients c_k such that the following approximation is exact for polynomials of degree $2n - 1$.

$$I_w(-1,1) \triangleq \int_{-1}^{1} f(x)w(x)\,dx \approx \sum_{k=1}^{n} c_k f(x_k) \tag{7.6.1}$$

Here the integrand has been generalized to include a *weighting function*, $w(x) > 0$. For now, it is sufficient to consider the case of uniform weighting, namely, $w(x) = 1$. Later, we examine non-uniform weighting functions that allow us to evaluate integrals with infinite limits of integration. To simplify the development of the Gauss quadrature method, we assume that the interval of integration is $[-1,1]$. This is not a limiting assumption because we can always transform or map $[-1,1]$ into a general finite interval $[a,b]$ by a change of variable. In particular, suppose $z = \alpha x + \beta$. We choose α and β such that

$$\alpha(-1) + \beta = a \tag{7.6.2a}$$

$$\alpha(1) + \beta = b \tag{7.6.2b}$$

Solving this linear algebraic system for α and β yields the following transformation, which maps the interval $[-1,1]$ into the interval $[a,b]$.

$$z = \overbrace{\left(\frac{b-a}{2}\right)}^{\alpha} x + \overbrace{\frac{b+a}{2}}^{\beta} \tag{7.6.3}$$

Given a function $f(z)$ where $a \le z \le b$, if z is replaced by $\alpha x + \beta$ and dz is replaced by αdx, then the range of integration is normalized to $[-1,1]$. Thus, if $\{x_k\}$ are evaluation points in the interval $[-1,1]$, then the general Gauss quadrature approximation becomes

$$\boxed{I_w(a,b) \approx \alpha \sum_{k=1}^{n} c_k f(\alpha x_k + \beta)} \tag{7.6.4}$$

Note that when $a = -1$ and $b = 1$, we have $\alpha = 1$ and $\beta = 0$, in which case Equation (7.6.4) reduces to Equation (7.6.1).

7.6.1 Legendre Polynomials

Our objective is to choose the n evaluation points x_k plus the n coefficients c_k such that the approximation in Equation (7.6.1) is exact when $f(x)$ is a polynomial of degree $2n - 1$. Suppose $f(x) = x^k$. Then the constraint that (7.6.1) be exact for $0 \le k < 2n$ generates the following system of $2n$ nonlinear algebraic equations.

$$c_1 x_1^k + c_2 x_2^k + \cdots + c_n x_n^k = \frac{1 - (-1)^{k+1}}{k + 1} \quad , \quad 0 \le k < 2n \tag{7.6.5}$$

For the simplest case $n = 1$, the first equation is $c_1 = 2$ and the second is $c_1 x_1 = 0$, which yields $x_1 = 0$. Thus when $n = 1$ we have $I_w(-1,1) = 2f(0)$, which is the midpoint rule.

The case $n = 2$ is more involved. Applying Equation (7.6.5), we get the following four-dimensional nonlinear algebraic system.

$$c_1 + c_2 = 2 \tag{7.6.6a}$$

$$c_1 x_1 + c_2 x_2 = 0 \tag{7.6.6b}$$

$$c_1 x_1^2 + c_2 x_2^2 = 2/3 \tag{7.6.6c}$$

$$c_1 x_1^3 + c_2 x_2^3 = 0 \tag{7.6.6d}$$

The system in (7.6.6) can be regarded as a system of the form $g(y) = 0$ where $y = [x_1, x_2, c_1, c_2]^T$, which can be solved using root finding techniques from Chapter 5. The solution in this case is $x_{1,2} = \pm 1/\sqrt{3}$ and $c_{1,2} = 1$. Although we can proceed in this manner for larger values of n, there is a way to decouple the problem into two smaller subproblems. It can be shown that the evaluation points, x_k, are the roots of a special family of polynomials called the *Legendre polynomials*. The first two Legendre polynomials are $P_0(x) = 1$ and $P_1(x) = x$. The remaining members of the sequence are generated by the following two-term recurrence relation.

$$\boxed{(k + 1)P_{k+1}(x) = (2k + 1)xP_k(x) - kP_{k-1}(x) \quad , \quad k \ge 1} \tag{7.6.7}$$

Thus $P_2(x) = (3x^2 - 1)/2$. Observe that when $n = 1$, the evaluation point $x_1 = 0$ is the root of $P_1(x) = x$. Similarly, when $n = 2$, the evaluation points $x_{1,2} = \pm 1/\sqrt{3}$ are the roots of $P_2(x) = (3x^2 - 1)/2$. For a general $n \ge 1$, the evaluation points are the roots of the nth Legendre polynomial, $P_n(x)$.

Once the evaluation points are available, the coefficients c_k can be determined. One way to do this is to apply Equation (7.6.5) for the case when k is even. This results in the following linear algebraic system of n equations in the n unknown coefficients.

$$x_1^{2j} c_1 + x_2^{2j} c_2 + \cdots + x_n^{2j} c_n = \frac{2}{2j + 1} \quad , \quad 0 \le j < n \tag{7.6.8}$$

Given the roots $\{x_1, x_2, \ldots, x_n\}$, this system can be solved for the $n \times 1$ coefficient vector c using the methods from Chapter 2. A summary of the roots x_k and the coefficients c_k for the first few Legendre polynomials is shown in Table 7.6.1. Additional values can be found in Beyer (1991). Note that all of the roots lie within the open interval $(-1,1)$, which makes the *Gauss-Legendre quadrature formulas* open integration for-

TABLE 7.6.1 Gauss-Legendre Quadrature: $w(x) = 1$

n	*Roots*	*Coefficients*
1	$x_1 = 0.00000000$	$c_1 = 2.00000000$
2	$x_{1,2} = \pm 0.57735027$	$c_{1,2} = 1.00000000$
3	$x_1 = 0.00000000$	$c_1 = 0.88888889$
	$x_{2,3} = \pm 0.77459667$	$c_{2,3} = 0.55555556$
4	$x_{1,2} = \pm 0.33998104$	$c_{1,2} = 0.65214515$
	$x_{3,4} = \pm 0.86113631$	$c_{3,4} = 0.34785485$
5	$x_1 = 0.00000000$	$c_1 = 0.56888889$
	$x_{2,3} = \pm 0.53846931$	$c_{2,3} = 0.47862867$
	$x_{3,4} = \pm 0.90617985$	$c_{4,5} = 0.23692689$
6	$x_{1,2} = \pm 0.23861919$	$c_{1,2} = 0.46791393$
	$x_{3,4} = \pm 0.66120939$	$c_{3,4} = 0.36076157$
	$x_{5,6} = \pm 0.93246951$	$c_{5,6} = 0.17132449$

mulas. One of the drawbacks of the Gauss quadrature method is that the formula parameters are typically irrational numbers; hence, they can never be represented exactly using finite precision.

The technique for finding the coefficients depicted in Equation (7.6.8) is straightforward, but it suffers from the drawback that the equations are ill-conditioned. Consequently, for large values of n, the numerical solution is not reliable. As an alternative, it can be shown that if $Q_k(x)$ is the kth Lagrange interpolating polynomial of degree n, as defined in (4.2.7), then the kth coefficient can be computed as follows (Hultquist, 1988).

$$c_k = \int_{-1}^{1} Q_k(x)\, dx \quad , \quad 1 \le k \le n \tag{7.6.9}$$

The Legendre polynomials have many interesting properties. One important property is that they form an *orthogonal* sequence on $[-1, 1]$ with respect to the weighting function $w(x) = 1$. That is,

$$\langle P_k, P_j \rangle \triangleq \int_{-1}^{1} P_k(x) P_j(x) w(x)\, dx = 0 \quad , \quad k \ne j \tag{7.6.10}$$

The integral $\langle P_k, P_j \rangle$ is referred to as the *inner product* of $P_k(x)$ with $P_j(x)$. An arbitrary polynomial can be written as a weighted sum of a complete set of orthogonal polynomials. In particular, if $g(x)$ is a polynomial of degree n, then

$$g(x) = \sum_{k=0}^{n} a_k P_k(x) \tag{7.6.11}$$

If we take the inner product of both sides of Equation (7.6.11) with $P_j(x)$ and then apply the orthogonality condition in (7.6.10), it is easy to show that the jth coefficient must be

$$a_j = \frac{\langle g, P_j \rangle}{\langle P_j, P_j \rangle} \quad , \quad 0 \le j \le n \tag{7.6.12}$$

It follows that the coefficients of an orthogonal series expansion can be computed *independently* from one another without solving a linear algebraic system. Instead, computing a_j requires performing numerical integration.

7.6.2 Chebyshev Polynomials

The Legendre polynomials can be generated by orthogonalizing the elementary polynomials $\{1, x, x^2, \ldots\}$ over the interval $[-1, 1]$ using uniform weighting $w(x) = 1$. If a different weighting function is used, a new set of orthogonal polynomials is generated. Consider, in particular, the following weighting function.

$$w(x) = \frac{1}{\sqrt{1 - x^2}} \tag{7.6.13}$$

When this weighting is used, the resulting orthogonal polynomials, $T_k(x)$, are called the *Chebyshev polynomials*. The first two Chebyshev polynomials are $T_0(x) = 1$ and $T_1(x) = x$. The remaining polynomials in the sequence are generated by the following two-term recurrence relation.

$$T_{k+1}(x) = 2xT_k(x) - T_{k-1}(x) \quad , \quad k \ge 1 \tag{7.6.14}$$

Recall that the Chebyshev polynomials were first encountered in Chapter 4 where they were used to reduce the oscillations in higher-order interpolating polynomials. If the roots of $T_n(x)$ are used as the evaluation points, x_k, the resulting integration formula is referred to as a *Gauss-Chebyshev quadrature formula*. Unlike the case of the Legendre polynomials, the roots of the Chebyshev polynomials do not have to be estimated numerically. Recall from Equation (4.4.21) that the n roots of $T_n(x)$ are

$$x_k = \cos\left[\frac{(2k - 1)\pi}{2n}\right] \quad , \quad 1 \le k \le n \tag{7.6.15}$$

The coefficients $\{c_k\}$ for the Gauss-Chebyshev quadrature formula are also available analytically. They have the interesting property that they are all equal.

$$c_k = \frac{\pi}{n} \quad , \quad 1 \le k \le n \tag{7.6.16}$$

A summary of the roots x_k and the coefficients c_k for the first few Chebyshev polynomials is shown in Table 7.6.2.

TABLE 7.6.2 Gauss-Chebyshev Quadrature: $w(x) = 1/\sqrt{1 - x^2}$

n	*Roots*	*Coefficients*
1	$x_1 = 0.00000000$	$c_1 = \pi$
2	$x_{1,2} = \pm 0.707106768$	$c_{1,2} = \pi/2$
3	$x_1 = 0.00000000$ $x_{2,3} = \pm 0.86602540$	$c_1 = \pi/3$ $c_{2,3} = \pi/3$
4	$x_{1,2} = \pm 0.92387953$ $x_{3,4} = \pm 0.38268343$	$c_{1,2} = \pi/4$ $c_{3,4} = \pi/4$
5	$x_1 = 0.00000000$ $x_{2,3} = \pm 0.95105552$ $x_{3,4} = \pm 0.58778525$	$c_1 = \pi/5$ $c_{2,3} = \pi/5$ $c_{4,5} = \pi/5$
6	$x_{1,2} = \pm 0.96592583$ $x_{3,4} = \pm 0.70710678$ $x_{5,6} = \pm 0.25881905$	$c_{1,2} = \pi/6$ $c_{3,4} = \pi/6$ $c_{5,6} = \pi/6$

7.6.3 Laguerre Polynomials

Since the Gauss quadrature approximations are open integration formulas, the values of the integrand at the two end points are not required. This feature can be exploited to extend one or both of the limits of integration to infinity. Consider the case when $a = 0$ and $b = \infty$. Of course, if a polynomial is integrated over the interval $[0, \infty)$, the result is unbounded. Consequently, we must consider a weighting function that goes to zero faster than polynomials go to infinity, namely,

$$w(x) = \exp(-x) \quad , \quad 0 \le x < \infty \tag{7.6.17}$$

When this exponential weighting function is used over the interval $[0, \infty)$, the resulting orthogonal polynomials, $L_k(x)$, are called the *Laguerre polynomials*. The first two Laguerre polynomials are $L_0(x) = 1$ and $L_1(x) = 1 - x$. The remaining polynomials in the sequence are generated by the following two-term recurrence relation.

$$(k + 1)L_{k+1}(x) = (2k + 1 - x)L_k(x) - kL_{k-1}(x) \quad , \quad k \ge 1 \tag{7.6.18}$$

If the roots of $L_n(x)$ are used as the evaluation points, x_k, then the resulting integration formula is referred to as a *Gauss-Laguerre quadrature formula*. There are no analytical expressions for the roots x_k or the coefficients c_k in this case. Instead the roots must be computed using the root finding techniques in Chapter 5. Once the roots are obtained, the coefficients can be determined by solving the following linear algebraic system:

$$x_1^{2k} c_1 + x_2^{2k} c_2 + \cdots + x_n^{2k} c_n = 2k! \quad , \quad 0 \le k < n \tag{7.6.19}$$

TABLE 7.6.3 Gauss-Laguerre Quadrature: $w(x) = \exp(-x)$

n	*Roots*	*Coefficients*
1	$x_1 = 1.00000000$	$c_1 = 1.00000000$
2	$x_1 = 0.58578644$	$c_1 = 0.85355339$
	$x_2 = 3.41421356$	$c_2 = 0.14644661$
3	$x_1 = 0.41577456$	$c_1 = 0.71109301$
	$x_2 = 2.29428036$	$c_2 = 0.27851773$
	$x_3 = 6.28994508$	$c_3 = 0.01038926$
4	$x_1 = 0.32254769$	$c_1 = 0.60315410$
	$x_2 = 1.74567110$	$c_2 = 0.35741869$
	$x_3 = 4.53662030$	$c_3 = 0.03888791$
	$x_4 = 9.39507091$	$c_4 = 0.00053929$

This system is analogous to Equation (7.6.8) and is obtained by applying the constraint that the approximation be exact when $f(x) = x^{2k}$ for $0 \le k < n$. A summary of the roots x_k and coefficients c_k for the first few Laguerre polynomials is shown in Table 7.6.3.

7.6.4 Hermite Polynomials

If both limits of integration are infinite, a weighting function must be chosen that goes to zero for both positive and negative values of x. One such function is the bell-shaped Gaussian distribution type function

$$w(x) = \exp(-x^2) \quad , \quad -\infty < x < \infty \tag{7.6.20}$$

When this weighting function is used over the interval $(-\infty, \infty)$, the resulting orthogonal polynomials, $H_k(x)$, are called the *Hermite polynomials*. The first two Hermite polynomials are $H_0(x) = 1$ and $H_1(x) = 2x$. The remaining polynomials are generated by the following two-term recurrence relation.

$$\boxed{H_{k+1}(x) = 2xH_k(x) - 2kH_{k-1}(x) \quad , \quad k \ge 1} \tag{7.6.21}$$

If the roots of $H_n(x)$ are used as the evaluation points, x_k, the resulting integration formula is referred to as a *Gauss-Hermite quadrature formula*. Again, there are no analytical expressions for the roots x_k, they must be found numerically. Once the roots are obtained, the coefficients can be determined by solving the following linear algebraic system.

$$\boxed{x_1^{2k}c_1 + x_2^{2k}c_2 + \cdots + x_n^{2k}c_n = \frac{1(3)5\cdots(2k-1)\sqrt{\pi}}{2^k} \quad , \quad 0 \le k < n} \tag{7.6.22}$$

This system of constraints is generated by making the approximation exact when $f(x) = x^{2k}$ for $0 \le k < n$. A summary of numerical values for the roots x_k and coefficients c_k for the first few Hermite polynomials is shown in Table 7.6.4.

TABLE 7.6.4 Gauss-Hermite Quadrature: $w(x) = \exp(-x^2)$

n	*Roots*	*Coefficients*
1	$x_1 = 0.00000000$	$c_1 = \sqrt{\pi}$
2	$x_{1,2} = \pm 0.70710678$	$c_{1,2} = 0.88622693$
3	$x_1 = 0.00000000$	$c_1 = 1.18163590$
	$x_{2,3} = \pm 1.22474487$	$c_{2,3} = 0.29540898$
4	$x_{1,2} = \pm 1.65068012$	$c_{1,2} = 0.08131284$
	$x_{3,4} = \pm 0.52464762$	$c_{3,4} = 0.80491409$

An example of an application of the Gauss-Legendre quadrature method can be found in Section 7.7, where an integral with infinite limits of integration is converted to two equivalent integrals with finite limits of integration.

7.7 IMPROPER INTEGRALS

In this section, we examine techniques that are applicable to integrals whose integrands contain singularities at one or both of the limits of integration. We also consider integrals for which the limits of integration are infinite.

$$I(a,b) = \int_a^b f(x)\,dx \tag{7.7.1}$$

Suppose $f(x)$ has a singularity at either a or b. For example, suppose $|f(x)| \to \infty$ as $x \to a$. The evaluation of $f(x)$ at the singularity can be avoided by using an *open* integration formula. Recall that the open integration formulas do not require the value of $f(x)$ at the end points. Examples of open integration formulas include the Gauss quadrature formulas and the midpoint rule formula. When an analytical expression for $f(x)$ is available, the midpoint rule integration formula can be reformulated in the following way where $h = (b - a)/n$.

$$I(a,b) \approx h \sum_{k=1}^{n} f\left[a + \frac{(2k-1)h}{2}\right] \tag{7.7.2}$$

In this case, the evaluation points are uniformly distributed with a step size h, and each of the end points is located within $h/2$ of a limit of integration as shown in Figure 7.7.1.

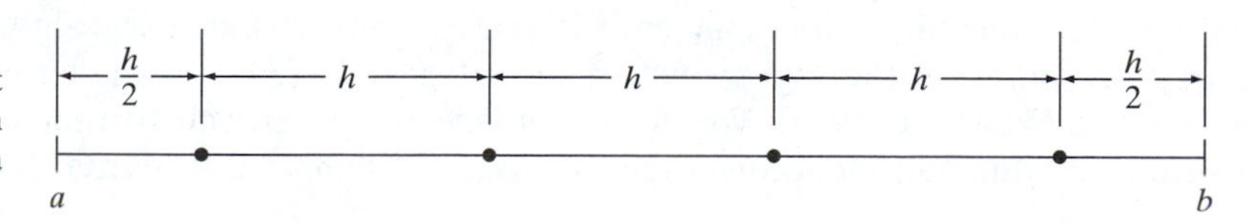

FIGURE 7.7.1 Midpoint Integration Formula

Unlike the original formulation of the midpoint method in Equation (7.4.13), the formulation in (7.7.2) uses all n points and does not require that n be even.

Not all functions with a singularity at a or b are integrable. For example, the integral of $f(x) = 1/x^p$ over the interval $[0, 1]$ does not exist when $p \geq 1$. When $I(a, b)$ does exist, an open integration formula such as Equation (7.7.2) can be used to approximate the integral while avoiding the singularity.

EXAMPLE 7.7.1 End Point Singularity

To illustrate the use of an open integration formula, consider the problem of numerically evaluating the following integral.

$$I(0,1) = \int_0^1 \frac{dx}{\sqrt{x}}$$

The integrand $f(x) = 1/\sqrt{x}$ is not defined at $x = 0$, yet the integral itself is well defined and has the value $I(0, 1) = 2$. Example program *e771* on the distribution CD applies the composite midpoint method in Equation (7.7.2) for various values of n with the results summarized in Table 7.7.1. The first column is the number of points n, the second is the estimated value of the integral, and the third is the percent error. The presence of the singularity makes this a challenging numerical problem with over one thousand points needed to achieve 1% accuracy in this case.

TABLE 7.7.1 Midpoint Integration of $1/\sqrt{x}$

n	$I(0,1)$	*Percent Error*
1	1.4142135	−29.2893231
2	1.5773503	−21.1324871
4	1.6988441	−15.0577962
8	1.7864610	−10.6769502
16	1.8488566	−7.5571716
32	1.8930882	−5.3455889
64	1.9243925	−3.7803769
128	1.9465351	−2.6732445
256	1.9621946	−1.8902719
512	1.9732664	−1.3366818
1024	1.9810950	−0.9452522

The slow convergence of the midpoint integration formula in Example 7.7.1 is due to the presence of the singularity at $a = 0$. Since the magnitude of the slope of $f(x)$ increases without bound as $x \to a$, the midpoint rule consistently underestimates the contribution to the integral that occurs to the left of the first point. One way to distribute the evaluation points more efficiently is to use an *adaptive* step size algorithm which places more points in the regions where the integrand $f(x)$ is changing most rapidly (Gerald and Wheatley, 1989). We do not pursue this approach further at this point because integrating a function is a special case of solving a differential equation, and adaptive step size methods for solving ordinary differential equations are considered in detail in Chapter 8.

Next, we examine the problem of evaluating integrals with infinite limits of integration. If the integration interval is $[0, \infty)$ and the integrand includes a factor $w(x) = \exp(-x)$, then Gauss-Laguerre quadrature is a natural choice. Similarly, if the integration interval is $(-\infty, \infty)$ and the integrand includes a factor $w(x) = \exp(-x^2)$, then Gauss-Hermite quadrature is a natural choice. Of course, the integrand $f(x)$ may not include either of these weighting functions. For more general integrands, the change of variable $x = 1/z$ may be of use. In this case, $dx = -\,dz/z^2$, and the integral can be recast as follows.

$$\int_a^b f(x)\,dx = \int_{1/b}^{1/a} \frac{1}{z^2} f\left(\frac{1}{z}\right) dz \quad , \quad ab > 0 \tag{7.7.3}$$

This formulation of $I(a,b)$ is valid for functions $f(x)$, which tend to zero at least as fast as $1/x^2$. It is applicable for the case when $a = -\infty$ and $b < 0$, and also for the case when $a > 0$ and $b = \infty$. If neither of these conditions on a and b is satisfied, we can always decompose the integral into parts that do satisfy them. For example, if the integration interval is $(-\infty, \infty)$ and we pick $\alpha < 0$ and $\beta > 0$, then the integral can be decomposed as follows.

$$\int_{-\infty}^{\infty} f(x)\,dx = \int_{-\infty}^{\alpha} f(x)\,dx + \int_{\alpha}^{\beta} f(x)\,dx + \int_{\beta}^{\infty} f(x)\,dx \tag{7.7.4}$$

The products of the limits of integration are positive for the first and last terms, $I(-\infty, \alpha)$, and $I(\beta, \infty)$, respectively. Hence, the transformation in Equation (7.7.3) can be applied to these two parts of the integral. After simplification, this leads to the following reformulation.

$$\boxed{\int_{-\infty}^{\infty} f(x)\,dx = \int_{1/\alpha}^{1/\beta} \frac{1}{z^2} f\left(\frac{1}{z}\right) dz + \int_{\alpha}^{\beta} f(x)\,dx} \tag{7.7.5}$$

Each of these integrals now involves finite limits. The integral $I(1/\alpha, 1/\beta)$ can be estimated using an open integration formula if needed, whereas $I(\alpha, \beta)$ can be approximated using either a closed or an open integration formula. The constants $\alpha < 0$ and $\beta > 0$ should be chosen sufficiently large in magnitude that the "tail" of the function $f(x)$ has begun to approach zero, at least as fast as $1/x^2$.

EXAMPLE 7.7.2 Infinite Limits

To illustrate the of use of a change of variable to evaluate an integral with an infinite limit of integration, consider the problem of numerically estimating the following integral.

$$I(0, \infty) = \int_0^{\infty} \frac{4}{\pi(4 + x^2)}\,dx$$

This integral can be decomposed at the point $\beta = 1$, thereby generating the following reformulation based on Equation (7.7.3).

$$\begin{aligned} I(0,\infty) &= \frac{4}{\pi}\left(\int_0^1 \frac{dx}{4+x^2} + \int_1^\infty \frac{dx}{4+x^2}\right) \\ &= \frac{4}{\pi}\left(\int_0^1 \frac{dx}{4+x^2} + \int_0^1 \frac{dz}{z^2(4+1/z^2)}\right) \\ &= \frac{4}{\pi}\left(\int_0^1 \frac{dx}{4+x^2} + \int_0^1 \frac{dz}{4z^2+1}\right) \end{aligned}$$

Example program *e772* on the distribution CD applies the Gauss-Legendre quadrature method to estimate these two integrals using increasing numbers of points n, with the results summarized in Table 7.7.2. The first column is the number of points, the second is the estimated value of the integral, and the third is the percent error with the exact value of the integral being $I(0,\infty) = 1$. It is evident that the Gauss-Legendre quadrature formula converges quite rapidly in this case with an accurate estimate obtained using relatively few function evaluations.

TABLE 7.7.2 Gauss-Legendre Quadrature

n	$I(0,\infty)$	*Percent Error*
1	0.9362055	−6.3794494
2	1.0177838	1.7783761
3	0.9999266	−0.0073433
4	0.9997398	−0.0260234
5	1.0000157	0.0015736
6	1.0000029	0.0002861
7	0.9999998	−0.0000238
8	0.9999850	−0.0015020
9	1.0000000	0.0000000

7.8 MULTIPLE INTEGRALS

All of the integration formulas examined thus far are applicable to one-dimensional integrals—that is, integrals of functions of a single variable. The problem of numerically estimating the value of integrals of functions of several variables is more challenging. We examine two approaches.

7.8.1 Parameterization Method

Let $I(D)$ denote the integral of a two-dimensional function $f(x,y)$ over a domain D in the xy plane.

$$I(D) = \iint_D f(x,y)\, dy\, dx \tag{7.8.1}$$

Suppose D is a region in the xy plane characterized by $a \le x \le b$ and $\alpha(x) \le y \le \beta(x)$ as in Figure 7.8.1. One way to integrate a function of two variables is to use a two-step approach. For each x in the interval $[a,b]$, let $g(x)$ be defined as follows.

$$g(x) \triangleq \int_{\alpha(x)}^{\beta(x)} f(x,y)\, dy \quad , \quad a \le x \le b \tag{7.8.2}$$

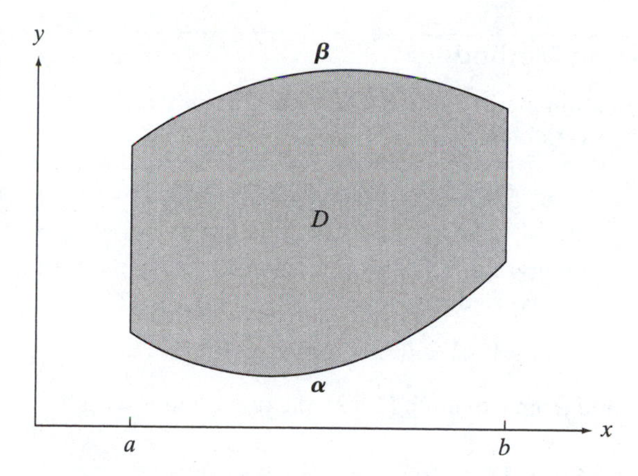

FIGURE 7.8.1 Domain of Integration D

Thus, $g(x)$ represents a family of one-dimensional integrals *parameterized* by x. Once $g(x)$ is determined, the original two-dimensional integral is simply

$$I(D) = \int_a^b g(x)\, dx \tag{7.8.3}$$

If the domain D can not be characterized as in Figure 7.8.1, it may be possible to reverse the roles of x and y with $a \le y \le b$ and $\alpha(y) \le x \le \beta(y)$.

The process of reformulating a two-dimensional integral as a parameterized sequence of one-dimensional integrals can be extended to integrals of functions of three or more variables. For example, consider the case of integrating a function of three variables.

$$I(D) = \iiint_D f(x, y, z)\, dz\, dy\, dx \tag{7.8.4}$$

In this case, the domain of integration D is a region in three-dimensional space. Suppose D can be characterized as $a \le x \le b$, $\alpha(x) \le y \le \beta(x)$, and $\gamma(x, y) \le z \le \delta(x, y)$. Then $I(D)$ can be computed by a three-step process that involves the following two intermediate functions:

$$h(x, y) \triangleq \int_{\gamma(x,y)}^{\delta(x,y)} f(x, y, z)\, dz \quad , \quad a \le x \le b, \quad \alpha(x) \le y \le \beta(x) \tag{7.8.5a}$$

$$g(x) \triangleq \int_{\alpha(x)}^{\beta(x)} h(x, y)\, dy \quad , \quad a \le x \le b \tag{7.8.5b}$$

Once $g(x)$ is determined, the original three-dimensional integral can be computed using Equation (7.8.3). The computational effort required to find a multi-dimensional integral is significant. If a numerical approximation to (7.8.3) requires evaluating $g(x)$ at n points, and each evaluation of $g(x)$ requires evaluating $f(x, y)$ at another n points, then the total computational effort for a two-dimensional integral grows as n^2. A similar analysis shows that the computational effort for evaluating the three-dimensional integral grows as n^3 where n is the number of points used to approximate a one-dimensional integral.

EXAMPLE 7.8.1 Parameterization Method

To illustrate the parameterization method for computing a multidimensional integral, consider the following function of two variables.

$$f(x, y) = \frac{\exp(-x^2)}{1 + y^4}$$

Suppose we are interested in integrating $f(x, y)$ over a region D that consists of the area above the x axis that is enclosed by the inverted parabola $y = 1 - x^2$.

$$D = \{(x, y) | -1 \le x \le 1, 0 \le y \le 1 - x^2\}$$

In this case, $[a, b] = [-1, 1]$, and from Equation (7.8.2), the one-dimensional function $g(x)$ is

$$g(x) = \int_0^{1-x^2} f(x, y)\, dy \quad , \quad -1 \le x \le 1$$

From (7.8.3), the two-dimensional integral in (7.8.1), can be evaluated as follows.

$$I(D) = \int_{-1}^{1} g(x)\, dx$$

Example program *e781* on the distribution CD applies Simpson's 1/3 rule to evaluate $g(x)$ at each x and the integral of $g(x)$. The results are summarized in Table 7.8.1. The first column is the number of panels (which must be even), the second column is the estimated integral $I(D)$, and the third column is the number of function evaluations required. In this case, for values of $n \ge 50$, the estimate of $I(D)$ does not change in the first seven decimal places.

TABLE 7.8.1 A Parameterized Double Integral

n	$I(D)$	*Evaluations*
10	1.0898604	100
20	1.0895975	400
30	1.0895830	900
40	1.0895807	1600
50	1.0895796	2500
60	1.0895796	3600

7.8.2 Monte Carlo Integration

An alternative approach to the numerical evaluation of integrals is based on the use of random numbers (recall Example 1.8.2). For two-dimensional integrals, the basic idea is to estimate the average value of $f(x, y)$ over the domain of integration D, and then multiply this estimate by the area of D. Suppose (x_k, y_k) for $1 \le k \le n$ denotes a set of n random points uniformly distributed over the region D. Assuming n is sufficiently large, the following can be used as an estimate of the average value of $f(x, y)$ over D.

$$f_{ave} \approx \frac{1}{n} \sum_{k=1}^{n} f(x_k, y_k) \qquad (7.8.6)$$

Next, let $\mu(D)$ denote the *area* of the domain of integration D in the xy plane. For example, if D is characterized as $a \le x \le b$ and $\alpha(x) \le y \le \beta(x)$, then

$$\boxed{\mu(D) = \int_a^b \int_{\alpha(x)}^{\beta(x)} dy\, dx} \tag{7.8.7}$$

That is, $\mu(D)$ is just the value of the integral when the integrand is $f(x, y) = 1$. Given the area of D and the estimate of the average value of $f(x, y)$ over D, the following random sequence can be constructed.

$$q_n \triangleq \frac{\mu(D)}{n} \sum_{k=1}^{n} f(x_k, y_k) \quad , \quad n \ge 1 \tag{7.8.8}$$

The random variable q_n has a number of interesting properties. The most important one is that the *mean* value of q_n is $I(D)$. Consequently, q_n can be used to estimate the value of the integral in Equation (7.8.1) as long as the value of n is sufficiently large. To determine how large a value of n is needed, we examine the standard deviation of q_n. It can be shown that the standard deviation, $\sigma(q_n)$, varies with n as follows.

$$\sigma(q_n) = \frac{c\mu(D)}{\sqrt{n}} \tag{7.8.9}$$

Here c is a constant that depends on $f(x, y)$. The main conclusion to draw from Equation (7.8.9) is that the standard deviation decreases with n, but very slowly. Increasing n by a factor of 100 decreases $\sigma(q_n)$ by a factor of only 10. If we assume that the random sequence q_n follows a normal distribution, then we can "roughly" model the estimation process as follows.

$$\boxed{I(D) \approx \frac{\mu(D)}{n} \sum_{k=1}^{n} f(x_k, y_k) + O\left(\frac{1}{\sqrt{n}}\right)} \tag{7.8.10}$$

This numerical approximation to a two-dimensional integral is referred to as *Monte Carlo integration* because of its reliance on random numbers. One of the attractive features of the Monte Carlo method is the fact that it can be easily extended to integrals of higher dimension. For example, for the three-dimensional case, suppose the domain of integration D is characterized by: $a \le x \le b$, $\alpha(x) \le y \le \beta(x)$, and $\gamma(x, y) \le z \le \beta(x, y)$. Let $\mu(D)$ be the *volume* of D.

$$\boxed{\mu(D) = \int_a^b \int_{\alpha(x)}^{\beta(x)} \int_{\gamma(x,y)}^{\delta(x,y)} dz\, dy\, dx} \tag{7.8.11}$$

If (x_k, y_k, z_k) for $1 \le k \le n$ denotes a set of n random points uniformly distributed throughout the region D, then the Monte Carlo approximation to the three-dimensional integral is

$$\boxed{I(D) \approx \frac{\mu(D)}{n} \sum_{k=1}^{n} f(x_k, y_k, z_k) + O\left(\frac{1}{\sqrt{n}}\right)} \tag{7.8.12}$$

It is evident that the integration formula for the three-dimensional case in Equation (7.8.12) is essentially no more complicated than it is for the two-dimensional case in (7.8.10). Note in particular that the estimation error (the standard deviation) is still of order $O(1/\sqrt{n})$.

Although the Monte Carlo method generalizes nicely to higher dimensional integrals, it does suffer from a number of drawbacks. One difficulty that can arise is in generating random points that are uniformly distributed throughout the region D. If the domain of integration has a complicated boundary, this may be a difficult task. However, this problem can be circumvented by enclosing D in a simpler region H.

$$D \subset H \tag{7.8.13}$$

It is helpful to choose H to be as small as possible. The random points are then uniformly distributed throughout H rather than D. In addition, the value of the integrand $f(x, y, z)$ is *defined* to be zero for points in H that are outside of D. Finally, $\mu(D)$ in Equation (7.8.12) is replaced by $\mu(H)$. This variation of the basic Monte Carlo formula is simpler to implement but tends to require larger values for n, particularly if the difference between the volumes of D and H is significant.

The main drawback of the Monte Carlo method is its very slow order of convergence, $O(1/\sqrt{n})$. In principle, the Monte Carlo method can be applied to one-dimensional integrals, but it rarely is because it does not compete favorably with the much faster methods such as Simpson's method, which is of order $O(1/n^4)$. The convergence rate of the Monte Carlo method can be increased to order $O(1/n)$ by using adaptive techniques. The basic idea is to increase the number of samples in subregions of D where the integrand changes most rapidly. A discussion of these techniques is included in (Press et al., 1992), and a treatment of Monte Carlo simulation methods in general can be found in Rubinstein (1981).

EXAMPLE 7.8.2 Monte Carlo Integration

To illustrate the Monte Carlo integration method, consider the problem of numerically estimating the value of the following two-dimensional integral.

$$I(D) = \int_0^1 \int_0^1 \frac{2x}{(1+y)^2}\, dy\, dx$$

In this case, $a = 0, b = 1, \alpha(x) = 0$, and $\beta(x) = 1$. Thus, the area of the domain of integration is $\mu(D) = 1$. Example program *e782* on the distribution CD applies the Monte Carlo method with the results summarized in Table 7.8.2. The first column is the number of random samples n, the

TABLE 7.8.2 Monte Carlo Integration

n	$I(D)$	*Percent Error*	$I(D)$	*Percent Error*
1	0.5765995	15.3198957	0.4691423	−6.1715307
10	0.6162403	23.2480526	0.5251005	5.0201058
100	0.5052300	1.0460019	0.5040472	0.8094311
1000	0.4964063	−0.7187486	0.5049201	0.9840250
10000	0.4995373	−0.0925362	0.5003672	0.0734329

second is an estimate of $I(D)$, the third is the percent error, the fourth is a second estimate of $I(D)$ using a different random sequence, and the last column is its percent error. Clearly, a large number of random samples are needed to reduce the error to less than 0.1% because of the slow rate of convergence. Given the probabilistic nature of the process, the error does *not* decrease monotonically. The exact value of the integral in this case is $I(D) = 0.5$.

7.9 APPLICATIONS

The following examples illustrate applications of numerical differentiation and integration techniques using both MATLAB and C. The relevant MATLAB functions in the NLIB toolbox are described in Section 1.8 of Appendix 1, while the corresponding C functions in the NLIB library are described in Section 2.11 of Appendix 2.

7.9.1 Change in Enthalpy: C

Chemical engineers use tabulated pressure-volume-temperature (pvT) data to analyze the characteristics of various liquids that change phases between solid, liquid, and gas. Recall from Section 4.7.1 that one such problem is to determine the change in enthalpy of saturated water as it is vaporized to steam. Consider a system consisting of one kg of saturated water. The appropriate mathematical model for computing the change in enthalpy is the Clapeyron equation (Moran and Shapiro, 1995).

$$\Delta h = T(v_g - v_f)\frac{dp}{dt} \tag{7.9.1}$$

Here, Δh is the change in enthalpy between the liquid and gas phase as the water vaporizes. The remaining parameters are the temperature T in degrees Kelvin, the specific volume v_g of the gas, and the specific volume v_f of the liquid. The factor dp/dT is the rate of change of pressure with respect to temperature. Pressure-temperature data points are available from standard steam tables and are summarized in Table 7.9.1.

TABLE 7.9.1 Pressure-Temperature Data for Saturated Water

T (°C)	p (bars)
50	0.1235
60	0.1994
70	0.3119
80	0.4739
90	0.7014
100	1.014
110	1.433
120	1.958
130	2.701
140	3.613
150	4.758

Values for the constants T, v_g, and v_f can also be obtained from the steam tables. For this problem $T = 373.15\text{K}$, $v_g = 1.63\ \text{m}^3/\text{kg}$, and $v_f = 0.00104\ \text{m}^3/\text{kg}$. To develop a numerical estimate for dp/dT, two distinct approaches are available. In Section 4.7.1, the method employed was to fit a curve to the data and then differentiate the curve. The other technique is to compute the numerical derivative directly from the data in Table 7.9.1. This is the approach taken in the following C program on the distribution CD.

```
/*----------------------------------------------------------------------*/
/* Example 7.9.1: Change in Enthalpy                                    */
/*----------------------------------------------------------------------*/

#include "c:\nlib\util.h"

float    g (float x)
{
   int             m = 11;
   float           y;
   matrix          D = mat (2,m,"50 60 70 80 90 100 110 120 130 140 150 "
                                ".1235 .1994 .3119 .4739 0.7014 1.014 "
                                "1.433 1.958 2.701 3.613 4.758");
   y = piece (x,D[1],D[2],m);
   freemat (D,2);
   return y;
}

int main (void)
{
   int             m = 11,                /* number of data points */
                   c = 0;                 /* central difference */
   float           f  = 1.e5f,            /* conversion factor */
                   vg = 1.673f,           /* specific volume (gas) */
                   vf = 0.00104f,         /* specific volume (liquid) */
                   dT = 10.0f,            /* temperature spacing */
                   T0 = 100.0f,           /* temperature point */
                   dh,                    /* change in enthalpy */
                   s;                     /* slope dp/DT */

/* Find numerical derivative, dp/dT, at T = 100 */

   printf ("\nExample 7.9.1: Change in Enthalpy\n");
   shownum ("3-point dp/dT",s = deriv (T0,dT,3,c,0,g));
   shownum ("5-point dp/dT",s = deriv (T0,dT,5,c,0,g));
   shownum ("3-point extrapolated dp/dT",s = deriv (T0,dT,3,c,1,g));
   shownum ("5-point extrapolated dp/dT",s = deriv (T0,dT,5,c,1,g));

/* Find change in enthalpy */

   dh = 373.15f*(vg - vf)*f*s;
   shownum ("Change in enthalpy (kJ/kg)",dh/1000);
```

```
    return 0;
}
/*-----------------------------------------------------------------*/
```

When program *e791.c* is executed, it uses central differences of various orders with and without extrapolation to estimate dp/dT at $T = 100°\text{C}$ using the data in Table 7.9.1. The resulting estimates of the slope are summarized in Table 7.9.2.

To obtain the change in enthalpy associated with the vaporization of water, we then use Clapeyron's equation (7.9.1). A conversion factor of 10^5 N/m^2 per bar is used to express the answer in units of J/kg. Using the five-point central-difference approximation with extrapolation, this results in the following estimate of the change in enthalpy.

$$\Delta h = 2276 \text{ kJ/kg} \tag{7.9.2}$$

Recall from Equation (4.7.4) that when a cubic least-squares polynomial was fitted to the data in Table 4.7.1 and then differentiated directly, this resulted in the estimate $\Delta h = 2239$. Thus, the two approaches yield generally similar results, with the percentage difference between the two estimates being 1.65 %.

A solution to Example 7.9.1 using MATLAB can be found in the file *e791.m* on the distribution CD.

7.9.2 Dam Design: MATLAB

Suppose a dam is placed across a river in order to prevent floods and generate hydroelectric power. When the dam is constructed, it causes a large pond to form behind the dam as the water backs up. Recall that this problem was first examined in Section 4.7.2. Using a 10×10 grid of depth measurements and bilinear interpolation, the size and shape of the pond was modeled as shown in Figure 7.9.1 where $y = 0$ corresponds to the dam location. The surface plot in Figure 7.9.1 represents the pond depth, which is approximated by the bilinear interpolation function

$$z = f(x, y) \tag{7.9.3}$$

Given a mathematical model of the pond depth, a number of useful engineering calculations can be made. First, consider the volume of water stored behind the dam. Suppose $-a/2 \le x \le a/2$ and $0 \le y \le b$ is a rectangular area that encloses the pond. Then the volume of water V stored behind the dam is

$$V = \int_{-a/2}^{a/2} \int_0^b f(x, y)\, dy\, dx \tag{7.9.4}$$

TABLE 7.9.2 Numerical Estimates of dp/dT at $T = 100°\text{C}$ Using Central Differences

Points	*Extrapolation*	*dp/dT (bars/°C)*
3	no	0.03658
5	no	0.03641
3	yes	0.03641
5	yes	0.03648

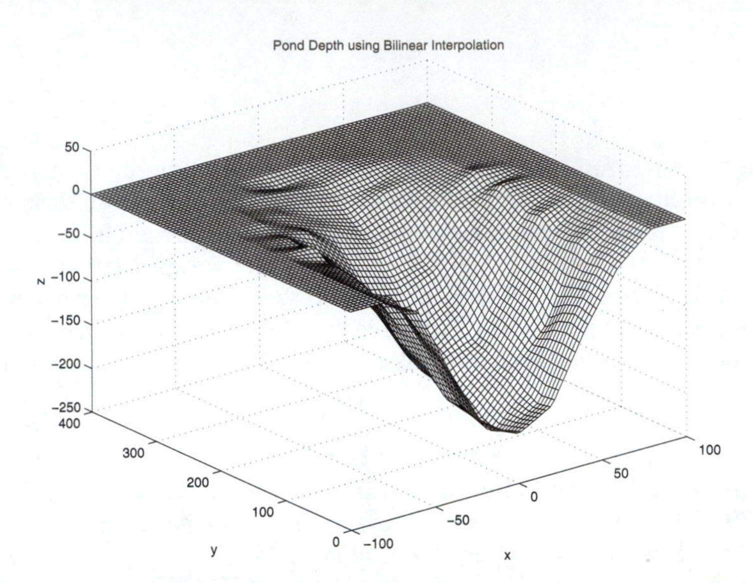

FIGURE 7.9.1 Pond Depth Using Bilinear Interpolation

Thus, a two-dimensional integral is used to determine V. Another important engineering calculation is the force that is exerted by the water on the dam. This will dictate the dam thickness, the material, and the construction methods used. Recall from physics that the change in pressure p exerted by water varies with the depth z of the water as follows.

$$p(z) = \rho g z \tag{7.9.5}$$

Here, $\rho = 1000 \text{ kg/m}^3$ is the density of the water, and $g = 9.81 \text{ m/s}^2$ is acceleration due to gravity. Once the pressure is known, the total force of the water acting on the dam is obtained by integrating over the area of the dam exposed to the water.

$$F = \int_{-a/2}^{a/2} \int_0^{f(x,0)} p(z)\, dz\, dx \tag{7.9.6}$$

Substituting the expression for $p(z)$ from Equation (7.9.5) into (7.9.6) and integrating with respect to z yields the following simplified one-dimensional integral for the force on the dam.

$$F = \frac{\rho g}{2} \int_{-a/2}^{a/2} f^2(x,0)\, dx \tag{7.9.7}$$

The following MATLAB script on the distribution CD can be used to solve this problem.

```
%-----------------------------------------------------------------
% Example 7.9.2: Dam Design
%-----------------------------------------------------------------

% Initialize

   clc                                  % clear screen
```

```
   clear                            % clear window
   global x y Z                     % used by funf792.m
   q = 40;                          % number of x plot points
   r = 80;                          % number of y plot points
   s = 300;                         % number of random points
   a = zeros (2,1);                 % limits on x (m)
   b = zeros (2,1);                 % limits on y (m)
   x1 = zeros (q,1);
   y1 = zeros (r,1);
   Z1 = zeros (q,r);
   randinit (1000)

% Get pond depth data: x,y,Z

   fprintf ('Example 7.9.2: Dam Design\n');
   data792                          % read in depth data x,y,Z
   [m,n] = size(Z);

% Plot surface

   for i = 1 : q
      x1(i) = x(1) + (x(m) - x(1))*(i-1)/(q - 1);
   end
   for j = 1 : r
      y1(j) = y(1) + (y(n) - y(1))*(j-1)/(r - 1);
   end
   Z1 = bilin (x,y,Z,x1,y1);
   plotxyz (x1,y1,Z1,'Pond Depth','x (m)','y (m)','z (m)')

% Compute volume of water in pond

   a(1) = x(1);
   a(2) = y(1);
   b(1) = x(m);
   b(2) = y(n);
   disp ('Integrating ... ')
   V = -monte(a,b,s,'funf792');
   show ('Volume of water (m^3))',V);

% Compute force on dam

   F = simpson (x(1),x(m),100,'fung792');
   show ('Force on dam (kg/(m-s^2))',F);


function s = funf792 (x0)
%----------------------------------------------------------------
% Volume integrand
%----------------------------------------------------------------
   global x y Z
   s = bilin (x,y,Z,x0(1),x0(2));
```

```
function y = fun792g (x)
%-----------------------------------------------------------------
% Force integrand
%-----------------------------------------------------------------
   rho   = 1000;                      % water density (kg/m^3)
   g     = 9.81;                      % gravity (m/s^2)
   f = funf792([x(1) 0]);
   y = rho*g*f^2/2;
%-----------------------------------------------------------------
```

When script *e792.m* is executed, it uses the pond depth measurements from file *data792.m* to construct the bilinear depth function $f(x, y)$. It then uses the Monte Carlo method to compute the water volume V using Equation (7.9.4) and Simpson's 1/3 rule to compute the water force F using (7.9.7). From Figure 7.9.1, the x and y limits used are $a = 180$ m and $b = 400$ m. This results in the following estimates for volume and force.

$$V = 2.29346 \times 10^6 \text{ m}^3 \tag{7.9.8a}$$

$$F = 1.13274 \times 10^{10} \text{ N} \tag{7.9.8b}$$

This is an example where two numerical techniques—interpolation and integration—are used together to perform useful engineering calculations.

A solution to Example 7.9.2 using C can be found in the file *e792.c* on the distribution CD.

7.9.3 RC Network: C

Linear electrical circuits can be easily miniaturized if they do not include inductors, which are relatively large and bulky. Instead, they consist of resistors, capacitors, and voltage and current sources. An example of a simple *RC* one-port network driven by a current source is shown in Figure 7.9.2.

When a current $u(t)$ is applied to the input port, a voltage $y(t)$ develops across the port terminals. To determine $y(t)$ we apply Kirchhoff's voltage law by setting the sum of voltage drops around the loop to zero, which yields

$$y(t) = Ru(t) + x(t) \tag{7.9.9}$$

The first term represents the voltage drop across the resistor R which is obtained from Ohm's law, while $x(t)$ denotes the voltage drop across the capacitor. The rate of change

FIGURE 7.9.2 An *RC* One-Port Network

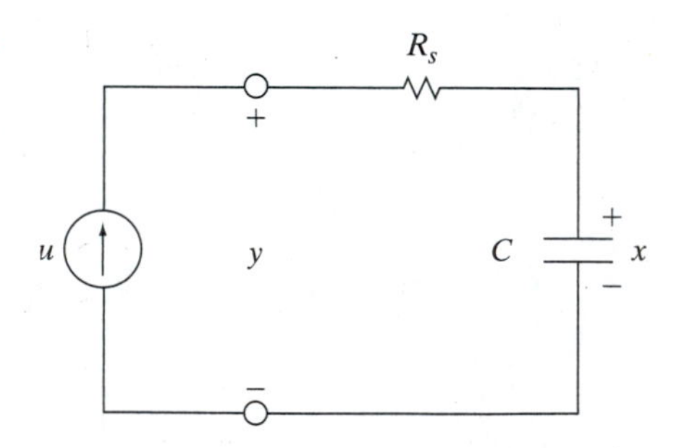

of the voltage across a capacitor is proportional to the current through the capacitor. More specifically, if C is the capacitance, then

$$C \frac{dx(t)}{dt} = u(t) \tag{7.9.10}$$

Suppose the capacitor starts out at time zero with an initial voltage of $x(0)$. Integrating both sides of Equation (7.9.10) and solving for $x(t)$ yields the following expression for the capacitor voltage.

$$x(t) = x(0) + \frac{1}{C} \int_0^t u(\tau)\, d\tau \tag{7.9.11}$$

The voltage developed across the input port terminals as a result of applying a current $u(t)$ is then obtained by substituting (7.9.11) into (7.9.9), which yields

$$y(t) = Ru(t) + x(0) + \frac{1}{C} \int_0^t u(\tau)\, d\tau \tag{7.9.12}$$

To make the problem specific, suppose the resistance is $R = 33\ \Omega$, the capacitance is $C = 470\ \mu$F, and the initial capacitor voltage is $x(0) = 2$ V. Consider the following current pulse.

$$u(t) = \frac{10t \exp(-t)}{1 + \cos^2(t)} \text{ mA} \tag{7.9.13}$$

The following C program on the distribution CD can be used to solve this problem.

```
/*----------------------------------------------------------------------*/
/* Example 7.9.3: RC Network                                            */
/*----------------------------------------------------------------------*/

#include "c:\nlib\util.h"

float f (float t)
{
   return (float) (t*exp(-t)/(1 + pow(cos(t),2))/100);
}

int main (void)
{
   int         q = 101,               /* number of plot points */
               n = 10,                /* number of panels */
               i;
   float       T = 10.0f,             /* upper limit */
               R = 33.0f,             /* resistance (ohm) */
               C = 470.e-6f;          /* capacitance (F) */
   vector      t = vec (q,""),        /* time (sec) */
               x = vec (q,""),        /* capacitor voltage (V) */
               y = vec (q,""),        /* input port voltage (V) */
               u = vec (q,"");
```

```
/* Compute capacitor and port voltage */

   printf ("\nExample 7.9.3: RC Network\n");
   x[1] = 2.0;
   for (i = 1; i <= q; i++)
   {
      t[i] = (i-1)*T/(q - 1);
      u[i] = 1000*f(t[i]);
      if (i > 1)
         x[i] = x[i-1] + (1/C)*trapint(t[i-1],t[i],n,f);
      y[i] = R*f(t[i]) + x[i];
   }

/* Plot results */

   graphxy (t,y,q,"","t (sec)","y (V)","fig793");
   shownum ("Final voltage (V)",y[q]);
   return 0;
}
/*------------------------------------------------------------------*/
```

When program *e793.c* is executed, it computes $y(t)$ for $0 \le t \le 10$. A plot of the voltage that develops across the input port is shown in Figure 7.9.3. Since $u(t) \to 0$ as $t \to \infty$, the steady-state value of $y(t)$ is the steady-state capacitor voltage x, which in this case is found to be

$$x = 17.61 \text{ V} \tag{7.9.14}$$

A solution to Example 7.9.3 using MATLAB can be found in the file *e793.m* on the distribution CD.

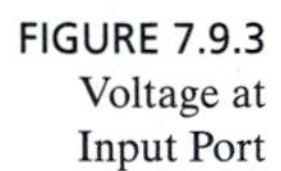

FIGURE 7.9.3 Voltage at Input Port

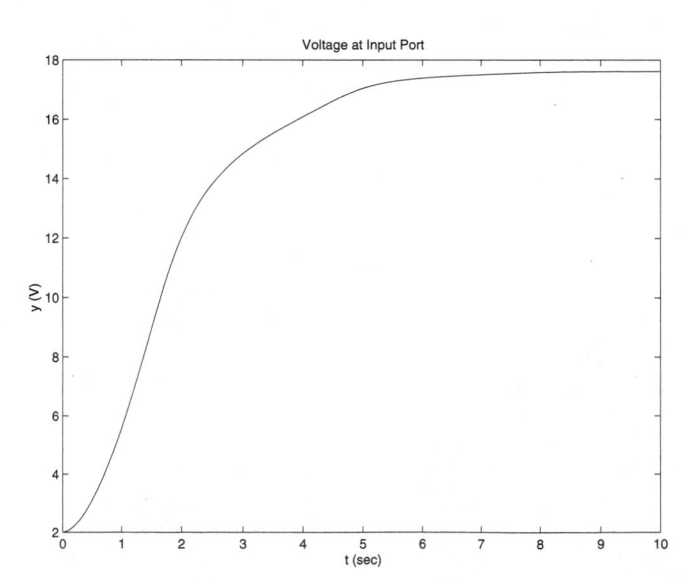

FIGURE 7.9.4 Link of a Robotic Arm

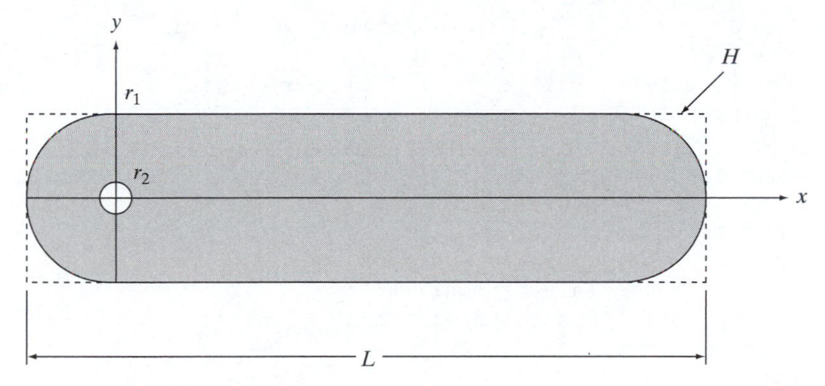

7.9.4 Link of Robotic Arm: MATLAB

A challenging yet practical example of a mechanical system is a robotic arm, which can be thought of as a chain of rigid links interconnected by flexible joints. To develop a complete dynamic model, we must first compute the moments of inertia of the links about the joint axes. A typical link is shown in Figure 7.9.4. The moment of inertia about the z axis (pointing out of the page) is as follows.

$$J = \rho w \int\int_D (x^2 + y^2)\, dy\, dx \tag{7.9.15}$$

Here, ρ is the density of the material used to fabricate the link, and w is the thickness of the link, which is assumed to be constant. The domain of integration D can be enclosed in a simpler rectangular region, H, which is shown by the dashed line in Figure 7.9.4.

$$H = \{(x, y): -r_1 \le x \le L - r_1, -r_1 \le y \le r_1\} \tag{7.9.16}$$

To make the problem specific, suppose the length of the link is $L = 0.9$ m, the radius at each end of the link is $r_1 = 0.1$ m, the radius of the hole about which the link rotates is $r_2 = 0.02$ m, the thickness of the link is $w = 0.1$ m, and the density of the material is $\rho = 100$ kg/m ^{3}. The following MATLAB script on the distribution CD can be used to solve this problem.

```
%-------------------------------------------------------------------
% Example 7.9.4: Link of Robotic Arm
%-------------------------------------------------------------------

% Initialize

   clc                          % clear screen
   clear                        % clear variables
   global L r                   % used by funf794.m
   L    = 0.9;                  % length of link
   r = [0.1 0.02];              % hole radii
   n = 2;                       % dimension of integral
   p = 10;                      % number of runs
   q = 500;                     % number of random points
```

```
   J = 0;                          % moment of inertia
   a = [-r(1)   ; -r(1)];          % limits on x (m)
   b = [L-r(1)  ;  r(1)];          % limits on y (m)
   randinit(1000);

% Compute moment of inertia of link use average of p runs

   fprintf ('Example 7.9.4: Link of Robotic Arm\n');
   for i = 1 : p
       y = monte (a,b,q,'funf794');
       fprintf ('Run %i of %i = %g\n',i,p,y);
       J = J + y;
   end
   J = J/p;
   show ('Moment of inertia J (kg-m^2)',J);


function y = funf794 (x)
%------------------------------------------------------------------
% Integrand for inertia of link of robotic arm
%------------------------------------------------------------------
   global L r
   rho = 100;                 % density
   w   = 0.1;                 % thickness
   z = norm (x,2);
   y = x(1) - (L - 2*r(1));
   if ((x(1) < 0) & (z > r(1))) | ...
       ((y > 0) & (y^2 + x(2)^2 > r(1)^2)) | ...
       (z < r(2))
      y = 0;
   else
      y = rho*w*z^2;
   end
%------------------------------------------------------------------
```

When script *e794.m* is executed, it uses the Monte Carlo method to compute the link moment of inertia. It includes a function which defines the integrand to be zero for those points in H that are outside of D. A total of 10 estimates of J are computed, each using 10000 random points. The results are averaged to produce the following final estimate of the moment of inertia of the link.

$$J = 0.326885 \text{ kg-m}^2 \tag{7.9.17}$$

A solution to Example 7.9.4 using C can be found in the file *e794.c* on the distribution CD.

7.10 SUMMARY

Numerical differentiation and integration methods are summarized in Table 7.10.1. The truncation error depends on the step size h, the number of samples n, the order of the derivative k, and the extrapolation level m.

TABLE 7.10.1 Differentiation and Integration Techniques

Method	*Operation*	*Truncation Error*
Finite difference	differentiation	$O(h^{n-k})$
Trapezoid rule	integration	$O(h^2)$
Simpson's rules	integration	$O(h^4)$
Midpoint rule	integration	$O(h^2)$
Romberg Method	integration	$O(h^{2m})$
Gauss quadrature	integration	—
Monte Carlo	integration	$O(1/\sqrt{n})$

The finite difference method is a differentiation technique that uses numerical differences of function values to estimate derivatives. It is assumed that the step size h between the samples is constant. If n is the number of points, then the truncation error for the kth derivative is of order $O(h^{n-k})$. The forward difference methods use points in front of the evaluation point, the backward difference methods use points in back of the evaluation point, and the central difference methods use points that are symmetrically distributed about the evaluation point. A central difference approximation of the first derivative requires only $n - 1$ function values because the coefficient of the central point is always zero. Richardson extrapolation can be used improve the accuracy of the estimated derivatives. Numerical differentiation is an inherently ill-conditioned process, which tends to amplify additive high-frequency noise that may be present in the data samples.

The next three methods in Table 7.10.1 are Newton-Cotes integration techniques for approximating the following integral, which has finite limits of integration.

$$I(a,b) = \int_a^b f(x)\,dx \tag{7.10.1}$$

The Newton-Cotes formulas are based on approximating the integrand, $f(x)$, with an interpolating polynomial. The trapezoid rule uses piecewise-linear interpolation, and has a truncation error of order $O(h^2)$. Improved performance is obtained with Simpson's 1/3 rule which uses piecewise-quadratic interpolation, and has a truncation error of order $O(h^4)$. Simpson's 3/8 rule uses piecewise-cubic interpolation and also has a truncation error of order $O(h^4)$. The trapezoid rule and Simpson's rules are closed integration formulas which require the values of $f(x)$ at the two end points. In contrast, the midpoint rule is an open integration formula which does not require $f(a)$ or $f(b)$. Open formulas are useful when $f(x)$ has a singularity at one or both ends of the integration interval. The midpoint rule uses piecewise-constant interpolation, and has a truncation error of order $O(h^2)$. In general, if the integrand is approximated with a polynomial of degree m, the truncation error will be of order $O(h^{m+1})$ when m is odd, and $O(h^{m+2})$ when m is even. Simpson's 1/3 rule and the midpoint rule require that the number of integration panels, n, be even. Integrals with infinite limits of integration often can be converted to integrals with finite limits by decomposition and a change of variables, $x = 1/z$.

The Romberg method is a variable-order integration method that is based on Richardson extrapolation. It computes $I(a,b)$ with the trapezoid rule using a sequence of step sizes that are successively reduced by a factor of two. An estimate of $I(a, b)$ based on m extrapolation levels requires a total of $m(m+1)/2$ trapezoid rule integrations, and results in a truncation error of order $O(h^{2m})$. Romberg's method adjusts the extrapolation level automatically so the magnitude of the estimated truncation error falls below a user-prescribed tolerance.

The Gauss quadrature method includes a family of integration techniques that are applicable when the sample spacing is variable and can be controlled. The Gauss quadrature methods apply to integrals which have a weighting function $w(x) > 0$.

$$I_w(a, b) = \int_a^b f(x)w(x)\,dx \tag{7.10.2}$$

There are four Gauss quadrature methods, each with its own integration interval and weighting function as summarized in Table 7.10.2. In each case the n points of evaluation are determined by finding the roots of an nth degree classical orthogonal polynomial that can be generated with a simple recurrence relation. The coefficients of the integration formula are then obtained by solving an n-dimensional linear algebraic system. By construction, the n-point Gauss quadrature integration formulas are exact for polynomials of degree $2n - 1$.

The Monte Carlo method is a statistical numerical integration technique that is useful for estimating higher-dimensional integrals. For example, if $f(x, y, z)$ is a function of three variables defined over a domain of integration D, then Monte Carlo integration can be used to estimate

$$I(D) = \iiint_D f(x, y, z)\,dz\,dy\,dx \tag{7.10.3}$$

The basic approach is to approximate the average value of the integrand by evaluating $f(x, y, z)$ at random points uniformly distributed throughout D. The integral $I(D)$ is then obtained by multiplying the average value of the integrand by the volume of D. It is important that the random sequence not be correlated in order to generate points in n-dimensional space, which are uniformly distributed. The Monte Carlo method produces a random sequence of estimates whose mean is $I(D)$ and whose standard deviation is of order $O(1/\sqrt{n})$ where n is the number of random points. Consequently, the Monte Carlo method is very slow. By using adaptive nonuniform sampling of D, the order of the method can be increased to $O(1/n)$.

TABLE 7.10.2 Gauss Quadrature Integration

Method	*Integration Interval*	*Weighting Function*	*Evaluation Points*
Gauss-Legendre	$[-1, 1]$	$w(x) = 1$	$P_n(x_k) = 0$
Gauss-Chebyshev	$[-1, 1]$	$w(x) = 1/\sqrt{1 - x^2}$	$T_n(x_k) = 0$
Gauss-Laguerre	$[0, \infty)$	$w(x) = \exp(-x)$	$L_n(x_k) = 0$
Gauss-Hermite	$(-\infty, \infty)$	$w(x) = \exp(-x^2)$	$H_n(x_k) = 0$

PROBLEMS

The problems are divided into Analysis problems, which can be solved by hand, and Computation problems, which require the use of MATLAB or C. Solutions to selected problems can be found in Appendix 4. Students are encouraged to use these problems, which are identified with an (S), as a check on their understanding of the material. Problems marked with a (P) are programming problems that require the student to implement one or more of the algorithms discussed in the chapter. The remaining Computation problems require the student to write a main program that uses one or more of the NLIB functions discussed in Appendix 1 or Appendix 2.

Analysis

7.1 Suppose we want to develop a finite difference formula for the first derivative that has a truncation error of order $O(h^q)$ where h is the step size. How many points are needed?

7.2 Why is the three-point central difference formula for the derivative more efficient than either the three-point forward difference formula or the three-point backward difference formula? Describe a practical application where only a backward difference formula can be used.

(S)7.3 Consider the function $f(x) = x \exp(-x)$. Fill in Table 7.11.1 with numerical estimates of the first derivative of $f(x)$ at $x = 0$ using a step size of $h = 0.1$.

7.4 Consider the function $f(x) = x^2 \cos(\pi x)$. Fill in Table 7.11.2 with numerical estimates of the second derivative of $f(x)$ at $x = 0$ using a step size of $h = 0.1$.

7.5 Consider the following function.

$$f(x) = \ln\left(\frac{1}{1 + x^2}\right)$$

(a) Find the exact value of the first derivative $f'(1)$.

(b) Estimate $f'(1)$ using a three-point central difference with $h_1 = 0.1$. What is the percent error?

(c) Estimate $f'(1)$ using a three-point central difference with $h_2 = 0.05$. What is the percent error?

(d) Estimate $f'(1)$ using Richardson extrapolation with $h_1 = 0.1$ and $h_2 = 0.05$. What is the percent error?

TABLE 7.11.1 Estimates of First Derivative

Type	*Points*	$f^{(1)}(0)$	*Percent Error*
Forward	3		
Backward	3		
Central	3		

TABLE 7.11.2 Estimates of Second Derivative

Type	*Points*	$f^{(2)}(0)$	*Percent Error*
Forward	3		
Backward	3		
Central	3		

(S)7.6 Suppose an integration formula with a truncation error of order $O(h^q)$ is used to compute

$$I(a,b) = \int_a^b f(x)\, dx$$

The magnitude of the truncation error is 0.01 when the number of integration panels is $n = 10$. What is the minimum value of n needed to ensure that the magnitude of the truncation error less that 0.0001? What other source of error contributes to the total error?

7.7 Which of the integration methods presented in this chapter are adaptive with respect to the step size h?

7.8 Develop a *semi-open* integration formula that is open at the left end and closed at the right end. Use the midpoint rule and the trapezoid rule. You can assume that the number of integration panels satisfies $n \geq 3$. What is the order of convergence of the truncation error with respect to n?

7.9 Develop a hybrid open integration formula that uses the midpoint rule at each end and Simpson's 1/3 rule in the interior. You can assume that the number of panels, n, is even and $n \geq 6$. What is the order of convergence of the truncation error with respect to n ?

7.10 Consider the following numerical integration problem.

$$I(0,1) = \int_0^1 \frac{1}{1 + x^2}\, dx$$

(a) Find the exact value of $I(0,1)$.
(b) Estimate $I(0,1)$ using the trapezoid rule with $n = 2$.
(c) Estimate $I(0,1)$ using the trapezoid rule with $n = 4$.
(d) Find the percent error when $n = 4$.

7.11 Use Gauss-Legendre quadrature to estimate the value of the following integral using $h = 1$ and $n = 3$ points. The exact answer is $I(0,2) = \ln(1 + \sqrt{2})$. Compute the percent error in each case.

$$I(0,2) = \int_0^2 \frac{1}{\sqrt{4 + x^2}}\, dx$$

7.12 Evaluate the following integral using Gauss-Laguerre quadrature using $n = 1$ and $n = 3$ points. The exact answer is $I(0, \infty) = 3!$. Compute the percent error in each case.

$$I(0,\infty) = \int_0^\infty x^3 \exp(-x)\, dx$$

7.13 Use a change of variable to write the integral in problem 7.12 as a sum of two integrals, each having finite limits of integration.

7.14 Evaluate the following integral using Gauss-Hermite quadrature using $n = 1$ and $n = 3$ points. The exact answer is $I(-\infty, \infty) = \sqrt{\pi}/2$. Compute the percent error in each case.

$$I(-\infty,\infty) = \int_{-\infty}^\infty x^2 \exp(-x^2)\, dx$$

7.15 Use a change of variable to write the integral in problem 7.14 as a sum of integrals, each having finite limits of integration.

Computation

7.16 Write a program that uses the NLIB function *deriv* to estimate the derivative of the following function. Use a three-point central difference and a five-point central difference. Plot the exact derivative and the two estimates on the same graph using $p = 30$ points equally spaced over $0 \le x \le 3$.

$$f(x) = \exp(-x)\cos(2\pi x)$$

(P)7.17 Suppose $f(x)$ is a function mapping R^n into R. The *gradient* of $f(x)$ is defined as the $n \times 1$ vector $\nabla f(x)$ of partial derivatives of $f(x)$ with respect to the components of x.

$$\nabla_k f(x) \triangleq \frac{\partial f(x)}{\partial x_k} \quad , \quad 1 \le k \le n$$

Write a function called *gradient* that estimates the value of the gradient vector using the three-point central difference approximation. The pseudo-prototype of *gradient* is

```
[vector g] = gradient (vector x,int n,function f);
```

On entry to *gradient*: the $n \times 1$ vector x is the evaluation point, n is the number of variables, and f is the name of a user-supplied function whose gradient is to be estimated. The pseudo-prototype of f is

```
[float y] = f (vector x,int n);
```

When f is called with the $n \times 1$ vector x, it must return the value of $y = f(x)$. On exit from *gradient:* the $n \times 1$ vector g contains the estimated value of the gradient. Test your function by writing a main program that computes and displays the gradient of the following function at $x = [1, -1, 2]^T$.

$$f(x) = \frac{\sin(x_1)\cos(2x_2)}{\sqrt{1 + x_3^2}}$$

(S)7.18 Write a program that uses the NLIB functions *midpoint*, *trapint* and *simpson* to estimate the value of the following integral. Print the estimates for the number of panels n ranging from 2 to 64 in powers of 2.

$$I(-1,1) = \int_{-1}^{1} \exp[\sin(\pi x)]\, dx$$

7.19 Write a program which uses the NLIB function *romberg* to evaluate the following integral. Use a truncation error upper bound of $tol = 10^{-6}$. Print the value of the integral and the number of extrapolation levels used.

$$I(0,1) = \int_{0}^{1} \sin(1 - x + x^2)\, dx$$

(P)7.20 Write a function called *quad1* that implements Gauss-Chebyshev quadrature to integrate a function over the interval $[-1,1]$. The pseudo-prototype for *quad1* is

```
[float z] = quad1 (int n,function f);
```

On entry to *quad1:* n is the number of evaluation points, and f is the name of a user-supplied function to be integrated. The pseudo-prototype of f is

```
[float y] = f (float x);
```

When f is called with scalar x, it must return the scalar $y = f(x)$. On exit from *quad1*, z is the estimate of the integral. Test your function by writing a main program that computes and displays the following integral for $1 \le n \le 10$ points.

$$I(-1,1) = \int_{-1}^{1} \frac{\ln(1 + x^2)}{\sqrt{1 - x^2}}\, dx$$

7.21 Suppose you are given the problem of computing the weight of the block of material D shown in Figure 7.11.1. The block is of uniform thickness, $c = 1$ m. The density of the material varies with the position as $\rho(x, y) = x \exp(y)$ kg/m^3. Thus, the total weight is

$$W = c \int\int_D \rho(x, y)\, dy\, dx \quad \text{kg}$$

(a) Find $a, b, \alpha(x)$ and $\beta(x)$ such that the domain of integration can be characterized:

$$D = \{(x, y) | a \le x \le b, \infty(x) \le y \le \beta(x)\}$$

(b) Use the parameterization method to compute the exact value of the weight W.

(c) Write a program that uses the NLIB function *simpson* and the parameterization method to estimate the weight W. Use $n = 2, 4, 6, 8, 10$ panels. Print the estimated weight and the percent error in each case.

7.22 Write a main program that uses the NLIB function *monte* to estimate the weight of the block of material described in problem 7.21. Print the estimated value of W and the percent error for $n = 1000, 2000, 4000$ random samples. Initialize the random number generator using a seed of 2000.

7.23 Write a program that uses NLIB functions to find the cross sectional area of the river in Figure 7.1.3. The program should display the depth profile $y(x)$ in Table 7.1.2 and the following cross sectional area profile $A(x)$ in graphical form. What is the total cross sectional area?

$$A(x) = \int_0^x y(\alpha) d\alpha$$

(S)7.24 Write a program that uses the NLIB function *romberg* to compute the value of the following integral using a truncation error bound of $tol = 10^{-6}$. The program should print the number of extrapolation levels used and the value of the integral.

$$I(0,1) = \int_0^1 \frac{x^3 \exp(-x^2)}{2 + \cos(\ln(1 + x^2))}\, dx$$

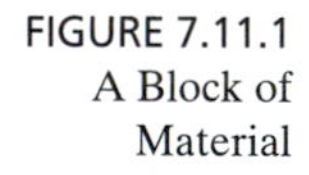
FIGURE 7.11.1 A Block of Material

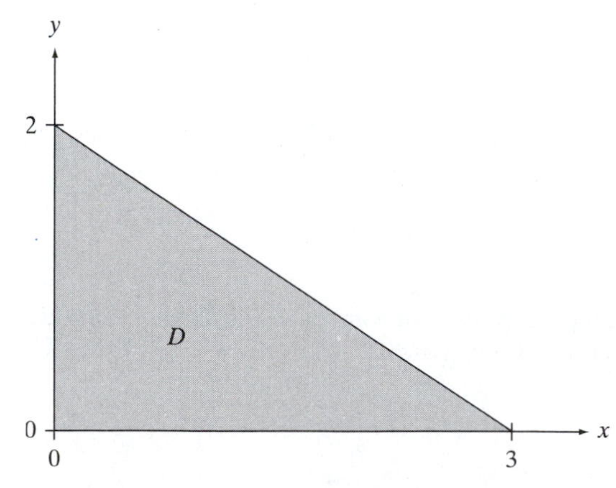

CHAPTER 8

Ordinary Differential Equations

Since the time of Isaac Newton, differential equations have been a useful tool for describing the behavior of a wide variety of *dynamic* physical systems. In this chapter, we investigate numerical techniques for solving a system of first-order ordinary differential equations of the following form.

$$\frac{dx}{dt} = f(t,x) \quad , \quad \alpha \le t \le \beta$$

The independent variable, t, typically denotes *time*, but it can also be used to represent another physical quantity, such as a spatial coordinate. The dependent variable or solution, $x(t)$, is an $n \times 1$ vector called the *state* of the system. If the function f on the right-hand side of the state equation is sufficiently smooth, there is a unique solution $x(t)$ over the interval $[\alpha, \beta]$ corresponding to a given *initial condition*:

$$x(\alpha) = a$$

The $n \times 1$ vector a is called the *initial state*, and finding a solution $x(t)$ that satisfies $x(\alpha) = a$ is called an *initial value problem*. More generally, it is possible to specify some components of $x(t)$ at $t = \alpha$ and the remaining components at $t = \beta$. Since this involves algebraic constraints at both ends of the solution interval, this is a referred to as a two-point *boundary value problem*.

There are a number of important special cases. If $f(t,x)$ does not depend *explicitly* on t, then the system is said to be *autonomous*, otherwise it is *nonautonomous*. For many mathematical models, $f(t,x) = Ax + u$ where A is an $n \times n$ coefficient matrix. In this case the system is said to be *linear*, otherwise it is *nonlinear*. Linear systems can be solved analytically, and a substantial body of theory is available for the analysis and design of linear systems (see, e.g., Chen, 1984). Although linear systems and autonomous systems occur frequently in applications, we focus on solving nonlinear nonautonomous systems because they are more general and the numerical techniques are essentially the same. No attempt is made to find the complete solution trajectory $x(t)$ for $\alpha \le t \le \beta$. Instead, we develop techniques to accurately and efficiently estimate $x(t)$ at a set of discrete values of t distributed over the interval $[\alpha, \beta]$.

In this chapter, we begin by converting higher-order ordinary differential equations into equivalent first-order systems of equations. We then proceed to pose some engineering problems that are formulated naturally in terms of differential equations. Next, we introduce a very simple technique called Euler's method for solving first-order systems. We follow with a discussion of a family of higher-order one-step techniques called the Runge-Kutta methods. The Runge-Kutta methods are made more efficient by allowing the step size to vary in an effort to control the estimated truncation error; this leads to the popular fifth-order Runge-Kutta-Fehlberg method. We then consider multistep methods in the form of the Adams-Bashforth-Moulton predictor-corrector formulas. We follow with a discussion of the Bulirsch-Stoer extrapolation technique based on the modified midpoint method. Next, we examine the question of stiff systems—systems whose solutions contain terms with time scales that differ by more than an order of magnitude. We develop a semi-implicit technique based on the extrapolation method, suitable for stiff systems. Then we discuss solution techniques that are applicable to two-point boundary value problems, including the shooting method and the finite difference method. The chapter concludes with a summary of ordinary differential equation solution techniques, and a presentation of engineering applications. The selected examples include a stirred tank chemical reactor, deflection of a cantilever beam, a phase-locked loop circuit for FM demodulation, and a model of turbulent fluid flow using a chaotic system.

8.1 MOTIVATION AND OBJECTIVES

Many physical phenomena can be modeled by higher-order ordinary differential equations. For example, the motion of a mass acted upon by a force can be modeled by a second-order differential equation using Newton's second law. In almost all cases of practical interest, a higher-order differential equation can be recast as a system of first-order differential equations by introducing new variables. Consider, in particular, the following nth-order nonlinear ordinary differential equation (ODE):

$$\frac{d^n y}{dt^n} = g\left(t, y, \frac{dy}{dt}, \ldots, \frac{d^{n-1}y}{dt^{n-1}}\right) \tag{8.1.1}$$

As is traditional, we leave the dependence of y on t *implicit* in order to streamline the notation. The nth order equation in (8.1.1) can be converted into an equivalent system of n first-order equations by defining the following $n \times 1$ vector of *state* variables.

$$x \triangleq \left[y, \frac{dy}{dt}, \ldots, \frac{d^{n-1}y}{dt^{n-1}}\right]^T \tag{8.1.2}$$

If we differentiate the kth component of x in Equation (8.1.2), and then substitute using (8.1.2), this yields

$$\begin{aligned}\frac{dx_k}{dt} &= \frac{d^k y}{dt^k} \\ &= x_{k+1} \quad , \quad 1 \le k < n\end{aligned} \tag{8.1.3}$$

Thus, the derivative of the kth state variable is just the $(k+1)$th state variable for $1 \le k < n$. To get the last state equation, we differentiate the final component of (8.1.2) and then substitute using (8.1.1) and (8.1.2), which yields

$$\begin{aligned}\frac{dx_n}{dt} &= \frac{d^n y}{dt^n} \\ &= g\left(t, y, \frac{dy}{dt}, \ldots, \frac{d^{n-1}y}{dt^{n-1}}\right) \\ &= g(t, x_1, x_2, \ldots, x_n)\end{aligned} \tag{8.1.4}$$

It follows that if the state vector is defined as in Equation (8.1.2), then the nth-order ODE in (8.1.1) can be cast in the standard form of an n-dimensional first-order system with

$$f(t,x) = \begin{bmatrix} x_2 \\ x_3 \\ \vdots \\ x_{n-1} \\ g(t, x_1, x_2, \ldots, x_n) \end{bmatrix} \tag{8.1.5}$$

8.1.1 Satellite Attitude Control

As an example of a task that is faced by engineers, consider the single-axis satellite attitude control problem shown in Figure 8.1.1. Using the moment equation (the rotational version of Newton's second law), the angular position, ϕ, of the satellite is governed by the following second-order ordinary differential equation.

$$J\frac{d^2\phi}{dt^2} = \tau \tag{8.1.6}$$

Here J is the mass moment of inertia about the axis of rotation, $d^2\phi/dt^2$ is the angular acceleration, and τ is the torque applied by the thrusters. A typical control problem is to determine the thruster control signal $\tau(t)$ so as to stop the satellite from spinning and reorient it to point the antenna toward a ground station. This is a second-order differential equation in the form of (8.1.1). Thus, if we define the state vector to be $x = [\phi, d\phi/dt]^T$ as in (8.1.2), then from (8.1.5) we can express the equation of motion as

$$\frac{dx}{dt} = \underbrace{\begin{bmatrix} 0 & 1 \\ 0 & 0 \end{bmatrix}}_{A} x + \underbrace{\begin{bmatrix} 0 \\ \tau/J \end{bmatrix}}_{u} \tag{8.1.7}$$

It follows that this is an example of a linear system of dimension $n = 2$. If the applied torque, τ, varies with t, then this is a nonautonomous linear system.

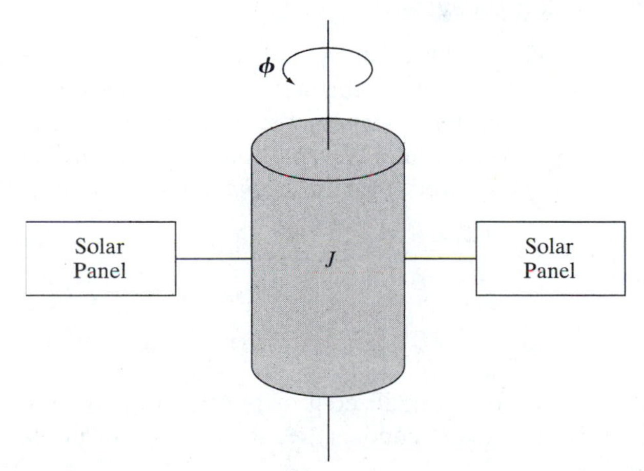

FIGURE 8.1.1 Satellite Attitude Control

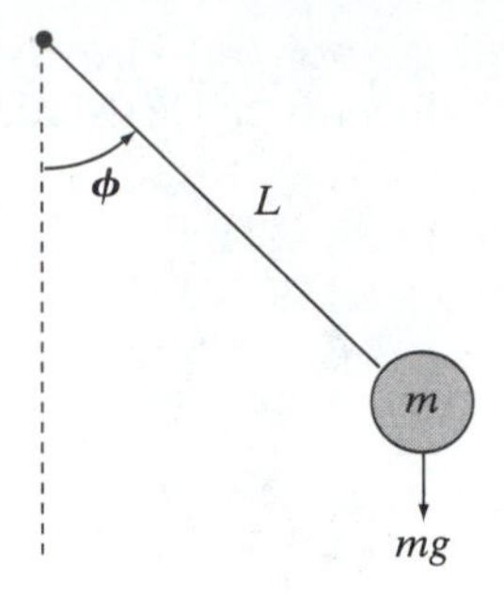

FIGURE 8.1.2
A Pendulum

8.1.2 Pendulum

As an example of a system that is nonlinear, consider the swinging pendulum shown in Figure 8.1.2. When the mass of the pendulum is small in comparison with the mass m at the end of the pendulum, the equation of motion of the pendulum is as follows.

$$mL^2\frac{d^2\phi}{dt^2} + \mu\frac{d\phi}{dt} + mLg \sin \phi = 0 \tag{8.1.8}$$

Here ϕ is the angular position of the pendulum measured relative to vertical, m is mass at the end of the pendulum, L is the length of the pendulum, μ is the coefficient of viscous friction, and g is the acceleration due to gravity. Again, this is a second-order equation in the form of (8.1.1). If we define the state vector to be $x = [\phi, d\phi/dt]^T$ as in (8.1.2), then from (8.1.5) we get the following first-order system.

$$\frac{dx_1}{dt} = x_2 \tag{8.1.9a}$$

$$\frac{dx_2}{dt} = -\left(\frac{g}{L}\right)\sin x_1 - \left(\frac{\mu}{mL^2}\right)x_2 \tag{8.1.9b}$$

This is an autonomous system because there is no explicit dependence on t. Furthermore, it is nonlinear due to the presence of the sin x_1 term.

8.1.3 Predator-Prey Ecological System

Dynamic systems occur in many fields of study. Consider, for example, the problem of modeling the population levels of a predator-prey pair of species. Let x_1 denote the population level of the *prey*, and let x_2 denote the population level of the *predator*. Suppose x_1 and x_2 are expressed in units of, say, thousands. The following simplified model of population growth is referred to as the *Lotka-Volterra* system (Pielou, 1969).

$$\frac{dx_1}{dt} = (b_1 - c_1x_2)x_1 \tag{8.1.10a}$$

$$\frac{dx_2}{dt} = (-b_2 + c_2x_1)x_2 \tag{8.1.10b}$$

Here the parameter b_1 denotes the normalized growth rate of the prey when the predator is not present ($x_2 = 0$). Similarly, b_2 denotes the rate at which the predator population decreases in the absence of prey ($x_1 = 0$). The term $-c_1x_1x_2$ represents the decrease in the prey population as a result of the actions by the predator, and the term $c_2x_1x_2$ represents the increase in the predator population as a result of the availability of prey. This is a nonlinear system due to the presence of the x_1x_2 product terms. As we shall see, it has a periodic solution in which the population levels of the predator and prey go through ecological cycles. It can be shown that amplitude of the cycle depends on the initial conditions, while the period of the cycle is

$$T = \frac{2\pi}{b_1 b_2} \tag{8.1.11}$$

8.1.4 Chapter Objectives

When you finish this chapter, you will be able to compute numerical solutions to systems of ordinary differential equations. You will know how to convert higher-order differential equations into equivalent systems of first-order differential equations using the notion of a state vector. You will know how to solve initial value problems using single-step methods ranging from Euler's method to the Runge-Kutta-Fehlberg method. You will be able to adjust the step size up and down to control the truncation error while minimizing the number of steps required. You will know how and when to apply multi-step solution techniques, including the Adams-Bashforth-Moulton predictor-corrector method. You will understand Richardson extrapolation and how it is used in explicit and implicit extrapolation methods. You will know how to use implicit methods to solve stiff systems of differential equations, including differential-algebraic systems. Finally, you will be able to solve boundary value problems using the shooting method and the finite difference method. These overall goals will be achieved by mastering the chapter objectives summarized below.

Objectives for Chapter 8

- Know how to convert a higher-order ordinary differential equation into an equivalent system of first-order equations.
- Understand the difference between initial and boundary conditions.
- Understand the relationship between local and global truncation error.
- Be able to apply the Runge-Kutta single-step solution methods.
- Know how to adjust the step size to control the local truncation error.
- Be able to apply the Adams-Bashforth-Moulton multi-step solution method.
- Be able to apply the Bulirsch-Stoer extrapolation solution method.
- Understand the difference between an implicit and an explicit method.
- Know what stiff differential equations are and what type of solution methods are most appropriate.
- Understand how differential-algebraic systems can be approximated as stiff systems of differential equations.
- Be able to solve boundary value problems by the shooting method.
- Be able to solve boundary value problems by the finite difference method.

- Understand how ordinary differential equation techniques can be used to solve practical engineering problems.
- Understand the relative strengths and weaknesses of each computational method and know which are most applicable for a given problem.

8.2 EULER'S METHOD

The examples of dynamic systems introduced earlier represent special cases of the following general first-order nonlinear system where $x(t)$ is an $n \times 1$ vector.

$$\boxed{\frac{dx}{dt} = f(t,x) \quad , \quad x(\alpha) = a} \tag{8.2.1}$$

We restrict our consideration to systems for which the right-hand side function is sufficiently smooth that Equation (8.2.1) has a unique solution satisfying the initial condition, $x(\alpha) = a$. Sufficient conditions on $f(t,x)$ to ensure the existence of a unique solution over $[\alpha, \beta]$ can be found in (Vidyasagar, 1978).

We are interested in estimating $x(t^k)$ for $0 \le k \le m$ where the t^k are equally spaced over the interval $[\alpha, \beta]$. That is, $t^k = \alpha + kh$ where the *step size* h is

$$\boxed{h \triangleq \frac{\beta - \alpha}{m}} \tag{8.2.2}$$

It should be pointed out that a *superscript* k is used to denote the kth discrete time step t^k *throughout* this chapter. Superscripts are used, instead of subscripts, because $x^k \approx x(t^k)$ is a vector and subscripts are reserved to denote *elements* of vectors (see, e.g., Section 8.1). Rather than use an awkward combination of subscripts for t and superscripts for x, we instead use superscripts throughout to indicate the time step. There should be no confusion about the meaning of t^k because there are no instances where the scalar t is raised to the kth power. Many authors use subscripts to denote the kth iteration or time step. Although this approach has its merits, it requires the use of a second subscript to distinguish between individual elements of the solution vector at a particular iteration, and double subscripts have been reserved to denote elements of a matrix. A more complete discussion of vector and matrix notation can be found in Appendix 3.

Suppose the value of $x(t^k)$ is known. This is certainly true for $k = 0$ because $x(t^0) = a$. To find $x(t^{k+1})$ in terms of $x(t^k)$ we multiply both sides of Equation (8.2.1) by dt and then integrate from t^k to t^{k+1}. This yields the following reformulation of (8.2.1) as an integral equation.

$$x(t^{k+1}) = x(t^k) + \int_{t^k}^{t^{k+1}} f[\tau, x(\tau)]\, d\tau \tag{8.2.3}$$

The problem with applying (8.2.3) directly is that we do not know the value of $x(\tau)$ for $t^k < \tau < t^{k+1}$, and without it we can not evaluate the integral. However, if the step size, h, is sufficiently small, we can approximate the integrand over the interval $[t^k, t^k + h]$, by its value $f[t^k, x(t^k)]$ at the start of the interval. In this case, the integral in (8.2.3) simplifies to $hf[t^k, x(t^k)]$. If $x^k \approx x(t^k)$ denotes the approximate solution obtained in this manner, this yields the following solution formula, which is called *Euler's method.*

$$\boxed{x^{k+1} = x^k + hf(t^k, x^k) \quad , \quad 0 \le k < m} \tag{8.2.4}$$

Clearly, Euler's method is very easy to implement, and this is one of its virtues. However, it is not particularly accurate except for very small step sizes. To see this, we expand the solution $x(t)$ into a Taylor series about $t = t^k$ and evaluate the result at $t = t^{k+1}$

$$x(t^{k+1}) = x(t^k) + x^{(1)}(t^k)h + \frac{x^{(2)}(\xi)h^2}{2} \tag{8.2.5}$$

Here $x^{(i)} = d^i x/dt^i$ for $i \ge 1$. Therefore, using Equation (8.2.1), we have $x^{(1)}(t^k) = f[t^k, x(t^k)]$. Comparing with Equation (8.2.4), it follows that Euler's method is simply a truncated Taylor series expansion of $x(t^{k+1})$ about $t = t^k$. The second and higher-order terms that are ignored are of order $O(h^2)$, which means that Euler's method has a *local* truncation error of order $O(h^2)$. Of more practical interest is the *global* truncation error, which is the accumulated truncation error in the final estimate x^m. It can be shown (see, e.g., Gerald and Wheatley, 1989) that the global truncation error is of order $O(h)$, which means that Euler's method is referred to as a *first-order* method. A simplified way to look at the relationship between local and global truncation errors is to observe that if the m local errors are added to approximate the global error, then the truncation error goes from order $O(h^2)$ to order $O(h)$ because $m = (\beta - \alpha)/h$. The relationship between the local and global truncation error for the scalar case, $n = 1$, is shown in Figure 8.2.1.

The global error at time t^{k+1} is the global error at time t^k, which can be thought of as an error in the initial condition $x^k \neq x(t^k)$, plus the local error generated in going from t^k to t^{k+1}. The total error at the end of the solution interval is the global truncation error plus the accumulated round-off error. If many steps are performed, and there are many computations per step, the accumulated round-off error can become significant.

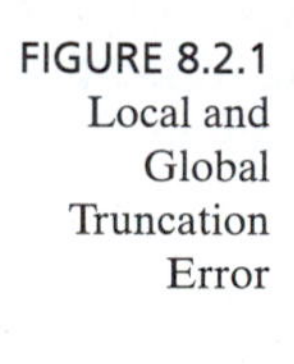

FIGURE 8.2.1 Local and Global Truncation Error

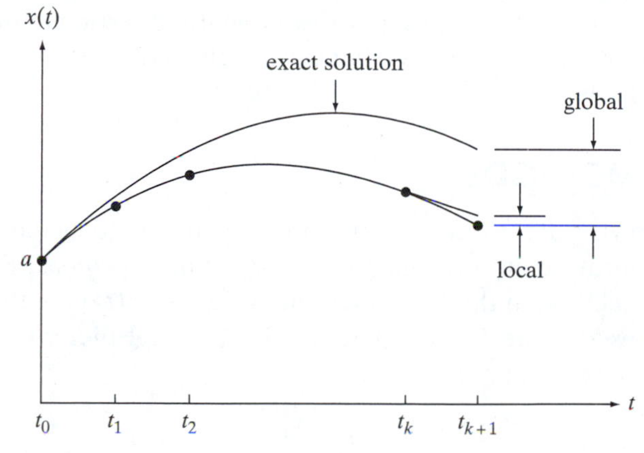

EXAMPLE 8.2.1 Euler's Method

To illustrate the use of Euler's method, consider the following simple one-dimensional first-order system.

$$\frac{dx}{dt} = -cx \quad , \quad x(0) = a$$

Here the constant $c > 0$. This is a one-dimensional linear system whose exact solution is $x(t) = a \exp(-ct)$. Applying Euler's method in Equation (8.2.4), we have

$$\begin{aligned} x^{k+1} &= x^k + hf(t^k, x^k) \\ &= x^k - hcx^k \\ &= (1 - hc)x^k \end{aligned}$$

This difference equation is simple enough that we can write a closed-form expression for the solution. If $x^0 = a$, then

$$x^k = (1 - hc)^k a \quad , \quad k \geq 0$$

Recall that the exact solution is a decaying exponential that approaches zero in the steady state. The Euler estimate of the solution will go to zero as k approaches infinity only if $|1 - hc| < 1$ or

$$0 < h < \frac{2}{c}$$

Consequently, we say that Euler's method *converges* for step sizes in the range $0 < h < 2/c$. When $h > 2/c$, the estimated solution diverges from the exact solution with $|x^k - x(t^k)|$ growing without bound as $k \to \infty$.

The solution to the simple linear system in Example 8.2.1 illustrates some important characteristics that are found in more general n-dimensional nonlinear systems. A numerical solution method *diverges* if the norm of the difference between the computed solution x^k and the exact solution $x(t^k)$ grows without bound with increasing time. In the case of Example 8.2.1, the numerical solution diverges when the step size is larger than $h = 2/c$. In general, the range of step sizes over which a method converges will depend on the details of the system being solved and also on the type of solution method used. As we shall see, the more sophisticated methods estimate the truncation error and adjust the step size accordingly. That way, relatively large steps can be taken without the numerical solution diverging from the true solution.

8.3 RUNGE-KUTTA METHODS

Euler's method for solving differential equations can be regarded as the first member of a family of higher-order methods called the *Runge-Kutta methods*. Each of these methods is a *one-step* method in the sense that only values of $f(t, x)$ in the integration interval $[t^k, t^{k+1}]$ are used. Note from Equation (8.2.4) that Euler's method can be rewritten as a system of two equations as follows.

$$q^1 = f(t^k, x^k) \tag{8.3.1a}$$

$$x^{k+1} = x^k + hq^1 \tag{8.3.1b}$$

Again, we use a superscript for q because it is an $n \times 1$ vector, and subscripts are reserved to denote components of vectors. To simplify the interpretation of the $n \times 1$ vector q^1, consider the case $n = 1$. Then q^1 is just the slope of $x(t)$ at $t = t^k$. In order to generalize this approach, suppose we use the average of the slope at $t = t^k$ and the slope at $t = t^{k+1}$. The slope at $t = t^{k+1}$ is not known because $x(t^{k+1})$ is not known. However, the slope at t^{k+1} can be approximated by using $f(t^{k+1}, x^{k+1})$, where x^{k+1} is computed as in (8.3.1b). This yields the following system of three equations called a *second-order Runge-Kutta* method.

$$q^1 = f(t^k, x^k) \tag{8.3.2a}$$

$$q^2 = f(t^k + h, x^k + hq^1) \tag{8.3.2b}$$

$$x^{k+1} = x^k + \frac{h}{2}(q^1 + q^2) \tag{8.3.2c}$$

Using a Taylor series expansion of $x(t^{k+1})$ and x^{k+1}, it is possible to show (see, e.g., Chapra and Canale, 1988) that the estimate in Equation (8.3.2c) has a local truncation error of order $O(h^3)$ and a global truncation error of order $O(h^2)$. Hence, this is a second-order method.

The constant q^1 in (8.3.2a) is the slope of $x(t)$ at $t = t^k$, while the constant q^2 in (8.3.2b) approximates the slope of $x(t)$ at $t = t^{k+1}$. Since these two constants are averaged in (8.3.2c) to produce the estimated change in x, we see that (8.3.2c) is equivalent to using the trapezoid rule to estimate the integral of $f[t, x(t)]$ over $[t^k, t^{k+1}]$. For the *third-order Runge-Kutta* method we use Simpson's rule, which leads to the following four equations.

$$q^1 = f(t^k, x^k) \tag{8.3.3a}$$

$$q^2 = f(t^k + h/2, x^k + hq^1/2) \tag{8.3.3b}$$

$$q^3 = f(t^k + h, x^k - hq^1 + 2hq^2) \tag{8.3.3c}$$

$$x^{k+1} = x^k + \frac{h}{6}(q^1 + 4q^2 + q^3) \tag{8.3.3d}$$

Again, q^1 is the slope of $x(t)$ at $t = t^k$. Next, q^2 represents the slope of $x(t)$ at the midpoint $t = t^k + h/2$. It is an estimate that uses $x^k + hq^1/2$ to approximate $x(t^k + h/2)$. Finally, q^3 represents the slope of $x(t)$ at the end point $t = t^k + h$. Again, this is an estimate that uses x^k plus a linear combination of q^1 and q^2 to approximate $x(t^k + h)$. The particular linear combination that is used is chosen to generate a local truncation error of order $O(h^4)$ and a global truncation error of order $O(h^3)$, hence a third-order method.

EXAMPLE 8.3.1 Runge-Kutta Methods

In order to compare the different Runge-Kutta methods and examine the effects of step size, consider the following differential equation.

$$\frac{d^2y}{dt^2} + 11\frac{dy}{dt} + 10y = 10u$$

If the forcing term is $u(t) = 1$, and the initial conditions are $y(0) = 0$ and $dy(t)/dt = 0$ at $t = 0$, then by direct substitution, one can verify that the following function of time is a solution to the equation satisfying the initial conditions.

$$y(t) = 1 - \frac{10}{9}\exp(-t) + \frac{1}{9}\exp(-10t)$$

Letting $x = [y, dy/dt]^T$ and using Equation (8.1.5), the second-order differential equation can be recast as the following equivalent first-order system of two equations.

$$\frac{dx_1}{dt} = x_2$$

$$\frac{dx_2}{dt} = 10u - 10x_1 - 11x_2$$

Example program *e831* on the distribution CD uses the first-, second-, and third-order Runge-Kutta methods to solve this system over the time interval $[\alpha, \beta] = [0, 1]$ with $u = 1$ and $x(\alpha) = 0$. For each of the methods, the following step sizes are used.

$$h = [0.00625, 0.0125, 0.025, 0.05, 0.1, 0.2]^T$$

To determine the effectiveness of each method, the following error was used where the maximum is found over the $m = (\beta - \alpha)/h$ solution points.

$$E = \max_{k=1}^{m}\{|x_1(kh) - y(kh)|\}$$

The results showing the effects of step size on the error E are shown in Figure 8.3.1. As expected the error increases as the step sizes increase, and as the order of the method decreases.

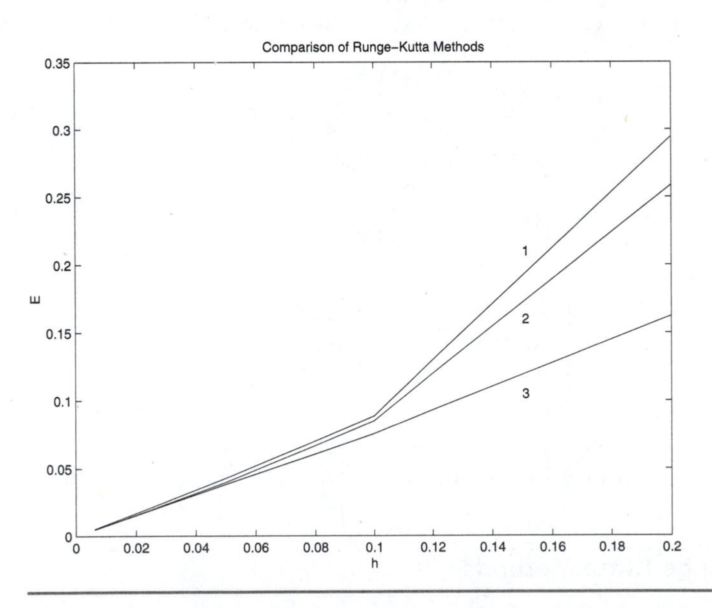

FIGURE 8.3.1 Comparison of Runge-Kutta Methods

The process used to develop the first three Runge-Kutta methods can be continued in a similar manner. The next iteration leads to one of the most popular methods for solving differential equations, namely, the *fourth-order Runge-Kutta* method:

$$q^1 = f(t^k, x^k) \tag{8.3.4a}$$
$$q^2 = f(t^k + h/2, x^k + hq^1/2) \tag{8.3.4b}$$
$$q^3 = f(t^k + h/2, x^k + hq^2/2) \tag{8.3.4c}$$
$$q^4 = f(t^k + h, x^k + hq^3) \tag{8.3.4d}$$
$$x^{k+1} = x^k + \frac{h}{6}(q^1 + 2q^2 + 2q^3 + q^4) \tag{8.3.4e}$$

The coefficients of the fourth-order Runge-Kutta method are chosen to ensure that its local truncation error is of order $O(h^5)$, and its global truncation error is of order $O(h^4)$. It is also possible to develop fifth- and higher-order Runge-Kutta methods. However, the fourth-order method is the highest-order Runge-Kutta method for which the number of function evaluations (four) equals the order of the method. For example, for the fifth-order Runge-Kutta method, six evaluations of the right-hand side function $f(t, x)$ are required. The number of function evaluations is often used as a measure of the computational effort required. It is for this reason that the "classical" fourth-order method is more popular than the higher-order methods.

Although the second- and higher-order Runge-Kutta methods tend to be more accurate than Euler's method, they still can become unstable if the step size h is too large. For example, for the one-dimensional linear system in Example 8.2.1, the fourth-order Runge-Kutta solution diverges from the exact solution for step sizes in the range $h > 2.785/c$ (Nakamura, 1993) as opposed to $h > 2/c$ for Euler's method.

EXAMPLE 8.3.2 Fourth-Order Runge-Kutta Method

As an illustration of the use of the fourth-order Runge-Kutta method, consider the predator-prey equations discussion in Section 8.1.3. For convenience, suppose the parameters of the system are $b = [1, 1]^T$ and $c = [1, 1]^T$. This yields the following system of equations.

$$\frac{dx_1}{dt} = (1 - x_2)x_1$$
$$\frac{dx_2}{dt} = (-1 + x_1)x_2$$

Recall that x_1 represents the prey population level and x_2 the predator population level, both in units of thousands. Example program *e832* on the distribution CD uses the fourth-order Runge-Kutta method to solve this system from $\alpha = 0$ to $\beta = 10$ using an initial condition of $x(0) = [0.5, 0.5]^T$. The number of points used was $m = 250$, which corresponds to a step size of $h = 0.04$. The resulting solution forms a closed trajectory in state space as can be seen in Figure 8.3.2. The direction of the trajectory can be deduced from physical considerations. When the predator population is low, the prey will flourish and grow. The increase in prey eventually brings about favorable conditions for the predator at which point it grows. The increase in the predator population generates a decrease in the prey as they are consumed by the predator. Excessive harvesting of the prey eventually leads to a decrease in the predator population. This brings the system back to the initial condition and the counter-clockwise ecological cycle of period 2π then repeats.

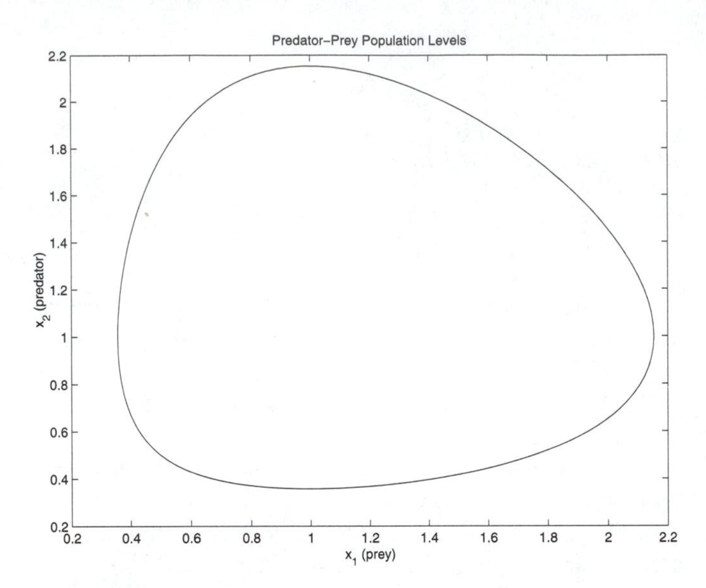

FIGURE 8.3.2 Predator-Prey Population Levels

The predator-prey equations have a number of constant solutions that are obtained by setting $dx/dt = 0$ and solving for x. Constant solutions are referred to as *equilibrium points*. The only equilibrium point for which both population levels are positive is $\overline{x} = [1, 1]^T$. Note that the periodic solution starting from $x(0) = [2, 0.5]^T$ in Figure 8.3.2 encircles this equilibrium point. Solutions that start from other initial conditions in the first quadrant are also periodic and encircle $\overline{x}$. Decreasing the solution step size in Figure 8.3.2 to $h = 0.01$ does not change the plot in any noticeable way. However, increasing the step size to $h = 1$ causes the numerical solution to diverge.

8.4 STEP SIZE CONTROL

The computational efficiency of the fourth-order Runge-Kutta method can be significantly improved by allowing the step size to vary as the solution progresses. For example, for certain values of x, the function $f(t, x)$ may change very rapidly, in which case small steps are needed. However, in other regions the function $f(t, x)$ might be quite smooth, even flat, in which case much larger steps are permitted. The key to implementing *adaptive* step size control is to develop an estimate of the local truncation error. The step size can then be decreased or increased as appropriate in order to maintain a desired level for the error. In this way, a prescribed truncation error can be maintained while minimizing the total number of function evaluations.

Before we examine techniques for step size control, recall that adaptive integration methods were not considered in Section 7.7; the discussion was deferred until now. This was done because evaluating an integral is equivalent to solving a simple differential equation. Indeed, suppose the right-hand side function $f(t, x)$ in Equation (8.2.1) depends only on t, and suppose the initial condition is $x(\alpha) = 0$. Multiplying both sides of (8.2.1) by dt and integrating from α to β then yields

$$x(\beta) = \int_{\alpha}^{\beta} f(t)\, dt \tag{8.4.1}$$

Consequently, methods used to solve differential equations can also be used to integrate functions by using a zero initial condition and evaluating the solution at the upper limit of integration.

8.4.1 Interval Halving

The simplest way to estimate the local truncation error is to use a Richardson extrapolation technique similar to the one used for Romberg integration in Section 7.5. That is, we estimate the solution using two different step sizes and then compare the two estimates. Let $x^{k+1}(h)$ denote the estimated solution at time t^{k+1} using one step of length h, and let $E(h)$ be the associated local truncation error. Similarly, let $x^{k+1}(h/2)$ denote the estimated solution at time t^{k+1} using two steps of length $h/2$, and let $E(h/2)$ represent its local truncation error. Again, superscripts are used to represent the estimated solution at different times, rather than subscripts, because x is a vector and subscripts are reserved to denote components of the vector. If t^k is taken as the initial time, and $x(t^k) = x^k$ is taken as the initial state, then the solution at time t^{k+1} is the estimated solution plus the local truncation error. Consequently, neglecting round-off error, we have

$$x^{k+1}(h) + E(h) = x^{k+1}(h/2) + E(h/2) \tag{8.4.2}$$

If a fourth-order Runge-Kutta method is used to compute x^{k+1}, then the local truncation error is of order $O(h^5)$. It follows that there exists an $n \times 1$ vector c such that for $|h| \ll 1$:

$$E(h) \approx ch^5 \tag{8.4.3a}$$

$$E(h/2) \approx 2c(h/2)^5 \tag{8.4.3b}$$

The factor two in Equation (8.4.3b) arises from the fact that two steps of length $h/2$ are needed. Comparing (8.4.3a) and (8.4.3b), we see that the two errors satisfy the relationship

$$E(h/2) \approx \frac{E(h)}{16} \tag{8.4.4}$$

Substituting the expression for $E(h)$ into (8.4.2), and solving the resulting equation for the half-step error yields

$$\boxed{E(h/2) \approx \frac{x^{k+1}(h/2) - x^{k+1}(h)}{15}} \tag{8.4.5}$$

The expression in Equation (8.4.5) is an estimate of the local truncation error for the fourth-order Runge-Kutta method using a step size of $h/2$. Consequently, it can be used to adjust the step size so as to keep the truncation error within a prescribed tolerance. Given the estimated truncation error $E(h/2)$, we can add it to the estimated solution $x^{k+1}(h/2)$ thereby producing the following *extrapolated* estimate of the solution.

$$\boxed{\bar{x}^{k+1} \triangleq \frac{16x^{k+1}(h/2) - x^{k+1}(h)}{15}} \tag{8.4.6}$$

The process of computing $\bar{x}^{k+1}$ in (8.4.6) is called *local extrapolation*. The extrapolated estimate has a local truncation error of order $O(h^6)$ and a global truncation error of order $O(h^5)$. However, the estimate of the truncation error in (8.4.5) applies to $x^{k+1}(h/2)$, rather that $\overline{x}^{k+1}$.

The computation of an extrapolated solution estimate comes at a cost of additional function evaluations. The required evaluations of $f(t,x)$ include four to compute $x^{k+1}(h)$ plus eight to compute the two steps needed for $x^{k+1}(h/2)$. Since there is a common starting point, the total number of function evaluations is 11. In comparison, eight evaluations are required to compute $x^{k+1}(h/2)$. It follows that there is a computational overhead of 37.5%that must be paid to estimate the local truncation error and generate the extrapolated estimate of the solution. This extra computational effort is more than worthwhile because it allows us to adjust the step size and thereby reduce the *number* of steps while maintaining a prescribed error tolerance.

8.4.2 Runge-Kutta-Fehlberg Method

Before we examine how to adjust the step size to maintain a prescribed truncation error, it is useful to consider an alternative approach to estimating the local truncation error. Rather than use estimates of the solution based on two different step sizes, we can instead use Runge-Kutta formulas of different orders to produce two estimates. These estimates can be compared to develop an expression for the local truncation error. Fehlberg has developed a fifth-order Runge-Kutta formula that requires six function evaluations, but has the additional benefit that it generates a separate fourth-order estimate using the same six evaluation points. By taking the difference between the fourth and fifth-order solution estimates it is possible to estimate the local truncation error as well. The following system of equations is the *Runge-Kutta-Fehlberg* method.

$$q^1 = f(t^k, x^k) \tag{8.4.7a}$$

$$q^2 = f\left(t^k + \frac{h}{4}, x^k + \frac{hq^1}{4}\right) \tag{8.4.7b}$$

$$q^3 = f\left(t^k + \frac{3h}{8}, x^k + h\left(\frac{3q^1}{32} + \frac{9q^2}{32}\right)\right) \tag{8.4.7c}$$

$$q^4 = f\left(t^k + \frac{12h}{13}, x^k + h\left(\frac{1932q^1}{2197} - \frac{7200q^2}{2197} + \frac{7296q^3}{2197}\right)\right) \tag{8.4.7d}$$

$$q^5 = f\left(t^k + h, x^k + h\left(\frac{439q^1}{216} - 8q^2 + \frac{3680q^3}{513} - \frac{845q^4}{4104}\right)\right) \tag{8.4.7e}$$

$$q^6 = f\left(t^k + \frac{h}{2}, x^k + h\left(-\frac{8q^1}{27} + 2q^2 - \frac{3544q^3}{2565} + \frac{1859q^4}{4104} - \frac{11q^5}{40}\right)\right) \tag{8.4.7f}$$

$$x^{k+1} = x^k + h\left(\frac{16q^1}{135} + \frac{6656q^3}{12825} + \frac{28561q^4}{56430} - \frac{9q^5}{50} + \frac{2q^6}{55}\right) \tag{8.4.7g}$$

$$E^{k+1} = h\left(\frac{q^1}{360} - \frac{128q^3}{4275} - \frac{2197q^4}{75240} + \frac{q^5}{50} + \frac{2q^6}{55}\right) \tag{8.4.7h}$$

Here E^{k+1} is an estimate of the local truncation error at time t^{k+1} obtained by taking the difference between the fourth- and fifth-order solution estimates. The solution estimate x^{k+1} in Equation (8.4.7g) is the fifth-order solution and therefore has a global truncation error of order $O(h^5)$. The beauty of the Runge-Kutta-Fehlberg method is that it only requires six function evaluations to obtain expressions for both the local truncation error and the estimated solution. This is in contrast to the eleven function evaluations required by the interval halving method. Thus we have achieved a reduction of 45% in the computational effort. Note that both methods produce a fifth-order solution and an estimate of the truncation error based on a fourth-order solution.

Verner has developed fifth- and sixth-order Runge-Kutta formulas that together use a total of eight function evaluations. They are implemented in the IMSL subroutine DVERK. It is also possible to apply the same technique to lower-order Runge-Kutta methods. For example, the MATLAB function *ode23* is an implementation that uses the second- and third-order Runge-Kutta formulas. For a more complete discussion of Runge-Kutta methods in general see, for example, Butcher (1987).

8.4.3 Step Size Adjustment

Suppose ε denotes a desired upper bound on the size of the local truncation error. Our objective is to choose a step size h that is as large as possible consistent with satisfying the local error criterion:

$$\|E^{k+1}\| \leq \varepsilon \tag{8.4.8}$$

Here the $n \times 1$ vector E^{k+1} is the estimated local truncation error at time t^{k+1}. Thus, E^{k+1} can be determined using either Equation (8.4.5) or (8.4.7h), the latter being the method of choice. Recall that the notation $\|E\|$ denotes the infinity norm, which is the magnitude of the largest component. That is, $\|E\| = \max\{|E_1|,|E_2|, \ldots, |E_n|\}$. Since the local truncation error is based on a fourth-order Runge-Kutta formula, it is of order $O(h^5)$. Therefore, there exists an $n \times 1$ vector c such that

$$E^{k+1} \approx ch^5 \quad , \quad |h| \ll 1 \tag{8.4.9}$$

Let h_0 denote the trial step size used over the interval $[t^k, t^{k+1}]$, and let h_1 denote the new step size. Since the truncation error is roughly proportional to the fifth power of the step size, we can compute the new step size as follows.

$$h_1 = \left(\frac{\varepsilon}{\|E^{k+1}\|} \right)^{1/5} h_0 \tag{8.4.10}$$

For example, if the estimated truncation error is too large by a factor of 32, the new step size will be one-half of the original step size. Of course, the expression for h_1 can also *lengthen* the next step when $\|E^{k+1}\| < \varepsilon$. It is this feature that potentially reduces the total number of steps.

The step size adjustment in Equation (8.4.10) is based on the absolute error. It is often more meaningful to work with the relative error. If ε denotes a relative error bound, then the error criterion in (8.4.8) generalizes as follows.

$$\boxed{\|E^{k+1}\| \leq \max\{\|x^{k+1}\|, 1\}\varepsilon} \tag{8.4.11}$$

This error criterion includes a combination of absolute and relative errors. The relative error is used except when the solution becomes "small" in the sense that $\|x^{k+1}\| < 1$. The threshold does not need to be unity. The point is that we want to guard against the relative error criterion becoming too stringent as $\|x^{k+1}\|$ approaches zero. We should also take into account the fact that (8.4.9) is only an approximation. To be conservative we can include a *safety factor* which makes the new step size somewhat smaller than that predicted to yield $\|E^{k+1}\| = \varepsilon$. Using a safety factor of say, $\gamma = 0.8$, this results in the following generalized update formula for the step size.

$$\boxed{h_1 = \gamma\left(\frac{\max\{\|x^{k+1}\|, 1\}\varepsilon}{\|E^{k+1}\|}\right)^{1/5} h_0} \tag{8.4.12}$$

A practical implementation of adaptive step size control might also put some overall bounds on the minimum and maximum step size. For example, if $\|E^{k+1}\|$ happens to come out exceedingly small, then the next step might be too large and subsequent time would be spent reducing it. At the other extreme, if h_1 is made smaller than $\varepsilon_M t^k$ where ε_M is the machine epsilon, then the solution will stop advancing because $t^{k+1} = t^k$ in this case.

The following algorithm is an implementation of the Runge-Kutta-Fehlberg method with adaptive step size control. The user specifies a relative error bound ε and a minimum step size, h_{min}.

Alg. 8.4.1 **Runge-Kutta-Fehlberg Method**

1. Pick $\varepsilon > 0$, and $h_{min} > 0$.
2. Set $\gamma = 0.8, H = 1, k = 0, x^0 = a, t^0 = \alpha, h = \beta - \alpha$.
3. Do

   {

   (a) Do

   {

   (1) Compute x^{k+1} and E^{k+1} using (8.4.7).

   (2) Compute

   $$\delta = \frac{\max\{\|x^{k+1}\|, 1\}\varepsilon}{\|E^{k+1}\|}$$

   (3) If $\delta \geq 1$, set $t^{k+1} = t^k + h$

   (4) Compute the new step size

   $$h = \gamma\delta^{1/5}h$$

   (5) If $h < h_{min}$, set $H = 0$ and exit.

   (6) If $h > \beta - t^k$, set $h = \beta - t^k$

   }

   (b) While $(\delta < 1)$

   (c) Set $k = k + 1$

}
4. While ($t^k < \beta$)
5. Set $x = x^k$

Alg. 8.4.1 solves the system in Equation (8.2.1) over the interval $[\alpha, \beta]$ subject to the initial condition $x(\alpha) = a$. When Alg. 8.4.1 terminates with $H = 1$, the vector x is the estimate of $x(\beta)$. If $H = 0$, then the error criterion could not be satisfied using the minimum step size. In this case, either h_{min} has to be decreased or ε has to be increased. The Runge-Kutta-Fehlberg method is a medium-accuracy method capable of achieving relative error bounds down to about $\varepsilon = 10^{-6}$ using single precision arithmetic.

EXAMPLE 8.4.1 Runge-Kutta-Fehlberg Method

As an illustration of adaptive step size control of the Runge-Kutta-Fehlberg method, consider the predator-prey equations discussed in Example 8.3.1.

$$\frac{dx_1}{dt} = (1 - x_2)x_1$$

$$\frac{dx_2}{dt} = (-1 + x_1)x_2$$

Example program *e841* on the distribution CD uses the Runge-Kutta-Fehlberg method to solve this system from $\alpha = 0$ to $\beta = 16$ using an initial condition of $x(0) = [0.5, 0.5]^T$ and a local truncation error bound, ε, ranging from 10^{-4} to 10^{-6}. The number of solution points used was $m = 100$. The step size selected at the start of each solution interval using Equation (8.4.12) is plotted in Figure 8.4.1. Recall that the solution of this system is periodic with a period of $T = 2\pi$. Note how the optimal step size varies throughout the cycle.

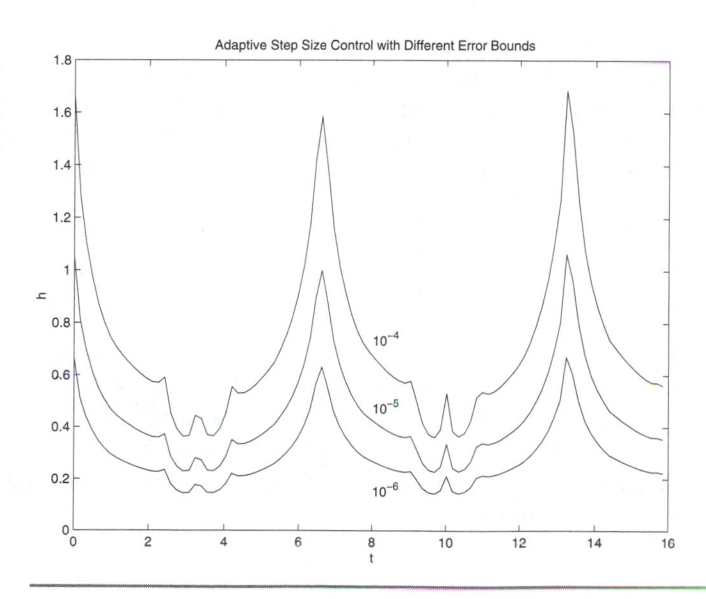

FIGURE 8.4.1 Adaptive Step Size Control with Different Error Bounds

8.5 MULTI-STEP METHODS

The Euler and Runge-Kutta methods are examples of *single-step* methods because they use only the most recent value, x^k, to compute the solution point x^{k+1}. Once the solution has progressed over several steps, it is possible to use past values of $x(t)$ to construct a polynomial approximation to $f[t, x(t)]$. This approximation can then be integrated over the interval $[t^k, t^{k+1}]$ and the result added to x^k as in Equation (8.2.3) to produce x^{k+1}. This is the basic idea behind the *multi-step* methods. Suppose a total of $m + 1$ solution points $\{x^k, x^{k-1}, \ldots, x^{k-m}\}$ are available to predict x^{k+1}. The general formula for computing the next point is as follows.

$$x^{k+1} = x^k + h \sum_{j=-1}^{m} b_j f(t^{k-j}, x^{k-j}) \qquad (8.5.1)$$

Notice that if $b_{-1} \neq 0$, then x^{k+1} appears on both sides of (8.5.1). In this case, (8.5.1) is referred to as an *implicit* method, whereas when $b_{-1} = 0$, it is called an *explicit* method. Implicit methods are difficult to implement because they involve solving a system of nonlinear algebraic equations for x^{k+1}. Root finding techniques from Chapter 5, such as Newton's vector method, must be used at each step.

8.5.1 Adams-Bashforth Predictor

To generate a multi-step method, we start by approximating $f[t, x(t)]$ using a Newton backward difference interpolating polynomial of degree m. For the case $m = 3$, this corresponds to the following cubic polynomial.

$$f[t, x(t)] \approx f^k + s\Delta f^{k-1} + \frac{s(s+1)}{2}\Delta^2 f^{k-2} + \frac{s(s+1)(s+2)}{6}\Delta^3 f^{k-3} \qquad (8.5.2)$$

Here $s = (t - t^k)/h$, and we have used the notational shorthand $f^j = f[t^j, x(t^j)]$. The term $\Delta^i f^{k-i}$ denotes the ith forward difference of $f[t, x(t)]$ evaluated at $t = t^{k-i}$. We can now substitute this approximation for $f[t, x(t)]$ into Equation (8.2.3) and evaluate the integral from t^k to t^{k+1}. Using the observations that $dt = hds$ and s goes from 0 to 1 as t goes from t^k to t^{k+1}, this results in the following expression for the *predicted* value.

$$x^{k+1} = x^k + h\left(f^k + \frac{1}{2}\Delta f^{k-1} + \frac{5}{12}\Delta^2 f^{k-2} + \frac{3}{8}\Delta^3 f^{k-3}\right) \qquad (8.5.3)$$

There is no need to evaluate the forward differences that appear in (8.5.3). Instead, the formulation can be recast directly in terms of $f^j = f(t^j, x^j)$ by expanding the forward difference terms. Recall from Table 7.4.1 that $\Delta f^{k-1} = f^k - f^{k-1}$, and $\Delta^2 f^{k-2} = f^k - 2f^{k-1} + f^{k-2}$, respectively. Making these substitutions into (8.5.3) and collecting coefficients results in the following formulation, which is called the *fourth-order Adams-Bashforth* formula.

$$x^{k+1} = x^k + \frac{h}{24}(55f^k - 59f^{k-1} + 37f^{k-2} - 9f^{k-3}) \tag{8.5.4}$$

The four-point estimate of $x(t^{k+1})$ in Equation (8.5.4) has a local truncation error of order $O(h^5)$ and a global truncation error of order $O(h^4)$. Note that in order to evaluate the predicted value x^{k+1}, the four previous solution points $\{x^k, x^{k-1}, x^{k-2}, x^{k-3}\}$ must be known. Therefore, the fourth-order Adams-Bashforth formula is not a *self-starting* method. A single-step technique, such as a Runge-Kutta method, must be used to compute the first four points $\{x^0, x^1, x^2, x^3\}$.

8.5.2 Adams-Moulton Corrector

The accuracy of the predicted value obtained from the fourth-order Adams-Bashforth method can be improved by using a second application of the multi-step formula in Equation (8.5.1). The basic idea is to use an explicit formula to predict x^{k+1} as in (8.5.4), and then use an implicit formula to refine or correct the solution estimate. To develop a corrector formula we again approximate $f[t, x(t)]$ with a cubic interpolating polynomial, but this time based on the points $\{x^{k+1}, x^k, x^{k-1}, x^{k-2}\}$ where it is understood that x^{k+1} comes from the predictor formula, (8.5.4). Using the four-point Newton backward difference interpolating polynomial yields

$$f[t, x(t)] \approx f^{k+1} + s\Delta f^k + \frac{s(s+1)}{2}\Delta^2 f^{k-1} + \frac{s(s+1)(s+2)}{6}\Delta^3 f^{k-2} \tag{8.5.5}$$

Here $s = (t - t^k)/h$, and $\Delta^i f^{k-i+1}$ denotes the ith forward difference of $f[t, x(t)]$ evaluated at $t = t^{k-i+1}$. If we now substitute this approximation for $f[t, x(t)]$ into (8.2.3) and evaluate the integral from t^k to t^{k+1}, the result is the following expression for the *corrected* value.

$$x^{k+1} = x^k + h\left(f^{k+1} - \frac{1}{2}\Delta f^k - \frac{1}{12}\Delta^2 f^{k-1} - \frac{1}{24}\Delta^3 f^{k-2}\right) \tag{8.5.6}$$

As before, the computation in (8.5.6) can be made more direct by expanding the forward differences and collecting terms. This yields the following simplified formulation, which is called the *fourth-order Adams-Moulton* formula.

$$x^{k+1} = x^k + \frac{h}{24}(9f^{k+1} + 19f^k - 5f^{k-1} + f^{k-2}) \tag{8.5.7}$$

The four-point estimate of $x(t^{k+1})$ in Equation (8.5.7) has a local truncation error of order $O(h^5)$ and a global truncation error of order $O(h^4)$. Equations (8.5.4) and (8.5.7) taken together form a *predictor-corrector* pair. The fourth-order Adams-Bashforth formula in (8.5.4) is the *predictor*, and the fourth-order Adams-Moulton formula in (8.5.7) is the *corrector*. Note that the predictor-corrector pair is actually an explicit method because the corrector is only applied once. It is possible to iterate (8.5.7) by taking the corrected solution point x^{k+1}, substituting it into the right-hand side, and recomputing x^{k+1}. However, this does not necessarily generate a more accurate solution, because the accuracy is ultimately limited by the order of the truncation error of the corrector formula.

The fourth-order Adams predictor-corrector method has the drawback that it is not a self-starting method because it needs four initial solution points. However, once a sufficient number of points are available, this multi-step method is quite efficient because it requires only two function evaluations per step. This is in contrast to the fourth-order Runge-Kutta method, which requires four function evaluations per step. The multi-step methods can generate solutions whose accuracy is superior to the Runge-Kutta methods when the right-hand side function $f[t, x(t)]$ is sufficiently smooth.

One of the useful features of a predictor-corrector method is that is it easy to estimate the local truncation error. Let x_p^{k+1} denote the fourth-order predictor solution in Equation (8.5.4), and let x_c^{k+1} denote the fourth-order corrector solution in (8.5.7). If we take t^k as the initial time and $x(t^k) = x^k$ as the initial state, the exact solution $x(t^{k+1})$ is the estimated solution plus the local truncation error. By examining the form of the first neglected term in the polynomial approximation, we find that the local truncation errors are as follows:

$$x(t^{k+1}) = x_p^{k+1} + \frac{251}{720} h^5 x^{(4)}(\xi_1) \tag{8.5.8a}$$

$$x(t^{k+1}) = x_c^{k+1} - \frac{19}{720} h^5 x^{(4)}(\xi_2) \tag{8.5.8b}$$

In this case, ξ_1 lies somewhere in the interval $[t^{k-3}, t^{k+1}]$, while ξ_2 lies in the overlapping interval $[t^{k-2}, t^{k+1}]$. Suppose the intervals are sufficiently small that $\xi_2 \approx \xi_1$. Then we can subtract (8.5.8b) from (8.5.8a) and solve for $h^5 x^{(4)}(\xi_2)$. Multiplying the solution by the factor $-19/720$ then gives us the following estimate for the local truncation error for the corrector in (8.5.8b).

$$\boxed{E^{k+1} \approx \frac{19}{270}(x_p^{k+1} - x_c^{k+1})} \tag{8.5.9}$$

Thus, the local truncation error of the fourth-order Adams predictor-corrector method can be estimated by simply comparing the predictor and corrector solutions. No additional function evaluations are required. Again, this is in contrast to the Runge-Kutta-Fehlberg method, where two additional function evaluations are needed to estimate E^{k+1}. The principal reason for estimating the truncation error is to adjust the step size so a prescribed error tolerance can be maintained while taking as few steps as possible. Unfortunately, adjusting the step size for a multi-step method is awkward and inefficient. If the step size changes, at least some of the past points have to be re-evaluated. Alternatively, a formulation based on variable spacing can be used, but this means the weighting coefficients in Equations (8.5.4) and (8.5.7) have to be recomputed every step.

The fourth-order predictor-corrector method can be initialized by generating the first four points with a single-step method such as a Runge-Kutta method. Another approach is to implement a variable-order predictor-corrector method. In this case, the technique can begin as a first-order method, which makes it self-starting. Press et al. (1992) make the observation that the multi-step methods may be losing their niche as they are squeezed out by the slow but dependable Runge-Kutta methods at one end and the fast, highly accurate extrapolation methods at the other. Extrapolation methods are presented in the next section.

EXAMPLE 8.5.1 Predictor-Corrector Method

As an illustration of the use of the fourth-order Adams predictor-corrector method, consider the following linear two-dimensional system.

$$\frac{dx_1}{dt} = x_2$$

$$\frac{dx_2}{dt} = -x_1 - 2x_2$$

It is easy to verify through direct substitution that this system has the exact solution $x(t) = \exp(-t)[t, 1-t]^T$ corresponding to the initial condition $x(0) = [0, 1]^T$. Example program *e851* on the distribution CD applies the predictor-corrector method to solve this system, using $m = 10$ steps equally spaced over the interval $[0, 2]$ with the results summarized in Table 8.5.1.

TABLE 8.5.1 Comparison of Solution Methods

	Euler	*Runge-Kutta-Fehlberg*		*Adams-Bashforth-Moulton*	
t	*error*	*local*	*global*	*local*	*global*
0.2	0.0549846	2.6766161e-06	3.5762787e-07	0.0000000e+00	0.0000000e+00
0.4	0.0821920	2.1186679e-06	5.9604645e-07	0.0000000e+00	0.0000000e+00
0.6	0.0915247	1.6763481e-06	7.1525574e-07	0.0000000e+00	0.0000000e+00
0.8	0.0898658	1.3235252e-06	7.6740980e-07	3.4644701e-05	3.0092895e-05
1.0	0.0819200	1.0443295e-06	7.6481371e-07	2.8009788e-05	4.6600908e-05
1.2	0.0708332	8.2270151e-07	7.3388219e-07	2.1389609e-05	5.5279583e-05
1.4	0.0586476	6.4684974e-07	6.7800283e-07	1.7128361e-05	5.8315694e-05
1.6	0.0466343	5.0802936e-07	6.1839819e-07	1.3455639e-05	5.7652593e-05
1.8	0.0355331	3.9788878e-07	5.3644180e-07	1.0567793e-05	5.4642558e-05
2.0	0.0257260	3.1104591e-07	4.7683716e-07	8.2818442e-06	5.0321221e-05

The first column lists the independent variable time, while the second column is the global error for Euler's method. The next two columns correspond to the fifth-order Runge-Kutta-Fehlberg method. Column three is the estimated local truncation error in Equation (8.4.15), while column four is the actual global error. The last two columns correspond to the fourth-order Adams predictor-corrector method. Column five is the estimated local truncation error in (8.5.9), while column six is the actual global error. To start the predictor-corrector method, exact values for the first four points were used. It is clear that both the Runge-Kutta-Fehlberg method and the Adams multi-step method offer superior accuracy to Euler's first-order method, as expected. Of course, the accuracy of all the methods can be improved by reducing the step size, which is $h = 0.2$ in this case. A plot of the solution trajectory for this two-dimensional system is shown in Figure 8.5.1.

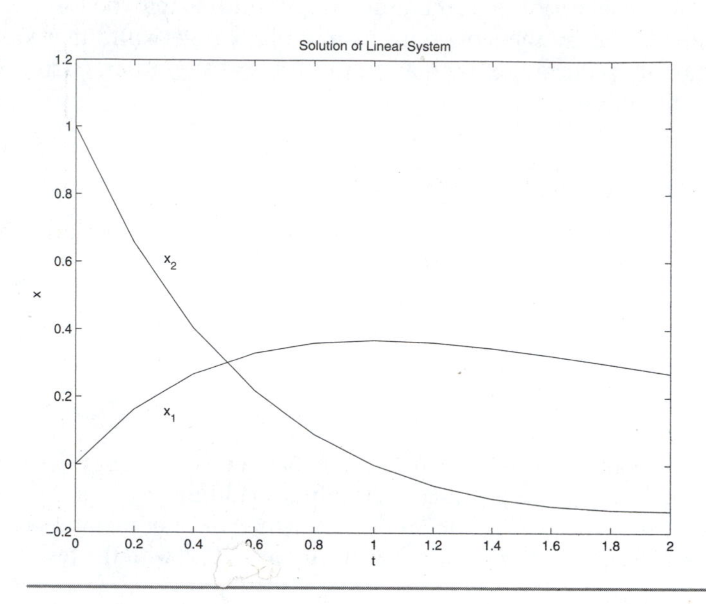

FIGURE 8.5.1 Solution of Linear System

8.6 BULIRSCH-STOER EXTRAPOLATION METHODS

Recall that Romberg's method is a highly accurate technique for numerical evaluation of integrals. In this section, we apply the idea behind Romberg integration, namely Richardson extrapolation, to a system of differential equations. The basic approach is to solve the differential equations with an elementary technique using different step sizes, and then use Richardson extrapolation to estimate what the solution would be in the limit as the step size approaches zero. This general approach is referred to as the *Bulirsch-Stoer extrapolation method* (Stoer and Bulirsch, 1980). Since extrapolation requires multiple solutions of the differential equations, it is important that a simple, efficient technique be used to obtain solutions. Euler's method is one possibility, but we can achieve faster convergence using the following approach.

8.6.1 Modified Midpoint Method

Suppose we are interested in solving Equation (8.2.1) over the interval $[\alpha, \beta]$ starting from the initial condition $x(\alpha) = a$. We partition the interval into m steps of equal length:

$$h \triangleq \frac{\beta - \alpha}{m} \tag{8.6.1}$$

Next, let $q^0 = a$ denote the initial state. Euler's method is used for the first step, which yields

$$q^1 = q^0 + hf(\alpha, q^0) \tag{8.6.2}$$

This gives us an estimate $q^1 \approx x(\alpha + h)$. The midpoint method is then used to obtain estimated solutions at the remaining points as follows.

$$q^{k+1} = q^{k-1} + 2hf(\alpha + kh, q^k) \quad , \quad 1 \leq k < m \tag{8.6.3}$$

Note that (8.6.3) uses two past solution points; hence, the need for one Euler step to start the process. Equation (8.6.3) can be interpreted as integrating over an interval of length $2h$ using the slope of $x(t)$ at the midpoint, $t = \alpha + kh$. To complete the procedure, we perform an end-point correction using the last two points as follows.

$$x^m = \frac{q^m}{2} + \frac{q^{m-1} + hf(\beta, q^m)}{2} \tag{8.6.4}$$

The point x^m is an estimate of $x(\beta)$. It is computed by taking the average of the midpoint estimate q^m and the estimate obtained by applying an implicit version of Euler's method to the last step. Equations (8.6.2) to (8.6.4) are referred to collectively as the *modified midpoint method*. Note that the modified midpoint method requires $m + 1$ function evaluations to compute x^m. The truncation error of estimate x^m can be shown to be of order $O(h^2)$. More important, the truncation error can be represented by the following *even* power series.

$$x(\beta) = x^m + \sum_{j=1}^{\infty} \gamma_j h^{2j} \tag{8.6.5}$$

Here the coefficients, γ_k, do not depend on h. By combining estimates associated with different step sizes, we can eliminate the terms in the power series one by one starting with the h^2 term. The importance of (8.6.5) lies in the fact that the odd terms are missing. Consequently, every time we eliminate a term in the power series, the order of the truncation error goes up by *two* rather than one. This makes the technique converge rapidly.

8.6.2 Richardson Extrapolation

The modified midpoint method in Equations (8.6.2) to (8.6.4) produces an estimate $x^m \approx x(\beta)$ based on m steps of length h. Suppose we compute a sequence of estimates of decreasing step size. In particular, let A_{k1} denote the midpoint estimate of $x(\beta)$ using $m_k = 2^k$ steps.

$$A_{k1} \triangleq x^{m_k} \quad , \quad 1 \leq k \leq p \tag{8.6.6}$$

The p estimates in (8.6.6) can be combined to form new higher-order extrapolated estimates. Let E_{k1} denote the truncation error of estimate A_{k1}. Since the exact solution (ignoring round-off error) is the estimated solution plus the truncation error, we have

$$A_{k1} + E_{k1} = A_{k-1,1} + E_{k-1,1} \tag{8.6.7}$$

Let $h_k = (\beta - \alpha)/2^k$ denote the step size for estimate A_{k1}. From (8.6.5), we see that the truncation error is of the form $E_{k1} \approx \gamma_1 h_k^2$ for $|h_k| \ll 1$. Since $h_k = h_{k-1}/2$, it follows that the relationship between successive truncation errors is

$$E_{k1} \approx \frac{E_{k-1,1}}{4} \tag{8.6.8}$$

Substituting the expression for $E_{k-1,1}$ from (8.6.8) into (8.6.7) and solving for E_{k1} yields

$$E_{k1} \approx \frac{A_{k1} - A_{k-1,1}}{3} \tag{8.6.9}$$

To construct a new estimate, A_{k2}, we take the original estimate, A_{k1}, and add the truncation error, E_{k1}, in Equation (8.6.9) to produce the following *extrapolated* estimate.

$$A_{k2} = \frac{4A_{k1} - A_{k-1,1}}{3} \quad , \quad 2 \le k \le p \tag{8.6.10}$$

This linear combination of A_{k1} and $A_{k-1,1}$ eliminates the $j = 1$ term in (8.6.5), thereby increasing the order of the new estimate to $O(h^4)$. The estimates in (8.6.10) can themselves be combined to eliminate the $j = 2$ term. This will produce $p - 2$ estimates labeled A_{k3}, each with a truncation error of order $O(h^6)$. This process can be continued until a final estimate A_{pp} of order $O(h^{2p})$ is obtained. To develop the general formula for computing extrapolated estimates of $x(\beta)$, let E_{kj} denote the truncation error associated with estimate A_{kj}. Then, expressing the exact solution as the estimate plus the truncation error (ignoring round-off error), we have

$$A_{k,j-1} + E_{k,j-1} = A_{k-1,j-1} + E_{k-1,j-1} \tag{8.6.11}$$

The truncation error for A_{kj} is of the form $E_{kj} \approx \gamma_j h_k^{2j}$ for $|h_k| \ll 1$. Since $h_k = h_{k-1}/2$, the general relationship between successive truncation errors is

$$E_{k,j-1} \approx \left(\frac{1}{2}\right)^{2(j-1)} E_{k-1,j-1} \tag{8.6.12}$$

Substituting the expression for $E_{k-1,j-1}$ from (8.6.12) into (8.6.11) and solving for $E_{k,j-1}$ results in the following error formula, which is a generalization of (8.6.9).

$$\boxed{E_{k,j-1} \approx \frac{A_{k,j-1} - A_{k-1,j-1}}{4^{j-1} - 1}} \tag{8.6.13}$$

To construct the new estimate, A_{kj}, we correct $A_{k,j-1}$ by adding the truncation error, $E_{k,j-1}$. After simplification, this yields the following generalization of Equation (8.6.10).

$$\boxed{A_{kj} = \frac{4^{j-1}A_{k,j-1} - A_{k-1,j-1}}{4^{j-1} - 1} \quad , \quad 2 \le j \le k} \tag{8.6.14}$$

As was the case with Romberg integration, the extrapolated estimates can be arranged into a lower-triangular extrapolation array A as shown in Figure 8.6.1. Note that each element in Figure 8.6.1 is in fact an $n \times 1$ vector where n is the number of differential equations. The first column represents increasingly accurate midpoint estimates with

FIGURE 8.6.1 Extrapolated Estimates of $x(\beta)$

$$\begin{bmatrix} A_{11} & & & & \\ A_{21} & A_{22} & & & \\ A_{31} & A_{32} & A_{33} & & \\ \vdots & \vdots & \vdots & \ddots & \\ A_{p1} & A_{p2} & A_{p3} & \cdots & A_{pp} \end{bmatrix}$$

the kth entry constructed using 2^k steps. Each of the remaining entries in the triangular array is formed from a linear combination of two adjacent estimates, the west estimate and the northwest estimate as indicated in Equation (8.6.14). A sequence of extrapolations of increasing order is then obtained from the diagonal elements $\{A_{11}, A_{22}, \ldots, A_{pp}\}$. The final estimate, A_{pp}, has a truncation error of order $O(h^{2p})$.

Although the diagonal estimates converge rapidly, the computational effort also grows quickly with the number of rows in the array. Computing A_{11} requires three function evaluations ($m = 2$), while computing A_{21} requires five function evaluations ($m = 4$). Thus, when the level of extrapolation is $p = 2$, a total of eight function evaluations are required. In general, the total number of function evaluations needed to construct the level p extrapolation shown in Figure 8.6.1 is

$$N(p) = p + \sum_{k=1}^{p} 2^k. \tag{8.6.15}$$

It is evident that the computational effort required to obtain a high-order estimate can be substantial. It should be kept in mind that A_{pp} solves the differential equation over the entire interval $[\alpha, \beta]$. If the solution is needed at several points within $[\alpha, \beta]$, the extrapolation technique must be applied over each subinterval. In this case the method is not as efficient, but because the subintervals are smaller, so is the required value for p.

It is clear from Equation (8.6.14) that the triangular array of extrapolated estimates can be computed one row at a time. Consequently, there is no need to know p ahead of time. Instead, the truncation error $E_{k,k-1}$ of estimate $A_{k,k-1}$ can be used as part of a termination criterion. The process can be terminated when $\|E_{k,k-1}\|$ falls below a given threshold or when a maximum extrapolation level, r, has been reached. If the latter occurs, it might be helpful to break the original interval $[\alpha, \beta]$ into subintervals and then apply the extrapolation method to each subinterval.

Note from (8.6.14) that in order to compute row k of A, only row $k - 1$ is needed. Consequently, it is only necessary to reserve storage for two rows of A. This is useful because the kth "row" of the extrapolation array A is in fact an $n \times k$ matrix where n is the dimension of the system in Equation (8.2.1). The following algorithm is an implementation of the extrapolation method.

Alg. 8.6.1 Extrapolation Method

1. Pick $\varepsilon > 0$ and $r \geq 2$.
2. Set $k = 0, q^0 = a$.

3. Do

{

(a) Set $k = k + 1, m = 2^k$ and $h = (\beta - \alpha)/m$.

(b) Compute

{

$$q^1 = q^0 + hf(\alpha, q^0)$$

$$q^{j+1} = q^{j-1} + 2hf(\alpha + jh, q^j) \quad , \quad 1 \le j < m$$

$$A_{k1} = \frac{q^m + q^{m-1} + hf(\beta, q^m)}{2}$$

}

(c) If $k > 1$ then compute

{

$$A_{kj} = \frac{4^{j-1} A_{k,j-1} - A_{k-1,j-1}}{4^{j-1} - 1} \quad , \quad 2 \le j \le k$$

$$E_{k,k-1} = \frac{A_{k,k-1} - A_{k-1,k-1}}{4^{k-1} - 1}$$

}

}

4. While $(k < 2)$ or $[(\|E_{k,k-1}\| > \max\{\|A_{kk}\|, 1\}\varepsilon)$ and $(k < r)]$.

When Alg. 8.6.1 terminates, A_{kk} is the estimated value of $x(\beta)$. If $k < r$, the convergence criterion $\|E_{k,k-1}\| \le \max\{\|A_{kk}\|, 1)\varepsilon$ was achieved. This is the combined relative and absolute error criterion developed in Equation (8.4.11).

EXAMPLE 8.6.1 Extrapolation Method

The extrapolation method converges quite rapidly. To illustrate just how fast, consider the following three-dimensional linear system.

$$\frac{dx_1}{dt} = x_2$$

$$\frac{dx_2}{dt} = x_3$$

$$\frac{dx_3}{dt} = -2x_1 - 5x_2 - 4x_3$$

Suppose the initial condition is $x(0) = [1, -3, 6]^T$. It is easy to verify by direct substitution that the following is the solution in this case:

$$x(t) = \exp(-t)\begin{bmatrix} -t \\ t - 1 \\ 2 - t \end{bmatrix} + \exp(-2t)\begin{bmatrix} 1 \\ -2 \\ 4 \end{bmatrix}$$

TABLE 8.6.1 Extrapolation Method

t	i	$\|E_{k,k-1}\|$	$\|E\|$
0.5	201	1.0518467e-09	2.3841858e-07
1.0	201	7.0123113e-10	2.3841858e-07
1.5	201	2.3374372e-10	1.7881393e-07
2.0	201	2.3374372e-10	8.9406967e-08
2.5	102	4.3846075e-08	4.4703484e-08
3.0	102	1.6970766e-08	2.9802322e-08
3.5	102	6.9183961e-09	3.7252903e-08
4.0	102	2.8383165e-09	3.7252903e-08

Example program *e861* on the distribution CD applies the extrapolation method to solve this system over the interval $[0, 4]$ using a step size of $h = 0.5$ and an error bound of $\varepsilon = 10^{-7}$ with the results summarized in Table 8.6.1. The first column is the independent variable time, the second is the number of scalar function evaluations performed i, the third is the norm of the estimated truncation error $E_{k,k-1}$, and the last is the norm of the actual global error E. In each case the infinity norm is used. A linear system with a closed-form solution was used so that the predicted error at each solution interval could be compared with the actual error.

There are a number of variations on the basic extrapolation method that have been investigated. Instead of having the step size cut in half with each new extrapolation level, one can have it decrease more slowly, such as $h_k = (\beta - \alpha)/(2k)$. This way, the number of function evaluations does not grow as rapidly. It is also possible to represent the truncation error with a more general rational polynomial approximation rather than with a polynomial. Implementations of these ideas can be found in Press et al. (1992).

8.7 STIFF DIFFERENTIAL EQUATIONS

In engineering applications, there are many instances of differential equations whose solutions contain terms that have widely differing *time* scales. Some transient components of the solution may decay to zero very quickly, while others change much more slowly. Examples of widely differing time scales occur naturally in chemical reactions, in the study of mechanical vibrations, in electrical circuits and in feedback control systems. Differential equations that have time constants that vary by more than an order of magnitude are often referred to as *stiff* systems. The name arises from mechanical springs where the compression or extension of a stiff spring causes a very rapid transient.

Stiff systems are difficult to solve numerically. On the one hand, the presence of the fast transient component dictates that a very small step must be used. However, the presence of a slow component means that the solution has to be evaluated over a long period of time. This can result in a very large number of very small steps, which means significant computational effort. It might be tempting to think that a variable step size method could take small steps until the fast component has almost vanished and then lengthen the steps to obtain an efficient solution for the remaining time. The problem with this approach is that the fast part of the solution, although small, is still present, and if the step length is made too large, it will cause the numerical solution to *diverge* from the exact solution and grow without bound.

Although it is possible for a single differential equation with a slowing-varying forcing term to be stiff, the adjective *stiff* is usually reserved for systems of two or more differential equations. Consider, for example, the following linear system of n differential equations.

$$\frac{dx}{dt} = Ax + u \tag{8.7.1}$$

Here A is an $n \times n$ matrix and $u(t)$ is an $n \times 1$ forcing term which drives the system. Let $\{\lambda_1, \ldots, \lambda_n\}$ denote the n eigenvalues of the coefficient matrix A. When the forcing term is removed, $u = 0$, and the n eigenvalues are distinct, the solution to (8.7.1) can be shown to be of the general form (Chen, 1984):

$$x(t) = \sum_{k=1}^{n} \exp(\lambda_k t)\, c^k \tag{8.7.2}$$

Here the constant vectors $\{c^k\}$ depend on the initial condition $x(0)$. It is clear from Equation (8.7.2) that the solution goes to zero as time approaches infinity if and only if the real part of each eigenvalue is negative. That is, the linear system is *stable* if and only if

$$\boxed{\mathrm{Re}(\lambda_k) < 0 \quad , \quad 1 \le k \le n} \tag{8.7.3}$$

Thus for stable linear systems, the eigenvalues are all located strictly in the left half of the complex plane. The stability criterion in (8.7.3) is valid even when the eigenvalues of A are not distinct.

The speed with which each term in (8.7.2) goes to zero depends on the size of the real part of the eigenvalue. If some eigenvalues have very large real parts and others have very small real parts, then the system is stiff. In particular, the following ratio can be used as a measure of the stiffness of this system:

$$\eta(A) \triangleq \frac{\max_{k=1}^{n}\{|\mathrm{Re}(\lambda_k)|\}}{\min_{k=1}^{n}\{|\mathrm{Re}(\lambda_k)|\}} \tag{8.7.4}$$

By construction, $\eta(A) \ge 1$. When $\eta(A) \gg 1$, the linear system is stiff because it has some transient terms that decay rapidly while others decay slowly. Thus, a linear system is stiff when its coefficient matrix has eigenvalues whose real parts differ by more than an order of magnitude. If the forcing term in Equation (8.7.1) is nonzero, or if the eigenvalues are not distinct, the solution $x(t)$ will contain some additional terms. It is also possible for nonlinear systems to be stiff, but the characterization is less clear. For nonlinear systems of the form $dx/dt = f(t, x)$, one can look at the eigenvalues of the $n \times n$ *Jacobian* matrix.

$$J_{kj}(t, x) \triangleq \frac{\partial f_k(t, x)}{\partial x_j} \quad , \quad 1 \le k, j \le n \tag{8.7.5}$$

In general, the eigenvalues of $J(t, x)$ vary with t and x, which means that the system can change back and forth between being stiff and nonstiff. Observe that the Jacobian matrix of the linear system in (8.7.1) is simply $J(t, x) = A$.

EXAMPLE 8.7.1 Stiff System

As an illustration of an engineering application where a stiff system of differential equations arises, consider the electrical circuit shown in Figure 8.7.1. The component values listed in Figure 8.7.1 correspond to typical off-the-shelf parts. Applying Kirchhoff's laws results in the following first-order linear system where the 2×1 state vector $x = [x_1, x_2]^T$ denotes the capacitor voltages.

$$\frac{dx}{dt} = \begin{bmatrix} -1 & -1 \\ -1 & -10001 \end{bmatrix} x + \begin{bmatrix} 1 \\ 1 \end{bmatrix} v$$

FIGURE 8.7.1 A Stiff Electrical Circuit

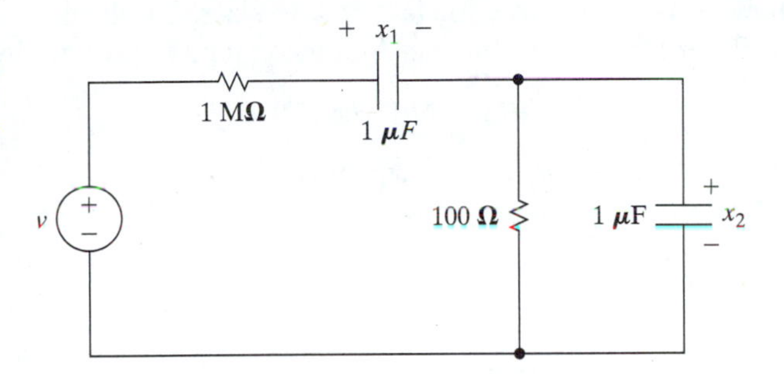

To find the eigenvalues of the coefficient matrix A, we factor the characteristic polynomial.

$$\begin{aligned}\Delta(\lambda) &= \det(\lambda I - A)\\ &= (\lambda + 1)(\lambda + 10001) - 1\\ &= \lambda^2 + 10002\lambda + 10000\\ &= (\lambda - \lambda_1)(\lambda - \lambda_2)\end{aligned}$$

Application of the quadratic formula yields eigenvalues of $\lambda_1 \approx -10001$ and $\lambda_2 \approx -1$. When the voltage source v is constant, the complete solution is of the form

$$x(t) = \exp(-10001t)a + \exp(-t)b + c$$

Here $c = [v, 0]^T$, while a and b depend on the initial condition $x(0)$. Applying (8.7.4), we see that $\eta(A) = 10001$ which makes this circuit a very stiff system.

8.7.1 Implicit Methods

Stiff equations are difficult to solve because they require a large number of very small integration steps. In order to obtain an efficient technique for solving stiff systems, we must develop a solution formula that does not become unstable when the step size is increased. It is helpful to begin by reexamining the characteristics of the following simple one-dimensional linear system.

$$\frac{dx}{dt} = -cx \tag{8.7.6}$$

If $c > 0$, this system is stable and has the exact solution $x(t) = \exp(-ct)x(0)$. Recall from Example 8.2.1 that the Euler estimates of the solution to this system diverge from the exact solution for step sizes in the range $h > 2/c$. Euler's method in (8.2.4) is an explicit integration formula because the solution estimate x^{k+1} appears only on the left-hand side. Consider the following implicit version of the Euler technique, which is called *Euler's backward method.*

$$x^{k+1} = x^k + hf(t^{k+1}, x^{k+1}) \tag{8.7.7}$$

In contrast to Euler's regular or forward method, the slope of $x(t)$ is evaluated at the end of the interval, t^{k+1}, rather than the beginning, t^k. Since x^{k+1} appears on both sides, Equation (8.7.7) is an *implicit* integration formula. Implicit formulas have the disadvantage that a system of nonlinear algebraic equations has to be solved at each step to obtain x^{k+1}. However, the extra computational effort is rewarded by the fact that implicit methods tend to remain stable for larger step sizes. To illustrate, suppose we apply the backward Euler formula to the one-dimensional linear system in (8.7.6).

$$x^{k+1} = x^k + h(-cx^{k+1}) \tag{8.7.8}$$

Since this system is linear, we can solve for x^{k+1} explicitly, which yields $x^{k+1} = x^k/(1 + ch)$. It follows that the solution estimate at time t^k is

$$x^k = \frac{x^0}{(1 + ch)^k}, \quad k \geq 0 \tag{8.7.9}$$

Since $c > 0$, we conclude that the solution estimates produced by the backward Euler method do not diverge for *any* step size $h > 0$. It is this relative insensitivity to step size that makes implicit methods attractive for solving stiff systems.

The work associated with solving (8.7.7) for x^{k+1}, say, by using Newton's vector method, can be substantial depending on the details of the function $f(t, x)$. One way to reduce the computational effort is to expand $f(t^{k+1}, x^{k+1})$ into a Taylor series about the point (t^k, x^k). Dropping the second- and higher-order terms then produces the following approximation to the backward Euler method.

$$x^{k+1} = x^k + h\left[f(t^k, x^k) + \frac{\partial f(t^k, x^k)}{\partial x}(x^{k+1} - x^k) + \frac{\partial f(t^k, x^k)}{\partial t}h\right] \tag{8.7.10}$$

Recall from (8.7.5) that $\partial f(t,x)/\partial x$ is the $n \times n$ Jacobian matrix $J(t,x)$. If $\Delta x^k = x^{k+1} - x^k$ denotes the kth forward difference of x, then Equation (8.7.10) can be reformulated in terms of $J(t^k, x^k)$ and Δx^k as follows.

$$\boxed{[I - hJ(t^k, x^k)]\Delta x^k = hf(t^k, x^k) + h^2 \frac{\partial f(t^k, x^k)}{\partial t}} \tag{8.7.11}$$

From (8.7.11), we see that, given x^k, we can obtain Δx^k by solving an n-dimensional linear algebraic system using the techniques from Chapter 2. Once Δx^k is known, we can compute $x^{k+1} = x^k + \Delta x^k$. In this way, x^{k+1} can be approximated without solving a nonlinear algebraic system. This is referred to as a *semi-implicit* method. Note that it is equivalent to using Newton's vector method to solve (8.7.7) but stopping after a single iteration.

Although (8.7.11) can be used to solve stiff systems of differential equations, it is only a first order method. Treatment of higher-order implicit methods for solving stiff differential equations based on backward differences can be found in Gear (1971).

8.7.2 Semi-Implicit Extrapolation Method

The extrapolation method is an example of a higher-order explicit method that is a good candidate for conversion to a semi-implicit method suitable for solving stiff differential equations (Press, et al., 1992). The basic approach is to formulate the modified midpoint method using implicit formulas, and then convert them to semi-implicit form. The extrapolations can then be computed just as before. Suppose we want to solve a stiff system over the interval $[\alpha, \beta]$ starting from the initial condition $x(\alpha) = a$. We begin by dividing the solution interval into m steps of equal length $h = (\beta - \alpha)/m$. The solution times are

$$t^j = \alpha + jh \quad , \quad 0 \le j \le m \tag{8.7.12}$$

To initiate the process, we compute the first solution point (and only the first) using Euler's implicit backward method. If $q^0 = a$, then to compute q^1 we solve

$$q^1 = q^0 + hf(t^1, q^1) \tag{8.7.13}$$

Rather than solve this implicit equation directly, we convert it to semi-implicit form. Expanding $f(t^1, q^1)$ into a Taylor series about (t^0, q^0) and dropping the second and higher-order terms yields

$$q^1 = q^0 + h\left[f(t^0, q^0) + \frac{\partial f(t^0, q^0)}{\partial x}(q^1 - q^0) + \frac{\partial f(t^0, q^0)}{\partial t}h\right] \tag{8.7.14}$$

Let $\Delta q^k = q^{k+1} - q^k$ be the kth forward difference of q. Recalling that $J(t, x) = \partial f(t, x)/\partial x$, and using $t^0 = \alpha$ and $q^0 = a$, we can rewrite Equation (8.7.14) as the following n dimensional linear algebraic system similar to (8.7.11).

$$\boxed{[I - hJ(\alpha, a)]\Delta q^0 = hf(\alpha, a) + h^2 \frac{\partial f(\alpha, a)}{\partial t}} \tag{8.7.15}$$

First, Δq^0 is found by solving (8.7.15), and then $q^1 \approx x(\alpha + h)$ is obtained using $q^1 = q^0 + \Delta q^0$. Once the points q^0 and q^1 are available, the following implicit version of the midpoint formula can be used to compute the remaining points.

$$q^{j+1} = q^{j-1} + 2hf\left(t^j, \frac{q^{j-1} + q^{j+1}}{2}\right) \quad , \quad 1 \le j < m \tag{8.7.16}$$

To convert (8.7.16) to semi-implicit form, we expand the last term into a Taylor series about (t^j, q^j). Dropping the second and higher-order terms of the series then yields the following approximation to (8.7.16).

$$q^{j+1} = q^{j-1} + 2h\left[f(t^j, q^j) + \frac{\partial f(t^j, q^j)}{\partial x}\left(\frac{q^{j-1} + q^{j+1}}{2} - q^j\right)\right] \tag{8.7.17}$$

Note that there is no $\partial f(t, x)/\partial t$ term because $f(t, x)$ is evaluated at the point about which we are expanding. Again, using the notation $J(t, x) = \partial f(t, x)/\partial x$ for the Jacobian matrix, we can rearrange the terms in (8.7.17) as

$$[I - hJ(t^j,q^j)]q^{j+1} = [I + hJ(t^j,q^j)]q^{j-1} + 2h[f(t^j,q^j) - J(t^j,q^j)q^j] \quad (8.7.18)$$

Next, to minimize the number of matrix multiplications, we can subtract $2hJ(t^j,q^j)q^{j-1}$ from the first term on the right-hand side of (8.7.18) and then add it to the last term. This allows us to rewrite the equation as follows.

$$\boxed{[I - hJ(t^j,q^j)](q^{j+1} - q^{j-1}) = 2h[f(t^j,q^j) - J(t^j,q^j)\Delta q^{j-1}] \quad , \quad 1 \le j < m} \quad (8.7.19)$$

This n-dimensional linear algebraic system can be solved for the difference vector $v = q^{j+1} - q^{j-1}$. The new point is then $q^{j+1} = q^{j-1} + v$.

To complete the procedure, we perform an end point correction. First, an implicit backward Euler step is used to compute q^{m+1}.

$$q^{m+1} = q^m + hf(t^{m+1},q^{m+1}) \quad (8.7.20)$$

Just as with the first step, we convert it to semi-implicit form by expanding $f(t^{m+1},q^{m+1})$ into a Taylor series about (t^m,q^m). After dropping the second and higher-order terms and recalling that $t^m = \beta$, this yields the following semi-implicit formula.

$$\boxed{[I - hJ(\beta,q^m)]\Delta q^m = hf(\beta,q^m) + h^2\frac{\partial f(\beta,q^m)}{\partial t}} \quad (8.7.21)$$

Solving Equation (8.7.21) for Δq^m then yields $q^{m+1} = q^m + \Delta q^m$. Finally, the smoothed estimate $x^m \approx x(\beta)$ is obtained by averaging q^{m-1} and q^{m+1}.

$$\boxed{x^m = \frac{q^{m-1}}{2} + \frac{q^m + \Delta q^m}{2}} \quad (8.7.22)$$

Whereas the regular extrapolation method requires $m + 1$ function evaluations to estimate $x(\beta)$, the semi-implicit method requires solving $m + 1$ linear algebraic systems. Furthermore, the formation of each of these linear algebraic systems requires evaluating the Jacobian matrix, which is roughly equivalent to n function evaluations. Of course, if the system is linear as in (8.7.1), the Jacobian matrix is constant. In this case, the Jacobian only has to be evaluated once, and the linear algebraic systems can be efficiently solved with the *LU* decomposition method. The number of function evaluations is $n + 2$ for the initial step, $(m - 1)(n + 1)$ for the midpoint steps, and $n + 2$ for the final step. Thus, the total number of (vector) function evaluations for one iteration of the semi-implicit midpoint method is

$$N(m) = (m + 1)(n + 1) + 2 \quad (8.7.23)$$

Bader and Dueffhard (1983) have shown that the truncation error of their semi-implicit extrapolation method is of order $O(h^2)$ and can be expressed as an even power series as in Equation (8.6.5). Thus, the same extrapolation techniques that were developed in (8.6.6), (8.6.13), and (8.6.14) can be used on the semi-implicit solution estimates without modification. The following algorithm summarizes the semi-implicit extrapolation method.

Alg. 8.7.1 **Semi-Implicit Extrapolation Method**

1. Pick $\varepsilon > 0$ and $r \geq 2$.
2. Set $k = 0, q^0 = a$.
3. Do

{

(a) Set $k = k + 1, m = 2^k$ and $h = (\beta - \alpha)/m$.

(b) For $j = 1$ to m, set $t^j = \alpha + jh$.

(c) Solve the following linear algebraic systems

{

$$[I - hJ(\alpha, a)]v = hf(\alpha, a) + h^2 \frac{\partial f(\alpha, a)}{\partial t}$$

$$q^1 = q^0 + v$$

$$[I - hJ(t^j, q^j)]v = 2h[f(t^j, q^j) - J(t^j, q^j)(q^j - q^{j-1})]$$

$$q^{j+1} = q^{j-1} + v \quad , \quad 1 \leq j < m$$

$$[I - hJ(\beta, q^m)]v = hf(\beta, q^m) + h^2 \frac{\partial f(\beta, q^m)}{\partial t}$$

$$A_{k1} = \frac{q^{m-1} + q^m + v}{2}$$

}

(d) If $k > 1$ then compute

{

$$A_{kj} = \frac{4^{j-1}A_{k,j-1} - A_{k-1,j-1}}{4^{j-1} - 1} \quad , \quad 2 \leq j \leq k$$

$$E_{k,k-1} = \frac{A_{k,k-1} - A_{k-1,k-1}}{4^{k-1} - 1}$$

}

}

4. While $(k < 2)$ or $[(\|E_{k,k-1}\| > \max\{\|A_{kk}\|, 1\}\varepsilon)$ and $(k < r)]$

When Alg. 8.6.1 terminates, A_{kk} is the estimated value of $x(\beta)$. If $k < r$, then the convergence criterion $\|E_{k,k-1}\| \leq \max\{\|A_{kk}\|, 1)\varepsilon$ was satisfied.

EXAMPLE 8.7.2 **Semi-Implicit Extrapolation Method**

To illustrate the use of the semi-implicit extrapolation method, consider the stiff circuit shown in Figure 8.7.1. Recall that the state equations for this circuit are

$$\frac{dx_1}{dt} = -x_1 - x_2$$

TABLE 8.7.1 Solution of a Stiff Electrical Circuit

	Semi-Implicit Extrapolation		*Numerical Derivatives*		*Runge-Kutta-Fehlberg*	
t^k	$\|E^k\|$	i	$\|E^k\|$	i	$\|E^k\|$	i
0.1	7.0984271e-07	412	5.2359104e-07	700	8.2488583e-07	3612
0.2	7.9472862e-08	108	7.5499216e-08	188	6.4949694e-07	3300
0.3	7.5499216e-08	108	7.5499216e-08	188	8.9182993e-07	3288
0.4	6.3578291e-08	108	6.3578291e-08	188	9.9269778e-07	3288
0.5	5.1657359e-08	108	5.1657359e-08	188	4.6162077e-07	3312
0.6	4.7683717e-08	108	4.7683717e-08	188	9.9745512e-07	3288
0.7	4.3710074e-08	108	4.3710074e-08	188	4.9031524e-07	3300
0.8	4.1723251e-08	108	4.1723251e-08	188	8.2533523e-07	3288
0.9	3.7749608e-08	108	3.9736431e-08	188	7.0422806e-07	3300
1.0	3.5762788e-08	108	3.5762788e-08	188	7.5404199e-07	3300
average	1.1866798e-07	138.4	9.9844137e-08	239.2	7.5919070e-07	3327.6

$$\frac{dx_2}{dt} = -x_1 - 10001x_2$$

Suppose the initial condition is $x(0) = [1, -1]^T$. Example program *e872* on the distribution CD applies the semi-implicit extrapolation method to solve this system over the interval $[0, 1]$ using a step size of $h = 0.1$ and an error bound of $\varepsilon = 10^{-6}$ with the results summarized in Table 8.7.1. The first column is the independent variable time. The next two columns correspond to solving the system with the semi-implicit extrapolation method. The second column is the truncation error, and the third is the number of scalar function evaluations, i. Columns four and five again correspond to solving the system with the semi-implicit extrapolation method, but this time using central differences to approximate the partial derivatives of $f(t, x)$ with respect to t and x. Column four is the norm of the truncation error, and column five is the number of scalar function evaluations. The last two columns correspond to solving the system with the variable step size Runge-Kutta-Fehlberg method. Again, column six is the truncation error, and column seven is the number of scalar function evaluations. Finally, the last row of Table 8.7.1 shows the average truncation error and the average number of scalar function evaluations for each integration interval.

For this stiff system, it is clear that the semi-implicit extrapolation method is not only more accurate, it is also substantially more efficient. Notice how, after the fast transient has vanished, the implicit extrapolation method takes larger steps and consequently fewer function evaluations. This is in contrast to the Runge-Kutta-Fehlberg method, which must take small steps throughout to avoid having the solution diverge. When central differences are used to approximate the partial derivatives, the number of function evaluations increases because two function evaluations are needed for each derivative.

8.7.3 Differential-Algebraic Systems

The first-order system in Equation (8.2.1) can be generalized to include both differential equations and algebraic equations. Suppose the system of n differential equations is augmented with m algebraic equations as follows.

$$\frac{dx}{dt} = g(t, x, y) \tag{8.7.24a}$$

$$0 = h(t, x, y) \tag{8.7.24b}$$

Systems that can be modeled by (8.7.24) are referred to as *differential-algebraic* systems. Under certain special conditions, differential-algebraic systems are relatively easy to solve. Suppose the nonlinear algebraic constraint (8.7.24b) can be solved for the vector y. That is, suppose there exists a function v such that (8.7.24b) can be rewritten as

$$y = v(t, x) \tag{8.7.25}$$

The expression for y in (8.7.25) can then be substituted directly into (8.7.24a). This results in first-order n-dimensional system in the remaining variable x. This system can be solved by any of the techniques covered thus far to obtain $x(t)$. Finally, the remaining unknowns are recovered by substituting the solution $x(t)$ into (8.7.25).

It is also possible to obtain a numerical estimate of the solution to a differential-algebraic system even when (8.7.24b) can not be solved for y. The approach in this case is to rewrite (8.7.24) as follows.

$$\frac{dx}{dt} = g(t, x, y) \tag{8.7.26a}$$

$$\mu \frac{dy}{dt} \approx h(t, x, y) \tag{8.7.26b}$$

When $\mu = 0$, the approximation in Equation (8.7.26b) is exact. Thus we take $|\mu| \ll 1$. The approximate system in (8.7.26) is referred to as a *singular perturbation* of the original system. Assuming $\mu \neq 0$, we can rewrite (8.7.26) as a standard first-order $(n + m)$-dimensional system.

$$\frac{dx}{dt} = g(t, x, y) \tag{8.7.27a}$$

$$\frac{dy}{dt} = \frac{1}{\mu} h(t, x, y) \tag{8.7.27b}$$

Since the *perturbation parameter* μ is very small, this means that y is likely to change much faster than x. That is, the complete solution contains a relatively slow component, $x(t)$, and a very fast component, $y(t)$. Consequently, a singular perturbation of a differential-algebraic system generates a stiff system of differential equations. As the accuracy of the approximation improves, the system becomes extremely stiff. Further discussion of singular perturbation systems can be found in Kokotovic and Khalil (1986).

8.8 BOUNDARY VALUE PROBLEMS

For the systems of differential equations that we have considered thus far, the $n \times 1$ solution vector $x(t)$ has been specified at a single value of the independent variable, $t = \alpha$. The algebraic constraint, $x(\alpha) = a$, is called an initial condition and finding a solution which satisfies it is called an *initial value problem*. There is another important class of problems for which some of the n algebraic constraints are specified at the initial value α, while the remaining constraints are specified at the final value β. Problems of this type are referred to as two-point *boundary value problems*. To illustrate the methods for solving a boundary value problem (BVP), consider the following nonlinear second-order differential equation, where the *independent variable* is a spatial coordinate x rather than t.

$$\frac{d^2y}{dx^2} = g\left(x, y, \frac{dy}{dx}\right) \tag{8.8.1}$$

If we specify y and dy/dx at $x = \alpha$, then this is an initial value problem that can be cast as a two-dimensional first-order system using the state variable $z = [y,\ dy/dx]^T$. Instead, we specify the value of the solution $y(x)$ at the two boundaries of the integration interval.

$$y(\alpha) = a \tag{8.8.2a}$$

$$y(\beta) = b \tag{8.8.2b}$$

Depending on g, there may or may not be a solution to Equation (8.8.1) satisfying the *boundary conditions* in (8.8.2). When a solution does exist, it is generally much harder to compute than for an initial value problem.

8.8.1 Shooting Method

The first technique for solving boundary value problems is based on a very simple idea. Suppose we are attempting to fire an artillery shell to hit a distant target as in Figure 8.8.1. Assuming the artillery piece is pointed in the direction of the target, the only variable available is the elevation angle of the barrel, which is the initial slope of the trajectory of the projectile. We can fire two rounds, each with a different barrel angle, and record where the projectile lands in each case. It should then be possible to come close to the target with the third round by interpolating between (or perhaps extrapolating from) the two previous trial attempts. The process can then be repeated as necessary.

This iterative technique can be applied directly to differential equations and is referred to as the *shooting method*. The barrel angle is the initial slope dy/dx at $x = \alpha$, shooting the gun is solving the differential equation, and hitting the target is matching the boundary condition $y(\beta) = b$. To develop the shooting method, first we convert the second-order equation (8.8.1) into an equivalent two-dimensional first-order system. Let the state vector be $z = [y,\ dy/dx]^T$. Then, proceeding as in (8.1.5), we have

$$\frac{dz_1}{dx} = z_2 \tag{8.8.3a}$$

$$\frac{dz_2}{dx} = g(x, z_1, z_2) \tag{8.8.3b}$$

The system of equations in (8.8.3) can be thought of as a parameterized family of initial value problems with initial condition $z(\alpha) = [a, q]^T$, where the *parameter* q remains to be determined. The objective is to find a value for $q = z_2(\alpha)$ such that the solution trajectory $z(t)$ satisfies the terminal boundary condition $z_1(\beta) = b$. Since the solution to (8.8.3) depends on the initial condition parameter q, suppose we denote the depen-

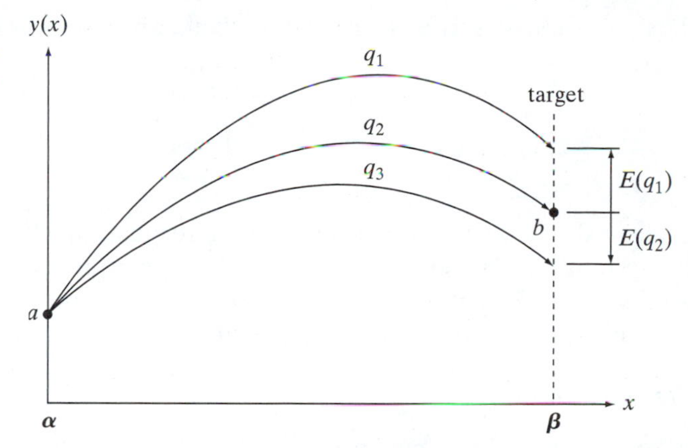

FIGURE 8.8.1 Shooting Method

dence explicitly as $z(x,q)$. Then the error in hitting the terminal boundary condition can be defined as follows.

$$E(q) \triangleq x(\beta, q) - b \tag{8.8.4}$$

Computing the error function $E(q)$ is rather complex because it involves solving the differential equation over the interval $[\alpha, \beta]$ subject to the initial condition $z(\alpha) = [a, q]^T$. In spite of this, $E(q)$ is an algebraic function of q, which means that the original boundary value problem has been converted to a simpler root finding problem. That is, we want to find q such that $E(q) = 0$. Recall from Section 5.6 that one of the most effective methods for finding roots is Newton's iterative method. If q_0 is an initial guess, then subsequent estimates of q are obtained as follows.

$$q_{k+1} = q_k - \left(\frac{dE(q_k)}{dq}\right)^{-1} E(q_k) \quad , \quad k \geq 0 \tag{8.8.5}$$

The problem with applying Newton's method directly is that there is no closed-form expression for $E(q)$, much less $dE(q)/dq$. However, we can approximate the derivative using a central difference. In particular, if Δq is a small variation in q, then

$$\frac{dE(q_k)}{dq} \approx \frac{E(q_k + \Delta q) - E(q_k - \Delta q)}{2\Delta q} \tag{8.8.6}$$

An important special case arises when the differential equation to be solved is *linear*. This occurs when the function $g(x, z_1, z_2)$ in Equation (8.8.3b) can be represented as a sum of three terms as follows.

$$g(x, z_1, z_2) = c(x)z_1 + d(x)z_2 + e(x) \tag{8.8.7}$$

If the coefficient functions $c(x)$, $d(x)$, and $e(x)$ are continuous and if $c(x) > 0$, then a unique solution to the boundary value problem exists. Although we can find that solution using Newton's method in (8.8.5), it is also possible to express it directly in this case. Let $u(x)$ denote the solution to the initial value problem $z(\alpha) = [a, 0]^T$, and let $v(x)$ denote the solution to the initial value problem $z(0) = [0, 1]^T$. Then it is possible to show by direct substitution (see, e.g., Faires and Burden, 1993) that the following linear combination of the two solutions solves the boundary value problem when the system is linear.

$$z(x) = u(x) + \left(\frac{b - u_1(\beta)}{v_1(\beta)}\right)v(x) \tag{8.8.8}$$

This sum of two initial-value solutions represents linear interpolation between the two solutions. When the system itself is linear, a single linear interpolation step produces the exact solution as pictured in Figure 8.8.1. For nonlinear systems, multiple iterations using Newton's method or some other root finding technique are needed.

8.8.2 Finite Difference Method

An alternative approach to solving boundary value problems is based on the idea of approximating the derivative terms with finite numerical differences. This generates a system of algebraic equations that can be solved to obtain estimates of the solution at discrete points. We begin by partitioning the solution interval $[\alpha, \beta]$ into $m + 1$ subintervals of equal length as shown in Figure 8.8.2 for the case $m = 7$. If $h = (\beta - \alpha)/(m + 1)$ denotes the grid size, then the $m + 2$ points at which the solution is estimated are as follows.

$$x_k = \alpha + kh \quad , \quad 0 \le j \le m + 1 \tag{8.8.9}$$

Next, we work directly with the original second-order differential equation in (8.8.1). Recall from Table 7.4.1 that the first and second derivatives can be approximated by the following central difference formulas, which use only values at the grid points.

$$\left.\frac{dy}{dx}\right|_{x=x_k} \approx \frac{y(x_{k+1}) - y(x_{k-1})}{2h} \tag{8.8.10a}$$

$$\left.\frac{d^2y}{dx^2}\right|_{x=x_k} \approx \frac{y(x_{k+1}) - 2y(x_k) + y(x_{k-1})}{h^2} \tag{8.8.10b}$$

To simplify the notation, let $y_k = y(x_k)$. Then, using the finite difference approximations in (8.8.10), we get the following discrete approximation to the second-order differential equation (8.8.1).

$$y_{k+1} - 2y_k + y_{k-1} = h^2 g\left(x_k, y_k, \frac{y_{k+1} - y_{k-1}}{2h}\right) \quad 1 < k < m \tag{8.8.11}$$

Note that (8.8.11) is only applied at the interior points of the integration interval. At the boundary points, we have $y_0 = a$ and $y_{m+1} = b$. Thus, the $k = 1$ and $k = m$ cases result in the following two boundary condition equations.

$$y_2 - 2y_1 + a = h^2 g\left(x_1, y_1, \frac{y_2 - a}{2h}\right) \tag{8.8.12a}$$

$$b - 2y_m + y_{m-1} = h^2 g\left(x_m, y_m, \frac{b - y_{m-1}}{2h}\right) \tag{8.8.12b}$$

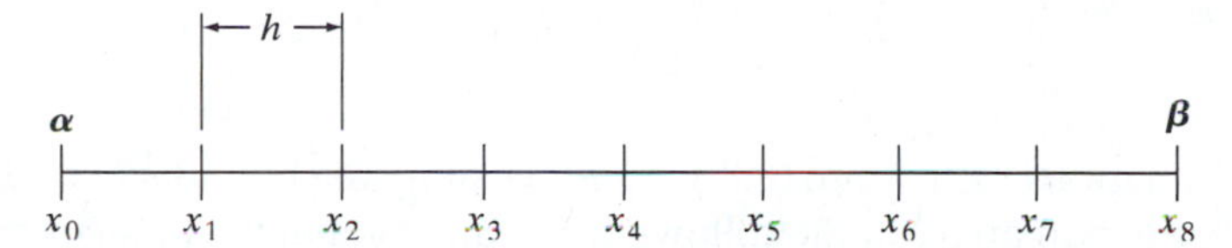

FIGURE 8.8.2 Finite-Difference Method with $m = 7$

Taken together, (8.8.11) and (8.8.12) constitute a system of m nonlinear algebraic equations in the vector of m unknowns $y = [y_1, y_2, \ldots, y_m]^T$. Solving the equations is equivalent to finding a root of the m-dimensional nonlinear algebraic system $E(y) = 0$ where

$$E_k(y) \triangleq y_{k+1} - 2y_k + y_{k-1} - h^2 g\left(x_k, y_k, \frac{y_{k+1} - y_{k-1}}{2h}\right) \quad , \quad 1 \le k \le m \tag{8.8.13}$$

Here it is understood that when $k = 1$, $y_0 = a$; and when $k = m$, $y_{m+1} = b$. The nonlinear algebraic system $E(y) = 0$ can be solved by Newton's vector method. If the $m \times 1$ vector y^0 denotes an initial guess, then subsequent estimates y^j can be computed as follows.

$$y^{j+1} = y^j - \left[\frac{\partial E(y^j)}{\partial y}\right]^{-1} E(y^j) \quad , \quad j \ge 0 \tag{8.8.14}$$

The $m \times m$ Jacobian matrix $J(y^j) = \partial E(y^j)/\partial y$ does not have to be inverted. Instead, it is more efficient to solve the following linear algebraic system for $\Delta y^j = y^{j+1} - y^j$ and then compute $y^{j+1} = y^j + \Delta y^j$.

$$J(y^j)\Delta y^j = -E(y^j) \tag{8.8.15}$$

Note that unlike with the shooting method, it may be possible to get a closed-form expression for the Jacobian matrix because it depends directly on the right-hand-side function g. In any event, we can always evaluate the Jacobian numerically, but this tends to be less accurate and it requires additional function evaluations.

If we restrict our attention to the case of a *linear* boundary value problem as in Equation (8.8.7), the system of equations to be solved becomes much simpler. Substituting the linear expression for g from (8.8.7) into (8.8.13), we get the following linear constraints.

$$y_{k+1} - 2y_k + y_{k-1} - h^2\left(c_k y_k + \frac{d_k(y_{k+1} - y_{k-1})}{2h} + e_k\right) = 0 \quad , \quad 1 \le k \le m \tag{8.8.16}$$

Here we have used the notational shorthand, $c_k = c(x_k)$, $d_k = d(x_k)$ and $e_k = e(x_k)$. Collecting coefficients of the y_j terms, we can rewrite (8.8.16) as

$$\left(1 + \frac{hd_k}{2}\right)y_{k-1} - (2 + h^2 c_k)y_k + \left(1 - \frac{hd_k}{2}\right)y_{k+1} = h^2 e_k \quad , \quad 1 \le k \le m \tag{8.8.17}$$

To simplify the final equations, we define the following intermediate variables:

$$u_k \triangleq 1 + \frac{hd_k}{2} \tag{8.8.18a}$$

$$v_k \triangleq -(2 + h^2 c_k) \tag{8.8.18b}$$

$$w_k \triangleq 1 - \frac{hd_k}{2} \tag{8.8.18c}$$

Substituting Equation (8.8.18) into (8.8.17) and recalling the boundary conditions $y_0 = a$ and $y_{m+1} = b$, we then arrive at the following m-dimensional linear algebraic system.

$$\begin{bmatrix} v_1 & w_1 & 0 & \cdots & 0 & 0 & 0 \\ u_2 & v_2 & w_2 & \cdots & 0 & 0 & 0 \\ \vdots & \vdots & \vdots & \ddots & \vdots & \vdots & \vdots \\ 0 & 0 & 0 & \cdots & u_{m-1} & v_{m-1} & w_{m-1} \\ 0 & 0 & 0 & \cdots & 0 & u_m & v_m \end{bmatrix} \begin{bmatrix} y_1 \\ y_2 \\ \vdots \\ y_{m-1} \\ y_m \end{bmatrix} = \begin{bmatrix} h^2 e_1 - u_1 a \\ h^2 e_2 \\ \vdots \\ h^2 e_{m-1} \\ h^2 e_m - w_m b \end{bmatrix} \tag{8.8.19}$$

Not only is (8.8.19) a linear system, but the coefficient matrix is *tridiagonal*. Recall from Section 2.4.3 that there are efficient techniques for storing tridiagonal matrices and solving tridiagonal systems. In practical terms, this means that linear boundary value problems can be *easily* solved with the finite difference method, even for relatively large values for m.

EXAMPLE 8.8.1 Boundary Value Problem

As an illustration of a numerical solution to a two-point boundary value problem, consider the following second-order differential equation:

$$\frac{d^2y}{dx^2} + 6\frac{dy}{dx} = \sin(\pi x)$$

Suppose the boundary conditions are $y(0) = 2$ and $y(1) = -1$. Example program *e881* on the distribution CD applies the finite difference method to this differential equation with the results shown in Figure 8.8.3. In this case, $m = 49$ was used, which resulted in a grid size of $h = 0.02$.

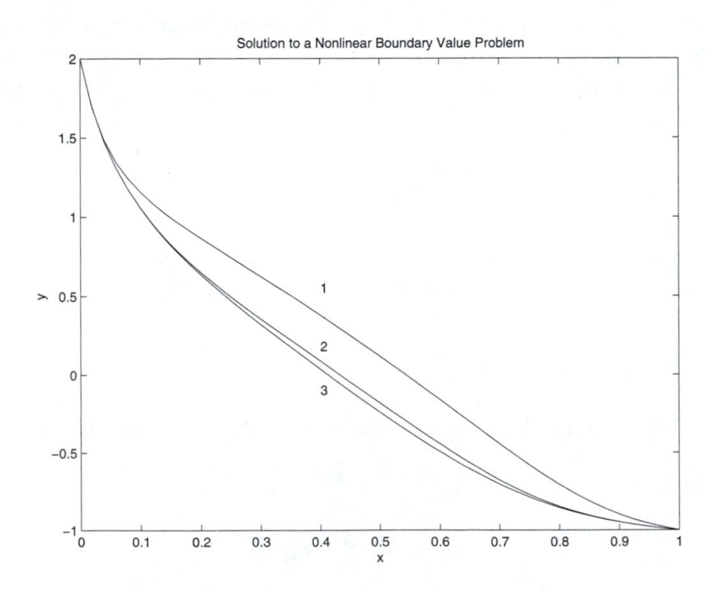

FIGURE 8.8.3 Three Iterations of Solution to a Boundary Value Problem

For clarity, only the first three iterations are plotted. Subsequent iterations are essentially the same as the third iteration.

There are many examples of engineering applications that can be formulated as boundary value problems. For example, if an object such as a heat dissipation fin is connected to a heat source, then the temperature as a function of position is characterized by an ordinary differential equation subject to boundary conditions. A discussion of this example and others, such as neutron diffusion, can be found in Nakamura (1993).

8.9 APPLICATIONS

The following examples illustrate applications of ordinary differential equation techniques using both MATLAB and C. The relevant MATLAB functions in the NLIB toolbox are described in Section 1.9 of Appendix 1, while the corresponding C functions in the NLIB library are described in Section 2.12 of Appendix 2.

8.9.1 Chemical Reactor: MATLAB

In the process control industry, chemical reactors are used to provide controlled conditions under which mixtures of two or more chemicals can react with one another to produce other chemicals. An example of a continuous-flow stirred tank chemical reactor is shown in Figure 8.9.1 (McClamroch, 1980). Here, two chemical reactants enter the tank. As they are mixed they react with one another to form a third chemical, which is drained from the tank. As an illustration, one of the reactants might be water (H_2O) and the other sulfur trioxide (SO_3). When mixed, they react to produce a third product chemical, sulfuric acid (H_2SO_4).

To develop a simplified mathematical model for this process, suppose the temperature in the tank is constant, and the volume of the mixture within the tank is also constant and equal to V. Let q be the constant flow rate into and out of the tank. Suppose the inflow or feed stream includes two reactants, one with a molar concentration of u_1

FIGURE 8.9.1 A Stirred Tank Chemical Reactor

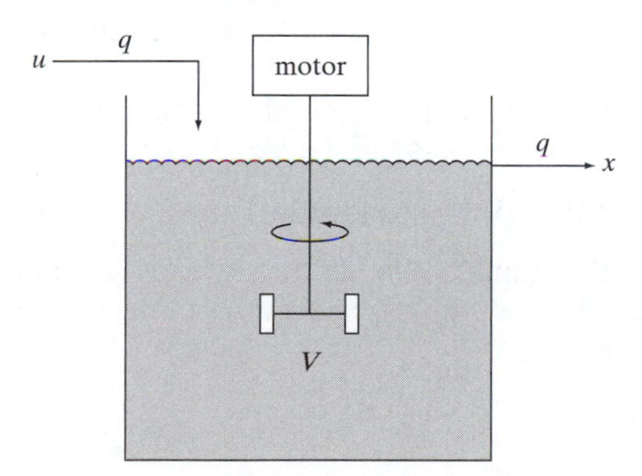

and the other with a molar concentration of u_2. The outflow or product stream consists of the mixture, which includes the original two reactants plus the product chemical formed from them. Let x_1, x_2 denote the molar concentrations of the two reactants within the stirred tank, and let x_3 denote the molar concentration of the product. Finally, let r denote the rate at which the two chemicals react. Applying a component balance to the three chemical species then results in the following dynamic model for the stirred tank system.

$$V\frac{dx_1}{dt} = qu_1 - qx_1 - Vr \tag{8.9.1a}$$

$$V\frac{dx_2}{dt} = qu_2 - qx_2 - Vr \tag{8.9.1b}$$

$$V\frac{dx_3}{dt} = -qx_3 + Vr \tag{8.9.1c}$$

The rate r at which the chemicals x_1 and x_2 react to form x_3 is assumed to be proportional to the product of the two molar concentrations. That is,

$$r = \alpha x_1 x_2 \tag{8.9.2}$$

for some constant $\alpha > 0$. Substituting the expression for r into (8.9.1) and dividing each equation by V, then yields the following mathematical model for the reactor vessel.

$$\frac{dx_1}{dt} = \frac{q}{V}(u_1 - x_1) - \alpha x_1 x_2 \tag{8.9.3a}$$

$$\frac{dx_2}{dt} = \frac{q}{V}(u_2 - x_2) - \alpha x_1 x_2 \tag{8.9.3b}$$

$$\frac{dx_3}{dt} = -\frac{q}{V}x_3 + \alpha x_1 x_2 \tag{8.9.3c}$$

Since x_3 does not appear in the first two equations, (8.9.3a) and (8.9.3b) can be thought of as a two-dimensional stand-alone subsystem that can be solved for $x_1(t)$ and $x_2(t)$. Once a solution is obtained, the product x_1x_2 serves as an input to the third equation, which can then be solved for $x_3(t)$. Of course, if a general purpose numerical technique is used, then all three equations can be solved simultaneously given the initial concentrations within the tank, $x(0)$, and the input stream concentrations $u(t)$.

To make the problem specific, suppose the initial concentrations of the mixture within the tank are

$$x(0) = [5 \;\; 3 \;\; 0]^T \text{ moles/m}^3 \tag{8.9.4}$$

This is equivalent to assuming that, initially, the mixture does not contain any product formed from the reaction. Next, suppose the concentrations of the two reactants in the feed stream are constant as follows.

$$u(t) = [3.2 \;\; 4.8]^T \text{ moles/m}^3 \tag{8.9.5}$$

Suppose the flow rate is $q = 10$ m ^{3}/min, the volume of the mixture is $V = 2$ m^3, and the reaction rate constant is $\alpha = 2.6$ m ^{3}/mole-min. The following MATLAB script on the distribution CD can be used to solve this problem.

```
%------------------------------------------------------------------------
% Example 8.9.1: Chemical Reactor
%------------------------------------------------------------------------

% Initialize

   clc                          % clear screen
   clear                        % clear variables
   n       = 3;                 % dimension of system
   m       = 100;               % number of solution points
   q       = 5000;              % maximum function evaluations
   alpha   = 0;                 % initial time
   beta    = 1.2;               % final time
   tol     = 1.e-6;             % error bound
   x0      = [5 3 0]';          % initial state

% Solve equations

   fprintf ('Example 8.9.1: Chemical Reactor\n');
   [t,X,e,k] = rkf (x0,alpha,beta,m,tol,q,'funf891');

% Display results

   show ('Number of scalar function evaluations',k);
   show ('Maximum truncation error',e);
   show ('Final product concentration (moles/m^3)',X(m,n));
   graphxy (t,X,'Concentrations','t (min)','x(t) (moles/m^3)')


function dx = funf891 (t,x)
%------------------------------------------------------------------------
% Description: Stirred tank reactor
%------------------------------------------------------------------------
   q     = 10;        % flow rate (m^3/min)
   V     = 2;         % mixture volume (m^3)
   alpha = 2.6;       % reaction rate (moles/min)
   u1    = 3.2;       % feed concentration 1 (moles/m^3)
   u2    = 4.8;       % feed concentration 2 (moles/m^3)

   b  = q/V;
   dx = [ b*(u1 - x(1)) - alpha*x(1)*x(2);
          b*(u2 - x(2)) - alpha*x(1)*x(2);
         -b*x(3) + alpha*x(1)*x(2) ];
%------------------------------------------------------------------------
```

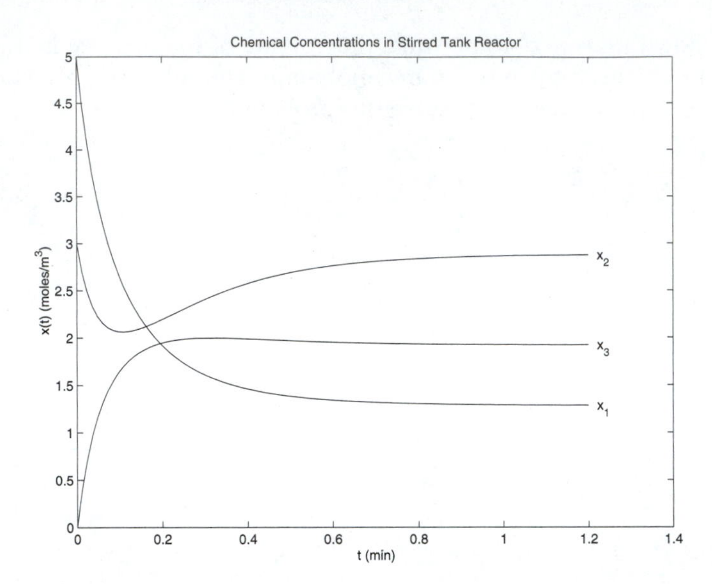

FIGURE 8.9.2 Chemical Concentrations in Stirred Tank Reactor

When script *e891.m* is executed, it uses the Runge-Kutta-Fehlberg method to compute $x(t)$ for $0 \le t \le 1.2$. A total of 1782 scalar function evaluations are required to compute $x(t)$ at 100 points uniformly distributed over $[0, 1.2]$. A plot of the resulting solution is shown in Figure 8.9.2, where it is seen that steady-state concentrations within the tank are achieved after about one min. The steady-state concentration of the product chemical in this case is

$$x_3 = 1.92074 \text{ moles/m}^3 \qquad (8.9.6)$$

A solution to Example 8.9.1 using C can be found in the file *e891.c* on the distribution CD.

8.9.2 Cantilever Beam: C

One of the problems that occurs in civil engineering is an analysis of the deflection of beams under various loads and supported in different ways. As an example, consider the cantilever beam shown in Figure 8.9.3. A cantilever beam is supported one end such that both its position y and its slope dy/dz are fixed at zero. For example, one end might be embedded in concrete or otherwise firmly secured.

Suppose the beam is of length L and is thin and uniform. The vertical deflection y of the beam due to its own weight can then be modeled as follows (Lastman and Sinha, 1989).

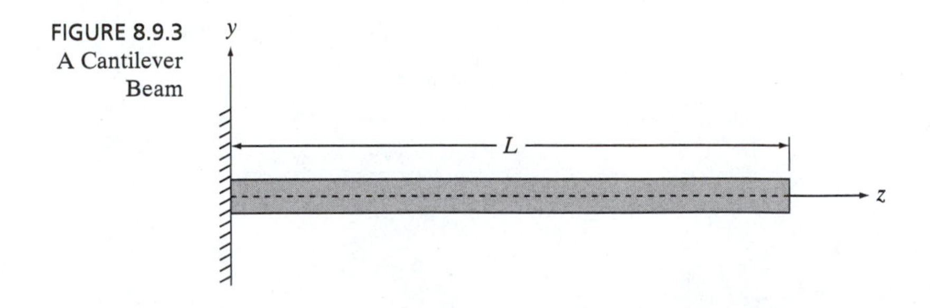

FIGURE 8.9.3 A Cantilever Beam

$$JE\frac{dy^2}{dz^2} = \rho g\left(1 + \frac{dy}{dz}\right)^{3/2}\left[z\left(z - \frac{L}{2}\right) + \frac{L^2}{2}\right] \qquad (8.9.7)$$

Here J is the moment of inertia of the beam cross section about its principal axis, and E is Young's modulus, which depends on the material from which the beam is made. The constant ρ is the linear mass density of the beam, and g is the acceleration due to gravity. Note that the independent variable in Equation (8.9.7) is the spatial variable z rather than the more typical time variable t. However, (8.9.7) represents an initial value problem because the boundary conditions are that y and dy/dz are both zero at $z = 0$. The second-order nonlinear beam deflection equation can be converted to a system of two first-order equations by introducing the state variables

$$x = [y \; dy/dz]^T \qquad (8.9.8)$$

Using (8.1.5), with t replaced by z, yields the following first-order system of ordinary differential equations.

$$\frac{dx_1}{dz} = x_2 \qquad (8.9.9a)$$

$$\frac{dx_2}{dz} = \frac{\rho g}{JE}(1 + x_2^2)^{3/2}\left[z\left(z - \frac{L}{2}\right) + \frac{L^2}{2}\right] \qquad (8.9.9b)$$

To make the problem specific, suppose the length of the beam is $L = 2$m, the linear mass density is $\rho = 10$ kg/m, and the moment of inertia times Young's modulus is $JE = 2400$ kg-m^3/s^2. The following C program on the distribution CD can be used to solve this problem.

```
/*----------------------------------------------------------------*/
/* Example 8.9.2: Cantilever Beam                                 */
/*----------------------------------------------------------------*/

#include "c:\nlib\util.h"

float           L = 2.0;        /* length of thin beam */

void beam (float z,vector x,int n,vector dx)
{
/* Description: Cantilever Beam */

   float       rho = 10.0f,     /* linear mass density (kg/m) */
               g = 9.81f,       /* acceleration due to gravity (m/s^2) */
               EJ = 2400.0f;    /* moment of inertia * Young's modulus */

   dx[1]  =  x[2];
   dx[2]  =  (float) (-(g*rho/EJ)*pow(1+x[2]*x[2],1./3.)*(z*(z-L/2)
             + L*L/2));
}

int main (void)
```

```
{
   int          n          = 2,              /* system dimension */
                m          = 100,            /* solution points */
                r;                           /* function evaluations */
   long         q = 10000;                   /* maximum evaluations */
   float        alpha      = 0.0f,           /* initial distance */
                beta       = L,              /* final distance */
                eps        = 1.e-6f,         /* error bound */
                E;                           /* estimated error */
   vector       z          = vec (m,""),     /* solution points */
                x0         = vec (n,"");     /* initial state */
   matrix       X          = mat (n,m,"");   /* solution points */

/* Solve equations */

   printf ("\nExample 8.9.2: Cantilever Beam\n");
   r = bsext (x0,alpha,beta,n,m,eps,q,z,X,&E,beam);
   shownum ("Number of scalar function evaluations",(float) r);
   shownum ("Maximum truncation error",E);
   shownum ("Maximum deflection (cm)",-100*X[1][m]);
   graphxy (z,X[1],m,"","z (m)","y (m)","fig894");
   return 0;
}
/*----------------------------------------------------------------*/
```

When program *e892.c* is executed, it uses the Bulirsch-Stoer extrapolation method, starting from the initial or end condition $x(0) = 0$. The method required 1584 scalar function evaluations to compute the beam deflection $y(z)$ at 100 point uniformly distributed over $[0, L]$. A plot of the resulting deflection, due to the effects of gravity, is shown in Figure 8.9.4. The maximum deflection occurs at the end of the beam and is found to be

$$y(L) = 16.38 \text{ cm} \tag{8.9.10}$$

A solution to Example 8.9.2 using MATLAB can be found in the file *e892.m* on the distribution CD.

8.9.3 Phase-Locked Loop: MATLAB

One of the most widely used circuits in the telecommunications industry is the phase-locked loop. A phase-locked loop (PLL) is a nonlinear feedback system that can be used to detect and track changes in the frequency of a periodic input signal. The most common practical example of a periodic signal whose frequency varies with time is a frequency-modulated or FM radio signal. For example, suppose the radio frequency to which the dial is tuned is f_0 Hz and $m(t)$ represents the music or speech that is "carried" by the radio wave. Then an FM radio wave that includes the audio information can be represented as follows.

$$z(t) = a \sin(2\pi[f_0 + m(t)]t) \tag{8.9.11}$$

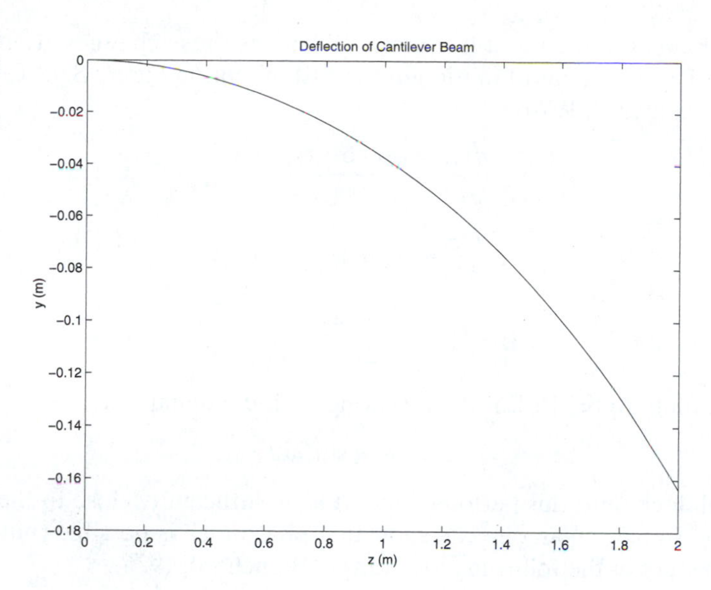

FIGURE 8.9.4 Deflection of Cantilever Beam

Here the frequency of the received signal is $f(t) = f_0 + m(t)$ Hz, and the amplitude is a. Typically, a can be very small if the transmitted signal travels a long distance. If $f(t)$ can be extracted from $z(t)$ through the process of frequency demodulation, the music or speech $m(t)$ can be recovered, amplified, and played back at the receiver. A block diagram of the basic PLL circuit is shown in Figure 8.9.5.

Here $u(t)$ is a periodic input to the PLL, and $y(t)$ is the output. The input $u(t)$ is multiplied by the feedback signal $v(t)$ and the product is then processed by a low-pass filter with gain α and time constant τ. The steady-state filter output, x_1, can be approximated as α times the constant part of $u(t)v(t)$. The filter output is then added to ω_c, which is called the center frequency of the PLL. The sum $x_1(t) + \omega_c$ is an estimate of the frequency of $u(t)$ with division by 2π converting this estimate from rad/s to Hz. The two components in the feedback path represent a harmonic oscillator whose frequency is $x_1(t) + \omega_c$. This oscillator speeds up or slows down as needed until its frequency exactly matches the frequency of the input $u(t)$. When $v(t)$ becomes synchronized with $u(t)$, the loop is said to be *locked* onto the input. It will

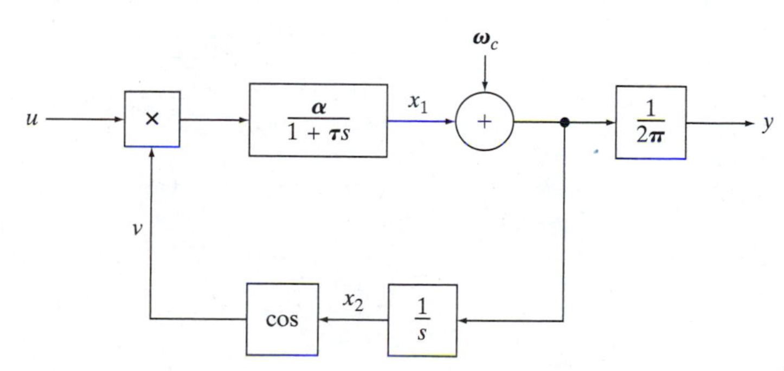

FIGURE 8.9.5 A Phase-Locked Loop Circuit

then track changes in the input frequency as long as those changes are not too large or too rapid. The equations of motion of the PLL circuit in Figure 8.9.5 can be written as follows (Blanchard, 1976).

$$\frac{dx_1}{dt} = \frac{\alpha u \cos(x_2) - x_1}{\tau} \tag{8.9.12a}$$

$$\frac{dx_2}{dt} = x_1 + \omega_c \tag{8.9.12b}$$

$$y = \frac{x_1 + \omega_c}{2\pi} \tag{8.9.12c}$$

Suppose the input to the PLL is the following periodic signal.

$$u(t) = a \sin(\omega_0 t) \tag{8.9.13}$$

The PLL will lock onto this periodic input if ω_0 is sufficiently close to the loop center frequency ω_c. In particular, one can show that phase-lock is possible (but not guaranteed) when ω_0 lies in the following lock range (Blanchard, 1976).

$$|\omega_0 - \omega_c| < a\alpha \tag{8.9.14}$$

Thus, the lock range depends on the input signal strength a, and the loop gain α. To illustrate the operation of the PLL circuit, suppose the loop parameters consist of a center frequency of $w_c = 2000\pi$ rad/s, a loop gain of $\alpha = 1600$, and a filter time constant of $\tau = 1.5$ msec. Suppose the input amplitude is $a = 0.5$, and the input frequency is $\omega_0 = 2100\pi$ rad/s. Finally, suppose the initial condition is $x(0) = 0$, which corresponds to an initial estimate of $y(0) = 1000$ Hz. The following MATLAB script on the distribution CD can be used to solve this problem.

```
%-----------------------------------------------------------------
% Example 8.9.3: Phase-Locked Loop
%-----------------------------------------------------------------

% Initialize

   clc                          % clear screen
   clear                        % clear variables
   global wc w0 alpha           % used by funu893.m, funp893.m
   wc     = 2*pi*1000;          % PLL center frequency (rad/s) */
   m      = 1000;               % solution points
   n      = 2;                  % dimension of system
   t1     = 0.025;              % final time
   tol    = 1.e-6;              % error bound
   q      = 40000;              % maximum function evaluations
   t      = zeros (m,1);        % solution points
   u      = zeros (m,1);        % input samples
   y      = zeros (m,1);        % output samples
   v      = zeros (m,1);        % feedback signal
```

```
   x0    = [0 0]';                  % initial state
   w0    = 2*pi*1050;               % input frequency
   alpha = 1600;

% Solve equations

   fprintf ('Example 8.9.3: Phase-Locked Loop\n');
   [t,X,e,k] = rkf (x0,0,t1,m,tol,q,'funp893');

% Compute input and output

   for i = 1 : m
      u(i) = funu893 (t(i));
      y(i) = (X(i,1) + wc)/(2*pi);
      v(i) = cos(X(i,2));
   end

% Display results

   t = 1000*t;
   show ('Number of scalar function evaluations',k);
   show ('Maximum truncation error',e);
   show ('Final estimate of frequency (Hz)',y(m));
   graphxy (u,v,'Oscillator','u','v')
   graphxy (t,y,'PLL Output','t (msec)','y (Hz)')


function dx = funp893 (t,x)
%------------------------------------------------------------------
% Description: Phase-locked loop
%------------------------------------------------------------------
   global wc w0 alpha
   tau = 3*pi/wc;
   u = funu893(t);
   dx = [ (alpha*u*cos(x(2))-x(1))/tau;
            x(1) + wc ];

function u = funu893 (t)
%------------------------------------------------------------------
% Periodic input
%------------------------------------------------------------------
   global wc w0 alpha
   u = 0.5*sin (w0*t);
%------------------------------------------------------------------
```

When script *e893.m* is executed, it uses the Runge-Kutta-Fehlberg method to solve the PLL equations at 1000 points uniformly distributed over the time interval $0 \le t \le 25$ msec. A total of 11,990 scalar function evaluations are required to produce

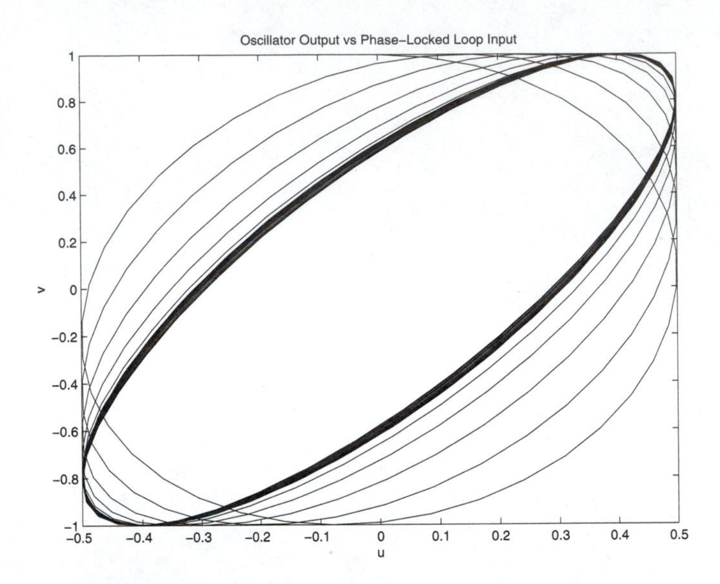

FIGURE 8.9.6 Oscillator Output vs. Phase-Locked Loop Input

the solution shown in Figure 8.9.6, which is a plot of the harmonic oscillator output $v(t)$ versus the PLL input $u(t)$. The presence of the dark ellipse indicates that the loop has achieved phase lock. In this case, the area of the ellipse specifies how much phase shift or delay there is between $u(t)$ and $v(t)$.

The locked condition can also be seen from the plot in Figure 8.9.7, which shows the PLL output $y(t)$. It is evident that phase lock is achieved after roughly 10 msec. Note that the final value of $y(t)$ yields the following estimate of the input frequency.

$$y = 1052.11 \text{ Hz} \tag{8.9.15}$$

This differs from the actual value of 1050 Hz by 0.19 percent. A significant improvement in the accuracy of the estimate can be obtained by post-processing the samples of $y(t)$ with a smoothing filter. Digital filtering operations are treated in detail in Chapter 10.

A solution to Example 8.9.3 using C can be found in the file *e893.c* on the distribution CD.

8.9.4 Turbulent Flow and Chaos: C

Mechanical and aeronautical engineers often focus on the study of fluid dynamics. For example, they might investigate the flow of air around an aircraft in order to estimate lift and drag. Turbulence is an important type of fluid flow that can occur under certain conditions. Turbulence is challenging to model in detail because of its random nature. Certain nonlinear systems produce solutions that behave as if they had a random component even though the equations themselves are completely deterministic. This mathematical phenomenon is known as *chaos* (Gleick, 1987). An example of a chaotic system that is used to model turbulent convection in fluids is the *Lorenz attractor* system (Cook, 1986).

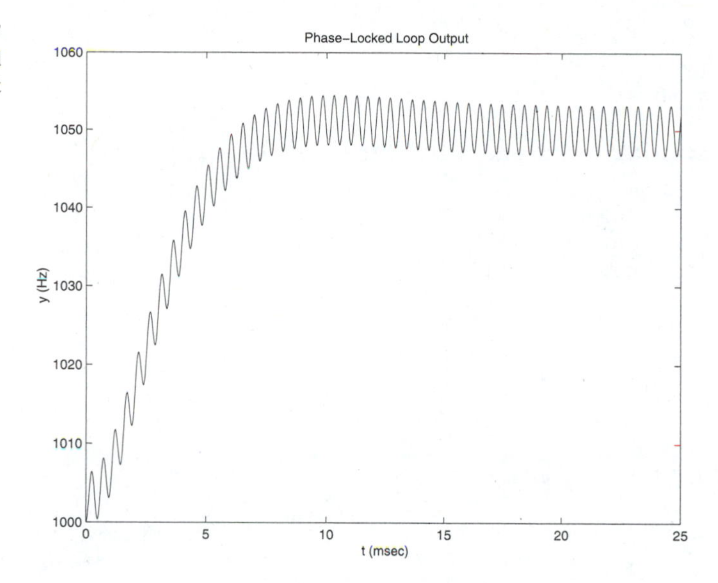

FIGURE 8.9.7 Phase-Locked Loop Output

$$\frac{dx_1}{dt} = \sigma(x_2 - x_1) \tag{8.9.16a}$$

$$\frac{dx_2}{dt} = (1 + \lambda - x_3)x_1 - x_2 \tag{8.9.16b}$$

$$\frac{dx_3}{dt} = x_1x_2 - \gamma x_3 \tag{8.9.16c}$$

Here the parameters σ, λ, and γ are all positive constants. This system has three equilibrium points, or constant solutions, namely

$$x^1 = \begin{bmatrix} 0 \\ 0 \\ 0 \end{bmatrix}, \quad x^2 = \begin{bmatrix} \sqrt{\gamma\lambda} \\ \sqrt{\gamma\lambda} \\ \lambda \end{bmatrix}, \quad x^3 = \begin{bmatrix} -\sqrt{\gamma\lambda} \\ -\sqrt{\gamma\lambda} \\ \lambda \end{bmatrix} \tag{8.9.17}$$

Chaotic behavior is observed when all of the equilibrium points are unstable. This occurs (see, e.g., Cook, 1986) when the parameters are chosen to satisfy

$$\sigma > \gamma + 1 \tag{8.9.18a}$$

$$\lambda > \frac{(\sigma + 1)(\sigma + \gamma + 1)}{\sigma - \gamma - 1} \tag{8.9.18b}$$

For example, one such choice of parameters is $\sigma = 10$, $\lambda = 24$ and $\gamma = 2$.

The following C program on the distribution CD can be used to solve this problem.

```
/*--------------------------------------------------------------------*/
/* Example 8.9.4: Turbulent Flow and Chaos                            */
/*--------------------------------------------------------------------*/
```

```
#include "c:\nlib\util.h"

void Lorenz (float t,vector x,int n,vector dx)
{
/* Description: The Lorenz attractor, a chaotic system. */

   float        sigma  = 10.0f,
                lambda = 24.0f,
                gamma  = 2.0f;

   dx[1] = sigma*(x[2] - x[1]);
   dx[2] = (1 + lambda - x[3])*x[1] - x[2];
   dx[3] = x[1]*x[2] - gamma*x[3];
}

int main (void)
{
   int          n   = 3,                      /* system dimension */
                m   = 4000;                   /* solution points */
   long         q   = 80000,                  /* maximum evaluations */
                r;                            /* function evaluations */
   float        t0  = 0.0f,                   /* initial time */
                t1  = 30.0f,                  /* final time */
                eps = 1.e-6f,                 /* error bound */
                E;                            /* estimated error */
   vector       t   = vec (m,""),             /* solution times */
                x0  = vec (n,"1 0 20");       /* initial state */
   matrix       X   = mat (n,m,"");           /* solution points */

/* Solve equations */

   printf ("\nExample 8.9.4: Turbulent Flow and Chaos\n");
   r = rkf (x0,t0,t1,n,m,eps,q,t,X,&E,Lorenz);
   shownum ("Number of scalar function evaluations",(float) r);
   shownum ("Maximum truncation error",E);
   graphxy (X[1],X[3],m,"","x_1","x_3","fig898");
   return 0;
}
/*------------------------------------------------------------*/
```

When program *e894.c* is executed, it uses the Runge-Kutta-Fehlberg method to compute a solution to the Lorenz attractor system over the time interval $[0, 30]$ starting from the initial state $x(0) = [1, 0, 20]^T$. The plot of x_3 versus x_1 produced by the program is shown in Figure 8.9.8. Notice how the solution wanders back and forth between regions which surround two of the equilibrium points. Plots of x_2 versus x_1 and x_3 versus x_2 show similar behavior, but from different perspectives. This example program required $71,980$ scalar function evaluations and produced a maximum truncation error of $E = 5.48 \times 10^{-7}$.

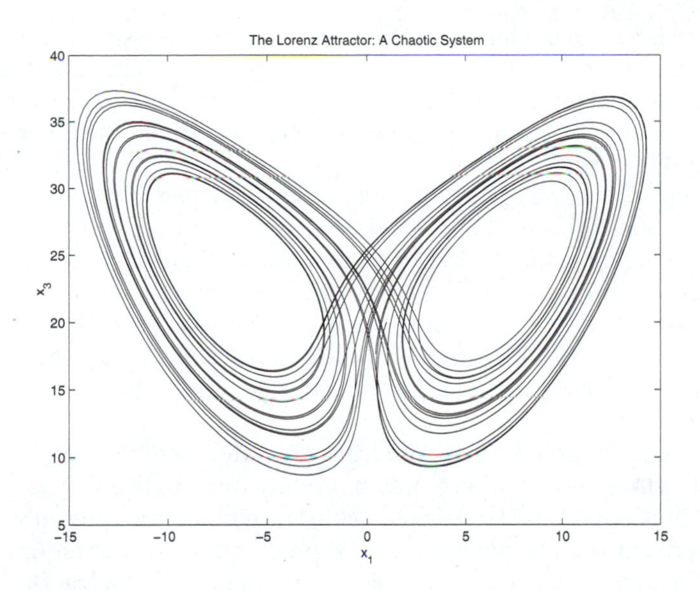

FIGURE 8.9.8 The Lorenz Attractor: A Chaotic System

A solution to Example 8.9.4 using MATLAB can be found in the file *e894.m* on the distribution CD.

8.10 SUMMARY

Numerical solution techniques for ordinary differential equations are summarized in Table 8.10.1. They are divided into methods that solve initial value problems and methods that solve boundary value problems. The order of the global truncation error depends on the step size h and the extrapolation level p.

The techniques applicable to initial value problems are designed to solve the following n-dimensional nonlinear first-order system of ordinary differential equations over the time interval $\alpha \le t \le \beta$.

$$\frac{dx}{dt} = f(t, x) \quad , \quad x(\alpha) = a \tag{8.10.1}$$

TABLE 8.10.1 Ordinary Differential Equation Solution Techniques

Method	*Problem*	*Truncation Error*
Euler	initial value	$O(h)$
Runge-Kutta-Fehlberg	initial value	$O(h^5)$
Adams multistep	initial value	$O(h^4)$
Bulirsch-Stoer extrapolation	initial value	$O(h^{2p})$
Semi-implicit extrapolation	stiff initial value	$O(h^{2p})$
Shooting	boundary value	—
Finite difference	boundary value	$O(h^2)$

Euler's method is a simple solution technique with a fixed step size h and a global truncation error of order $O(h)$. Euler's method is very easy to implement and requires only one function evaluation per step. However, it is necessary to use very small steps to achieve reasonable accuracy and avoid divergence between the computed numerical solution and the exact solution.

The Runge-Kutta-Fehlberg method is a popular dependable one-step method. It produces a solution estimate that has a global truncation error of order $O(h^5)$. The Runge-Kutta-Fehlberg method requires six function evaluations per step. The step size is adjusted automatically to ensure that the magnitude of the estimated truncation error in Equation (8.4.7h) stays below a user-specified threshold. This allows the method to take small steps over difficult parts of the solution interval and larger steps otherwise. The Runge-Kutta-Fehlberg method is an effective general technique that is applicable when a medium-accuracy solution is required.

The Adams-Bashforth-Moulton predictor-corrector method is a multi-step method that extrapolates from past solution points. It produces an estimate that has a global truncation error of order $O(h^4)$. The Adams multi-step method requires only two function evaluations per step, but it is not a self-starting method because it requires four initial values. These can be obtained by using another method for initialization or by using a variable-order family of multi-step methods. Multi-step methods are most effective when the right-hand side function, $f(t, x)$, is smooth.

The Bulirsch-Stoer extrapolation method is a variable-order technique that is based on a modified version of the midpoint method. The system in Equation (8.10.1) is solved using successively smaller step sizes, and the solutions are combined using Richardson extrapolation. The global truncation error of the pth level extrapolation is of order $O(h^{2p})$. To construct a level p extrapolation, a total $p(p+1)/2$ midpoint method solutions must be computed. Consequently, the number of function evaluations required grows rapidly with the extrapolation level and is given in (8.6.15). The extrapolation method is an effective general method that is particularly useful when $f(t, x)$ is relatively smooth and a high-accuracy solution is required.

The semi-implicit extrapolation method is a modified version of the Bulirsch-Stoer extrapolation method designed to solve stiff differential equations, equations whose solutions contain terms with time constants that differ by more than an order of magnitude. An n-dimensional linear algebraic system, based on the Jacobian matrix $J(t, x) = \partial f(t, x)/\partial x$, must be solved to obtain each midpoint method solution. A total of $p(p+1)$ midpoint method solutions are needed to generate the level p extrapolation. For stiff systems, the semi-implicit extrapolation method is more efficient than the previously discussed explicit methods because it can solve the system on the slow time scale without the numerical solution diverging.

The last two methods in Table 8.10.1 are applicable to the following nonlinear second-order ordinary differential equation subject to a pair of boundary conditions where the independent variable is a spatial variable x:

$$\frac{d^2y}{dx^2} = g\left(x, y, \frac{dy}{dx}\right) \quad , \quad y(\alpha) = a, \; y(\beta) = b \tag{8.10.2}$$

The shooting method is a simple iterative technique that solves a series of initial value problems parameterized by the initial slope $dy(\alpha)/dx$. Newton's method is used to converge to an initial value $dy(\alpha)/dx$, which generates the terminal boundary condition $y(\beta) = b$. If the function g is linear, the method converges to the exact solution in a single iteration.

The finite difference method solves the boundary value problem in Equation (8.10.2) along a grid of $m + 2$ points equally spaced over the interval $[\alpha, \beta]$. Using central difference approximations for the first and second derivatives, (8.10.2) is converted to an m-dimensional nonlinear algebraic system. The approximate solution is obtained iteratively using Newton's vector method. When the function g is linear, the nonlinear algebraic system reduces to a simple linear tridiagonal system. Consequently, it can be solved very efficiently for large values of m in this case. The finite difference approach is a general technique that is easily extended to partial differential equations, a topic that is treated in detail in Chapter 9.

PROBLEMS

The problems are divided into Analysis problems, that can be solved by hand, and Computation problems, that require the use of MATLAB or C. Solutions to selected problems can be found in Appendix 4. Students are encouraged to use these problems, which are identified with an (S), as a check on their understanding of the material. Problems marked with a (P) are programming problems that require the student to implement one or more of the algorithms discussed in the chapter. The remaining Computation problems require the student to write a main program that uses one or more of the NLIB functions discussed in Appendix 1 or Appendix 2.

Analysis

8.1 Write the following differential equation as a first-order system using the state variables $x = [y, y^{(1)}, y^{(2)}, y^{(3)}]^T$. Is this system linear?

$$y^{(4)} + 5(1 - y)y^{(3)} + 2y^{(2)} + 3y^{(1)} + y^3 = 10\sin(\pi t)$$

8.2 Consider the DC motor shown in Figure 8.11.1. The motion of this electro-mechananical device is modeled by a mechanical equation obtained from Newton's second law, an electrical

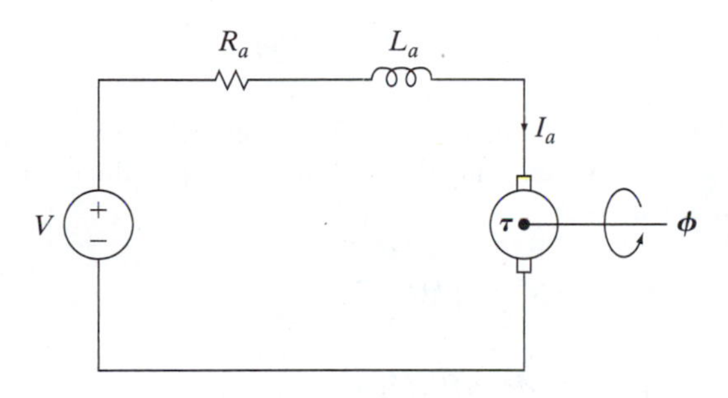

FIGURE 8.11.1 A DC Motor

equation obtained from Kirchhoff's voltage law, and an algebraic electro-mechanical coupling equation.

$$J\frac{d^2\phi}{dt^2} + \mu\frac{d\phi}{dt} = \tau$$

$$L_a\frac{dI_a}{dt} + R_a I_a + k_b\frac{d\phi}{dt} = V$$

$$\tau = k_t I_a$$

The system variables are the motor shaft angle ϕ, the torque developed at the motor shaft τ, the armature winding current I_a, and the applied armature voltage V. The remaining parameters are all positive constants. They include the moment of inertia about the motor shaft J, the coefficient of viscous friction μ, the armature winding inductance L_a, the armature winding resistance R_a, the back emf constant k_b, and the torque constant k_t. Write the equations of motion of this DC motor as a first-order system using the state variables $x = [\phi, d\phi/dt, I_a]^T$. Is this system linear?

8.3 Suppose $x^{k+1}(h)$ is the estimate of $x(t^{k+1})$ obtained by starting at $x(t^k)$ and taking one Euler step of length h. Similarly, let $x^{k+1}(h/2)$ be the estimate of $x(t^{k+1})$ obtained by starting from $x(t^k)$ and taking two Euler steps of length $h/2$. Finally, let $E(h)$ and $E(h/2)$ denote the truncation error of $x^{k+1}(h)$ and $x^{k+1}(h/2)$, respectively.

(a) What is the relationship between $E(h)$ and $E(h/2)$ when $|h| \ll 1$?

(b) What linear combination of $x^{k+1}(h)$ and $x^{k+1}(h/2)$ produces an extrapolated estimate that has a global truncation error of order $O(h^2)$?

8.4 Suppose we have developed a differential equation solution technique that has a local truncation error of order $O(h^p)$, and suppose $E(h)$ and $E(h/2)$ denote local truncation error estimates using step sizes of length h and $h/2$, respectively. What is the relationship between $E(h)$ and $E(h/2)$ when $|h| \ll 1$?

8.5 Consider the following first-order system of two equations in three variables:

$$\frac{dx_1}{dt} = x_3 - 2x_1$$

$$\frac{dx_2}{dt} = 3x_3 - x_2$$

Suppose the solution to this system is subject to the following algebraic constraint

$$x_1^2 + x_2^2 + x_3^2 = 1$$

Write this differential-algebraic system as a three-dimensional singular perturbation system using perturbation parameter μ.

Computation

(P)8.6 Euler's method for solving the initial value problem in Equation (8.10.1) can be generalized by taking an Euler step of length $h/2$ and then using the estimated slope at the midpoint to go back and take a full step of length h. That is,

$$q^1 = x^k + \frac{h}{2}f(t^k, x^k)$$

$$x^{k+1} = x^k + hf\left(t^k + \frac{h}{2}, q^1\right)$$

Write a function called *euler2* that implements this solution formula. The pseudo-prototype for *euler2* is as follows:

```
[vector x1, int ev] = euler2 (vector x0,float t0,float t1,int n,
                              function f);
```

On entry to *euler2:* the $n \times 1$ vector $x0$ is the initial state, $t0$ is the initial time, $t1$ is the final time, $n \geq 1$ is the number of equations, and f is the name of a user-supplied function that defines the equations to be solved. The pseudo-prototype of f is

```
[vector dx] = f (float t,vector x,int n);
```

When f is called with the scalar t and the $n \times 1$ vector x, it must compute the right-hand side function $f(t, x)$ in (8.10.1) and return the result in the $n \times 1$ vector dx. On exit from *euler2*: the $n \times 1$ vector $x1$ is the solution at time $t1$, and ev is the number of scalar function evaluations performed. Test *euler2* by solving the following system at $m = 200$ points uniformly distributed over the time interval $[0, 1]$ starting from the initial conditions $x(0) = [4, 0]^T$ and $x(0) = [-3, 4]^T$ Plot x_2 versus x_1 in each case. What do you predict the plotted solution will look like if $x(0) = [0, -6]^T$?

$$\frac{dx_1}{dt} = 9x_2$$

$$\frac{dx_2}{dt} = -9x_1$$

8.7 The following nonlinear differential equation is known as the *Van der Pol oscillator*. When the parameter $\mu = 0$, this is a linear harmonic oscillator similar to the system in problem 8.6. If $\mu > 0$, the steady-state solution continues to be periodic, but the shape of the dy/dt vs. y plot is no longer circular.

$$\frac{d^2y}{dt} - \mu(1 - y^2)\frac{dy}{dt} + y = 0$$

(a) Write the Van der Pol equation as a first-order system using the state variables $x = [y, dy/dt]^T$.

(b) Write a program that uses the NLIB function *rkf* to solve this system at $m = 250$ points uniformly distributed over the time interval $[0, 12]$ using $\mu = 2$ and the initial conditions $x(0) = [0.5, 0]^T$ and $x(0) = [-1, 2]^T$. Plot x_2 vs. x_1 in each case.

8.8 One of the simplest models for population growth is the *Verhulst-Pearl* logistic equation where the scalar $x(t)$ denotes the population level at time t.

$$\frac{dx}{dt} = x(\gamma - \delta x)$$

The parameters γ and δ are positive constants. The parameter γ is the normalized rate of population growth when the population level is small, and γ/δ is the carrying capacity of the environment. Although this one-dimensional system is nonlinear, the solution can be expressed in closed-form (Pielou, 1969) as

$$x(t) = \frac{\gamma}{\delta + (1/\lambda)\exp(-\gamma t)}$$

$$\lambda \triangleq \frac{-x(0)}{\delta x(0) - \gamma}$$

Suppose $\gamma = 1$ and $\delta = 0.001$. Write a program that uses the NLIB functions *rkf* and *bsext* to solve this system at 16 points equally spaced over the time interval $[0, 15]$ using a truncation error bound of $tol = 10^{-6}$ and starting from the initial condition $x(0) = 2$. Plot x versus t. Compare the two methods by printing a table that shows the actual error at each solution point.

8.9 Techniques used to solve the differential equation in (8.10.1) can also be used to compute numerical integrals. Suppose $I(\alpha, \beta)$ denotes the integral of $f(t)$ from α to β. Then $I(\alpha, \beta) = x(\beta)$ where $x(t)$ is the solution of the following first-order system.

$$\frac{dx}{dt} = f(t) \quad , \quad x(\alpha) = 0$$

Write a function called *rkfint* that uses the NLIB function *rkf* to compute a variable step size integral of $f(t)$. The pseudo-prototype of *rkfint* is as follows.

```
[float x, float e] = rkfint (float alpha,float beta,float tol,
                             int q,function f);
```

On entry to *rkfint: alpha* is the lower limit of integration, *beta* is the upper limit of integration, $tol \geq 0$ is an upper bound on the local truncation error, $q \geq 1$ is the maximum number of function evaluations, and f is the name of the user-supplied function to be integrated. The pseudo-prototype of f is

```
[float y] = f (float t, vector x);
```

When f is called with scalar input t, it must return the value $y = f(t)$. Note that the vector argument x is present (for compatatbility with rkf), but not used in this case. On exit from *rkfint: x* is the value of the integral and *e* is the magnitude of the maximum local truncation error. Test *rkfint* by computing the following integral. Plot the integrand over $[-1, 1]$. Print $I(-1, 1)$ and the magnitude of the maximum truncation error.

$$I(-1, 1) = \int_{-1}^{1} \frac{2^t \exp[t \sin(\pi t)]}{1 + t^4 + \ln(1 + t^2)} \, dt$$

8.10 In an industrial high-temperature oven, heat is added to the oven through the heating element and is simultaneously lost to the environment due to the temperature drop across the oven walls. Let x denote the temperature inside the oven, which is assumed to be uniform, and let T denote the ambient temperature of the air outside the oven. The rate of change of the oven temperature is proportional to the rate of heat entering the oven from the heating element minus the heat leaving the oven due to convection and radiation. Based on experimental observations, this process can be modeled as a nonlinear equation of the following general form (McClamroch, 1980).

$$\frac{dx}{dt} = \gamma(T - x) + \delta(T^4 - x^4) + \lambda u^2$$

Here γ, δ, and λ are positive constants that depend on the thermal capacity of the oven and the convective and radiative parameters, and the variable u represents the heat transfer rate of the heating element. For many manufacturing processes such as annealing, the temperature x must be carefully controlled. Suppose the heat transfer rate is constant with $u = 21$, and suppose $T = 293.15$, $\gamma = 50$, $\delta = 2 \times 10^{-7}$, and $\lambda = 800$. Write a program that uses the NLIB function *bsext* to solve this system at 101 points uniformly distributed over the time interval $[0, 0.1]$ using a truncation error bound of $tol = 10^{-6}$ and starting from the initial condition $x(0) = 25$. Plot x versus t. What is the steady-state temperature inside the oven?

8.11 Consider the problem of modeling the effects of vaccination on the spread of an epidemic. Let x_1 denote the number of individuals who are *susceptible* to the disease, and let x_2 denote the number of individuals who have contracted the disease and are therefore *infectious*. Infected individuals eventually die from the disease or become immune. The following nonlinear system can be used to model the spread of the epidemic (McClamroch, 1980).

$$\frac{dx_1}{dt} = -\gamma x_1 x_2 - \delta x_1$$

$$\frac{dx_2}{dt} = \gamma x_1 x_2 - \lambda x_2$$

The term $\gamma x_1 x_2$ on the right-hand side represents the rate at which susceptible individuals catch the disease due to contact with infected individuals. Where $\gamma > 0$ depends on how contagious the disease is. The term δx_1 represents the level of vaccination. Finally, the term λx_2 represents the rate at which infected individuals either become immune or die from the disease. Thus the constant $1/\lambda$ is a measure of the duration of the disease. Write a program that uses the NLIB function *rkf* to solve this system at $m = 200$ points uniformly distributed over the time interval $[0, 15]$. Use $\gamma = 0.1, \lambda = 2$ and three levels of vaccination: $\delta = 0, 4, 8$. Start from the initial condition $x(0) = [90, 10]^T$ and plot x_2 vs. x_1 for all three cases on the same graph. Print the maximum number infected in each case.

8.12 The motion of a satellite tumbling in space can be modeled by the following nonlinear Euler equations.

$$J_1 \frac{dx_1}{dt} = (J_2 - J_1)x_2 x_3 + \tau_1$$

$$J_2 \frac{dx_2}{dt} = (J_3 - J_1)x_3 x_1 + \tau_2$$

$$J_3 \frac{dx_3}{dt} = (J_1 - J_2)x_1 x_2 + \tau_3$$

Here x_k denotes the angular velocity, J_k denotes the moment of inertia, and τ_k denotes the applied torque about axis k for $1 \le k \le 3$. Suppose the satellite is non-symmetrical with $J_1 = 1$, $J_2 = 3$, and $J_3 = 2$. Write a program that uses the NLIB functions to solve this system over the time interval $[0, 10]$ when the applied torques are $\tau = \exp(-t)[2, 0, t]^T$. Start from the initial condition $x(0) = [1, 0, -1]$, and plot $x(t)$ versus t.

8.13 Solve the singular-perturbation system in problem 8.5 by writing a program that uses the NLIB function *stiff* to solve this system at $m = 51$ points uniformly distributed over the interval $[0, 1]$. Use the initial condition $x(0) = [1, 0, 0]^T$ and a truncation error bound of $tol = 10^{-5}$. Plot $x(t)$ versus t for the cases: $\mu = 0.1, 0.01, 0.001$. Also plot, on the same graph, $e(t) = |x_1^2(t) + x_2^2(t) + x_3^2(t) - 1|$ versus t.

(P)8.14 Consider the following linear second-order differential equation together with two boundary conditions.

$$\frac{d^2y}{dx^2} + c_1 \frac{dy}{dx} + c_2 y = r$$

$$y(\alpha) = a$$

$$y(\beta) = b$$

Here c_1 and c_2 are constants, but the forcing term u can vary with x. Write a function called *bvplin* that uses the shooting method and NLIB functions to solve this linear two-point boundary value problem. The pseudo-prototype of *bvplin* is as follows.

```
[vector x,vector y] = bvplin (float alpha,float beta,float a,float b,
                              vector c,float tol,int m,int q,function r);
```

On entry to *bvplin: alpha* is the initial value of the independent variable x, *beta* is the final value of x, a is the initial value of the dependent variable y, b is the final value of y, the 2×1 vector c specifies the equation coefficients, $tol \geq 0$ is an upper bound on the truncation error, $m \geq 1$ is the number of interior solution points, $q \geq 1$ is the maximum number of function evaluations and r is the name of a user-supplied function which specifies the forcing term. The pseudo-prototype for r is

```
[float y] = r (float x);
```

When r is called with scalar input x, it must return the scalar $y = r(x)$. On exit from *bvplin:* the $(m + 2) \times 1$ vector x contains the uniformly spaced points at which the equation is solved, and the $(m + 2) \times 1$ vector y contains the solution points. Test *bvplin* by solving the system with $c = [2, 1]^T$, $\alpha = 0$, $\beta = 3$, $a = 4$, $b = 7$, and $r(x) = x \exp(-x)$. Use $m = 29$ points, and plot $y(x)$ versus x.

(S)8.15 Consider the circular membrane shown in Figure 8.11.2. When the membrane is under constant pressure, p, the displacement y (out of the page) as a function of radial position r can be modeled as follows (Nakamura, 1993).

$$\frac{d^2y}{dr^2} + \left(\frac{1}{r}\right)\frac{dy}{dr} + \frac{p}{\tau} = 0$$

Here the parameter $\tau > 0$ is the membrane tension. The boundary conditions at the inner and outer radius of the annular membrane are $y(\alpha) = 0$ and $y(\beta) = 0$. Write a program that uses the NLIB function *bvp* to find the membrane displacement profile $y(r)$. Use $p = 200$, $\tau = 80$,

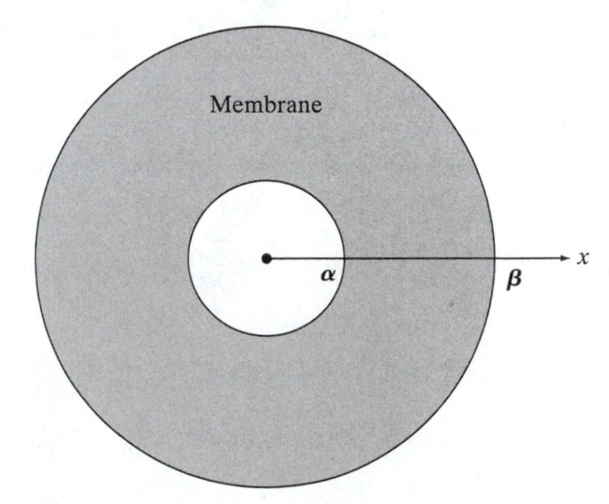

FIGURE 8.11.2 Displacement of a Circular Membrane

$\alpha = 0.5$, and $\beta = 1.5$. Solve the system for $m = 19$ points using an error tolerance of $tol = 10^{-4}$. Print the number iterations, and plot the cross-section $y(r)$ versus r.

8.16 Write a program that uses the NLIB functions to solve the linear system in problem 8.2. Use the parameter values $J = 2, \mu = 3, L_a = 1, R_a = 10, k_b = 0.5$, and $k_t = 1$. Use the following applied armature voltage:

$$V(t) = 10t \exp(-t)$$

Solve the system at $m = 100$ points uniformly distributed over the time interval $[0, 10]$ using a truncation error bound of $tol = 10^{-5}$ and an initial condition of $x(0) = 0$. The program should graph the solution. How far does the motor shaft turn in degrees?

CHAPTER 9

Partial Differential Equations

Dynamic physical systems that have more than one independent variable are modeled using *partial differential equations*. Partial differential equations are substantially more difficult to solve than ordinary differential equations. As such, we restrict our attention to some simple, but popular, numerical methods for solving partial differential equations or PDEs. To illustrate the solution techniques, we focus our attention on linear second-order equations of the following general form.

$$\alpha \frac{\partial^2 u}{\partial t^2} + \beta \frac{\partial u}{\partial t} + \gamma \left(\frac{\partial^2 u}{\partial x^2} + \frac{\partial^2 u}{\partial y^2} \right) = f$$

Here the scalars t, x, and y denote the *independent* variables with t representing time and x and y representing spatial coordinates. The coefficients $\{\alpha, \beta, \gamma\}$ and the forcing term f can depend on the independent variables. However, in many cases of practical interest they are constant, often zero or one. The solution $u(t, x, y)$ depends on both time and space. Consequently, to completely specify the problem we must impose both initial conditions and boundary conditions on $u(t, x, y)$. Detailed numerical treatments of more general partial differential equations can be found, for example, in Nakamura (1993) and Gerald and Wheatley (1989). Although the solution of partial differential equations requires substantial computational effort, one is rewarded by the fact that there are numerous applications in engineering. Application areas include steady-state temperature distribution, electric field patterns, heat flow, fluid dynamics, mechanical vibrations, and the spread of populations.

We begin this chapter by posing a number of engineering problems that require the solution of partial differential equations. We include Laplace's equation, the heat equation, and the wave equation. We then classify second-order linear partial differential equations into elliptic, parabolic, and hyperbolic forms. Laplace's equation is an example of an elliptic equation; the heat or diffusion equation is an example of a parabolic equation; and the wave equation is an example of a hyperbolic equation. We then examine numerical solution techniques based on finite difference methods for each separate class of equations. For elliptic partial differential equations, we develop an iterative central difference method using successive overrelaxation. One- and two-dimensional parabolic equations are then considered. The techniques for the one-dimensional heat equation include the explicit forward Euler method, the implicit backward Euler method, and the Crank-Nicolson method. The latter two methods are unconditionally stable, with the Crank-Nicolson method being second-order accurate in both time and space. We then develop the alternating direction implicit (ADI) method for the two-dimensional heat equation. Next, we investigate one- and two-dimensional hyperbolic equations. We present an elegant analytical solution technique for the one-dimensional wave equation called the d'Alembert method. We follow with discussions of one- and two-dimensional versions of the explicit central difference method. The chapter concludes with a summary of partial differential equation solution techniques and a presentation of engineering applications. The selected examples include heat flow in a metal rod, deflection of a rectangular plate, electrostatic field of a charge dipole, and torsion within a twisted bar.

9.1 MOTIVATION AND OBJECTIVES

Many physical phenomena can be modeled by partial differential equations. In this section, we examine a number of representative examples.

9.1.1 Laplace's Equation

As an illustration of an engineering application of a simple partial differential equation, consider the problem of modeling the steady-state distribution of temperature in the rectangular metal plate of width a and length b shown in Figure 9.1.1. Suppose the temperature at each point is denoted $u(x, y)$ with the temperature along the boundary of the plate held constant. If the plate is thin and insulated from above and below, the heat flow is in the xy plane. Under these conditions, it can be shown that the steady-state distribution of temperature is a solution of the following partial differential equation (see, e.g., Gerald and Wheatley, 1989).

$$\frac{\partial^2 u}{\partial x^2} + \frac{\partial^2 u}{\partial y^2} = 0 \tag{9.1.1}$$

Since it is the steady-state temperature that is of interest, there is no dependence on time. A forcing term, $f(x, y)$, is not present because there are no heat sources or sinks within the plate. Equation (9.1.1) is called *Laplace's equation*. Laplace's equation can also be used to model the electrostatic potential generated by a charge distribution. In this context, it is sometimes referred to as the *potential equation*. A compact way to

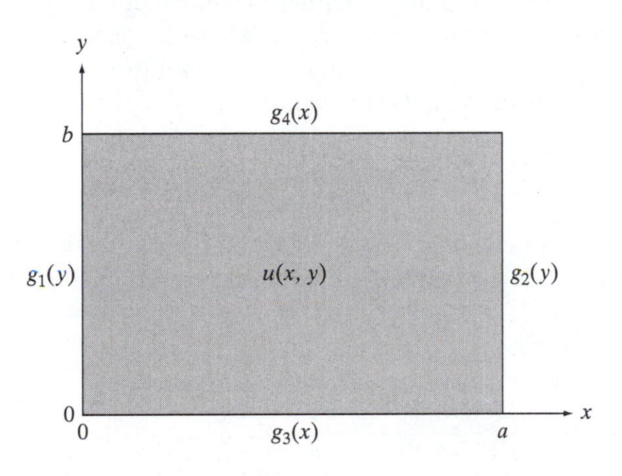

FIGURE 9.1.1 Temperature Distribution in a Rectangular Plate

write Laplace's equation is to use the following operator, called the two-dimensional *Laplacian operator*.

$$\boxed{\nabla^2 \triangleq \frac{\partial^2}{\partial x^2} + \frac{\partial^2}{\partial y^2}} \tag{9.1.2}$$

We can define a three-dimensional Laplacian operator involving x, y, and z in a similar manner. Using the Laplacian operator, (9.1.1) can be written compactly as

$$\nabla^2 u = 0 \tag{9.1.3}$$

There are many possible solutions to Laplace's equation. For example, it is easy to verify using direct substitution that both $u(x, y) = x^2 - y^2$ and $u(x, y) = \exp(x)\sin(y)$ are solutions. To obtain a unique solution, we must impose additional constraints in the form of boundary conditions. For example, suppose the temperature along the boundary of the plate in Figure 9.1.1 is as follows.

$$u(0, y) = g_1(y) \tag{9.1.4a}$$

$$u(a, y) = g_2(y) \tag{9.1.4b}$$

$$u(x, 0) = g_3(x) \tag{9.1.4c}$$

$$u(x, b) = g_4(x) \tag{9.1.4d}$$

Boundary conditions of this type are referred to as *Dirichlet* boundary conditions. It is also possible to instead specify the first derivative of u along all or part of the boundary. Derivative conditions are referred to as *Neumann* boundary conditions. They are useful, for example, when the boundary is insulated (zero derivative) or when there is a steady loss of heat by radiation and conduction.

9.1.2 Heat Equation

Transients associated with the *flow* of heat are often of interest. Consider, for example, the thin metal rod of length a shown in Figure 9.1.2. Suppose the rod is made of a homogeneous material and is enclosed in an insulating jacket except at its end points. Let $u(t, x)$ denote the temperature at time t and point x. Using the laws of physics, one can show (see, e.g., Gerald and Wheatley, 1989) that $u(t, x)$ is a solution to the following partial differential equation known as the *heat equation*.

$$\frac{\partial u}{\partial t} = \beta \frac{\partial^2 u}{\partial x^2} \tag{9.1.5}$$

The parameter β is a positive constant that depends on the properties of the material, specifically the thermal conductivity, the specific heat, and the density. The heat con-

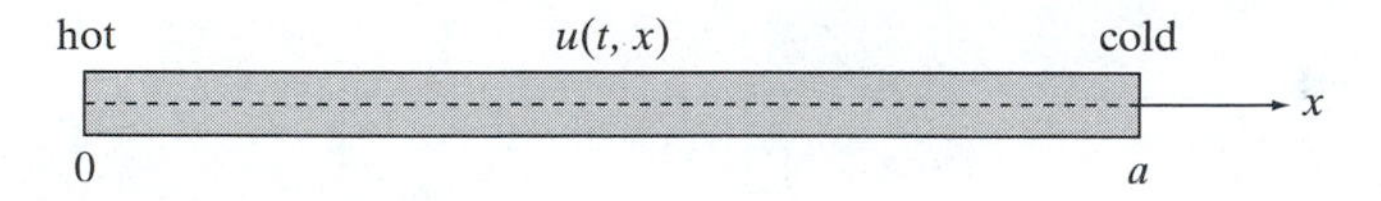

FIGURE 9.1.2 Heat Flow Along a Thin Rod

duction equation can be generalized to two dimensions by replacing $\partial^2 u/\partial x^2$ in Equation (9.1.5) with $\nabla^2 u$ where ∇^2 is the Laplacian operator in (9.1.2). The heat equation can also be used to model the diffusion process and is therefore sometimes referred to as the *diffusion equation*.

As was the case with Laplace's equation, there are many solutions to the heat equation. To make the solution unique, we must impose both an initial condition and boundary conditions. The initial condition specifies the distribution of temperature along the rod at time zero.

$$u(0, x) = f(x) \tag{9.1.6}$$

The boundary conditions at the two ends of the rod can be of different types. For the simplest case, the temperature at each end of the rod is held constant, which corresponds to two Dirichlet boundary conditions.

$$u(t, 0) = b_1 \tag{9.1.7a}$$

$$u(t, a) = b_2 \tag{9.1.7b}$$

For these boundary conditions, the temperature approaches the following steady-state condition after all transients have died out.

$$u(t, x) \to b_1 + \left(\frac{x}{a}\right)(b_2 - b_1) \quad \text{as} \quad t \to \infty \tag{9.1.8}$$

Alternatively, one or both ends of the bar may be insulated, in which case $\partial u/\partial x = 0$. Similarly, there may be a heat source or a heat sink on either end, which can be modeled with nonzero derivatives. Note that when the heat flow or the diffusion process reaches the steady state condition, u is no longer changing with time in which case $\partial u/\partial t = 0$. It follows that the two-dimensional version of the heat equation reduces to Laplace's equation, $\nabla^2 u = 0$, in the steady state. That is, the steady state solution to the two-dimensional heat equation is the solution to Laplace's equation.

9.1.3 Wave Equation

Consider a taut elastic string of length a fixed at both ends as shown in Figure 9.1.3. Suppose the string is homogeneous and of uniform diameter. Let $u(t, x)$ denote the vertical displacement of the string at time t and point x. We assume that the initial displacement is small and that the effects of gravity can be ignored. Using Newton's second law, one can show (see, e.g., Gerald and Wheatley, 1989) that the string dis-

FIGURE 9.1.3 Vibration of a String

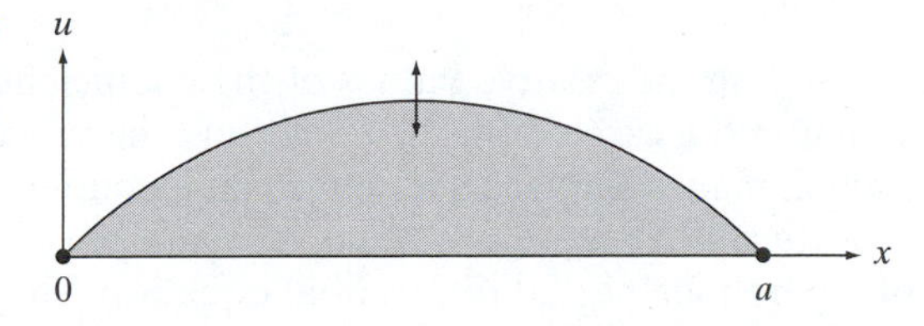

placement is a solution to the following partial differential equation called the *wave equation*.

$$\frac{\partial^2 u}{\partial t^2} = \beta \frac{\partial^2 u}{\partial x^2} \tag{9.1.9}$$

The parameter β is a positive constant that depends on the string tension and density. It has units of $(m/s)^2$, with $\sqrt{\beta}$ being the *velocity* of wave propagation. The wave equation can be generalized to two dimensions by replacing $\partial^2 u/\partial x^2$ in Equation (9.1.9) with $\nabla^2 u$ where ∇^2 is the Laplacian operator in (9.1.2). The two-dimensional formulation is useful, for example, in modeling the vibration of an elastic membrane.

To make the solution to the wave equation unique, we must again impose both initial conditions and boundary conditions. Since the wave equation is second order in time, two initial conditions are required.

$$u(0, x) = f(x) \tag{9.1.10a}$$

$$u_t(0, x) = g(x) \tag{9.1.10b}$$

Here, to make subsequent formulations more concise, we have adopted the standard subscript notation for partial derivatives:

$$u_t(t, x) \triangleq \frac{\partial u(t, x)}{\partial t} \tag{9.1.11}$$

Thus, an independent variable used as a subscript denotes *partial differentiation* with respect to that variable. If the string is fixed at both ends then the two Dirichlet boundary conditions are simply

$$u(t, 0) = 0 \tag{9.1.12a}$$

$$u(t, a) = 0 \tag{9.1.12b}$$

More generally, we can impose a constraint on a linear combination of u and $u_x = \partial u/\partial x$ at each end of the string. The boundary conditions in Equation (9.1.12), for example, might correspond to a string on a musical instrument with the initial conditions in (9.1.10) created by plucking the string.

9.1.4 Equation Classification

For the purposes of developing numerical solution techniques, it is useful to classify partial differential equations into three groups. A general linear second-order partial differential equation involving the variables t and x can be written as follows.

$$c_1 \frac{\partial^2 u}{\partial t^2} + c_2 \frac{\partial^2 u}{\partial t\, \partial x} + c_3 \frac{\partial^2 u}{\partial x^2} = d_1 u + d_2 \frac{\partial u}{\partial t} + d_3 \frac{\partial u}{\partial x} \tag{9.1.13}$$

The equations are classified based on the relative values of the coefficients c_k as summarized in Table 9.1.1. Note that in the case of elliptic equations, the variable t can be replaced by the variable y, which makes Laplace's equation an example of an elliptic partial differential equation. Since neither the c_1 nor the c_2 term are present in the heat equation, it is an example of a parabolic partial differential equation. The wave equation is an example of a hyperbolic partial differential equation.

TABLE 9.1.1 Linear Partial Differential Equations

Condition	*Equation Type*	*Example*
$c_2^2 < 4c_1c_3$	elliptic	Laplace's equation
$c_2^2 = 4c_1c_3$	parabolic	heat equation
$c_2^2 > 4c_1c_3$	hyperbolic	wave equation

Each of the classes of partial differential equations has its own numerical solution techniques based on the idea of approximating derivatives with finite differences as discussed in Section 7.2. There are also techniques for solving partial differential equations that are not based on finite difference methods, notably the finite element method. The finite element method is a powerful group of techniques that are readily applicable to regions with curved or irregular boundaries. However, a detailed treatment of the finite element method is outside the scope of this work. Instead, the interested reader is referred to Allaire (1985) and Burnett (1987).

9.1.5 Chapter Objectives

When you finish this chapter, you will be able to compute numerical solutions to linear second-order partial differential equations subject to initial and boundary conditions. You will know how to classify partial differential equations into elliptic, parabolic, and hyperbolic categories. You will be able to solve elliptic equations, including Laplace's equation and Poisson's equation, using the central difference method. You will know how to solve the one-dimensional heat equation, using the explicit and implicit Euler methods and the Crank-Nicolson method, and the two-dimensional heat equation using the ADI method. You will be able to construct an analytical solution to the one-dimensional wave equation, using the d'Alembert technique, and a numerical solution, using the explicit central difference method. You will also be able to compute a numerical solution to the two-dimensional wave equation using the two-dimensional central difference method. These overall goals will be achieved by mastering the following chapter objectives.

Objectives for Chapter 9

- Know how to classify linear partial differential equations into elliptic, parabolic, and hyperbolic equations.
- Understand the difference between Dirichlet and Neumann boundary conditions.
- Understand what a computational template is and how to use central differences to approximate Poisson's equation with a linear algebraic system.
- Understand why direct methods are not suitable for solving Poisson's equation.
- Know how to compute the optimal relaxation parameter for the successive over-relaxation method.
- Be able to solve the one-dimensional heat equation using the forward Euler, backward Euler, and Crank-Nicolson methods.
- Understand the conditions under which the parabolic methods are stable.

- Be able to solve the two-dimensional heat equation using the ADI method.
- Know how to solve the one-dimensional wave equation using d'Alembert's explicit method.
- Be able to solve the one- and two-dimensional wave equation using the explicit central difference method.
- Understand the conditions under which the hyperbolic methods are stable.
- Understand how partial differential equation techniques can be used to solve practical engineering problems.
- Understand the relative strengths and weaknesses of each computational method and know which are most applicable for a given problem.

9.2 ELLIPTIC EQUATIONS

In this section, we examine numerical techniques for finding a solution to a partial differential equation of the following general form, which is called *Poisson's equation*.

$$\boxed{\frac{\partial^2 u}{\partial x^2} + \frac{\partial^2 u}{\partial y^2} = f} \tag{9.2.1}$$

The forcing term, f, can depend on x and y. Note that when $f = 0$, Poisson's equation reduces to Laplace's equation, also called the potential equation. Suppose we are interested in solving (9.2.1) over the following rectangular region R in the xy plane.

$$R \triangleq \{(x, y) \mid 0 < x < a, 0 < y < b\} \tag{9.2.2}$$

To further specify the problem, we must place constraints on the solution along the *boundary* δR of R. Suppose the following Dirichlet conditions are applied.

$$u(x, y) = g(x, y) \quad , \quad (x, y) \in \delta R \tag{9.2.3}$$

9.2.1 Central Difference Method

The simplest way to solve Equation (9.2.1) is to convert it to an equivalent system of difference equations using a finite difference approximation. Recall that this was one of the techniques used to solve boundary value problems in Section 8.8.2. Suppose we are interested in estimating the solution $u(x, y)$ on a uniform grid consisting of $m + 2$ values of x and $n + 2$ values of y as shown in Figure 9.2.1. Let Δx and Δy denote the step size in the x and y directions, respectively.

$$\Delta x \triangleq \frac{a}{m + 1} \tag{9.2.4a}$$

$$\Delta y \triangleq \frac{b}{n + 1} \tag{9.2.4b}$$

To simplify subsequent equations, let x_k and y_j denote the values of x and y at the grid points. That is,

$$x_k = k\Delta x \quad , \quad 0 \le k \le (m + 1) \tag{9.2.5a}$$

$$y_j = j\Delta y \quad , \quad 0 \le j \le (n + 1) \tag{9.2.5b}$$

FIGURE 9.2.1 Uniform Solution Grid on the Region R

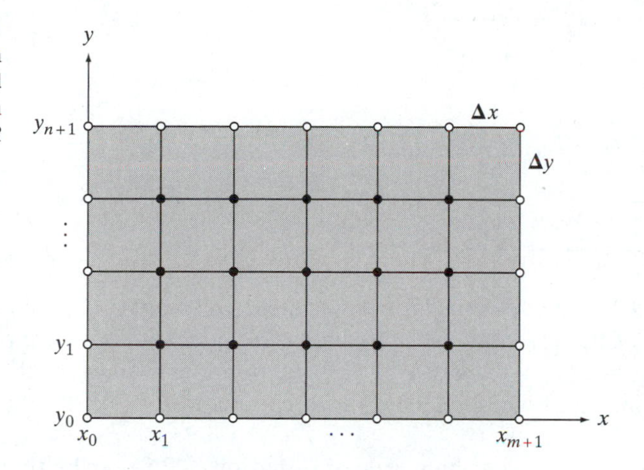

Note that x_0 and x_{m+1} correspond to the boundaries in the x direction, and y_0 and y_{n+1} correspond to the boundaries in the y direction. The boundary points are indicated with empty circles and the interior points with filled circles in Figure 9.2.1. Next, let u_{kj} denote the computed value of $u(x_k, y_j)$. To convert (9.2.1) to a difference equation, we use finite-difference approximations for the second derivative terms. Recall from (7.2.15) that the three-point central difference approximations for the second derivative are

$$u_{xx}(x_k, y_j) = \frac{u_{k+1,j} - 2u_{k,j} + u_{k-1,j}}{\Delta x^2} + O(\Delta x^2) \tag{9.2.6a}$$

$$u_{yy}(x_k, y_j) = \frac{u_{k,j+1} - 2u_{k,j} + u_{k,j-1}}{\Delta y^2} + O(\Delta y^2) \tag{9.2.6b}$$

Here we have extended the single-subscript notation for partial derivatives, first introduced in Equation (9.1.11), to a double-subscript notation for second-order partial derivatives:

$$u_{xx}(x, y) \triangleq \frac{\partial^2 u(x, y)}{\partial x^2} \tag{9.2.7a}$$

$$u_{yy}(x, y) \triangleq \frac{\partial^2 u(x, y)}{\partial y^2} \tag{9.2.7b}$$

Note that mixed partial derivatives such as $u_{xy} = \partial^2 u/\partial x \partial y$ can be accommodated as well. Next, let $f_{kj} = f(x_k, y_j)$. Substituting (9.2.6) into (9.2.1) and ignoring the higher-order terms, we get the following system of difference equations for the *interior* grid points, $1 \le k \le m$ and $1 \le j \le n$.

$$\boxed{\frac{u_{k+1,j} - 2u_{k,j} + u_{k-1,j}}{\Delta x^2} + \frac{u_{k,j+1} - 2u_{k,j} + u_{k,j-1}}{\Delta y^2} = f_{kj}} \tag{9.2.8}$$

FIGURE 9.2.2 Interior Computational Template

$$\begin{array}{ccc} 0 & \dfrac{1}{\Delta y^2} & 0 \\[2ex] \dfrac{1}{\Delta x^2} & -2\left(\dfrac{1}{\Delta x^2} + \dfrac{1}{\Delta y^2}\right) & \dfrac{1}{\Delta x^2} \\[2ex] 0 & \dfrac{1}{\Delta y^2} & 0 \end{array}$$

We refer to (9.2.8) as the *central difference method*. In order to evaluate (9.2.8) at the interior points of the grid, values for u_{kj} along the grid boundary are required. For convenience, let $g_{kj} = g(x_k, y_j)$ where $g(x, y)$ is the boundary value function in (9.2.3). Then, $u_{kj} = g_{kj}$ along the boundary, which corresponds to $k = 0,\ k = m + 1,\ j = 0,$ and $j = n + 1$.

The computation on the left-hand side of Equation (9.2.8) can be thought of as an operator acting on point u_{kj} and its neighbors. Specifically, the operator forms a linear combination of u_{kj} and its four neighbors to the north, south, east, and west using the 3×3 weighting coefficient *template* shown in Figure 9.2.2.

The five-point computation in Figure 9.2.2 has a truncation error of order $O(\Delta x^2 + \Delta y^2)$, assuming that u is sufficiently smooth. By using a 3×3 template that has nine nonzero entries, it is possible to develop an approximation that is of order $O(\Delta x^6 + \Delta y^6)$, again assuming u is sufficiently smooth (Gerald and Wheatley, 1989).

EXAMPLE 9.2.1 Central Difference Equations

The grid elements can often be taken to be square with $\Delta y = \Delta x = h$. This results in the following simplification of the central difference equations in (9.2.8).

$$u_{k+1,j} + u_{k-1,j} - 4u_{kj} + u_{k,j+1} + u_{k,j-1} = h^2 f_{kj}$$

Suppose the values of u_{kj} are arranged as an $mn \times 1$ vector $v = [u_{11}, u_{12}, \ldots, u_{mn}]^T$ constructed by placing the rows of u end to end. The difference equations can then be cast as a linear algebraic system of dimension mn in the unknown vector v. For the case $m = 3$ and $n = 3$, this results in the following system.

$$\left[\begin{array}{ccc|ccc|ccc} -4 & 1 & 0 & 1 & 0 & 0 & 0 & 0 & 0 \\ 1 & -4 & 1 & 0 & 1 & 0 & 0 & 0 & 0 \\ 0 & 1 & -4 & 0 & 0 & 1 & 0 & 0 & 0 \\ \hline 1 & 0 & 0 & -4 & 1 & 0 & 1 & 0 & 0 \\ 0 & 1 & 0 & 1 & -4 & 1 & 0 & 1 & 0 \\ 0 & 0 & 1 & 0 & 1 & -4 & 0 & 0 & 1 \\ \hline 0 & 0 & 0 & 1 & 0 & 0 & -4 & 1 & 0 \\ 0 & 0 & 0 & 0 & 1 & 0 & -1 & -4 & 1 \\ 0 & 0 & 0 & 0 & 0 & 1 & 0 & 1 & -4 \end{array}\right] \begin{bmatrix} u_{11} \\ u_{12} \\ u_{13} \\ u_{21} \\ u_{22} \\ u_{23} \\ u_{31} \\ u_{32} \\ u_{33} \end{bmatrix} = \begin{bmatrix} h^2 f_{11} - g_{01} - g_{10} \\ h^2 f_{12} - g_{02} \\ h^2 f_{13} - g_{03} - g_{14} \\ h^2 f_{21} - g_{20} \\ h^2 f_{22} \\ h^2 f_{23} - g_{24} \\ h^2 f_{31} - g_{41} - g_{30} \\ h^2 f_{32} - g_{42} \\ h^2 f_{33} - g_{43} - g_{34} \end{bmatrix}$$

Note that the $mn \times mn$ coefficient matrix is symmetric and that the boundary values appear in the right-hand-side vector.

For modest values of m and n such as $mn \leq 100$, the linear algebraic system in Example 9.2.1 can be solved by the direct elimination methods discussed in Chapter 2. However, as m and n increase to more realistic values, a direct solution becomes unfeasible because of the excessive computational time, storage requirements, and accumulated round-off error. Fortunately, the coefficient matrix in Example 9.2.1 is quite *sparse,* as there are, at most, five nonzero elements in each row. Furthermore, the matrix is diagonally dominant, with the magnitude of each diagonal element at least as large as the sum of the magnitudes of the remaining elements in each row. These two features make this linear algebraic system an excellent candidate for the iterative solution methods discussed in Section 2.6. Before we examine this solution technique in more detail, we first consider the question of how to incorporate additional types of boundary conditions.

9.2.2 Boundary Conditions

The formulation of the finite difference method in Equation (9.2.8) is valid for the Dirichlet boundary conditions in (9.2.3). Another common type of boundary condition involves constraints on the value of the *derivative* of the solution at the boundary. For the case of a rectangular region R, it is helpful to consider each edge separately.

$$-u_x(0, y) = g_1(y) \tag{9.2.9a}$$

$$u_x(a, y) = g_2(y) \tag{9.2.9b}$$

$$-u_y(x, 0) = g_3(x) \tag{9.2.9c}$$

$$u_y(x, b) = g_4(x) \tag{9.2.9d}$$

The negative signs in (9.2.9a) and (9.2.9c) are included so that in each case the partial derivative is in the direction of the *outward normal* to the boundary. Recall that constraints of this form are called Neumann boundary conditions. To illustrate how the basic finite-difference equations must be modified, suppose the left edge ($x = 0$) and the right edge ($x = a$) must satisfy Neumann boundary conditions. First, we evaluate (9.2.8) along the two edges. This yields the following supplementary difference equations for $1 \leq j \leq n$.

$$\frac{u_{1j} - 2u_{0j} + u_{-1,j}}{\Delta x^2} + \frac{u_{0,j+1} - 2u_{0j} + u_{0,j-1}}{\Delta y^2} = f_{0j} \tag{9.2.10a}$$

$$\frac{u_{n+2,j} - 2u_{n+1,j} + u_{nj}}{\Delta x^2} + \frac{u_{n+1,j+1} - 2u_{n+1,j} + u_{n+1,j-1}}{\Delta y^2} = f_{n+1,j} \tag{9.2.10b}$$

Supplementary equations (9.2.10) contain some new *fictitious* points $u_{-1,j}$ and $u_{n+2,j}$ that lie just outside the boundary. To obtain estimated values for these points, we approximate the first derivatives in (9.2.9a) and (9.2.9b) using three-point central difference approximations. This results in the following difference equations for $1 \leq j \leq n$.

$$\frac{-(u_{1j} - u_{-1,j})}{2\Delta x} = g_1(y_j) \tag{9.2.11a}$$

$$\frac{(u_{n+2,j} - u_{nj})}{2\Delta x} = g_2(y_j) \tag{9.2.11b}$$

We can now solve (9.2.11a) and (9.2.11b), respectively, for the fictitious points $u_{-1,j}$ and $u_{n+2,j}$. Substituting the results into (9.2.10) results in the following difference equations along the left and right edges of the region R.

$$\frac{2u_{1j} - 2u_{0j}}{\Delta x^2} + \frac{u_{0,j+1} - 2u_{0j} + u_{0,j-1}}{\Delta y^2} = f_{0j} - \frac{2g_1(y_j)}{\Delta x} \tag{9.2.12a}$$

$$\frac{-2u_{n+1,j} + 2u_{nj}}{\Delta x^2} + \frac{u_{n+1,j+1} - 2u_{n+1,j} + u_{n+1,j-1}}{\Delta y^2} = f_{n+1,j} - \frac{2g_2(y_j)}{\Delta x} \tag{9.2.12b}$$

Note that along the boundaries we now use a four-point computation rather than a five-point computation because one neighbor is not present. In addition, there is an extra term on the right-hand side due to the presence of a nonzero derivative. By applying the same analysis to the bottom and top edges of R using the boundary conditions in (9.2.9c) and (9.2.9d), respectively, we arrive at the following additional difference equations.

$$\frac{u_{k+1,0} - 2u_{k0} + u_{k-1,0}}{\Delta x^2} + \frac{2u_{k1} - 2u_{k0}}{\Delta y^2} = f_{k0} - \frac{2g_3(x_k)}{\Delta y} \tag{9.2.13a}$$

$$\frac{u_{k+1,m+1} - 2u_{k,m+1} + u_{k-1,m+1}}{\Delta x^2} + \frac{-2u_{k,m+1} + 2u_{km}}{\Delta y^2} = f_{k,m+1} - \frac{2g_4(x_k)}{\Delta y} \tag{9.2.13b}$$

Each of the four corner points of the solution grid is at the intersection of two edges. The easiest way to include the corner points is to apply one of the edge constraints in Equation (9.2.9). For example, if corner point $(k, j) = (0, 0)$ is included with the left edge, then (9.2.9a) is used, in which case (9.2.12a) is applicable with $j = 0$. The four-point computations along the boundary are conveniently summarized with the computational templates shown in Figure 9.2.3. Note that in each case the point that would have been outside the boundary is added to the closest interior point, thereby doubling its weight.

When derivative or Neumann boundary conditions are applied to the edges, the number of equations that must be solved increases. There can be a combination of Dirichlet conditions along some edges and Neumann conditions along the remaining edges.

9.2.3 Iterative Solution Methods

Direct numerical solution of the central difference equations in Equation (9.2.8) using elimination methods can be a very challenging task. Since there are mn internal grid points, the coefficient matrix contains $(mn)^2$ elements. Solving a linear algebraic system of dimension mn using the most efficient of the direct methods requires approximately $(mn)^3/3$ FLOPs. The prohibitive growth in both the storage requirements (at four bytes/float) and the computational effort is summarized in Table 9.2.1, which illustrates the special case $m = n$.

Fortunately, the iterative methods introduced in Section 2.6 are ideally suited for the solution of finite difference equations. For the iterative methods, approximately mn elements must be stored for large values of m and n. For the case $m = n = 80$, this reduces to 25,600 bytes, which represents a reduction of 99.7% over the direct method!

FIGURE 9.2.3 Computational Templates for Neumann Boundary Conditions

$\frac{1}{\Delta x^2}$	$-2\left(\frac{1}{\Delta x^2}+\frac{1}{\Delta y^2}\right)$	$\frac{1}{\Delta x^2}$
0	$\frac{2}{\Delta y^2}$	0
0	0	0

$\frac{1}{\Delta y^2}$	0	0
$-2\left(\frac{1}{\Delta x^2}+\frac{1}{\Delta y^2}\right)$	$\frac{2}{\Delta x^2}$	0
$\frac{1}{\Delta y^2}$	0	0

0	0	$\frac{1}{\Delta y^2}$
0	$\frac{2}{\Delta x^2}$	$-2\left(\frac{1}{\Delta x^2}+\frac{1}{\Delta y^2}\right)$
0	0	$\frac{1}{\Delta y^2}$

0	0	0
0	$\frac{2}{\Delta y^2}$	0
$\frac{1}{\Delta x^2}$	$-2\left(\frac{1}{\Delta x^2}+\frac{1}{\Delta y^2}\right)$	$\frac{1}{\Delta x^2}$

TABLE 9.2.1 Storage and Computational Costs of Direct Methods

$m = n$	*Storage (Bytes)*	*Operations (FLOPs)*
5	2,500	5,208
10	40,000	333,333
20	640,000	2.13×10^7
40	1.02×10^7	1.37×10^9
80	1.64×10^8	8.75×10^{10}

The computational time can also be substantially reduced, but this depends on the required accuracy and the initial guess.

To apply an iterative solution method to the difference equations in (9.2.8) subject to the Dirichlet boundary conditions in (9.2.3), we isolate u_{kj} on the left-hand side to produce the new estimate as follows, where it is understood that $1 \leq k \leq m$ and $1 \leq j \leq n$.

$$u_{kj}^{i+1} = \frac{\Delta y^2(u_{k+1,j}^i + u_{k-1,j}^i) + \Delta x^2(u_{k,j+1}^i + u_{k,j-1}^i) - \Delta x^2 \Delta y^2 f_{kj}}{2(\Delta x^2 + \Delta y^2)} \qquad (9.2.14)$$

Here, u_{kj}^i denotes the ith estimate of the solution at point (x_k, y_j). To start the process, we need an initial guess u_{kj}^0. Suppose u_{kj} represents temperature. In this case, the solution will be no colder than the coldest boundary point and no warmer than the warmest boundary point. Therefore, in the absence of more detailed knowledge about the solution, a reasonable initial guess is to set u_{kj} equal to the *average* of the boundary values.

$$u_{kj}^0 = \frac{1}{mn} \sum_{p=1}^{m} \sum_{q=1}^{n} g_{pq} \tag{9.2.15}$$

Once initial values for u_{kj} are chosen, a new generation of estimates can be computed using Equation (9.2.14). This process can then be repeated to yield successively more accurate estimates. Recall from Section 2.6.1 that when the $(i + 1)$th estimate of u_{kj} is computed using only the values u_{kj} from the ith estimate, as in (9.2.14), the technique is called the Jacobi iterative method.

Improved convergence rates can be obtained by using the new values of u_{kj} on the right-hand side of (9.2.14) as soon as they become available. This results in the following reformulation of (9.2.14), called the Gauss-Seidel iterative method.

$$u_{kj}^{i+1} = \frac{\Delta y^2(u_{k+1,j}^i + u_{k-1,j}^{i+1}) + \Delta x^2(u_{k,j+1}^i + u_{k,j-1}^{i+1}) - \Delta x^2 \Delta y^2 f_{kj}}{2(\Delta x^2 + \Delta y^2)} \tag{9.2.16}$$

When the Gauss-Seidel iterative method is used to solve the Poisson equation in (9.2.1), it is called *Liebmann's method*. Still faster convergence can be obtained by using the *successive over-relaxation* or *SOR* method, which includes the Gauss-Seidel method as a special case. Recall from Section 2.6.3 that the *SOR* method is implemented by adding and subtracting u_{kj}^i from the right-hand side and then scaling all but the first term by a relaxation factor α. Combining terms involving u_{kj} then results in the following iterative method where v_{kj}^i denotes the right-hand side of (9.2.16).

$$u_{kj}^{i+1} = (1 - \alpha)u_{kj}^i + \alpha v_{kj}^i \tag{9.2.17}$$

By changing the relaxation parameter α, the rate of convergence can be controlled. For a diagonally dominant linear system, the optimal value of α lies in the interval $1 \le \alpha \le 2$. Note that when $\alpha = 1$, the *SOR* method in Equation (9.2.17) reduces to the Gauss-Seidel method in (9.2.16).

Recall from Section 2.6.4 that the rate of convergence of a linear iterative method depends on the spectral radius, which is the magnitude of the largest eigenvalue of the coefficient matrix. For the case of a rectangular region R and Dirichlet boundary conditions, the optimal value α_{opt} is as follows (Press et al., 1992).

$$\alpha_{opt} = \frac{2}{1 + \sqrt{1 - \rho_J^2}} \tag{9.2.18a}$$

$$\rho_J = \frac{\cos(\pi/m) + (\Delta x/\Delta y)^2 \cos(\pi/n)}{1 + (\Delta x/\Delta y)^2} \tag{9.2.18b}$$

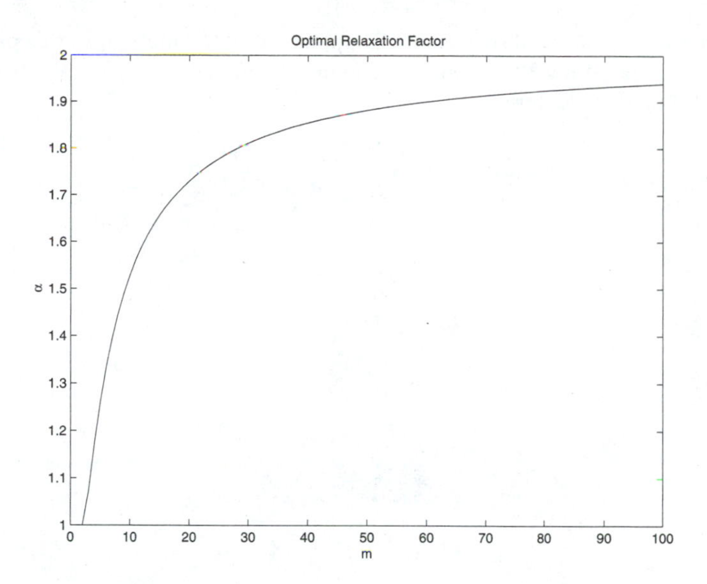

FIGURE 9.2.4 Optimal Relaxation Parameter for Rectangular R with $m = n$

Here ρ_J is the spectral radius of the coefficient matrix associated with the Jacobi method in Equation (9.2.14). A plot of the optimal relaxation parameter versus the grid size m for the case $m = n$ is shown in Figure 9.2.4. It is apparent that for a very coarse grid $\alpha_{opt} \approx 1$, but as m increases, so does α_{opt} until $\alpha_{opt} \approx 2$ for a very fine grid.

Each of the iterative solution methods requires a termination criterion to decide when to stop the computations. Since the exact solution $u(x_k, y_j)$ is not known, a technique that is often used is to stop when the *change* between successive estimates falls below a user-specified threshold of $\varepsilon > 0$. That is, the iterations continue until the following criterion is satisfied or a maximum number of iterations is performed.

$$\max_{k=1}^{m} \max_{j=1}^{n} \{|u_{kj}^{i+1} - u_{kj}^{i}|\} < \varepsilon \tag{9.2.19}$$

EXAMPLE 9.2.2 Steady-State Temperature

As an illustration of the use of the *SOR* iterative method to solve an elliptic partial differential equation, consider the problem of determining the steady-state temperature distribution over a square plate four units on a side. Recall from Section 9.1.1 that the steady-state temperature $u(x, y)$ is a solution of Laplace's equation, which is Poisson's equation with $f(x, y) = 0$. For the square plate, $a = b = 4$. Let the size of the grid be $m = n = 39$. Note that this produces a linear algebraic system of dimension 1521, which indicates that a direct solution method is clearly not practical! Instead we use the *SOR* iterative method. From Equation (9.2.18), the optimal relaxation parameter in this case is

$$\alpha_{opt} = 1.851$$

To obtain a unique solution, we must also specify the boundary conditions. Suppose the Dirichlet conditions in (9.2.3) are used with the following boundary value function.

$$g(x, y) = \exp(y)\cos(x) - \exp(x)\cos(y)$$

Example program *e922* on the distribution CD applies the *SOR* method to this problem, with the results shown in Figure 9.2.5. In this case, an error bound of $\varepsilon = 10^{-4}$ was used, and convergence was obtained in 90 iterations.

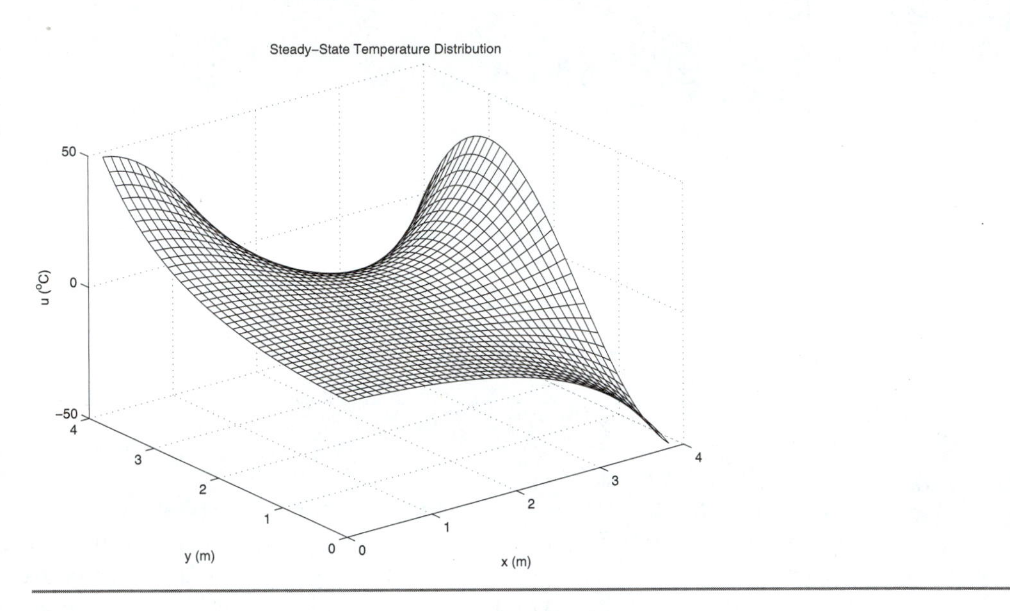

FIGURE 9.2.5 Steady-State Temperature Distribution

All of the analysis of the Poisson equation in (9.2.1) has been based on the assumption that the domain of the independent variables is a rectangular region R in the xy plane. Regions with curved boundaries also occur in applications, but the solution can become considerably more involved. There are a number of approaches available for handling regions with curved boundaries. Sometimes a change of variables can be helpful. For example, if the region R is a disk of radius a centered at the origin, then polar coordinates (r, θ) are useful because the boundary is just $r = a$. The two-dimensional Laplacian operator in polar coordinates is

$$\nabla^2 = \frac{\partial^2}{\partial r^2} + \left(\frac{1}{r}\right)\frac{\partial}{\partial r} + \left(\frac{1}{r^2}\right)\frac{\partial^2}{\partial \theta^2} \tag{9.2.20}$$

If the boundary is irregular, the rectangular coordinate formulation in (9.2.4) can be used, but the difference equations must be modified to accommodate a nonuniform grid near the boundary. The intersections of the curved boundary with the uniform grid produce new grid points at distances $\alpha\Delta x$ and $\beta\Delta y$ from their neighbors where $0 \le \alpha \le 1$ and $0 \le \beta \le 1$. The central difference approximations for the second derivatives must then be generalized to include the parameters α and β (see Problem 9.2). Another approach for curved boundaries is to simply choose the uniform grid points that are closest to the boundary and assign the boundary values to these points. Although this is only an approximation to the actual boundary, the error in the approximation can be made small if the grid is sufficiently fine. Still another technique is the finite element method (Allaire, 1985). The finite element method is attractive for irregular regions because it partitions the domain R into simple subregions, such as triangles, which span R.

9.3 ONE-DIMENSIONAL PARABOLIC EQUATIONS

Solutions to partial differential equations often involve time as well as space. A partial differential equation that is first-order in time, t, and second-order in a spatial coordinate, x, is a *parabolic* equation of the following form.

$$\boxed{\frac{\partial u}{\partial t} = \beta \frac{\partial^2 u}{\partial x^2}} \tag{9.3.1}$$

Recall that this is the one-dimensional heat conduction equation, also called the diffusion equation. The coefficient β is a positive constant. Since $u(t,x)$ depends on both time and space, we must specify both an initial condition and boundary conditions. Suppose we are interested in solving the heat equation over the interval $0 < x < a$. Then the *initial condition* will be of the following form for some initial value function $f(x)$.

$$u(0,x) = f(x) \quad , \quad 0 < x < a \tag{9.3.2}$$

We can formulate the boundary conditions in a general manner by placing constraints on the following linear combination of u and $u_x = \partial u/\partial x$.

$$(1 - c_1)u(t,0) - c_1 u_x(t,0) = g_1(t) \tag{9.3.3a}$$

$$(1 - c_2)u(t,a) + c_2 u_x(t,a) = g_2(t) \tag{9.3.3b}$$

In this characterization of the boundary conditions, the components of the 2×1 coefficient vector c take on values in the range $0 \le c_k \le 1$. For example, if $c_1 = 0$, then Equation (9.3.3a) becomes the Dirichlet boundary condition $u(t,0) = g_1(t)$. Similarly, if $c_2 = 1$, then (9.3.3b) reduces to the Neumann boundary condition $u_x(t,a) = g_2(t)$. More generally, $0 < c_k < 1$ produces a *mixed* boundary condition.

9.3.1 Explicit Forward Euler Method

Suppose we are interested in solving the heat equation over the time interval $0 \le t \le T$. We compute a numerical solution by estimating $u(t,x)$ over a uniform grid consisting of $m + 1$ values of t and $n + 2$ values of x as shown in Figure 9.3.1. Let Δt and Δx denote the step sizes of the variables t and x, respectively.

$$\Delta t \triangleq \frac{T}{m} \tag{9.3.4a}$$

$$\Delta x \triangleq \frac{a}{n+1} \tag{9.3.4b}$$

To simplify the final equations, let t_k and x_j denote the values of t and x at the grid points. That is,

$$t_k = k\Delta t \quad , \quad 0 \le k \le m \tag{9.3.5a}$$

$$x_j = j\Delta x \quad , \quad 0 \le j \le (n+1) \tag{9.3.5b}$$

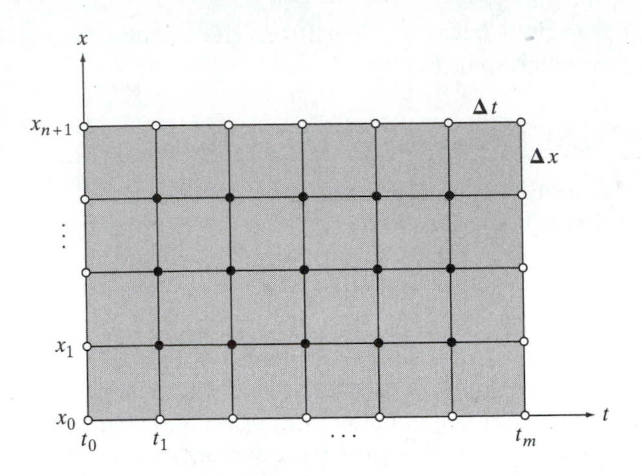

FIGURE 9.3.1 Uniform Solution Grid for Heat Equation

Next, let u_j^k denote the computed value of $u(t_k, x_j)$. Note that the superscript specifies the time step, while the subscript specifies the space step. To convert the heat equation in (9.3.1) to a difference equation, we again use finite difference approximations from Section 7.2 for the derivative terms. The first-order time derivative can be approximated with a two-point forward Euler difference, while the second-order space derivative can be approximated with a three-point central difference.

$$u_t(t_k, x_j) = \frac{u_j^{k+1} - u_j^k}{\Delta t} + O(\Delta t) \tag{9.3.6a}$$

$$u_{xx}(t_k, x_j) = \frac{u_{j+1}^k - 2u_j^k + u_{j-1}^k}{\Delta x^2} + O(\Delta x^2) \tag{9.3.6b}$$

The approximation of $u_t = \partial u/\partial t$ in (9.3.6a) has a truncation error of order $O(\Delta t)$, while the approximation of $u_{xx} = \partial^2 u/\partial x^2$ in (9.3.6b) is more accurate for the same step size because it has a truncation error of order $O(\Delta x^2)$. If we substitute (9.3.6) into (9.3.1), and ignore the higher order terms, this yields the following system of difference equations for the interior grid points, $0 \le k \le m$ and $1 \le j \le n$.

$$\frac{u_j^{k+1} - u_j^k}{\Delta t} = \beta\left(\frac{u_{j+1}^k - 2u_j^k + u_{j-1}^k}{\Delta x^2}\right) \tag{9.3.7}$$

The formulation in Equation (9.3.7) is referred to as the *explicit forward Euler method* because the time derivative is represented with a forward Euler approximation, and the solution at time t_{k+1} can be solved for explicitly as follows:

$$\boxed{u_j^{k+1} = \gamma u_{j-1}^k + (1 - 2\gamma)u_j^k + \gamma u_{j+1}^k} \tag{9.3.8}$$

where the *gain* parameter $\gamma > 0$ is defined

$$\boxed{\gamma \triangleq \frac{\beta \Delta t}{\Delta x^2}} \tag{9.3.9}$$

In order to evaluate (9.3.8) at $j = 1$ and $j = n$, the boundary values u_0^k and u_{n+1}^k are needed. To develop expressions for these values, we can use a two-point forward difference for the derivative in (9.3.3a) and a two-point backward difference for the derivative in (9.3.3b). This converts the boundary condition constraints into the following difference equations.

$$(1 - c_1)u_0^k - c_1\left(\frac{u_1^k - u_0^k}{\Delta x}\right) = g_1(t_k) \tag{9.3.10a}$$

$$(1 - c_2)u_{n+1}^k + c_2\left(\frac{u_{n+1}^k - u_n^k}{\Delta x}\right) = g_2(t_k) \tag{9.3.10b}$$

Solving (9.3.10) for the boundary values then yields

$$\boxed{u_0^k = \frac{g_1(t_k)\Delta x + c_1 u_1^k}{c_1 + (1 - c_1)\Delta x}} \tag{9.3.11a}$$

$$\boxed{u_{n+1}^k = \frac{g_2(t_k)\Delta x + c_2 u_n^k}{c_2 + (1 - c_2)\Delta x}} \tag{9.3.11b}$$

In contrast to the previous treatment of Neumann boundary conditions for the Poisson equation, the formulations in (9.3.11) for the heat equation do *not* require increasing the dimension of the system. However, the use of two-point forward and backward differences in place of three-point central differences means that the order of the truncation error at the boundary is only $O(\Delta x)$, rather than $O(\Delta x^2)$, except when Dirichlet boundary conditions are used.

EXAMPLE 9.3.1 Explicit Forward Euler Method

The forward Euler equations can be expressed in vector form by letting $u^k = [u_1^k, u_2^k, \ldots, u_n^k]^T$ denote the solution at time t_k for $0 \le k \le m$. Suppose the lower boundary has a Dirichlet constraint ($c_1 = 0$), while the upper boundary has a Neumann constraint ($c_2 = 1$). For the case $n = 5$, this results in the following explicit solution.

$$u^{k+1} = \begin{bmatrix} 1-2\gamma & \gamma & 0 & 0 & 0 \\ \gamma & 1-2\gamma & \gamma & 0 & 0 \\ 0 & \gamma & 1-2\gamma & \gamma & 0 \\ 0 & 0 & \gamma & 1-2\gamma & \gamma \\ 0 & 0 & 0 & \gamma & 1-\gamma \end{bmatrix} u_k + \begin{bmatrix} \gamma g_1(t_k) \\ 0 \\ 0 \\ 0 \\ \gamma \Delta x g_2(t_k) \end{bmatrix}$$

An important consideration for any iterative formula is its sensitivity to errors in the initial conditions. For example, suppose there is a small round-off error, $\Delta f(x)$, in representing the initial condition function $f(x)$ in Equation (9.3.2). We wish to determine under what conditions the effects of this error grow as the solution to (9.3.8) progresses in time. If the effects of initial errors do not increase with time, an iterative formula is said to be *stable*. The stability of the forward Euler method depends on the value of the gain parameter γ defined in (9.3.9). To see this, we begin by considering the following candidate for a solution to (9.3.8).

$$u_j^k = (\lambda)^k \exp\left(\frac{ij\pi}{p}\right) \tag{9.3.12}$$

Here, $i = \sqrt{-1}$, and p is any nonzero integer. The value λ is called the *amplitude factor* because the solution grows with k when $|\lambda| > 1$. To determine λ, we substitute the expression for u_j^k from (9.3.12) into (9.3.8) and solve for λ. Canceling $(\lambda)^k \exp(ij\pi/p)$ from both sides, and using the trigonometric identity $\exp(i\pi/p) + \exp(-i\pi/p) = 2\cos(\pi/p)$, we arrive at the following expression for the amplitude factor.

$$\lambda = 1 + 2\gamma[\cos(\pi/p) - 1] \tag{9.3.13}$$

To ensure stability, we must constrain γ such that $|\lambda| \le 1$. Since $|\cos(\pi/p)| \le 1$, it follows that $(1 - 4\gamma) \le \lambda \le 1$. Therefore, setting $1 - 4\gamma \ge -1$ and solving for γ yields $\gamma \le 1/2$. Recalling the definition of γ in (9.3.9), this yields the following stability criterion for the explicit forward Euler method.

$$\boxed{\Delta t \le \frac{\Delta x^2}{2\beta}} \tag{9.3.14}$$

From Equation (9.3.14), we see that as the space step Δx gets smaller, so must the time step Δt in order to maintain stability. This is a drawback of the forward Euler method. It is a very efficient method because an explicit formula is available for u^{k+1}, but the two step sizes can not be chosen independent of one another.

9.3.2 Implicit Backward Euler Method

A simple way to improve the stability characteristics of the forward Euler method is to evaluate the three-point central difference approximation for u_{xx} at t_{k+1} rather than t_k.

$$u_{xx}(t_{k+1}, x_j) = \frac{u_{j+1}^{k+1} - 2u_j^{k+1} + u_{j-1}^{k+1}}{\Delta x^2} + O(\Delta x^2) \tag{9.3.15}$$

Using t_{k+1} as the reference time in place of t_k has the effect of making the Euler approximation of u_t in (9.3.6a) a backward difference rather than a forward difference. Substituting (9.3.6a) and (9.3.15) into (9.3.1), and ignoring the higher-order terms, we get the following alternative system of difference equations for the interior grid points, $0 \le k \le m$ and $1 \le j \le n$.

$$\frac{u_j^{k+1} - u_j^k}{\Delta t} = \beta\left(\frac{u_{j+1}^{k+1} - 2u_j^{k+1} + u_{j-1}^{k+1}}{\Delta x^2}\right) \tag{9.3.16}$$

Notice that, unlike (9.3.7), this equation can not be solved explicitly for u_j^{k+1}. Instead, there are three terms that involve the solution at time t_{k+1}. Grouping them together on the left-hand side and recalling the definition of the gain constant γ in (9.3.9) yields

$$\boxed{-\gamma u_{j-1}^{k+1} + (1 + 2\gamma)u_j^{k+1} - \gamma u_{j+1}^{k+1} = u_j^k} \tag{9.3.17}$$

The formulation in (9.3.17) is referred to as the *implicit backward Euler method*. At each time step, a tridiagonal linear algebraic system of n equations must be solved. Note that to solve (9.3.17), the boundary values u_0^k and u_{n+1}^k in (9.3.11) must be used.

EXAMPLE 9.3.2 Implicit Backward Euler Method

The backward Euler equations can be expressed in compact vector form. Let $u^k = [u_1^k, u_2^k, \ldots, u_n^k]^T$ denote the solution at time t_k for $0 \le k \le m$. Suppose the lower boundary has a Neumann constraint ($c_1 = 1$), while the upper boundary has a Dirichlet constraint ($c_2 = 0$). For the case $n = 5$, this results in the following implicit linear algebraic system.

$$\begin{bmatrix} 1+\gamma & -\gamma & 0 & 0 & 0 \\ \hline -\gamma & 1+2\gamma & -\gamma & 0 & 0 \\ 0 & -\gamma & 1+2\gamma & -\gamma & 0 \\ 0 & 0 & -\gamma & 1+2\gamma & -\gamma \\ \hline 0 & 0 & 0 & -\gamma & 1+2\gamma \end{bmatrix} u^{k+1} = u^k + \begin{bmatrix} \gamma \Delta x g_1(t_{k+1}) \\ \hline 0 \\ 0 \\ 0 \\ \hline \gamma g_2(t_{k+1}) \end{bmatrix}$$

The tridiagonal nature of the $n \times n$ coefficient matrix of the backward Euler equations is apparent from Example 9.3.2. Since $\gamma > 0$, the coefficient matrix is diagonally dominant. Furthermore, the same coefficient matrix is used at each time step, so the backward Euler equations can be solved very efficiently using the tridiagonal *LU* decomposition method in Section 2.4.3.

Although the backward Euler method in Equation (9.3.17) is not as efficient as the explicit forward Euler method in (9.3.8), it does have improved stability characteristics. To see this, again consider the following candidate for a solution to (9.3.17).

$$u_j^k = (\lambda)^k \exp\left(\frac{ij\pi}{p}\right) \tag{9.3.18}$$

Here, $i = \sqrt{-1}$, and p is any nonzero integer. To determine the amplitude factor λ, we substitute this expression for u_j^k into (9.3.17) and solve for λ. Simplification, using the identity $\exp(i\pi/p) + \exp(-i\pi/p) = 2\cos(\pi/p)$, yields the following expression for the amplitude factor.

$$\lambda = \frac{1}{1 + 2\gamma[1 - \cos(\pi/p)]} \tag{9.3.19}$$

Since $1 - \cos(\pi/p) \ge 0$ and $\gamma > 0$, it follows that $|\lambda| \le 1$ for *all* values of γ. Therefore, the backward Euler method is *unconditionally stable*. Unlike the forward Euler method, there is no constraint similar to (9.3.14) on the relative values of Δt and Δx.

9.3.3 Crank-Nicolson Method

Although the implicit backward Euler method is unconditionally stable, it can be improved upon in an important way. Recall that both the forward Euler and the backward Euler methods are first-order accurate in time with a truncation error of order $O(\Delta t)$, but second-order accurate in space (except possibly at the boundary) with a truncation error of order $O(\Delta x^2)$. Observe that the two-point approximation of u_t in Equation (9.3.6a) can be viewed as a three-point central difference approximation about the midpoint, $(t_k + t_{k+1})/2$, using a step of size $\Delta t/2$. To make this second-order interpretation consistent, the central difference approximation to u_{xx} must also be evaluated at the midpoint. This can be achieved by simply averaging $u_{xx}(t_k, x_j)$ with $u_{xx}(t_{k+1}, x_j)$. This yields the following system of difference equations for the interior grid points $0 \le k \le m$ and $1 \le j \le n$.

$$\frac{u_j^{k+1} - u_j^k}{\Delta t} = \frac{\beta}{2}\left(\frac{u_{j+1}^{k+1} - 2u_j^{k+1} + u_{j+1}^{k+1}}{\Delta x^2} + \frac{u_{j+1}^k - 2u_j^k + u_{j-1}^k}{\Delta x^2}\right) \tag{9.3.20}$$

The formulation in (9.3.20) is second-order accurate in both time and space and is referred to as the *Crank-Nicolson method*. Observe that the Crank-Nicolson equations can be obtained by simply averaging the explicit forward Euler equations in (9.3.7) with the implicit backward Euler equations in (9.3.16). We can simplify the Crank-Nicolson equations somewhat by grouping terms with similar time indices together and by using the gain parameter γ defined in (9.3.9). This results in the following more compact formulation.

$$\boxed{-\gamma u_{j+1}^{k+1} + 2(1+\gamma)u_j^{k+1} - \gamma u_{j-1}^{k+1} = \gamma u_{j+1}^k + 2(1-\gamma)u_j^k + \gamma u_{j-1}^k} \tag{9.3.21}$$

It is evident that the Crank-Nicolson method is also an implicit method because at each time step, a tridiagonal system of n equations must be solved. In order to solve (9.3.21) at $j = 1$ and $j = n$, the boundary values u_0^k and u_{n+1}^k in (9.3.11) must be used. Alternatively, fictitious points can be added as in Section 9.2, in which case there are $n + 2$ equations that must be solved when Neumann boundary conditions are used.

EXAMPLE 9.3.3 Crank-Nicolson Method

The Crank-Nicolson equations can be expressed in vector form by letting $u^k = [u_1^k, u_2^k, \ldots, u_n^k]^T$ denote the solution at time t_k for $0 \le k \le m$. Suppose both boundary conditions are of the Dirichlet type, which means $c = 0$ in Equation (9.3.3). For the case $n = 5$, this results in the following implicit linear algebraic system.

$$\begin{bmatrix} 2(1+\gamma) & -\gamma & 0 & 0 & 0 \\ -\gamma & 2(1+\gamma) & -\gamma & 0 & 0 \\ 0 & -\gamma & 2(1+\gamma) & -\gamma & 0 \\ 0 & 0 & -\gamma & 2(1+\gamma) & -\gamma \\ 0 & 0 & 0 & -\gamma & 2(1+\gamma) \end{bmatrix} u^{k+1} =$$

$$\begin{bmatrix} 2(1-\gamma) & \gamma & 0 & 0 & 0 \\ \gamma & 2(1-\gamma) & \gamma & 0 & 0 \\ 0 & \gamma & 2(1-\gamma) & \gamma & 0 \\ 0 & 0 & \gamma & 2(1-\gamma) & \gamma \\ 0 & 0 & 0 & \gamma & 2(1-\gamma) \end{bmatrix} u^k + \begin{bmatrix} \gamma[g_1(t^k) + g_1(t_{k+1})] \\ 0 \\ 0 \\ 0 \\ \gamma[g_2(t_k) + g_2(t_{k+1})] \end{bmatrix}$$

Since the tridiagonal $n \times n$ coefficient matrix of u^{k+1} in Example 9.3.3 does not depend on time, the Crank-Nicolson equations can be solved efficiently by using the tridiagonal *LU* decomposition method. The Crank-Nicolson method is an implicit method that has stability characteristics similar to those of the implicit backward Euler method. In particular, suppose we use the solution candidate in (9.3.18). To determine the amplitude factor λ, we substitute this expression for u_j^k from (9.3.18) into (9.3.21) and solve for λ. After simplification, this results in the following expression for the amplitude factor.

$$\lambda = \frac{1 - \gamma[1 - \cos(\pi/p)]}{1 + \gamma[1 - \cos(\pi/p)]} \tag{9.3.22}$$

Since $1 - \cos(\pi/p) \geq 0$ and $\gamma > 0$, it follows that $|\lambda| \leq 1$ for *all* values of γ. Consequently, the Crank-Nicolson method is also *unconditionally stable*. However, because the Crank-Nicolson method has a second-order truncation error in both time and space, it tends to be more accurate than the backward Euler method using identical step sizes. Although the Crank-Nicolson method takes a little more work per step than the backward Euler method, one can solve the system using larger time steps. For this reason, the Crank-Nicolson method is the method of choice for solving the one-dimensional heat equation.

EXAMPLE 9.3.4 Heat Flow

As an illustration of the use of the Crank-Nicolson method to solve a parabolic partial differential equation, consider the problem of determining the temperature along a thin rod of length $a = 1$. Suppose the thermal diffusivity is $\beta = 10^{-5}$. From Section 9.1.2, the temperature $u(t, x)$ at time t and position x is a solution to the following heat equation.

$$\frac{\partial u}{\partial t} = (10^{-5})\frac{\partial^2 u}{\partial x^2}$$

Suppose the time interval is $T = 6000$. Let the size of the grid be $m = 50$ and $n = 39$. Note from Equation (9.3.9) that this produces a gain parameter of

$$\gamma = \frac{(10^{-5})(6000)(40)^2}{50} = 1.92$$

Since $\gamma > 0.5$, the forward Euler method would be unstable for this choice of m and n. However, the Crank-Nicolson method is stable for all $\gamma > 0$. As initial and boundary conditions, suppose that

$$u(0, x) = 2x + \sin(2\pi x)$$

$$u(t,0) = 0$$
$$u(t,1) = 2$$

Thus, the boundary conditions are constant conditions of the Dirichlet type. Example program *e934* on the distribution CD applies the Crank-Nicolson method to this problem with the results shown in Figure 9.3.2. Note how the solution approaches the following function in the steady state.

$$u(t,x) \to 2x \quad \text{as} \quad t \to \infty$$

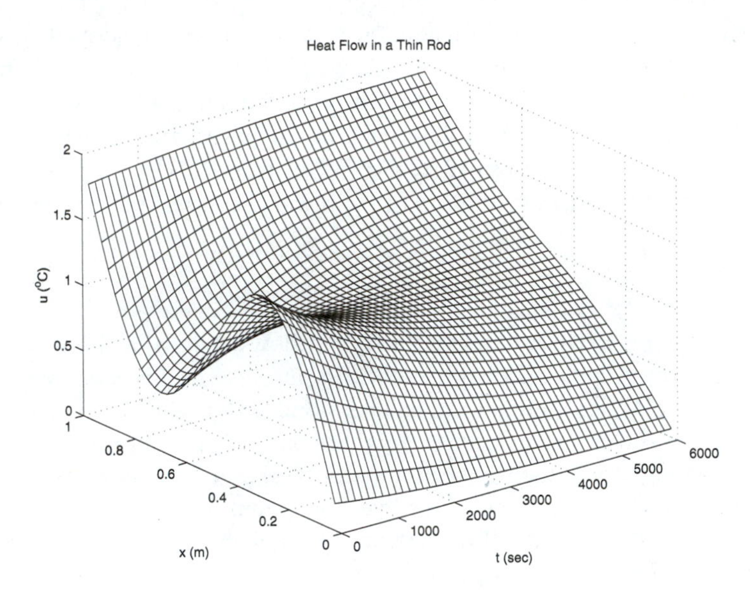

FIGURE 9.3.2 Heat Flow Along a Thin Rod

It is a simple matter to verify through direct substitution that the steady-state linear interpolation function, $u(t,x) = 2x$, satisfies the heat equation and the boundary conditions.

9.4 TWO-DIMENSIONAL PARABOLIC EQUATIONS

In engineering applications, there are instances where we want to simulate heat flow or diffusion in two or three spatial dimensions. In this section, we examine a numerical technique for solving the following two-dimensional version of the heat equation.

$$\frac{\partial u}{\partial t} = \beta\left(\frac{\partial^2 u}{\partial x^2} + \frac{\partial^2 u}{\partial y^2}\right) \tag{9.4.1}$$

Here the solution, $u(t,x,y)$, depends on three independent variables. Note that in the steady-state when $u_t = 0$, the two-dimensional heat equation reduces to Laplace's equation. As a consequence, techniques for solving (9.4.1) provide us with an alternative way to solve Laplace's equation. Of course, if only the steady-state solution is of interest, then it may be more efficient to solve Laplace's equation directly.

Suppose we are interested in solving (9.4.1) over the following rectangular region in the xy plane.

$$R = \{(x, y) \mid 0 < x < a, 0 < y < b\} \tag{9.4.2}$$

To obtain a unique solution, again we must specify both an initial condition and boundary conditions. In this case, the *initial condition* will be of the following form for some initial condition function $f(x, y)$.

$$u(0, x, y) = f(x, y) \quad , \quad (x, y) \in R \tag{9.4.3}$$

To keep the discussion simple, we illustrate the technique by considering only Dirichlet boundary conditions. In particular, if δR denotes the boundary of the region R, then we place the following constraints on the solution along the boundary.

$$u(t, x, y) = g(t, x, y) \quad , \quad (x, y) \in \delta R \tag{9.4.4}$$

The one-dimensional methods discussed in the last section can all be generalized to the two-dimensional case. The generalization of the explicit forward Euler method is efficient, but it is stable only for sufficiently small values of the time step. Indeed, the following stability criterion for the two-dimension case (Gerald and Wheatley, 1989) is somewhat stricter than the one-dimensional criterion.

$$\Delta t \le \frac{\Delta x^2 + \Delta y^2}{8\beta} \tag{9.4.5}$$

The backward Euler and Crank-Nicolson methods can also be generalized to apply to Equation (9.4.1), but they lose their attractiveness because the systems of linear equations that must be solved at each time step are much larger, and moreover, they are no longer tridiagonal.

Instead, we examine an alternative finite difference method specifically tailored to the two-dimensional case. Suppose we are interested in solving the heat equation over the time interval $0 \le t \le T$. We compute numerical estimates of $u(t, x, y)$ over a uniform grid consisting of $m + 1$ values of t, $n + 2$ values of x, and $p + 2$ values of y. If Δt, Δx and Δy denote the step sizes of the variables t, x, and y, respectively, then

$$\Delta t \triangleq \frac{T}{m} \tag{9.4.6a}$$

$$\Delta x \triangleq \frac{a}{n + 1} \tag{9.4.6b}$$

$$\Delta y \triangleq \frac{b}{p + 1} \tag{9.4.6c}$$

To simplify the final equations, we again let t_k, x_i, and y_j denote the values of t, x, and y at the grid points. That is,

$$t_k = k\Delta t \quad , \quad 0 \le k \le m \tag{9.4.7a}$$

$$x_i = i\Delta x \quad , \quad 0 \le i \le (n + 1) \tag{9.4.7b}$$

$$y_j = j\Delta y \quad , \quad 0 \le j \le (p + 1) \tag{9.4.7c}$$

Next, let u_{ij}^k denote the computed value of $u(t_k, x_i, y_j)$. Here the superscript specifies the time step, while the subscripts specify the space steps. To convert the two-dimensional heat equation in (9.4.1) to a difference equation, we use finite difference approximations for the derivative terms. The first-order time derivative can be approximated with a three-point central difference about the midpoint, $(t_k + t_{k+1})/2$, just as it was with the Crank-Nicolson method. However, in this case we evaluate one of the second-order derivative terms, u_{xx}, at time t_k, and the other second-order derivative term, u_{yy}, at time t_{k+1}. This yields the following system of difference equations at the interior points $0 \le k \le m, 1 \le i \le n$, and $1 \le j \le p$.

$$\frac{u_{ij}^{k+1} - u_{ij}^k}{\Delta t} = \beta\left(\frac{u_{i+1,j}^k - 2u_{ij}^k + u_{i-1,j}^k}{\Delta x^2} + \frac{u_{i,j+1}^{k+1} - 2u_{ij}^{k+1} + u_{i,j-1}^{k+1}}{\Delta y^2}\right) \tag{9.4.8}$$

The formula in (9.4.8) has the virtue that the coefficient matrix of u^{k+1} is tridiagonal. However, this is achieved by adding some bias, which evaluates u_{xx} at an earlier time than u_{yy}. This bias can be compensated for by adding a second equation for u^{k+2}, which reverses the roles of x and y by evaluating u_{xx} at time t_{k+2} and u_{yy} at time t_{k+1}.

$$\frac{u_{ij}^{k+2} - u_{ij}^{k+1}}{\Delta t} = \beta\left(\frac{u_{i+1,j}^{k+2} - 2u_{ij}^{k+2} + u_{i-1,j}^{k+2}}{\Delta x^2} + \frac{u_{i,j+1}^{k+1} - 2u_{ij}^{k+1} + u_{i,j-1}^{k+1}}{\Delta y^2}\right) \tag{9.4.9}$$

When Equations (9.4.8) and (9.4.9) are taken together, we see that the derivatives with respect to x are computed at the beginning of the interval $[t_k, t_{k+1}]$ and then at the end of the interval $[t_{k+1}, t_{k+2}]$, while the derivatives with respect to y are computed at the end of the interval $[t_k, t_{k+1}]$ and then at the beginning of interval $[t_{k+1}, t_{k+2}]$. The formulation in (9.4.8) and (9.4.9) is referred to as the *alternating direction implicit method* or *ADI* method (Peaceman and Rachford, 1955). The ADI formulas can be simplified by introducing the following *gain* parameters.

$$\boxed{\gamma_x \triangleq \frac{\beta \Delta t}{\Delta x^2}} \tag{9.4.10a}$$

$$\boxed{\gamma_y \triangleq \frac{\beta \Delta t}{\Delta y^2}} \tag{9.4.10b}$$

Using γ_x and γ_y, we can then recast (9.4.8) and (9.4.9) more compactly as follows.

$$\boxed{-\gamma_y(u_{i,j-1}^{k+1} + u_{i,j+1}^{k+1}) + (1 + 2\gamma_y)u_{ij}^{k+1} = \gamma_x(u_{i-1,j}^k + u_{i+1,j}^k) + (1 - 2\gamma_x)u_{ij}^k} \tag{9.4.11a}$$

$$\boxed{-\gamma_x(u_{i-1,j}^{k+2} + u_{i+1,j}^{k+2}) + (1 + 2\gamma_x)u_{ij}^{k+2} = \gamma_y(u_{i,j-1}^{k+1} + u_{i,j+1}^{k+1}) + (1 - 2\gamma_y)u_{ij}^{k+1}} \tag{9.4.11b}$$

Note that first (9.4.11a) is solved for the ith row of the $n \times p$ matrix u^{k+1} for $1 \le i \le n$. This involves solving n tridiagonal systems, each of dimension p. However, they all have the same coefficient matrix, so *LU* decomposition can be put to good use. Then (9.4.11b) is solved for the jth column of the $n \times p$ matrix u^{k+2} for $1 \le j \le p$. This involves solving p tridiagonal systems, each of dimension n. Again the *LU* decomposition method can be used because they all have the same coefficient matrix.

Given the bias that is inherent in the individual equations of (9.4.11), they should always be used as a pair. That is, only u^{k+2} should be used, not the less accurate intermediate value u^{k+1}. Consequently, the ADI method produces the solution points $\{u^0, u^2, u^4, \ldots\}$. This is not a problem because Δt can always be set sufficiently small that $2\Delta t$ is the final desired time resolution. That is, (9.4.11) can represent two half-steps.

Like the backward Euler and Crank-Nicolson methods, the two-dimensional ADI method is an implicit method that is unconditionally stable. However, for all three methods, if the gain parameters are too large, the accuracy of the estimated solution can suffer.

EXAMPLE 9.4.1 Two-Dimensional Heat Flow

As an illustration of the use of the ADI method to solve a two-dimensional parabolic partial differential equation, consider the problem of determining the temperature over a square plate that is four units on a side so that $a = b = 4$. Suppose the thermal diffusivity is $\beta = 10^{-4}$. The temperature $u(t, x, y)$ at time t and position (x, y) is a solution of the following two-dimensional heat equation.

$$\frac{\partial u}{\partial t} = (10^{-4})\left(\frac{\partial^2 u}{\partial x^2} + \frac{\partial^2 u}{\partial y^2}\right)$$

Suppose the time interval is $T = 5000$. Let the size of the grid be $m = 50$, and $n = p = 39$. Using Equation (9.4.10) yields the following gain parameters.

$$\gamma_x = \gamma_y = 0.625$$

As initial and boundary conditions, suppose that

$$u(0, x, y) = 0$$
$$u(t, x, y) = \exp(y)\cos(x) - \exp(x)\cos(y)$$

Example program *e941* on the distribution CD applies the ADI method to this problem. Plots of the resulting solution at times $t = 500$ and $t = 5000$ are shown in Figure 9.4.1 and Figure 9.4.2, respectively. When $t = 5000$, sufficient time has elapsed that the system is close to

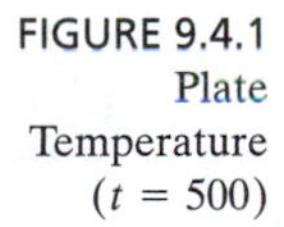

FIGURE 9.4.1 Plate Temperature ($t = 500$)

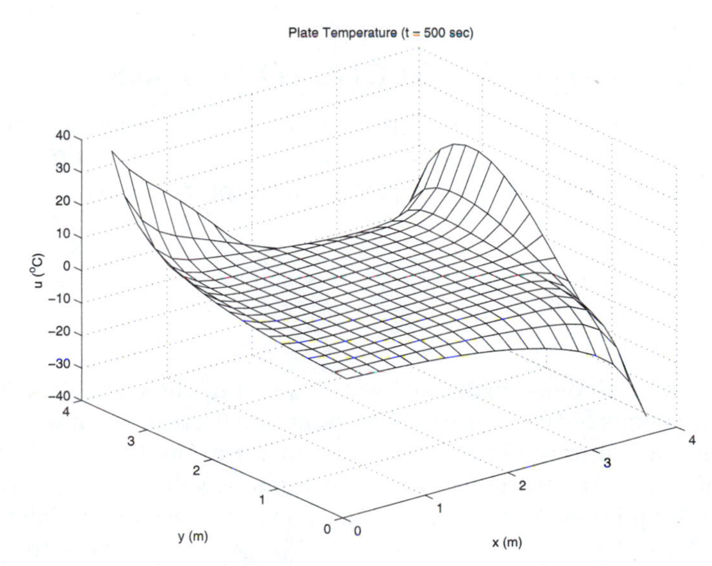

its steady-state condition. This can be verified by comparing Figure 9.4.2 with Figure 9.2.5. Recall that Figure 9.2.5 displays the steady-state plate temperature obtained by solving Laplace's equation with the SOR method.

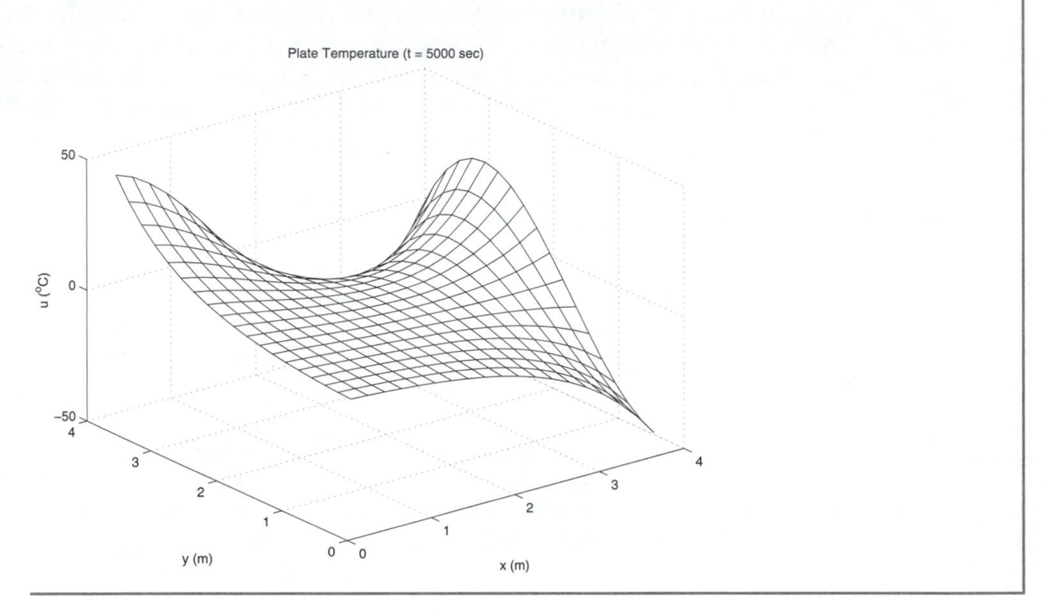

FIGURE 9.4.2 Plate Temperature ($t = 5000$)

The ADI method can also be applied directly to elliptic partial differential equations including Poisson's equation, although it is more difficult to program. One virtue of the ADI method is that it is amenable to parallel (vector) processing on a super computer (Kershaw, 1982). A discussion of the ADI method applied to elliptic systems can be found in (Birkhoff et al., 1962).

9.5 ONE-DIMENSIONAL HYPERBOLIC EQUATIONS

In this section, we examine solution techniques for *hyperbolic* partial differential equations. An important example of a hyperbolic equation is a partial differential equation in which the second derivative with respect to time is proportional to the second derivative with respect to space.

$$\boxed{\frac{\partial^2 u}{\partial t^2} = \beta \frac{\partial^2 u}{\partial x^2}} \tag{9.5.1}$$

Recall that this is the one-dimensional wave equation. The coefficient β is a positive constant having units of $(m/s)^2$ with $\sqrt{\beta}$ representing the *velocity* of wave propagation. The one-dimensional wave equation can be used, for example, to model the vibration of a string on a musical instrument. To obtain a unique solution, $u(t, x)$, we must specify both initial conditions and boundary conditions. Suppose we are interested in solving the wave equation over the interval $0 < x < a$. If a string of length a is fixed at both ends, the boundary conditions are

$$u(t,0) = 0 \tag{9.5.2a}$$

$$u(t,a) = 0 \tag{9.5.2b}$$

Since the wave equation is second-order in time, there are also two initial conditions that must be specified. In particular, we can specify both u and its time derivative u_t at the initial time.

$$u(0,x) = f(x) \quad , \quad 0 < x < a \qquad 9.5.3a$$

$$u_t(0,x) = g(x) \quad , \quad 0 < x < a \qquad 9.5.3b$$

Note that in order for $u(t,x)$ to satisfy the boundary conditions, it is necessary that $f(0) = 0$ and $f(a) = 0$.

9.5.1 d'Alembert's Solution

Before we examine numerical methods for solving the wave equation, it is useful to consider an elegant analytical technique called *d'Alembert's method*. In certain cases, the solutions obtained in this manner generate simple algebraic expressions that can be used as a check on the accuracy of numerical solutions.

The basic idea behind d'Alembert's method is to use a change of variables. For convenience, let $\alpha = \sqrt{\beta}$ where β is the wave equation coefficient in Equation (9.5.1). Next, consider the following set of new variables.

$$v \triangleq x + \alpha t \tag{9.5.4a}$$

$$w \triangleq x - \alpha t \tag{9.5.4b}$$

Since $t = (v - w)/(2\alpha)$ and $x = (v + w)/2$, the solution u can be regarded as a function of v and w. From (9.5.4), we have $v_x = w_x = 1, v_t = \alpha$ and $w_t = -\alpha$. Using the chain rule, the first derivative of u with respect to t can be expressed as

$$\begin{aligned} u_t &= u_v v_t + u_w w_t \\ &= \alpha(u_v - u_w) \end{aligned} \tag{9.5.5}$$

Using (9.5.5), the second derivative of u with respect to t can be expressed in terms of v and w as

$$\begin{aligned} u_{tt} &= \alpha(u_{vv}v_t + u_{vw}w_t - u_{wv}v_t - u_{ww}w_t) \\ &= \alpha^2(u_{vv} - (u_{wv} + u_{vw}) + u_{ww}) \\ &= \alpha^2(u_{vv} - 2u_{vw} + u_{ww}) \end{aligned} \tag{9.5.6}$$

Applying a similar analysis to evaluate the second derivative of u with respect to x yields

$$u_{xx} = u_{vv} + 2u_{vw} + u_{ww} \tag{9.5.7}$$

Note that the wave equation in (9.5.1) can be expressed $u_{tt} = \beta u_{xx}$. Substituting (9.5.6) and (9.5.7) into (9.5.1), and recalling that $\alpha^2 = \beta$, we find that most of the terms cancel and we are left with $u_{vw} = 0$ or

$$\frac{\partial^2 u}{\partial v \partial w} = 0 \tag{9.5.8}$$

Thus, the wave equation has a very simple form when expressed in terms of the independent variables v and w. Suppose h_1 and h_2 are two arbitrary differentiable functions. Then direct substitution into (9.5.8) verifies that the following expression is a solution.

$$u = h_1(v) + h_2(w) \tag{9.5.9}$$

The definitions of v and w in (9.5.4) yield a generic wave equation solution of the following form.

$$u(t, x) = h_1(x + \alpha t) + h_2(x - \alpha t) \tag{9.5.10}$$

The unknown functions h_1 and h_2 are determined from the initial and boundary conditions. We first consider the special case when $g(x) = 0$ in (9.5.3b). From the initial condition on $u(t, x)$ in (9.5.3a), we have

$$h_1(x) + h_2(x) = f(x) \tag{9.5.11}$$

Next, applying the initial condition on u_t in (9.5.3b) with $g(x) = 0$ yields

$$\alpha[h_1'(x) - h_2'(x)] = 0 \tag{9.5.12}$$

Here h_1' and h_2' denote the derivatives of h_1 and h_2 with respect to their arguments. If we integrate both sides of (9.5.12), we find that $h_2(x) = h_1(x) + c$ where c is a constant of integration. Substituting this expression for $h_2(x)$ into (9.5.11) and solving for $h_1(x)$ then yields

$$h_1(x) = \frac{f(x) - c}{2} \tag{9.5.13a}$$

$$h_2(x) = \frac{f(x) + c}{2} \tag{9.5.13b}$$

Interestingly enough, when the expressions in (9.5.13) are substituted into the generic wave equation solution in (9.5.10), the unknown constant terms cancel, and the result is

$$u_1(t, x) = \frac{f(x + \alpha t) + f(x - \alpha t)}{2} \tag{9.5.14}$$

The formulation for $u_1(t, x)$ in (9.5.14) is a solution to the wave equation for the initial conditions $f \neq 0$ and $g = 0$. Next, we consider the problem of finding a wave equation solution that satisfies the complementary initial conditions $f = 0$ and $g \neq 0$. Consider the following candidate.

$$u_2(t, x) = \frac{1}{2\alpha} \int_{x-\alpha t}^{x+\alpha t} g(z)\, dz \tag{9.5.15}$$

By inspection, we see that $u(0, x) = 0$. The derivative of $u(t, x)$ with respect to t is

$$u_t(t, x) = \frac{g(x + \alpha t) + g(x - \alpha t)}{2} \tag{9.5.16}$$

It follows that $u_t(0,x) = g(x)$. It remains to show that $u(t,x)$ satisfies the wave equation. Differentiating both sides of (9.5.16) with respect to time, we find that $u_{tt}(t,x) = 0$. A similar analysis shows that $u_{xx}(t,x) = 0$. Therefore, $u(t,x)$ is indeed a solution to the wave equation in (9.5.1). Finally, because the wave equation is linear, we can add the solution in (9.5.14) to the solution in (9.5.15) to get a general solution valid for $f \neq 0$ and $g \neq 0$.

$$\boxed{u(t,x) = \frac{f(x+\alpha t) + f(x-\alpha t)}{2} + \frac{1}{2\alpha}\int_{x-\alpha t}^{x+\alpha t} g(z)\,dz} \qquad (9.5.17)$$

Note that Equation (9.5.17) assumes that the domains of the functions f and g are the *entire* real line. To satisfy the boundary conditions, there are some additional constraints that must be placed on f and g. Applying the boundary condition $u(t,0) = 0$ to (9.5.17), we see that f and g must be *odd* functions.

$$f(-x) = -f(x) \qquad (9.5.18a)$$

$$g(-x) = -g(x) \qquad (9.5.18b)$$

Applying the second boundary condition $u(t,a) = 0$ to (9.5.17) and using the odd symmetry, we find after a change of variables that f and g must be periodic with period $2a$.

$$f(x + 2a) = f(x) \qquad (9.5.19a)$$

$$g(x + 2a) = g(x) \qquad (9.5.19b)$$

Thus, f and g must be odd periodic functions with periods of $2a$. This in turn implies that $f(0) = f(a) = 0$ and $g(0) = g(a) = 0$.

EXAMPLE 9.5.1 d'Alembert's Solution

To illustrate the use of d'Alembert's method to find an analytic solution, consider the following wave equation.

$$\frac{\partial^2 u}{\partial t^2} = 4\left(\frac{\partial^2 u}{\partial x^2}\right)$$

In this case, the velocity of wave propagation is $\alpha = 2$. Suppose the length of the solution interval is $a = 1$. Consider the following initial conditions where the 2×1 vector c is arbitrary.

$$u(0,x) = c_1 \sin(\pi x) \quad , \quad 0 < x < 1$$

$$u_t(0,x) = 2\pi c_2 \sin(\pi x) \quad , \quad 0 < x < 1$$

Note that the two initial condition functions are odd and periodic with a period of 2 as required. The boundary conditions are $u(t,0) = u(t,1) = 0$. Using (9.5.17), the solution to this wave equation satisfying these boundary conditions is

$$\begin{aligned} u(t,x) &= \frac{c_1}{2}(\sin[\pi(x+2t)] + \sin[\pi(x-2t)]) + \frac{c_2\pi}{2}\int_{2-2t}^{x+2t} \sin(\pi z)\,dz \\ &= \sin(\pi x)[c_1\cos(2\pi t) + c_2\sin(2\pi t)] \end{aligned}$$

9.5.2 Explicit Central Difference Method

Next, we examine a simple numerical technique for solving the wave equation over the time interval $0 \le t \le T$. Numerical estimates of $u(t, x)$ are computed over a uniform grid consisting of $m + 1$ values of t and $n + 2$ values of x as shown previously for the heat equation in Figure 9.3.1. Let Δt and Δx denote the step sizes of the variables t and x, respectively.

$$\Delta t \triangleq \frac{T}{m} \tag{9.5.20a}$$

$$\Delta x \triangleq \frac{a}{n+1} \tag{9.5.20b}$$

To simplify the final equations, let t_k and x_j denote the values of t and x at the grid points.

$$t_k = k\Delta t \quad , \quad 0 \le k \le m \tag{9.5.21a}$$

$$x_j = j\Delta x \quad , \quad 0 \le j \le (n+1) \tag{9.5.21b}$$

Similarly, let u_j^k denote the computed value of $u(t_k, x_j)$ where the superscript specifies the time step and the subscript specifies the space step. To convert the wave equation in (9.5.1) to a difference equation, three-point central difference approximations are used for the two derivative terms. This leads to the following system of difference equations for the interior grid points, $0 \le k \le m$ and $1 \le j \le n$.

$$\frac{u_j^{k+1} - 2u_j^k + u_j^{k-1}}{\Delta t^2} = \beta\left(\frac{u_{j+1}^k - 2u_j^k + u_{j-1}^k}{\Delta x^2}\right) \tag{9.5.22}$$

The formulation in (9.5.22) is referred to as the *explicit central difference method* because the derivative terms are represented with central differences. Given that three-point differences are used for both the time derivative and the space derivative, this is a second-order method with a truncation error of order $O(\Delta t^2 + \Delta x^2)$. The explicit nature of the method arises from the fact that there is only one term that involves the solution at time t_{k+1}. Solving for this term yields

$$\boxed{u_j^{k+1} = \gamma u_{j-1}^k + 2(1-\gamma)u_j^k + \gamma u_{j+1}^k - u_j^{k-1}} \tag{9.5.23}$$

where the *gain* parameter $\gamma > 0$ is defined

$$\boxed{\gamma \triangleq \frac{\beta \Delta t^2}{\Delta x^2}} \tag{9.5.24}$$

Note that the gain γ in Equation (9.5.24) is not the same as the gain used in the heat equation in (9.3.9) due to the presence of a second factor Δt. In order to evaluate (9.5.23) at $j = 1$ and $j = n$, the boundary values $u_0^k = 0$ and $u_{n+1}^k = 0$ are used. In addition, computing u^{k+1} requires values of u at *two* previous time steps. This presents a problem when $k = 0$ because the last term in (9.5.23) is then u_j^{-1}. To estimate the solution at time $t_{-1} = -\Delta t$, we use the initial condition constraint in (9.5.3b). Replacing the first derivative term with a three-point central difference evaluated at t_0 yields the following difference equation.

$$\frac{u_j^1 - u_j^{-1}}{2\Delta t} = g(x_j) \quad , \quad 1 \le j \le n \tag{9.5.25}$$

This equation can be solved for the fictitious point u_j^{-1}.

$$u_j^{-1} = u_j^1 - 2\Delta t g(x_j) \quad , \quad 1 \le j \le n \tag{9.5.26}$$

Finally, the expression for u_j^{-1} can be substituted into (9.5.23) for the case when $k = 0$. Recalling from (9.5.3a) that $u_j^0 = f(x_j)$, this yields the following formulation for starting the computation.

$$\boxed{u_j^1 = \frac{1}{2}[\gamma f_{j-1} + 2(1-\gamma)f_j + \gamma f_{j+1}] + \Delta t g_j} \tag{9.5.27}$$

Here we have used the discrete notation $f_j = f(x_j)$ and $g_j = g(x_j)$ for the boundary condition functions in (9.5.3). Thus, we use (9.5.3a) to compute u^0, (9.5.27) to compute u^1, and then (9.5.23) to compute u^{k+1} for $1 \le k < m$.

EXAMPLE 9.5.2 Explicit Central Difference Method

The central difference equations can be expressed in compact vector form. In this case, let $u^k = [u_1^k, u_2^k, \ldots, u_n^k]^T$ denote the solution at time t_k for $0 \le k \le m$. First, consider the initialization step. Using Equation (9.5.27) with $n = 4$ results in the following explicit system of equations.

$$u^1 = \frac{1}{2}\begin{bmatrix} 2(1-\gamma) & \gamma & 0 & 0 \\ \gamma & 2(1-\gamma) & \gamma & 0 \\ 0 & \gamma & 2(1-\gamma) & \gamma \\ 0 & 0 & \gamma & 2(1-\gamma) \end{bmatrix}\begin{bmatrix} f_1 \\ f_2 \\ f_3 \\ f_4 \end{bmatrix} + \Delta t \begin{bmatrix} g_1 \\ g_2 \\ g_3 \\ g_4 \end{bmatrix}$$

Here we have used the boundary conditions $f(x_0) = f(x_{n+1}) = 0$. For $1 \le k < m$, the equations in (9.5.23) simplify to

$$u^{k+1} = \begin{bmatrix} 2(1-\gamma) & \gamma & 0 & 0 \\ \gamma & 2(1-\gamma) & \gamma & 0 \\ 0 & \gamma & 2(1-\gamma) & \gamma \\ 0 & 0 & \gamma & 2(1-\gamma) \end{bmatrix} u^k - u^{k-1}$$

Since the central difference method in Equation (9.5.23) is an explicit method, it is stable only for certain values of the gain parameter γ defined in (9.5.24). To determine the stable range, consider the following candidate for a solution to (9.5.23).

$$u_j^k = (\lambda)^k \exp\left(\frac{ij\pi}{p}\right) \tag{9.5.28}$$

Here $i = \sqrt{-1}$ and p is any nonzero integer. The amplitude factor λ depends on γ. To determine this dependence, we substitute the expression for u_j^k from (9.5.28) into (9.5.23) and solve for λ. Canceling $(\lambda)^{k-1} \exp(ij\pi/p)$ from both sides and using the trigonometric identity $\exp(i\pi/p) + \exp(-i\pi/p) = 2\cos(\pi/p)$, we arrive at the following quadratic expression for the amplitude factor.

$$\lambda^2 + 2(\gamma[1 - \cos(\pi/p)] - 1)\lambda + 1 = 0 \tag{9.5.29}$$

To ensure that the solution in (9.5.28) remains bounded, we must constrain γ such that $|\lambda| \leq 1$. The simplest way to determine if the roots of a polynomial, such as (9.5.29), have magnitudes less than or equal to unity is to apply the Jury test (Franklin et al., 1990). The Jury test yields the following constraint on γ.

$$\gamma \leq \frac{2}{1 - \cos(\pi/p)} \tag{9.5.30}$$

Recall that p is any nonzero integer. The smallest value for the right-hand side of (9.5.30) occurs when $|p| = 1$, which yields $\gamma \leq 1$. Using the definition of γ from (9.5.24), this results in the following stability criterion for the explicit central difference method.

$$\boxed{\Delta t^2 \leq \frac{\Delta x^2}{\beta}} \tag{9.5.31}$$

It follows that as the space step Δx gets smaller, so must the time step Δt in order to maintain stability. As Δx approaches zero, the accuracy improves as long as Δt satisfies (9.5.31). Curiously enough, when Δt is set equal to its maximum value in (9.5.31), it can be shown that the discrete estimates, u_j^k, are exact when infinite precision arithmetic is used (Gerald and Wheatley, 1989).

EXAMPLE 9.5.3 Vibrating String

As an illustration of the use of the central difference method to solve a hyperbolic partial differential equation, consider the problem of determining the motion of a taut elastic string of length $a = 2$ fixed at both ends. Suppose the string tension and density are such that the wave equation coefficient is $\beta = 4$. From Section 9.1.3, the string displacement $u(t, x)$ at time t and position x is a solution of the following wave equation.

$$\frac{\partial^2 u}{\partial t^2} = 4\frac{\partial^2 u}{\partial x^2}$$

Suppose the time interval is $T = 1.0$, and the size of the grid is $m = 80$ and $n = 39$. Using (9.5.24) results in a gain parameter of

$$\gamma = \frac{4(1/80)^2}{(2/40)^2} = 0.25$$

Thus, $\gamma < 1$ and the central difference method is stable for this combination of solution parameters. Since the string is fixed at both ends, the boundary conditions are

$$u(t,0) = 0$$
$$u(t,2) = 0$$

To start the string vibrating, suppose the following initial conditions are applied.

$$u(0,x) = 0.1\sin(\pi x) \quad , \quad 0 < x < 2$$
$$u_t(0,x) = 0.2\pi\sin(\pi x) \quad , \quad 0 < x < 2$$

Example program *e953* on the distribution CD applies the explicit central difference method to this problem with the results shown in Figure 9.5.1. Note how the string vibrates in time. The horizontal line in the plot indicates that the string is motionless at $x = 1$ while it oscillates up and down on both sides of this *stationary* point. If the solution displayed in Figure 9.5.1 is compared with the analytic solution obtained by d'Alembert's method in Example 9.5.1, the maximum error over the grid is found to be $E = 7.276 \times 10^{-4}$ in this case.

FIGURE 9.5.1 Vibration of a String

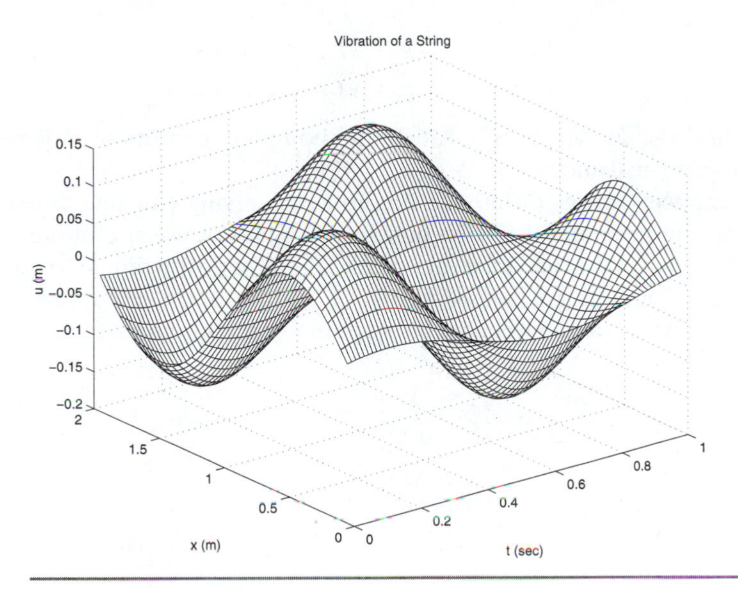

9.6 TWO-DIMENSIONAL HYPERBOLIC EQUATIONS

Mechanical vibrations in two spatial dimensions are also common. For example, a thin elastic membrane stretched across an opening will vibrate when displaced from its equilibrium position. In this section, we examine a numerical technique for solving the following two-dimensional version of the wave equation.

$$\boxed{\frac{\partial^2 u}{\partial t^2} = \beta\left(\frac{\partial^2 u}{\partial x^2} + \frac{\partial^2 u}{\partial y^2}\right)} \tag{9.6.1}$$

Note that the solution, $u(t, x, y)$, depends on three independent variables. Suppose we are interested in solving Equation (9.6.1) over a rectangle of width a and length b in the xy plane.

$$R = \{(x, y) \,|\, 0 < x < a,\ 0 < y < b\} \tag{9.6.2}$$

If we were instead interested in solving the system over a circular region such as the face of a drum, it would make more sense to reformulate (9.6.1) in polar coordinates using the expression for the Laplacian in (9.2.20). To obtain a unique solution, we must specify both initial conditions and boundary conditions. If the membrane is fixed along the boundary of R, then the boundary conditions are

$$u(t, 0, y) = 0 \tag{9.6.3a}$$

$$u(t, a, y) = 0 \tag{9.6.3b}$$

$$u(t, x, 0) = 0 \tag{9.6.3c}$$

$$u(t, x, b) = 0 \tag{9.6.3d}$$

Two initial conditions must also be specified. They take the form of constraints on u and u_t at $t = 0$.

$$u(0, x, y) = f(x, y) \quad , \quad (x, y) \in R \tag{9.6.4a}$$

$$u_t(0, x, y) = g(x, y) \quad , \quad (x, y) \in R \tag{9.6.4b}$$

Note that in order for $u(t, x, y)$ to satisfy the boundary conditions, it is necessary that $f(0, y) = f(a, y) = 0$ and $f(x, 0) = f(x, b) = 0$.

The explicit central difference method for solving the one-dimensional wave equation can be generalized to two dimensions. Suppose we are interested in solving Equation (9.6.1) over the time interval $0 \le t \le T$. Numerical estimates of $u(t, x, y)$ will be computed over a uniform grid consisting of $m + 1$ values of t, $n + 2$ values of x, and $p + 2$ values of y. Let $\Delta t, \Delta x$, and Δy denote the step sizes of the variables t, x, and y, respectively.

$$\Delta t \triangleq \frac{T}{m} \tag{9.6.5a}$$

$$\Delta x \triangleq \frac{a}{n + 1} \tag{9.6.5b}$$

$$\Delta y \triangleq \frac{b}{p + 1} \tag{9.6.5c}$$

To simplify the final equations, let t_k, x_i, and y_j denote the discrete values of t, x, and y on the grid.

$$t_k = k\Delta t \quad , \quad 0 \le k \le m \tag{9.6.6a}$$

$$x_i = i\Delta x \quad , \quad 0 \le j \le (n + 1) \tag{9.6.6b}$$

$$y_j = j\Delta y \quad , \quad 0 \le j \le (p + 1) \tag{9.6.6c}$$

Similarly, let u_{ij}^k denote the computed value of $u(t_k, x_i, y_j)$ where the superscript specifies the time step and the subscripts specify the space steps. To convert the wave equation in (9.6.1) to a difference equation, three-point central difference approximations are used for the three derivative terms. This leads to the following system of difference equations for the interior grid points, $0 \le k \le m, 1 \le i \le n$, and $1 \le j \le p$.

$$\frac{u_{ij}^{k+1} - 2u_{ij}^k + u_{ij}^{k-1}}{\Delta t^2} = \beta\left(\frac{u_{i+1,j}^k - 2u_{ij}^k + u_{i-1,j}^k}{\Delta x^2} + \frac{u_{i,j+1}^k - 2u_{ij}^k + u_{i,j-1}^k}{\Delta y^2}\right) \tag{9.6.7}$$

The formulation in Equation (9.6.7) is the *two-dimensional central difference method*. Given that three-point differences are used for the time derivative and the two space derivatives, this is a second-order method with a truncation error of order

$O(\Delta t^2 + \Delta x^2 + \Delta y^2)$. It is also an explicit method because there is only one term that involves the solution at time t_{k+1}. Solving for this term yields

$$u_{ij}^{k+1} = \gamma_x(u_{i-1,j}^k + u_{i+1,j}^k) + 2(1 - \gamma_x - \gamma_y)u_{ij}^k + \gamma_y(u_{i,j-1}^k + u_{i,j+1}^k) - u_{ij}^{k-1} \tag{9.6.8}$$

In this case, there are two *gain* parameters $\gamma_x > 0$ and $\gamma_y > 0$, defined as follows:

$$\gamma_x \triangleq \frac{\beta \Delta t^2}{\Delta x^2} \tag{9.6.9a}$$

$$\gamma_y \triangleq \frac{\beta \Delta t^2}{\Delta y^2} \tag{9.6.9b}$$

The gains γ_x and γ_y are different from the gains used in the two-dimensional heat equation due to the presence of a second factor Δt. To evaluate (9.6.8) at $i = 1$, $i = n$, $j = 1$, and $j = p$, the boundary values in (9.6.3) must be used. Observe that computing u^{k+1} requires values of u at *two* previous time steps. Consequently, special consideration must be given to the initial case, $k = 0$, because the last term in (9.6.8) becomes u_{ij}^{-1}. An estimate of the solution at time $t_{-1} = -\Delta t$ can be developed by using the initial condition constraint in (9.6.4b). Replacing the first derivative term with a three-point central difference evaluated at t_0 results in the following difference equation for $1 \le i \le n$ and $1 \le j \le p$.

$$\frac{u_{ij}^1 - u_{ij}^{-1}}{2\Delta t} = g(x_i, y_j) \tag{9.6.10}$$

This equation can be solved for the fictitious point u_{ij}^{-1}. The expression for u_j^{-1} can then be substituted into (9.6.8) for the case when $k = 0$. Using (9.6.4a) then yields the following formulation for starting the computation.

$$u_{ij}^1 = \frac{1}{2}[\gamma_x(f_{i-1,j} + f_{i+1,j}) + 2(1 - \gamma_x - \gamma_y)f_{i,j} + \gamma_y(f_{i,j-1} + f_{i,j+1})] + \Delta t g_{ij} \tag{9.6.11}$$

Here we have used the discrete notation $f_{ij} = f(x_i, y_j)$ and $g_{ij} = g(x_i, y_j)$ for the boundary condition functions in (9.6.4). To summarize, we use (9.6.4a) to compute u^0, (9.6.11) to compute u^1, and then (9.6.8) to compute u^{k+1} for $1 \le k < m$.

The two-dimensional central difference method in Equation (9.6.8) is an explicit method that is stable only for certain values of the gains γ_x and γ_y. It can be shown that the stability criterion for the one-dimensional case in (9.5.31) generalizes to $\gamma_x\gamma_y/(\gamma_x + \gamma_y) \le 1/4$. Consequently, using (9.6.9), we find that stability requires that

$$\Delta t^2 \le \frac{\Delta x^2 + \Delta y^2}{4\beta} \tag{9.6.12}$$

Observe that when the grid elements are square ($\Delta y = \Delta x$), this reduces to $\lambda_x = \lambda_y \le 1/2$. This is more stringent than the one-dimensional case by a factor of two. As the space steps Δx and Δy get smaller, so must the time step Δt in accordance with (9.6.12) in order to maintain stability.

EXAMPLE 9.6.1 Vibrating Membrane

As an illustration of the use of the two-dimensional central difference method to solve a hyperbolic partial differential equation, consider the problem of determining the displacement of a thin elastic membrane stretched over a rectangular opening of width $a = 2$ and length $b = 2$. Suppose the membrane tension and density are such that the wave equation coefficient is $\beta = 1/4$. The membrane displacement $u(t, x, y)$ at time t and coordinates (x, y) is a solution of the following wave equation.

$$\frac{\partial^2 u}{\partial t^2} = \frac{1}{4}\left(\frac{\partial^2 u}{\partial x^2} + \frac{\partial^2 u}{\partial y^2}\right)$$

Suppose the item interval is $T = 2.0$, and the size of the grid is $m = 40$ and $n = p = 39$. In this case $\Delta t = 1/20, \Delta x = 1/20$, and $\Delta y = 1/20$. Hence $(\Delta x^2 + \Delta y^2)/4\beta = 1/200$. Since $\Delta t^2 = 1/400$, it follows that the stability criterion in Equation (9.6.12) is satisfied by this time step. Since the membrane is fixed along the rectangular boundary, the boundary conditions are

$$u(t,0,y) = 0$$
$$u(t,2,y) = 0$$
$$u(t,x,0) = 0$$
$$u(t,x,2) = 0$$

To start the membrane vibrating, suppose the following initial conditions are applied for $0 < x < 2$ and $0 < y < 2$.

$$u(0,x,y) = 0.1 \sin(\pi x) \sin(\pi y/2)$$
$$u_t(0,x,y) = 0$$

Example program *e961* on the distribution CD applies the two-dimensional central difference method to solve this problem. Plots of the resulting solution showing the membrane dis-

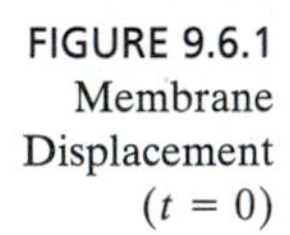
FIGURE 9.6.1 Membrane Displacement $(t = 0)$

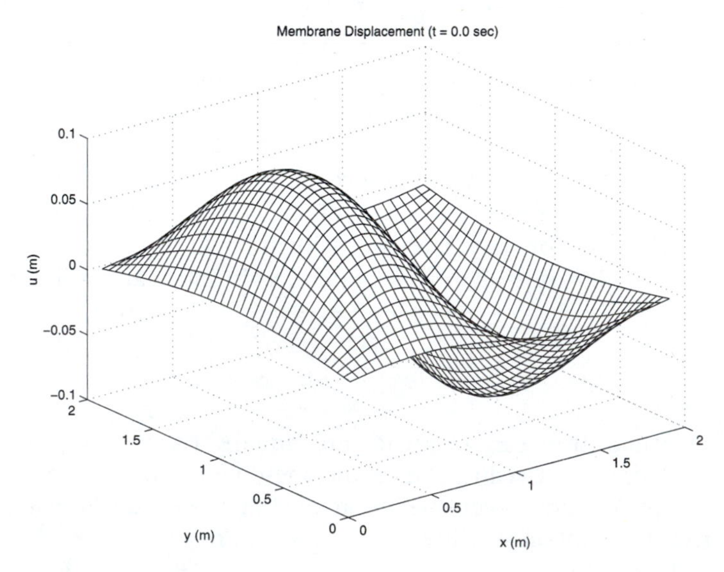

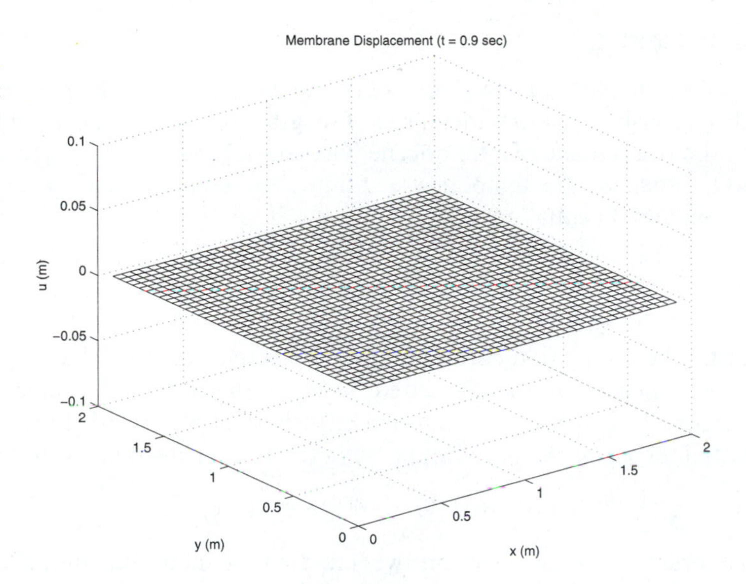

FIGURE 9.6.2 Membrane Displacement ($t = 0.9$)

placements at $t = 0, t = 0.9$, and $t = 1.8$ are shown in Figure 9.6.1 through Figure 9.6.3, respectively. The membrane is essentially flat at the snap shot, $t = 0.9$, in Figure 9.6.2, but it is clearly vibrating, as can be seen in the other two figures.

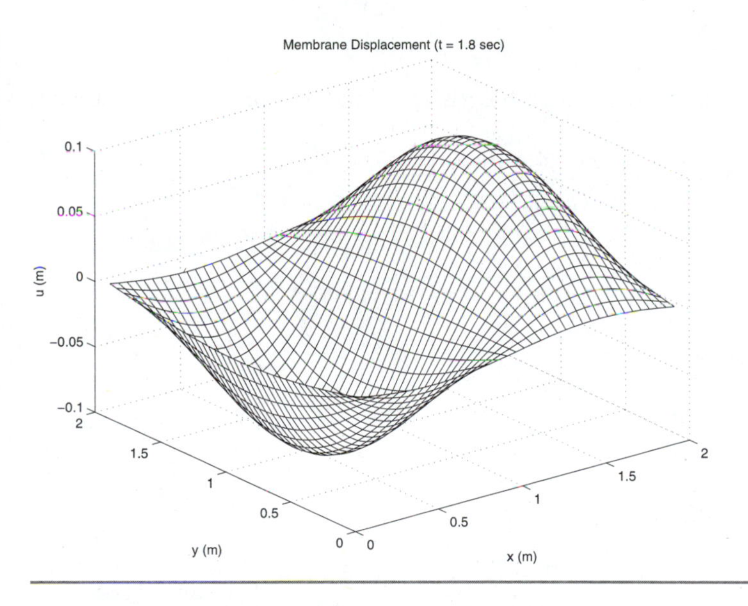

FIGURE 9.6.3 Membrane Displacement ($t = 1.8$)

9.7 APPLICATIONS

The following examples illustrate applications of partial differential equation techniques using both MATLAB and C. The relevant MATLAB functions in the NLIB toolbox are described in Section 1.10 of Appendix 1, while the corresponding C functions in the NLIB library are described in Section 2.13 of Appendix 2.

9.7.1 Heated Rod: C

Thermodynamics, the flow of heat, is one of the areas of focus for chemical engineers. As an illustration, consider the thin metal rod of length a previously shown in Figure 9.1.2. We assume the rod is made of a homogeneous material and is enclosed in an insulating jacket. If $u(t,x)$ denotes the temperature at time t and point x, then $u(t,x)$ is a solution to the heat conduction equation:

$$\frac{\partial u}{\partial t} = \beta \frac{\partial^2 u}{\partial x^2} \tag{9.7.1}$$

Recall that the heat equation coefficient β is a positive constant that depends on the material from which the rod is fabricated. To make the problem specific, suppose the rod is of length $a = 2$ m and has a thermal diffusivity of $\beta = 0.001$. Furthermore, suppose the initial temperature distribution along the rod is sinusoidal as follows.

$$u(0,x) = 30 + 10[1 - \cos(\pi x)]°\text{C} \tag{9.7.2}$$

To generate a unique solution, we must supplement the initial condition in Equation (9.7.2) with a pair of boundary conditions, one at each end. Suppose the two ends are insulated, which means that there is no heat loss—the temperature gradient is $\partial u/\partial x = 0$. This corresponds to the pair of Neumann boundary conditions:

$$u_x(t,0) = 0 \tag{9.7.3a}$$

$$u_x(t,a) = 0 \tag{9.7.3b}$$

The following C program on the distribution CD can be used to solve this problem.

```
/*------------------------------------------------------------------*/
/* Example 9.7.1: Heated Rod                                        */
/*------------------------------------------------------------------*/

#include "c:\nlib\util.h"

float f (float x)
{
   return (float) (30 + 10*(1 - cos(PI*x)));
}

int main (void)
{
  int           m = 49,                 /* t precision */
                n = 40;                 /* x precision */
  float         T = 120.0f,             /* maximum t */
                a = 2.0f,               /* maximum x */
                beta = 0.005f,          /* thermal diffusivity */
                c1 = 1.0f,              /* Neumann constraint */
                c2 = 1.0f,              /* Neumann constraint */
                dt,dx;
```

```
   vector        t = vec (m+1,""),          /* t grid values */
                 x = vec (n,"");            /* x grid values */
   matrix        U = mat (m+1,n,"");        /* solution */

/* Initialize */

   printf ("\nExample 9.7.1: Heated Rod\n");
   dt = T/m;
   dx = a/(n+1);
   shownum ("gamma",beta*dt/(dx*dx));
   heat1 (T,a,m,n,beta,c1,c2,t,x,U,f,NULL);
   shownum ("Steady-state temperature (deg C)",U[m+1][n]);
   plotxyz (t,x,U,m+1,n,"","t (sec)","x (m)","u (^oC)","fig971");
   return 0;
}
/*-----------------------------------------------------------------*/
```

When program *e971.c* is executed, it uses the Crank-Nicolson method to solve the one-dimensional heat conduction equation. The solution is computed at 50 values of t uniformly distributed over 120 sec. and 40 values of x uniformly distributed over the length of the rod. The resulting temperature profile is plotted as a surface and is shown in Figure 9.7.1. Here, time increases from left to right while the front corresponds to one end of the rod and the back corresponds to the other end. Notice that as time increases, the initial sinusoidal temperature distribution along the rod evens out. Because no heat is gained or lost, the temperature will eventually approach a constant steady-state value of $u = 40°$ C, which corresponds to the average of $u(0, x)$ in Equation (9.7.2).

A MATLAB solution to Example 9.7.1 can be found in the file *e971.m* on the distribution CD.

FIGURE 9.7.1 Heat Flow Along a Thin Rod Insulated at Both Ends

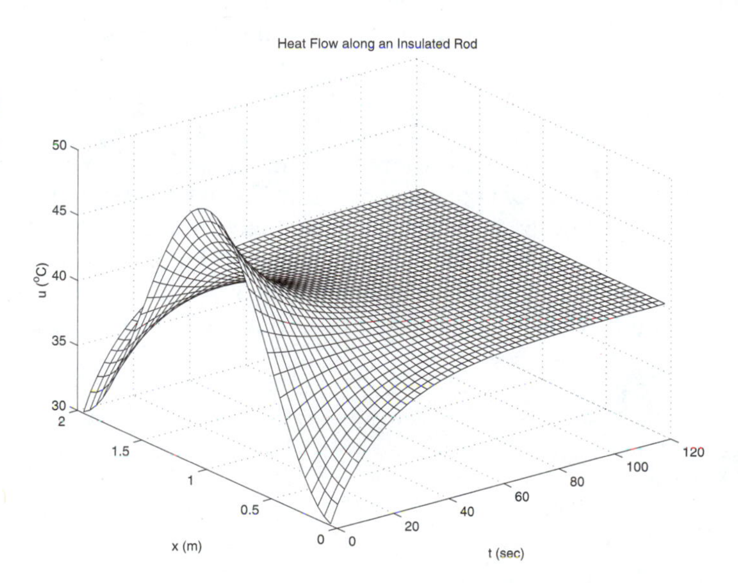

9.7.2 Plate Deflection: MATLAB

Civil engineers are sometimes asked to determine the deflection of mechanical structures under various loads. As an illustration, consider the problem of computing the deflection of a rectangular plate that is subject to a distributed area load f. Suppose the plate is simply supported along the edges, which means that both the deflection and the slope of the plate are zero along the edges. If $u(x, y)$ denotes the deflection of the plate at coordinates (x, y), the plate can be described by the following fourth-order partial differential equation (Carnahan et al., 1969).

$$\frac{\partial^4 u}{\partial x^4} + 2\frac{\partial^4 u}{\partial x^2\,\partial y^2} + \frac{\partial^4 u}{\partial y^4} = \frac{f}{\alpha} \tag{9.7.4}$$

Here the parameter α, called the flexural rigidity of the plate, depends on such things as the plate thickness and the modulus of elasticity of the material.

The plate deflection equation in (9.7.4) is roughly similar to Poisson's equation, but it differs in that it is of higher order and it includes a mixed partial derivative term. One way to approach its solution would be to develop finite difference formulas suitable for this class of equations. However, because of the special perfect-square structure of (9.7.4), an alternative approach is available. In particular, (9.7.4) can be reformulated as a sequence of two Poisson problems (Chapra and Canale, 1988). Let $v(x, y)$ denote a new intermediate variable defined as

$$v \triangleq \nabla^2 u \tag{9.7.5}$$

where ∇^2 is the Laplacian operator in Equation (9.1.2). Applying the Laplacian operator to both sides of (9.7.5), we find that (9.7.4) can be recast in terms of v as

$$\frac{\partial^2 v}{\partial x^2} + \frac{\partial^2 v}{\partial y^2} = \frac{f}{\alpha} \tag{9.7.6}$$

We find $v(x, y)$ by solving (9.7.6) subject to the appropriate boundary conditions. Then, using (9.7.5), the deflection $u(x, y)$ can be found by solving

$$\frac{\partial^2 u}{\partial x^2} + \frac{\partial^2 u}{\partial y^2} = v \tag{9.7.7}$$

In this way, the original fourth-order problem in (9.7.4) is transformed into a sequence of two second-order Poisson problems. To make the problem specific, suppose the plate is $a = 3$m by $b = 3$m, the area load on the plate is $f = 2000$ N/m^2 and the flexural rigidity is $\beta = 1.5 \times 10^4$ m$/N$. The boundary conditions for a plate that is simply supported along the edges are $u(x, y) = v(x, y) = 0$ along the boundary. The following MATLAB script on the distribution CD can be used to solve this problem.

```
%-----------------------------------------------------------------
% Example 9.7.2: Plate Deflection
%-----------------------------------------------------------------

% Initialize
```

```
   clc                       % clear screen
   clear                     % clear variables
   global xg yg V            % used in funq972.m
   m = 15;                   % number of x grid points
   n = 15;                   % number or y grid points
   q   = 500;                % maximum number of iterations
   a   = 3;                  % plate width (m)
   b   = 3;                  % plate length (m)
   tol = 1.e-4;              % error tolerance

% Solve first Poisson system

   fprintf ('Example 9.7.2: Plate Deflection\n');
   [alpha,r,xg,yg,V] = poisson (a,b,m,n,q,tol,'funp972','');
   show ('Number of iterations',r)
   show ('Relaxation parameter alpha',alpha)

% Solve second Poisson system

   [alpha,r,xg,yg,U] = poisson (a,b,m,n,q,tol,'funq972','');
   show ('Number of iterations',r)
   show ('Maximum deflection (m)',max(max(U)))
   plotxyz (xg,yg,U,'Plate Deflection','x (m)','y (m)','u (m)')


function z = funp972 (x,y)
%--------------------------------------------------------------------
% Description: Plate deflection
%--------------------------------------------------------------------
   f = 2000;                % area load (N/m^2)
   beta = 1.5e4;            % flexural rigidity
   z = f/beta;

function z = funq972 (x,y)
%--------------------------------------------------------------------
% Description: Plate deflection
%--------------------------------------------------------------------
   global xg yg V
   z = bilin (xg,yg,V,x,y);
%--------------------------------------------------------------------
```

When script *e972.m* is executed, it solves this problem on a uniformly distributed 20-by-20 grid of interior points. From (9.2.18), the optimal relaxation parameter for the *SOR* method is

$$\alpha = 1.729 \tag{9.7.8}$$

Using an error tolerance of $\varepsilon = 10^{-5}$, the PDE in (9.7.6) required 43 iterations to converge, while the PDE in (9.7.7) required 39 iterations. The resulting deflection of the plate is shown in Figure 9.7.2. In this case, the maximum deflection occurs at the center of the plate and is

$$u_{max} = 4.36 \text{ cm}$$

A solution to Example 9.7.2 using C can be found in the file *e972.c* on the distribution CD.

9.7.3 Electrostatic Field: C

In electronic devices, electric fields form due to the presence of electrical charge. Consider the case when there are two point charges of equal strength, but opposite sign, located near one another. This is referred to as a charge dipole. The electrostatic potential of the field surrounding the dipole can be modeled by the Poisson equation

$$\frac{\partial^2 u}{\partial x^2} + \frac{\partial^2 u}{\partial y^2} = -\frac{\rho}{\varepsilon_p} \tag{9.7.10}$$

Here $u(x, y)$ is the electrostatic potential, in volts, at coordinates (x, y). The parameter ε_p is the permitivity of the material, and $\rho(x, y)$ is the charge density. Suppose the electrostatic potential is to be determined over a rectangular region of width $a = 2$ cm and length $b = 2$ cm. Let the charge density be zero everywhere except near the point $(x_1, y_1) = (0.8, 0.8)$,where there is a positive charge, and the point $(x_2, y_2) = (1.2, 1.2)$, where there is an equal negative charge.

$$\rho(0.8, 0.8) = -\rho(1.2, 1.2) = 10^{-8} \text{ coul/m}^2 \tag{9.7.11}$$

To obtain a unique solution to Poisson's equation, we also need boundary conditions. Suppose the voltage along the boundary of the rectangular region is zero. Finally, suppose the permitivity is $\varepsilon_p = 8.9 \times 10^{-12} \text{ coul}^2/N\text{-}m^2$, which corresponds to air. The following C program on the distribution CD can be used to solve this problem.

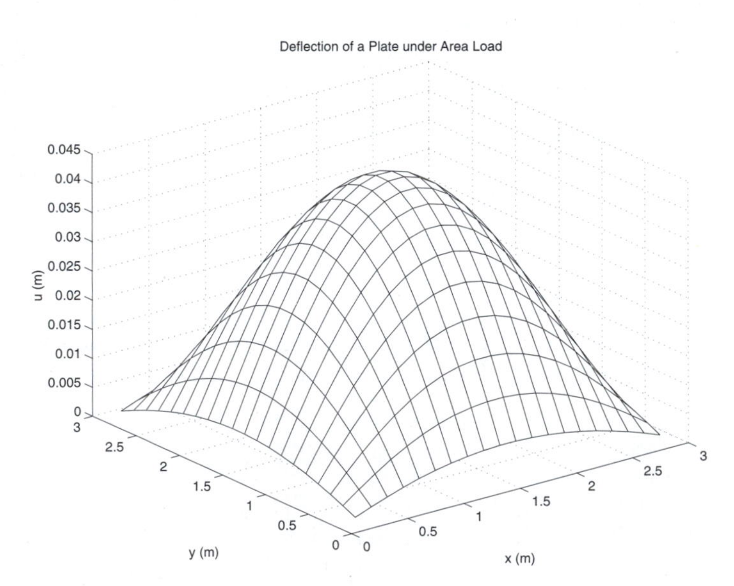

FIGURE 9.7.2 Deflection of a Simply Supported Rectangular Plate under an Area Load

```
/*----------------------------------------------------------------------*/
/* Example 9.7.3: Electrostatic Field                                   */
/*----------------------------------------------------------------------*/

#include "c:\nlib\util.h"

float            a = 1.0f,                  /* width (m) */
                 b = 1.0f;                  /* length (m) */

float f (float x,float y)
{
   float         ep = 8.9e-12f,             /* permitivity of air */
                 rho = 1.e-8f,              /* charge */
                 z = 0.03f;

   if      ((fabs(x-0.4*a) < z) && (fabs(y - 0.4*b) < z))
      return -rho/ep;
   else if ((fabs(x-0.6*a) < z) && (fabs(y - 0.6*b) < z))
      return rho/ep;
   else
      return 0;
}

int main (void)
{
   int           m = 30,                    /* number of x grid points */
                 n = 30,                    /* number or y grid points */
                 q = 500,                   /* maximum iterations */
                 r,                         /* number of iterations */
                 i;
   float         eps = 1.e-5f,              /* error tolerance */
                 alpha;                     /* relaxation parameter */
   vector        x = vec (m,""),            /* x axis grid points (m) */
                 y = vec (n,"");            /* y axis grid points (m) */
   matrix        U = mat (m,n,"");          /* potential (V) */

/* Solve Poisson equation */

   printf ("\nExample 9.7.3: Electrostatic Field\n");
   plotfun (0,a,0,b,"Charge Density","x","y","-f",f,"figf");
   r = poisson (a,b,m,n,q,eps,&alpha,x,y,U,f,NULL);
   shownum ("Relaxation parameter alpha",alpha);
   shownum ("Number of iterations",(float) r);
   plotxyz (x,y,U,m,n,"","x (cm)","y (cm)","u (V)","fig973");
   return 0;
}
/*----------------------------------------------------------------------*/
```

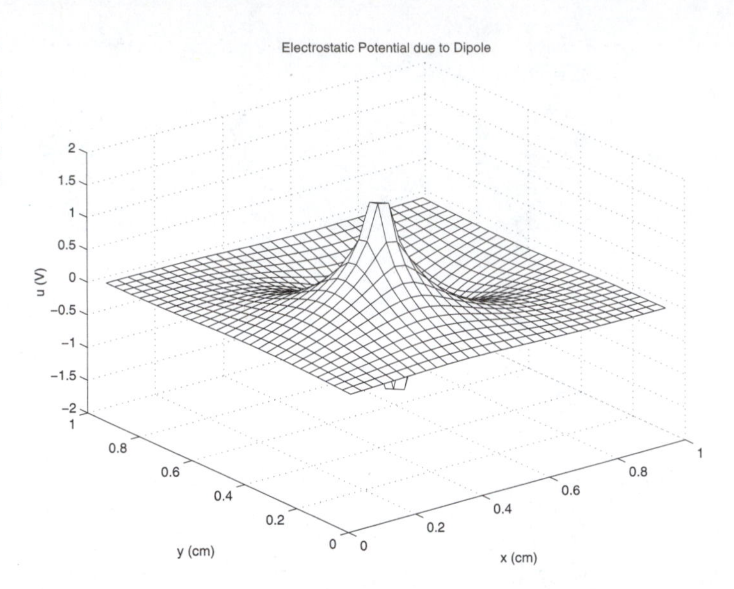

FIGURE 9.7.3 Electrostatic Potential Due to Charge Dipole

When program *e973.c* is executed, it uses the *SOR* method to solve this system over a 30 by 30 grid of interior points. The following optimal value for the relaxation parameter is obtained from (9.2.18).

$$\alpha = 1.811 \tag{9.7.12}$$

Using an error tolerance of $\varepsilon = 10^{-5}$, the algorithm converges in 71 iterations. The resulting electrostatic field is shown in Figure 9.7.3, where the presence of the two charges is apparent.

A solution to Example 9.7.3 using MATLAB can be found in the file *e973.m* on the distribution CD.

9.7.4 Twisted Bar: MATLAB

Consider the problem of analyzing the torsion in a rectangular prismatic bar that is subject to twisting. For example, the bar might be part of a support for a structure that is subject to disturbances from the wind. Suppose the bar has a constant rectangular cross section that is of width a and depth b. The torsion function ϕ for the bar is a solution to the following partial differential equation:

$$\frac{\partial^2 \phi}{\partial x^2} + \frac{\partial^2 \phi}{\partial y^2} = -2 \tag{9.7.13}$$

The torsional function $\phi(x, y)$ is found by solving the Poisson equation in (9.7.13) subject to the boundary condition that $\phi(x, y) = 0$. The tangential stresses within the bar are then proportional to the partial derivatives of $\phi(x, y)$. To make the problem specific, suppose the cross section of the rectangular bar is $a = 3$ cm wide by $b = 5$ cm deep. The following MATLAB script on the distribution CD can be used to solve this problem.

```
%-----------------------------------------------------------------------
% Example 9.7.4: Twisted Bar
%-----------------------------------------------------------------------

% Initialize

   clc                        % clear screen
   clear                      % clear variables
   a = 3;                     % width (cm)
   b = 5;                     % length (cm)
   m = 25;                    % number of x grid points
   n = 25;                    % number or y grid points
   q = 100;                   % maximum number of iterations
   tol = 5.e-5;               % error tolerance

% Right-hand side function

   f = inline ('-2','x','y');

% Solve Poisson equation

   fprintf ('Example 9.7.4: Twisted Bar\n');
   [alpha,r,x,y,U] = poisson (a,b,m,n,q,tol,f,'');
   show ('Relaxation parameter alpha',alpha)
   show ('Number of iterations',r)
   plotxyz (x,y,U,'Twisted Bar','x (cm)','y (cm)','u (cm^2)')
%-----------------------------------------------------------------------
```

When script *e974.m* is executed, it computes the torsion function over a 25-by-25 grid of interior points. The following optimal value for the relaxation parameter (9.2.18) is used.

$$\alpha = 1.777 \tag{9.7.14}$$

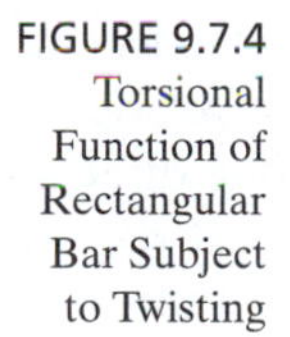

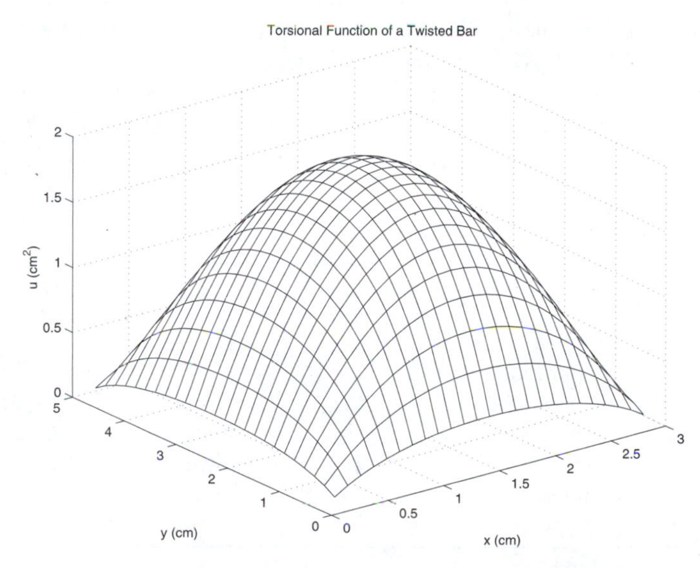

FIGURE 9.7.4 Torsional Function of Rectangular Bar Subject to Twisting

Using an error tolerance of $\varepsilon = 10^{-5}$, the *SOR* algorithm converges in 68 iterations. The resulting torsion function is shown in Figure 9.7.4. Observe that the gradient of the surface is zero at the center and largest near the boundary. Consequently, the tangential stresses will be smaller in the middle of the bar and larger at the edges.

A solution to Example 9.7.4 using C can be found in the file *e974.c* on the distribution CD.

9.8 SUMMARY

Numerical methods for solving partial differential equations using finite difference methods are summarized in Table 9.8.1. The dimension column in Table 9.8.1 refers to the number of spatial dimensions. All of the solution methods are applicable to a rectangular region R in the space of independent variables.

There are three solution techniques applicable to elliptic partial differential equations, including Poisson's equation and Laplace's equation, all based on central difference equations. The Gauss-Seidel method is faster than Jacobi's method, while the *SOR* method is still faster and includes the Gauss-Seidel method as a special case. Consequently, the *SOR* method is the method of choice. An optimal value for the relaxation parameter α_{opt} is obtained using Equation (9.2.18).

Four techniques are applicable to parabolic partial differential equations, which include the heat conduction equation or diffusion equation. The forward Euler method is an explicit method that is fast, but it is stable only when the gain parameter in (9.3.9) satisfies $\gamma \leq 0.5$. Therefore, the time step must be bounded as follows.

$$\Delta t \leq \frac{\Delta x^2}{2\beta} \tag{9.8.1}$$

The forward Euler method has a truncation error that is first-order accurate in time and second-order in space. The backward Euler and Crank-Nicolson methods are stable implicit methods that require the solution of tridiagonal linear algebraic systems at each time step. These tridiagonal systems can be solved very efficiently using *LU* decomposition. The backward Euler method is first-order accurate in time and second-order accurate in space, while the Crank-Nicolson method is second-order accurate in both variables. Consequently, the Crank-Nicolson method is preferred. The stable alternating direction implicit (ADI) method is applicable to two-dimensional parabolic equations. It requires solving several tridiagonal systems at each time step, and it is second-order accurate.

TABLE 9.8.1 Partial Differential Equation Techniques

Method	*Equation*	*Dimension*	*Stable*	*Truncation Error*
central-difference	elliptic	2	always	$O(\Delta x^2 + \Delta y^2)$
forward Euler	parabolic	1	$\gamma \leq 0.5$	$O(\Delta t + \Delta x^2)$
backward Euler	parabolic	1	always	$O(\Delta t + \Delta x^2)$
Crank-Nicolson	parabolic	1	always	$O(\Delta t^2 + \Delta x^2)$
ADI	parabolic	2	always	$O(\Delta t^2 + \Delta x^2 + \Delta y^2)$
central difference	hyperbolic	1	$\gamma \leq 1$	$O(\Delta t^2 + \Delta x^2)$
central difference	hyperbolic	2	$\gamma \leq 0.5$	$O(\Delta t^2 + \Delta x^2 + \Delta y^2)$

Two versions of the explicit central difference method are applicable to hyperbolic partial differential equations that include the wave equation. The method for one-dimensional systems is a fast second-order method that is stable when the gain parameter in (9.5.24) satisfies $\gamma \leq 1$. This requires that the time step be bounded as follows.

$$\Delta t^2 \leq \frac{\Delta x^2}{\beta} \tag{9.8.2}$$

The method for two-dimensional systems is also a second-order method. When the aspect ratio of the grid elements is unity ($\Delta y = \Delta x$), this method is stable for gain parameters satisfying $\gamma \leq 0.5$. More generally, stability requires that the time step be bounded in the following manner.

$$\Delta t^2 \leq \frac{\Delta x^2 + \Delta y^2}{4\beta} \tag{9.8.3}$$

PROBLEMS

The problems are divided into Analysis problems, which can be solved by hand, and Computation problems, which require the use of MATLAB or C. Solutions to selected problems can be found in Appendix 4. Students are encouraged to use these problems, which are identified with an (S), as a check on their understanding of the material. Problems marked with a (P) are programming problems that require the student to implement one or more of the algorithms discussed in the chapter. The remaining Computation problems require the student to write a main program that uses one or more of the NLIB functions discussed in Appendix 1 or Appendix 2.

Analysis

9.1 Show by direct substitution that the following are all solutions to Laplace's equation. Also, show that if u_1 and u_2 are solutions, then so is $\alpha u_1 + \beta u_2$ for arbitrary constants α and β

$$\frac{\partial^2 u}{\partial x^2} + \frac{\partial^2 u}{\partial y^2} = 0$$

(a) $u(x, y) = x^2 - y^2$

(b) $u(x, y) = \exp(x)\cos(y)$

(c) $u(x, y) = \exp(y)\sin(x)$

(d) $u(x, y) = \ln(x^2 + y^2)$

9.2 Consider the problem of solving Poisson's equation over a region R that has a curved boundary. At the boundary of R, new grid points can be generated by taking the intersection of the boundary with the uniformly-spaced grid lines as shown in Figure 9.9.1. This is equivalent to having a *variable* grid size with the horizontal grid size being $\alpha_1 \Delta x$ to the west, $\alpha_2 \Delta x$ to the east, $\beta_1 \Delta y$ to the south, and $\beta_2 \Delta y$ to the north.

(a) Find approximations to $u_x = \partial u/\partial x$ at $(x, y) = (x_k - \alpha_1 \Delta x/2, y_j)$ and $(x, y) = (x_k + \alpha_2 \Delta x/2, y_j)$ using two-point forward differences.

(b) Use the results of part (a) to find a three-point central difference approximation of $u_{xx} = \partial^2 u/\partial x^2$. Simplify your final answer as much as possible. What do you get when $\alpha_1 = \alpha_2 = 1$?

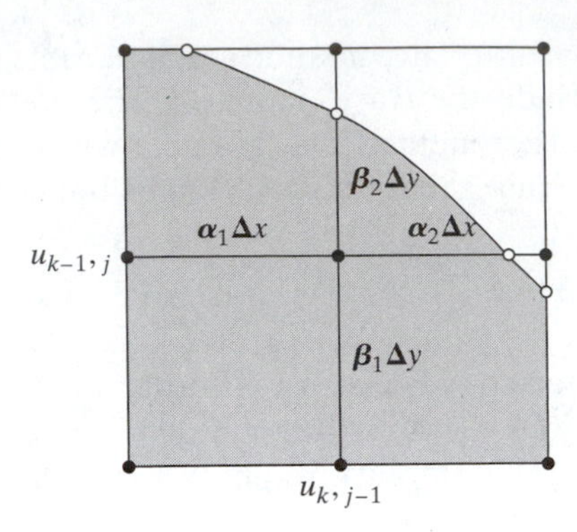

FIGURE 9.9.1 Non-Uniform Grid

(c) Find approximations to $u_y = \partial u/\partial y$ at $(x, y) = (x_k, y_j - \beta_1\Delta y/2)$ and $(x, y) = (x_k, y_j + \beta_2\Delta y/2)$ using two-point forward differences.

(d) Use the results of part (c) to find a three-point central difference approximation of $u_{yy} = \partial^2 u/\partial y^2$. Simplify your final answer as much as possible. What do you get when $\beta_1 = \beta_2 = 1$?

(e) Use the results of parts (b) and (d) to write a variable grid difference equation for Poisson's equation.

(S)9.3 Fisher developed the following model for the spread of genes in a population where p denotes a probability of occurrence a specific allele and α and β are positive constants (Edelstein-Keshet, 1988).

$$\frac{\partial p}{\partial t} = \beta\left(\frac{\partial^2 p}{\partial x^2}\right) + \alpha p(1 - p)$$

Let p_j^k denote $p(t_k, x_j)$ with $t_k = k\Delta t$ and $x_j = j\Delta x$ where $\Delta t = T/m$ and $\Delta x = a/(n + 1)$. Develop an explicit difference equation approximation for this *nonlinear* partial differential equation using a forward difference in time and a central difference in space.

9.4 Consider the following one-dimensional wave equation.

$$\frac{\partial^2 u}{\partial t^2} = \beta\frac{\partial^2 u}{\partial x^2}$$

Suppose we try a *separation of variables* solution of the form

$$u(t, x) = v(t)w(x)$$

Suppose v and w are solutions to the following second-order *ordinary* differential equations where c is an arbitrary constant. Show that $u(t, x)$ is a solution to the one-dimensional wave equation.

$$\frac{d^2 v(t)}{dt^2} = c\beta v(t)$$

$$\frac{d^2 w(x)}{dx^2} = cw(x)$$

9.5 There is an alternative first-order version of the wave equation that is often used to model the transport of matter.

$$\frac{\partial u}{\partial t} + \alpha\left(\frac{\partial u}{\partial x}\right) = f$$

To simplify the discussion, suppose $\alpha > 0$ is *constant,* and the forcing term is $f = 0$. Suppose we are interested in solving this equation over the rectangular region

$$R = \{(t, x) \mid 0 \le t \le T, 0 < x < a\}$$

Let $\Delta t = T/m$ and $\Delta x = a/(n + 1)$ be the step size in t and x, respectively, and let u_j^k denote the estimate of $u(t_k, x_j)$ where $t_k = k\Delta t$ and $x_j = j\Delta x$. Suppose a two-point forward difference is used to approximate the time derivative u_t and a two-point backward difference is used to approximate the space derivative u_x.

$$u_t \approx \frac{u_j^{k+1} - u_j^k}{\Delta t}$$

$$u_x \approx \frac{u_j^k - u_{j-1}^k}{\Delta x}$$

Show that the resulting forward-time backward-space (FTBS) method generates the following explicit difference equation where γ is called the *Courant number*. Find γ.

$$u_j^{k+1} = (1 - \gamma)u_j^k + \gamma u_{j-1}^k$$

9.6 Consider the FTBS method from problem 9.5. As a solution candidate, consider

$$u_j^k = (\lambda)^k \exp\left(\frac{ij\pi}{p}\right)$$

where $i = \sqrt{-1}$ and p is any nonzero integer. Show that this is a solution by substituting it into the difference equation and finding an expression for the amplitude factor λ in terms of γ. For what values of the Courant number γ is the FTBS method stable? That is, for what values of γ is $|\lambda| \le 1$?

9.7 For the first-order wave equation in problem 9.5 with $f = 0$, suppose the time derivative, u_t, is approximated with a two-point forward difference, and the space derivative, u_x, is approximated with a three-point central difference.

$$u_t \approx \frac{u_j^{k+1} - u_j^k}{\Delta t}$$

$$u_x \approx \frac{u_{j+1}^k - u_{j-1}^k}{2\Delta x}$$

Show that the resulting forward-time central-space (FTCS) method generates an explicit difference equation of the following form. Find γ.

$$u_j^{k+1} = u_j^k - \frac{\gamma(u_{j+1}^k - u_{j-1}^k)}{2}$$

9.8 Find the amplitude factor λ for the FTCS method in problem 9.7 using a candidate solution of the form $u_j^k = (\lambda)^k \exp(ij\pi/p)$ where $i = \sqrt{-1}$ and p is any nonzero integer. For what values of γ is the FTCS method stable?

9.9 For the first-order wave equation in problem 9.5 with $f = 0$, suppose the time derivative, u_t, is approximated with a two-point backward difference, and the space derivative, u_x, is approximated with a three-point central difference. This can be achieved by evaluating the central difference of u at t_{k+1} rather than t_k.

$$u_t \approx \frac{u_j^{k+1} - u_j^k}{\Delta t}$$

$$u_x \approx \frac{u_{j+1}^{k+1} - u_{j-1}^{k+1}}{2\Delta x}$$

Show that the resulting backward-time central-space (BTCS) method generates an implicit tridiagonal difference equation of the following form. Find γ.

$$\gamma u_{j+1}^{k+1} + 2u_j^{k+1} - \gamma u_{j-1}^{k+1} = 2u_j^k$$

9.10 Find the amplitude factor λ for the BTCS method in problem 9.9 using a candidate solution of the form $u_j^k = (\lambda)^k \exp(ij\pi/p)$ where $i = \sqrt{-1}$ and p is any nonzero integer. For what values of γ is the BTCS method stable?

Computation

9.11 Consider the problem of finding the electrostatic potential, $v(x, y)$, over a plate of width $a = 3$ and length $b = 3$. If ε_p denotes the permitivity of the material and $\rho(x, y)$ denotes the charge density, then Poisson's equation can be used to model the distribution of electrostatic potential as follows.

$$\frac{\partial^2 v}{\partial x^2} + \frac{\partial^2 v}{\partial y^2} = \frac{-\rho}{\varepsilon_p} \tag{9.9.1}$$

Suppose the plate is free of charge, which means $\rho = 0$. Let the boundary conditions be the following voltages along the edges.

$$v(0, y) = 0$$

$$v(a, y) = \left(\frac{y}{b}\right)^2$$

$$v(x, 0) = 0$$

$$v(x, b) = \sqrt{\left(\frac{x}{a}\right)}$$

Write a program that uses the NLIB function *poisson* to solve this system on a *25* $\times$ *25* grid. Plot the solution as a surface. Print the number of iterations, the optimal relaxation factor, and the maximum potential

(S)9.12 Random motion of particles at the molecular level causes movement of matter from regions of high concentration to regions of low concentration through the process of *diffusion*. Suppose a substance is constrained to move in one dimension. If $c(t, x)$ denotes the concentration of the substance at time t and position x, then using *Fick's law* we can model the change in concentration due to diffusion as follows.

$$\frac{\partial c}{\partial t} = D\left(\frac{\partial^2 c}{\partial x^2}\right)$$

Here $D > 0$ is the diffusion coefficient. If the substance is oxygen and the medium is air at 20° C, then the diffusion coefficient is $D = 0.201$ (Leyton, 1975). Suppose we are interested in solving this system for $0 \le t \le 30$ and $0 < x < 5$. The initial and boundary conditions are as follows:

$$u(0, x) = (x-2.5)^2$$

$$u(t, 0) = 1$$

$$u_x(t, 5) = 0$$

Write a program that uses the NLIB function *heat1* to solve this system. Print γ and plot the solution as a surface.

9.13 Write a program that uses the NLIB function *heat2* to find the electrostatic potential of the plate in problem 9.11 using $\beta = 1$. You can assume that the initial potential is

$$v(0, x, y) = 0$$

Plot the solution surface on a *25* $\times$ *25* grid at $t = 0.1$ and $t = 1.0$, and compare it with the surface generated in problem 9.11. Print γ_x and γ_y in each case.

9.14 Consider the problem of determining the displacement of a taut elastic string of length $a = 4$. Suppose the tension and density of the string are such that the wave equation coefficient is $\beta = 4$. Thus, the string displacement is a solution of

$$\frac{\partial^2 u}{\partial t^2} = 4\left(\frac{\partial^2 u}{\partial x^2}\right)$$

The string is fixed at both ends, and the initial conditions are

$$u(0, x) = \sin\left(\frac{\pi x}{4}\right)$$

$$u_t(0, x) = 0$$

Write a program that uses the NLIB function *wave1* to find the string displacement $u(t, x)$. Print the value of γ and plot the solution as a surface over $0 < t < 4$ and $0 < x < 4$ using a *40* $\times$ *19* grid.

9.15 Write a function called *wave0* that implements d'Alembert's solution in (9.5.17) to solve the following one-dimensional wave equation:

$$\frac{\partial^2 u}{\partial t^2} = \beta\left(\frac{\partial^2 u}{\partial x^2}\right)$$

over the rectangular region

$$R = \{(t, x) \mid 0 \le x \le T, 0 < x < a\}$$

subject to the zero boundary condition $u(t, x) = 0$, and the following initial conditions for $0 < x < a$.

$$u(0, x) = f(x)$$

$$u_t(0, x) = g(x)$$

The pseudo-prototype for *wave0* is as follows:

```
[float t,vector x,matrix U] = wave0 (float T,float a,int m,
                                     int n,float beta,
                                     function f,function g)
```

On entry to *wave0:* $T > 0$ is the maximum value of t, $a > 0$ is the maximum value of $x, m \ge 1$ is the number of steps in $t, n \ge 1$ is the number of steps in x, *beta* > 0 is the wave equation coefficient, and f and g are the names of user-supplied functions that specify the initial conditions on u and its time derivative u_t, respectively. The pseudo-prototypes for f and g are

```
[float y] = f (float x);
[float z] = g (float x);
```

When f is called, it must return the initial value $u(0,x)$, and when g is called, it must return the initial value $u_t(0,x)$ where $u_t = \partial u/\partial t$. On exit from *wave0:* the $(m+1) \times 1$ vector t contains the t grid points, the $n \times 1$ vector x contains the x grid points, and the $(m+1) \times n$ matrix U contains the solution, $u(t,x)$, evaluated at the $(m+1)n$ interior grid points. That is, for $1 \le k \le (m+1)$ and $1 \le j \le n$,

$$U_{kj} = u(t_k, x_j)$$

$$t_k = \frac{(k-1)T}{m}$$

$$x_j = \frac{ja}{n+1}$$

Test *wave0* by solving the system in problem 9.14. Compare your solution to that obtained from problem 9.14 by computing the maximum of the magnitude of the difference between the two solutions over the $(m+1)n$ interior grid points.

9.16 Consider the problem of determining the displacement of a thin elastic membrane stretched over a square opening of width $a = 2$ and length $b = 2$. Suppose the tension and density of the membrane are such that the wave equation coefficient is $\beta = 0.5$. Thus, the membrane displacement is a solution of

$$\frac{\partial^2 u}{\partial t^2} = 0.5\left(\frac{\partial^2 u}{\partial x^2} + \frac{\partial^2 u}{\partial y^2}\right)$$

The membrane is fixed along the edges, and the initial conditions are

$$u(0,x,y) = 0$$

$$u_t(0,x,y) = [1 - \cos(\pi x)][(1 - \cos(\pi y)]$$

Write a program that uses the NLIB function *wave2* to find the membrane displacement $u(t,x,y)$. Plot the membrane displacement as a surface on a *19 × 19* grid at $t = 1.5, t = 2.0$ and $t = 2.5$.

9.17 Consider the problem of determining the heat flow along a thin rod of length $a = 2$. Let the heat equation coefficient for the rod be $\beta = 0.5$.

$$\frac{\partial u}{\partial t} = 0.5\left(\frac{\partial^2 u}{\partial x^2}\right)$$

Suppose the rod is insulated at each end, which means that $u_x(t,x) = 0$ at $x = 0$ and $x = a$. Let the initial temperature distribution along the rod be

$$u(0,x) = \frac{\sin(\pi x)}{1+x} \quad , \quad 0 < x < 2$$

Suppose we are interested in the temperature along the rod over the time interval $0 \le t \le 1.5$. Write a program that uses the NLIB function *heat1* to solve this problem using a grid size of $m = 50$ and $n = 39$. Plot the solution as a surface. What is the steady-state solution in this case?

CHAPTER 10

Digital Signal Processing

An analog signal is a physical variable whose value varies with time or space. For the purpose of discussion we assume that the signal $x_a(t)$ varies with time t, where $t \geq 0$. Practical signal processing techniques are applicable to discrete samples of analog signals that can be represented as

$$\boxed{x(k) \triangleq x_a(kT) \quad , \quad k = 0,1,2,\ldots}$$

where T denotes the sampling interval. When finite precision is used, the resulting sequence $x(k)$ is called a *digital signal*. Since $x(k)$ is a scalar, it also is possible to use the notation x_k for the kth sample of $x_a(t)$. However, in order to maintain compatibility with the digital signal processing (DSP) literature, the traditional notation $x(k)$ will be used throughout this chapter. Digital signal processing techniques have widespread applications, and they play an increasingly important role in modern engineering practice. Application areas include the detection of targets with radar, the filtering of audio and video signals, image processing, communications system analysis, and the identification of dynamic systems from input and output measurements. Digital signal processing is an extensive subject area with entire books devoted to the topic (see, e.g., Proakis and Manolakis, 1988).

We begin this chapter by posing a number of problems from engineering that require digital signal processing. We then introduce the Fourier transform of analog signals in order to provide a theoretical foundation for the numerical methods that follow. The first of these is a high-speed implementation of the discrete Fourier transform called the fast Fourier transform or FFT. The FFT is a fundamental tool that is useful for determining the frequency content or spectra of digital signals. It also provides an effective means for computing the cross correlation and the convolution of pairs of signals, topics which we examine next. We use cross correlation to detect the presence of weak signals buried within random noise, and we use convolution to compute the output of a linear discrete-time dynamic system. The next topic is digital filter design, including the design of linear-phase FIR filters with prescribed frequency response characteristics. We then examine a two-dimensional version of the FFT that is useful for image processing. We follow with an investigation of system identification techniques which focus on the problem of determining the parameters of linear discrete-time dynamic systems from input and output measurements. We also examine adaptive filters using the popular least-mean square (LMS) method. We conclude the chapter with a summary of digital signal processing techniques and a presentation of engineering applications. The selected examples include the frequency response of a heat exchanger, a dynamic model of flagpole motion, band pass filter design, and spectral analysis of helicopter noise.

10.1 MOTIVATION AND OBJECTIVES

With the widespread use of digital computers, digital signal processing is now commonplace throughout engineering. The following examples illustrate some typical applications.

10.1.1 Harmonic Distortion

Engineers who develop audio and video equipment strive to design amplifiers that boost signal strength without distorting the shape of the input signal. For the amplifier shown in Figure 10.1.1, suppose the input signal $u(t)$ is a pure tone of amplitude α and frequency f_0 Hz.

$$u(t) = \alpha \sin(2\pi f_0 t) \quad , \quad t \geq 0 \tag{10.1.1}$$

An ideal amplifier produces an output signal $y(t)$ that is a scaled and delayed version of the input signal. For example, if the scale factor or amplifier *gain* is K and the delay is T_d, then

$$y(t) = Ku(t - T_d) \quad , \quad t \geq T_d \tag{10.1.2}$$

In a practical amplifier, the transformation from input to output is only approximately linear, so some additional terms are present in the actual output y_a.

$$y_a(t) = \sum_{k=1}^{n-1} a_k \sin(2\pi k f_0 t + \phi_k) \quad , \quad t \geq 0 \tag{10.1.3}$$

The presence of the additional harmonics indicates that there is *distortion* in the amplified signal due to nonlinearities within the amplifier. For example, if the amplifier is driven with an input whose amplitude α is too large, then the amplifier will saturate with the result that the output is a "clipped" sine wave that sounds distorted when played through a speaker. To quantify the amount of distortion, the following measure called the *total harmonic distortion* is often used.

$$\beta \triangleq \frac{a_2^2 + \cdots + a_{n-1}^2}{a_1^2 + \cdots + a_{n-1}^2} \tag{10.1.4}$$

The average power contained in the kth harmonic is a_k^2. Consequently, the total harmonic distortion is the power in the second and higher harmonics normalized by the total signal power. For an ideal amplifier, we have $a_k = 0$ for $2 \leq k < n$ and

$$a_1 = K\alpha \tag{10.1.5}$$

$$\phi_1 = -2\pi f_0 T_d \tag{10.1.6}$$

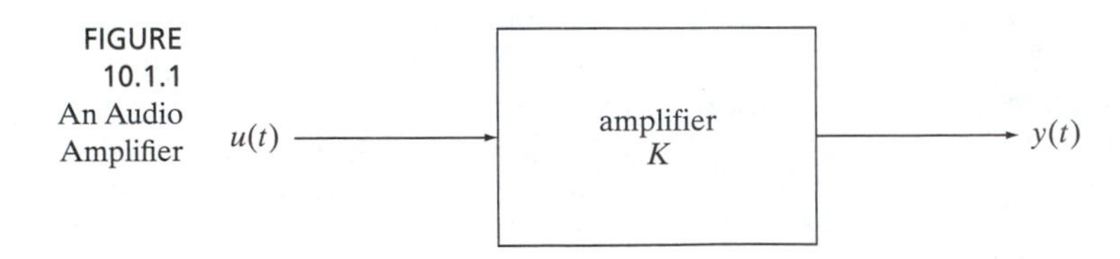

FIGURE 10.1.1 An Audio Amplifier

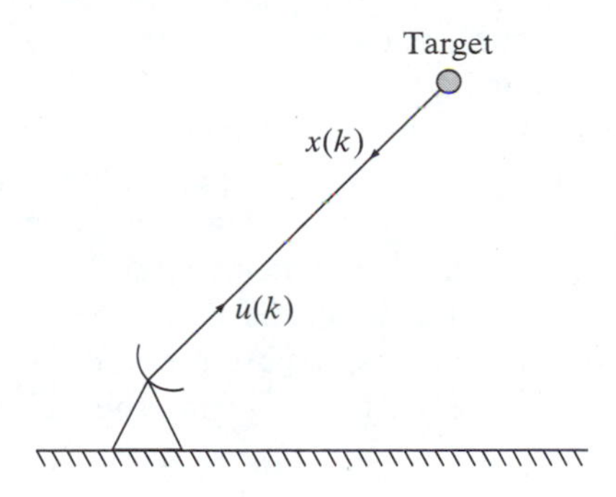

FIGURE 10.1.2 A Radar Installation

Thus, for a good design, $\beta \ll 1$, and when no distortion is present, $\beta = 0$. Suppose the amplifier output is sampled to produce the following digital signal of length $2n$.

$$y(k) = y_a(kT) \quad , \quad 0 \leq k < 2n \tag{10.1.7}$$

If the sampling interval is set to $T = 1/(2nf_0)$, by processing the digital signal $y(k)$ with the fast Fourier transform, it is possible to determine a_k and ϕ_k for $1 \leq k < n$. In this way, the total harmonic distortion can be measured.

10.1.2 Radar

Suppose a radar installation scans a region of space by transmitting an electromagnetic signal, $u(k)$, from its antenna as shown in Figure 10.1.2. If a target is in the path of the transmitted signal, it will be *illuminated* by the radar. In this case, some of the signal bounces off the target and returns to the radar station, where it is detected as an echo by the receiver. Typically, the received signal $x(k)$ contains two components as follows:

$$x(k) = \alpha u(k - d) + v(k) \tag{10.1.8}$$

The first term in $x(k)$ represents the *echo*, which consists of a delayed and scaled version of the transmitted signal. The delay, d, is a measure of the time of flight of the signal, while the scaling $0 < \alpha \ll 1$ represents an attenuation of the signal due to dispersion and energy dissipation. The second term in the received signal represents random atmospheric and electrical noise, $v(k)$, that is picked up and amplified by the receiver. The problem is to detect the presence of an echo hidden or buried within the noise of the received signal. Once an echo is identified, the *range* of the target that produced it can be estimated from the delay d and the propagation speed of the signal.

10.1.3 Chapter Objectives

When you finish this chapter, you will be able apply digital signal processing techniques to extract information contained in the samples of discrete-time signals. You will know how to compute the energy contained in a signal over a range of frequencies and how to reconstruct a band-limited continuous-time signal from its samples. You will be able to compute the fast Fourier transform (FFT) of a signal. You will know how to compute the normalized cross correlation of two signals and how to use this to detect the presence of one signal in another. You will know how to compute the discrete convolution of two signals and how to find the output of a linear system from the pulse response and the input. You will be able to design a linear-phase finite impulse response (FIR) digital filter with a prescribed magnitude response. You will know how to compute a two-dimensional FFT and how to analyze the spatial frequency content of images. Finally, you will be able to identify the parameters of a linear discrete-time dynamic system from input and output measurements using the least-square method and the adaptive LMS method. These overall goals will be achieved by mastering the following chapter objectives.

Objectives for Chapter 10

- Understand the difference between a continuous-time signal and a discrete-time signal.
- Know what the Fourier transform is and how to use it to find the magnitude spectrum of a signal.
- Understand why aliasing occurs when the sampling rate is too low.
- Know what a band-limited signal is and how to reconstruct a band-limited signal from its samples.
- Be able to compute the fast Fourier transform (FFT) of a signal.
- Know how many floating-point operations are required to compute an FFT.
- Be able to compute the cross correlation of two signals and use it to perform signal detection.
- Be able to compute the discrete convolution of two signals and use it to find the output of a linear discrete-time system.
- Know how to design a digital filter using the Fourier series and windowing.
- Be able to compute the two-dimensional FFT of a spatial signal.
- Understand the difference between MA, AR, and ARMA models and know which one is always stable.
- Be able to perform least-squares system identification from input and output data using an ARMA model.
- Be able to perform adaptive LMS system identification using a MA model.
- Understand how digital signal processing techniques can be used to solve practical engineering problems.
- Understand the relative strengths and weaknesses of each computational method and know which are most applicable for a given problem.

10.2 FOURIER TRANSFORM

In communications systems, signals carry *information* that can be extracted by appropriate processing. One of the most useful signal processing tools is the Fourier transform, which is a transformation that converts an analog signal $x(t)$ for $-\infty < t < \infty$ into a second signal $X(f)$ for $-\infty < f < \infty$ as follows.

$$X(f) \triangleq \int_{-\infty}^{\infty} x(t) \exp(-j2\pi ft)\, dt \qquad (10.2.1)$$

The new signal $X(f)$ is called the *Fourier transform* of $x(t)$. Since the exponent in (10.2.1) contains the imaginary number $j = \sqrt{-1}$, the Fourier transform $X(f)$ is complex-valued. That is, X is a function that maps the real line R into the complex plane C. The independent variable f ranges from $-\infty$ to ∞ and is called the *frequency* variable. It is easy to show that the function $X(f)$ is finite for all f if the time signal $x(t)$ is absolutely integrable:

$$\int_{-\infty}^{\infty} |x(t)|\, dt < \infty \qquad (10.2.2)$$

The Fourier transform is a one-to-one mapping of real-valued functions of time onto complex-valued functions of frequency. Therefore, given the Fourier transform of a signal, we can recover the original signal $x(t)$ from its Fourier transform $X(f)$ by applying the *inverse Fourier transform* as follows.

$$x(t) = \int_{-\infty}^{\infty} X(f) \exp(j2\pi ft)\, df \qquad (10.2.3)$$

The Fourier transform satisfies a number of symmetry properties, one of which is that if $x(t)$ is real, then $X(-f) = X^*(f)$ where X^* denotes the complex-conjugate of $X(f)$. In effect, this means that all of the essential information about $x(t)$ is contained in the non-negative frequency band, $f \geq 0$.

To illustrate the use of the Fourier transform, suppose the time signal $x(t)$ is a pure sinusoidal *tone* of the form $x(t) = \sin(2\pi f_0 t)$. It is possible to show that the Fourier transform of this signal is zero everywhere except at $f = \pm f_0$ where it contains an impulse. Consequently, one can conclude that *all* of the energy of the time signal $x(t)$ is concentrated at the frequency $f = f_0$ Hz. The function $|X(f)|^2$ is called the *energy density spectrum* of $x(t)$. It can be used to determine the amount of energy a signal contains in a given band of frequencies $[f_0, f_1]$ as follows.

$$E(f_0, f_1) = 2 \int_{f_0}^{f_1} |X(f)|^2\, df \qquad (10.2.4)$$

Letting $f_0 = 0$ and $f_1 = \infty$, it follows that the area under the energy density spectrum is the *total energy* of the signal. The factor two in Equation (10.2.4) accounts for the fact that the range of integration includes only non-negative frequencies. A signal $x(t)$ is said to be *band limited* to frequency B Hz if $X(f) = 0$ for $|f| \geq B$.

Practical signal processing methods do not operate directly on the analog signal $x(t)$. Instead, they are applied to a sequence of discrete *samples* of $x(t)$. If T denotes the *sampling interval*, the samples of $x(t)$ can be represented by the sequence

$$x(0), x(T), x(2T), \ldots \qquad (10.2.5)$$

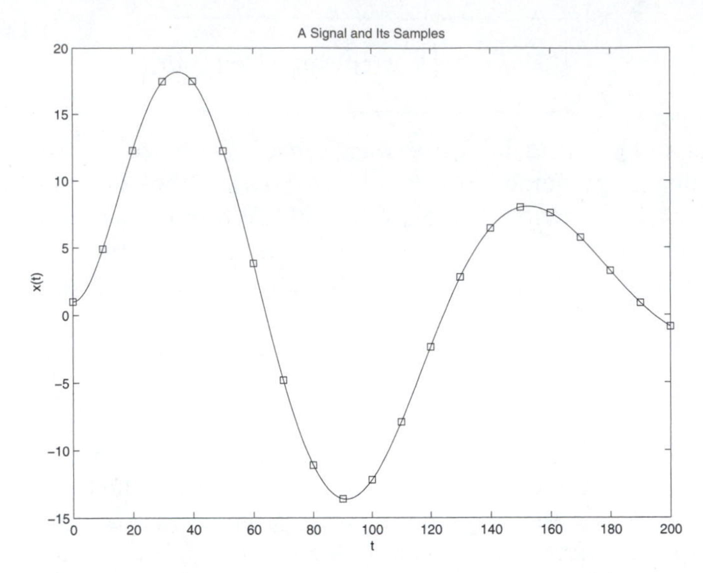

FIGURE 10.2.1 A Signal and Its Samples

A section of a typical signal and its samples is shown in Figure 10.2.1. The rate at which the samples are taken is called the *sampling frequency* and is denoted f_s.

$$\boxed{f_s \triangleq \frac{1}{T}} \tag{10.2.6}$$

If a signal $x(t)$ is band limited to B Hz, it is possible to reconstruct $x(t)$ from its samples as long as the sampling rate f_s is sufficiently high. In particular, the time signal must be sampled at a rate that is higher than twice the highest frequency present in the signal (Shannon, 1949). If $X(f) = 0$ for $f \geq B$, this means that the sampling frequency must satisfy

$$f_s \geq 2B \tag{10.2.7}$$

The frequency $f_N = f_s/2$, is called the *Nyquist rate*. If (10.2.7) is satisfied, all of the information needed to reconstruct the value of $x(t)$ between the samples is contained in the samples themselves. In particular, using the *sampling theorem,* we can interpolate between the samples as follows (see, e.g., Ahmed and Natarajan, 1983).

$$\boxed{x(t) = T \sum_{k=-\infty}^{\infty} \frac{x(kT) \sin[\pi f_s(t - kT)]}{\pi f_s(t - kT)}} \tag{10.2.8}$$

10.3 FAST FOURIER TRANSFORM (FFT)

The Fourier transform of a signal $x_a(t)$ can be approximated by applying a discrete version of the transform to N samples of the signal.

$$x(k) = x_a(kT) \quad , \quad 0 \leq k < N \tag{10.3.1}$$

Note that to simplify subsequent equations the sampling interval, T, is left implicit. To develop an efficient discrete version of the Fourier transform, it is helpful to introduce the following exponential scalar:

$$W_N \triangleq \exp\left(\frac{-j2\pi}{N}\right) \tag{10.3.2}$$

The scalar W_N satisfies $W_N^N = 1$, so W_N can be thought of as the Nth root of unity. The Fourier transform integral in Equation (10.2.1) can be approximated at N discrete frequencies by the following sum.

$$X(m) \triangleq \sum_{k=0}^{N-1} x(k)W_N^{mk} \quad , \quad 0 \le m < N \tag{10.3.3}$$

The expression in (10.3.3) is called the *discrete Fourier transform* of $x(k)$ or, simply, the *DFT*. As we shall see, the DFT is a powerful tool for analyzing the frequency content of digital signals. It is also useful for designing digital filters that modify the spectrum of a signal by enhancing certain frequencies and removing others.

Just as with the Fourier transform, the discrete Fourier transform has an inverse that has an almost identical structure. A digital signal is recovered from its DFT as follows.

$$x(k) = \frac{1}{N} \sum_{m=0}^{N-1} X(m)W_N^{-km} \quad , \quad 0 \le k < N \tag{10.3.4}$$

The expression in (10.3.4) is the *inverse DFT* of $X(m)$. Note that, apart from scaling by $1/N$, the inverse DFT is just the DFT with W_N replaced by its complex conjugate, $W_N^* = W_N^{-1}$. Consequently, any algorithm that computes the DFT can be used, with minor modification, to compute the inverse DFT as well.

Unfortunately, the DFT can be relatively expensive to compute. This presents a problem for long signals and also for real-time applications where computations must be performed on line. Note from Equation (10.3.3) that to compute $X(m)$ for $0 \le m < N$, it is necessary to perform N^2 floating-point multiplications and additions. Additional multiplications are needed to form the powers of W_N. Thus, a DFT applied to a thousand-point sequence requires in excess of one million FLOPs. However, there is substantial *symmetry* in the DFT, and this can be exploited to drastically reduce the computational effort (Cooley and Tukey, 1965). To see this, suppose that N is a power of two. We can then break the sum in (10.3.3) into two parts as follows.

$$X(m) = \sum_{k=0}^{N/2-1} x(k)W_N^{mk} + \sum_{k=N/2}^{N-1} x(k)W_N^{mk} \tag{10.3.5}$$

The second sum in (10.3.5) can be rewritten using the change of variable $k = n + N/2$. This makes the range of indices for the two sums identical and results in the following reformulation.

$$X(m) = \sum_{k=0}^{N/2-1} x(k)W_N^{mk} + \left(\sum_{n=0}^{N/2-1} x(n + N/2)W_N^{mn}\right)W_N^{mN/2} \tag{10.3.6}$$

Note from (10.3.2) that $W_N^{N/2} = \exp(-j\pi)$. Applying Euler's identity, this means that $W_N^{mN/2} = (-1)^m$. Using this observation and combining the two sums in (10.3.6), we arrive at the following simplification.

$$X(m) = \sum_{k=0}^{N/2-1} [x(k) + (-1)^m x(k + N/2)]W_N^{mk} \tag{10.3.7}$$

The formulation in (10.3.7) reduces the number of multiplications by approximately a factor of two. However, substantially more savings can be obtained if we now consider the even and odd cases separately.

$$X(2m) = \sum_{k=0}^{N/2-1} [x(k) + x(k + N/2)]W_N^{2mk} \tag{10.3.8a}$$

$$X(2m + 1) = \sum_{k=0}^{N/2-1} [x(k) - x(k + N/2)]W_N^{2mk}W_N^k \tag{10.3.8b}$$

This, in effect, removes the variable m from the expression within the square brackets. Next, define two new sequences of length $N/2$ as follows:

$$a(k) \triangleq x(k) + x(k + N/2) \tag{10.3.9a}$$

$$b(k) \triangleq [x(k) - x(k + N/2)]W_N^k \tag{10.3.9b}$$

These sequences can be substituted in (10.3.8a) and (10.3.8b), respectively. Finally, note from (10.3.2) that $W_N^{2mk} = W_{N/2}^{mk}$. Thus, (10.3.8) can be written as

$$X(2m) = \sum_{k=0}^{N/2-1} a(k)W_{N/2}^{mk} \tag{10.3.10a}$$

$$X(2m + 1) = \sum_{k=0}^{N/2-1} b(k)W_{N/2}^{mk} \tag{10.3.10b}$$

Comparing (10.3.10) with (10.3.3), we see that the even and odd parts of $X(m)$ can be computed *separately*, using DFTs of length $N/2$. The beauty of this decomposition technique is that it can be applied recursively. That is, if N is a power of two, then the two $N/2$ point DFTs in (10.3.10) can each be decomposed into a pair of $N/4$ point DFTs. These DFTs can be further decomposed into pairs of $N/8$ point DFTs, etc., until we arrive at a collection of two-point DFTs, which are computed very simply as

$$X(0) = x(0) + x(1) \tag{10.3.11a}$$

$$X(1) = x(0) - x(1) \tag{10.3.11b}$$

The resulting algorithm is called the *fast Fourier transform* or, simply, the *FFT*. Since an N point DFT requires approximately N^2 FLOPs, decomposing it into two $N/2$ point DFTs results in a 50% savings in computational effort: $2(N/2)^2 = N^2/2$. Therefore, each iteration reduces the remaining computational effort by a factor of two. Assuming N is a power of two, the total number of iterations is $q = \log_2 N$. It can be shown that the overall computational effort is reduced from N^2 complex FLOPs to

$$\boxed{p = 0.5N \log_2 N} \tag{10.3.12}$$

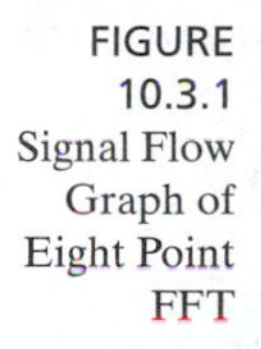

FIGURE 10.3.1 Signal Flow Graph of Eight Point FFT

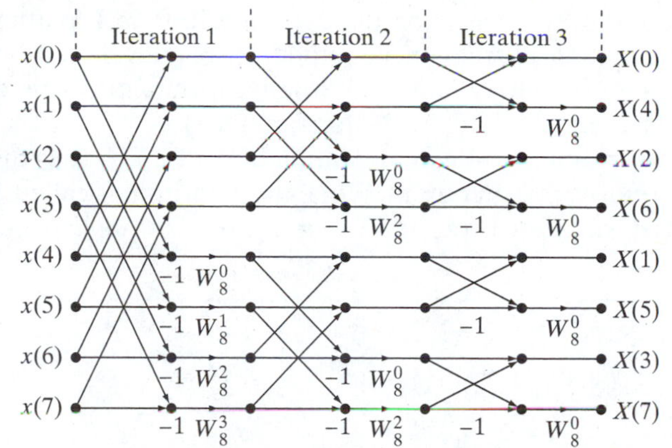

This is a *substantial* saving for large values of N. For example, it is not uncommon to transform signals of length $N = 1024$. Using the DFT, this requires approximately 1,048,575 FLOPs, while the FFT requires only 5120 FLOPs, a reduction of over 99.5 percent! A signal flow graph that summarizes the computations required to perform the three iterations of an eight-point FFT is shown in Figure 10.3.1. Here signals are added at the nodes and multiplied by the scalars used to label the arcs.

One difficulty with the FFT is that the order of the computed points gets *scrambled* with each iteration. For example, for the eight-point DFT in Figure 10.3.1, the order after the first iteration is the even-odd decomposition $\{X(0), X(2), X(4), X(6), X(1), X(3), X(5), X(7)\}$. The next iteration scrambles the order further, treating each run of four points as having even and odd components, thereby yielding $\{X(0), X(4), X(2), X(6), X(1), X(5), X(3), X(7)\}$. There is no additional scrambling in the last iteration, which is composed of two-point DFTs. To identify the pattern of the scrambled DFT values, it is helpful to examine the *binary* representations of the indices or subscripts as shown in Table 10.3.1.

Comparing the binary representations of the original DFT ordering with the FFT ordering, we observe that the original DFT order can be recovered by a simple *bit reversal* process. That is, if the binary representation of the FFT index is scanned from right to left instead of left to right, it yields the binary representation of the DFT index. Thus, the FFT points can be sorted and returned to the original order using a bit reversal procedure.

TABLE 10.3.1 FFT Ordering When $N = 8$

DFT	*Binary*	*FFT*	*Binary*
$X(0)$	000	$X(0)$	000
$X(1)$	001	$X(4)$	100
$X(2)$	010	$X(2)$	010
$X(3)$	011	$X(6)$	110
$X(4)$	100	$X(1)$	001
$X(5)$	101	$X(5)$	101
$X(6)$	110	$X(3)$	011
$X(7)$	111	$X(7)$	111

The derivation of the FFT based on the decomposition in Equation (10.3.10) is referred to as the decimation in frequency method. It was developed by Cooley and Tukey (1965). It is also possible to obtain an alternative formulation called the decimation in time method (see, e.g., Ahmed and Natarajan, 1983).

The following algorithm is an implementation of the FFT. It uses the $N \times 1$ vector z as an input and returns z as an output. When the algorithm is called, it is also supplied a *direction code* d with $d = 1$ for the FFT or $d = -1$ for the inverse FFT.

Alg. 10.3.1 FFT

1. Set $\theta = -2\pi d/N$ and $r = N/2$.
2. For $i = 1$ to $N - 1$ do
   {
   (a) Set $w = \cos(i\theta) + j\sin(i\theta)$.
   (b) For $k = 0$ to $N - 1$ do
   {
   (1) Set $u = 1$
   (2) For $m = 0$ to $r - 1$ do
   {

$$t = z(k + m) - z(k + m + r)$$
$$z(k + m) = z(k + m) + z(k + m + r)$$
$$z(k + m + r) = tu$$
$$u = wu$$

   }
   (3) Set $k = k + 2r$
   }
   (c) Set $i = 2i$ and $r = r/2$
   }
3. For $i = 0$ to $N - 1$ do
   {
   (a) Set $r = i$ and $k = 0$.
   (b) For $m = 1$ to $N - 1$ do
   {

$$k = 2k + (r \% 2)$$
$$r = r/2$$
$$m = 2m$$

   }
   (c) If $k > i$ do
   {

$$t = z(i)$$
$$z(i) = z(k)$$
$$z(k) = t$$

```
            }
    }
4. If d < 0 then for i = 0 to N - 1 do
   {
                    z(i) = z(i)/N
   }
```

On entry to Alg. 10.3.1, the vector z contains the samples of the complex digital signal to be transformed, and on exit, it contains the transformed signal. Thus, Alg 10.3.1 is an *in place* implementation, which does not require separate storage for the signal and its transform. When Alg 10.3.1 is called with $d = 1$, it computes the forward FFT of z, and when $d = -1$, it computes the inverse FFT of z. The actual FFT calculations are performed in step (2). Sorting of the results using bit reversal is performed in step (3), and normalization of the inverse FFT is performed in step (4).

Suppose a digital signal $x(k)$ is obtained by generating N equally-spaced samples of analog signal $x_a(t)$ using a sampling frequency of $f_s = 1/T$ samples per second as in (10.3.1). The FFT of $x(k)$ can then be used to find the Fourier transform of $x_a(t)$ at $N/2$ discrete frequencies. If $x_a(t)$ is band limited to $0.5/T$ Hz, then the discrete frequencies at which the Fourier transform is computed are

$$f_k = \frac{kf_s}{N} \quad , \quad 0 \le k < N/2 \tag{10.3.13}$$

In this case, $X(m)$ is the Fourier transform of $x_a(t)$ evaluated at $f = f_m$. Note that even though the FFT consists of N points, only the first $N/2$ points provide essential information about the Fourier transform of $x_a(t)$. The remaining points of the FFT correspond to negative values of f and add no new information because $X(-f) = X^*(f)$ when $x_a(t)$ is real. Since $X(m)$ is complex, it can be expressed as $X(m) = A(m)\exp[j\phi(m)]$ where the magnitude and phase components are defined as follows:

$$A(m) \triangleq |X(m)| \tag{10.3.14a}$$

$$\phi(m) \triangleq \arctan\left\{\frac{\text{Im}\,[X(m)]}{\text{Re}\,[X(m)]}\right\} \tag{10.3.14b}$$

We refer to $A(m)$ as the *magnitude spectrum* of $x(k)$ and to $\phi(m)$ as the *phase spectrum* of $x(k)$. The square of the magnitude spectrum is the energy density spectrum.

EXAMPLE 10.3.1 FFT

To illustrate the use of the FFT, suppose an experiment is performed in which measurements produce the following digital signal.

$$x(k) = \cos(400\pi kT) + 0.5\sin(120\pi kT) + v(kT) \quad , \quad 0 \le k < N$$

Here the first term in $x(k)$ represents the underlying signal whose presence we would like to detect. The second term is a *disturbance* that is caused by the 60 Hz commercial AC power supply used by the measurement instrument. Finally, the last term represents random *measurement noise*, which is modeled using a Gaussian distribution with zero mean and a standard deviation of 0.5. A plot of the digital signal $x(k)$ produced by example program *e1031* on the distribution CD is shown in Figure 10.3.2. Here a sampling frequency of $f_s = 600$ Hz was used to generate $N = 512$ samples.

FIGURE 10.3.2 A Discrete-Time Signal $x(k)$

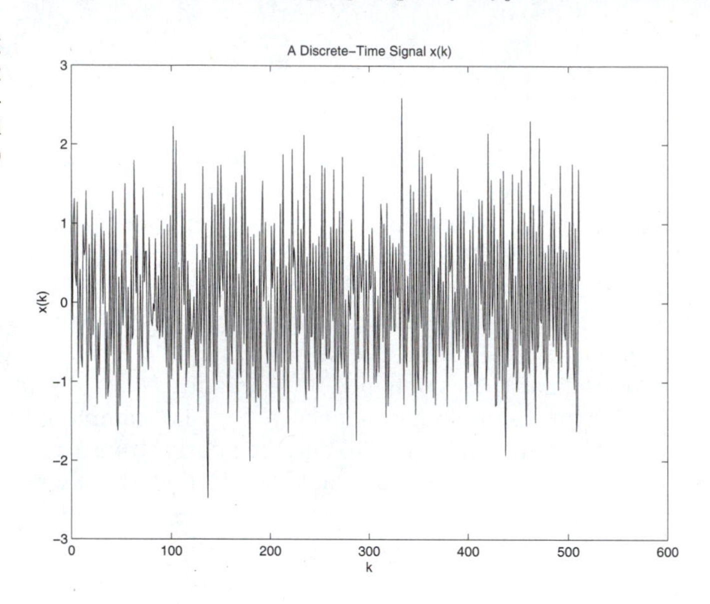

The presence of the 200 Hz signal and the 60 Hz disturbance is not obvious from direct inspection of $x(k)$ due to the presence of the additive noise $v(k)$. To isolate these terms from the noise, it is helpful to compute the FFT of the signal. The magnitude spectrum of $x(k)$ is shown in Figure 10.3.3. Note the large peak centered at $f = 200$ Hz indicating the presence and

FIGURE 10.3.3 The Magnitude Spectrum of $x(k)$

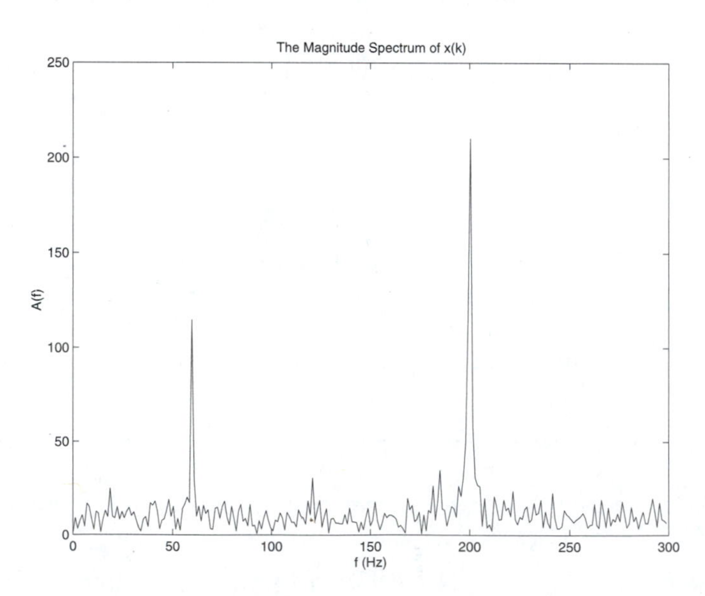

strength of the underlying signal. There is also a smaller peak centered at $f = 60$ Hz that corresponds to the disturbance term. Finally, the distribution of energy over the remaining frequencies indicates the presence of the broad band random noise $v(k)$.

10.4 CORRELATION

In engineering applications, there is often a need to measure the *similarity* between two digital signals. As an illustration, consider the radar target detection problem posed in Section 10.1.2. In this case, the received signal $x(k)$ consists of a weak echo contaminated with random noise as follows.

$$x(k) = \alpha u(k - d) + v(k) \tag{10.4.1}$$

The first term in $x(k)$ represents a delayed and attenuated version of the transmitted signal $u(k)$, and the second term represents random atmospheric noise. By examining the similarity between the transmitted signal $u(k)$ and the received signal $x(k)$, we can detect whether or not an echo is present. If an echo is detected, the range to the target that produced it can be estimated from the delay d and the propagation speed of the signal.

To solve the echo detection problem, suppose $x(k)$ is a signal of length N and $y(k)$ is a signal of length M where $M \leq N$. When $M < N$, it simplifies the discussion to *pad* the shorter signal $y(k)$ with $N - M$ zeros at the end so that the new $y(k)$ is also of length N. The *cross correlation* of $x(k)$ with $y(k)$ is then defined as follows.

$$r_{xy}(i) \triangleq \sum_{k=i}^{N-1} x(k)y(k-i) \quad , \quad 0 \leq i < N \tag{10.4.2}$$

Note that $r_{xy}(i)$ is computed by sliding $y(k)$ to the right by i samples and then taking the dot product of the overlapping components of $x(k)$ with $y(k)$ as illustrated in Figure 10.4.1. If $x(k)$ is a random signal with zero mean and $y(k)$ is $x(k)$ delayed by d samples, the maximum value of $r_{xy}(i)$ will occur at $i = d$ where all of the product terms match. The larger the number of samples, N, the more pronounced the difference will be between $r_{xy}(d)$ and the other values of the cross correlation. To detect a match that indicates the presence of an echo, we look for a peak in the cross correlation function.

It is also possible to compute the cross correlation of a signal with itself. This is referred to as the *auto correlation* of the signal $x(k)$ and denoted $r_{xx}(i)$.

$$r_{xx}(i) \triangleq \sum_{k=i}^{N-1} x(k)x(k-i) \tag{10.4.3}$$

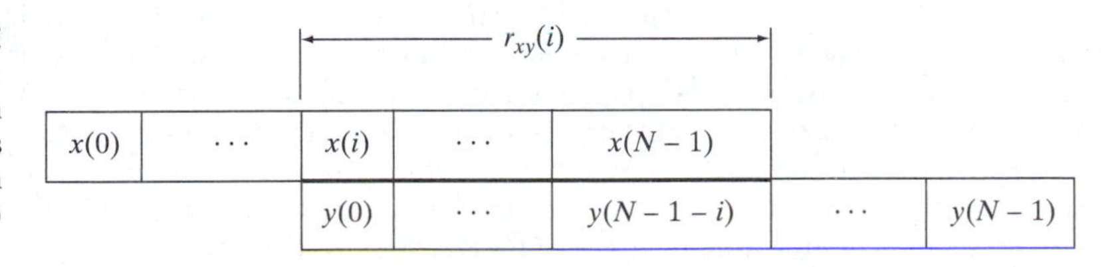

FIGURE 10.4.1 Evaluation of Cross Correlation $r_{xy}(i)$

The auto correlation can be used to measure the correlation between successive samples of a random signal. If the samples of a random sequence with zero mean are uncorrelated from one another, then $r_{xx}(i) \approx 0$ for $i \neq 0$ assuming N is sufficiently large. Uncorrelated random sequences are useful for generating uniformly distributed random vectors such as those used by the Monte Carlo integration techniques in Section 7.8.2. When $i = 0$ the value of the auto correlation function is the *energy* of the signal. That is,

$$E_x \triangleq \sum_{k=0}^{N-1} x^2(k) = r_{xx}(0) \tag{10.4.4}$$

It can be shown that the magnitude of the auto correlation is bounded from above by the signal energy. There is also an analogous upper bound that can be placed on the magnitude of the cross correlation.

$$|r_{xx}(i)| \leq E_x \tag{10.4.5a}$$

$$|r_{xy}(i)| \leq \sqrt{E_x E_y} \tag{10.4.5b}$$

Given these upper limits on the values of the auto correlation and the cross correlation, it is possible to define normalized versions of these two functions. The *normalized auto correlation* of $x(k)$ is denoted $\rho_{xx}(i)$, and the *normalized cross correlation* of $x(k)$ with $y(k)$ is denoted $\rho_{xy}(i)$.

$$\boxed{\rho_{xx}(i) \triangleq \frac{r_{xx}(i)}{r_{xx}(0)}} \tag{10.4.6a}$$

$$\boxed{\rho_{xy}(i) \triangleq \frac{r_{xy}(i)}{\sqrt{r_{xx}(0) r_{yy}(0)}}} \tag{10.4.6b}$$

The utility of the normalized auto correlation and cross correlation is that they return values that are restricted to the interval $[-1, 1]$. Consequently, it is easier to make a judgment as to whether or not a match has occurred because it is known that the largest possible peak has unit amplitude.

EXAMPLE 10.4.1 Target Detection

To illustrate the use of the normalized cross correlation, consider the radar installation problem described in Section 10.1.2. Suppose, for example, that a Gaussian random signal $u(k)$ with zero mean and unit standard deviation is transmitted from the radar station shown in Figure 10.1.2. The number of points in the transmitted signal is $M = 256$. The received signal $x(k)$ in Equation (10.4.1) consists of $n = 512$ points and includes an echo term that is delayed by $d = 200$ samples and attenuated by a gain of $\alpha = 0.05$. Also included in the received signal is a second random noise term, $v(k)$, that consists of Gaussian noise with zero mean and standard deviation $\sigma = 0.05$. A plot of the received signal produced by example program *e1041* on the distribution CD is shown in Figure 10.4.2. From direct inspection of $x(k)$, it is not at all apparent that an echo is buried within the noise. However, the presence of an echo can be confirmed,

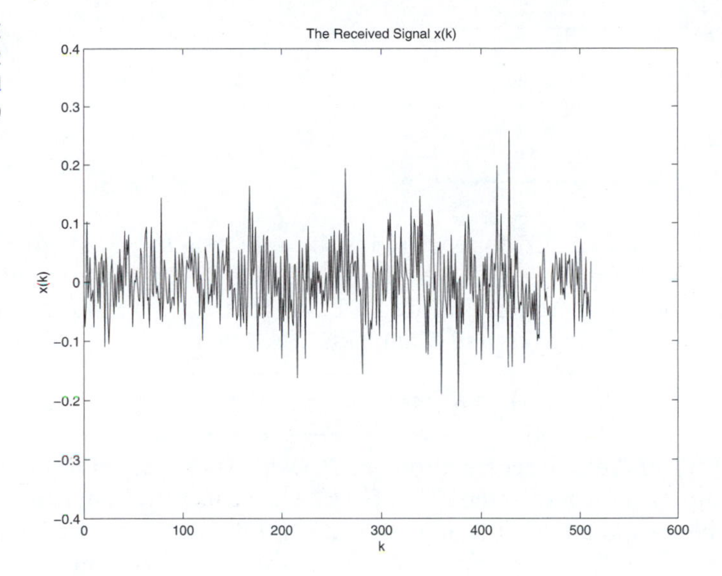

FIGURE 10.4.2 The Received Signal $x(k)$

and its location identified, if we examine the normalized cross correlation of $x(k)$ with the transmitted signal $u(k)$ as is shown in Figure 10.4.3.

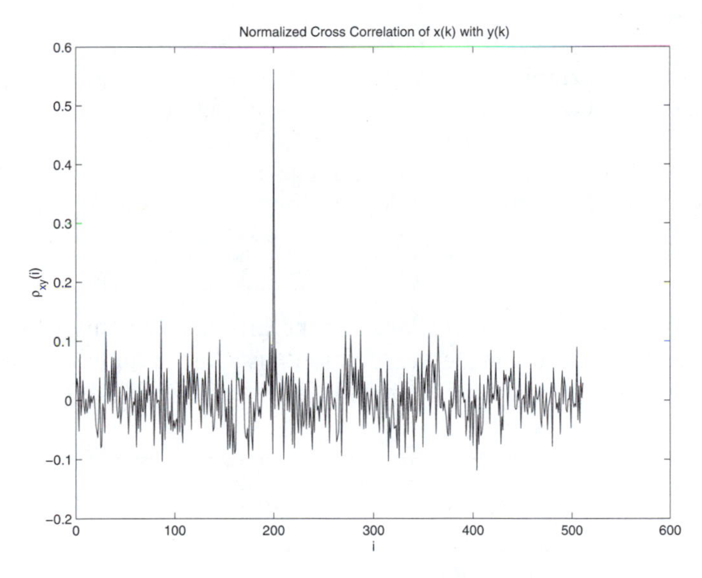

FIGURE 10.4.3 Normalized Cross Correlation of $x(k)$ with $y(k)$

The clear spike at $i = 200$ in Figure 10.4.3 indicates that an echo was received with a delay of $d = 200$ samples. To convert this to an estimate of the target range, suppose the time between samples is T sec. and the propagation speed of the signal is c m/sec. Since the delay d represents a round trip, the estimated range to the target is

$$r = \frac{cdT}{2} \quad \text{m}$$

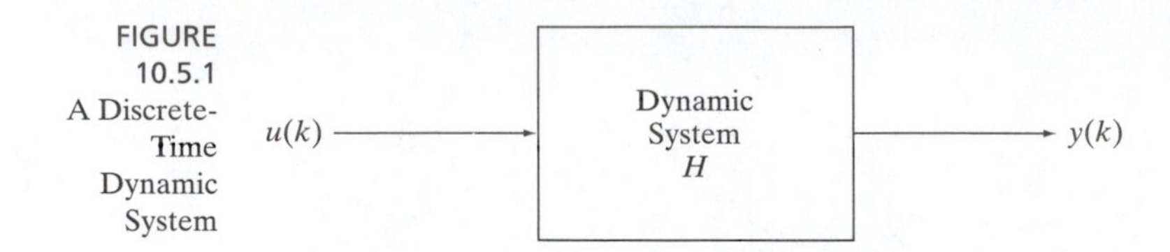

FIGURE 10.5.1 A Discrete-Time Dynamic System

The computational effort associated with evaluating the cross correlation function $r_{xy}(i)$ in (10.4.2) for $0 \le i < N$ requires $N^2/2$ FLOPs. This can be reduced by making use of the following property of the FFT. Suppose $R_{xy}(m) = \text{FFT}\{r_{xy}(i)\}$. Then it can be shown (see, e.g., Proakis and Manolakis, 1988) that

$$R_{xy}(m) = X(m)Y^*(m) \quad , \quad 0 \le m < N \tag{10.4.7}$$

That is, the FFT of the cross correlation of $x(k)$ with $y(k)$ is equal to the FFT of $x(k)$ times the complex conjugate of the FFT of $y(k)$. Consequently, from Equation (10.3.2), $R_{xy}(m)$ can be computed using $N \log_2 N$ FLOPs. Taking the inverse FFT of (10.4.7) then results in $r_{xy}(i)$. Therefore, the cross correlation of two N-point sequences can be computed with the FFT, using $1.5N \log_2 N$ FLOPs. This computational technique is referred to as *fast cross correlation.*

10.5 CONVOLUTION

Physical systems that exhibit memory are called *dynamic* systems. A block diagram of a discrete-time dynamic system with *input* or excitation $u(k)$ and *output* or response $y(k)$ is shown in Figure 10.5.1. The system H is dynamic if the current output $y(k)$ depends not only on the current input $u(k)$, but also on *past* values of the input and output. For example, if the current output depends on the current and past m values of the input, and the past n values of the output, then

$$y(k) = f[y(k-1), \ldots, y(k-n), u(k), \ldots, u(k-m)] \quad , \quad k \ge 0 \tag{10.5.1}$$

The function f in Equation (10.5.1) specifies the internal structure of the system. An important special case occurs when the function f is a linear combination, or weighted sum, of its arguments. In this case, the resulting system is called a *linear* discrete-time dynamic system.

$$y(k) + \sum_{i=1}^{n} a_i y(k-i) = \sum_{i=0}^{m} b_i u(k-i) \quad , \quad k \ge 0 \tag{10.5.2}$$

EXAMPLE 10.5.1 Running Average Filter

Suppose it is desired to compute the average value the last p samples of an input sequence $u(k)$. This can be performed by setting $a_i = 0$ in (10.5.2) and setting $b_i = 1/p$ for $0 \le i < p$.

$$y(k) = \frac{1}{p} \sum_{i=0}^{p-1} u(k-i)$$

This is referred to a *running average* filter of length p. It tends to "smooth" and delay the input sequence, particularly for large values of p.

10.5.1 Pulse Response

Next, we consider the problem of finding the output of the linear discrete-time system in Equation (10.5.2) given an input sequence $u(k)$. To evaluate the output starting at $y(0)$, it is necessary to know $u(0)$ and the $m + n$ past values $\{u(-1), \ldots, u(-m)\}$ and $\{y(-1), \ldots, y(-n)\}$. However, if the input is a *causal* signal, then $u(k) = 0$ for $k < 0$. Hence, for causal inputs, it is sufficient to know $u(0)$ and the n *initial conditions*: $\{y(-1), \ldots, y(-n)\}$. For the purpose of this discussion, suppose the initial conditions are zero. The resulting output $y(k)$ generated by a causal input $u(k)$ can then be expressed in terms of the system response to a special input called the *unit pulse* input $\delta(k)$, which is defined as follows.

$$\delta(k) \triangleq \begin{cases} 1 & , \quad k = 0 \\ 0 & , \quad k \neq 0 \end{cases} \tag{10.5.3}$$

Thus, the unit pulse, $\delta(k)$, is a digital signal that is zero everywhere except at $k = 0$, where it takes on the value one. The output of a linear system, H, generated by the unit pulse input $\delta(k)$ when the initial conditions are zero, is called the *pulse response* of the system and is denoted $h(k)$.

$$h(k) \triangleq H[\delta(k)] \tag{10.5.4}$$

The pulse response $h(k)$ of a linear system can be computed by applying Equation (10.5.2) iteratively with $u(k) = \delta(k)$. Once the pulse response is determined, the response to *any* input can be easily computed. To see this, first note that an arbitrary input $u(k)$ can be expressed as a weighted sum of pulses as follows.

$$u(k) = \sum_{i=0}^{k} u(i)\delta(k - i) \quad , \quad k \geq 0 \tag{10.5.5}$$

Here $u(i)\delta(k - i)$ is a pulse of amplitude $u(i)$ occurring at time $k = i$. Thus, (10.5.5) expresses $u(k)$ as a sum of scaled and shifted pulses. For a linear system, as in (10.5.2), the response to a sum of inputs is just the sum of the responses to the individual inputs. Furthermore, if an input is delayed by i samples and scaled by α, the corresponding output is also delayed by i samples and scaled by α. Thus, from (10.5.4), the output corresponding to $u(i)\delta(k - i)$ is $u(i)h(k - i)$. Using this observation and the representation of $u(k)$ in (10.5.5), it follows that the response to an arbitrary input, assuming zero initial conditions, can be written as

$$y(k) = \sum_{i=0}^{k} h(k - i)u(i) \quad , \quad k \geq 0 \tag{10.5.6}$$

The formulation in (10.5.6) is referred to as the *convolution* of $h(k)$ with $u(k)$. We see from (10.5.6) that the output produced by an input $u(k)$ can be expressed as the convolution of the pulse response $h(k)$ with the input $u(k)$.

Note that the convolution operation is closely related to the cross correlation function in (10.4.2). The computation of $y(k)$ for $0 \le k < N$ requires approximately $N^2/2$ FLOPs for large values of N. However, like the cross correlation function, the computation of the convolution function can be made more efficient by using the FFT. In particular, it can be shown that if $H(m) = \text{FFT}\{h(k)\}$ and $U(m) = \text{FFT}\{u(k)\}$, then $Y(m) = \text{FFT}\{y(k)\}$ can be expressed as follows (Ahmed and Natarajan, 1983).

$$\boxed{Y(m) = H(m)U(m)} \qquad (10.5.7)$$

That is, convolution in the time domain becomes multiplication in the FFT domain. Comparing (10.5.7) with (10.4.7), we see that the only difference between cross correlation and convolution is the use of the complex conjugate. Taking the inverse FFT of both sides of (10.5.7) yields $y(k)$. Using the FFT to perform a convolution of two N-point signals reduces the number of FLOPs from $N^2/2$ to $1.5N \log_2 N$. The implementation of convolution based on (10.5.7) is called *fast convolution*.

EXAMPLE 10.5.2 Convolution

Consider the case of the running average filter of length p in Example 10.5.1. Direct computation shows that the pulse response is a pulse of amplitude $1/p$ and width p starting at $k = 0$.

$$h(k) = \begin{cases} 1/p \ , & 0 \le k < p \\ 0 \ , & k \ge p \end{cases}$$

Suppose $u(k)$ is an input signal with $N = 512$ points that includes a damped sine wave plus Gaussian random noise, $v(k)$, with zero mean and unit standard deviation.

$$u(k) = \exp(-3k/q)\sin(6\pi k/q) + v(k)$$

A plot of $u(k)$ produced by example program *e1052* on the distribution CD is shown in Figure 10.5.2. Let $y(k)$ be the filtered version of $u(k)$ using a running average filter of length $p = 50$. The output $y(k)$ computed using fast convolution is shown in Figure 10.5.3. It is clear from

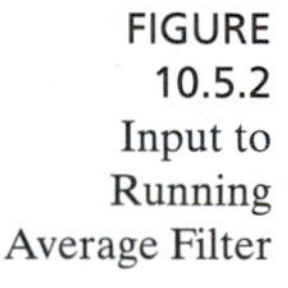

FIGURE 10.5.2 Input to Running Average Filter

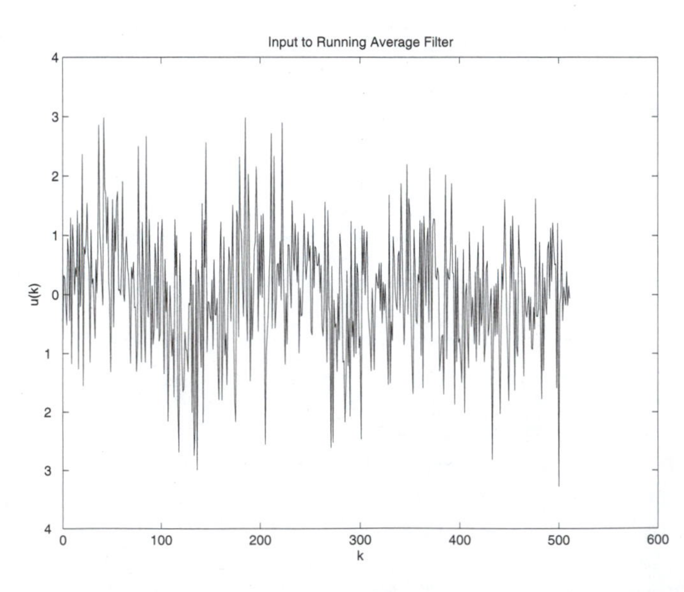

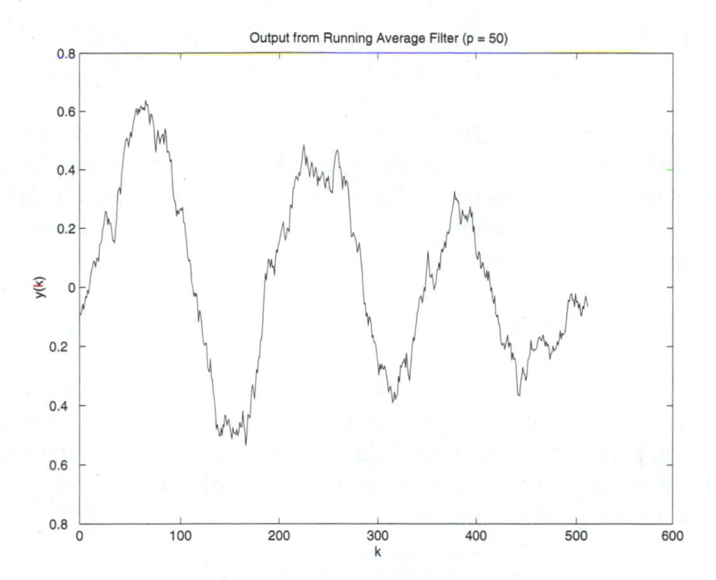

FIGURE 10.5.3 Output from Running Average Filter ($p = 50$)

inspection that $y(k)$ is noticeably smoother than $u(k)$, and the running average attempts to track the underlying lower frequency-damped sine wave that is embedded within the noise.

A discrete-time system whose pulse response goes to zero in a finite number of samples is referred to as a *finite impulse response* or *FIR* filter. Otherwise, it is called an *infinite impulse response* or *IIR* filter. The running average filter in Example 10.5.1 is an example of an FIR filter. In general, FIR filters are discrete-time systems as in Equation (10.5.2), but with $a_i = 0$ for $1 \le i \le n$. Using a direct calculation, one can show that the pulse response of a linear FIR filter can be specified in terms of its coefficients as follows.

$$h(k) = \begin{cases} b(k) & , \quad 0 \le k \le m \\ 0 & , \quad k > m \end{cases} \tag{10.5.8}$$

10.5.2 Stability

There is an important qualitative characteristic that is shared by all FIR filters and some IIR filters. To describe it, first note that a signal $x(k)$ is *bounded* if and only if there exists a constant $B > 0$ called a *bound* such that

$$|x(k)| \le B \quad , \quad k \ge 0 \tag{10.5.9}$$

Thus, a bounded signal is a signal whose magnitude does not grow arbitrarily large with increasing time. A discrete-time system is *stable* if and only if every bounded input signal $u(k)$ generates a bounded output signal $y(k)$. Otherwise, the system is *unstable*. Stability is an important practical feature because without it, the output can grow arbitrarily large and generate an overflow condition.

There are two ways to determine the stability of a linear discrete-time system. Starting with the convolution equation in (10.5.6), it is not difficult to show that if $|u(k)| \le B$, then $|y(k)| \le \alpha B$ where α is the following constant.

$$\alpha = \sum_{k=0}^{\infty} |h(k)| \tag{10.5.10}$$

If the constant α in (10.5.10) is finite, we say that the pulse response of the system is *absolutely summable*. A linear discrete-time system is stable if and only if $h(k)$ is absolutely summable.

The stability criterion in (10.5.10) is somewhat indirect because it requires the computation of the pulse response. A more direct way to determine the stability of a linear discrete-time system is to use the coefficients of the equation in (10.5.2). In particular, given the $n \times 1$ coefficient vector a, we can form the following polynomial called the *characteristic polynomial* of the system.

$$\Delta(z) \triangleq z^n + a_1 z^{n-1} + \cdots + a_{n-1} z + a_n \tag{10.5.11}$$

The characteristic polynomial has n roots in the complex plane, which are denoted $\{p_1, p_2, \ldots, p_n\}$. The roots of $\Delta(z)$ are called the *poles* of the system. Using an analytical tool called the Z-transform, it is possible to show that the system in (10.5.2) is stable if and only if the poles of the system all lie inside the unit circle of the complex plane (Ahmed and Natarajan, 1983). That is,

$$|p_k| < 1 \quad , \quad 1 \le k \le n \tag{10.5.12}$$

By using either the criterion $\alpha < \infty$ in Equation (10.5.10) or the criterion in (10.5.12), one can show that FIR filters are *always* stable. By definition, an FIR filter has a pulse response that goes to zero in a finite number of samples. Hence $h(k)$ is always absolutely summable. Alternatively, for an FIR filter, $a_k = 0$ for $1 \le k \le n$. Thus, the characteristic polynomial in (10.5.11) is $\Delta(z) = z^n$, which means that all n poles are at $z = 0$, well within the unit circle.

EXAMPLE 10.5.3 Stability

Suppose you open a savings account that pays an annual interest rate of α percent compounded daily. Let $y(k)$ denote the balance at the end of day k, and let $u(k)$ denote the amount deposited (if positive) or withdrawn (if negative) on day k. The balance at the end of day k is the balance at the end of day $k - 1$, plus the interest earned on that balance, plus the amount deposited.

$$\begin{aligned} y(k) &= y(k-1) + \frac{\alpha y(k-1)}{100(365)} + u(k) \\ &= \left(1 + \frac{\alpha}{36500}\right) y(k-1) + u(k) \end{aligned}$$

Thus, the accumulation of interest can be modeled with a simple discrete-time dynamic system with $a_1 = -(1 + \alpha/36500)$ and $b_0 = 1$. Since $a_1 \ne 0$, this is an IIR filter. From (10.5.11), its characteristic polynomial is

$$\Delta(z) = z - \left(1 + \frac{\alpha}{36500}\right)$$

The characteristic polynomial has a single root at $z = 1 + \alpha/36500$, which is outside the unit circle. It follows that this dynamic system is unstable. That is, there is a bounded input that generates an unbounded output. Indeed, if the initial balance when the account is opened is

$y(0) = \beta$ and $u(k) = 0$ for $k \geq 0$, then the daily balance grows without bound (albeit slowly) as follows.

$$y(k) = \left(1 + \frac{\alpha}{36500}\right)^k \beta \quad , \quad k \geq 0$$

10.6 DIGITAL FILTERS

Perhaps the most widespread practical application of discrete-time dynamic systems is in the area of digital filtering. If the parameters of the linear discrete-time system in Equation (10.5.2) are selected with care, the frequency content, or spectrum, of the input signal can be reshaped or modified in a desired manner as the signal is processed by the system. In this case, we refer to the discrete-time system as a *digital filter*.

10.6.1 Frequency Response

To see how the spectrum of the input can be modified, we start with the relationship between the DFT of the output and the DFT of the input from Equation (10.5.7), which is repeated here for convenient reference.

$$Y(m) = H(m)U(m) \quad , \quad 0 \leq m < N \tag{10.6.1}$$

Recall that $H(m)$ is the DFT (or equivalently the FFT) of the pulse response of the filter. In general, $U(m)$, $H(m)$, and $Y(m)$ are complex valued. Consequently, the DFT of the input and output signals can be represented in polar form as

$$U(m) = A_u(m) \exp[j\phi_u(m)] \tag{10.6.2a}$$

$$Y(m) = A_y(m) \exp[j\phi_y(m)] \tag{10.6.2b}$$

Here $A_u(m)$ and $\phi_u(m)$ denote the magnitude spectrum of $u(k)$ and the phase spectrum of $u(k)$, respectively. Similarly, $A_y(m)$ and $\phi_y(m)$ are the magnitude and phase spectra of $y(k)$. The DFT of the pulse response of a digital filter is called the *frequency response* of the filter. Again, it can be represented in polar form as

$$\boxed{H(m) = A(m) \exp[j\phi(m)]} \tag{10.6.3}$$

The magnitude spectrum, $A(m)$, is called the *magnitude response* of the filter, and the phase spectrum, $\phi(m)$, is called the *phase response* of the filter. Thus, the filter frequency response is made up of the magnitude response and the phase response. To see how the magnitude and phase spectra of an input signal get altered by the filter, we recast (10.6.1) in polar coordinates.

$$\begin{aligned} A_y(m) \exp[j\phi_y(m)] &= Y(m) \\ &= H(m)U(m) \\ &= A(m) \exp[j\phi(m)]A_u(m) \exp[j\phi_u(m)] \\ &= A(m)A_u(m) \exp\{j[\phi(m) + \phi_u(m)]\} \end{aligned} \tag{10.6.4}$$

FIGURE 10.6.1 Common Idealized Filters

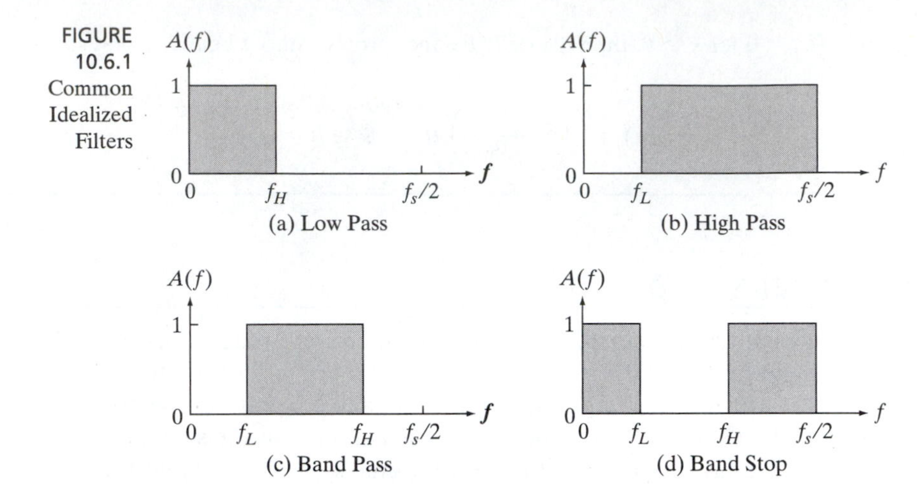

It is apparent that the magnitude spectrum of the output signal is the magnitude spectrum of the input signal *times* the magnitude response of the filter.

$$\boxed{A_y(m) = A_u(m)A(m)} \tag{10.6.5}$$

Similarly, the phase spectrum of the output signal is the phase spectrum of the input signal *plus* the phase response of the filter.

$$\boxed{\phi_y(m) = \phi_u(m) + \phi(m)} \tag{10.6.6}$$

To reshape the spectrum of $u(k)$, it is therefore necessary to design a filter with an appropriate magnitude and/or phase response. The four most common types of magnitude responses that occur in engineering applications are shown in Figure 10.6.1. Recall that $H(m)$ denotes the frequency response at $f_m = mf_s/N$ for $0 \le m < N/2$ where $f_s = 1/T$ is the sampling frequency. Therefore, the highest frequency that the filter can process is the Nquist frequency, $f_s/2$. Although the idealized plots of $A(f)$ for $0 \le f \le f_s/2$ in Figure 10.6.1 *cannot* be realized exactly with a filter of finite length, they can be approximated arbitrarily closely.

Figure 10.6.1(a) shows a *low-pass* filter, a filter that passes frequencies in the range 0 to f_H but completely attenuates or stops the higher frequencies in the range f_H to $f_s/2$. The corner frequency f_H is called the high-frequency cutoff of the filter. Low-pass filters are often used to smooth signals by removing high frequency noise. The opposite of a low-pass filter is a *high-pass* filter as shown in Figure 10.6.1(b). Here all frequencies above a low frequency cutoff, f_L, are passed, while frequencies below f_L are completely eliminated. A high-frequency filter can be used, for example, to turn a general signal into a signal with zero mean because the average value of the signal can be thought of as the component with frequency $f = 0$.

A more general *band-pass* filter is shown in Figure 10.6.1(c). Here all frequencies between a low frequency cutoff, f_L, and a high-frequency cutoff, f_H, are passed by the filter, while frequencies outside this pass band are rejected. A band-pass filter can be used to extract a particular frequency component from a signal. Finally, the opposite of a band pass filter is a *band-stop* filter or notch filter as shown in Figure 10.6.1(d). Here all frequencies between a low-frequency cutoff, f_L, and a high-frequency cutoff, f_H, are rejected by the filter, while the remaining frequencies outside this stop band are passed. Band-stop filters can be used to remove a particular frequency component from a signal.

It is less common to design a filter based on a desired phase response, $\phi(f)$. However, there is one case of particular interest called a *linear-phase* filter, which has a phase response of the form

$$\phi(f) = -2\pi\tau_d f \tag{10.6.7}$$

The virtue of having a phase response that is proportional to the frequency, f, lies in the fact that all frequency components take the same amount of time to pass through the filter. For example, if an input signal, $u(k) = \sin(2\pi f_0 k)$, has a phase shift of $\phi = -2\pi\tau_d f_0$ radians, then the input, $u(k) = \sin(4\pi f_0 k)$, has a phase shift of $\phi = -4\pi\tau_d f_0$ radians. But if doubling the frequency produces twice the phase shift, that means that the two signals are delayed by exactly the same amount. More generally, the *delay* for each frequency component is defined as

$$\tau(f) \triangleq -\frac{1}{2\pi}\frac{d\phi(f)}{df} \tag{10.6.8}$$

For a linear-phase filter, the delay function is the constant $\tau(f) = \tau_d$. Linear-phase or constant-delay filters are particularly useful for applications like speech processing, where a variable delay would cause a distortion in the signal processed by the filter.

10.6.2 FIR Filter Design

A linear-phase digital filter with a desired magnitude response can be constructed by a Fourier series method using a FIR discrete-time system of the form

$$y(k) = \sum_{i=0}^{2p} b_i u(k-i) \tag{10.6.9}$$

The magnitude response of a digital filter is periodic with a period of $f_s = 1/T$. The first step in determining values for the filter parameters, b_i, is to approximate the desired magnitude response $A(f)$ with a truncated Fourier series.

$$A(f) \approx \sum_{i=-p}^{p} c_i \exp\left(\frac{2\pi ijf}{f_s}\right) \tag{10.6.10}$$

For digital filters with real coefficients, $A(f)$ will be an *even* function of f. Consequently, the Fourier coefficients can be expressed in terms of cosines as follows:

$$\boxed{c_i = \frac{2}{f_s}\int_0^{f_s/2} A(f)\cos\left(\frac{2\pi if}{f_s}\right) df \quad , \quad -p \le i \le p} \tag{10.6.11}$$

TABLE 10.6.1 Fourier Coefficients of Common Filters

Filter	c_0	c_i
Low pass	$\frac{2f_H}{f_s}$	$\frac{1}{\pi i}\sin\left(\frac{2\pi i f_H}{f_s}\right)$
High pass	$\frac{f_s - 2f_L}{f_s}$	$-\frac{1}{\pi i}\sin\left(\frac{2\pi i f_L}{f_s}\right)$
Band pass	$\frac{2(f_H - f_L)}{f_s}$	$\frac{1}{\pi i}\left[\sin\left(\frac{2\pi i f_H}{f_s}\right) - \sin\left(\frac{2\pi i f_L}{f_s}\right)\right]$
Band stop	$\frac{f_s - 2(f_H - f_L)}{f_s}$	$\frac{1}{\pi i}\left[\sin\left(\frac{2\pi i f_L}{f_s}\right) - \sin\left(\frac{2\pi i f_H}{f_s}\right)\right]$

Note that for the case $i = 0$, the Fourier coefficient is just the average value of $A(f)$ for $-f_s/ \le f \le f_s/2$. The Fourier coefficients of the four common idealized filters in Figure 10.6.1 are summarized in Table 10.6.1.

It is instructive to examine what happens to the Fourier series approximation of $A(f)$ for large but finite values of p. The case of a low pass filter with cutoff frequency $f_H = f_s/4$, $p = 20$, and $f_s = 200$ Hz is shown in Figure 10.6.2.

Although the approximation in Figure 10.6.2 is quite good, there is a noticeable *ringing* or oscillation, particularly on both sides of the cutoff frequency, $f_H = 50$ Hz. If the number of terms p is increased, the approximation will improve, but there continues to be a significant oscillation in the neighborhood of the cutoff frequency. This characteristic of the truncated Fourier series occurs whenever there is a jump discontinuity in the function being approximated, and it is called the *Gibbs phenomenon*. The

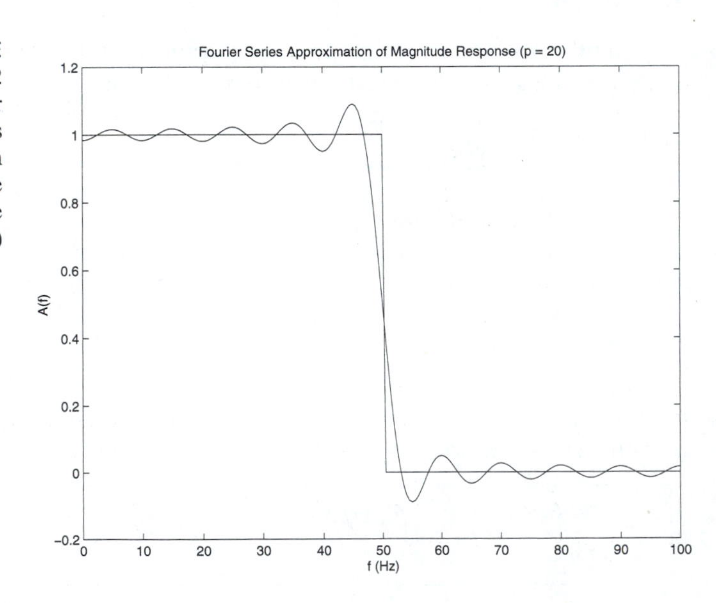

FIGURE 10.6.2 Fourier Series Approximation of Magnitude Response ($p = 20$)

TABLE 10.6.2 Fourier Series Window Sequences

Window	$w(i)$
Bartlett	$1 - \frac{\lvert i \rvert}{p}$
Welch	$1 - \left(\frac{\lvert i \rvert}{p}\right)^2$
Hanning	$0.5\left[1 + \cos\left(\frac{i\pi}{p}\right)\right]$
Hamming	$0.54 + 0.46\cos\left(\frac{i\pi}{p}\right)$

amplitude of the oscillation can be reduced by modifying the Fourier coefficients slightly. Notice that truncating a Fourier series after the pth term is equivalent to multiplying the sequence of Fourier coefficients $\{c_i\}$ by a *rectangular window* sequence $w(i) = 1$ for $-p \leq i \leq p$ and $w(i) = 0$ for $|i| > p$. To reduce the oscillation, we instead multiply the Fourier coefficients by an alternative window that goes to zero less abruptly. Some popular alternatives to the rectangular window are summarized in Table 10.6.2.

A graphical comparison of the Fourier series window sequences for the case $p = 100$ is shown in Figure 10.6.3. All of these offer improvements over the rectangular window. The use of the Hamming window is illustrated in Figure 10.6.4, which depicts the same low-pass approximation as in Figure 10.6.2, but with the ith Fourier coefficient multiplied by $w(i)$. Comparing Figure 10.6.4 with Figure 10.6.2, it is clear that the oscillations

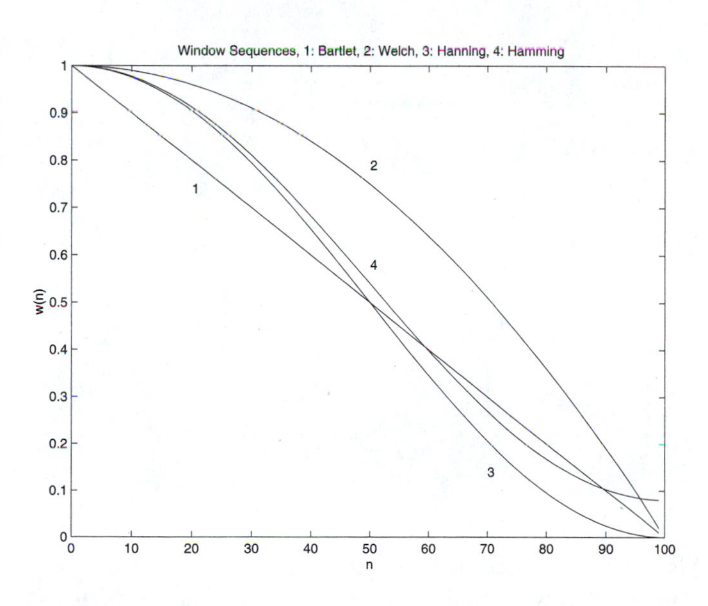

FIGURE 10.6.3 Window Sequences: 1: Bartlett, 2: Welch, 3: Hanning, 4: Hamming

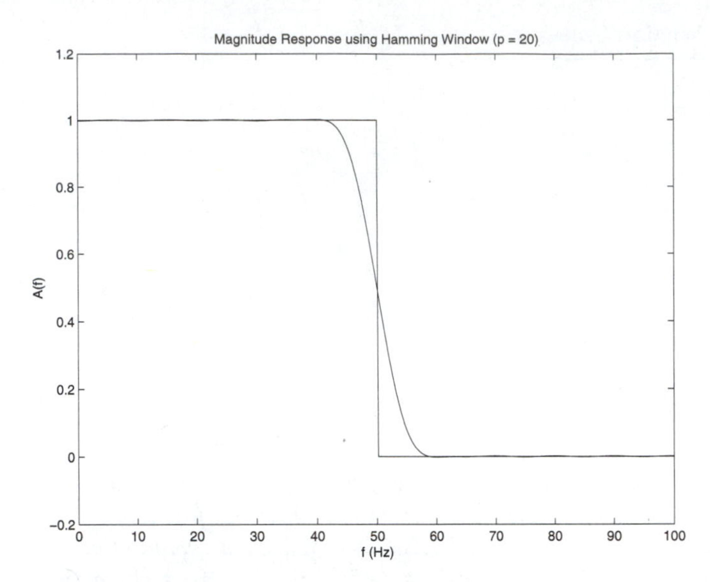

FIGURE 10.6.4 Magnitude Response Using Hamming Window ($p = 20$)

have been significantly reduced. It should also be noted that this is accomplished at the expense of a more gradual transition between the pass band and the stop band.

The windowed versions of the Fourier coefficients of $A(f)$ can now be used to construct the parameters of the FIR filter in Equation (10.6.9). The steps are summarized in the following algorithm. Details of the derivation can be found, for example, in Ahmed and Natarajan (1983).

Alg. 10.6.1 FIR Filter

1. Pick $p > 0$.
2. For $i = 0$ to p compute

{

$$c_i = (2/f_s) \int_0^{f_s/2} A(f) \cos(2\pi i f/f_s)\, df$$

$$b_{p+i} = w(i) c_i$$

$$b_{p-i} = b_{p+i}$$

}

3. Set

$$y(k) = \sum_{i=0}^{2p} b_i u(k-i)$$

When Alg. 10.6.1 is called, it is passed a desired magnitude response $A(f)$ for $0 \leq f \leq f_s/2$ and a window sequence $w(i)$ from Table 10.6.2. It returns the coefficients

b_i for $0 \le i \le 2p$ for the FIR filter. The magnitude response of the FIR filter approximates $A(f)$ when p is sufficiently large. Furthermore, it can be shown that the FIR filter is a linear-phase filter with a constant delay of p samples. That is,

$$\phi(f) = -2\pi pTf \tag{10.6.12}$$

EXAMPLE 10.6.1 White Noise

To illustrate the use of Alg. 10.6.1 to perform digital filtering, let $u(k)$ denote a random Gaussian signal with zero mean and unit standard deviation. Suppose there are $N = 1024$ samples of $u(k)$ corresponding to a sampling rate of $f_s = 2000$ Hz. Example program *e1061* on the distribution CD applies Alg. 10.6.1 to construct a filter. The magnitude spectrum of $u(k)$ is shown in Figure 10.6.5. Note that $A_u(f)$ is approximately flat over the entire range of frequencies, $0 \le f \le f_s/2$. A random signal that contains energy uniformly distributed over all frequencies is referred to as a *white noise* signal. That is, white noise has a constant or flat magnitude spectrum. Like light, the noise is "white" because it contains energy at all frequencies.

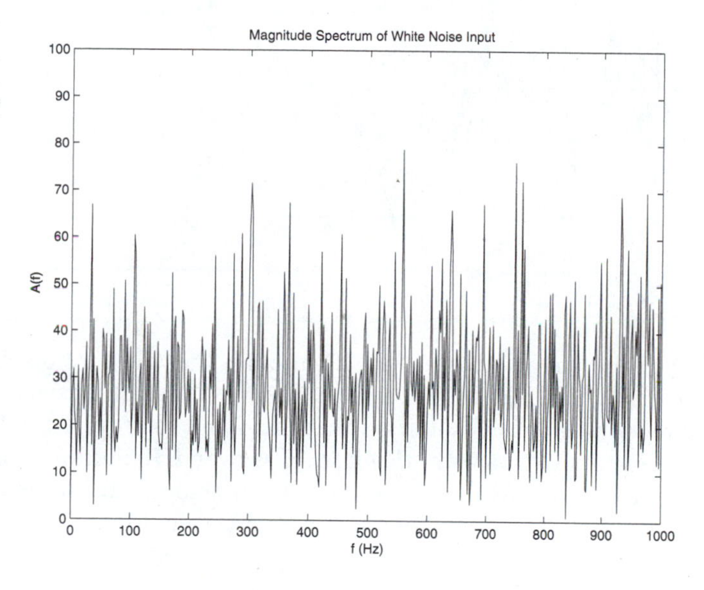

FIGURE 10.6.5 Magnitude Spectrum of White Noise Input

When white noise is passed through a digital filter, it becomes "colored" in the sense that the spectrum is no longer flat with energy uniformly distributed over all frequencies. For example, consider the problem of designing a filter that passes only frequencies in the interval [200, 600] Hz. Since $f_s = 2000$ Hz, this corresponds to a band-pass filter with lower cutoff $f_L = 0.1f_s$ and upper cutoff $f_H = 0.3f_s$. From Table 10.6.1, the Fourier series coefficients are $c_0 = 0.4$ and

$$c_i = \frac{\sin(0.6\pi i) - \sin(0.2\pi i)}{\pi i} \quad , \quad -p \le i \le p$$

Suppose the Hamming window in Table 10.6.2 is used. From Alg. 10.6.1, the coefficients of the FIR filter of length $2p + 1$ are then

$$b_{p+i} = b_{p-i} = \frac{[0.54 + 0.46 \cos(i\pi/p)]c_i}{2}, \quad 0 \le i \le p$$

The case $p = 20$ is shown in Figure 10.6.6, which plots the magnitude spectrum of the output $y(k)$ corresponding to the white noise input in Figure 10.6.5. It is clear that frequencies outside the pass band $[200, 600]$ Hz have been attenuated by the filter.

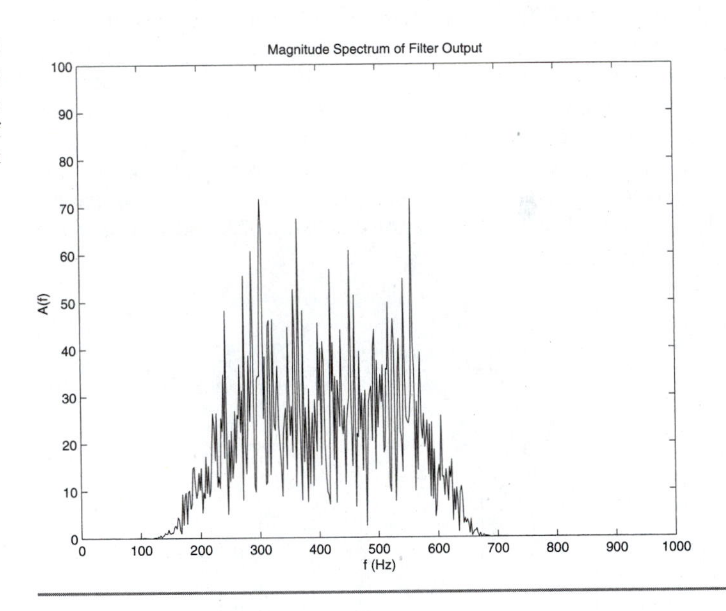

FIGURE 10.6.6 Magnitude Spectrum of Filter Output

One of the inherent advantages of FIR filters is that they are guaranteed to be *stable*. Consequently, if the input signal $u(k)$ is bounded, the output signal $y(k)$ will be bounded. A disadvantage of FIR filters is that they tend to require a relatively large number of terms in comparison with IIR filters. However, IIR filters are not always stable. One way to design an IIR filter is to convert an analog filter to its discrete-time equivalent. The interested reader is referred to Proakis and Manolakis (1988).

10.7 TWO-DIMENSIONAL FFT

All of the digital signals we have examined thus far are one-dimensional signals or sampled functions of time. It is also possible to apply signal processing techniques to two-dimensional signals or sampled functions of space. Two-dimensional signals arise naturally in the representation of *images*. For example, suppose $z_a(x, y)$ denotes the intensity or gray level of an image at coordinates (x, y). Suppose the image is partitioned into a grid of MN picture elements or *pixels* as shown in Figure 10.7.1 where

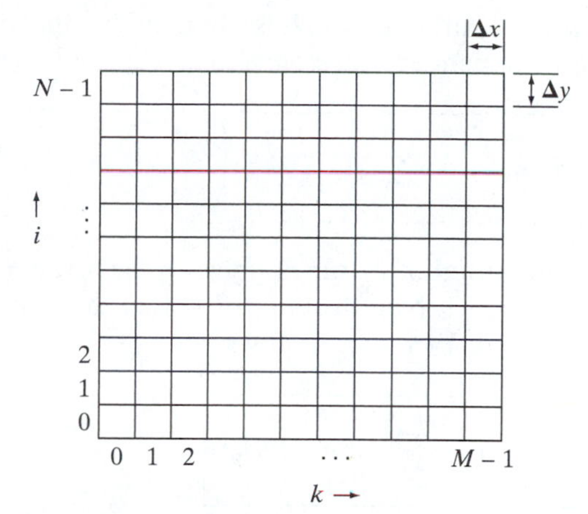

FIGURE 10.7.1 A Digital Image with MN Pixels

each pixel is of width Δx and of height Δy. In this case, the image can be represented by a two-dimensional signal where $z(k,i)$ denotes the intensity of the pixel located at $x = k\Delta x$ and $y = i\Delta y$.

Just as the discrete Fourier transform or DFT is useful for analyzing the frequency content of a one-dimensional signal, $x(k)$, a two-dimensional version of the DFT is useful for analyzing the spatial frequency content of a two-dimensional signal, $z(k,i)$. The DFT in Equation (10.3.3) is generalized to two dimensions in the following manner. Let $W_P = \exp(-j2\pi/P)$ be the Pth root of unity for $P = M$ and $P = N$, and define

$$Z(m,n) \triangleq \sum_{k=0}^{M-1} \sum_{i=0}^{N-1} z(k,i) W_N^{ni} W_M^{mk} \quad , \quad 0 \leq m < M, \ 0 \leq n < N \tag{10.7.1}$$

We call $Z(m,n)$ the *two-dimensional DFT* of $z(k,i)$ evaluated at discrete spatial frequencies m and n. The inverse of the two-dimensional DFT is obtained by replacing W_M and W_N by their complex conjugates, W_M^{-1} and W_N^{-1}, and normalizing by MN. That is, to recover the original signal $z(k,i)$ from its DFT in (10.7.1), we use

$$z(k,i) = \frac{1}{MN} \sum_{k=0}^{M-1} \sum_{i=0}^{N-1} Z(m,n) W_N^{-ni} W_M^{-mk} \quad , \quad 0 \leq k < M, \ 0 \leq i < N \tag{10.7.2}$$

It is clear that any technique used to compute the two-dimensional DFT can also be used, with minor modification, to compute the two-dimensional inverse DFT. The computational effort required to compute the two-dimensional DFT is considerable, as can be seen from Equation (10.7.1) where $(MN)^2$ FLOPs are required to find the MN values of $Z(m,n)$. This does not include the multiplications needed to form powers of W_N and W_M. We can reduce the computational effort substantially by using the one-dimensional fast Fourier transform or FFT developed in Section 10.3. To that end, consider the following intermediate variable, which is obtained by applying the FFT to $z(k,i)$ while treating i as the independent variable and k as a fixed parameter.

$$Y(n,k) \triangleq \sum_{i=0}^{N-1} z(k,i)W_N^{ni} \tag{10.7.3}$$

This FFT operation is performed a total of M times, once for each value of the parameter k, which ranges from 0 to $M - 1$. Note from (10.7.1) that the two-dimensional DFT can then be rewritten in terms of $Y(n,k)$ as follows.

$$Z(m,n) = \sum_{k=0}^{M-1} Y(n,k)W_M^{mk} \tag{10.7.4}$$

That is, $Z(m,n)$ can be obtained by taking the FFT of $Y(n,k)$, this time treating k as the independent variable and n as a parameter. This FFT is performed a total of N times, once for each value of the parameter n, which ranges from 0 to $N - 1$. It follows that a total of M FFTs of length N plus N FFTs of length M must be performed to compute a two-dimensional FFT in this manner. Therefore, the number of FLOPs required is $0.5MN(\log_2 M + \log_2 N)$. Often, $M = N$, in which case the computation effort is $N^2 \log_2 N$ FLOPs in comparison with N^4 FLOPs using the brute force approach in (10.7.1). It is clear that the savings are very substantial, even for moderate values of N. The following algorithm is an implementation of the two-dimensional FFT of $z(k,i)$ where $0 \le k < M$ and $0 \le i < N$.

Alg. 10.7.1 Two-Dimensional FFT

1. For $k = 0$ to $M - 1$ apply Alg. 10.3.1 to compute

   {

   $$Y(n,k) = \sum_{i=0}^{N-1} z(k,i)W_N^{ni} \quad , \quad 0 \le n < N$$

   }

2. For $n = 0$ to $N - 1$ apply Alg. 10.3.1 to compute

   {

   $$Z(m,n) = \sum_{k=0}^{M-1} Y(n,k)W_M^{mk} \quad , \quad 0 \le m < M$$

   }

Note that Alg 10.7.1 is not an in-place procedure because it requires an $N \times M$ matrix Y of intermediate variables. This is in contrast to the one-dimensional FFT, which requires no extra memory. Since $Z(m,n)$ is a complex-valued function, it can be expressed in polar coordinates in terms of a *magnitude* $A(m,n)$ and *phase* $\phi(m,n)$ as follows.

$$A(m,n) \triangleq |Z(m,n)| \tag{10.7.5a}$$

$$\phi(m,n) \triangleq \arctan\left\{\frac{\text{Im}\,[Z(m,n)]}{\text{Re}\,[Z(m,n)]}\right\} \tag{10.7.5b}$$

Symmetry arguments similar to those used for the one-dimension FFT can be used to show that for a real signal, all of the important information about the two-dimensional FFT is contained in the first quadrant: $0 \le m < M/2$, $0 \le n < N/2$. If Δx is the sampling interval in the x dimension and Δy is the sampling interval in the y dimension, then $Z(m,n)$ specifies the frequency response at spatial frequencies

$$f_{xm} = \frac{m}{M\Delta x} \tag{10.7.6a}$$

$$f_{yn} = \frac{n}{N\Delta y} \tag{10.7.6b}$$

EXAMPLE 10.7.1 Two-Dimensional FFT

As a simple illustration of the two-dimensional FFT, consider the following two-dimensional analog signal.

$$z_a(x,y) = \exp(-y)\sin(\pi x)$$

Suppose this signal is sampled with $M = 64$ samples in the x dimension using a sampling interval of $\Delta x = 1/10$, and $N = 64$ samples in the y dimension using a sampling interval of $\Delta y = 1/8$. A plot of the two-dimensional signal $z(x,y)$ produced by example program *e1071* on the distribution CD is shown in Figure 10.7.2. Notice that the surface is periodic in the x dimension with

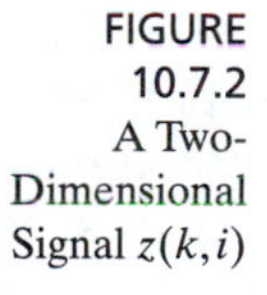
FIGURE 10.7.2 A Two-Dimensional Signal $z(k,i)$

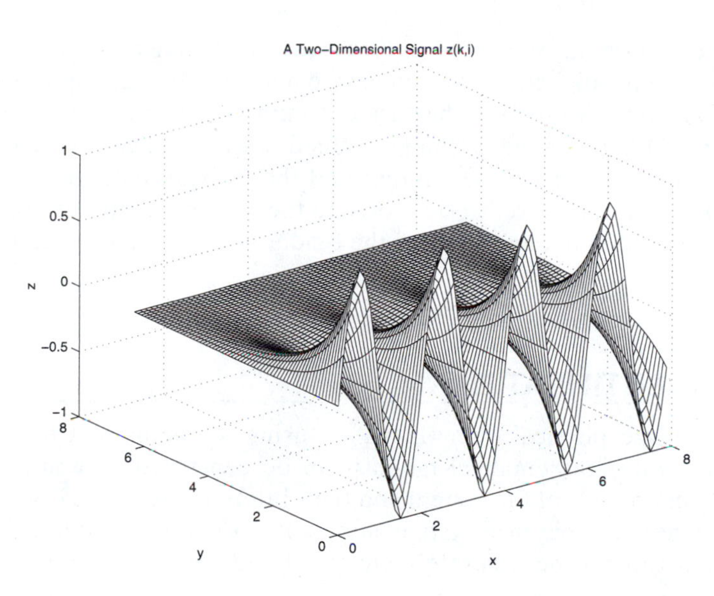

a spatial frequency of $f_x = 0.5$ cycles per unit length. However, it is not periodic in the y dimension; instead, it contains energy at all spatial frequencies. This becomes evident when we examine the plot of the two-dimensional magnitude spectrum, $A(m,n)$, shown in Figure 10.7.3. The concentration of energy at one spatial frequency in the x dimension is indicated by the ridge centered at $f_x = 0.5$ cycles per unit length. Similarly, the distribution of energy in the y dimension is evident from the fact that the ridge decays gradually with the higher spatial frequencies having less energy than the lower spatial frequencies.

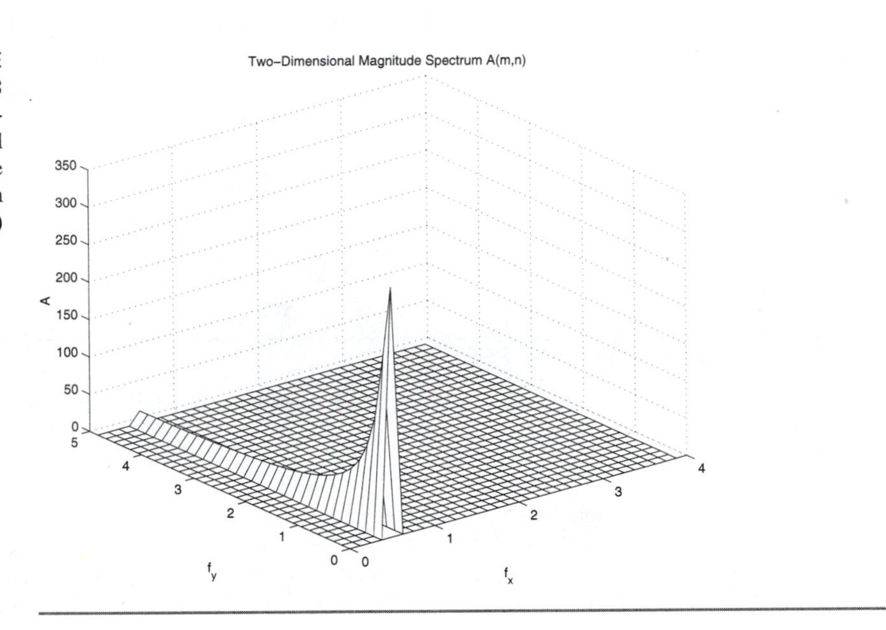

FIGURE 10.7.3 Two-Dimensional Magnitude Spectrum $A(m,n)$

All of the basic signal processing techniques, including correlation, convolution, and filtering, can be applied to two-dimensional signals. For example, low-pass filtering of an image tends to smooth the variations in intensity and consequently remove the fine detail. Other techniques can be applied to detect edges in an image or separate objects in a scene from one another and the background, a process called segmentation. Entire books have been devoted to the topic of computer vision and image processing. For additional information, the reader is referred, for example, to Ballard and Brown (1982).

10.8 SYSTEM IDENTIFICATION

Engineers analyze physical phenomena by using mathematical models. In many instances, accurate mathematical models can be constructed by applying physical laws to the components of the system and their interconnections. However, there are other cases where the physical system is too complex or the individual components of the system can not be accurately characterized, perhaps because they are not

FIGURE 10.8.1 A Physical System

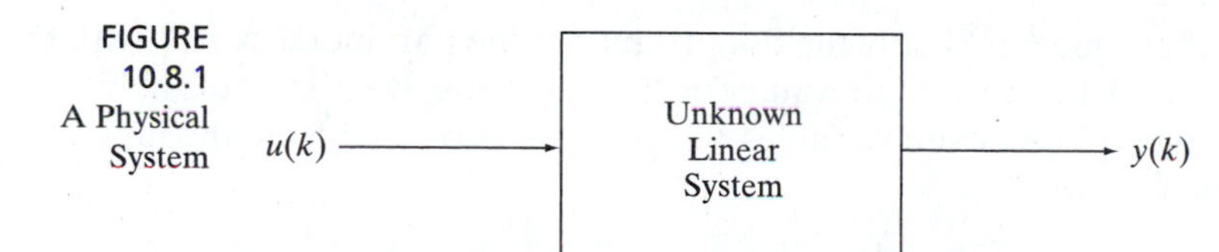

accessible. In these cases, it is useful to develop a mathematical model based solely on measurements of the input $u(k)$ and output $y(k)$ of the system as shown in Figure 10.8.1. This "black box" approach to mathematical modeling is referred to as *system identification*.

10.8.1 Least-Squares Method

To mathematically model a dynamic discrete-time system using measurements of the input and output, we assume that it is a *linear* system that can be characterized by an input/output relationship of the form

$$y(k) + \sum_{i=1}^{n} a_i y(k-i) = \sum_{i=0}^{m} b_i u(k-i) \tag{10.8.1}$$

A system for which the $a_i = 0$ for $1 \le i \le n$ is called a *moving average* or MA model because the output is a weighted sum of the present and past inputs. When $b_i = 0$ for $1 \le i \le m$, the system in Equation (10.8.1) is called an *auto-regressive* or AR model. The general case in (10.8.1) is called an *auto-regressive moving average* or ARMA model. For the general ARMA model, the system identification problem reduces to one of finding appropriate values for the components of the following $(n + m + 1) \times 1$ *parameter* vector.

$$\theta \triangleq [a_1, \ldots, a_n, b_0, \ldots, b_m]^T \tag{10.8.2}$$

The formulation of the output, $y(k)$, can be made more concise by introducing an $(n + m + 1) \times 1$ *state* vector, $x(k)$, that summarizes the past inputs and outputs as follows.

$$x(k) \triangleq [-y(k-1), \ldots, -y(k-n), u(k), \ldots, u(k-m)]^T \tag{10.8.3}$$

Thus, the state at time k contains the recent history of the system input and the system output. Solving for $y(k)$ in (10.8.1) and writing the result in terms of the state vector and the parameter vector yields

$$y(k) = \theta^T x(k) \quad , \quad k \ge 0 \tag{10.8.4}$$

That is, the output at time k is simply the dot product of the parameter vector with the state at time k. To find an appropriate value for the parameter vector θ, it is helpful to introduce the following composite variables, which are constructed from the input and output measurements.

$$X \triangleq \begin{bmatrix} x^T(0) \\ x^T(1) \\ \vdots \\ x^T(N-1) \end{bmatrix}, \quad y \triangleq \begin{bmatrix} y(0) \\ y(1) \\ \vdots \\ y(N-1) \end{bmatrix} \tag{10.8.5}$$

Here X is an $N \times (n + m + 1)$ matrix whose kth row is the system state at time $k - 1$, while y is an $N \times 1$ vector whose kth component is the system output at time $k - 1$. Note that to construct the first few rows of X, it is necessary to know the values of $u(k)$ and $y(k)$ for $k < 0$. We will assume that the input signal is causal (which means $u(k) = 0$ for $k < 0$) and that the initial conditions of the system are zero (which means that $y(k) = 0$ for $k < 0$).

If the system in Figure 10.8.1 can be represented by the ARMA model in Equation (10.8.4) for some $\theta = \bar{\theta}$, then the vector of system outputs can be written compactly as

$$y = X\bar{\theta} \tag{10.8.6}$$

For a general parameter vector θ, the accuracy of the fit can be measured by using the following *error* vector:

$$e(\theta) \triangleq y - X\theta \tag{10.8.7}$$

The objective is to compute a parameter vector that minimizes the error. A convenient way to measure the overall fit of the model to the data is to use the following *least-squares* error criterion.

$$\begin{aligned} J(\theta) &\triangleq e^T(\theta)e(\theta) \\ &= (y - X\theta)^T(y - X\theta) \\ &= (y^T - \theta^T X^T)(y - X\theta) \\ &= y^T y - y^T X\theta - \theta^T X^T y + \theta^T X^T X\theta \\ &= y^T y - 2y^T X\theta + \theta^T X^T X\theta \end{aligned} \tag{10.8.8}$$

To find a value of θ that minimizes $J(\theta)$, we apply the necessary condition $\partial J(\theta)/\partial\theta = 0$. By performing the partial derivatives component by component, it is possible to show that $\partial(c^T\theta)/\partial\theta = c^T$ and $\partial(\theta^T B\theta)/\partial\theta = 2\theta^T B$. Using these observations, we can compute the derivative of the expression for $J(\theta)$ in (10.8.8) and set the result to zero. Taking the transpose of both sides then results in the following system of *normal equations,* which the optimal parameter vector must satisfy.

$$X^T X\theta = X^T y \tag{10.8.9}$$

Thus, to find the parameter vector θ, we must solve a linear algebraic system of dimension $n + m + 1$. Consequently, the techniques discussed in Chapter 2 can be applied. Note that a unique solution exists if and only if the $(n + m + 1) \times (n + m + 1)$ coefficient matrix $X^T X$ is nonsingular. In this case, the solution can be written as

$$\boxed{\theta = (X^T X)^{-1} X^T y} \tag{10.8.10}$$

The matrix $X^T X$ will be invertible if two conditions hold. First, the number of samples, N, must be at least as great as the number of unknown parameters:

$$N \geq n + m + 1 \tag{10.8.11}$$

Typically, a model is identified using substantially more points than the minimum number in (10.8.11) because this tends to improve the accuracy of the model for general inputs. The second condition for $X^T X$ to be invertible is that the input $u(k)$ used for identification have sufficiently "rich" frequency content, particularly in the range of frequencies over which the ARMA model is to be used. More specifically, it is necessary that the magnitude spectrum of $u(k)$ be nonzero at $q \geq n + m + 1$ frequencies. For example, a white noise signal might be used to identify the system.

A useful generalization of the least-squares ARMA model arises when we consider the possibility that some of the samples may be more significant than others in determining the best fit. For example, perhaps recent data are more important than data collected in the remote past. This can be accounted for by introducing a diagonal *weighting matrix*, W, with positive diagonal elements. The new error criterion to be minimized is then

$$\boxed{J(\theta) = e^T(\theta) W e(\theta)} \tag{10.8.12}$$

Using the same procedure as before, one can show that the parameter vector for the weighted least-squares fit must satisfy the following system of normal equations.

$$X^T W X \theta = X^T W y \tag{10.8.13}$$

It is clear that this reduces to the original case in (10.8.9) when *uniform weighting*, $W = I$, is used. In the general case, the weighted least-squares solution is

$$\boxed{\theta = (X^T W X)^{-1} X^T W y} \tag{10.8.14}$$

EXAMPLE 10.8.1 ARMA Model

To illustrate the process of system identification, consider the following linear discrete-time dynamic system.

$$y(k) = 0.5y(k-1) - 0.6y(k-2) + 0.3y(k-3) + u(k) - 2u(k-1) + 8u(k-2) - 4u(k-3)$$

Example program *e1081* on the distribution CD identifies this system using a 512 -point white noise input uniformly distributed over the interval $[-1, 1]$, and $n = m = 3$. The estimated values for the components of the parameter vector θ are summarized in Table 10.8.1.

TABLE 10.8.1 Estimated Parameter Vector θ

k	*Actual* $\theta(k)$	*Estimated* $\theta(k)$	*Percent Error*
1	−0.5000000	−0.5002529	0.0505805
2	0.6000000	0.5999916	−0.0014007
3	−0.3000000	−0.3001547	0.0515679
4	1.0000000	0.9999979	−0.0002146
5	−2.0000000	−2.0002496	0.0124812
6	8.0000000	8.0003767	0.0047088
7	−4.0000000	−4.0018291	0.0457287

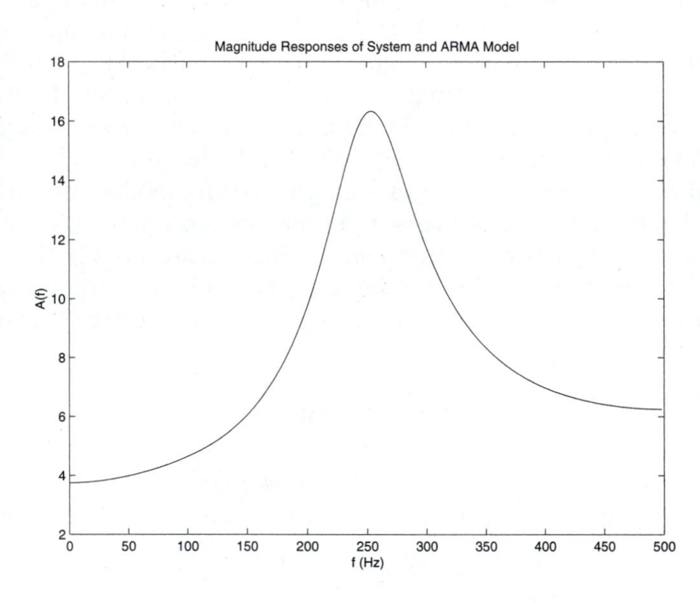

FIGURE 10.8.2 Magnitude Responses of System and ARMA Model

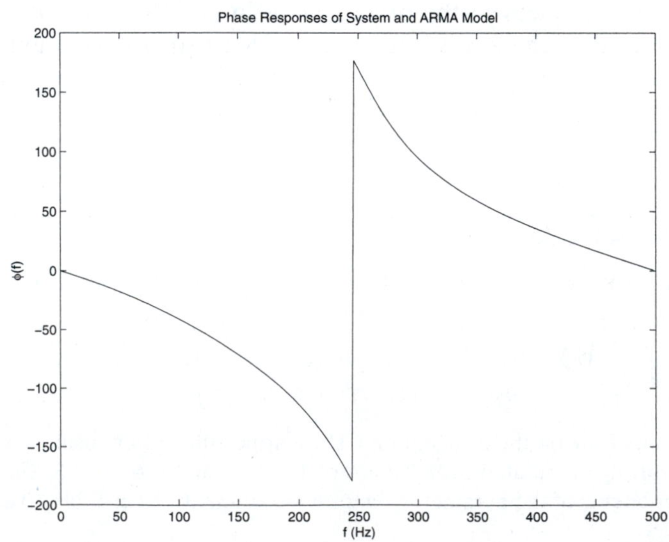

FIGURE 10.8.3 Phase Responses of System and ARMA Model

Another way to compare the original system with the ARMA model is to compute the frequency responses of the two systems. Using a sampling frequency of $f_s = 1000$ Hz, the magnitude responses are shown in Figure 10.8.2, and the phase responses are shown in Figure 10.8.3. In both cases, the two curves are essentially indistinguishable.

The least-squares identification technique in Equations (10.8.10) and (10.8.14) is referred to as a *batch* calculation because all of the input and output data have to be known ahead of time to construct the matrix X and the vector y. It is possible to reformulate the least-squares method in a *recursive* fashion where the estimate of θ is updated in real time as each new sample becomes available (Franklin et al., 1990). Closely related to this is the problem of using input and output measurements to estimate the state of a general linear discrete-time system in the presence of noise. A popular recursive technique for computing an optimal estimate of the state is the *Kalman filter*. The interested reader is referred to Anderson and Moore (1979).

10.8.2 Adaptive LMS Method

Real-time identification of the parameters of a linear discrete-time dynamic system is useful in situations where the characteristics of the system change with time and it is necessary to *adapt* to these changes. A simple and elegant adaptive method can be developed if we restrict our attention to the case of a moving-average or MA model of the form:

$$\boxed{y(k) = \sum_{i=0}^{m} b_k u(k - i)} \tag{10.8.15}$$

In this case, the $(m + 1) \times 1$ parameter vector simplifies to $\theta = b$, and the $(m + 1) \times 1$ state vector is

$$x(k) = [u(k), u(k - 1), \ldots, u(k - m)]^T \tag{10.8.16}$$

The output $y(k)$ is the dot product of θ with $x(k)$ as in Equation (10.8.4). To develop an adaptive technique for determining an optimal value for θ, we use the arrangement shown in Figure 10.8.4. Let $\overline{y}(k)$ denote the actual output of the system being identified, and let $y(k)$ be the output of the MA model. Then the *error* at time k can be computed as follows:

$$\begin{aligned} e(k) &\triangleq \overline{y}(k) - y(k) \\ &= \overline{y}(k) - \theta^T x(k) \end{aligned} \tag{10.8.17}$$

To develop a strategy for finding an optimal value of θ, suppose we attempt to minimize the objective $e^2(k)$. The basic idea is to change each component θ in a direction that reduces the value of $e^2(k)$. Using (10.8.17), the partial derivative of $e^2(k)$ with respect to θ_i is

$$\frac{\partial e^2(k)}{\partial \theta_i} = -2e(k)x_i(k) \quad , \quad 1 \le i \le m + 1 \tag{10.8.18}$$

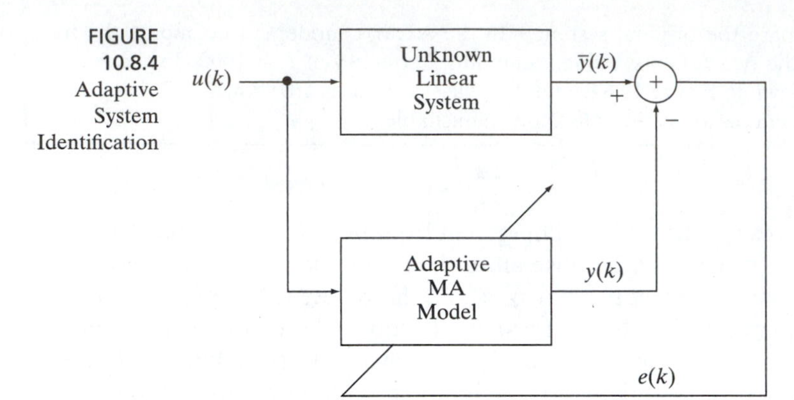

FIGURE 10.8.4 Adaptive System Identification

Recall from Section 6.4 that the simplest iterative technique for finding an optimal value of θ is the method of steepest descent. That is, we search in the direction opposite to the gradient of the objective function. This yields the following simple adaptive update formula for θ, which is called the *least-mean square* or LMS method (Widrow and Stearns, 1985).

$$\theta(k+1) = \theta(k) + 2\mu e(k)x(k) \quad , \quad k \geq 0 \tag{10.8.19}$$

Here $\mu > 0$ is the step size for the search. Rather than use a variable step size as in Chapter 6, we use a small fixed step size because all the computations must be done in real time, so computational speed is at a premium. A simple upper bound can be placed on the step size μ. Recall from (10.4.4) that the energy of a finite signal $u(k)$ is the sum of the squares of the samples. Dividing energy by time yields units of power. The *average power* of a bounded discrete-time signal $u(k)$ for $k \geq 0$ is defined:

$$P(u) \triangleq \lim_{n\to\infty} \frac{1}{n} \sum_{k=0}^{n} u^2(k) \tag{10.8.20}$$

Under very general conditions on a random white noise input $u(k)$, the LMS adaptive method in Equation (10.8.19) converges to an optimal parameter vector if the step size is confined to the following range (Widrow and Stearns, 1985):

$$0 < \mu < \frac{1}{(m+1)P(u)} \tag{10.8.21}$$

The criterion in (10.8.21) is quite easy to apply. For example, if $u(k)$ is a random sequence that is uniformly distributed over the interval $[-1, 1]$, then clearly $P(u) < 1$. Hence, a conservative upper bound in this case is $\mu = 1/(m+1)$.

EXAMPLE 10.8.2 LMS Method

To illustrate the LMS method, we again consider the linear discrete-time system in Example 10.8.1 where $n = m = 3$. Even though $n > 0$, this system can be modeled effectively with an adaptive MA model as long as the number of terms is sufficiently large. Suppose a MA model with

$m = 32$ is used. If the input signal $u(k)$ is random white noise uniformly distributed over the interval $[-1, 1]$, then $P(u) < 1$, and from (10.8.21), we can use the following step size:

$$\mu = \frac{1}{33}$$

Example program *e1082* on the distribution CD applies the LMS method to identify this system starting with an initial guess of $\theta = 0$. The resulting error signal $e(k)$ indicating the difference between the actual output and the estimated output is shown in Figure 10.8.5. It is clear that LMS method converges to a very small error after about 300 samples. The accuracy of the MA model can be examined by applying a second random white noise input. A graph showing the actual output and the predicted output using the θ identified by the LMS method is shown in Figure 10.8.6. It is apparent that the two outputs are essentially identical in this case. Thus,

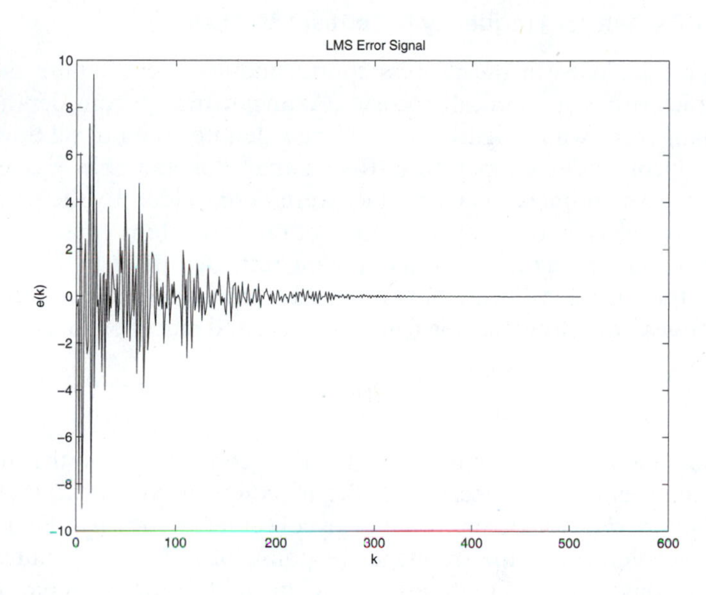

FIGURE 10.8.5 LMS Error Signal

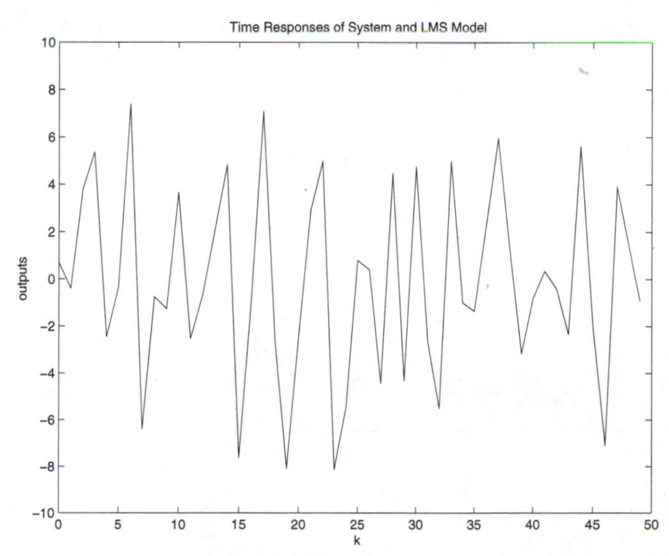

FIGURE 10.8.6 Time Responses of System and LMS Model

an MA model can be used to model a more general stable linear system if the number of terms m is sufficiently large.

10.9 APPLICATIONS

The following examples illustrate applications of digital signal processing techniques using both MATLAB and C. The relevant MATLAB functions in the NLIB toolbox are described in Section 1.11 of Appendix 1, while the corresponding C functions in the NLIB library are described in Section 2.14 of Appendix 2.

10.9.1 Heat Exchanger Frequency Response: MATLAB

DSP techniques are used in the process control industry to compute the frequency response characteristics of chemical processes. As an illustration, consider the glycol–hot oil heat exchanger shown in Figure 10.9.1. Here r denotes the hot oil flow rate and T denotes the glycol outlet temperature. Recall that this example was examined in Section 6.9.1, where optimization techniques were used to identify the parameters of a mathematical model. Suppose that when the hot oil flow rate is constant and equal to $r(t) = \alpha$, the corresponding constant steady-state outlet temperature is $T(t) = \beta$. Let u and y denote the variations in the flow rate and temperature, respectively, from their steady state values. Then, from Section 6.9.1, the linearized model of the heat exchanger is

$$\frac{dy(t)}{dt} = \frac{1}{\tau}[au(t - T_d) - y(t)] \tag{10.9.1}$$

Here the model parameters are the gain a, the time constant τ, and the time delay T_d. Recall that a time delay occurs because the thermocouple used to measure temperature is placed a distance d downstream from the glycol outlet as shown in Figure 10.9.1.

In order to determine the frequency response of this process numerically, it is helpful to first convert (10.9.1) into an equivalent discrete-time system or difference equation. Let T denote the sampling interval. Recall from Section 7.2 that if $t = kT$

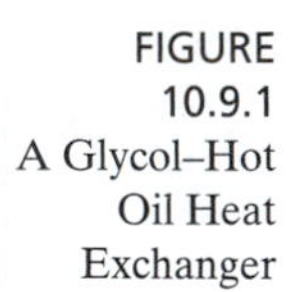
FIGURE 10.9.1 A Glycol–Hot Oil Heat Exchanger

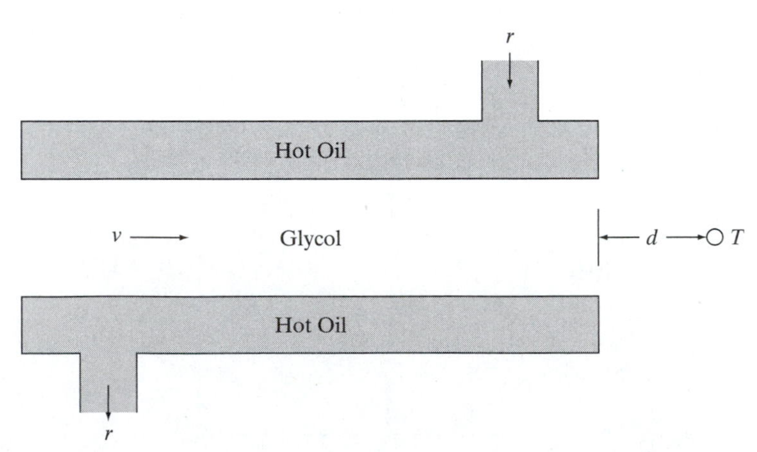

denotes the current time and $y(k)$ is $y(t)$ at $t = kT$, then the derivative in (10.9.1) can be approximated with a first-order backwards difference operator:

$$\frac{dy(t)}{dt} \approx \frac{y(k) - y(k-1)}{T} \tag{10.9.2}$$

Using this backward Euler approximation for the derivative, the continuous-time model in (10.9.1) can be converted to the following discrete-time model:

$$y(k+1) = y(k) + \frac{T}{\tau}[au(k-m) - y(k)] \tag{10.9.3}$$

In this case, the integer delay m is in units of sampling intervals. That is, m is computed as

$$m = \text{round}\left\{\frac{T_d}{T}\right\} \tag{10.9.4}$$

If the sampling interval T is chosen to be an integer submultiple of the time delay T_d, then no rounding is needed, and the time delay can be represented exactly in discrete time. To make the problem specific, suppose the three parameters are a gain of $a = 10\,^\circ\text{C-m}^3/\text{min}$, a time constant of $\tau = 4.5$ min, and a time delay of $T_d = 1.8$ min. If the sampling interval is $T = 0.3$ min, or one sample every 18 sec, the time delay is $m = 6$ samples. The following MATLAB script on the distribution CD can be used to solve this problem.

```
%------------------------------------------------------------------
% Example 10.9.1: Heat Exchanger Frequency Response
%------------------------------------------------------------------

% Initialize

    clc                        % clear screen
    clear                      % clear variables
    n   = 256;                 % number of samples
    T   = 0.25;                % sampling interval (min)
    fs  = 1/T;                 % sampling rate (samples/min)
    a   = 10;                  % gain (deg C-m^3/min)
    tau = 4.5;                 % time constant (min)
    Td  = 1.5;                 % time delay (min)
    u   = zeros (n,1);         % unit pulse input
    y   = zeros (n,1);         % pulse response
    t   = zeros (n,1);         % discrete times

% Compute pulse response of heat exchanger

    fprintf ('Example 10.9.1: Heat Exchanger Frequency Response\n');
    u(1) = 1;
    m   = round (Td/T);
    for k = 1 : n
       t(k) = (k-1)*T;
```

```
      if k > max(m+1,1)
         y(k) = y(k-1) + (T/tau)*(a*u(k-1-m) - y(k-1));
      end
   end

% Compute and display frequency response

[A,phi,f] = freqrsp (u,y,fs);
graphxy(f(1:n/2),A(1:n/2),'Magnitude Response','f (samples/min)','A')
%------------------------------------------------------------------------
```

When script *e1091.m* is executed, it computes the frequency response of the glycol–hot oil heat exchanger. This is achieved by driving the system in Equation (10.9.3) with a unit pulse input and finding the resulting pulse response $y(k)$. The frequency response is then obtained by taking the FFT of the $y(k)$. It includes the magnitude response $A(f)$ shown in Figure 10.9.2, which reveals a low-pass characteristic.

A solution to Example 10.9.1 using C can be found in the file *e1091.c* on the distribution CD.

10.9.2 Flagpole Motion: C

Tall structures such as flagpoles, masts on sailboats, and even buildings, deflect from their nominal vertical positions when they are exposed to sufficiently strong gusts of wind. Consider the problem of developing an empirical dynamic model for the motion of the top of a flagpole in the presence of a variable wind. Let $u(t)$ denote the wind speed, which is assumed to be uniform over the length of the flagpole, and let $y(t)$

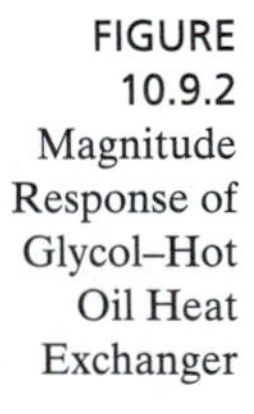
FIGURE 10.9.2 Magnitude Response of Glycol–Hot Oil Heat Exchanger

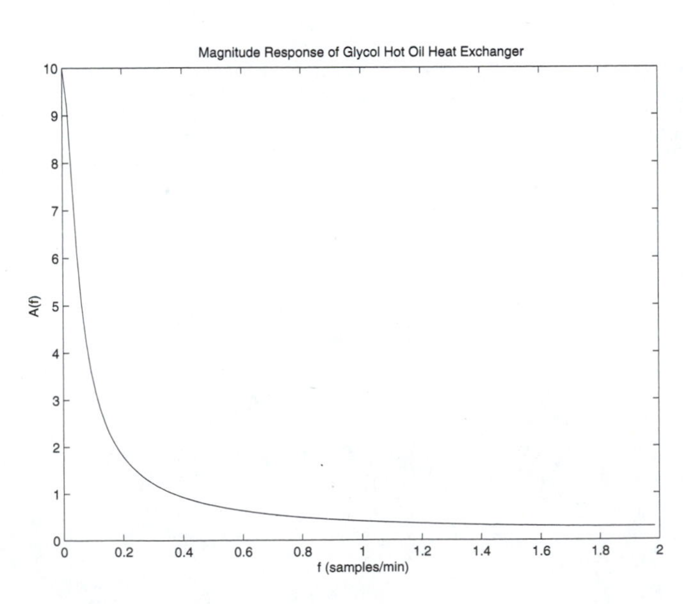

denote the deflection of the top of the flagpole from the vertical position as shown in Figure 10.9.3.

A complete dynamic model of the flagpole, which specifies the deflection along each point of the pole, is quite complicated and involves a nonlinear partial differential equation (see, e.g., Chapra and Canale, 1988). Suppose the wind speed is not too large, so the deflections remain small. Under these conditions, the motion of the top of the flagpole can be modeled empirically with a lumped linear discrete-time model. In particular, suppose the wind speed $u(k)$ and deflection $y(k)$ are measured every $T = 0.5$ sec and result in the data summarized in Table 10.9.1.

FIGURE 10.9.3 Deflection of a Flagpole Due to Wind

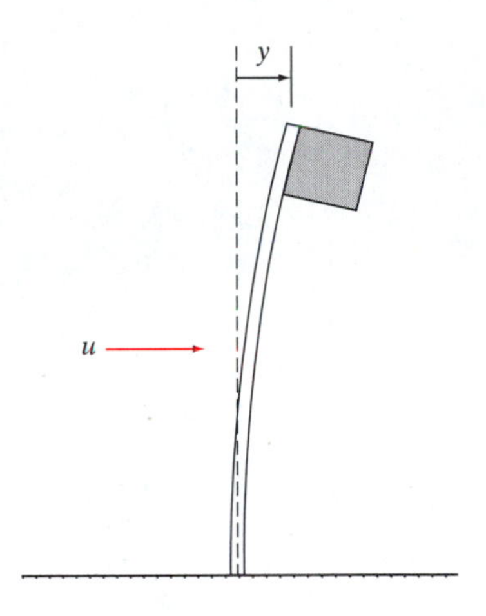

TABLE 10.9.1 Deflection Measurements of Top of Flagpole

t (sec)	$u(t)$ (MPH)	$y(t)$ (cm)	t (sec)	$u(t)$ (MPH)	$y(t)$ (cm)
0.0	−5.4	−3.21	8.0	0.6	2.37
0.5	12.1	2.03	8.5	2.1	14.95
1.0	−12.3	0.22	9.0	7.9	22.55
1.5	−12.8	−10.58	9.5	10.4	23.72
2.0	−7.9	−16.36	10.0	0.5	12.29
2.5	−10.0	−15.07	10.5	5.1	−2.65
3.0	−0.2	−6.66	11.0	5.4	−11.80
3.5	2.7	8.48	11.5	−9.0	−19.40
4.0	−12.2	11.14	12.0	0.9	−18.66
4.5	7.1	10.79	12.5	6.3	−3.68
5.0	13.0	15.65	13.0	−7.8	6.43
5.5	−1.0	10.46	13.5	−6.8	7.16
6.0	−5.2	−4.59	14.0	−0.3	6.65
6.5	−3.5	−17.06	14.5	0.4	4.26
7.0	−8.4	−22.96	15.0	14.5	8.78
7.5	5.7	−14.07	15.5	−10.9	2.75

System identification techniques can be used to fit a model to these data. In particular, consider the following moving-average (MA) model of order $m + 1$.

$$y(k) = \sum_{i=0}^{m} b_i u(k - i) \tag{10.9.5}$$

The following C program on the distribution CD can be used to solve this problem.

```
/*----------------------------------------------------------------*/
/* Example 10.9.2: Flagpole Motion                                 */
/*----------------------------------------------------------------*/

#include "c:\nlib\util.h"

int main (void)
{
   int         p,                        /* number of samples */
               n = 0,                    /* number of past outputs */
               m = 20,                   /* number of past inputs */
               r = n + m + 1,
               i,j;
   vector      theta = vec (r,""),       /* parameters */
               x     = vec (r,""),       /* states */
               t,                        /* time data */
               u,                        /* input data */
               y;                        /* output data */
   matrix      Y;                        /* outputs */
   FILE        *f;

/* Get input/output data */

   printf ("\nExample 10.9.2: Flag Pole Motion\n");
   f = fopen ("data1092.m","r");
   t = getvec (f,&p);
   u = getvec (f,&p);
   y = getvec (f,&p);
   fclose (f);
   Y = mat (2,p,"");
   Y[1] = y;

/* Identify MA model and compare responses */

   getarma(u,y,p,n,m,theta);
   showvec ("theta",theta,r,0);
   for (i = 1; i <= p; i++)
      Y[2][i] = arma (u[i],theta,x,n,m);
   graphxm (t,Y,2,p,"","t (sec)","y (cm)","fig1094");

/* Write theta to file */
```

```
   f = fopen ("theta.dat","w");
   for (i = 1; i <= r/2; i++)
   {
      j = r/2 + 1 + i;
      fprintf (f,"%i & %.3f & %i & %.3f \\\\ \n",
               i,theta[i],j,theta[j]);
   }
   fprintf (f,"%i & %.3f & & \\\\ \n",r/2+1,theta[r/2+1]);
   return 0;
}
/*-----------------------------------------------------------------*/
```

When program *e1092.c* is executed, it computes a least-squares MA model to find the optimal coefficients for the case $m = 20$. This results in the coefficient vector shown in Table 10.9.2.

To determine how well the model fits the data, the output of the model is compared to the data in Figure 10.9.4. Although the responses are not identical, they are quite close. Differences between the two can be attributed to such factors as unmodeled dynamics and measurement noise that is present in the data in Table 10.9.1.

A solution to Example 10.9.2 using MATLAB can be found in the file *e1092.m* on the distribution CD.

10.9.3 Band Pass Filter: MATLAB

Suppose a received electrical signal is sampled at a rate of $f_s = 1000$ Hz. The signal is corrupted with random white noise, and the information embedded within the signal is known to have a magnitude spectrum that lies in the frequency range from 200 Hz to 400 Hz. In order to remove as much of the white noise as possible without losing any of the information contained in the signal, we process the signal with a band pass filter with gain $a = 1$ and cutoff frequencies

$$[f_L, f_H] = [200, 400] \text{ Hz} \tag{10.9.6}$$

TABLE 10.9.2 Coefficient Vector for Least-Squares MA Model

k	b_k	k	b_k
1	0.492	12	0.187
2	0.691	13	−0.165
3	0.316	14	−0.356
4	−0.154	15	−0.290
5	−0.422	16	−0.102
6	−0.570	17	0.035
7	−0.373	18	0.205
8	−0.037	19	0.191
9	0.307	20	0.065
10	0.419	21	−0.070
11	0.389		

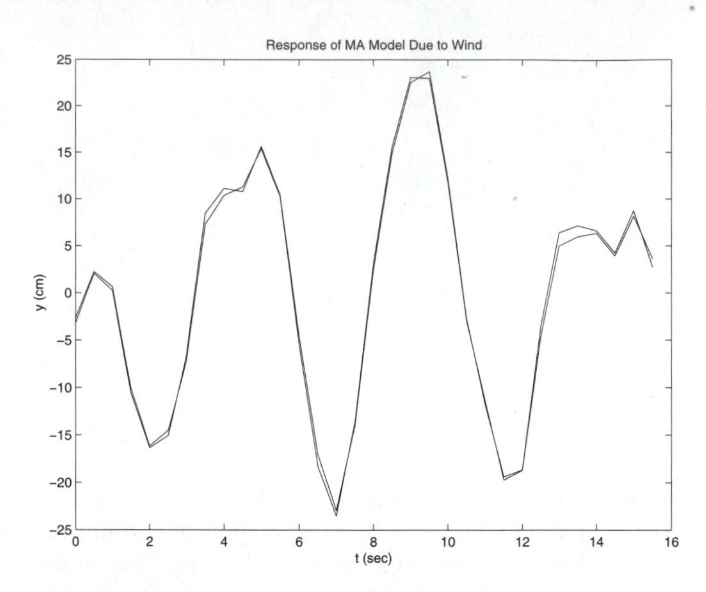

FIGURE 10.9.4 Response of MA Model Due to Wind

Recall that since the sampling frequency is $f_s = 1000$ Hz, the highest frequency the digital filter can process is the Nyquist frequency $f_s/2 = 500$ Hz. Therefore, the normalized low-frequency cutoff in this case is $f_0 = 0.4$, and the normalized high-frequency cutoff is $f_1 = 0.8$. Suppose a linear-phase FIR filter of order $2p + 1$ is to be designed:

$$y(k) = \sum_{i=0}^{2p} b_i u(k - i) \tag{10.9.7}$$

The following MATLAB script on the distribution CD can be used to solve this problem.

```
%------------------------------------------------------------------
% Example 10.9.3: Band Pass Filter
%------------------------------------------------------------------

% Initialize

   clc                       % clear screen
   clear                     % clear variables
   p = 25;                   % select filter order
   r = 2*p + 1;              % number of terms
   n = 512;                  % number of samples
   fs = 1000;                % sampling frequencies
   x   = zeros (r,1);        % filter state
   u   = zeros (n,1);        % unit pulse input
   y   = zeros (n,1);        % pulse response

% Find coefficients
```

```
    fprintf ('Example 10.9.3: Band Pass Filter\n');
    b = fir ('funf1093',p);
    show ('Filter coefficients',b)

% Compute pulse response and magnitude response

    u(1) = 1;
    for i = 1 : n
       [x,y(i)] = arma (u(i),b,x,0,r-1);
    end
    [A,phi,f] = freqrsp (u,y,fs);
    graphxy (f(1:n/2),A(1:n/2),'Filter Magnitude Response','f (Hz)','A')
```

```
function a = funf1093 (f)
%----------------------------------------------------------------------
% Description: Band pass magnitude response
%----------------------------------------------------------------------
    if (f >= 0.4) & (f <= 0.8)
       a =  1;
    else
       a = 0;
    end
%----------------------------------------------------------------------
```

When script *e1093.m* is executed, it designs a FIR band pass filter of order $2p + 1 = 51$ using a Hamming window. A plot of the magnitude response of the filter is shown in Figure 10.9.5. Observe that the cutoff frequencies are at $f_L = 200$ Hz and

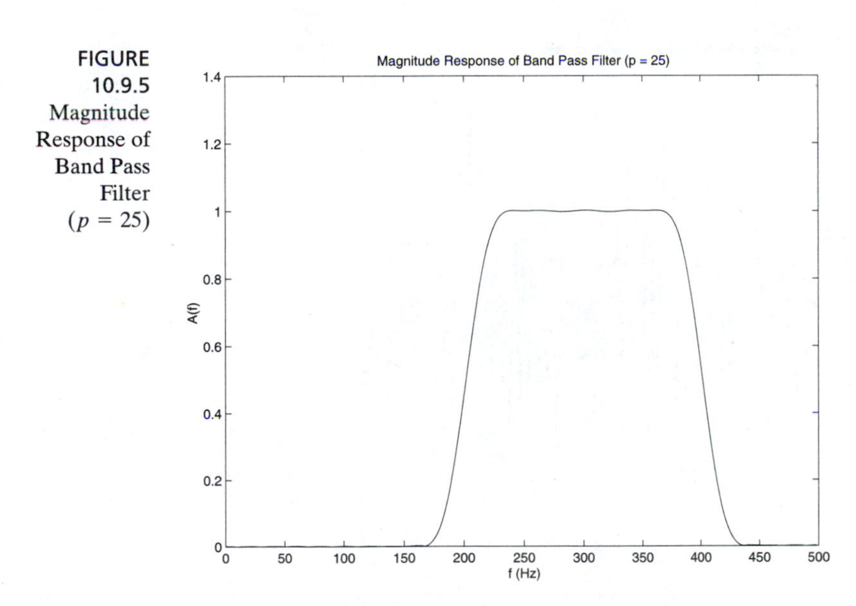

FIGURE 10.9.5 Magnitude Response of Band Pass Filter ($p = 25$)

$f_H = 400$ Hz per the specifications, the pass band gain is one, and there is very little ripple in the magnitude response. Because this is an MA filter, it is guaranteed to be a stable discrete-time system.

A solution to Example 10.9.3 using C can be found in the file *e1093.c* on the distribution CD.

10.9.4 Helicopter Noise: C

To illustrate the application of DSP techniques in aeronautical engineering, suppose the rotor noise from an overhead helicopter is sampled at a rate of $f_s = 512$ Hz and recorded as shown in Figure 10.9.6. The problem is to determine the rate at which the helicopter rotor is turning.

The microphone samples $x(k)$ include a narrow-band periodic component whose fundamental frequency is determined by the rotor speed, plus several harmonics. It is apparent from inspection of Figure 10.9.6 that there is also a significant random component due to such factors as ambient noise in the environment and noise created by turbulent air flow.

To estimate the speed at which the helicopter blades are rotating, we compute the magnitude spectrum of the time samples in Figure 10.9.6. The following C program on the distribution CD can be used to solve this problem.

```
/*----------------------------------------------------------------*/
/* Example 10.9.4: Helicopter Noise                                */
/*----------------------------------------------------------------*/

#include "c:\nlib\util.h"

int main (void) {
```

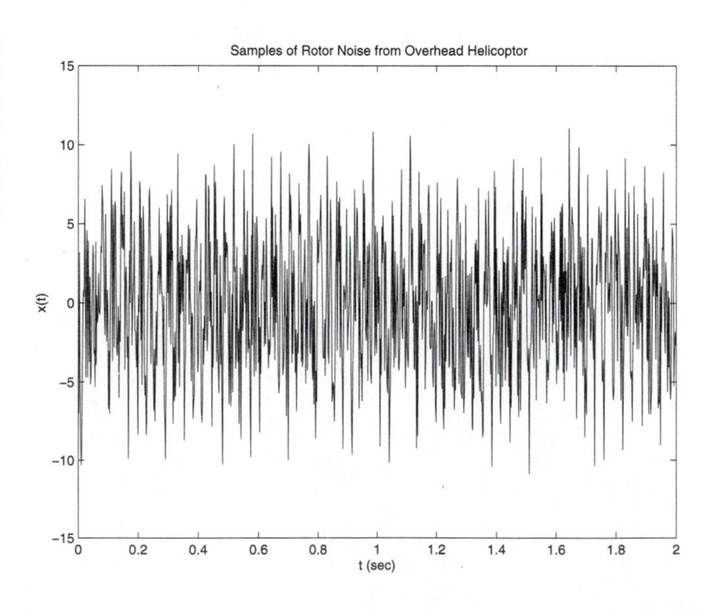

FIGURE 10.9.6 Samples of Rotor Noise from Overhead Helicopter

```
   int          n = 1024,           /* number of samples */
                m = 3,              /* number of harmonics */
                i,j;
   float        fs = 512.0f,        /* sampling frequency (Hz) */
                T = 1/fs,           /* sampling interval (sec) */
                blade = 32.0f,      /* blade passage frequency (Hz) */
                r   = 6.0f,         /* random noise magnitude */
                df = fs/n,
                b;
   vector       t   = vec (n,""),              /* time */
                x   = vec (n,""),              /* noise */
                A   = vec (n,""),              /* magnitude */
                phi = vec (n,""),              /* phase */
                f   = vec (n,""),              /* frequency */
                a   = vec (m,"3.2 2.1 1.3"),   /* magnitudes */
                p   = vec (m,"1.8 -0.7 2.1");  /* phase angles */

/* Generate noise */

   printf ("\nExample 10.9.4: Helicoptor Noise\n");
   for (i = 1; i <= n; i++)
   {
      t[i] = (i-1)*T;
      x[i] = randu(-r,r);
      for (j = 1; j <= m; j++)
      {
          b = j*2*PI*blade*(i-1)*T;
         x[i] += (float) (a[j]*cos(b + p[j]));
      }
   }
   graphxy (t,x,n,"","t (sec)","x(t)","fig1096");

/* Convert from dB sound measurements to amplitude */

   for (i = 1; i <= n; i++)
      x[i] = (float) pow(10,x[i]/20);

/* Compute spectrum */

   spectra (x,n,fs,A,phi,f);
   A[1] = 0;                         /* remove average value */
   graphxy (f,A,n/2,"","f (Hz)","A","fig1097");
   return 0;
}
/*----------------------------------------------------------------*/
```

When program *e1094.c* is executed, it computes the FFT of the 1024 samples in $x(k)$. The magnitude response $A(f)$ is then computed directly from the FFT and is shown in Figure 10.9.7. In order to view the results clearly, the DC component, $A(0)$, has been

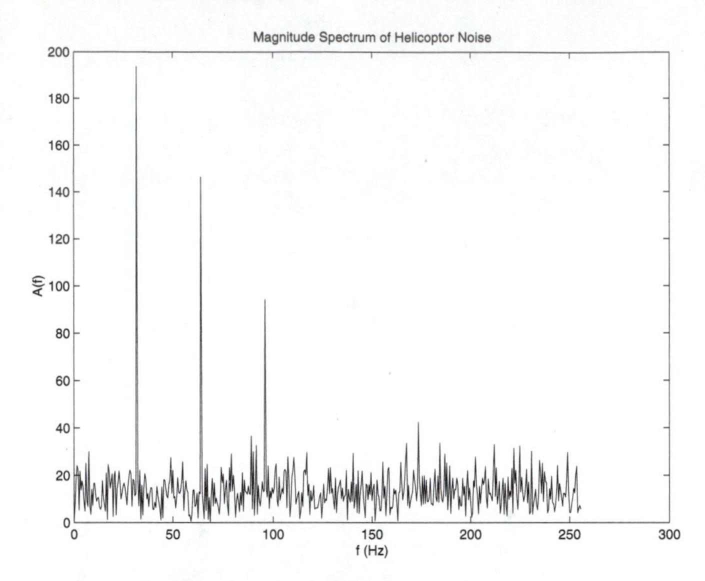

FIGURE 10.9.7 Magnitude Spectrum of Helicopter Noise

removed before plotting. It is apparent from inspection of Figure 10.9.7 that there is a fundamental frequency component at 32 Hz and two harmonics embedded within the white noise associated with turbulent air flow. Thus, the blade rotation frequency in this case is

$$f_{blade} = 32 \text{ Hz} \tag{10.9.8}$$

A solution to Example 10.9.4 using MATLAB can be found in the file *e1094.m* on the distribution CD.

10.10 SUMMARY

Numerical methods for digital signal processing are summarized in Table 10.10.1. The dimension column in Table 10.10.1 refers to the number of independent variables.

The fast Fourier transform or FFT is a highly efficient technique for computing the discrete Fourier transform of a signal $x(k)$ as in Equation (10.3.3). With minor mod-

TABLE 10.10.1 Digital Signal Processing Techniques

Method	*Dimension*	*FLOPs*
FFT	1	$0.5N \log_2 N$
Correlation	1	$1.5N \log_2 N$
Convolution	1	$1.5N \log_2 N$
Frequency response	1	$0.5N \log_2 N$
FIR filter	1	$2p + 1$
Two-dimensional FFT	2	$N^2 \log_2 N$
Least-squares identification	1	$4(n + m + 1)^3/3$
LMS identification	1	$m + 3$

ifications, it can also be used to compute the inverse FFT in (10.3.4). The FFT requires that the number of samples N be a positive power of two. The number of complex FLOPs required to compute an N-point FFT is $0.5N \log_2 N$. The FFT is a fundamental analytical tool that is useful for determining the frequency content or spectrum of a discrete-time signal.

The normalized cross correlation of two signals $x(k)$ and $y(k)$ is a function, with values in the interval $[-1, 1]$, that can be used to determine the similarity between $x(k)$ and $y(k)$. For example, cross correlation can be used to detect the presence and location of a signal that is embedded within noise. The cross correlation of a signal with itself is called the auto correlation. The FFT can be used to efficiently compute the cross correlation of two N-point signals using $1.5N \log_2 N$ FLOPs.

The convolution of two signals $h(k)$ and $u(k)$ is a third signal, $y(k)$. If $h(k)$ is the pulse response of a linear discrete-time dynamic system, subject to zero initial conditions, then the output $y(k)$ produced by an input $u(k)$ is the convolution of $h(k)$ with $u(k)$ as in (10.5.6). Again, the FFT can be used to efficiently compute the convolution of two N-point signals using $1.5N \log_2 N$ FLOPs.

The frequency response $H(m)$ of a discrete-time dynamic system or digital filter is obtained by taking the FFT of the pulse response $h(k)$ of the system. More generally, the frequency response is the FFT of the system output $y(k)$ divided by the FFT of the system input $u(k)$ assuming zero initial conditions.

$$H(m) = \frac{Y(m)}{U(m)} \tag{10.10.1}$$

The complex frequency response $H(m)$ specifies how the mth discrete frequency, $f_m = mf_s/N$, is scaled in amplitude and shifted in phase by the linear system. Here N is the number of samples, $f_s = 1/T$ is the sampling frequency, and $0 \leq m < N/2$. The magnitude of $H(m)$ is called the magnitude response, and phase angle of $H(m)$ is called the phase response of the linear system. If the pulse response is available, the frequency response of a linear discrete-time system can be computed using $0.5N \log_2 N$ FLOPs.

A finite impulse response or FIR filter is a linear-phase moving-average (MA) filter that is designed using a truncated Fourier series of the desired magnitude response. Frequency responses over the range $0 \leq f \leq f_s/2$ can be specified where $f_s = 1/T$ is the sampling frequency. Ringing or oscillations in the magnitude response are reduced by using window functions, which gradually taper the magnitudes of the Fourier coefficients. FIR filters are guaranteed to be stable. A filter with $2p + 1$ terms is a linear-phase filter with a time delay of $\tau = pT$ where T is the sampling interval.

The two-dimensional FFT is an efficient procedure for computing the discrete Fourier transform of a two-dimensional signal or image $z(k, i)$ as in (10.7.1). The two-dimensional FFT is formulated as a sequence of one-dimensional FFTs. If the number of samples in each dimension is N, the number of FLOPs required is $N^2 \log_2 N$. The two-dimensional FFT is useful for image-processing applications, where it can be used to analyze the spatial frequency content of images.

Least-squares identification is used to find optimal parameter values for the auto-regressive moving average (ARMA) model in (10.8.1) using input and output measurements. If n is the number of AR terms and $m + 1$ is the number of MA terms, then at least $n + m + 1$ samples must be used. Furthermore, the input signal used for identification must have a magnitude spectrum that is nonzero at $r \geq n + m + 1$ frequencies. The number of FLOPs needed to formulate and solve the normal equations for the $(n + m + 1) \times 1$ parameter vector θ is approximately $4(n + m + 1)^3/3$.

Adaptive least-mean square (LMS) identification is a real-time technique for identifying the parameters of an MA model. The $(m + 1) \times 1$ parameter vector θ in (10.8.19) is updated as each new sample becomes available. The technique applies the steepest descent method to the square of the current error signal using a fixed step size. To ensure convergence, the step size μ must be restricted as in (10.8.21), where $P(u)$ is the average power of the input. The number of FLOPs per iteration is $m + 3$.

PROBLEMS

The problems are divided into Analysis problems, which can be solved by hand, and Computation problems, which require the use of MATLAB or C. Solutions to selected problems can be found in Appendix 4. Students are encouraged to use these problems, which are identified with an (S), as a check on their understanding of the material. Problems marked with a (P) are programming problems that require the student to implement one or more of the algorithms discussed in the chapter. The remaining Computation problems require the student to write a main program that uses one or more of the NLIB functions discussed in Appendix 1 or Appendix 2. Some of the Computation problems are also marked with a (D). These problems require the use of a data file that can be found in the data subfolder within the NLIB folder or directory. To read these plain text files from MATLAB, simply enter the file name. To read them from a C program, use the appropriate sequence of getvec and getmat NLIB functions as described in Appendix 2.

Analysis

10.1 Consider the following analog signal where $\alpha > 0$.

$$x_a(t) = \begin{cases} \exp(-\alpha t) & , \quad t \geq 0 \\ 0 & , \quad t < 0 \end{cases}$$

(a) Find the Fourier transform, $X(f)$, of this signal.

(b) Find the energy density spectrum, $|X(f)|^2$, of this signal.

(c) What percentage of the total energy of $x_a(t)$ is contained in the frequency band $[0, 100]$ Hz?

(d) Is this signal band limited? If so, what is its bandwidth B ?

(S)10.2 Consider the analog signal $x_a(t) = \cos(2\pi f_0 t)$. The Fourier transform of this signal is as follows where $\delta(f)$ denotes the unit impulse, which is zero for $f \neq 0$.

$$X(f) = \frac{\delta(f + f_0) + \delta(f - f_0)}{2}$$

For what range of sampling intervals, T, can this signal be reconstructed from its samples?

10.3 If an analog signal is not sampled fast enough to reconstruct it from its samples, then a phenomenon called *aliasing* occurs. Suppose the following two signals are sampled with a sampling rate of $f_s = 10$ Hz. Show that the samples of $x_2(t)$ are identical to the samples of $x_1(t)$. How fast must $x_2(t)$ be sampled to avoid aliasing?

$$x_1(t) = \cos(10\pi t)$$
$$x_2(t) = \cos(30\pi t)$$

10.4 Consider the following linear discrete-time system. The pulse response of this system, $h(k)$, is a sequence of numbers called the *Fibonacci sequence.*

$$y(k) = u(k) + y(k-1) + y(k-2) \quad , \quad k \geq 0$$

It can be shown that $h(k)/h(k-1) \rightarrow \gamma$ as $k \rightarrow \infty$, where the ancient Greeks referred to the constant γ as the *golden ratio*. Find the first 10 samples of $h(k)$, and estimate the golden ratio.

10.5 Let $u(k) = u_a(kT)$ for $k \geq 0$. The following linear discrete-time system can be used to approximate the integral of $u_a(t)$ over $0 \leq t \leq kT$:

$$y(k) = y(k-1) + \frac{T}{2}[u(k) + u(k-1)] \quad , \quad k \geq 1$$

Recall from Chapter 7 that if $y(0) = 0$, this is the trapezoid rule integrator.

(a) Show that the trapezoid integrator is an ARMA model by identifying appropriate values for n, m, a, and b.

(b) Show that the trapezoid integrator is unstable.

(c) Find a bounded input that generates an unbounded output.

(S)10.6 Consider the following discrete-time system. For what values of α is this system stable?

$$y(k) = u(k) - 1.6y(k-1) - (0.64 + \alpha^2)y(k-2)$$

10.7 Each semester, a college evaluates the courses that are taught and assigns a numerical score between 0 and 10 to rate the overall performance of the instructor. Suppose you are in charge of record-keeping, and you want to keep track of running averages of the course evaluation scores so that current instructors can be compared with previous instructors. Let $y(k)$ denote the running average of the evaluation score assuming the course has been taught k times, and let $u(k)$ be the score for the kth semester for $k \geq 1$. Show that you only need to know k, $y(k-1)$ and $u(k)$ in order to compute $y(k)$.

Computation

(D)10.8 Suppose a message is sent over the digital communications channel shown in Figure 10.11.1. The received signal, $x(k)$, consists of an information part, which may or may not be present, plus a random noise component, $n(k)$.

$$x(k) = \alpha y(k-d) + n(k) \quad , \quad 0 \leq k < N$$

The receiver has a *key* consisting of two signals, $y_a(k)$ and $y_b(k)$. The sequence $y_a(k)$ for $0 \leq k < M$ is the code for the message *yes*, and the sequence $y_b(k)$ for $0 \leq k < M$ is the code for the message *no*. Typically, $M \ll N$, so several yes and no symbols can be included in one signal, $y(k)$. Examples of $y_a(k)$, $y_b(k)$ and $x(k)$ (in that order) have been stored in the data file *data108.m*. Write a program that reads the signals in *data108.m* and uses the NLIB function *crosscor* to determine the sequence of yes and no messages contained in $x(k)$. Your program should plot the following variables:

(a) The received signal $x(k)$

(b) The normalized cross correlation of $x(k)$ with $y_a(k)$

(c) The normalized cross correlation of $x(k)$ with $y_b(k)$

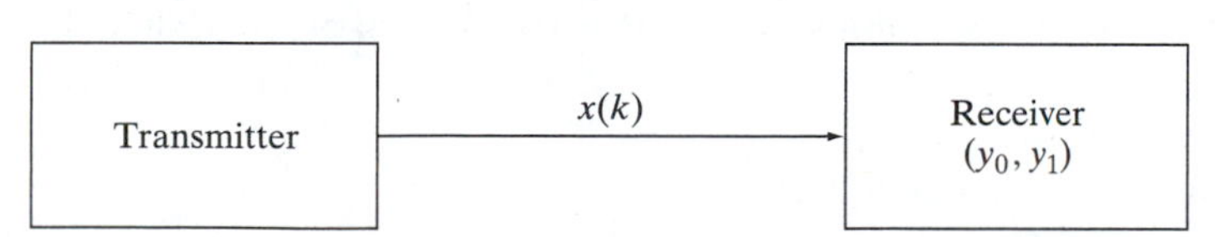

FIGURE 10.11.1 A Digital Communications Channel

(P)10.9 Write a function called *notch* that computes the coefficients of a MA band-stop filter. The pseudo-prototype of *notch* is as follows:

```
[vector b] = notch (float f0,float f1,float a,int p);
```

On entry to *notch:* $f0$ specifies the lower cutoff frequency, $f1$ specifies the upper cutoff frequency, a specifies the pass band gain, and p selects the filter order, which is $2p + 1$. The cutoff frequencies are expressed as fractions of the Nyquist frequency, $f_N = f_s/2$, which means that $0 \le f0 < 1$ and $f0 < f1 \le 1$. On exit from *filter:* the $(2p + 1)$ vector b contains the coefficients of the filter. Use the Hanning window in Table 10.6.2 to smooth the ripples in the magnitude response of the filter. The output of the filter can be computed by calling the NLIB function *arma* with $n = 0$, $m = 2p$, and $theta = b$ or by using *convolve*. Test your function by writing a main program that filters a 512-point white noise signal using $f_0 = 0.2$, $f_1 = 0.4$, $a = 1$, $p = 30$, and $f_s = 10000$ Hz. Your program should plot the following variables.

(a) The magnitude response of the filter

(b) The magnitude spectrum of the white noise input

(c) The magnitude spectrum of the filter output

(D)10.10 The file *data1010.m* contains the samples of a noise-corrupted periodic signal x. Write a program that uses the NLIB function *spectra* to find the fundamental frequency of x, assuming the sampling rate is $f_s = 500$ Hz. Your program should plot the following variables.

(a) The signal $x(k)$

(b) The magnitude spectrum of $x(k)$

(S)10.11 Consider the following discrete-time linear system. Write a program that uses the NLIB function *arma* to compute and plot the pulse response, $h(k)$, for $0 \le k < N$ where $N = 64$.

$$y(k) = u(k - 1) - 0.81y(k - 2)$$

10.12 Write a program that uses the NLIB functions to find the frequency response of the following discrete-time system using a sampling frequency of $f_s = 1000$ Hz. Your program should plot both the magnitude response and the phase response.

$$y(k) = u(k) + u(k - 1) - 0.49y(k - 2)$$

10.13 Repeat problem 10.12, but use the moving average filter in Example 10.5.1 with $p = 9$.

(D)10.14 The data file *data1014.m* contains an input signal $u(k)$ and an output signal $y(k)$ from a linear discrete-time system. Write a program that identifies an ARMA model for this system using $n = m = 5$ and the NLIB function *getarma*. Your program should display the parameter vector θ, and it should compare the system and ARMA model outputs by plotting them on the same graph.

10.15 Write a program that uses the NLIB function *roots* (see Section 2.7 of Appendix 2 for the C version) to find the poles of the following discrete-time system. Is this system stable?

$$y(k) = u(k) - \sum_{i=1}^{10} \frac{y(k - i)}{i}$$

(D)10.16 The data file *data1016.m* contains a noise-corrupted two-dimensional signal $x(k,i)$ that has some small sinusoidal terms buried within it. Verify this by writing a program that uses the NLIB function *dft2* to find the magnitude spectrum of $x(k,i)$. Your program should plot the following variables:

(a) The signal $x(k,i)$

(b) The magnitude spectrum of $x(k,i)$

(P)10.17 Write a program that uses the NLIB function *fir* to design a linear-phase filter with the following magnitude response:

$$g(f) = \frac{\sin(\pi f)}{1 + f^2}$$

Use $p = 9$ and $f_s = 1000$. Plot the actual and desired magnitude responses on the same graph for $0 \le f \le f_s/2$.

10.18 Write a program that uses the NLIB function *fir* to design a low-pass filter with cutoff frequency 600 Hz, $p = 20$, and a sampling frequency of $f_s = 2000$ Hz. The program should graph the magnitude response. What is the phase response of this filter?

(S)10.19 Write a program that uses the NLIB function *arma* to find the first 20 samples of the Fibonacci sequence by computing the pulse response of the system in Problem 10.4. The script should graph the ratio $h(k)/h(k-1)$ and print the final estimate of the golden ratio.

(D)10.20 Consider the radar system in Figure 10.1.2. The data file *data1020.m* contains the $n \times 1$ received signal $x(k)$, followed by the $m \times 1$ transmitted signal $u(k)$. Write a program that uses NLIB functions to determine if the noise-corrupted signal $x(k)$ contains an echo of the transmitted signal $u(k)$. Suppose the propagation speed of the signals is $c = 10^7$ m/sec, and the sampling frequency is $f_s = 10^5$ Hz. What is the range to the target?

References and Further Reading

Acton, F. S. 1970. *Numerical methods that work*. 2d ed. Washington: Mathematical Association of America.

Ahmed, N., and T. Natarajan. 1983. *Discrete-time signals and sytems*. Reston, VA: Reston.

Allaire, P. 1985. *Basic of the finite element method*. Dubuque, IA: Brown.

Allen, D. N. 1954. *Relaxation methods*. New York: McGraw-Hill.

Anderson, B. D. O., and J. B. Moore. 1979. *Optimal filtering*. Englewood Cliffs, NJ: Prentice Hall.

Atkinson, K. E. 1978. *An introduction to numerical analysis*. New York: Wiley.

Backus, J. 1979. The history of Fortran I, II, and III. *Annals of the History of Computing*. 1:21–37.

Bader, G., and P. A. Deuffhard. 1983. A semi-implicit mid-point rule for stiff systems of ordinary differential equations. *Numerical Mathematics* 41:373–398.

Ballard, D. H., and C. M. Brown. 1982. *Computer vision*. Englewood Cliffs, NJ: Prentice Hall.

Ben-Israel, A., and T. N. E. Grenville. 1974. *Generalized inverses: Theory and application*. New York: Wiley-Interscience.

Beyer, W. H. 1991. *CRC standard mathematic tables and formulae*. 29th ed. Boca Raton, FL: CRC Press.

Birkhoff, G., R. Varga, and D. Young. 1962. Alternating direction implicit methods. *Advances in Computers* 3:187–273.

Blanchard, A. 1976. *Phase-locked loops*. New York: Wiley.

Borse, G. J. 1997. *Numerical methods with MATLAB*. Boston: PWS.

Brent, R. P. 1973. *Algorithms for minimization without derivatives*. Englewood Cliffs, NJ: Prentice Hall.

Broyden, C. G. 1965. A class of methods for solving nonlinear simultaneous equations. *Mathematics of Computation* 19:577–593.

Bulirsch, R., and J. Stoer. 1966. Numerical treatment of ordinary differential equations by extrapolation methods. *Numerical Mathematics* 8:1–13.

Burnett, D. 1987. *Finite element analysis: From concepts to applications*. Reading, MA: Addison-Wesley.

Butcher, J. 1987. *The numerical analysis of ordinary differential equations*. New York: Wiley.

Carnahan, B., H. A. Luther, and J. O. Wilkes. 1969. *Applied numerical methods*. New York: Wiley.

Chapman, S. J. 1998. *Fortran 90/95 for scientists and engineers*. New York: McGraw-Hill.

Chapra, S. C., and R. P. Canale. 1988. *Numerical methods for engineers*. 2d ed. New York: McGraw-Hill.

Chen, C.-T. 1984. *Linear system theory and design*. New York: Holt, Rinehart and Winston.

Cline, A. K., C. B. Moler, G. W. Stewart, and J. H. Wilkinson. 1979. An estimate for the condition number of a matrix. *SIAM Journal of Numerical Analysis* 16(2):368–375.

Colquhoun, W. (Ed.). 1971. *Biological rhythms and human performance*. New York: Academic Press.

Cook, P. A. 1986. *Nonlinear dynamical systems*. Englewood Cliffs, NJ: Prentice Hall International.

Cooley, J. W., and R. W. Tuke. 1965. An algorithm for machine computation of complex Fourier series. *Mathematics of Computation* 19:297–301.

Dantzig, G. B. 1951. Maximization of a linear function of variables subject to linear inequalities. Chapter XXI in *Cowles commission monograph 13*, edited by T. C. Koopmans. New York: Wiley.

Edelstein-Keshet, L. 1988. *Mathematical models in biology*. New York: Random House.

Fadeeva, V. N. 1955. *Computational methods of linear algebra*. New York: Dover.

Faires, J. D., and R. L. Burden. 1993. *Numerical methods*. Boston: PWS-Kent.

Forsythe, G. E., M. A. Malcolm, and C. B. Moler. 1977. *Computer methods for mathematic computations*. Englewood Cliffs, NJ: Prentice Hall.

Franklin, G. E., J. D. Powell, and M. L. Workman. 1990. *Digital control of dynamic systems*. 2d ed. Reading, MA: Addison-Wesley.

Gear, C. W. 1971. *Numerical initial value problems in ordinary differential equations*. Englewood Cliffs, NJ: Prentice Hall.

Geman, S., and D. Geman. 1984. Stochastic relaxation, Gibbs distributions and Baysian restoration of images. *IEEE Transactions on Pattern Analysis and Machine Intelligence* 6:721–741.

Gerald, C. F., and P. O. Wheatley. 1989. *Applied numerical analysis*. 4th ed. Reading, MA: Addison-Wesley.

Gleick, J. 1987. *Chaos*. New York: Viking.

Griffiths, D. V., and I. M. Smith. 1991. *Numerical methods for engineers*. Boca Raton, FL: CRC Press.

Halliday, D., and R. Resnick. 1965. *Physics for students of science and engineering*. New York: Wiley.

Hanselman, D., and B. Littlefield. 1998. *Mastering MATLAB 5*. Upper Saddle River, NJ: Prentice Hall.

Hultquist, P. F. 1988. *Numerical methods for engineers and computer scientists*. Menlo Park, CA: Benjamin/Cummings.

Kahaner, D., C. Moler, and S. Nash. 1989. *Numerical methods and software*. Englewood Cliffs, NJ: Prentice Hall.

Kelley, A., and I. Pohl. 1990. *A book on C*. 2d ed. Redwood City, CA: Benjamin/Cummings.

Kernighan, B. W., and D. M. Ritchie. 1988. *The C programming language*. 2d ed. Englewood Cliffs, NJ: Prentice Hall.

Kershaw, D. 1982. Solution of single tridiagonal linear systems and vectorization of the ICCG algorithm on the Cray-1. In *Parallel Computations*, edited by G. Rodrigue. New York: Academic Press.

Kincaid, D., J. Respess, D. Young, and R. Grimes. 1982. Itpack 2C: A Fortran package for solving large sparse linear systems by adaptive accelerated iterative methods. *ACM Transactions on Mathematical Software* 8:302–322.

Knuth, D. E. 1981. *Seminumerical algorithms*. 2d ed. Vol. 2 of *The art of computer programming*. Reading, MA: Addison-Wesley.

Kokotivic, P. V., and H. K. Khalil. 1986. *Singular perturbation in systems and control*. New York: IEEE Press.

Lastman, G. J., and N. K. Sinha. 1989. *Microcomputer-based numerical methods for science and engineering*. New York: Saunders.

Leyton, L. 1975. *Fluid behavior in biological systems*. New York: Clarendon Press.

Luenberger, D. G. 1965. *Introduction to linear and nonlinear programming*. Reading, MA: Addison-Wesley.

McClamroch, N. H. 1980. *State models for dynamic systems*. New York: Springer-Verlag.

McLean, W. G., and E. W. Nelson. 1978. *Theory and problems of engineering mechanics*. 3d ed. New York: McGraw-Hill.

Metropolis, N., A. Rosenbluth, M. Rosenbluth, A. Teller, and E. Teller. 1953. Equations of state calculations by fast computing machines. *Journal of Chemistry and Physics* 21:1087–1091.

Milman, J., and C. C. Halkias. 1967. *Electronic devices and circuits*. New York: McGraw-Hill.

Moran, M. J., and H. N. Shapiro. 1995. *Fundamentals of engineering thermodynamics*. New York: Wiley.

Moses, J. 1971. Symbolic integration: The stormy decade. *Communication of the ACM* 14:548–460.

Nakamura, S. 1993. *Applied numerical methods in C*. Englewood Cliffs, NJ: Prentice Hall.

Nelson, L. C., and E. F. Obert. 1954. Generalized compressibility charts. *Chemical Engineering*. 61:203.

Noble, B. 1969. *Applied linear algebra*. Englewood Cliffs, NJ: Prentice Hall.

Ogata, K. 1997. *Modern control engineering*. 3d ed. Englewood Cliffs, NJ: Prentice Hall.

Palmer, J. 1976. *An introduction to biological rhythms*. New York: Academic Press.

Park, S. K., and K. W. Miller. 1988. Random number generators: Good ones are hard to find. *Communication of the ACM* 31:1192–1201.

Peaceman, D. W., and H. H. Rachford. 1955. The numerical solution of parabolic and elliptic differential equations. *Journal of Society for Industrial and Applied Mathematics* 3:28–41.

Pielou, E. C. 1969. *An introduction to mathematical ecology*. New York: Wiley-Interscience.

Polak, E. 1971. *Computational methods in optimization*. New York: Academic Press.

Press, W. H., B. P. Flannery, S. A. Teukolski, and W. T. Vetterling. 1992. *Numerical recipes in C*. 2d ed. Cambridge, UK: Cambridge University Press.

Proakis, J. G., and D. G. Manolakis. 1988. *Introduction to digital signal processing*. New York: Macmillan.

Ralston A., and P. Rabinowitz. 1978. *A first course in numerical analysis*. 2d ed. New York: McGraw-Hill.

Rao, S. S. 1990. *Mechanical vibrations*. Reading, MA: Addison-Wesley.

Rice, J. 1993. *Numerical methods, software, and analysis*. 2d ed. Boston: Academic Press.

Rubinstein, R. 1981. *Simulation and the Monte Carlo method*. New York: Wiley.

Schilling, R. J. 1990. *Fundamentals of robotics: Analysis and control*. Englewood Cliffs, NJ: Prentice Hall.

Schrage, L. 1979. *ACM transactions on mathematical software*. 5:132–138.

Seborg, D. E., T. F. Edgar, and D. A. Mellichamp. 1989. *Process dynamics and control*. New York: Wiley.

Shannon, C. E. 1949. Communication in the process of noise. *Proceedings of the IRE* 33(1): 10–21.

Stoer, J., and R. Bulirsch. 1980. *Introduction to numerical analysis*. New York: Springer-Verlag.

Szu, H., and R. Hartley. 1987. Fast simulated annealing. *Physics Letters* 1222:157–162.

Varga, R. S. 1962. *Matrix iterative analysis*. Englewood Cliffs, NJ: Prentice Hall.

Vidyasagar, M. 1978. *Nonlinear systems analysis*. Englewood Cliffs, NJ: Prentice Hall.

Wasserman, P. D. 1989. *Neural computing: Theory and practice*. New York: Van Nostrand Reinhold.

Widrow, B., and S. D. Stearns. 1985. *Adaptive signal processing*. Englewood Cliffs, NJ: Prentice Hall.

Wilkinson, J. H. 1963. *Rounding error in algebraic process*. Englewood Cliffs, NJ: Prentice Hall.

Wilkinson, J. H. 1965. *The algebraic eigenvalue problem*. Oxford: Oxford University Press.

Wilkinson, J. H., and C. Reinsch. 1971. *Linear algebra*. Vol. II of *Handbook for automatic computation*. New York: Springer-Verlag.

Wong, K. T., and R. Luus. 1980. Model reduction of high-order multistage systems by the method of orthogonal collocation. *Canadian Journal of Chemical Engineering* 58:382.

Young, D. M. 1971. *Iterative solution of large linear systems*. New York: Academic Press.

Zangwill, W. I. 1969. *Nonlinear programming: A unified approach*. Englewood Cliffs, NJ: Prentice Hall.

APPENDIX 1

NLIB Using MATLAB

This appendix describes the MATLAB version of *NLIB,* a numerical toolbox that implements the algorithms developed in the text. It is assumed that the reader is familiar with the fundamentals of MATLAB (see, e.g., Hanselman and Littlefield, 1998). The NLIB toolbox consists of a collection of functions directly executable from the MATLAB command window. Several of the algorithms developed in the text are already implemented as standard built-in MATLAB functions. They are supplemented, as needed, by customized numerical functions whose source code can be found in the directory *nlib\toolbox* on the distribution CD.

A block diagram that displays the modules or groups of functions in NLIB is shown in Figure 1.1.1. There is one main-program support module called *display* plus nine application modules. The module *display* contains tabular display functions designed to provide for display of scalars, vectors, and matrices on the console screen. It also contains graphical display functions for conveniently generating graphs and surface plots. Random number generation functions and other low-level utility functions are also included.

The NLIB toolbox features nine application modules corresponding to Chapters 2 through Chapter 10 of the text. The linear algebraic systems module, *linear,* includes direct and iterative methods for solving linear algebraic systems of equations. The eigenvalue and eigenvector module, *eigen,* focuses on computing eigenvalues and eigenvectors of square matrices. It also computes singular values of nonsquare matrices. Problems of interpolation, extrapolation, and least-squares curve fitting are covered in the curve fitting module, *curves*. The root finding module, *roots,* contains iterative solution techniques for nonlinear equations. The more general problem of finding the minimum of

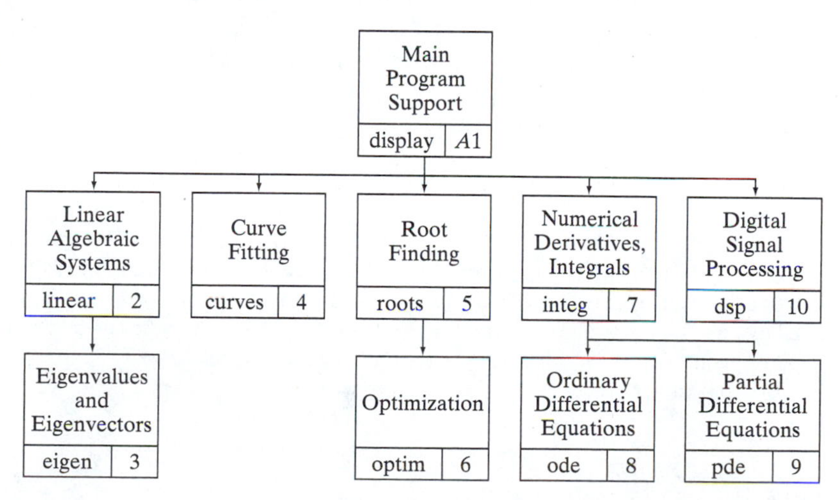

FIGURE 1.1.1 NLIB Structure (MATLAB Version)

an objective function subject to equality and inequality constraints is addressed in the optimization module, *optim*. Numerical differentiation and integration are the focus of the derivatives and integrals module, *integ*. The problem of solving systems of ordinary differential equations subject to initial or boundary conditions is treated in the ordinary differential equations module, *ode*. The partial differential equations module, *pde*, includes techniques for solving various classes of partial differential equations, including hyperbolic, parabolic, and elliptic equations. Finally, the digital signal processing module, *dsp*, focuses on numerical techniques for processing discrete-time signals, including spectral analysis, digital filtering, and system identification.

All of the functions in the NLIB toolbox can be executed using only the *student* edition of MATLAB; there is no need to have access to special toolboxes or the professional edition. NLIB was developed using MATLAB Version 5.2.

1.1 A NUMERICAL TOOLBOX: NLIB

1.1.1 Toolbox Installation

The numerical toolbox NLIB can be installed on a PC by executing the following command from the Windows Start/Run menu, assuming the distribution CD is in drive D.

```
D:\setup
```

More detailed instructions for installing the NLIB toolbox for both PC and non-PC users can be found by viewing the plain text file *user_mat.txt* on the distribution CD.

Access to the NLIB toolbox is obtained by appending the appropriate NLIB folders to the MATLAB search path. The simplest way to do this is through the startup file, *startup.m*, which is stored in the MATLAB *bin* folder and executed whenever MATLAB is launched. The following example startup file, included on the distribution CD, is installed automatically by the setup program.

```
%----------------------------------------------------------------------
% Usage:        startup
%
% Description: Add the NLIB folders to the MATLAB path.
%
% Note:         This file should be added to the MATLAB folder.
%----------------------------------------------------------------------
path(path,'c:\nlib\toolbox')     % add NLIB toolbox to path
path(path,'c:\nlib\exam_mat')    % add NLIB examples path
path(path,'c:\nlib\prob_mat')    % add NLIB problems to path
path(path,'c:\nlib\user_mat')    % add NLIB user folder to path
format compact                   % compact display
cd c:\nlib\user_mat              % working folder
clc                              % clear screen
what                             % list executables
%----------------------------------------------------------------------
```

TABLE 1.1.1 NLIB Modules

Module	*Description*	*Location*
display	main-program support	App. 1
linear	linear algebraic systems	Chap. 2
eigen	eigenvalues and eigenvectors	Chap. 3
curves	curve fitting	Chap. 4
roots	root finding	Chap. 5
optim	optimization	Chap. 6
integ	derivatives and integrals	Chap. 7
ode	ordinary differential equations	Chap. 8
pde	partial differential equations	Chap. 9
dsp	digital signal processing	Chap. 10

For illustration purposes, it is assumed that the NLIB files are stored in the folder *c:\nlib*. The remaining statements in *startup.m* control the display format, select a working folder called *user_mat,* and display its contents. As an alternative to replacing *startup.m,* the path statements can be appended to an existing version of *startup.m*.

1.1.2 NLIB Example Browser

NLIB is composed of the modules summarized in Table 1.1.1. The last column indicates the chapter or appendix where the algorithms implemented in the module are discussed.

The most convenient way to view and run all of the NLIB software is to use the *NLIB Example Browser,* which is launched by entering the following command from the MATLAB command prompt:

```
browse
```

The file *browse.m* is an easy-to-use menu-based program that allows the user to select from the options listed in Table 1.1.2. The *run* option allows the user to execute all of the computational examples discussed in the text, as well as solutions to selected problems discussed in Appendix 4. The *view* option is similar to *run,* but it allows the user to view the complete source code of the examples and the selected problems including data files. The *user* option provides a simple means of configuring *browse* to

TABLE 1.1.2 Menu Options of NLIB Example Browser

Menu Option	*Selections*
Run	examples, selected problems
View	examples, selected problems
User	create script, run, view, add, remove
Help	NLIB toolbox, software updates
Exit	terminate program

include user scripts. Selection are provided to create, run, and view user scripts as well as add them to or remove them from the browser menus. In this way, the user can employ *browse* as an environment for program development. Finally, the *help* option allows the user to view documentation on all of the functions in the NLIB toolbox and learn how to obtain software upgrades as they become available. The NLIB Example Browser is a fast and easy way to access all of the MATLAB software available on the distribution CD, and it also serves as an environment for user program development.

An alternative way to obtain online documentation of the NLIB toolbox functions is to use the MATLAB *help* command. To display a list of the names of all of the functions in the NLIB toolbox, arranged by category, enter the following command from the MATLAB command prompt:

```
help nlib
```

To obtain user documentation on a specific NLIB function, replace the operand *nlib* with the appropriate function name.

```
help funcname
```

There are over one hundred functions in the NLIB toolbox, and space does not permit all of them to be described in this appendix. However, user documentation on *all* of the NLIB functions, including the lower-level functions not discussed here, can be obtained through the *browse/help* menu option or with the *help* command.

1.2 MAIN-PROGRAM SUPPORT

Our goal is to develop a toolbox of MATLAB functions that can be used to solve a variety of numerical problems in engineering. There are a number of general-purpose low-level functions that can be introduced at this point to facilitate the development and testing of the higher-level functions. The toolbox is initially seeded with these low-level utility functions, which are referred to collectively as the *display* module. The files for the NLIB toolbox can be found in the *nlib\toolbox* directory on the distribution CD. Many of the built-in MATLAB functions have alternative calling sequences. No attempt is made to describe all of the calling sequences here. Instead, the reader can enter *help* followed by the function name to get a more complete description of function usage.

1.2.1 Tabular Display

Scalars, vectors, and matrices can be displayed from the MATLAB command prompt by simply entering the variable name. However, a more refined interactive control of the display can be achieved by using the main-program support functions in Table 1.2.1.

show The NLIB function *show* is used to display a scalar, vector, or matrix. All items are labeled, and vectors and matrices are enclosed in brackets. On entry to *show:*

TABLE 1.2.1 Tabular Display Functions

Usage	*Description*	*Toolbox*
$show(s, A)$	display $m \times n$ matrix A	nlib
$x = prompt(s, a, b, df)$	prompt for number in $[a, b]$	nlib
wait	pause until a key is pressed	nlib

s is a string used to label the matrix, and A is the matrix to be displayed. MATLAB handles operations with complex numbers automatically because the fundamental internal data type is a complex double-precision matrix. When complex numbers are displayed, the format is as follows, where $i = \sqrt{-1}$.

$$u = x + yi \tag{1.2.1}$$

In many instances, vectors and matrices are too large to fit on one screen. In these cases, one slice of the vector or matrix is shown at a time, and rows and columns are labeled. Once a display is complete, the *show* function displays a menu of disposition options as shown in Table 1.2.2.

The Save option is used to save the variable to a file. The user is prompted for the file name, and a *.m* extension is automatically added so that the file is executable from within MATLAB. The Append option is similar, but instead of overwriting it can be used repeatedly to save several variables (scalars, vectors, and matrices) in the same file. These variables can be subsequently recovered and placed in the MATLAB workspace by simply entering the file name from the MATLAB command prompt. When the Save or Append options are used, the string s used to label the matrix should consist of the variable name. If it is a long string that is not a legal variable name, then when the matrix is saved the label will be truncated, as needed, to create a variable name. The normal way to exit from the *show* function is to press the Enter key, which resumes program execution by returning to the calling program. Experience has shown that the *show* functions can be highly useful for program development in that they provide a convenient way to examine program variables and intermediate results from within the user program.

prompt The NLIB function *prompt* prompts the user for a number within a specified range. On entry to *prompt*: s is the prompt string to be displayed, a is the lower limit of the response, b is the upper limit, and df is the default response. The function *prompt* continues to prompt for responses until a number in the interval $[a, b]$ is entered. It then returns the number.

TABLE 1.2.2 *show* Options

Option	*Description*
1	save to a file
2	append to a file
3	continue

wait The NLIB function *wait* displays the message "Press any key to continue . . . " and then pauses until a key is pressed. This is useful, for example, when a user program displays multiple graphs or surface plots.

EXAMPLE 1.2.1 Tabular Display

The following MATLAB script prompts the user for some dimensions, creates a random vector and a random matrix using the the built-in MATLAB function *rand,* and then displays them on the console screen. To view the output, execute script *a121* on the distribution CD.

```
%------------------------------------------------------------------
% Example a1.2.1: Tabular Display
%------------------------------------------------------------------
clc          % clear command window
clear        % clear variables

fprintf ('Example a1.2.1: Tabular Display\n');
m = prompt ('Enter number of rows',1,20,10);
n = prompt ('Enter number of columns',1,20,10);
x = rand (m,1);
A = randn (m,n);
show ('Random uniform vector',x);
show ('Random Gaussian matrix',A);
%------------------------------------------------------------------
```

1.2.2 Graphical Display

Descriptions of the main-program support functions that display families of curves are summarized in Table 1.2.3.

graphmat A simple way to graphically display a matrix is to use the NLIB function *graphmat,* which generates a family of curves by graphing the n columns of the matrix $m \times n$ matrix Y using the row subscript as the independent variable. Thus, the kth curve is a graph of the column vector $Y(:,k)$. On entry to *graphmat:* the matrix Y contains the data to be graphed, *title* is the graph title, *xaxis* is the x-axis label, and *yaxis* is the y-axis label. To graph the rows Y instead of the columns, just replace Y by its transpose Y'. When $m = 1$ or $n = 1$, *graphmat* can be used to graph a vector.

TABLE 1.2.3 Graph Functions

Usage	*Description*	*Toolbox*
$graphmat(Y, title, xaxis, yaxis)$	graph $m \times n$ matrix Y	nlib
$graphfun(a, b, title, xaxis, yaxis, f)$	graph function f over $[a, b]$	nlib
$graphxy(X, Y, title, xaxis, yaxis, pts)$	graph matrix Y vs matrix X	nlib

graphfun A function $f(x)$ can be graphed using the NLIB function *graphfun*. On entry to *graphfun*: a is the lower limit of x, b is the upper limit of x, *title* is the graph title, *xaxis* is the x-axis label, *yaxis* is the y-axis label, and f is a string containing the name of the function to be graphed. The user-supplied function f must be stored in an M-file of the same name, or it must be declared as an inline function. The form of the function f is

```
function y = f(x)
```

When f is called with the scalar x, it must return the scalar $y = f(x)$. The function *graphfun* graphs the curve $f(x)$ over the interval $[a, b]$.

graphxy The NLIB function *graphxy* is the most general graph function. It graphs the columns of the $m \times n$ matrix Y versus the columns of the $m \times n$ matrix X. Consequently, the kth curve is a graph of $Y(:, k)$ vs. $X(:, k)$ with each curve having its own independent variable. A curve with fewer than m points can be graphed by repeating the last point as needed to fill the corresponding columns of X and Y. The family of curves is labeled using *title* for the title, *xaxis* for the x-axis label, and *y-axis* for the y-axis label. Normally, each column of Y is plotted by connecting the points with solid straight line segments. However, if the *optional* argument *pts* is included, the last column of Y is graphed using *isolated* points. In this case, *pts* is a one-character string specifying the plot symbol, such as "s" for square, "d" for diamond and "o" for circle. See *help plot* for other symbols.

EXAMPLE 1.2.2 Family of Curves

The following script generates a matrix of data and then uses *graphxy* to display the family of curves shown in Figure 1.2.2.

```
%----------------------------------------------------------------------
% Example a1.2.2: Family of Curves
%----------------------------------------------------------------------
clc          % clear command window
clear        % clear variables
m = 50;      % number of points
n = 6;       % number of curves

fprintf ('Example a1.2.2: Family of Curves\n');
a = 5;
x = linspace(0,a,m)';
for j = 1 : n
   for i = 1 : m
      Y(i,j) = j*x(i)*exp(-n*x(i)/j);
   end
end
graphxy (x,Y,'A Family of Curves','x','y')
%----------------------------------------------------------------------
```

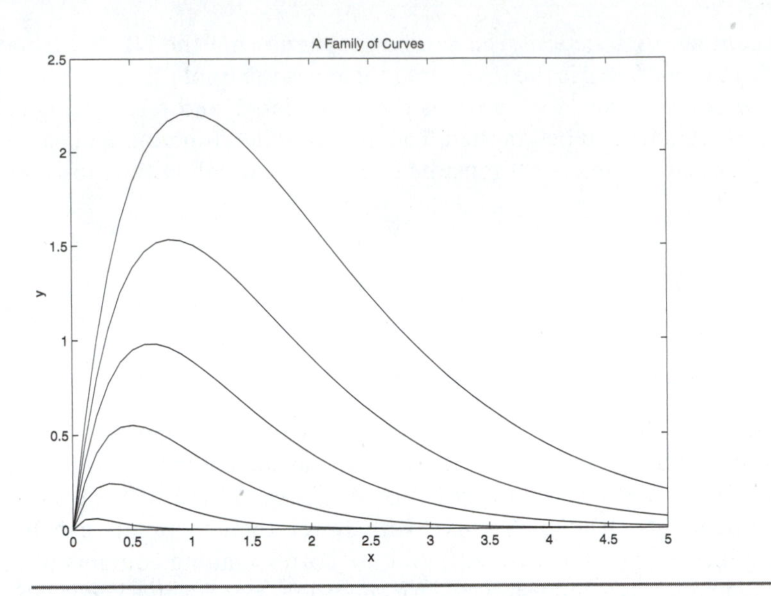

FIGURE 1.2.2 Output from Example 1.2.2

An alternative way to plot a family of curves is to treat the curve number as a second independent variable and thereby plot the curves "side by side" to generate a surface. The main-program support functions designed to display surface plots are summarized in Table 1.2.4.

plotmat A simple way to graphically display a matrix as a surface is to use the NLIB function *plotmat*, which plots the $m \times n$ matrix Z using the row subscript as one independent variable and the column subscript as the other independent variable. Matrices with a special form (e.g., symmetric, triangular, diagonal, sparse) have characteristic surface plots. The string *title* is the plot title, and the strings *xaxis, yaxis,* and *zaxis* are the labels for the x, y, and z axes, respectively.

plotfun A function $f(x, y)$ can be plotted as a surface using the NLIB function *plotfun*. On entry to *plotfun:* $x0$ is the minimum value of x, $x1$ is the maximum value of x, $y0$ is the minimum value of y, $y1$ is the maximum value of y, *title* is the plot title, and *xaxis, yaxis,* and *zaxis* are the axis labels for the x, y, and z axes, respectively. Finally, f is a string containing the name of the function to be plotted. The user-supplied function f must be stored in an M-file of the same name or declared as an inline function. The form of the function f is

```
function z = f(x,y)
```

TABLE 1.2.4 Surface Plot Functions

Usage	*Description*	*Toolbox*
$plotmat(Z, title, xaxis, yaxis, zaxis)$	plot $m \times n$ matrix Z	nlib
$plotfun(x0, x1, y0, y1, title, xaxis, yaxis, zaxis, f)$	plot function f	nlib
$plotxyz(x, y, Z, title, xaxis, yaxis, zaxis)$	plot $m \times n$ matrix Z vs x and y	nlib

When f is called with the scalars x and y, it must return the scalar $z = f(x, y)$. The function *plotfun* plots the surface $f(x, y)$ over the rectangular region $[x0, x1] \times [y0, y1]$.

plotxyz To obtain more sophisticated surface plots, one can use the NLIB function *plotxyz,* which plots the $m \times n$ matrix Z as a surface using the $m \times 1$ vector x as the first independent variable, and the $n \times 1$ vector y as the second independent variable. The plot is labeled using *title* for the title, *xaxis* for the x-axis label, *yaxis* for the y-axis label, and *zaxis* for the z-axis label.

EXAMPLE 1.2.3 Surface Plot

The following script constructs a surface using the two-dimensional *sinc* function and then plots it using *plotxyz* as shown in Figure 1.2.3.

```
%----------------------------------------------------------------
% Example a1.2.3: Surface Plot
%----------------------------------------------------------------
clc                                % clear command window
clear                              % clear variables
d = 10;                            % plot range
n = 90;                            % number of points

fprintf ('Example a1.2.3: Surface Plot\n');
x = linspace(-d,d,n)';
y = x;
for i = 1 : n
   for j = 1 : n
      r = sqrt (x(i)^2 + y(j)^2) + eps;
      Z(i,j) = sin(r)/r;
```

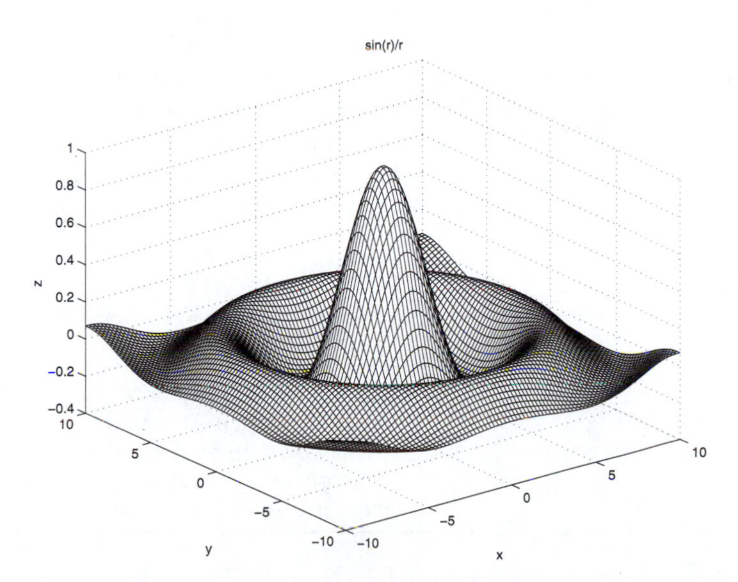

FIGURE 1.2.3 Output from Example 1.2.3

```
    end
end
plotxyz (x,y,Z,'sin(r)/r','x','y','z')
%------------------------------    ------------------------------
```

1.2.3 Utility Functions

Finally, the *display* module contains a few low-level general-purpose utility functions that can be useful for developing and testing user programs. The general purpose utility functions are summarized in Table 1.2.5.

randinit Random numbers are useful for testing numerical methods in general, and they are also required for specific applications such as Monte Carlo integration and system identification. The NLIB function *randinit* is used to *initialize* the random number generator. On entry to *randinit*, s is the seed for the random number sequence. Each seed generates a different random sequence. If *randinit* is called with $s \leq 0$, then a random seed is generated based on the time of day. In this way, different random sequences can be generated each time a program is executed.

randu The NLIB function *randu* is used to produce a matrix of random numbers *uniformly* distributed over an interval $[a, b]$. On entry to *randu:* $m \geq 1$ is the number of rows, $n \geq 1$ is the number of columns, a is the lower bound on the interval, and b is the upper bound. On exit, A is a $m \times n$ matrix whose components are random numbers uniformly distributed in the interval $[a, b]$. In theory, pseudo-random sequences up to length 2^{1492} can be generated. The probability density of the random numbers produced by successive calls to *randu* is

$$p(x) = \begin{cases} \dfrac{1}{b - a} & , \quad a \leq x \leq b \\ 0 & , \quad \text{otherwise} \end{cases} \tag{1.2.2}$$

randg The NLIB function *randg* is used to produce a matrix of random numbers with a *Gaussian* or normal distribution. On entry to *randg:* $m \geq 1$ is the number of rows, $n \geq 1$ is the number of columns, *mu* is the *mean* of the distribution, and *sigma* is the standard deviation. On exit, A is an $m \times n$ matrix whose components are Gaussian random numbers with mean *mu* and standard deviation *sigma*. The probability density

TABLE 1.2.5 Utility Functions

Usage	*Description*	*Toolbox*
randinit (s)	select random sequence	nlib
A = *randu* (m, n, a, b)	uniform random $m \times n$ matrix	nlib
A = *randg* $(m, n, mu, sigma)$	Gaussian random $m \times n$ matrix	nlib
x = *norm* (A, p)	pth norm of $m \times n$ matrix A	built-in
y = *args* (x, a, b, k, f)	check user calling arguments	nlib

of the random numbers produced by successive calls to *randg* is as follows where $\mu = mu$ and $\sigma = sigma$.

$$p(x) = \frac{1}{\sigma\sqrt{2\pi}} \exp\left[\frac{-(x-\mu)^2}{2\sigma^2}\right] \tag{1.2.3}$$

norm The function *norm* is a built-in MATLAB function that computes the *p*th Holder norm of a matrix or a vector. If x is a vector, the *p*th *Holder* norm is defined:

$$\|x\|_p \triangleq \left(\sum_{k=1}^{m} |x_k|^p\right)^{1/p} \quad p \geq 1 \tag{1.2.4}$$

The limiting special case as $p \to \infty$ is obtained by setting the second argument to *inf*, which yields the *infinity* norm:

$$\|x\| \triangleq \max_{k=1}^{m} \{|x_k|\} \tag{1.2.5}$$

On entry to *norm:* A is an $m \times n$ matrix, and $p \in \{1, 2, inf\}$ is the desired norm. The value returned by *norm* is $\|A\|_p$. When A is a matrix, the cases $p = 1, 2, inf$ correspond to the column-sum norm of A, the spectral radius of A, and the row-sum norm of A, respectively.

args The NLIB function *args* is a utility that can be used within user functions to check that scalar calling arguments are within range. On entry to *args:* x is the scalar calling argument whose value is to be checked, a is the lower limit for x, b is the upper limits for x, k is a positive integer indicating the position of x in the list of calling arguments, and f is a string containing the name of the function whose arguments are being checked. On exit from *args:* y is equal to x saturated to the interval $[a, b]$. If $x < a$ or $x > b$, an error message is printed by *args* to warn the user that x was out of range.

EXAMPLE 1.2.4 Utility Functions

The following script illustrates the use of some of the utility functions. To view the output, execute script *a124* on the distribution CD.

```
%------------------------------------------------------------
% Example a1.2.4: Utility Functions
%------------------------------------------------------------
clc                          % clear command window
clear                        % clear variables
randinit (100)               % select random sequence

fprintf ('Example a1.2.4: Utility Functions\n');
A = randu (3,4,-1,1);
x = randg (6,1,10,2);
show ('machine epsilon',eps)
show ('A',A);
show ('row sum norm of A',norm(A,inf))
```

```
show ('x',x);
show ('infinity norm of x',norm(x,inf))
%------------------------------------------------------------------
```

1.3 LINEAR ALGEBRAIC SYSTEMS

The module *linear* is an application module that implements the linear algebraic system techniques discussed in Chapter 2. The linear algebraic system functions are summarized in Table 1.3.1. Because MATLAB is designed specifically to work with matrices, a number of the linear algebraic system functions are already built into MATLAB itself.

rank A set of p vectors $X = \{x^1, x^2, \ldots, x^p\}$ is *linearly dependent* if and only if there exists a nonzero $p \times 1$ vector c such that

$$c_1 x^1 + c_2 x^2 + \cdots + c_p x^p = 0 \tag{1.3.1}$$

Otherwise, the set X is *linearly independent*. For a linearly dependent set X, at least one vector can be expressed as a weighted sum, or linear combination, of the other vectors. The *rank* of an $m \times n$ matrix A is the number of linearly independent columns. Since $\text{rank}(A^T) = \text{rank}(A)$, the rank is also equal to the number of linearly independent rows. The built-in MATLAB function *rank* computes the rank of a matrix. On entry to *rank*: A is an $m \times n$ matrix. The value returned by *rank* is the rank of A. The rank can never be larger than the number of rows or the number of columns. Consequently

$$0 \le \text{rank}(A) \le \min\{m, n\} \tag{1.3.2}$$

A matrix A for which $\text{rank}(A) = \min\{m, n\}$ is said to be of *full rank*. A square ($m = n$) matrix is nonsingular or invertible if and only if it is of full rank.

TABLE 1.3.1 Linear Algebraic System Functions

Usage	*Description*	*Toolbox*
$q = rank(A)$	rank of A	built-in
$x = gauss(A,b)$	solve $Ax = b$ by Gaussian elimination	nlib
$[L,U,P,Delta] = lufac(A,dm)$	LU factorization of A	nlib
$[x,Delta] = lusub\ (L,U,P,b)$	LU substitution	nlib
$[x,Delta] = ludec\ (A,b,eq)$	solve $Ax = b$ by LU decomposition	nlib
$d = det(A)$	determinant of A	built-in
$B = inv(A)$	inverse of A	built-in
$K = condnum(A,m)$	estimated condition number of A	nlib
$rho = residual(A,b,x)$	residual error of x	nlib
$[Q,Delta] = trifac\ (T)$	tridiagonal LU factorization of T	nlib
$[x,Delta] = trisub\ (Q,b)$	tridiagonal LU substitution	nlib
$[x,Delta] = tridec\ (T,b)$	solution of tridiagonal system	nlib
$[x,k] = sr\ (x,A,b,alpha,tol,m)$	solve $Ax = b$ using successive relaxation	nlib

gauss The NLIB function *gauss* uses Gaussian elimination, as implemented by the MATLAB left division operator, to solve the following linear algebraic system of equations.

$$Ax = b \tag{1.3.3}$$

On entry to *gauss:* A is an $m \times n$ coefficient matrix, and b is an $m \times 1$ right-hand-side vector. On exit, the $n \times 1$ vector x contains the solution. If A is of full rank, *gauss* finds a solution x that minimizes the Euclidean norm of the residual error vector:

$$r \triangleq b - Ax \tag{1.3.4}$$

Among all vectors x that minimize $\|r\|_2 = \sqrt{r^T r}$, *gauss* finds the x whose norm $\|x\|_2$ is smallest. This is called the *minimum least-squares* solution. When $m = n$, the minimum least-squares solution is the solution $x = A^{-1}b$. For large values of n, computing x requires approximately $n^3/3$ floating-point multiplications and divisions, or FLOPs.

lufac The NLIB function *lufac* factors the matrix A into lower- and upper-triangular factors. On entry to *lufac:* A is an $n \times n$ matrix, and dm is an optional display mode. If dm is present, intermediate results are displayed. On exit from *lufac:* the $n \times n$ matrix L contains the lower-triangular factor, the $n \times n$ matrix U contains the upper-triangular factor with ones on the diagonal, the $n \times n$ matrix P is a row permutation matrix, and *Delta* is the determinant of A. If *Delta* $\neq 0$, then

$$LU = PA \tag{1.3.5}$$

lusub The NLIB function *lusub* is used in conjunction with *lufac*. It uses the triangular factors of A to efficiently solve Equation (1.3.3) for different right-hand-side vectors b using forward and back substitution. On entry to *lufac:* L is an $n \times n$ lower-triangular matrix, U is an $n \times n$ upper-triangular matrix with ones on the diagonal, P is an $n \times n$ row permutation matrix, and b is an $n \times 1$ right-hand-side vector. On exit from *lusub:* the $n \times 1$ vector x contains the solution of

$$LUx = Pb \tag{1.3.6}$$

and *Delta* is the determinant of LU. If *Delta* $\neq 0$, then x is a valid solution vector. The inputs (L, U, P) are obtained by a single call to *lufac*. Then *lusub* can be called repeatedly for each new right-hand-side vector b.

ludec The NLIB function *ludec* uses LU decomposition to solve the system (1.3.3) by calling *lufac* and *lusub*. On entry to *ludec:* A is an $n \times n$ coefficient matrix, b is an $n \times 1$ right-hand-side vector, and eq is an optional scaling mode. If eq is present, then the equations are first prescaled using equilibration to reduce the condition number of A. On exit from *ludec:* the $n \times 1$ vector x contains the solution of (1.3.3), and *Delta* is the determinant of A. If *Delta* $\neq 0$, then x is a valid solution.

det The built-in MATLAB function *det* computes the *determinant* of the matrix A. On entry to *det:* A is an $n \times n$ matrix. The value returned by *det* is the determinant of A. For large values of n, this requires approximately $n^3/3$ FLOPs.

inv The built-in MATLAB function *inv* computes the inverse A^{-1} of the matrix A. On entry to *inv:* A is an $n \times n$ matrix. On exit, the $n \times n$ matrix B contains the inverse of A. If I is the $n \times n$ identity matrix, then the inverse satisfies the equation

$$A^{-1}A = AA^{-1} = I \tag{1.3.7}$$

For large values of n, computing A^{-1} requires approximately $4n^3/3$ FLOPs.

condnum The NLIB function *condnum* computes the estimated condition number $K(A)$ of the matrix A as follows using the infinity norm for the vectors.

$$K(A) \approx \|A\| \cdot \max_{k=1}^{m} \left\{ \frac{\|x^k\|}{\|b^k\|} \right\} \tag{1.3.8}$$

The row-sum matrix norm defined in Equation (2.5.5) of Chapter 2 is used for A. On entry to *condnum:* A is an $n \times n$ matrix, and $m \geq 0$ is the number of random right-hand-side vectors to use. The value returned by *condnum* is the estimated condition number. If $m = 0$, then the following "exact" expression for the condition number is used.

$$K(A) = \|A\| \cdot \|A^{-1}\| \tag{1.3.9}$$

Large values of $K(A)$ indicate that the linear algebraic system in (1.3.3) is an ill-conditioned system.

residual The NLIB function *residual* computes the norm of the residual error vector, r, in (1.3.4). On entry to *residual:* A is an $m \times n$ coefficient matrix, b is an $m \times 1$ right-hand-side vector, and x is an $n \times 1$ solution estimate. The value returned by *residual* is $rho = \|r\|$ where the infinity norm is used. A vector x is a solution of (1.3.3) if and only if $rho = 0$. Therefore, *rho* is a rough measure of the accuracy of the solution.

trifac The NLIB function *trifac* is analogous to *lufac* except that it factors a *tridiagonal* $n \times n$ matrix A into lower- and upper-triangular factors. To economize on storage, the nonzero entries of A are stored in a $3 \times n$ matrix T as follows. Note that the last elements in the first and third rows can be arbitrary.

$$T = \begin{bmatrix} A_{12} & A_{23} & \cdots & A_{n-1,n} & - \\ A_{11} & A_{22} & \cdots & A_{n-1,n-1} & A_{nn} \\ A_{21} & A_{32} & \cdots & A_{n,n-1} & - \end{bmatrix} \tag{1.3.10}$$

On entry to *trifac:* T is a $3 \times n$ matrix containing the nonzero elements of A. On exit from *trifac:* the $3 \times n$ matrix Q contains the nonzero elements of the lower- and upper-triangular factors of A, and *Delta* is the determinant of A. The matrix Q is used by *trisub*.

trisub The NLIB function *trisub* is used in conjunction with *trifac*. Specifically, it uses the triangular factors of the tridiagonal matrix A in Equation (1.3.10) to very efficiently solve (1.3.3) for different right-hand-side vectors b using forward and back substitution. On entry to *trisub:* Q is a $3 \times n$ matrix containing the nonzero entries of the lower- and upper-triangular factors of A, and b is an $n \times 1$ right-hand-side vector. On exit from *trisub:* the $n \times 1$ vector x contains the solution of (1.3.3), and *Delta* is the determinant of A. If $Delta \neq 0$, then x is a valid solution vector. The input Q is obtained by a single call to *trifac*. Then *trisub* can be called repeatedly for each new right-hand-side vector b.

tridec The NLIB function *tridec* uses LU decomposition to solve the system (1.3.3) when A is a tridiagonal matrix by calling *trifac* and *trisub*. On entry to *tridec:* T is a $3 \times n$ coefficient matrix containing the nonzero entries of A as in (1.3.10), and b is an $n \times 1$ right-hand-side vector. On exit from *tridec:* the $n \times 1$ vector x contains the solution of (1.3.3), and *Delta* is the determinant of A. If $Delta \neq 0$, then x is a valid solution. Computing x requires only $5n - 4$ FLOPS.

sr The function *sr* computes an iterative solution to Equation (1.3.3) using the successive relaxation or SR iterative method. This is particularly useful for large sparse systems. On entry to *sr:* x is an $n \times 1$ initial guess, A is an $n \times n$ coefficient matrix, b is an $n \times 1$ right-hand-side vector, $0 < alpha < 2$ is the relaxation factor, $tol \geq 0$ is an error tolerance used to terminate the search, and $m \geq 1$ is an upper bound on the number of iterations. On exit from *sr:* the $n \times 1$ vector x contains the solution estimate, and k is the number of iterations performed. The diagonal elements of A must be nonzero. Otherwise, *sr* returns with $k = 0$, and the equations must be reordered. If *sr* returns with $0 < k < m$, then the following convergence criterion was satisfied where r is the residual error vector in (1.3.4).

$$\|r\| \leq tol \tag{1.3.11}$$

EXAMPLE 1.3.1 Linear Algebraic Systems

As an illustration of the use of the linear algebraic system functions in Table 1.3.1, consider the problem of solving a random system of dimension $n = 5$. The following script solves this system using both the Gaussian elimination method and the SR iterative method.

```
%-----------------------------------------------------------------
% Example a1.3.1: Linear Algebraic Systems
%-----------------------------------------------------------------

% Initialize

clc                           % clear command window
clear                         % clear variables
randinit(1000)                % select random sequence
n = 5;                        % number of variables
m = 250;                      % maximum iterations
tol = 1.e-6;                  % error tolerance
alpha = 1.5;                  % relaxation parameter
B = randu (n,n,-1,1);
b = randu (n,1,-1,1);
A = B'*B;

% Solve system

fprintf ('Example a1.3.1: Linear Algebraic Systems\n');
show ('A',A)
show ('b',b)
show ('det(A)',det(A))
```

```
show ('inv(A)',inv(A))
show ('K(A)',condnum(A,0))
x = gauss (A,b);
show ('Gaussian elimination solution',x)
show ('||r||',residual(A,b,x))
x = zeros(n,1);
[x,k] = sr (x,A,b,alpha,tol,m);
show ('Number of iterations',k);
show ('Successive relaxation solution',x)
show ('||r||',residual(A,b,x))
%-----------------------------------------------------------------
```

When this script was executed, it produced the following random matrix and right-hand-side vector.

$$A = \begin{bmatrix} 0.5225 & 0.1448 & 0.2886 & -0.4402 & -0.0815 \\ 0.1448 & 2.4261 & 0.8815 & -0.3972 & -0.9537 \\ 0.2886 & 0.8815 & 1.1927 & -0.5393 & -0.3444 \\ -0.4402 & -0.3972 & -0.5393 & 1.1097 & 0.9351 \\ -0.0815 & -0.9537 & -0.3444 & 0.9351 & 1.4763 \end{bmatrix}, \quad b = \begin{bmatrix} -0.4502 \\ 0.3391 \\ -0.2122 \\ -0.7052 \\ 0.2231 \end{bmatrix}$$

The determinant of A was found to be $\det(A) = 0.130065$, and the inverse of A was

$$A^{-1} = \begin{bmatrix} 5.6205 & -1.1957 & 0.8563 & 5.2299 & -3.5750 \\ -1.1957 & 1.2085 & -1.0050 & -2.0049 & 1.7501 \\ 0.8563 & -1.0050 & 1.9353 & 2.2461 & -1.5731 \\ 5.2299 & -2.0049 & 2.2461 & 8.0554 & -5.5847 \\ -3.5750 & 1.7501 & -1.5731 & -5.5847 & 4.7809 \end{bmatrix}$$

The condition number of A was determined to be $K(A) = 65.6866$. When Gaussian elimination was applied, it produced the following solution:

$$x = \begin{bmatrix} -7.6031 \\ 2.9657 \\ -3.0718 \\ -10.4376 \\ 7.5417 \end{bmatrix}$$

This resulted in a residual error of $\|r\| = 1.554 \times 10^{-15}$. When the successive relaxation iterative method was applied with a relaxation factor of $\alpha = 1.5$, it produced a similar solution in 33 iterations. The residual error in this case was $\|r\| = 6.674 \times 10^{-7}$.

1.4 EIGENVALUES AND EIGENVECTORS

The module *eigen* is an application module that implements the eigenvalue and eigenvector techniques discussed in Chapter 3. The eigenvalue and eigenvector functions are summarized in Table 1.4.1.

TABLE 1.4.1 Eigenvalue and Eigenvector Functions

Usage	*Description*	*Toolbox*
$a = poly(A)$	characteristic polynomial of A	built-in
$[c, x, k] = powereig(A, tol, m, dir)$	dominant eigenvalue and eigenvector of A	nlib
$[B, Q] = house\ (A)$	Householder transformation of A	nlib
$[c, X, r] = jacobi\ (A), tol, m$	Jacobi's method for symmetric A	nlib
$[Q, R] = qr\ (A)$	orthogonal-triangular factorization of A	built-in
$[X, C] = eig\ (A)$	eigenvectors and eigenvalues of A	built-in
$[U, S, V] = svd\ (A)$	singular-value decomposition of A	built-in
$t = trace\ (A)$	trace of A	built-in

poly The built-in MATLAB function *poly* computes the coefficients of the characteristic polynomial of a matrix A. On entry to *poly:* A is an $n \times n$ matrix. On exit, the $1 \times (n + 1)$ *row* vector a contains the coefficients of the characteristic polynomial of A.

$$\Delta(\lambda) = \lambda^n + a_2\lambda^{n-1} + \cdots + a_n\lambda + a_{n+1} \tag{1.4.1}$$

Note that the characteristic polynomial, $\Delta(\lambda) = \det(\lambda I - A)$, is a monic polynomial, which means $a_1 = 1$.

powereig The NLIB function *powereig* computes the dominant eigenvalue and its eigenvector using the power method in Alg. 3.3.1 or the least-dominant eigenvalue and its eigenvector using the inverse power method in Alg. 3.3.2. On entry to *powereig:* A is an $n \times n$ matrix, $tol \geq 0$ is an error tolerance used to terminate the search, $m \geq 1$ is an upper bound on the number of iterations, and *dir* is a direction code that selects either the power method or the inverse power method. It is assumed that the eigenvalues of A can be ordered as follows.

$$|\lambda|_1 > |\lambda|_2 \geq \cdots \geq |\lambda_{n-1}| > |\lambda_n| \tag{1.4.2}$$

If $dir \geq 0$, then the power method is used to estimate the dominant eigenvalue, λ_1, and its eigenvector, x^1. When $dir < 0$, the inverse power method is used to estimate the least dominant eigenvalue, λ_n, and its eigenvector, x^n. When the inverse power method is selected, the matrix A must be nonsingular. A random initial guess is used for x. On exit from *powereig:* c is an estimate of the selected eigenvalue, the $n \times 1$ vector x is an estimate of its eigenvector, and k is the number of iterations performed. If $k < m$, the following convergence criterion was satisfied where $r = (\lambda I - A)x$ is the residual error vector.

$$\|r\| < tol \tag{1.4.3}$$

The function *powereig* should converge if A is diagonalizable, if (1.4.2) is satisfied, and if the random initial guess for x is not orthogonal to the selected eigenvector. If $dir \geq 0$, then on exit from *powereig,* the spectral radius of A is $\rho(A) = |c|$. The dominant eigenvalue and its eigenvector can also be computed using the built-in MATLAB function *eigs*.

house The NLIB function *house* implements the Householder transformation to convert a matrix to upper-Hessenberg form using Alg. 3.5.1. On entry to *house:* A is an $n \times n$ matrix. On exit from *house:* the $n \times n$ matrix B contains the upper-Hessenberg form of A, and the $n \times n$ matrix Q is an orthogonal transformation matrix such that $B = Q^T AQ$. Since $Q^{-1} = Q^T$, the eigenvalues of B are equal to the eigenvalues of A. When A is symmetric, the upper-Hessenberg form is a tridiagonal matrix.The function *house* is used as a preprocessing step to increase the rates of convergence of the iterative methods.

jacobi The NLIB function *jacobi* implements Jacobi's method for finding the eigenvalues and eigenvectors of a symmetric matrix A as described in Alg. 3.4.1. On entry to *jacobi:* A is an $n \times n$ symmetric matrix, $tol \geq 0$ is an error tolerance used to terminate the search, and $m \geq 1$ is an upper bound on the number of iterations. The eigenvalues and eigenvectors of a symmetric matrix are real. On exit from *jacobi:* the $n \times 1$ vector c contains the estimated eigenvalues of A, the $n \times n$ matrix X contains the estimated eigenvectors, and r is the number of iterations. If $r < m$, then convergence to a diagonal matrix with *zero* elements having magnitudes less than *tol* was achieved. The kth column of X is the eigenvector x^k associated with the kth eigenvalue, λ_k.

$$Ax^k = \lambda_k x^k \quad , \quad 1 \leq k \leq n \tag{1.4.4}$$

The function *jacobi* uses *house* to convert A to tridiagonal form to improve the convergence rate. If *tol* and m are sufficiently large, *jacobi* should always converge.

qr The built-in MATLAB function *qr* computes an orthogonal-triangular QR factorization of the $n \times n$ matrix A.

$$A = QR \tag{1.4.5}$$

Here the $n \times n$ factor Q is an orthogonal matrix, $Q^{-1} = Q^T$, and the $n \times n$ factor R is an upper-triangular matrix. On entry to *qr:* A is an $n \times n$ matrix. On exit from *qr:* the $n \times n$ matrix Q contains the orthogonal factor, and the $n \times n$ matrix R contains the upper-triangular factor. The QR factorization is used to find eigenvalues using the QR method.

eig The built-in MATLAB function *eig* computes the eigenvalues, $\{\lambda_k\}$, and eigenvectors, $\{x^k\}$, of an $n \times n$ matrix A using the QR method. On entry to *eig:* A is an $n \times n$ matrix. On exit from *eig:* the $n \times n$ complex matrix $X = [x^1, x^2, \ldots, x^n]$ contains the eigenvectors with the kth eigenvector stored in the kth column of X, and the $n \times n$ matrix $C = \text{diag}\{\lambda_1, \lambda_2, \ldots, \lambda_n\}$ is a complex diagonal matrix with the eigenvalues of A along the diagonal. Accuracy of the eigenvalues is improved by first balancing the matrix A with a diagonal similarity transformation, which makes the norms of the row and the column vectors approximately equal. An alternative way to call *eig* is $c = \text{eig}(A)$. This returns an $n \times 1$ vector of eigenvalues $c_k = \lambda_k$.

svd The built-in MATLAB function *svd* computes the *singular value decomposition* of a nonsquare $m \times n$ matrix A.

$$A = USV^T \tag{1.4.6}$$

Here U is an $m \times m$ orthogonal matrix whose rows are the eigenvectors of AA^T. Similarly, V is an $n \times n$ orthogonal matrix whose rows are the eigenvectors of $A^T A$. Finally, S is an $m \times n$ matrix of zeros except for q positive *singular values* along the diagonal where $q = \text{rank}(A)$. The singular values, $\{\sigma_1, \sigma_2, \ldots, \sigma_q\}$, are the square roots of the nonzero eigenvalues of the AA^T. On entry to *svd:* A is an $m \times n$ matrix. On exit from *svd:* U is the $m \times m$ orthogonal factor, S is the $m \times n$ matrix containing the singular values along the diagonal, and V is the $n \times n$ orthogonal factor.

trace The built-in MATLAB function *trace* computes the trace of an $n \times n$ matrix, which is the sum of the diagonal elements.

$$\text{trace}(A) \triangleq A_{11} + A_{22} + \cdots + A_{nn} \tag{1.4.7}$$

On entry to *trace:* A is an $n \times n$ matrix. The value returned by *trace* is the trace of A. The trace of a matrix is also equal to the sum of the eigenvalues. This can be used as an inexpensive partial check on the accuracy of the estimated eigenvalues of A.

EXAMPLE 1.4.1 Eigenvalues and Eigenvectors

As an illustration of computing eigenvalues and eigenvectors, the following script generates a random $n \times n$ matrix, computes its characteristic polynomial with the function *poly*, and its eigenvalues and eigenvectors with the function *eig*.

```
%------------------------------------------------------------------
% Example a1.4.1: Eigenvalues and Eigenvectors
%------------------------------------------------------------------

% Initialize

clc                         % clear command window
clear                       % clear variables
n = 4;                      % size of matrix
randinit(2000)              % select random sequence
A = randu (n,n,-1,1);

% Find eigenvalues and eigenvectors

fprintf ('Example a1.4.1: Eigenvalues and Eigenvectors\n');
show ('A',A)
a = poly(A);
show ('characteristic polynomial coefficients',a')
c = eig (A);
show ('eigenvalues',c)
[X,C] = eig (A);
show ('eigenvectors',X)
t = abs(trace(A) - sum(c));
show ('trace check',t)
%------------------------------------------------------------------
```

When this script was executed, it produced the following random matrix.

$$A = \begin{bmatrix} -0.5617 & 0.3187 & 0.7462 & -0.7563 \\ -0.4126 & 0.7709 & 0.1176 & -0.0623 \\ 0.1940 & 0.4465 & 0.1764 & -0.9232 \\ 0.3860 & -0.5150 & -0.5516 & 0.8227 \end{bmatrix}$$

The characteristic polynomial in this case was found to be

$$\Delta(\lambda) = \lambda^4 - 1.2082\lambda^3 - 0.3941\lambda^2 + 0.7451\lambda - 0.0114$$

The QR method in *eig* produced the following eigenvalues and eigenvectors of A:

$$c = \begin{bmatrix} -0.7352 + 0.0000j \\ 0.9640 + 0.2750j \\ 0.9640 - 0.2750j \\ 0.0154 + 0.0000j \end{bmatrix}$$

$$X = \left[\begin{array}{c|c|c|c} -0.7211 + 0.0000j & 0.4825 - 0.1741j & 0.4825 + 0.1741j & 0.5715 + 0.0000j \\ -0.2299 + 0.0000j & -0.0699 + 0.3141j & -0.0699 - 0.3141j & 0.2320 + 0.0000j \\ 0.5769 + 0.0000j & 0.5840 + 0.1994j & 0.5840 + 0.1994j & 0.7026 + 0.0000j \\ 0.3070 + 0.0000j & -0.4900 + 0.1115j & -0.4900 - 0.1115j & 0.3548 + 0.0000j \end{array}\right]$$

When the trace check was applied to the eigenvalues, the result was

$$\left| \text{trace}(A) - \sum_{k=1}^{4} \lambda_k \right| = 2.66454 \times 10^{-15}$$

1.5 CURVE FITTING

The module *curves* is an application module that implements the curve fitting techniques discussed in Chapter 4. The curve fitting functions are summarized in Table 1.5.1.

interp1 The built-in MATLAB function *interp1* implements one-dimensional piecewise-linear interpolation using the following data set.

$$D = \{(x_k, y_k) | 1 \le k \le n\} \tag{1.5.1}$$

On entry to *interp1:* x is an $n \times 1$ vector of strictly increasing independent variables, y is an $n \times 1$ vector of dependent variables, and a is an $m \times 1$ vector of evaluation points.

TABLE 1.5.1 Curve Fitting Functions

Usage	*Description*	*Toolbox*
$b = interp1\,(x, y, a)$	piecewise-linear fit	built-in
$b = spline\,(x, y, a)$	cubic spline fit	built-in
$a = polyfit\,(x, y, m)$	least-squares polynomial fit	built-in
$y = polyval\,(c, x)$	polynomial evaluation at x	built-in
$y = polynom\,(c, x)$	polynomial evaluation at x	nlib
$C = bilin\,(x, y, Z, a, b)$	bilinear fit	nlib

On exit from *interp1*, the $m \times 1$ vector b contains the piecewise-linear interpolation values evaluated at the points in a. The evaluation points must lie inside the range specified by x.

spline The built-in MATLAB function *spline* constructs a natural cubic spline function and evaluates it at a vector of points a. On entry to *spline:* x is an $n \times 1$ vector of strictly increasing independent variables, y is an $n \times 1$ vector of dependent variables, and a is an $m \times 1$ vector of evaluation points. On exit from *spline:* the $m \times 1$ vector b contains the cubic spline function evaluated at the vector of points a.

polyfit One measure of the closeness of the fit of a polynomial $f(x)$ to the discrete data D in Equation (1.5.1) is the sum of the squares of the errors at each point.

$$E = \sum_{k=1}^{n} [f(x_k) - y_k]^2 \tag{1.5.2}$$

The *least-squares* polynomial is the polynomial $f(x)$ that minimizes the error E. The built-in MATLAB function *polyfit* computes the coefficient vector of the least-squares polynomial of degree m. On entry to *polyfit:* x is an $n \times 1$ vector of strictly increasing independent variables, y is an $n \times 1$ vector of dependent variables, and $0 \le m < n$ is the polynomial degree. On exit from *polyfit:* the $1 \times (m + 1)$ *row* vector a contains the coefficients of the least-squares polynomial

$$f(x) = a_1 x^m + a_2 x^{m-1} + \cdots + a_{m+1} \tag{1.5.3}$$

When $m = n - 1$, the least-squares polynomial is the *interpolating* polynomial:

$$f(x_k) = y_k \quad , \quad 1 \le k \le n \tag{1.5.4}$$

polyval The built-in MATLAB function *polyval* evaluates a polynomial at a point. On entry to *polyval:* a is a $1 \times (m + 1)$ coefficient vector, and x is the evaluation point. On exit from *polyval:* y is the polynomial evaluated at x as in (1.5.3).

polynom The NLIB function *polynom* is similar to *polyval*, but it operates on a coefficient vector whose elements are reversed. On entry to *polynom:* a is an $(m + 1) \times 1$ coefficient vector, and x is the evaluation point. The value returned by *polyval* is $y = g(x)$ where

$$g(x) = a_1 + a_2 x + \cdots + a_{m+1} x^m \tag{1.5.5}$$

bilin The NLIB function *bilin* implements two-dimensional bilinear interpolation. It is based on the following data set.

$$D = \{(x_k, y_j, Z_{kj}) \mid 1 \le k \le m, 1 \le j \le n\} \tag{1.5.6}$$

On entry to *bilin:* x is an $m \times 1$ vector of strictly increasing independent variables, y is an $n \times 1$ vector of strictly increasing independent variables, Z is an $m \times n$ matrix of dependent variables, a is a $p \times 1$ vector of x coordinates for the evaluation points, and b is a $q \times 1$ vector of y coordinates for the evaluation points. On exit from *bilin:* the $p \times q$ matrix C contains bilinear function $C(i,j) = f[a(i), b(j)]$ for $1 \le i \le p$ and $1 \le j \le q$. The evaluation points must lie within the bounds of data range, $x \times y$.

EXAMPLE 1.5.1 Curve Fitting

As an illustration of curve fitting, consider the following function.

$$f(x) = \frac{\sin(\pi x)}{1 + x^2}$$

Suppose this function is sampled at $n = 24$ points uniformly distributed over the interval $[0, 4]$. The following script fits a set of three least-squares polynomials of increasing order to these samples using the function *polyfit* in Table 1.5.1.

```
%--------------------------------------------------------------------
% Example a1.5.1: Curve Fitting
%--------------------------------------------------------------------

% Initialize

clc                          % clear command window
clear                        % clear variables
n = 24;                      % number of data points
p = 10*n;                    % number of plot points
b = 4;                       % plot range
x = linspace (0,b,n)';
y = sin(pi*x) ./ (1 + x.*x);

% Compute least-squares fits

fprintf ('Example a1.5.1: Curve Fitting\n');
for i = 1 : 3
   m = 3*i - 1;
   a = polyfit (x,y,m);
   X(:,i) = linspace (0,b,p)';
   Y(:,i) = polyval (a,X(:,i));
end

% Add data points

for j = 1 : p
   k = min ([n,j]);
   X(j,4) = x(k);
   Y(j,4) = y(k);
end

% Graph

graphxy (X,Y,'Least Squares Polynomials','x','y','s')
%--------------------------------------------------------------------
```

When this script is executed, it produces the graph shown in Figure 1.5.1. The least-squares polynomial of degree $m = 2$ is clearly inadequate, and the polynomial of degree $m = 5$ begins to resem-

FIGURE 1.5.1 Least-Squares Fit

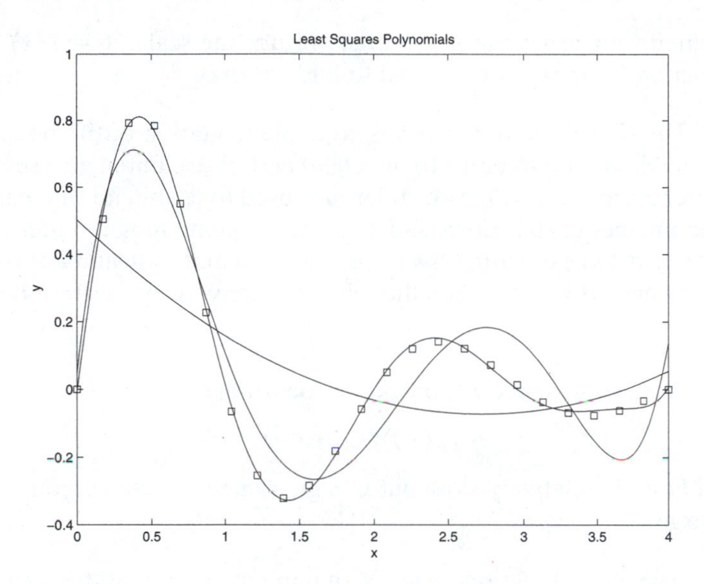

ble the pattern of data, but it is out of phase, while the polynomial of degree $m = 8$ is a reasonable fit, with some error evident near the end of the data interval.

1.6 ROOT FINDING

The module *roots* is an application module that implements the root finding techniques discussed in Chapter 5. The root finding functions are summarized in Table 1.6.1.

The first three functions in Table 1.6.1 attempt to find a root or solution of the following equation.

$$f(x) = 0 \tag{1.6.1}$$

The function f must be stored in an M-file of the same name or declared as an inline function, and it should be of the following form:

```
function y = f(x)
```

TABLE 1.6.1 Root Finding Functions

Usage	*Description*	*Toolbox*
$[x, k] = bisect(x0, x1, tol, m, f)$	bisection method	nlib
$[x, k] = secant(x0, x1, tol, m, f)$	secant method	nlib
$[x, k] = muller(x0, x1, x2, tol, m, f)$	Muller's method	nlib
$J = jacobian(x, f)$	Jacobian matrix	nlib
$[x, k] = newton(x0, tol, m, f)$	Newton's vector method	nlib
$x = roots(c)$	roots of polynomial	built-in

When f is called with scalar input x, it must return the scalar $y = f(x)$. The built-in MATLAB function *fzero* can also be used to find a root of $f(x)$.

bisect The NLIB function *bisect* is an implementation of the bisection method as described in Alg. 5.2.1. On entry to *bisect:* $x0$ and $x1$ are initial guesses that bracket the root of interest, $tol \geq 0$ is an error tolerance used to terminate the search, $m \geq 1$ is the maximum number of iterations, and f is a string containing the name of the function to be searched. On exit from *bisect:* x is the estimated root, and k is the number of iterations performed. If $k < m$, then the following convergence criterion was satisfied.

$$|f(x)| < tol \tag{1.6.2}$$

A pair of initial guesses, $x0$ and $x1$, bracket a root of $f(x)$ if

$$f(x0)f(x1) < 0. \tag{1.6.3}$$

The function *bisect* is relatively slow, but it is guaranteed to converge if m and tol are sufficiently large.

secant The NLIB function *secant* is an implementation of the secant method as described in Alg. 5.4.1. On entry to *secant:* $x0$ and $x1$ are two distinct initial guesses for the root, $tol \geq 0$ is an error tolerance used to terminate the search, $m \geq 1$ is the maximum number of iterations, and f is a string containing the name of the function to be searched. On exit from *secant:* x is an estimate of a root, and k is the number of iterations performed. If $k < m$, the convergence criterion in (1.6.2) was satisfied.

muller The NLIB function *muller* is an implementation of Muller's method as described in Alg. 5.5.1. On entry to *muller:* $x0, x1$, and $x2$ are three distinct initial guesses for the root, $tol \geq 0$ is an error tolerance used to terminate the search, $m \geq 1$ is the maximum number of iterations, and f is a string containing the name of the function to be searched. Both x and $f(x)$ can be complex in this case. On exit from *muller:* x is an estimate of a root, and k is the number of iterations performed. If $k < m$, then the convergence criterion in (1.6.2) was satisfied. The initial guesses can be real even if the root of interest is complex.

jacobian The function *jacobian* computes the Jacobian matrix $J(x) = \partial f(c)/\partial x$ of partial derivatives of $f(x)$ with respect to the components of x using a central difference approximation.

$$J_{kj}(x) \triangleq \frac{\partial f_k(x)}{\partial x_j} \quad , \quad 1 \leq k \leq m, 1 \leq j \leq n \tag{1.6.4}$$

On entry to *jacobian*: the $n \times 1$ vector x specifies the point at which the Jacobian matrix is to be evaluated, and the string f contains the name of the function to be differentiated. The function f must be stored in an M-file of the same name or declared as an inline function and must be of the following form.

```
function y = f(x)
```

When f is called, it must take the $n \times 1$ vector x and compute the $m \times 1$ vector $y = f(x)$. The function *jacobian* is used by the function *newton* to solve the system of equations $f(x) = 0$.

newton The NLIB function *newton* is an implementation of Newton's vector method as described in Alg. 5.8.1. On entry to *newton:* the $n \times 1$ vector $x0$ contains an initial guess, $tol \geq 0$ is an error tolerance used to terminate the search, $m \geq 1$ is the maximum number of iterations, and the string f contains the name of the function to be searched. The function f must be stored in an M-file of the same name or declared as an inline function and must be of the following form.

```
function y = f(x)
```

When f is called, it must take the $n \times 1$ vector x and compute the $n \times 1$ vector $y = f(x)$. On exit from *newton:* the $n \times 1$ vector x contains an estimate of the solution of $f(x) = 0$, and the integer k is the number of iterations performed. If $k < m$, the following convergence criterion was satisfied.

$$\|f(x)\| < tol \tag{1.6.5}$$

roots The built-in MATLAB function *roots* computes the roots of a polynomial. On entry to *roots:* a is a $1 \times (n + 1)$ row vector containing the coefficients of the polynomial. On exit from *roots:* the $n \times 1$ complex vector x contains the roots of the polynomial

$$f(x) = a_1 x^n + a_2 x^{n-1} + \cdots + a_{n+1} \tag{1.6.6}$$

EXAMPLE 1.6.1 Root Finding

To illustrate the root finding functions in Table 1.6.1, consider the liquid level control system introduced in Chapter 5. The steady-state liquid level in the tank is given by the solution of (5.1.6). For convenience, suppose $\alpha = 1, \beta = 1, \gamma = 1$, and $h = 1$ m. The nonlinear equation for the liquid level then reduces to

$$f(x) = 1 - x - \sqrt{x} = 0$$

This particular equation can be solved directly as formulated using *bisect*. It can also be solved using *roots* by using a change of variable $x = y^2$. This converts it to the following polynomial root solving problem.

$$y^2 + y - 1 = 0$$

Since the height of the tank is $h = 1$ m, the steady-state solution is known to lie in the interval $[a, b] = [0, 1]$. The following script uses both *bisect* and *roots* to estimate a root. The *inline* function f describes the equation to be solved.

```
%--------------------------------------------------------------------
% Example a1.6.1: Root Finding
%--------------------------------------------------------------------
```

```
% Initialize

clc                              % clear command window
clear                            % clear variables
a = 0;                           % lower limit
b = 1;                           % upper limit
c = [1 1 -1];                    % polynomial coefficients
tol = 1.e-4;                     % error tolerance
m = 100;                         % maximum iterations
f = inline ('1 - x - sqrt(x)','x');

% Find roots

fprintf ('Example a1.6.1: Root Finding\n');
y = 100*bisect(a,b,tol,m,f);
show ('Liquid level in cm = ',y)
x = 100*roots(c).^2;
show ('roots in cm',x)
%--------------------------------------------------------------------
```

When this script is executed, we find that the steady-state height of the liquid in the tank is $x = 38.1966$ cm. When both roots are computed, the second root at $x = 261.8034$ cm is not physically meaningful because it corresponds to an overflow of the tank.

1.7 OPTIMIZATION

The module *optim* is an application module that implements the optimization techniques discussed in Chapter 6. The optimization functions are summarized in Table 1.7.1.

conjgrad The NLIB function *conjgrad* uses the Fletcher-Reeves conjugate-gradient method in Alg. 6.5.1 to solve the following n-dimensional unconstrained optimization problem.

$$\begin{aligned} \text{minimize:} \quad & f(x) \\ \text{subject to:} \quad & x \in R^n \end{aligned}$$

TABLE 1.7.1 Optimization Functions

Usage	*Description*	*Toolbox*
$[x, ev, k] = conjgrad\ (x0, tol, v, m, f)$	conjugate gradient method	nlib
$[x, ev, k] = dfp\ (x0, tol, v, m, f)$	Davidon-Fletcher-Powell method	nlib
$[x, ev, k] = penalty\ (x0, mu, tol, m, f, p, q)$	constrained minimization using penalty functions	nlib
$[x, ev, k] = anneal\ (x0, f0, f1, tol, eta, seed, gamma, mu, m, a, b, f)$	global minimization using simulated annealing	nlib

If we are interested in *maximizing* $f(x)$ instead, we simply minimize $-f(x)$. On entry to *conjgrad:* the $n \times 1$ vector $x0$ is an initial guess that serves as a starting point for the search, $tol \geq 0$ is an error tolerance used to terminate the search, $1 \leq v \leq n$ is the desired *order* of the method, $m \geq 1$ is an upper bound on the number of iterations, and the string f contains the name of the function to be minimized. Normally, *conjgrad* is called with $v = n$, which corresponds to the full conjugate-gradient method. When $v = 1$, the conjugate-gradient method reduces to the steepest descent method. The function f must be stored in an M-file of the same name or declared an inline function, and it should be of the following form:

```
function y = f(x)
```

When f is called with the $n \times 1$ vector x, it must return the scalar $y = f(x)$. On exit from *conjgrad:* the $n \times 1$ vector x is an updated estimate of the optimal vector, *ev* is the number of scalar function evaluations performed, and k is the number of iterations. If $k < m$, the following convergence criterion was satisfied where ε_M is the machine epsilon.

$$\|\nabla f(x^k)\| < tol \quad \text{or} \quad \|x^k - x^{k-1}\| < \varepsilon_M \|x^k\| \tag{1.7.1}$$

dfp The NLIB function *dfp* uses the Davidon-Fletcher-Powell quasi-Newton method in Alg. 6.6.1 to solve the n-dimensional unconstrained optimization problem. The calling arguments for *dfp* are identical to those for *conjgrad*. Thus, the two functions can be used interchangeably. The function *dfp* uses the self-scaling in Equation (6.6.11) of Chapter 6. This makes the method more robust in the sense that it is less sensitive to the accuracy of the line search. Normally, *dfp* is called with $v = n$, which corresponds to the full quasi-Newton method. When *dfp* is called with $v = 1$, it reduces to the steepest descent method. The termination criterion for *DFP* is identical to that for *conjgrad* and is given in (1.7.1).

penalty The NLIB function *penalty* uses the penalty function method to solve the following n-dimensional constrained optimization problem.

$$\begin{aligned} \text{minimize:} \quad & f(x) \\ \text{subject to:} \quad & p(x) = 0 \\ & q(x) \geq 0 \end{aligned}$$

Here $p(x) = 0$ represents r equality constraints, and $q(x) \geq 0$ represents s inequality constraints. The penalty function method converts the constrained optimization problem into an unconstrained problem using the following generalized objective function.

$$F(x) \triangleq f(x) + \mu[P(x) + Q(x)] \tag{1.7.2}$$

The function *penalty* uses the function *dfp* to perform an unconstrained minimization of $F(x)$. Here $\mu > 0$ is a penalty parameter that controls the cost of violating the constraints. The term $P(x)$ is a penalty function for the equality constraints and is defined in (6.7.1), while the term $Q(x)$ is a penalty function for the inequality constraints and is defined in (6.7.2).

On entry to *penalty:* the $n \times 1$ vector $x0$ is an initial guess, $mu > 0$ is the penalty parameter, $tol \geq 0$ is an error tolerance used to terminate the search, $m \geq 1$ is an upper bound on the number of iterations, and f, p, and q are strings containing the names of functions that specify the objective, the equality constraints, and the inequality constraints, respectively. They must be stored in M-files of the same names or declared as inline functions, and they should be of the following form.

```
function y = f(x)
function u = p(x)
function v = q(x)
```

When f is called with the $n \times 1$ vector x, it must return the objective value $f(x)$. When p is called, it must compute the $r \times 1$ equality constraint function $u = p(x)$. When q is called, it must compute the $s \times 1$ inequality constraint function $v = q(x)$. The functions p and q are *optional* in the following sense. If *penalty* is called with the argument p replaced by the empty string '', a user-supplied function is *not* required. A similar remark holds for the argument q. Since both p and q can be replaced by the empty string, *penalty* can be used to solve unconstrained problems, problems with only equality constraints, and problems with only inequality constraints, as well as general problems. Larger values for the penalty parameter *mu* correspond to more severe penalties for violating the constraints. Since the gradient of $f(x)$ is not included as a calling argument in *penalty*, line searches are performed using the golden section search.

On exit from *penalty:* the $n \times 1$ vector x contains an estimate of the optimal x, *ev* is the number of scalar function evaluations, and k is the number of iterations performed. If $k < m$, then the following convergence criterion was satisfied where ε_M is the machine epsilon.

$$\|\nabla F(x^k)\| < tol \quad \text{or} \quad \|x^k - x^{k-1}\| < \varepsilon_M \|x^k\| \tag{1.7.3}$$

anneal The NLIB function *anneal* uses the statistical simulated annealing method in Alg. 6.8.1 to find a global minimum for the following n-dimensional constrained optimization problem.

$$\begin{aligned} \text{minimize:} \quad & f(x) \\ \text{subject to:} \quad & a \le x \le b \end{aligned} \tag{1.7.3}$$

On entry to *anneal:* the $n \times 1$ vector $x0$ is an initial guess, $f0$ is a lower bound on $f(x)$, $f1$ is an upper bound on $f(x)$, $tol \ge 0$ is an error tolerance used to terminate the random global search, $eta \ge 0$ is an error tolerance used to terminate the local search, *seed* is an integer used to initialize the random number generator, $0 < gamma < 1$ is a localization parameter that controls how fast the random search becomes localized, $mu > 0$ is the penalty parameter, $m \ge 1$ is an upper bound on the number of iterations, the $n \times 1$ vector a contains the lower limits for the components of x, the $n \times 1$ vector b contains the upper limits, and f is a string containing the name of the objective function. The function f must be stored in an M-file of the same name or declared as an inline function, and it should be of the following form.

```
function y = f(x)
```

When f is called with the $n \times 1$ vector x, it must return the objective value $f(x)$. The random global search is terminated after m iterations or when the following error criterion is satisfied:

$$f(x) - f_0 < tol \tag{1.7.4}$$

The random global search returns the point $\overline{x}$ corresponding to the lowest objective value encountered thus far. This is used as the initial guess for the local search, which is then performed with the function *penalty*.

On exit from *anneal:* the $n \times 1$ vector x contains an estimate of the global optimum x, ev is the number of scalar function evaluations, and k is the number of iterations performed. If $k < m$, the global search satisfied (1.7.4), and the following convergence criterion was satisfied by the local search where ε_M is the machine epsilon and $F(x)$ is the generalized objective function in (1.7.2).

$$\|\nabla F(x^k)\| < eta \quad \text{or} \quad \|x^k - x^{k-1}\| < \varepsilon_M \|x\| \tag{1.7.5}$$

EXAMPLE 1.7.1 Optimization

As an illustration of an optimization using the functions in Table 1.7.1, consider the problem of fitting a curve to the following experimental measurements.

$$D = \{(t_k, y_k) \mid 1 \le k \le p\}$$

Suppose a decaying exponential function is used to model the data.

$$y_k \approx x_1 \exp(-x_2 t_k) + x_3$$

Recall from Chapter 6 that to find suitable values for the parameters, x, we perform an unconstrained optimization using the following nonlinear objective function.

$$f(x) = \frac{1}{p} \sum_{k=1}^{p} [x_1 \exp(-x_2 t_k) + x_3 - y_k]^2$$

The following script uses *conjgrad* to solve this problem with $p = 100$ data points and starting from an initial guess of $x^0 = 0$. In this case, global variables are used to pass the data points (t, y) to the objective function f.

```
%---------------------------------------------------------------
% Example a1.7.1: Optimization
%---------------------------------------------------------------

% Initialize

clc                        % clear command window
clear                      % clear variables
global t y                 % used by funa171.m
q = [6 .5 2];              % nominal parameters
p = 100;                   % number of points
n = 3;                     % number of variables
t1 = 5;                    % range of t
```

```
noise = 0.25;               % measurement noise
x0 = [0 0 0]';              % initial guess
tol = 1.e-4;                % error tolerance
m = 200;                    % maximum iterations
v = n;                      % order of method
g = inline ('x(1)*exp(-x(2)*t) + x(3)','t','x');

% Construct data

fprintf ('Example a1.7.1: Optimization\n');
randinit (1000)
t = [0 : t1/(p-1) : t1]';
y = g(t,q) + randu (p,1,-noise,noise);

% Fit optimal curve

disp ('Searching for optimal x ...')
[x,ev,k] = conjgrad (x0,tol,v,m,'funa171');
show ('Optimal x',x)
show ('Optimal f(x)',funa171(x))
show ('Iterations',k);
show ('Function evaluations',ev);

% Display optimal curve and data

Y(:,1) = g(t,x);
Y(:,2) = y;
graphxy (t,Y,'Unconstrained Optimization','t','y')

function E = funa171(x)

% Description: Objective function

global t y
a = max([-10,min([10,x(2)])]);
E = sum((x(1)*exp(-a*t)+ x(3) - y).^2)/length(t);
%-----------------------------------------------------------------
```

The exponent in the objective function is bracketed by $[-10, 10]$ to eliminate possible floating-point underflow and overflow conditions during the search. The function *conjgrad* used 47 iterations and 2549 scalar function evaluations to produce the following curve parameters and objective value:

$$x = \begin{bmatrix} 6.016 \\ 0.4874 \\ 0.1941 \end{bmatrix}$$

$$f(x) = 0.0207986$$

A graph showing the original data and the fitted exponential curve is shown in Figure 1.7.1.

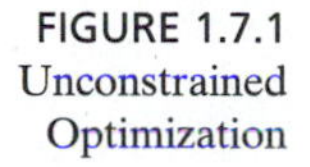

FIGURE 1.7.1 Unconstrained Optimization

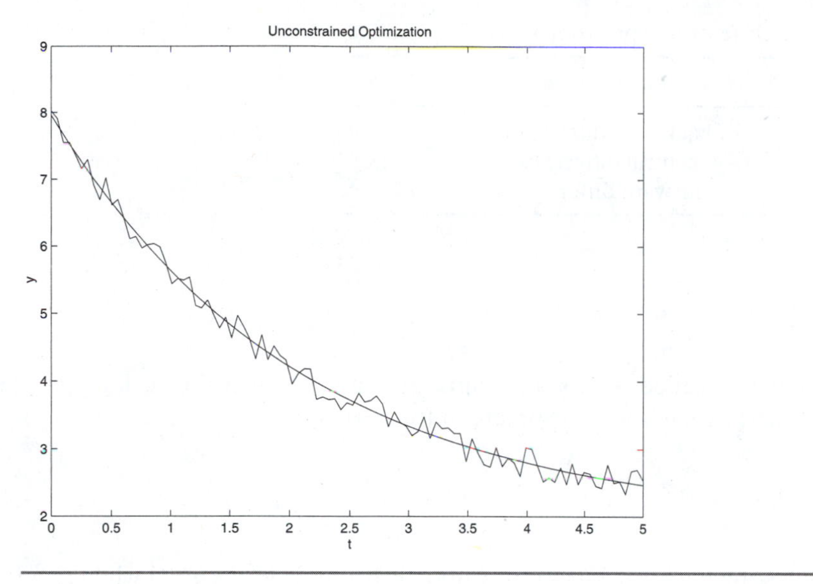

1.8 DIFFERENTIATION AND INTEGRATION

The module *integ* is an application module that implements the numerical differentiation and integration techniques discussed in Chapter 7. The differentiation and integration functions are summarized in Table 1.8.1.

deriv The NLIB function *deriv* estimates the derivative of a function using the finite difference method. On entry to *deriv:* a is the evaluation point, $h > 0$ is the step size, n is the number of points, and m selects the type of approximation as summarized in Table 1.8.2. The remaining calling arguments are e, which controls optional extrapolation, and f, which is a string containing the name of the function to be differentiated. If the extrapolation parameter e is nonzero, then Richardson extrapolation based on the step sizes $2h$ and h is used. The user-supplied function f must be stored in an M-file of the same name or declared as an inline function, and it should be of the following form.

TABLE 1.8.1 Numerical Differentiation and Integration Functions

Usage	*Description*	*Toolbox*
$dy = deriv(a,h,n,m,e,f)$	numerical derivatives	nlib
$y = trapint(a,b,n,f)$	trapezoid rule integration	nlib
$y = simpson(a,b,n,f)$	Simpson's rule integration	nlib
$y = midpoint(a,b,n,f)$	midpoint rule integration	nlib
$[y,r] = romberg(a,b,tol,m,f)$	Romberg integration	nlib
$y = gausquad(a,b,n,f)$	Gauss quadrature integration	nlib
$z = monte(a,b,r,f)$	Monte Carlo multi-dimensional integration	nlib

TABLE 1.8.2 Finite Difference Approximations of Derivative

m	*Approximation*	n
-1	backward difference	2,3,4
0	central difference	3,5
1	forward difference	2,3,4

```
function y = f(x)
```

When f is called with scalar input x, it must return the scalar $y = f(x)$. The value returned by *deriv* is the estimate of the derivative.

$$f'(a) = \left.\frac{df(x)}{dx}\right|_{x=a} \tag{1.8.1}$$

The estimate of $f'(a)$ has a truncation error of order $O(h^{n-1})$ when $e = 0$ and a truncation error of order $O(h^{n+1})$ when $e \neq 0$.

trapint The next four functions in Table 1.8.1 are used to numerically estimate the value of the integral

$$I(a,b) = \int_a^b f(x)\,dx \tag{1.8.2}$$

The NLIB function *trapint* implements the trapezoid rule integration formula in (7.4.6) to approximate $I(a,b)$. On entry to *trapint:* a is the lower limit of integration, b is the upper limit of integration, $n \geq 1$ is the number of integration panels, and the string f contains the name of the function to be integrated. The function f must be stored in an M-file of the same name or declared as an inline function, and it should be of the following form.

```
function y = f(x)
```

When f is called with scalar input x, it must return the scalar $y = f(x)$. On exit from *trapint:* y is an estimate of $I(a,b)$. The truncation error of y is of order $O(h^2)$ where

$$h = \frac{b-a}{n} \tag{1.8.3}$$

simpson The NLIB function *simpson* implements the Simpson's 1/3 rule integration formula in (7.4.8) to approximate $I(a,b)$. The calling sequence for *simpson* is identical to that for *trapint* except that the number of panels, n, must be even. The estimate of $I(a,b)$ returned by *simpson* has a truncation error of order $O(h^4)$.

midpoint The NLIB function *midpoint* implements the midpoint rule integration formula in (7.4.11) to approximate $I(a,b)$. The calling sequence for *midpoint* is identical to that for *trapint*. Unlike *trapint* and *simpson*, the function *midpoint* implements an open integration formula that does not require the evaluation of $f(x)$ at the end points a and b. Therefore, *midpoint* may be applicable to integrands, which contain singularities at the end points. The estimate of $I(a,b)$ returned by *midpoint* has a truncation error of order $O(h^2)$.

romberg The NLIB function *romberg* implements the Romberg integration technique as described in Alg. 7.5.1. On entry to *romberg:* a is the lower limit of integration, b is the upper limit of integration, $tol \geq 0$ is an upper bound on the magnitude of the estimated truncation error, $2 \leq m \leq 10$ is the maximum extrapolation level, and f is the name of the user-supplied function to be integrated as described in *trapint*. On exit from *romberg:* y is the estimate of the integral $I(a,b)$, and r is the extrapolation level used. If $r < m$, the following termination criterion was satisfied where E is the estimated truncation error.

$$|E| < tol \tag{1.8.4}$$

gausquad The NLIB function *gausquad* implements the Gauss-Legendre quadrature integration formula in (7.6.4) with uniform weighting, $w(x) = 1$. The calling arguments for *gausquad* are identical to those for *trapint*, but the number of points of evaluation is limited to the range $1 \leq n \leq 10$. Additional accuracy can be achieved by partitioning the integration interval $[a,b]$ into subintervals and performing Gauss-Legendre integration on each subinterval. It is also possible to compute $I(a,b)$ in (1.8.2) using the built-in MATLAB functions *quad* and *quad8*.

monte The NLIB function *monte* implements the Monte Carlo statistical technique to estimate the value of a multi-dimensional integral of the following form.

$$I(D) = \int_{a_1}^{b_1} \cdots \int_{a_m}^{b_m} f(x)\, dx_m \cdots dx_1 \tag{1.8.5}$$

On entry to *monte:* the $m \times 1$ vector a contains the lower limits of integration, the $m \times 1$ vector b contains the upper limits of integration, $n \geq 1$ is the number of random samples, and the string f contains the name of the function to be integrated. The function f must be stored in an M-file of the same name or declared as an inline function, and it should be of the following form.

```
function y = f(x)
```

When f is called with the $m \times 1$ vector x, it must return the scalar $y = f(x)$. The domain of integration D is the following rectangular region:

$$D = \{x \in R^m \mid a_k \leq x_k \leq b_k, 1 \leq k \leq m\} \tag{1.8.6}$$

On exit from *monte*, y is an estimate of $I(D)$. The standard deviation of the estimate is proportional to $1/\sqrt{n}$.

EXAMPLE 1.8.1 Numerical Integration

Suppose a sequence of random numbers has a Gaussian distribution with mean $\mu = 0$ and standard deviation $\sigma = 1$. Then the probability density function is

$$f(x) = \frac{1}{\sqrt{2\pi}} \exp\left(\frac{-x^2}{2}\right)$$

The probability that a number lies in the interval $[a, b]$ is then

$$I(a,b) = \int_a^b f(x)\, dx$$

There is no closed-form expression for the integral $I(a, b)$, so it must be computed numerically. The following program estimates the value of $I(-1, 1)$ using *simpson* and *trapint*. As a crude check, it also estimates $I(-1, 1)$ directly by generating a random sequence and computing the fraction of numbers that lie within $[-1, 1]$.

```
%----------------------------------------------------------------------
% Example a1.8.1: Numerical Integration
%----------------------------------------------------------------------

% Initialize

clc                       % clear command window
clear                     % clear variables
a = -1;                   % lower limit
b = 1;                    % upper limit
n = 100;                  % number of panels
randinit (500);           % select random sequence
f = inline ('exp(-x.*x/2) ./ sqrt(2*pi)','x');

% Simpson's method

fprintf ('Example a1.8.1: Numerical Integration\n\n');
y = simpson (a,b,n,f);
fprintf ('Simpson: I(-1,1) = %.6f\n',y)

% Direct method

disp('Computing by direct method ...')
c = 0;
for k = 1 : 1000
   x = randg(1,1,0,1);
   if (x >= a) & (x <= b)
      c = c + 1;
   end
   if (k == 10) | (k == 100) | (k == 1000)
      fprintf ('direct, p = %5i  %.6f \n',k,c/k);
   end
end
```

```
% Trapezoid rule

y = trapint(a,b,n,f);
fprintf ('Trapezoid rule: I(-1,1) = %.6f\n',y)
wait
%----------------------------------------------------------------------
```

The output from Example 1.8.1 is summarized in Table 1.8.2. The middle rows of Table 1.8.2 are direct estimates of $I(-1,1)$ based on increasing numbers of random samples p.

TABLE 1.8.2 Numerical Estimates of $I(-1,1)$

Method	*(−1, 1)*
simpson	0.682689
direct, p = 10	0.600000
direct, p = 100	0.720000
direct, p = 1000	0.691000
trapezoid	0.682673

1.9 ORDINARY DIFFERENTIAL EQUATIONS

The module *ode* is an application module that implements the ordinary differential equations techniques discussed in Chapter 8. The ordinary differential equations functions are summarized in Table 1.9.1.

The first three functions in Table 1.9.1 compute solutions to the following n-dimensional initial value problem:

$$\frac{dx}{dt} = f(t,x) \quad , \quad x(\alpha) = a \tag{1.9.1}$$

rkf The NLIB function *rkf* implements the variable step size Runge-Kutta-Fehlberg method as summarized in Alg. 8.4.1. It computes a solution on a set of m points equally spaced over the interval $[alpha, beta]$. On entry to *rkf:* the $n \times 1$ vector $x0$ is the initial state, *alpha* is the initial time, *beta* is the final time, $m \geq 2$ is the number of solution points, $tol \geq 0$ is an upper bound on the norm of the local truncation error, $q \geq 1$ is the maximum number of scalar function evaluations, and f is a string

TABLE 1.9.1 Ordinary Differential Equation Functions

Usage	*Description*	*Toolbox*
$[t, X, e, k] = rkf(x0, alpha, beta, m, tol, q, f)$	Runge-Kutta-Fehlberg method	nlib
$[t, X, e, k] = bsext(x0, alpha, beta, m, tol, q, f)$	Bulirsch-Stoer extrapolation method	nlib
$[t, X, e, k] = stiff(x0, alpha, beta, m, tol, q, f, fx, ft)$	semi-implicit stiff ODE solver	nlib
$[t, y, r] = bvp(alpha, beta, a, b, tol, m, p, q)$	two-point boundary value problem	nlib

containing the name of the function that specifies the equations to be solved. The function f must be stored in an M-file of the same name or declared as an inline function and should have the following form:

```
function dx = f(t,x)
```

When f is called with the scalar t and the $n \times 1$ vector x, it must evaluate the right-hand-side function $f(t,x)$ in Equation (1.9.1) and return the result in the $n \times 1$ *column* vector dx. On exit from *rkf*: the $m \times 1$ vector t contains the equally spaced solution times, the $m \times n$ matrix X contains the solution points, e is the infinity norm of the maximum local truncation error, and k is the number of scalar function evaluations performed. The ith row of X contains the solution at time $t(i)$. Thus, column vector $X(:,j)$ is the solution to the jth differential equation. If $k < q$, then the following termination criterion was satisfied.

$$\|E\| < \max\{\|x\|, 1\} tol \tag{1.9.2}$$

Here E is the estimated local truncation error computed in (8.4.7h). Note that when the solution x is large in the sense that $\|x\| > 1$, *tol* becomes a *relative* error bound; otherwise, it is an *absolute* error bound. The built-in MATLAB function *ode45* can also be used to solve (1.9.1).

bsext The NLIB function *bsext* is similar to *rkf* but uses the Bulirsch-Stoer extrapolation method as summarized in Alg. 8.6.1. It computes a solution to (1.9.1) at a set of m points equally spaced over the time interval $[alpha, beta]$. The calling sequence for *bsext* is identical to that for *rkf*, so the two functions can be used interchangeably. Typically, *bsext* is used when $f(t,x)$ is relatively smooth and high accuracy is required.

stiff The NLIB function *stiff* is designed for stiff systems of differential equations—equations whose solutions contain terms with time scales that differ by more than an order of magnitude. The function *stiff* implements the semi-implicit Bulirsch-Stoer extrapolation method summarized in Alg. 8.7.1. It computes a solution to (1.9.1) at a set of m points equally spaced over the time interval $[alpha, beta]$. On entry to *stiff:* the $n \times 1$ vector $x0$ is the initial state, *alpha* is the initial time, *beta* is the final time, $m \geq 2$ is the number of solution points, $tol \geq 0$ is an upper bound on the norm of the local truncation error, q is the maximum number of scalar function evaluations, f is a string containing the name of the function that specifies the equations to be solved, fx is a string containing the name of an optional function that computes the $n \times n$ Jacobian matrix $J(t,x) = \partial f(t,x)/\partial x$ in (8.7.5), and ft is a string containing the name of an optional function that computes the $n \times 1$ vector $\partial f(t,x)/\partial t$. The functions f, fx, and ft must be stored in M-files of the same names or declared as an inline functions and should have the following forms:

```
function dx   = f(t,x)
function dfdx = f(t,x)
function dfdt = f(t,x)
```

When f is called with the scalar t and the $n \times 1$ vector x, it must evaluate the right-hand-side function $f(t,x)$ in (1.9.1) and return the result in the $n \times 1$ *column* vector dx.

When fx is called, it must evaluate the Jacobian matrix defined in Equation (8.7.5) and return the result in the $n \times n$ matrix $dfdx$. Finally, when ft is called, it must evaluate $\partial f(t, x)/\partial t$ and return the result in the $n \times 1$ vector $dfdt$. The user-supplied functions fx and ft are *optional* in the sense that *stiff* can be called with either of these arguments set to the empty string, ''. When this is done, the partial derivatives are evaluated numerically using central differences. Thus, the user does *not* have to supply the functions fx and ft in this case. Numerical approximations to partial derivatives are provided as a convenience, but they can be somewhat less accurate and require additional function evaluations.

On exit from *stiff:* the $m \times 1$ vector t contains the equally spaced solution times, the $m \times n$ matrix X contains the solution points, e is the norm of the maximum local truncation error, and k is the number of scalar function evaluations performed. The ith row of X contains the solution at time $t(i)$. Consequently, column vector $X(:, j)$ is the solution to the jth differential equation. If $k < q$, the termination criterion in (1.9.2) was satisfied. The built-in MATLAB function *ode15s* can also be used to solve stiff systems of differential equations.

bvp The NLIB function *bvp* uses the finite difference method to solve the following second-order nonlinear two-point boundary value problem:

$$\frac{d^2y}{dx^2} = g\left(x, y, \frac{dy}{dx}\right) \quad , \quad y(\alpha) = a, y(\beta) = b \tag{1.9.3}$$

On entry to *bvp: alpha* is the initial value of the independent variable x, *beta* is the final value of x, a is the initial value of the dependent variable y, b is the final value of y, $tol \geq 0$ is an upper bound on the norm of the difference between successive solution estimates, $m \geq 1$ is the number of interior solution points, $p \geq 1$ is the maximum number of iterations, and g is a string containing the name of a function that describes the equation to be solved. The function g must be stored in an M-file of the same name or declared as an inline function. The form of the function g is

```
function y2 = g(x,y,y1)
```

When g is called, it must return the value $y2 = d^2y/dx^2$ as in Equation (1.9.3) where $y1 = dy/dx$. On exit from *bvp:* the $(m + 2) \times 1$ vector x contains the equally spaced values of the independent variable, the $(m + 2) \times 1$ vector y contains the estimated solution with $y(1) = a$ and $y(m + 2) = b$, and r is the number of iterations performed. If $r < p$, then the infinity norm of the difference between the last two estimates of the vector y was less than tol.

EXAMPLE 1.9.1 Ordinary Differential Equations

Some systems of nonlinear differential equations produce periodic steady-state solutions when started from *any* initial condition. One such system is the *van der Pol* oscillator:

$$\frac{dx_1}{dt} = x_2$$

$$\frac{dx_2}{dt} = -x_1 + \mu(1 - x_1^2)x_2$$

The parameter $\mu \geq 0$ controls the degree of nonlinearity. When $\mu = 0$, this is a linear harmonic oscillator whose steady-state solution is sinusoidal with the amplitude and phase of the solution dependent on the initial condition $x(0)$. When $\mu > 0$, the steady-state solution is periodic but is no longer a pure sinusoid. Furthermore, the periodic steady-state solution, called a *limit cycle*, is independent of the initial condition when $\mu > 0$ and $x(0) \neq 0$. The following script uses *rkf* to compute a solution to the van der Pol oscillator.

```
%---------------------------------------------------------------------
% Example a1.9.1: Ordinary Differential Equations
%---------------------------------------------------------------------

   clc                       % clear command window
   clear                     % clear variables
   n = 2;                    % number of equations
   m = 500;                  % number of solution points
   b = 20;                   % final time
   tol = 1.e-5;              % error bound
   q = 12000;                % maximum function evaluations
   x0 = [0.5 0]';            % initial state

% Solve system from x = x0 using rkf

   fprintf ('Example a1.9.1: Ordinary Differential Equations\n');
   [t,X,e,k] = rkf (x0,0,b,m,tol,q,'funa191');
   graphxy (t,X,'Solution Trajectories','t','x')
   graphxy (X(:,1),X(:,2),'Limit Cycle','x_1','x_2')
%---------------------------------------------------------------------
```

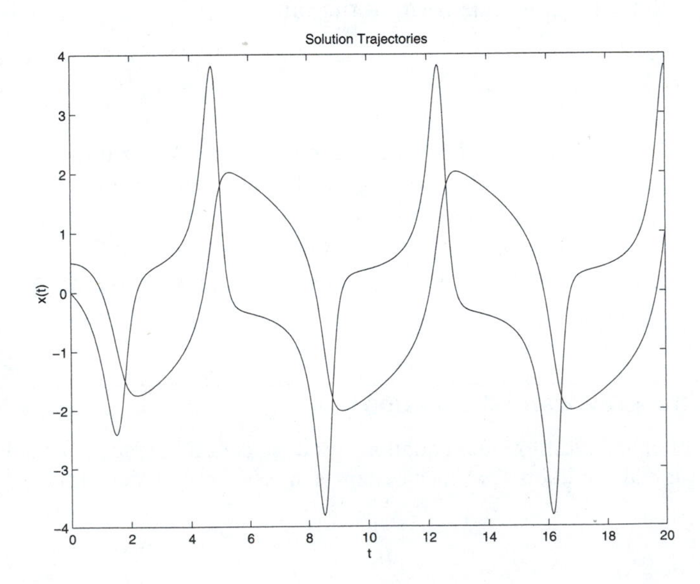

FIGURE 1.9.1 The van der Pol Oscillator

FIGURE 1.9.2 Limit Cycle of the van der Pol Oscillator

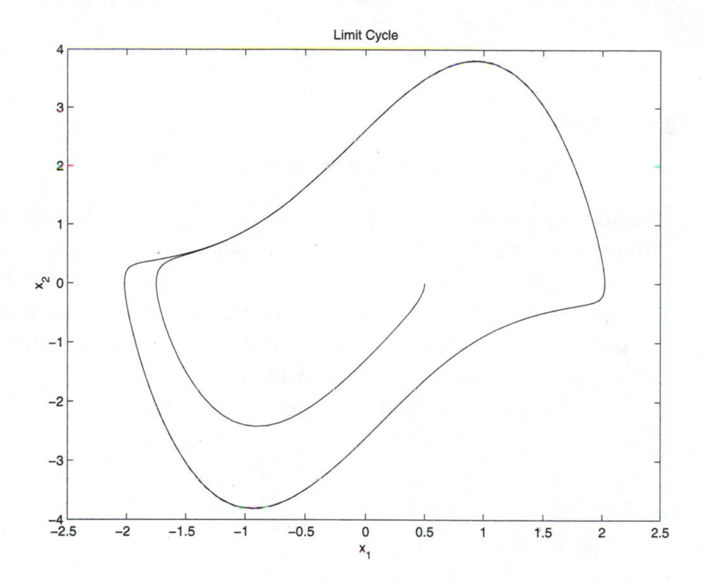

The first plot in Figure 1.9.1 displays the two elements of the solution vector $x(t)$ versus t. The limit cycle is clearly evident in the plot of x_2 versus x_1 shown in Figure 1.9.2.

1.10 PARTIAL DIFFERENTIAL EQUATIONS

The module *pde* is an application module that implements the partial differential equations techniques discussed in Chapter 9. The partial differential equations functions are summarized in Table 1.10.1.

Each of the functions in Table 1.10.1 contains arguments, f and g, which are strings containing the names of user-supplied M-file or inline functions. These functions allow the user to specify such things as forcing terms, initial values, and boundary values. In each case, the arguments f and g are *optional* in the sense that either can be replaced by the empty string, ''. When this is done, the corresponding function is assumed to be zero, which means that a user-supplied function is *not* required in this case.

poisson The NLIB function *poisson* uses central difference approximations and the *SOR* method to solve Poisson's equation:

TABLE 1.10.1 Partial Differential Equation Functions

Usage	*Description*	*Toolbox*
$[alpha, r, x, y, U] = poisson(a, b, m, n, q, \ tol, f, g)$	Poisson's equation	nlib
$[t, x, U] = heat1(T, a, m, n, beta, f, g)$	one-dimensional heat equation	nlib
$[x, y, U] = heat2(T, a, b, m, n, p, beta, f, g)$	two-dimensional heat equation	nlib
$[t, x, U] = wave1(T, a, m, n, beta, f, g)$	one-dimensional wave equation	nlib
$[x, y, U] = wave2(T, a, b, m, n, p, beta, f, g)$	two-dimensional wave equation	nlib

$$\frac{\partial^2 u}{\partial x^2} + \frac{\partial^2 u}{\partial y^2} = f \tag{1.10.1}$$

over the rectangular region

$$R = \{(x, y) \mid 0 < x < a, 0 < y < b\} \tag{1.10.2}$$

subject to the Dirichlet boundary condition, $u(x, y) = g(x, y)$. On entry to *poisson*: $a > 0$ is the maximum value of x, $b > 0$ is the maximum value of y, $m \geq 1$ is the number of steps in x, $n \geq 1$ is the number of steps in y, $q \geq 1$ is the maximum number of iterations, $tol \geq 0$ is an upper bound on the norm of the difference between successive solution estimates, and the strings f and g contain the names of user-supplied functions that specify the forcing term and boundary conditions, respectively. The functions f and g should be stored in M-files of the same name or declared as inline functions and should be of the form

```
function z = f(x,y)
function z = g(x,y)
```

When f is called with scalar inputs x and y, it must return the value of the forcing term $z = f(x, y)$ on the right-hand side of Equation (1.10.1). When g is called with (x, y) on the boundary of R, it must return the desired boundary value $z = u(x, y)$. Note that when the argument f is replaced with the empty string, '', Poisson's equation reduces to *Laplace's* equation, also called the potential equation. On exit from *poisson: alpha* is the optimal relaxation parameter computed using Equation (9.2.18), r is the number of iterations performed, the $m \times 1$ vector x contains the x grid points, the $n \times 1$ vector y contains the y grid points, and the $m \times n$ matrix U contains the solution, $u(x, y)$, evaluated at the mn interior grid points. That is, for $1 \leq k \leq m$ and $1 \leq j \leq n$,

$$U(k, j) = u[x(k), y(j)] \tag{1.10.3a}$$

$$x(k) = \frac{ka}{m + 1} \tag{1.10.3b}$$

$$y(j) = \frac{jb}{n + 1} \tag{1.10.3c}$$

If the number of iterations satisfies $r < q$, the following termination criterion was satisfied where U^i is the ith estimate of the solution.

$$\max_{k=1}^{m} \max_{j=1}^{n} \{|U^i(k, j) - U^{i-1}(k, j)|\} > tol \tag{1.10.4}$$

The estimate, U^0, used to start the *SOR* iterative procedure is the average value of $u(x, y)$ along the boundary of R.

heat1 The next two functions in Table 1.10.1 are used to solve the heat or diffusion equation. The NLIB function *heat1* uses the Crank-Nicolson method to solve the one-dimensional heat equation:

$$\frac{\partial u}{\partial t} = \beta\left(\frac{\partial^2 u}{\partial x^2}\right) \tag{1.10.5}$$

over the rectangular region

$$R = \{(t, x) \mid 0 \le x \le T, 0 < x < a\} \tag{1.10.6}$$

subject to a user-selectable combination of Dirichlet and Neumann boundary conditions and the following initial condition.

$$u(0, x) = f(x) \quad , \quad 0 < x < a \tag{1.10.7}$$

On entry to *heat1:* $T > 0$ is the maximum value of t, $a > 0$ is the maximum value of x, $m \ge 1$ is the number of steps in t, $n \ge 1$ is the number of steps in x, $beta \ge 0$ is the heat equation coefficient, the 1×2 vector $c = [c_1, c_2]$ contains parameters in the range $[0, 1]$ that select the type of boundary condition, and f and g are strings containing the names of user-supplied functions that specify the initial condition and boundary conditions, respectively. The functions f and g should be stored in M-files of the same name or declared as inline functions and should be of the form

```
function y = f(x);
function y = g(t,k);
```

When f is called with input x, it must return the initial value $y = u(0, x)$. The function g is used to specify the right-hand side of the following two general boundary conditions.

$$(1 - c_1)u(t, 0) - c_1 u_x(t, 0) = g(t, 1) \tag{1.10.8a}$$

$$(1 - c_2)u(t, a) + c_2 u_x(t, a) = g(t, 2) \tag{1.10.8b}$$

Here $u_x = \partial u / \partial x$. Note that when $c_1 = 0$, (1.10.8a) reduces to the Dirichlet boundary condition $u(t, 0) = g(t, 1)$. Similarly, when $c_2 = 1$, (1.10.8b) reduces to the Neumann boundary condition $u_x(t, a) = g(t, 2)$. On exit from *heat1:* the $(m + 1) \times 1$ vector t contains the t grid points, the $n \times 1$ vector x contains the x grid points, and the $(m + 1) \times n$ matrix U contains the solution, $u(t, x)$, evaluated at the $(m + 1)n$ interior grid points. That is, for $1 \le k \le (m + 1)$ and $1 \le j \le n$,

$$U(k, j) = u[t(k), x(j)] \tag{1.10.9a}$$

$$t(k) = \frac{(k - 1)T}{m} \tag{1.10.9b}$$

$$x(j) = \frac{ja}{n + 1} \tag{1.10.9c}$$

The Crank-Nicolson method is unconditionally stable and second-order accurate in both t and x. To obtain good accuracy, the gain parameter γ in Equation (9.3.9) should not be too large. It is recommended that the calling parameters be selected such that $\gamma \le 1$ where

$$\gamma = \frac{\beta(n + 1)^2 T}{ma^2} \tag{1.10.10}$$

heat2 The NLIB function *heat2* uses the alternating direction implicit (ADI) method to solve the two-dimensional heat or diffusion equation:

$$\frac{\partial u}{\partial t} = \beta\left(\frac{\partial^2 u}{\partial x^2} + \frac{\partial^2 u}{\partial y^2}\right) \tag{1.10.11}$$

over the rectangular region

$$R = \{(t, x, y) \mid 0 \le t \le T, 0 < x < a, 0 < y < b\} \tag{1.10.12}$$

subject to the Dirichlet boundary condition $u(t, x, y) = g(t, x, y)$ and the following initial condition where $0 < x < a$ and $0 < y < b$.

$$u(0, x, y) = f(x, y) \tag{1.10.13}$$

On entry to *heat2:* $T > 0$ is the maximum value of t, $a > 0$ is the maximum value of x, $b > 0$ is the maximum value of y, $m \ge 2$ is the number of steps in t, $n \ge 1$ is the number of steps in x, $p \ge 1$ is the number of steps in y, *beta* ≥ 0 is the heat equation coefficient, and the strings f and g contain the names of user-supplied functions that specify the initial conditions and boundary conditions, respectively. The functions f and g should be stored in M-files of the same name or declared as inline functions and should be of the form

```
function z = f(x,y)
function z = g(t,x,y)
```

When f is called with scalar inputs x and y, it must return the initial value $z = u(0, x, y)$. When g is called with (x, y) on the boundary of R, it must return the desired boundary value $z = u(t, x, y)$. On exit from *heat2:* the $n \times 1$ vector x contains the x grid points, the $p \times 1$ vector y contains the y grid points, and the $n \times p$ matrix U contains the final solution, $u(T, x, y)$, at the np interior grid points. That is, for $1 \le i \le n$ and $1 \le j \le p$,

$$U(i,j) = u[T, x(i), y(j)] \tag{1.10.14a}$$

$$x(i) = \frac{ia}{n+1} \tag{1.10.14b}$$

$$y(j) = \frac{jb}{p+1} \tag{1.10.14c}$$

The ADI method is an unconditionally stable implicit method. However, to obtain accurate results, the gain parameters γ_x and γ_y in Equation (9.4.10) should not be too large. It is recommended that the calling parameters be selected such that $\gamma_x \le 1$ and $\gamma_y \le 1$ where

$$\gamma_x = \frac{\beta(n+1)^2 T}{ma^2} \tag{1.10.15a}$$

$$\gamma_y = \frac{\beta(p+1)^2 T}{mb^2} \tag{1.10.15b}$$

wave1 The last two functions in Table 1.10.1 are used to solve the wave equation. The NLIB function *wave1* uses the explicit central difference method to solve the one-dimensional wave equation:

$$\frac{\partial^2 u}{\partial t^2} = \beta\left(\frac{\partial^2 u}{\partial x^2}\right) \tag{1.10.16}$$

over the rectangular region

$$R = \{(t,x) \mid 0 \le x \le T, 0 < x < a\} \tag{1.10.17}$$

subject to the zero boundary condition $u(t,x) = 0$ and the following initial conditions for $0 < x < a$ where $u_t = \partial u/\partial t$.

$$u(0,x) = f(x) \tag{1.10.18a}$$

$$u_t(0,x) = g(x) \tag{1.10.18b}$$

On entry to *wave1:* $T > 0$ is the maximum value of $t, a > 0$ is the maximum value of $x, m \ge 1$ is the number of steps in $t, n \ge 1$ is the number of steps in $x, beta > 0$ is the wave equation coefficient, and the strings f and g contain the names of user-supplied functions that specify the initial conditions on u and its time derivative u_t, respectively. The functions f and g should be stored in M-files of the same name or declared as inline functions and should be of the form

```
function y = f(x)
function y = g(x)
```

When f is called with input x, it must return the initial value $y = u(0,x)$, and when g is called with input x, it must return the initial value $y = u_t(0,x)$ where $u_t = \partial u/\partial t$. On exit from *wave1:* the $(m + 1) \times 1$ vector t contains the t grid points, the $n \times 1$ vector x contains the x grid points, and the $(m + 1) \times n$ matrix U contains the solution, $u(t,x)$, evaluated at the $(m + 1)n$ interior grid points. That is, for $1 \le k \le (m + 1)$ and $1 \le j \le n$,

$$U(k,j) = u[t(k), x(j)] \tag{1.10.19a}$$

$$t(k) = \frac{(k - 1)T}{m} \tag{1.10.19b}$$

$$x(j) = \frac{ja}{n + 1} \tag{1.10.19c}$$

The function *wave1* is an explicit method that is stable only when the gain parameter γ in Equation (9.5.24) satisfies $\gamma \le 1$. Therefore, it is the *responsibility* of the user to select the calling arguments to *wave1* such that $\gamma \le 1$ where

$$\gamma = \frac{\beta(n + 1)^2 T^2}{m^2 a^2} \tag{1.10.20}$$

wave2 The NLIB function *wave2* uses the explicit central difference method to solve the two-dimensional wave equation:

$$\frac{\partial^2 u}{\partial t^2} = \beta\left(\frac{\partial^2 u}{\partial x^2} + \frac{\partial^2 u}{\partial y^2}\right) \tag{1.10.21}$$

over the rectangular region

$$R = \{(t, x, y) \mid 0 \le t \le T,\ 0 < x < a,\ 0 < y < b\} \tag{1.10.22}$$

subject to the zero boundary condition $u(t, x, y) = 0$ and the following initial conditions where $0 < x < a, 0 < y < b$, and $u_t = \partial u/\partial t$.

$$u(0, x, y) = f(x, y) \tag{1.10.23a}$$

$$u_t(0, x, y) = g(x, y) \tag{1.10.23b}$$

On entry to *wave2:* $T > 0$ is the maximum value of t, $a > 0$ is the maximum value of x, $b > 0$ is the maximum value of y, $m \ge 2$ is the number of steps in t, $n \ge 1$ is the number of steps in x, $p \ge 1$ is the number of steps in y, *beta* > 0 is the wave equation coefficient, and the strings f and g contain the names of user-supplied functions that specify the initial conditions on u and its time derivative u_t, respectively. The functions f and g should be stored in M-files of the same name or declared as inline functions and should be of the form

```
function z = f(x,y)
function z = g(x,y)
```

When f is called with scalar inputs x and y, it must return the initial value $z = u(0, x, y)$, and when g is called with inputs x and y, it must return the initial value $z = u_t(0, x, y)$ where $u_t = \partial u/\partial t$. On exit from *wave2:* the $n \times 1$ vector x contains the x grid points, the $p \times 1$ vector y contains the y grid points, and the $n \times p$ matrix U contains the final solution, $u(T, x, y)$, at the np interior grid points. That is, for $1 \le i \le n$ and $1 \le j \le p$,

$$U(i,j) = u[T, x(i), y(j)] \tag{1.10.24a}$$

$$x(i) = \frac{ia}{n + 1} \tag{1.10.24b}$$

$$y(j) = \frac{jb}{p + 1} \tag{1.10.24c}$$

The function *wave2* is an explicit method that is stable only when the criterion in Equation (9.6.12) is satisfied. Therefore, it is the *responsibility* of the user to select the calling arguments to *wave2* such that

$$\frac{T^2}{m^2} \le \frac{1}{4\beta}\left[\frac{a^2}{(n + 1)^2} + \frac{b^2}{(p + 1)^2}\right] \tag{1.10.25}$$

EXAMPLE 1.10.1 Partial Differential Equation

As an illustration of the use of the functions in Table 1.10.1 to solve a partial differential equation, consider the problem of finding the steady-state temperature distribution in a metal plate of width $a = 2$ and length $b = 2$. This requires a solution to Laplace's equation:

$$\frac{\partial^2 u}{\partial x^2} + \frac{\partial^2 u}{\partial y^2} = 0$$

Suppose the temperature along the edges of the plate satisfies the following boundary condition.

$$u(x, y) = \cos[\pi(x + y)]$$

The following script uses the function *poisson* from Table 1.10.1 to compute the steady-state temperature using a grid size of $m = n = 20$.

```
%----------------------------------------------------------------------
% Example a1.10.1: Partial Differential Equations
%----------------------------------------------------------------------

% Initialize

clc                     % clear command window
clear                   % clear variables
a = 2;                  % maximum x
b = 2;                  % maximum y
tol = 1.e-4;            % error tolerance
m = 20;                 % x precision
n = 20;                 % y precision
q = 250;                % maximum SOR iterations
g = inline ('cos(pi*(x+y))','x','y');

% Compute solution

fprintf ('Example a1.10.1: Partial Differential Equations\n');
[alpha,r,x,y,U] = poisson (a,b,m,n,q,tol, '' ,g);

% Display results

show ('Optimal relaxation parameter',alpha)
show ('Number of SOR iterations',r)
plotxyz (x,y,U,'Steady-State Temperature','x','y','u')
%----------------------------------------------------------------------
```

Note how the empty string '' is used in place of the calling argument f, thereby converting Poisson's equation to Laplace's equation. Using $tol = 10^{-4}$, the SOR method converged in $r = 45$ iterations. The optimal value for the relaxation parameter was $\alpha = 1.72945$. A plot of the resulting steady-state temperature profile is shown in Figure 1.10.1.

FIGURE 1.10.1 Steady-State Temperature

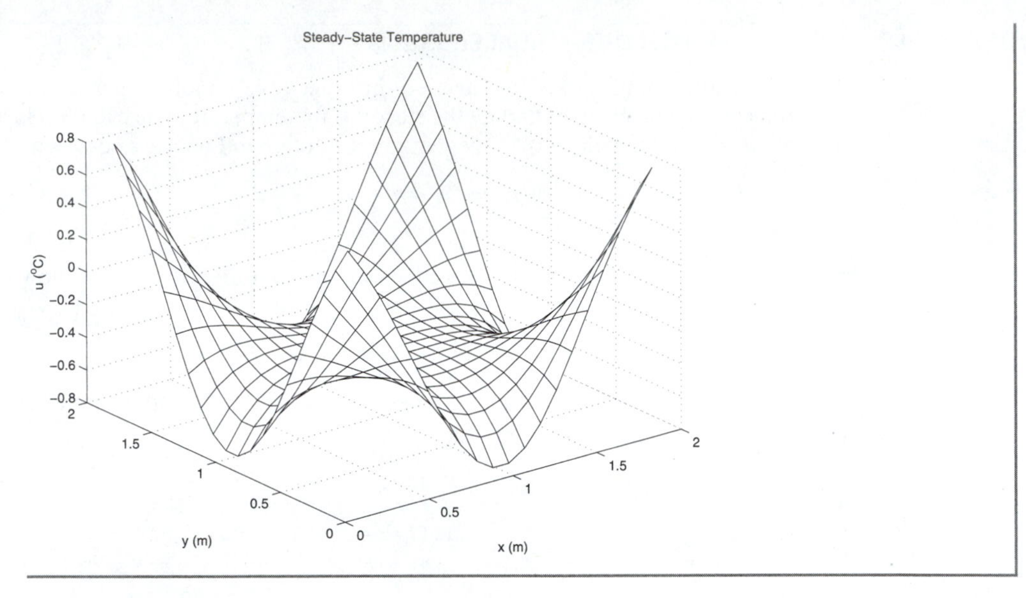

1.11 DIGITAL SIGNAL PROCESSING

The module *dsp* is an application module that implements the digital signal processing techniques discussed in Chapter 10. The digital signal processing (DSP) functions are summarized in Table 1.11.1. For high-speed signal processing, the number of samples of the signals should be a power of two.

dft The NLIB function *dft* computes the one-dimensional *discrete Fourier transform* or DFT. If $x(k)$ denotes the kth sample of the discrete-time input for $0 \le k < n$ and $j = \sqrt{-1}$, then the DFT of $x(k)$ is defined:

$$X(m) \triangleq \sum_{k=0}^{n-1} x(k) \exp\left(\frac{-jmk2\pi}{n}\right) \quad , \quad 0 \le m < n \tag{1.11.1}$$

TABLE 1.11.1 Digital Signal Processing Functions

Usage	*Description*	*Toolbox*
$y = dft(x, dir)$	fast Fourier transform	nlib
$Y = dft2(X, dir)$	two-dimensional fast Fourier transform	nlib
$[a, phi, f] = spectra(x, fs)$	magnitude and phase spectra	nlib
$z = convolve(x, y)$	convolution of x with y	nlib
$z = crosscor(x, y)$	normalized cross correlation of x with y	nlib
$b = fir(p, g)$	FIR filter design	nlib
$[a, phi, f] = freqrsp(u, y, fs)$	frequency response	nlib
$[x, y] = arma(u, theta, x, n, m)$	evaluate ARMA model	nlib
$theta = getarma(u, y, n, m)$	least-squares identification: ARMA	nlib
$[theta, x, e] = lms(u, y, x, theta, mu, m)$	adaptive LMS identification: MA	nlib

On entry to *dft:* the $n \times 1$ vector x contains the samples to be transformed, and the scalar *dir* is a direction code. If $dir \geq 0$, the forward DFT of x is computed j; otherwise, the inverse DFT of x is computed. On exit from *dft:* the $n \times 1$ vector y contains the transformed samples. When the number of samples n is a power of two, the highly efficient *fast Fourier transform* or FFT is used, which only requires $n \log_2(n)/2$ FLOPS.

dft2 The NLIB function *dft2* computes the two-dimensional *discrete Fourier transform* or DFT. If $z(k,i)$ denotes the (k,i)th sample of the input for $0 \leq k < m$, and $0 \leq i < n$ and $j = \sqrt{-1}$, then the two-dimensional DFT of $x(k,i)$ is defined:

$$X(p,q) \triangleq \sum_{k=0}^{m-1} \sum_{i=0}^{n-1} x(k,i) \exp\left[-j2\pi\left(\frac{pk}{m} + \frac{qi}{n}\right)\right] \quad , \quad 0 \leq p < m, 0 \leq q < n \quad (1.11.2)$$

On entry to *dft2:* the $m \times n$ matrix X contains the samples to be transformed, and *dir* is a direction code. If $dir \geq 0$, the forward DFT of X is computed; otherwise, the inverse DFT of X is computed. On exit from *dft2:* the $m \times n$ matrix Y contains the transformed samples. When *dft2* is called with both m and n being powers of two, the highly efficient *fast Fourier transform* or FFT is used.

spectra The NLIB function *spectra* computes the magnitude and phase spectra of a discrete-time signal $x(k)$. Suppose $X(m) = DFT\{x(k)\}$ denotes the discrete Fourier transform of $x(k)$. Then the *magnitude* spectrum $A(m)$ and *phase spectrum* $\phi(m)$ are defined by the relationship

$$X(m) = A(m) \exp[j\phi(m)] \quad , \quad 0 \leq m < n \quad (1.11.3)$$

On entry to *spectra:* the $n \times 1$ vector x contains the samples of the signal, and $fs > 0$ is the sampling frequency in samples/sec or Hz. On exit from *spectra:* the $n \times 1$ vector a contains the magnitude spectrum of x, the $n \times 1$ vector *phi* contains the phase spectrum of x in degrees, and the $n \times 1$ vector f contains the discrete frequencies at which the spectra are evaluated. Here $a(k)$ and $phi(k)$ specify the magnitude and phase, respectively, at discrete frequency

$$f(k) = \frac{(k-1)fs}{n} \quad , \quad 1 \leq k \leq n \quad (1.11.4)$$

Since values of $k > n/2$ correspond to negative frequencies, $a(k)$ and $phi(k)$ exhibit symmetry about the point $k = n/2$ when x is real.

convolve The NLIB function *convolve* computes the *convolution* of two discrete-time signals. If $x(k)$ and $y(k)$ are the two signals, then their convolution is defined:

$$z(i) \triangleq \sum_{k=0}^{i} x(i-k)y(k) \quad , \quad 0 \leq i < n \quad (1.11.5)$$

On entry to *convolve:* the $n \times 1$ vector x contains the samples of the first signal, and the $m \times 1$ vector y contains the samples of the second signal with $m \leq n$. If $m < n$, the signal y is padded with $n - m$ zeros. On exit from *convolve:* the $n \times 1$ vector z contains the convolution of x with y. If x is the pulse response of a linear discrete-time system subject to zero initial conditions, and y is the system input, then z is the system output.

crosscor The NLIB function *crosscor* computes the *normalized cross correlation* of two discrete-time signals. If $x(k)$ and $y(k)$ denote two discrete-time signals, the cross correlation of $x(k)$ with $y(k)$ is defined:

$$r_{xy}(i) \triangleq \sum_{k=i}^{n-1} x(k)y(k-i) \quad , \quad 1 \le i < n \tag{1.11.6}$$

The normalized cross correlation of $x(k)$ with $y(k)$ is found by dividing $r_{xy}(i)$ by its maximum value as follows:

$$\rho_{xy}(i) = \frac{r_{xy}(i)}{\sqrt{(r_{xx}(0)r_{yy}(0)}} \quad , \quad 0 \le i < n \tag{1.11.7}$$

On entry to *crosscor:* the $n \times 1$ vector x contains the samples of the first signal, and the $m \times 1$ vector y contains the samples of the second signal with $m \le n$. On exit from *crosscor:* the $n \times 1$ vector z contains the normalized cross correlation of x with y. The vector z is normalized to the range $|z(k)| \le 1$. If *crosscor* is called with $y = x$, then z contains the *normalized auto correlation* of x.

fir The NLIB function *fir* computes the coefficients of the following finite impulse response (FIR) digital filter using the Fourier series method described in Alg. 10.6.1.

$$y(k) = \sum_{i=0}^{2p} b_i u(k-i) \tag{1.11.8}$$

On entry to *fir:* the integer p selects the filter order (which is $2p + 1$), and the string g contains the name of a function that specifies the desired magnitude response. The user-supplied function g must be stored in an M-file of the same name or declared an inline function and should be of the form

```
function a = g(f)
```

When g is called with input frequency f, it must return the desired filter gain $a = g(f)$. The input frequency is expressed as a fraction of the Nyquist frequency f_N, which means that f is normalized to the interval $0 \le f \le 1$. On exit from *fir:* the $(2p + 1)$ vector b contains the coefficients of the filter. The Hamming window in Table 10.6.2 is used to smooth the ripples in the magnitude response of the filter. Depending on the shape of g, *fir* can be used to design low pass, high pass, band pass, band stop, or more general filters. The filter constructed by *fir* is a linear-phase or constant-delay filter with a delay of p samples. The output of the filter can be computed by calling the function *arma* with $n = 0$, $m = 2p$, and $theta = b$ or by using *convolve.*

freqrsp The NLIB function *freqrsp* is used to compute the *frequency response* of a discrete-time linear system from its input and output measurements. If $u(k)$ denotes the system input and $y(k)$ denotes the system output, the frequency response is $H(m) = \text{DFT}\,\{y(k)\}/\text{DFT}\,\{u(k)\}$ or

$$H(m) = A(m)\exp[j\phi(m)] \quad , \quad 0 \le m < n \tag{1.11.9}$$

On entry to *freqrsp:* the $n \times 1$ vector u contains the samples of the input signal, the $n \times 1$ vector y contains the samples of the output signal, and $fs > 0$ is the sampling frequency. On exit from *freqrsp:* the $n \times 1$ vector a contains the *magnitude response* of the system, the $n \times 1$ vector *phi* contains the phase response of the system in degrees, and the $n \times 1$ vector f contains the frequencies at which the frequency response is evaluated as in Equation (1.11.4). For efficient evaluation of $H(m)$, n should be a power of two.

arma The NLIB function *arma* is used to compute the output of the following auto-regressive moving-average (ARMA) model:

$$y(k) + \sum_{i=1}^{n} \theta_i y(k-i) = \sum_{i=0}^{m} \theta_{n+1+i} u(k-i) \tag{1.11.10}$$

On entry to *arma:* u is the current value of the system input, the $(n + m + 1) \times 1$ vector *theta* is the parameter vector, and the $(n + m + 1) \times 1$ vector x is the state vector defined as follows.

$$x(k) = [-y(k-1), \ldots, -y(k-n), u(k), \ldots, u(k-m)]^T \tag{1.11.11}$$

The remaining parameters are the AR order $n \geq 0$, and the MA order $m \geq 0$. Notice that the current output of the ARMA model, $y(k)$, can be expressed in terms of the parameter vector θ and the current state $x(k)$ using a dot product as follows:

$$y(k) = \theta^T x(k) \tag{1.11.12}$$

On exit from *arma:* the $(n + m + 1) \times 1$ vector x contains the updated state, and the scalar y is the current value of the system output. When *arma* is called, the value of $x(n + 1)$ does not have to be specified because it is replaced by the current input u. The function *arma* computes the output of a linear discrete-time system in real time. When $n = 0$, it evaluates a moving-average or MA filter, a filter that is always stable.

getarma The NLIB function *getarma* is used to identify parameter vector, θ, of the ARMA model in (1.11.10) using a least-squares fit. On entry to *getarma:* the $p \times 1$ vector u contains the samples of the input, the $p \times 1$ vector y contains the samples of the output, p is the number of samples, $n \geq 0$ is the AR order, and $m \geq 0$ is the MA order. The number of samples p must satisfy $p \geq n + m + 1$ and should be larger for a reliable fit. On exit from *getarma,* the $(n + m + 1) \times 1$ vector *theta* contains the parameters of the optimal least-squares ARMA model.

lms The NLIB function *lms* is used to perform real-time identification of the parameters of a MA model using the least-mean square (LMS) method in Equation (10.8.19). The MA model is described by the following equation, which corresponds to the ARMA model in (1.11.10) with $n = 0$.

$$y(k) = \sum_{i=0}^{m} \theta_{1+i} u(k-i) \tag{1.11.13}$$

The current state of the MA model is an $(m + 1) \times 1$ vector that corresponds to the state vector in (1.11.11), but with $n = 0$.

$$x(k) = [u(k), \ldots, u(k-m)]^T \tag{1.11.14}$$

The output of the MA model is then given by (1.11.12). On entry to *lms*: u is the current input, y is the current output, the $(m + 1) \times 1$ vector x is the current state as defined in (1.11.14), the $(m + 1) \times 1$ vector *theta* contains the parameters of the MA model, $mu > 0$ is the step size for the steepest descent search, and $m \geq 0$ specifies the order of the MA model.

On exit from *lms*: the $(m + 1) \times 1$ vector *theta* contains the updated parameter estimates, the $(m + 1) \times 1$ vector x contains the updated state in (1.11.14), and the scalar e is the current error in (10.8.17). When *lms* is called, $x(1)$ does not have to be specified because it is replaced by the current input u. The function *lms* is called each time a new sample becomes available. To ensure convergence, the step size *mu* must satisfy the following constraint where $P(u)$ is the average power of u.

$$0 < mu < \frac{1}{(m + 1)P(u)}$$

EXAMPLE 1.11.1 Digital Signal Processing

Consider the problem of designing a band stop or notch filter with a sampling frequency of $f_s = 1000$ Hz, a pass band gain of one, a low-frequency cutoff of $f_0 = 250$ Hz, and a high-frequency cutoff of $f_1 = 350$ Hz. The following script uses the functions in Table 10.8.2 to perform this task using an FIR filter of order of $2p + 1 = 61$.

```
%--------------------------------------------------------------------
% Example a1.11.1: Digital Signal Processing
%--------------------------------------------------------------------

% Initialize

clc                     % clear command window
clear                   % clear variables
p = 30;                 % select order
r = 2*p + 1;            % number of terms
n = 256;                % number of samples
fs = 1000;              % sampling frequency
u = zeros(n,1);         % input signal
h = zeros(n,1);         % pulse response
x = zeros(r,1);         % initial state vector

% Compute filter coefficients

fprintf ('Example a1.11.1: Digital Signal Processing\n\n');
b = fir ('funa1111',p);

% Compute pulse response

disp('Computing pulse response ... ')
t = [0 : 1/fs : (n-1)/fs]';
u(1) = 1;
for k = 1 : n
   [x,h(k)] = arma (u(k),b,x,0,2*p);
end
```

```
graphxy (t(1:100),h(1:100),'Pulse Response','t (sec)','h(t)')

% Compute frequency response

[a,phi,f] = freqrsp (u,h,fs);
graphxy (f(1:n/2),a(1:n/2),'Magnitude Response','f (Hz)','a');
%---------------------------------------------------------------------
```

A plot of the pulse response of the filter is shown in Figure 1.11.1. Note that it is bounded and goes to zero after 61 samples. All FIR filters are stable and have a pulse response that goes to zero in a finite number of samples. A plot of the magnitude response of the filter is shown in Figure 1.11.2. Observe that the cutoff frequencies are at $f_L = 250$ Hz and $f_H = 350$ Hz per the specifications, the gain is one, and there is very little ripple in the magnitude response.

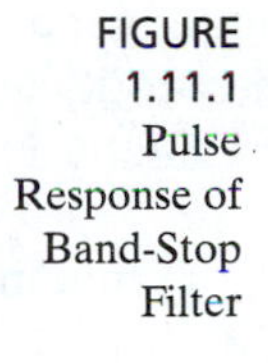

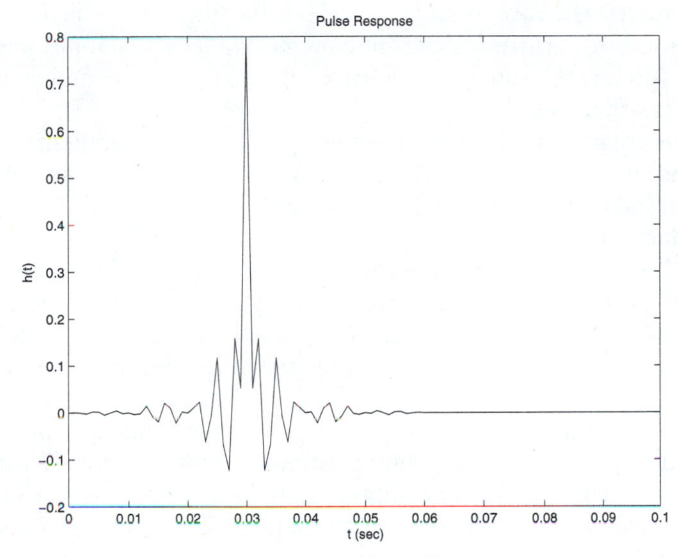

FIGURE 1.11.1 Pulse Response of Band-Stop Filter

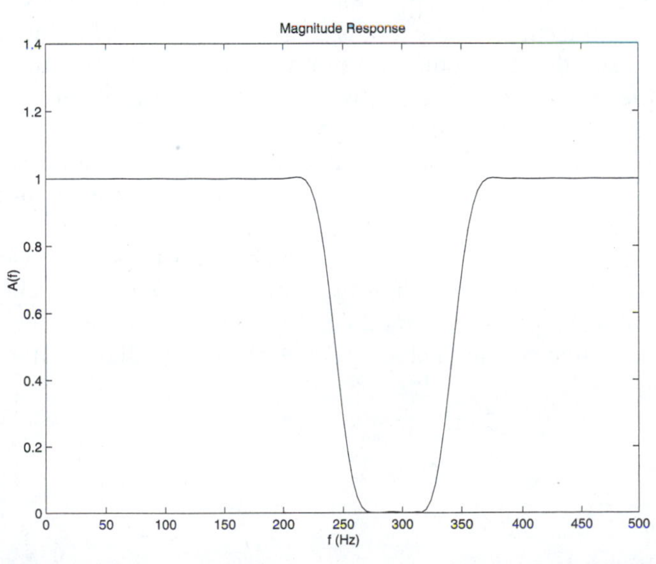

FIGURE 1.11.2 Magnitude Response of Band-Stop Filter

APPENDIX 2

NLIB Using C

This appendix describes the C version of *NLIB*, a *N*umerical *LIB*rary that implements the algorithms developed in the text. It is assumed that the reader is familiar with the fundamentals of C or C++ (see, e.g., Kelley and Pohl, 1990). In an effort to facilitate *portability* between programming environments, NLIB has been implemented using the ANSI C standard (Kernighan and Ritchie, 1988). NLIB is based on extensive use of vectors and matrices. The core module of NLIB is the file *nlib.c*, which includes general purpose functions for performing dynamic memory allocation, input and output, and algebraic operations with vectors and matrices. It also includes complex arithmetic and random number generation.

Two main-program support modules are provided as a supplement. The module *show.c* contains tabular display functions designed to provide for convenient display of scalars, vectors, and matrices on the console screen and the printer. The module *draw.c* contains graphical display functions that generate script files that are directly executable from the MATLAB command prompt. This two-step procedure for graphical output ensures portability between programming environments. The main-program support modules are not used by the remaining NLIB modules, and as such they are not part of NLIB proper. Instead, they are a supplement intended to facilitate the development and testing of user programs.

NLIB features nine application modules corresponding to Chapter 2 through Chapter 10 of the text. The linear algebraic systems module, *linear.c*, includes direct and iterative methods for solving linear algebraic systems of equations. The eigenvalue and eigenvector module, *eigen.c*, focuses on computing eigenvalues and eigenvectors of square matrices. Problems of interpolation, extrapolation, and least-squares curve fitting are covered in the curve fitting module, *curves.c*. The root finding module, *roots.c*, contains iterative solution techniques for nonlinear equations. The more general problem of finding the minimum of an objective function subject to equality and inequality constraints is covered in the optimization module, *optim.c*. Numerical differentiation and integration are the focus of the derivatives and integrals module, *integ.c*. The problem of solving systems of ordinary differential equations subject to initial or boundary conditions is treated in the ordinary differential equations module, *ode.c*. The partial differential equations module, *pde.c*, includes techniques for solving various classes of partial differential equations including hyperbolic, parabolic, and elliptic equations. Finally, the digital signal processing module, *dsp.c*, focuses on numerical techniques for processing discrete-time signals, including spectral analysis, digital filtering, and system identification.

2.1 A NUMERICAL LIBRARY: NLIB

2.2.1 NLIB Installation

The numerical library NLIB can be installed on a PC by executing the following command from the Windows Start/Run menu, assuming the distribution CD is in drive D.

```
D:\setup
```

More detailed instructions for installing the NLIB library for both PC and non-PC users can be found by viewing the plain text file *user_c.txt* on the distribution CD. Convenient access to the NLIB batch files, can be achieved by adding the NLIB directory or folder to the path in the *autoexec.bat* file. This is accomplished automatically when NLIB is installed, with a backup copy of *autoexec.bat* being created.

NLIB is composed of the modules summarized in Figure 2.1.1. The number in the lower right corner of each box indicates the chapter or appendix where the algorithms implemented in the module are discussed.

2.1.2 Library Usage

Every C program that uses functions in NLIB should have *one* of the following compiler directives placed near the top of the program:

```
#include "c:\nlib\nlib.h"          /* use compiled library */
#include "c:\nlib\util.h"          /* use source library */
```

FIGURE 2.1.1 NLIB Structure (C Version)

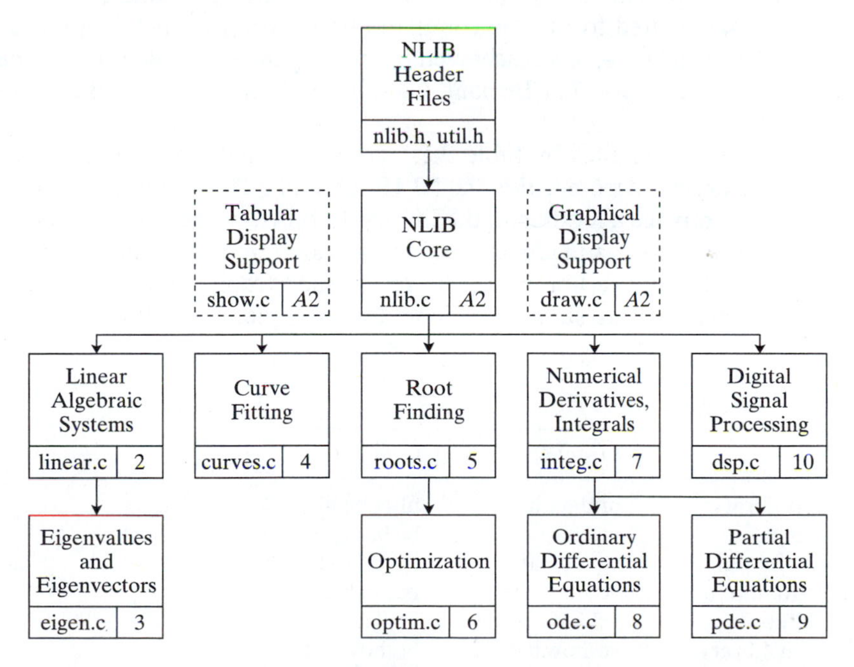

Here it is assumed that the NLIB folder is *c:\nlib*. The NLIB header file *nlib.h* contains symbolic constants, data types, macros, and prototypes of the functions in NLIB. The source for *nlib.h* and all of the NLIB modules can be found on the distribution CD. If the modules in NLIB are first precompiled into a library or *.lib* file, the only item that needs to be included at the top of a user program is the header file *nlib.h*. This is the most efficient way to use NLIB because it leads to smaller executable files, and it also reduces the compile time somewhat.

Many users may prefer to directly include the NLIB source files and compile them with their program. In this case, the auxiliary header file *util.h* should be used instead. The file *util.h* includes *nlib.h* and the NLIB source modules. This *simplified* approach is used in the examples included in the *exam_c* directory on the distribution CD. Although it is somewhat less efficient, it minimizes the amount of configuration required to build programs from the command line or an integrated environment. Furthermore, since the NLIB source is written in ANSI C, this approach should work with any ANSI C compatible compiler assuming sufficient memory is available.

For users interested in using a precompiled library, there are two precompiled versions of the NLIB library available on distribution CD, one for Microsoft Visual C++, Version 5.0, called *nlib_vc.lib,* and one for Borland C++, Version 4.5, called *nlib_bc.lib*. Several batch file commands are available to facilitate the maintenance and usage of the NLIB library. For the batch files summarized in Table 2.1.1, the user simply has to enter the file name to obtain instructions for usage.

To compile a C program that uses NLIB from the command line, the batch file commands *vc* and *bc* should be used. The program name is supplied as a command line argument. The batch files *vc* and *bc* can be executed from any folder once the NLIB folder is added to the path. This change to *autoexec.bat* will be made automatically (with confirmation from the user) when NLIB is installed. It is assumed that the environment has been configured to run the command-line version of the compiler. In the case of Microsoft Visual C++, this is achieved by running the batch file *vcvars32.bat* that is supplied with Visual C++. For Borland C++, it involves adding the Borland *bin* folder to the path.

The last four files in Table 2.1.1 are concerned with library maintenance. All should be executed from *within* the NLIB directory. The files *vcadd* and *bcadd* are used to insert or replace a module in the library. In this way, the user can add new functions or new features to existing functions. Modules are removed from the library using *vcsub* and *bcsub*. The entire library can be recompiled from scratch by using the *vclib* and *bclib* commands. This should be done, for example, if the PRECISION constant in

TABLE 2.1.1 NLIB Library and Batch Files

File	*Visual C++*	*Borland C++*
NLIB library	nlib_vc.lib	nlib_bc.lib
compile	vc.bat	bc.bat
insert module	vcadd.bat	bcadd.bat
remove module	vcsub.bat	bcsub.bat
build library	vclib.bat	bclib.bat
display library	vcshow.bat	bcshow.bat

nlib.h is changed from SINGLE to DOUBLE. The contents of the library are displayed with the batch files *vcshow* and *bcshow*.

It is also possible to use NLIB from an integrated environment. In this case, the path to the appropriate NLIB library file must be specified as part of the project configuration. Alternatively, the NLIB source files can be directly included in the user program using the directive: #include "c:\nlib\util.h". In this case, there is no need to access a compiled version of the library. This is the most portable way to use NLIB.

2.1.3 NLIB Example Browser

For PC users running Windows, a convenient way to view and run all of the NLIB software is to use the *NLIB Example Browser,* which is launched by entering the following command from the Windows Start/Run menu:

```
browse
```

The program *browse.exe* is an easy-to-use, menu-based program that allows the user to select from the options listed in Table 2.1.2. The *run* option allows the user to execute precompiled versions of all of the computational examples discussed in the text. The *view* option is similar to run, but it allows the user to view the complete source code of the examples, including data files. The *draw* option provides a simple means of displaying the figures generated by the examples. The *Compile* option compiles the examples. Next the *settings* option allows the user to select an editor and a compiler as well as run from the CD. Finally, the *help* option allows the user to view documentation on all of the functions in the NLIB library and learn how to obtain software upgrades as they become available. Thus, the NLIB Example Browser is a fast and easy way for PC users to access all of the C software on the distribution CD.

An alternative way to obtain online documentation of the NLIB library functions is to use the *help* utility. To display a list of the names of all of the functions in the NLIB library, arranged by category, enter the following command from the Windows Start/Run menu:

```
help nlib
```

To obtain user documentation on a specific NLIB module, replace the operand *nlib* with the appropriate module name.

TABLE 2.1.2 NLIB Example Browser: C Version

Menu Option	*Selections*
Run	examples, exit
View	examples
Draw	figures
Compile	examples
Settings	editor, compiler
Help	library, updates
Exit	terminate program

```
help modname
```

There are over one hundred functions in the NLIB library, and space does not permit all of them to be described in this appendix. However, user documentation on *all* of the NLIB functions, including the lower-level functions not discussed here, can be obtained through the *browse/help* menu option or with the NLIB *help* utility.

2.2 NLIB DATA TYPES

One of the appealing features of C is its facility to handle user-defined data types. A careful selection of data types results in a simplified user interface that corresponds directly to the underlying mathematics.

2.2.1 Scalars

Custom data types used by NLIB functions are all defined in the header file *nlib.h*. The most basic data type is the scalar data type *complex,* which is defined as follows.

```
/*-------------------------------------------------------------------*/
 typedef struct {
    float    x;                                /*      real part   */
    float    y;                                /* imaginary part   */
 } complex;
/*-------------------------------------------------------------------*/
```

The data type *complex* is a structure, composed of two scalar members, that is used to represent complex numbers. The first member is a float, x, which specifies the *real part* of the complex number, and the second member is a float, y, which specifies the *imaginary part* of the complex number. If z is declared to be a variable of type *complex*, then $z.x$ denotes the real part and $z.y$ denotes the imaginary part. Similarly, if z is declared to be a pointer to type complex, then $z->x$ denotes the real part of the complex number and $z->y$ denotes the imaginary part. Note that if C++ is used in place of C, the built-in data type complex is already available.

2.2.2 Vectors

Numerical methods make extensive use of vectors and matrices. Vectors are often represented with one-dimensional arrays, which can be thought of as fixed pointers whose memory is allocated at the time they are declared. However, it is more efficient to use variable pointers to represent vectors. The data type *vector* is defined in *nlib.h* as follows.

```
/*-------------------------------------------------------------------*/
 typedef vector float*                            /* real vector */
/*-------------------------------------------------------------------*/
```

The data type *vector* is a pointer to type *float*. Consequently, if x is declared to be a variable of type vector, then $x[k]$ denotes the kth component of the vector x. For vectors used in mathematics, it is traditional for subscripts to range from 1 to n, particularly in linear algebra applications. However, there are some instances where it is more nat-

FIGURE 2.2.1 Vector Storage

ural for subscripts to range from 0 to n (e.g., the coefficients of an nth degree polynomial) or from 0 to $n - 1$ (e.g., the samples of a discrete-time signal). Normally, when the memory for a variable of type *vector* is allocated, it is done in such a way that the subscripts range from 1 to n as illustrated in Figure 2.2.1. The special cases of subscripts ranging from 0 to n and from 0 to $n - 1$ can be accommodated using pointer arithmetic, as will be shown in the next section.

Rather than point to a float, it is also possible to have a variable that points to a structure of type *complex*. This is the basis for constructing vectors of complex numbers. In particular, the NLIB data type *cvector* is defined in *nlib.h* as follows.

```
/*-----------------------------------------------------------------------*/
 typedef cvector complex*                          /* complex vector */
/*-----------------------------------------------------------------------*/
```

The data type *cvector* is a pointer to type *complex*. Consequently, if z is declared to be a variable of type *cvector*, then $z[k]$ denotes the kth component of the complex vector z. In this case, $z[k]$ is a complex number with $z[k].x$ denoting the real part and $z[k].y$ denoting the imaginary part. Again, when a complex vector is allocated, it is normally done in such a way that the subscripts k range from 1 to n.

2.2.3 Matrices

Numerical applications not only make extensive use of real and complex vectors, they also use real and complex matrices. The NLIB data type for a real matrix is a variable of type *matrix,* which is defined in *nlib.h* as follows.

```
/*-----------------------------------------------------------------------*/
 typedef matrix vector*                               /* real matrix */
/*-----------------------------------------------------------------------*/
```

The data type *matrix* is a pointer to type *vector*. That is, if A, is declared to be a variable of type *matrix*, then $A[k]$ is a vector denoting the kth *row* of A, and $A[k][j]$ is a scalar denoting the element in the kth row and jth column. In effect, A is a double pointer, or a pointer to a pointer. The first pointer selects a row of the matrix, and the second pointer selects a column within that row. Matrices are allocated in such a way that the row and column subscripts start at *one*. For example, the first subscript ranges from 1 to m, and the second subscript ranges from 1 to n as shown in Figure 2.2.2.

FIGURE 2.2.2 Matrix Storage

$A[1]$	→	$A[1][1]$	$A[1][2]$	...	$A[1][n]$
$A[2]$	→	$A[2][1]$	$A[2][2]$	...	$A[2][n]$
⋮	⋮	⋮	⋮	⋰	⋮
$A[m]$	→	$A[m][1]$	$A[m][2]$	...	$A[m][n]$

Since $A[k]$ is in fact a vector, it can be used in any NLIB function where a vector argument is required. This feature is useful when one is trying to assemble a matrix of data for tabular or graphical display through repeated calls to a function.

The last custom data type in NLIB is the type *cmatrix,* which is used to represent a complex matrix as follows.

```
/*--------------------------------------------------------------*/
 typedef cmatrix cvector*                    /* complex matrix */
/*--------------------------------------------------------------*/
```

The data type *cmatrix* is a pointer to type *cvector*. Consequently, if A is declared to be a variable of type *cmatrix*, then $A[k]$ is a complex vector denoting the kth *row* of A, and $A[k][j]$ is a complex scalar denoting the element in the kth row and jth column. As with real matrices, A is a double pointer with the first pointer selecting a row and the second pointer selecting a column within the row. Again, the row and column subscripts start at *one*.

By using the custom data types *complex, vector, matrix, cvector,* and *cmatrix*, the meaning of calling arguments for NLIB functions is made more clear. In addition, usage of memory is optimized in the sense that only the memory actually needed is allocated, and only when it is needed. Furthermore, memory can be released as soon as the need for it has expired.

2.2.4 Precision

The custom data types defined in the header file *nlib.h* are summarized in Table 2.2.1. Since the NLIB functions make use of these data types, the file *nlib.h* (or *util.h*) must be included near the top of any C program that uses NLIB.

All of the custom data types in Table 2.2.1 use floats and therefore represent single precision data types (seven digits). They can be converted to double precision (sixteen digits) by changing the value of the symbolic constant PRECISION near the top of the NLIB header file *nlib.h*.

```
/*--------------------------------------------------------------*/
#define PRECISION DOUBLE
/*--------------------------------------------------------------*/
```

TABLE 2.2.1 NLIB Data Types

Data Type	*Description*
complex	complex number
vector	real vector
cvector	complex vector
matrix	real matrix
cmatrix	complex matrix

PRECISION can take on the values SINGLE or DOUBLE. If PRECISION is given the value DOUBLE, all occurrences of type float are replaced by type double. Whenever the precision is changed, the NLIB library should be recompiled to make the changes take effect.

When higher precision is selected, the numerical accuracy of the computations is increased at the expense of increased memory, because scalars of type float typically require four bytes of storage, whereas scalars of type double require eight bytes. For some applications—for example, finding eigenvalues and eigenvectors of a relatively large matrix—the increased accuracy is worth the trade-off in storage and, to some extent, speed.

2.3 NLIB CORE

Our goal is to develop a customized library of C functions that can be used to solve a variety of numerical problems in engineering. There are a number of general-purpose, low-level functions that can be introduced at this point to facilitate the development and testing of the higher-level functions. The numerical library is initially seeded with these low-level utility functions, which are referred to collectively as the NLIB *core*. Their source code is contained in the file *nlib.c* on the distribution CD.

2.3.1 Vector and Matrix Allocation

Vectors and matrices are used extensively in numerical methods applications. It is therefore important to handle them efficiently and in a manner that is convenient for the user. Numerical library functions should be able to work with vectors and matrices of any size. This is an area where C excels by allocating memory dynamically. Since the required size does not have to be known until execution time, general-purpose functions that efficiently use all of the memory available can be developed. NLIB includes a set of vector allocation functions whose prototypes are summarized in Table 2.3.1.

vec The functions in Table 2.3.1 create and manipulate vectors. The function *vec* allocates memory for an $n \times 1$ real vector and initializes it using the string s. For example, if x has been declared to be a variable of type *vector* as defined in Section 2.2, then

TABLE 2.3.1 Vector Allocation Functions

Function Prototype			*Function*
Type	*Name*	*Arguments*	*Description*
vector	*vec*	(int n, char $*s$)	allocate real vector
vector	*randvec*	(int n, float a, float b)	allocate random vector
cvector	*cvec*	(int n, char $*s$)	allocate complex vector
void	*freevec*	(vector x)	release real vector
void	*freecvec*	(cvector x)	release complex vector
void	*zerovec*	(vector x, int n)	zero real vector
void	*copyvec*	(vector x, int n, vector y)	copy real vector

$x = \text{vec}(n, "\,")$ creates a vector with elements, $x[1], \ldots, x[n]$ as shown in the following code fragment.

```
/*------------------------------------------------------------------*/
 vector      x;
 int         n;
                               /* initialize n */
 x = vec (n,"");
                               /* use x */
 freevec (x);
/*------------------------------------------------------------------*/
```

Note that the value of n does not have to be known at compile time. Instead, n is an integer variable whose value can be assigned at program execution time. For example, n might be initialized by an assignment statement, by reading a data file, or by prompting the user for the desired number of elements. The string s contains the initial values to assign to the elements of x. The values in s are separated by blank space or commas. If less than n values are specified, the remaining values are set zero. Consequently, using the empty string for s initializes all elements of x to zero.

randvec Another way to assign values to an allocated vector is to use *randvec*. The function *randvec* allocates memory for an $n \times 1$ real vector and initializes it with pseudo-random numbers uniformly distributed over the interval $[a, b]$.

cvec The function *cvec* allocates memory for an $n \times 1$ complex vector and initializes it using the string s with the initial values separated by blank space or commas. The format is real part, imaginary part, real part, imaginary part, etc. If less than n complex values are specified, the remaining values are set to zero.

freevec, freecvec There are often instances in numerical applications where *temporary* vectors are needed to hold intermediate results. Once a real vector is no longer needed, the memory it occupies can be released for subsequent use with the function *freevec*. In a similar way, the function *freecvec* releases the memory occupied by a complex vector. The memory occupied by a vector x can also be released using the statement free $(x + 1)$, where *free* is the ANSI C library function in *<stdlib.h>*.

zerovec, copyvec The function *zerovec* zeros the contents of the $n \times 1$ vector x, while the function *copyvec* copies the contents of the $n \times 1$ vector x into the $n \times 1$ vector y.

On occasion, it is natural to work with vectors whose subscripts start at 0 rather than 1. For example, $a[0], \ldots, a[n]$ might be used to represent the coefficients of a polynomial of degree n. To allocate a vector with subscripts ranging from 0 to n, one can use the following code fragment where it is assumed that n is a positive integer and a has been declared to be a variable of type vector.

```
/*------------------------------------------------------------------*/
 a = vec(n+1,"") + 1;                        /* allocate a[0],...a[n] */
/*------------------------------------------------------------------*/
```

TABLE 2.3.2 Matrix Allocation Functions

Function Prototype			*Function*
Type	*Name*	*Arguments*	*Description*
matrix	*mat*	(int m, int n, char $*s$)	allocate real matrix
matrix	*randmat*	(int m, int n, float a, float b)	allocate random matrix
cmatrix	*cmat*	(int m, int n, char $*s$)	allocate complex matrix
void	*freemat*	(matrix A, int m)	release real matrix
void	*freecmat*	(cmatrix A, int m)	release complex matrix
void	*zeromat*	(matrix A, int m, int n)	zero real matrix
void	*copymat*	(matrix A, int m, int n, matrix B)	copy real matrix

Although this use of pointer arithmetic allocates a vector with subscripts starting at 0, it should be emphasized that all of the subsequent low-level NLIB functions that work with vectors process only components $a[1], \ldots, a[n]$ unless otherwise noted.

mat A second group of functions, whose prototypes are summarized in Table 2.3.2, is used to create and manipulate matrices. The function *mat* allocates memory for an $m \times n$ real matrix and initializes it using the string s. For example, if A has been declared to be a variable of type *matrix*, as defined in Section 2.2, then $A = \text{mat}(m, n, "")$ creates a matrix with elements $A[1][1], A[1][2], \ldots, A[m][n]$ as illustrated by the following code fragment.

```
/*-----------------------------------------------------------------*/
matrix       A;
int          m,n;
                            /* initialize m and n */
A = mat (m,n,"");
                            /* use A */
freemat (A,m);
/*-----------------------------------------------------------------*/
```

Again, the values of m and n do not have to be known until execution time because the matrix is allocated dynamically. Here $A[k]$ is a pointer that points to the start of the kth row of A. The jth element of this vector is $A[k][j]$, the element appearing in the kth row and jth column of the matrix A.

As before, the initial values of A are specified with the string s and are separated by blank space or commas. The values are stored in row-wise order (second index changes faster than first index). Note that this is in contrast to two-dimensional FORTRAN arrays, which are stored in column-wise order. If less than mn values are specified in s, then the remaining components of A are set to zero. Consequently, when the empty string is used for s, all components of A are initialized to zero.

randmat The function *randmat* allocates memory for an $m \times n$ real matrix and initializes it with pseudo-random numbers uniformly distributed over the interval $[a, b]$.

cmat The function *cmat* allocates memory for an $m \times n$ complex matrix and initializes it using the string s. The format for the string of initial values is real part, imaginary part, real part, imaginary part, etc., in row-wise order. If less than mn complex values are specified, the remaining values are set to zero.

freemat, freecmat To release the memory allocated to an $m \times n$ real matrix, the function *freemat* should be used. Note that the number of rows must be specified in the call, but not the number of columns. Similarly, *freecmat* is used to release the memory allocated by *cmat*.

zeromat, copymat The function *zeromat* zeros the contents of the $m \times n$ matrix A, while the function *copymat* copies the contents of the $m \times n$ matrix A into the $m \times n$ matrix B.

EXAMPLE 2.3.1 Random Vector

As an illustration of the use of the functions in Table 2.3.1, the following example dynamically allocates a vector and initializes it with pseudo-random numbers uniformly distributed over the interval $[-1,1]$.

```
/* ------------------------------------------------------------ */
/* Example a2.3.1: Random vector                                */
/* ------------------------------------------------------------ */
#include "c:\nlib\util.h"
int main (void)
{
   int          n = 12,
                i;
   vector       x = randvec (n,-1,1);
   for (i = 1; i <= n; i++)
      printf ("x[%i] = %7.4g\n",i,x[i]);
   return 0;
}
/* ------------------------------------------------------------ */
```

Observe that the first line in Example 2.3.1 following the comments is an include directive, that includes the header file *util.h,* which is assumed to be located in the directory *c:\nlib*.

2.3.2 Vector and Matrix Input/Output

Since vectors and matrices are used extensively in numerical methods applications, it is useful to have a convenient way to transfer data back and forth between vectors and matrices, on the one hand, and files and devices on the other. Prototypes of the NLIB functions for vector input and output operations are summarized in Table 2.3.3.

putvec The function *putvec* is used to output the $n \times 1$ vector x to the file *fname*. The string *vname* specifies the name used for the vector x (e.g., *vname* = "x"), and the

TABLE 2.3.3 Vector Input/Output Functions

Function Prototype			*Function Description*
Type	*Name*	*Arguments*	
void	*putvec*	(vector x, int n, char **vname*, char **fname*, int *mode*)	output real vector
vector	*getvec*	(FILE *f, int *n)	input real vector
void	*putcvec*	(cvector x, int n, char **vname*, char **fname*, int *mode*)	output complex vector
cvector	*getcvec*	(FILE *f, int *n)	input complex vector

parameter *mode* controls the type of write operation. If *mode* is zero, then file *fname* is appended to; otherwise, it is overwritten. In this way, more than one vector can be written to the same file with successive calls to *putvec*.

The storage format used by *putvec* is shown in Table 2.3.4. The function *putvec* creates a plain text file using a format that is directly executable from within the MATLAB environment. Consequently, the variables stored in *fname* can be placed directly into the MATLAB environment by simply entering the file name from the MATLAB command prompt. When *fname* is saved, it is automatically given the extension *.m* so that it will be recognized as a MATLAB script file. Consequently, the user should *not* supply an extension with *fname*. If an extension is used, it will be replaced by *.m* to ensure MATLAB compatibility.

getvec If x is stored in a file with *putvec*, it can be recovered and read directly into a C program by using *getvec*. The function *getvec* is the inverse of *putvec*. It allocates memory for and inputs an $n \times 1$ real vector from the stream f. It is assumed that the FILE pointer f has already been assigned to a file that has been opened for reading with the C library function *fopen*. The functions *putcvec* and *getcvec* are analogous to *putvec* and *getvec*, respectively, except that they operate on complex vectors. The following program fragment illustrates the use of *getvec* to read a vector from a data file named *vector.m*.

TABLE 2.3.4 Storage Format for *putvec*

Line	*Contents*
1	n;
2	$vname = [$
3	$x[1]$
4	$x[2]$
$\vdots$	$\vdots$
$2 + n$	$x[n]$
$3 + n$	];

```
/*-------------------------------------------------------------------*/
 int         n;
 vector      b;
 FILE        *f = fopen ("vector.m","r");
 if (f == NULL)
    printf ("File not found!\n");
 else
    b = getvec (f,&n);
/*-------------------------------------------------------------------*/
```

This code fragment inputs the value n, allocates memory for the $n \times 1$ vector b, and then inputs the n values $b[1], \ldots, b[n]$.

When n is relatively large, the function *getvec* is the preferred way to initialize vectors. Smaller vectors can be initialized with the initialization string s in the *vec* function in Table 2.3.1. The functions *putvec* and *getvec* are compatible in the sense that a file written with *putvec* can later be read with *getvec*. Files written with *putvec* are also directly executable from within MATLAB.

If a data file is created directly from a text editor for use by *getvec*, then MATLAB-style comments (lines with % in the first column) can be included at the start of the file. More generally, files containing multiple vectors can include comments before each vector. Blank lines are also interpreted as comments by *getvec*. In this way, the data file can be made more readable for the user.

A second group of functions, whose prototypes are listed in Table 2.3.5, is designed to perform input/output operations with matrices.

putmat The function *putmat* is used to output the $m \times n$ matrix A to the file *fname*. The string *vname* specifies the name used for the matrix A (e.g., *vname* = "A"), and the parameter *mode* controls the type of write operation as in *putvec*. That is, if *mode* is zero, then file *fname* is appended to; otherwise, it is overwritten. In this way, combinations of vectors and matrices can be written to the same file with successive calls.

TABLE 2.3.5 Matrix Input/Output Functions

Function Prototype			
Type	*Name*	*Arguments*	*Function Description*
void	*putmat*	(matrix A, int m, int n, char $*vname$, char $*fname$, int $mode$)	output real matrix
matrix	*getmat*	(FILE $*f$, int $*m$, int $*n$)	input real matrix
void	*putcmat*	(cmatrix A, int m, int n, char $*vname$, char $*fname$, int $mode$)	output complex matrix
cmatrix	*getcmat*	(FILE $*f$, int $*m$, int $*n$)	input complex matrix
void	*putcol*	(matrix A, int m, int n, int c, vector x);	insert column
void	*getcol*	(matrix A, int m, int n, int c, vector x);	extract column

TABLE 2.3.6 Storage Format for *putmat*

Line	*Contents*
1	m ;
2	n ;
3	*vname* = [
4	$A[1][1] A[1][2] \cdots A[1][n]$
5	$A[2][1] A[2][2] \cdots A[2][n]$
⋮	⋮
$3+m$	$A[m][1] A[m][2] \cdots A[m][n]$
$4+m$	];

The storage format used by *putmat* is shown in Table 2.3.6. The function *putmat* creates a plain text file using a format that is directly executable from within the MATLAB environment. Consequently, the variables stored in *fname* can be placed directly into the MATLAB environment by simply entering the file name from the MATLAB command prompt. The storage format shown in Table 2.3.6 corresponds to "small" matrices. When the matrix width n is too large to fit on a single screen, long rows are continued to the next line using the MATLAB line continuation symbol, . . . , to improve readability.

getmat If A is stored in a file with *putmat,* it can be recovered and read directly into a C program by using *getmat*. The function *getmat* is the inverse of *putmat*. It allocates memory for and inputs an $m \times n$ real matrix from the stream f. It is assumed that the FILE pointer f has already been assigned to a file that has been opened for reading with the C library function *fopen*. The function *getmat* ignores MATLAB style comments (% in column one) and blank lines placed before each matrix. The functions *putcmat* and *getcmat* are analogous to *putmat* and *getmat,* respectively, except that they operate on complex matrices. The following program fragment illustrates the use of *getmat* to read a matrix from a data file named *matrix.m*.

```
/*--------------------------------------------------------------------*/
 int          m,n;
 matrix       A;
 FILE         *f = fopen ("matrix.m","r");
 if (f == NULL)
    printf ("File not found!\n");
 else
    A = getmat (f,&m,&n);
/*--------------------------------------------------------------------*/
```

This code fragment inputs the values m and n, allocates memory for the $m \times n$ matrix A and then inputs the mn values of A in the row-wise order $A[1][1], A[1][2], \ldots,$ $A[m][n]$.

When mn is relatively large, the function *getmat* is the preferred way to initialize matrices. Smaller matrices can be initialized by using the initialization string s in the *mat* function in Table 2.3.2. The functions *putmat* and *getmat* are compatible in the sense that a file written with *putmat* can later be read with *getmat*. Files written with *putmat* are also directly executable from within MATLAB.

putcol, getcol The next two functions in Table 2.3.5 are used to transfer data between vectors and columns of matrices. The function *putcol* copies the $m \times 1$ vector x into the cth column of the $m \times n$ matrix A. The function *getcol* transfers data in the opposite direction. It copies the cth column of the $m \times n$ matrix A into the $m \times 1$ vector x. Note that there is no need for analogous *putrow* and *getrow* functions because $A[r]$ is the rth row of the matrix A. Thus, the function *copyvec* in Table 2.3.1 can be used to perform *putrow* and *getrow* type of operations.

EXAMPLE 2.3.2 Matrix Output

The following program creates a random matrix and saves it in a data file called *mat232.m* using *putmat*. The matrix A can be recovered with *getmat* (see Example 2.4.2). Alternatively, the matrix A can be added to the MATLAB workspace by simply entering *mat232* from the MATLAB command prompt.

```
/* ----------------------------------------------------------------- */
/* Example a2.3.2: Matrix output                                     */
/* ----------------------------------------------------------------- */
#include "c:\nlib\util.h"
int main (void)
{
   char          *fname = "mat232";
   int           m = 10,
                 n = 6;
   matrix        A = randmat (m,n,-1,1);
   printf ("\nThis program creates the data file %s.m\n",fname);
   putmat (A,m,n,"A",fname,1);
   return 0;
}
/* ----------------------------------------------------------------- */
```

2.3.3 Matrix Algebra

There are certain routine arithmetic operations that are performed with vectors and matrices. For convenience, these common operations are implemented with NLIB functions whose prototypes are summarized in Table 2.3.7.

transpose The function *transpose* computes the transpose of an $m \times n$ matrix A and stores the result in the $n \times m$ matrix B. The transpose of A, denoted A^T, is obtained by interchanging the rows with the columns.

$$A^T_{kj} = A_{jk} \quad , \quad 1 \le k \le n \quad , \quad 1 \le j \le m \tag{2.3.1}$$

TABLE 2.3.7 Matrix Algebra Functions

Function Prototype			*Function Description*
Type	*Name*	*Arguments*	
void	*transpose*	(matrix A, int m, int n, matrix B)	$B = A^T$
float	*inner*	(vector x, vector y, int n)	$x^T y$
void	*outer*	(vector x, vector y, int m, int n, matrix A)	$A = xy^T$
void	*matvec*	(matrix A, vector x, int m, int n, vector b)	$b = Ax$
void	*vecmat*	(vector x, matrix A, int m, int n, vector b)	$b = x^T A$
void	*matmat*	(matrix A, matrix B, int m, int n, int p, matrix C)	$C = AB$
float	*minvec*	(vector x, int n, int $*k$);	vector minimum
float	*maxvec*	(vector x, int n, int $*k$);	vector maximum
float	*minmat*	(matrix A, int m, int n, int $*k$, int $*j$);	matrix minimum
float	*maxmat*	(matrix A, int m, int n, int $*k$, int $*j$);	matrix maximum

When $m = n$, the function *transpose* can be called with $B = A$ to save memory in the calling program.

inner The function *inner* multiplies the transpose of an $n \times 1$ vector x times the $n \times 1$ vector y. The result is a 1×1 scalar called the *inner product* of x with y. Thus, *inner* returns the value

$$x^T y = \sum_{k=1}^{n} x_k y_k \tag{2.3.2}$$

outer The function *outer* is the opposite of *inner*. It multiplies the $m \times 1$ vector x times the transpose of the $n \times 1$ vector y and stores the result in the $m \times n$ matrix A. The matrix A is called the *outer product* of x with y. Thus, $A_{kj} = x_k y_j$, or in vector notation,

$$A = xy^T \tag{2.3.3}$$

matvec The next function, *matvec*, multiplies the $m \times n$ matrix A times the $n \times 1$ vector x and stores the result in the $m \times 1$ vector b. Consequently, the kth element of b is the inner product of the kth row of A with x.

$$b = Ax \tag{2.3.4}$$

vecmat The function *vecmat* is the opposite of *matvec*. It multiplies the transpose of the $m \times 1$ vector x times the $m \times n$ matrix A and stores the resulting $1 \times n$ vector in b. Thus, the kth element of b is the inner product of x with the kth column of A.

$$b = x^T A \tag{2.3.5}$$

matmat The function *matmat* multiplies the $m \times n$ matrix A times the $n \times p$ matrix B and stores the result in the $m \times p$ matrix C. In this case, the element of C in the kth row and jth column is the inner product of the kth row of A with the jth column of B.

$$C = AB \tag{2.3.6}$$

When $m = n = p$, the function *matmat* can be called with $C = A$ or $C = B$. This saves memory in the calling program, but it destroys the original A or B, respectively.

minvec, maxvec The functions *minvec* and *maxvec* return the minimum value and the maximum value, respectively, of the $n \times 1$ vector x. On exit, k points to the subscript of the minimum or maximum element, respectively.

minmat, maxmat In a similar manner, the functions *minmat* and *maxmat* return the minimum value and the maximum value, respectively, of the $m \times n$ matrix A. On exit, the row and column subscripts of the minimum or maximum element are pointed to by k and j, respectively.

EXAMPLE 2.3.3 Vector Range

The following example creates and initializes a vector and then computes and prints its minimum and maximum values.

```
/* -------------------------------------------------------------- */
/* Example a2.3.3: Vector Range                                   */
/* -------------------------------------------------------------- */
#include "c:\nlib\util.h"
int main (void)
{
   int          n = 5,
                i,j;
   vector       x = vec (n,"3.7,-4.5,8.1,2.2,-1.9");
   printf ("\nx ranges from %g to %g",minvec(x,n,&i),maxvec(x,n,&j));
   printf ("\nThe minimum is x[%i] = %g",i,x[i]);
   printf ("\nThe maximum is x[%i] = %g",j,x[j]);
   return 0;
}
/* -------------------------------------------------------------- */
```

2.3.4 Complex Arithmetic

Numerical computations that involve complex numbers make use of the data type *complex* defined in Section 2.2. To facilitate complex arithmetic, the low-level functions summarized in Table 2.3.8 are provided. It should be pointed out that if C++ is used in place of C, then many of the functions in Table 2.3.8 are not needed because C++ has a built-in complex data type and the capability to process complex numbers by *overloading* the basic math operators and functions.

TABLE 2.3.8 Complex Arithmetic Functions

Function Prototype			*Function Description*
Type	*Name*	*Arguments*	
complex	*cnum*	(float x, float y)	Return $x + jy$
complex	*cadd*	(complex u, complex v)	Return $u + v$
complex	*csub*	(complex u, complex v)	Return $u - v$
complex	*cmul*	(complex u, complex v)	Return uv
complex	*cdiv*	(complex u, complex v)	Return u/v
float	*cmag*	(complex u)	Return $\|u\|$
float	*cphi*	(complex u)	Return $\arctan(u.y/u.x)$
complex	*csqrt*	(complex u)	Return $\sqrt{u}$
complex	*cexp*	(complex u)	Return $\exp(u)$
complex	*cpoly*	(complex u, vector a, int n)	Evaluate polynomial
void	*csort*	(cvector z, int n)	Sort complex vector

cnum The function *cnum* is used to *create* a complex number $u = x + jy$ where

$$j \triangleq \sqrt{-1} \tag{2.3.7}$$

cadd, csub, cmul, cdiv The next four functions—*cadd, csub, cmul, and cdiv*—perform the four basic operations of complex addition, complex subtraction, complex multiplication, and complex division, respectively. The function *cdiv* generates an error message when $v = 0$.

cmag, cphi The next two functions perform a transformation from rectangular coordinates $u = x + jy$ to *polar* coordinates.

$$u = a \exp(j\phi) \tag{2.3.8}$$

Here $a = \sqrt{u.x^2 + u.y^2}$ denotes the *magnitude* of u and $\phi = \tan^{-1}(u.y/u.x)$ denotes the *phase angle* of u. The function *cmag* returns the magnitude, a, and the function *cphi* returns the phase, ϕ, in radians. The real and imaginary parts of u can then be recovered from a and ϕ as follows, using Euler's identity.

$$u.x = a\cos(\phi) \tag{2.3.9a}$$

$$u.y = a\sin(\phi) \tag{2.3.9b}$$

csqrt, cexp The next two entries in Table 2.3.8 compute common functions of complex numbers. The function *csqrt* computes the complex *square root*, and the function *cexp* computes the complex *exponential*.

cpoly The function *cpoly* is used to evaluate a *polynomial* of degree n at u. On entry to *cpoly*: the $(n + 1) \times 1$ vector a contains the coefficients of the polynomial as follows.

$$p = a_1 + a_2 u + \cdots + a_{n+1}u^n \tag{2.3.10}$$

The value returned by *cpoly* is p.

csort The function *csort* is used to *sort* an $n \times 1$ vector z of complex numbers. On exit from *csort:* the complex numbers in z are arranged in order of decreasing magnitude.

EXAMPLE 2.3.4 Complex Numbers

The following example computes the magnitude, phase, and square root of the complex number $j = \sqrt{-1}$.

```
/* ---------------------------------------------------------------- */
/* Example a2.3.4: Complex numbers                                  */
/* ---------------------------------------------------------------- */
#include "c:\nlib\util.h"
int main (void)
{
   complex      j = cnum (0,1),
                u = csqrt (j);
   printf ("\nMagnitude of j:   a   = %g",cmag(j));
   printf ("\nPhase angle of j: phi = %g",cphi(j));
   printf ("\nSquare root of j: u   = %g + j%g\n",u.x,u.y);
   return 0;
}
/* ---------------------------------------------------------------- */
```

2.3.5 Random Number Generation

Random numbers are useful for testing numerical methods in general, and they are also needed for specific applications such as Monte Carlo integration and system identification. The random number generation functions are listed in Table 2.3.9.

randinit The function *randinit* is used to initialize the random number generator. On entry to *randinit:* s is the seed for the random number sequence. Each seed generates a different random sequence. The default seed is $s = 1$. If *randinit* is called with $s \leq 0$, then a random seed is generated based on the time of day. This way, different random sequences can be generated each time a program is executed.

TABLE 2.3.9 Random Number Generation Functions

Function Prototype			*Function Description*
Type	*Name*	*Arguments*	
void	*randinit*	(long s)	initialize random sequence
float	*randu*	(float a, float b)	uniform random number
float	*randg*	(float m, float s)	Gaussian random number

randu The function *randu* is used to produce a sequence of random floats *uniformly* distributed over an interval $[a, b]$. It uses the multiplicative congruential method with a multiplier of $\alpha = 16807$ and a modulus of $\gamma = 2147483647$.

$$u(k+1) = \alpha u(k) \quad \% \quad \gamma \tag{2.3.11}$$

This produces a pseudo-random sequence with a period of $p = 2^{31}-2$, which corresponds to over 2.147 billion random numbers. Serial correlation between successive random numbers is reduced by using the random shuffle method described in Alg. 1.6.1. On entry to *randu: a* is the lower bound on the interval, and *b* is the upper bound. The value returned by *randu* is the next random number in the sequence. The probability density of the random numbers produced by successive calls to *randu* is

$$p(x) = \begin{cases} \dfrac{1}{b-a} & , \quad a \le x \le b \\ 0 & , \quad \text{otherwise} \end{cases} \tag{2.3.12}$$

randg The function *randg* is used to produce a sequence of random floats with a *Gaussian* or normal distribution. The Box-Muller method is used to convert a uniform distribution over $[0, 1]$ into a Gaussian distribution. On entry to *randg: m* is the *mean* of the distribution and *s* is the standard deviation. The value returned by *randg* is the next random number in the sequence. The probability density of the random numbers produced by successive calls to *randg* is

$$p(x) = \frac{1}{s\sqrt{2\pi}} \exp\left[\frac{-(x-m)^2}{2s^2}\right] \tag{2.3.13}$$

EXAMPLE 2.3.5 Random Numbers

The following program constructs and displays a Gaussian random sequence with a mean of $m = 10$ and a standard deviation of $s = 2$.

```
/* -------------------------------------------------------------- */
/* Example a2.3.5: Random numbers                                  */
/* -------------------------------------------------------------- */
#include "c:\nlib\util.h"
int main (void)
{
   int          n = 20,
                i;
   float        m = 10,          /* mean */
                s = 2;           /* standard deviation */
   randinit (0);
   for (i = 1; i <= n; i++)
      printf ("\n%g",randg(m,s));
   return 0;
}
/* -------------------------------------------------------------- */
```

TABLE 2.3.10 Utility Functions

Function Prototype			*Function*
Type	*Name*	*Arguments*	*Description*
float	*args*	(float x, float a, float b, int k, char $*f$)	argument check
float	*epsilon*	(void)	machine epsilon ε_M
void	*dec2bin*	(unsigned x, char $*b$)	decimal to binary
float	*polynom*	(float x, vector a, int n)	evaluate polynomial
float	*vecnorm*	(vector x, int n)	infinity vector norm

2.3.6 Utility Functions

Finally, the NLIB core module contains a few low-level, general-purpose utility functions that can be useful for developing and testing user functions. The prototypes of the utility functions are summarized in Table 2.3.10.

args The function *args* is designed to verify that a scalar calling argument to a user function is within its proper range. On entry to *args:* x is the value of the argument to be checked, a is the lower limit of x, b is the upper limit of x, k is the argument number, and f is the function name. The value returned by *args* is x saturated or clipped to the interval $[a, b]$. If x lies outside this interval, an error message is generated, and the user is given the choice to exit or proceed. This simple check can uncover faulty calls to user functions.

epsilon The function *epsilon* returns the machine epsilon, ε_M, using the procedure in Alg. 1.3.1. If NLIB is converted to double precision as discussed in Section 2.2, the machine epsilon for the double precision data type will be returned by *epsilon*. Recall that ε_M is the smallest positive number that can be added to one and produce a result that is greater than one.

dec2bin The function *dec2bin* uses Alg. 1.2.1 to convert the two-byte unsigned integer x to a string of 16 zeros and ones corresponding to the binary equivalent of the decimal value x.

polynom The function *polynom* uses Horner's rule in Alg. 4.2.1 to evaluate the following polynomial of degree n.

$$f(x) = a_1 + a_2 x + \cdots + a_{n+1} x^n \qquad (2.3.14)$$

On entry to *polynom:* x is the evaluation point, a is an $(n + 1) \times 1$ coefficient vector, and $n \geq 0$ is the polynomial degree. The value returned by *polynom* is $f(x)$. Note that the subscripts of a range from 1 to $n + 1$. Horner's rule reduces the number of floating-point multiplications by about one half in comparison with a direct evaluation.

vecnorm The function *vecnorm* computes the *infinity* norm of the vector x, which is defined as follows.

$$\|x\| \triangleq \max_{k=1}^{n}\{|x_k|\} \tag{2.3.15}$$

On entry to *vecnorm:* x is an $n \times 1$ vector, and $n \geq 1$ is the number of elements of x. The value returned by *vecnorm* is $\|x\|$. Another popular vector norm is the Euclidean norm, which can be computed using the square root of the function *inner* defined in Table 2.3.7.

2.4. TABULAR DISPLAY

The module *show.c* is a supplementary main-program support module that is designed to facilitate user program development by providing a convenient and useful way to display vectors and matrices on the console screen and the printer.

2.4.1 Screen

shownum, showvec, showmat The first group of main-program support functions in Table 2.4.1 displays numerical data on the *stdout* device, the console screen. The most elementary function, *shownum*, displays a scalar q on the screen and labels it with the string s. The function *showvec* displays an $n \times 1$ vector x on the screen by enclosing it in brackets and labeling it with the string s. If t is nonzero, *showvec* displays the transposed vector instead. Still more general is the function *showmat,* which is used to display an $m \times n$ matrix of A. The matrix is enclosed in brackets and labeled with the string s. If t is nonzero, *showmat* displays the transposed matrix.

showcnum, showcvec, showcmat The functions *showcnum, showcvec,* and *showcmat* are analogous to *shownum, showvec,* and *showmat,* respectively, except that they display complex data. When complex numbers are displayed, the format is as follows where $j = \sqrt{-1}$.

$$u = x + yj \tag{2.4.1}$$

In many instances, vectors and matrices are too large to fit on one screen. In these cases, one slice of the vector or matrix is shown at a time and the rows and columns are

TABLE 2.4.1 Console and Keyboard Functions

Function Prototype			*Function Description*
Type	*Name*	*Arguments*	
void	*shownum*	(char *s, float q)	display real scalar
void	*showvec*	(char *s, vector x, int n, int t)	display real vector
void	*showmat*	(char *s, matrix A, int m, int n, int t)	display real matrix
void	*showcnum*	(char *s, complex u)	display complex scalar
void	*showcvec*	(char *s, cvector x, int n, int t)	display complex vector
void	*showcmat*	(char *s, cmatrix A, int m, int n, int t)	display complex matrix
float	*prompt*	(char *s, float a, float b, float df)	prompt for number
void	*wait*	(void)	wait for keystroke

TABLE 2.4.2 *Show* Options

Option	*Description*
Save	Save to a file
Append	Append to a file
Print	Send to printer
e**X**it	Exit to operating system

labeled. Once a display is complete, the *show* function lists the menu of options summarized in Table 2.4.2.

The **S**ave option is used to save the displayed numerical data to a new file, while the **A**ppend option is used to append the displayed numerical data to an existing file. The **P**rint option is used to obtain a hard copy of the item being displayed. If the user wants to exit the program and return to the operating system, the e**X**it option should be used. The normal way to return from a *show* function is to press the Enter key, which resumes program execution by returning to the calling program. Experience has shown that the *show* functions can be highly useful for program development; they provide a convenient way to examine program variables and intermediate results from within a user program.

2.4.2 Keyboard

prompt The second group of functions in Table 2.4.1 reads the *stdin* device or keyboard. The function *prompt* prompts the user for a number within a specified range. On entry to *prompt*: s is the prompt string to be displayed, a is the lower limit, b is the upper limit, and df is the default response. The function *prompt* continues to prompt for responses until a number in the interval $[a, b]$ is entered. It then returns the number. The lower and upper limits are optional in the sense that a can be replaced by $-$INFINITY or b can be replaced by INFINITY. The symbolic constant INFINITY, defined in *nlib.h,* represents the maximum real number.

wait The function *wait* suspends program execution and displays the message "Press Enter to continue . . . " on the console screen. Pressing the Enter key causes program execution to continue.

EXAMPLE 2.4.1 Console Display

The following program prompts the user for some dimensions, creates a random vector and a random matrix, and displays them on the console screen. Typical output for the responses $m = 12$ and $n = 4$ is shown in Figure 2.4.1.

```
/* ---------------------------------------------------------------- */
/* Example a2.4.1: Screen display                                   */
/* ---------------------------------------------------------------- */
#include "c:\nlib\util.h"
```

```
int main (void)
{
   int         m,n;
   vector      x;
   matrix      A;
   m = prompt ("Enter number of rows",1,20);
   n = prompt ("Enter number of columns",1,20);
   x = randvec (m,-1,1);
   A = randmat (m,n,0,10);
   showvec ("x",x,m,0);
   showmat ("A",A,m,n,0);
   return 0;
}
/*-------------------------------------------------------------------*/
```

$$x = \begin{bmatrix} -0.9789 \\ -0.9921 \\ -0.3297 \\ -0.9335 \\ -0.2886 \\ -0.5656 \\ 0.07395 \\ -0.6084 \\ 0.4007 \\ 0.8998 \\ -0.4504 \\ -0.1114 \end{bmatrix} \qquad A = \begin{bmatrix} -7.82 & 3.965 & 1.287 & -9.17 \\ -6.697 & 6.31 & 3.711 & 5.288 \\ 6.553 & 9.191 & -5.611 & -1.464 \\ 9.052 & 6.791 & 8.465 & 6.218 \\ -0.9793 & 2.095 & 3.233 & 1.994 \\ 0.9836 & 4.4 & -7.721 & -1.876 \\ -7.571 & 3.425 & -0.502 & -0.1779 \\ 1.285 & -3.11 & 7.375 & -4.712 \\ -6.404 & -1.538 & 3.896 & -6.726 \\ 0.7663 & 2.91 & 2.468 & -9.939 \\ 5.744 & -4.621 & -0.7785 & -2.269 \\ -2.464 & 1.623 & 2.069 & -4.404 \end{bmatrix}$$

FIGURE 2.4.1 Output from Example 2.4.1

2.4.3 Printer

listnum, listvec, listmat It is also useful to display vectors and matrices in hard-copy form on the printer. Prototypes of functions that display numerical output on the *stdprn* device or printer are summarized in Table 2.4.3. The function *listnum* is analogous to *shownum*, but it sends the scalar data to the printer. The function *listvec* prints the $n \times 1$ vector x. Again, it encloses the elements in brackets and labels them with the string s. If t is nonzero, *listvec* prints the transposed vector. The function *listmat* prints the $m \times n$ matrix A in a manner analogous to *showmat*.

Hard copy can also be obtained by using the *print* option of the *show* functions. This is equivalent to calling the corresponding *list* function. By calling the *list* functions directly, the need for user interaction during program execution is eliminated.

listcnum, listcvec, listcmat The functions *listcnum, listcvec,* and *listcmat,* are analogous to *listnum, listvec,* and *listmat,* respectively, except that they display complex data on the printer.

TABLE 2.4.3 Printer Functions

Function Prototype			*Function*
Typ/e	*Name*	*Arguments*	*Description*
void	*listnum*	(char *s, float q)	print real scalar
void	*listvec*	(char *s, vector x, int n, int t)	print real vector
void	*listmat*	(char *s, matrix A, int m, int n, int t)	print real matrix
void	*listcnum*	(char *s, complex u)	print complex scalar
void	*listcvec*	(char *s, cvector x, int n, int t)	print complex vector
void	*listcmat*	(char *s, cmatrix A, int m, int n, int t)	print complex matrix

EXAMPLE 2.4.2 Hard Copy

The following program reads the data file *mat232.m* produced by Example 2.3.2 using the *getmat* function. It then displays it on the screen. Hard copy is obtained by selecting the *print* option.

```
/* ---------------------------------------------------------------- */
/* Example a2.4.2: Hard copy                                        */
/* ---------------------------------------------------------------- */
#include "c:\nlib\util.h"
int main (void)
{
   char          *fname = "mat232.m";
   int           m,n;
   matrix        A;
   FILE          *f = fopen (fname,"r");
   if (f != NULL)
   {
      A = getmat (f,&m,&n);
      showmat ("A",A,m,n,0);
   }
   else
      printf ("\nThe file %s does not exist.\n",fname);
   return 0;
}
/* ---------------------------------------------------------------- */
```

2.5 GRAPHICAL DISPLAY

The module *draw.c* is a supplementary main-program support module that facilitates the *interpretation* of numerical results by displaying data in graphical form. This can be useful, for example, when the results can be expressed as a curve, a family of curves, or a surface.

2.5.1 Curves

Prototypes of the main-program support functions that display families of curves are summarized in Table 2.5.1. The functions in Table 2.5.1 do not directly produce graphics on the console screen. Instead, they generate a script file that contains the data to be graphed and the graph commands. The script file is a plain text file that is formatted in such a way that it is directly executable from within the MATLAB environment. A graph is then produced by simply entering the name of the script file from the MATLAB command prompt. By using this two-step procedure, the main-program support module *draw.c* conforms to the ANSI C standard. In this way, high-quality graphical output can be obtained on the wide variety of programming platforms that are supported by MATLAB (Hanselman and Littlefield, 1998). Whenever any of the graphical display functions are called, they produce a message on the console that specifies the name of the MATLAB script file that has been created. The file that is produced is automatically given the *.m* extension so that it can be recognized as a MATLAB script file.

graphvec The easiest way to graphically display a vector is to use the function *graphvec,* which graphs the $n \times 1$ vector y using the subscript of y as the independent variable. On entry to *graphvec:* y is the vector to be graphed, $n \geq 1$ is the number of points, *title* is the graph title, *xaxis* is the x-axis label, *yaxis* is the y-axis label, and *mfile* is the name of the MATLAB script file that will contain the data to be graphed and the graph commands. The file name *mfile* should be supplied *without* an extension because it is automatically be given a *.m* extension identifying it as a MATLAB script file.

graphmat A simple way to graphically display a matrix is to use the function *graphmat,* which generates a family of curves by graphing the m rows of the $m \times n$

TABLE 2.5.1 Graph Functions

Function Prototype			*Function Description*
Type	*Name*	*Arguments*	
void	*graphvec*	(vector y, int n, char **title*, char **xaxis* char **yaxis*, char **mfile*)	graph vector x
void	*graphmat*	(matrix Y, int m, int n, char **title*, char **xaxis*, char **yaxis*, char **mfile*)	graph matrix Y
void	*graphfun*	(float a, float b, char **title*, char **xaxis*, char **yaxis*, float (*f)(float x),char **mfile*)	graph function f
void	*graphxy*	(vector x, vector y, int n, char **title*, char **xaxis*, char **yaxis*, char **mfile*)	graph y vs. x
void	*graphxm*	(vector x, matrix Y, int m, int n, char **title*, char **xaxis*, char **yaxis*, char **mfile*)	graph Y vs. x
void	*graphmm*	(matrix X, matrix Y, int m, int n, char **title*, char **xaxis*, char **yaxis*, char **mfile*, int *dm*)	graph Y vs. X

matrix Y using the column subscript as the independent variable. Thus, the kth curve is a graph of the vector $Y[k]$. On entry to *graphmat:* Y is the $m \times n$ data matrix, $m \geq 1$ is the number rows, and $n \geq 2$ is the number of columns. The graph title and axis labels are supplied through the strings *title, xaxis,* and *yaxis,* respectively. Finally, the string *mfile* is the name of the MATLAB script file that will be produced. It will be given a *.m* extension so that the family of curves can be generated by executing the script from within MATLAB. To graph the n columns of Y instead of the rows, first transpose Y using the function *transpose* from Table 2.3.7.

graphfun A function $f(x)$ can be graphed using the function *graphfun*. On entry to *graphfun*: a is the minimum value of x, b is the maximum value of x, *title* is the graph title, *xaxis* is the x-axis label, *yaxis* is the y-axis label, f is the name of the function to be graphed, and *mfile* is the name of the MATLAB script file that will contain the graph commands and the data. The prototype of the function f is

```
float f (float x);
```

When f is called with input x, it must return the value $f(x)$. The function *graphfun* creates a script that graphs the curve $f(x)$ over the interval $[a, b]$.

graphxy A more flexible way to generate a single curve is to call *graphxy,* which graphs the $n \times 1$ vector y versus the $n \times 1$ vector x. The graph is labeled using *title* for the title, *xaxis* for the x-axis label, and *yaxis* for the y-axis label. The data to be graphed is stored in the MATLAB script file *mfile,* which will be given a *.m* extension. Unlike the function *graphvec,* for the function *graphxy,* the user specifies both the abscissa and the ordinate.

graphxm A family of curves can be generated by using the function *graphxm,* which graphs the m rows of the $m \times n$ matrix Y versus the $n \times 1$ vector x. Again, the family of curves is labeled using *title* for the title, *xaxis* for the x-axis label, and *yaxis* for the y-axis label. The graph commands and data are stored in the MATLAB script file *mfile*. In this case, the curves share the same user-specified independent variable x.

graphmm The function *graphmm* is the most general graph function. It graphs the m rows of the $m \times n$ matrix Y versus the m rows of the $m \times n$ matrix X. Consequently, the kth curve is a graph of $Y[k]$ versus $X[k]$, with each curve having its own independent variable. A curve with fewer than n points can be graphed by repeating the last point as needed to fill to n points. The family of curves is labeled using *title* for the title, *xaxis* for the x-axis label, and *yaxis* for the y-axis label. The graph commands and data are stored in the script file *mfile* so that the graph can be directly generated from within MATLAB. Normally, each row of Y is plotted by connecting the points with solid straight line segments. However, if the display mode argument dm is nonzero, the last row of Y is graphed using *isolated* plot symbols. Of course, the user can always edit the plot command near the end of the script file created by *graphmm* to change options such as line styles and colors.

EXAMPLE 2.5.1 Graph

The following program generates a matrix of data and then uses *graphxm* to display a family of curves. The call to *graphxm* produces the script file *fig251.m*. When the command *fig251* is entered from the MATLAB command prompt, it produces the graph shown in Figure 2.5.1.

```
/* ---------------------------------------------------------------- */
/* Example a2.5.1: Graph Family of Curves                           */
/* ---------------------------------------------------------------- */
#include "c:\nlib\util.h"
int main (void)
{
   int          n = 50,          /* points */
                m = 6,           /* curves */
                i,j;
   vector       x = vec (n,"");
   matrix       Y = mat (m,n,"");
   for (j = 1; j <= n; j++)
   {
      x[j] = 5.0*(j-1)/(n-1);
      for (i = 1; i <= m; i++)
         Y[i][j] = i*x[j]*exp(-m*x[j]/i);
   }
   graphxm  (x,Y,m,n,"A Family of Curves","x","y","fig251");
   return 0;
}
/* ---------------------------------------------------------------- */
```

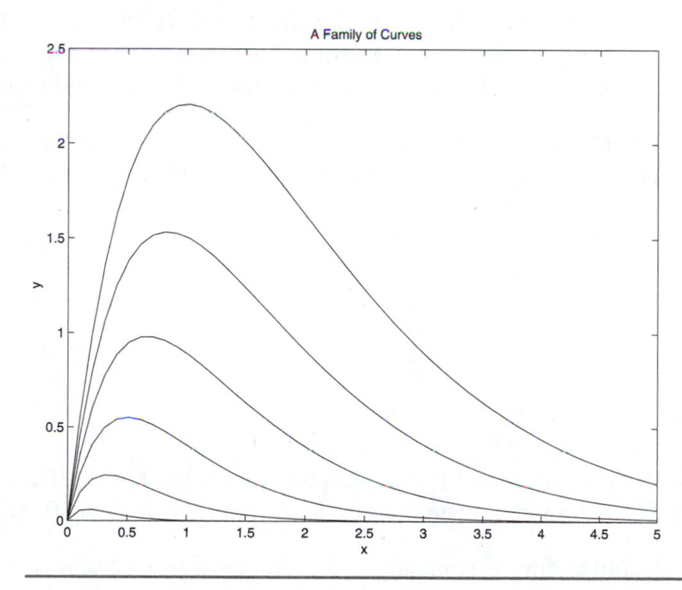

FIGURE 2.5.1 Graphical Output from *fig251.m*

TABLE 2.5.4 Surface Plot Functions

Function Prototype			*Function*
Type	*Name*	*Arguments*	*Description*
void	*plotmat*	(matrix Z, int m, int n, char *$title$, char *$xaxis$, char *$yaxis$, char *$zaxis$, char *$mfile$)	plot matrix Z
void	*plotfun*	(float $x0$, float $x1$, float $y0$, float $y1$, char *$title$, char *$xaxis$, char *$yaxis$, char *$zaxis$, float (*f)(float x, float y),char *$mfile$)	plot function f
void	*plotxyz*	(vector x, vector y, matrix Z, int m, int n, char *$title$, char *$xaxis$, char *$yaxis$, char *$zaxis$, char *$mfile$)	plot Z vs. x, y

2.5.2 Surfaces

An alternative way to plot a family of curves is to treat the curve number as a second independent variable and thereby plot the curves "side by side" to generate a surface. Prototypes of the main-program support functions designed to display surface plots are summarized in Table 2.5.4.

plotmat The easiest way to graphically display a matrix as a surface is to use the function *plotmat*, which plots the $m \times n$ matrix Z using the row subscript as one independent variable and the column subscript as the other independent variable. Matrices with a special form (e.g., symmetric, triangular, diagonal, sparse) have characteristic surface plots. The string *title* is the plot title, and the strings $xaxis$, $yaxis$, and $zaxis$ are the x-, y-, and z-axis labels, respectively. Finally, the string *mfile* is the name of the MATLAB script file that will contain the plot commands and the data. It will be given a *.m* extension so that the plot can be generated directly from the MATLAB command prompt.

plotfun A function of two variables, $f(x, y)$, can be plotted as a surface using the function *plotfun*. On entry to *plotfun:* $x0$ is the minimum value of x, $x1$ is the maximum value of x, $y0$ is the minimum value of y, $y1$ is the maximum value of y, *title* is the plot title, and $xaxis$, $yaxis$, and $zaxis$ are the axis labels for the x-, y-, and z-axes, respectively. Finally, f is the name of the user-supplied function to be plotted, and *mfile* is the name of the MATLAB script file that will hold the plot commands and the data. The prototype of f is

```
float f (float x,float y);
```

When f is called with inputs x and y, it must return the value $f(x, y)$. The function *plotfun* plots the surface $f(x, y)$ over the rectangular region $[x0, x1] \times [y0, y1]$.

plotxyz To obtain more sophisticated surface plots, one can use the function *plotxyz,* which plots the $m \times n$ matrix Z as a surface using the $m \times 1$ vector x as the first independent variable and the $n \times 1$ vector y as the second independent variable. The plot is labeled using *title* for the title, *xaxis* for the x-axis label, *yaxis* for the y-axis label, and *zaxis* for the z-axis label. The plot commands and data needed to generate the plot from within Matlab are stored in the script file *mfile,* which will be given a *.m* extension.

EXAMPLE 2.5.2 Plot

The following example uses *plotfun* to plot a surface. When *plotfun* is called, it generates the script file *fig252.m*. Executing the command *fig252* from the MATLAB command prompt then produces the plot displayed in Figure 2.5.2.

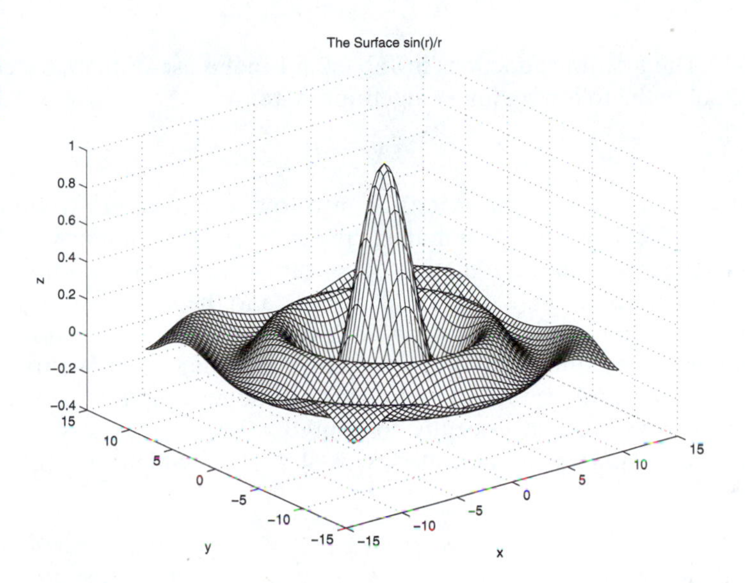

FIGURE 2.5.2 Graphical Output from *fig252.m*

```
/* ---------------------------------------------------------------- */
/* Example a2.5.2: Surface Plot                                      */
/* ---------------------------------------------------------------- */
#include "c:\nlib\util.h"

float f(float x,float y)
{
   float      eps=1.e-6,
              r;
   r = sqrt(x*x + y*y) + eps;
   return sin(r)/r;
}

int main (void)
{
   float      d = 12;
   plotfun (-d,d,-d,d,"sin(r)/r","x","y",
```

```
                "z",f,"fig252");
    return 0;
}
/* ------------------------------------------------------------------- */
```

2.6 LINEAR ALGEBRAIC SYSTEMS

The module *linear.c* is an application module that implements the linear algebraic system techniques discussed in Chapter 2. The function prototypes are summarized in Table 2.6.1, and the source code for these functions can be found in the file *linear.c* on the distribution CD.

lufac The first four functions in Table 2.6.1 make use of the LU decomposition method to solve the following linear algebraic system.

$$Ax = b \tag{2.6.1}$$

The function *lufac* factors the matrix A into lower- and upper-triangular factors. On entry to *lufac:* A is an $n \times n$ matrix, and $n \geq 1$ is the dimension of A. On exit from *lufac:* the $n \times n$ storage matrix Q contains the lower- and upper-triangular factors of A as in Equation (2.4.13) of Chapter 2, and the $n \times 1$ vector p is a row permutation vector that is initialized to $p_k = k$ for $1 \leq k \leq n$. Each time a row interchange is performed during pivoting, the corresponding components of p are also interchanged. The value returned by *lufac* is the determinant of A. If it is nonzero, then Q and p have been successfully computed. Otherwise, A was found to be singular. The function *lufac* can be called with $Q = A$, in which case the original A is destroyed.

TABLE 2.6.1 Linear Algebraic System Functions

Function Prototype			*Function Description*
Type	*Name*	*Arguments*	
float	*lufac*	(matrix A, int n, matrix Q, vector p);	LU factorization
int	*lusub*	(matrix Q, vector p, vector b, int n, vector x);	LU substitution
float	*ludec*	(matrix A, vector b, int n, vector x);	LU decomposition
float	*det*	(matrix A, int n);	matrix determinant
float	*inv*	(matrix A, int n, matrix B);	matrix inverse
float	*condnum*	(matrix A, int n, int m);	condition number
float	*residual*	(matrix A, vector b, vector x, int n);	residual error
int	*sr*	(vector x, matrix A, vector b, float *alpha*, float *tol*, int n, int m);	successive relaxation
float	*tridec*	(matrix T, vector b, int n, vector x)	tridiagonal system
int	*gensys*	(matrix A, vector b, int m, int n, vector x);	least-squares solution

lusub The function *lusub* implements the back substitution part of the *LU* decomposition method. The function *lusub* is used in conjunction with the companion function *lufac*. On entry to *lusub:* Q is an $n \times n$ storage matrix containing the factors L and U as in Equation (2.4.13) of Chapter 2, p is an $n \times 1$ row permutation vector, b is an $n \times 1$ right-hand-side vector, and $n \geq 1$ is the number of unknowns. The matrix Q and the vector p are computed with a preliminary call to the function *lufac*. On exit from *lusub:* the $n \times 1$ vector x is the solution. The value returned by *lusub* is zero if a singular matrix was detected. The functions *lufac* and *lusub* should be used whenever the system in (2.6.1) must be solved multiple times using different right-hand-side vectors b. In this case, *lufac* is called once to initialize Q and p, and then *lusub* is called for each right-hand-side vector.

ludec The linear algebraic system in Equation (2.6.1) is typically solved only once. When this is the case, a single call to the *LU* decomposition function, *ludec*, can be used. On entry to *ludec:* A is an $n \times n$ coefficient matrix, b is an $n \times 1$ right-hand-side vector, and $n \geq 1$ is the number of unknowns. On exit from *ludec:* the $n \times 1$ vector x contains the solution. The value returned by *ludec* is the determinant of A. If it is nonzero, the solution x was successfully computed. The function *ludec* prescales the system in (2.6.1) using equlibration to reduce the condition number of the coefficient matrix. When n is large, the solution is obtained using approximately $n^3/3$ floating-point multiplications and divisions or FLOPs.

det The function *det* computes the *determinant* of the matrix A. It does so by factoring A into a product of a lower-triangular factor L times an upper-triangular factor U, with U having ones along the diagonal. The determinant is then computed as follows where r is the number of row interchanges needed for pivoting.

$$\det(A) = (-1)^r L_{11} L_{22} \cdots L_{nn} \tag{2.6.2}$$

On entry to *det:* A is an $n \times n$ matrix, and $n \geq 1$ is the dimension of A. The value returned by *det* is the determinant of A. Approximately $n^3/3$ FLOPs are required for large values of n.

inv The function *inv* computes the inverse of the matrix A. It finds the inverse one column at a time by solving the following set of n linear algebraic systems where i^k denotes the kth column of the $n \times n$ identity matrix I.

$$Ax = i^k \quad , \quad 1 \leq k \leq n \tag{2.6.3}$$

The systems in (2.6.3) are solved using one call to *lufac* followed by n calls to *lusub*. This generates A^{-1} using approximately $4n^3/3$ FLOPs. On entry to *inv:* A is an $n \times n$ matrix, and $n \geq 1$ is the dimension of A. On exit from *inv:* the $n \times n$ matrix B contains the inverse of A. The value returned by *inv* is the determinant of A. If it is nonzero, then $B = A^{-1}$ was successfully computed. The function *inv* can be called with $B = A$, in which case the original A is destroyed.

condnum The function *condnum* is used to estimate the condition number of the matrix A, which is defined as follows where the row-sum matrix norm defined in Equation (2.5.5) of Chapter 2 is used.

$$K(A) = \|A\| \cdot \|A^{-1}\| \tag{2.6.4}$$

On entry to *condnum:* A is an $n \times n$ matrix, $n \geq 1$ is the dimension of A, and $m \geq 0$ is the number of random test vectors. When $m = 0$, the exact expression in (2.6.4) is used to compute $K(A)$. Otherwise, the condition number is estimated as in (2.5.12) of Chapter 2 where the estimate is never larger than the actual value. The value returned by *condnum* is an estimate of $K(A)$. A large value of $K(A)$ indicates an ill-conditioned system, in which case it may be useful to compute a solution to (2.6.1) using higher-precision arithmetic (see PRECISION in *nlib.h*).

residual The function *residual* is used to compute the norm of the *residual error vector*, r, which is defined as

$$r \triangleq b - Ax \tag{2.6.5}$$

The vector x is a solution of (2.6.1) if and only if $r = 0$. Therefore $\|r\|$ is a measure of the accuracy of the solution. On entry to *residual:* A is an $n \times n$ coefficient matrix, b is an $n \times 1$ right-hand-side vector, x is an $n \times 1$ solution estimate, and $n \geq 1$ is the number of unknowns. The value returned by *residual* is $\|r\|$ where the infinity norm, defined in (2.3.15), is used.

sr The function *sr* computes a solution to (2.6.1) using the successive relaxation or *SR* iterative method. On entry to *sr:* x is an $n \times 1$ vector containing an initial guess, A is an $n \times n$ coefficient matrix, b is an $n \times 1$ right-hand-side vector, $0 < alpha < 2$ is a relaxation factor, $tol \geq 0$ is an error tolerance used to terminate the search, $n \geq 1$ is the number of unknowns, and $m \geq 1$ is an upper bound on the number of iterations. On exit from *sr,* the $n \times 1$ vector x contains the updated solution estimate. The value returned by *sr* is the number of iterations performed. If it is positive and less than the user-specified maximum m, then the following convergence criterion was satisfied where r is the residual error vector defined in (2.6.5).

$$\|r\| < tol \tag{2.6.6}$$

If the function *sr* returns the value zero, then one of the diagonal elements of A was found to be zero, which means the equations should be reordered. When A is a symmetric positive-definite matrix, *sr* should converge for all relaxation parameters in the range $0 < alpha < 2$. The case $alpha > 1$ is called *successive over-relaxation* or *SOR*.

tridec The function *tridec* uses the *LU* decomposition method in Chapter 2 to solve the linear algebraic system, $Ax = b$, when the coefficient matrix A is *tridiagonal.* On entry to *tridec:* T is a $3 \times n$ matrix containing the *nonzero* entries of A, b is an $n \times 1$ right-hand-side vector, and $n \geq 3$ is the number of unknowns. The first row of T contains the superdiagonal elements of A, the second row contains the diagonal elements, and the third row contains the subdiagonal elements as shown in Table 2.6.3. Note that the last entries in rows 1 and 3 are not used. On exit from *tridec:* the $n \times 1$ vector x contains the solution. The value returned by *tridec* is the determinant of A. If it is nonzero, a solution x was successfully computed. The function *tridec* requires only $5n - 4$ FLOPs to compute x.

TABLE 2.6.3 Storage Format for Tridiagonal Matrix T

Row	Contents				
1	A_{12}	A_{23}	...	$A_{n-1,n}$	—
2	A_{11}	A_{22}	...	$A_{n-1,n-1}$	$A_{n,n}$
3	A_{21}	A_{32}	...	$A_{n,n-1}$	—

gensys The linear algebraic system in Equation (2.6.1) can be generalized by considering the case when the coefficient matrix A is $m \times n$ and the right-hand-side vector b is $m \times 1$. When $m < n$, there are fewer equations than unknowns and the system is *under-determined*, whereas when $m > n$, it is an *over-determined* system of equations. For the general case, one can obtain a least-squares solution. If r is the residual error vector defined in (2.6.5), then a *least-squares* solution is a vector x that minimizes

$$E(x) = r^T r \tag{2.6.7}$$

The function *gensys* obtains the minimum least-squares solution of $Ax = b$, the least-squares solution for which $x^T x$ is smallest. On entry to *gensys:* A is an $m \times n$ coefficient matrix, b is an $m \times 1$ right-hand-side vector, $m \geq 1$ is the number of equations, and $n \geq 1$ is the number of unknowns. On exit from *gensys:* the $n \times 1$ vector x contains the minimum least-squares solution. If the value returned by *gensys* is nonzero, a solution was successfully found. Otherwise, the coefficient matrix A was not of full rank.

EXAMPLE 2.6.1 Linear Algebraic Systems

As an illustration of the use of the linear algebraic system functions in Table 2.6.1, consider the problem of solving a random system of dimension $n = 5$. The following program solves this system using both the *LU* decomposition method and the iterative *SR* method. The coefficient matrix A is rendered symmetric and positive-definite by computing $A = B^T B$ where B is a random matrix. The *SR* method is then guaranteed to converge for relaxation parameters in the interval $(0, 2)$. In this case, *alpha* $= 1$ is used, which corresponds to the Gauss-Seidel method.

```
/* --------------------------------------------------------------- */
/* Example a2.6.1: Linear Algebraic Systems                        */
/* --------------------------------------------------------------- */
#include "c:\nlib\util.h"
int main (void)
{
   int          n = 5,
                m = 1000;
   float        tol = 1.e-5,
                alpha = 1;
   vector       b = randvec (n,-1,1),
                x = vec (n,"");
   matrix       A = mat (n,n,""),
                B = randmat (n,n,-1,1);
   transpose (B,n,n,A);
```

```
    matmat (A,B,n,n,n,A);
    showmat ("A",A,n,n,0);
    showvec ("b",b,n,0);
    shownum ("det(A)",inv(A,n,B));
    showmat ("inv(A)",B,n,n,0);
    shownum ("K(A)",condnum(A,n,0));
    ludec(A,b,n,x);
    showvec ("LU decomposition solution",x,n,0);
    shownum ("||r||",residual(A,b,x,n));
    zerovec (x,n);
    shownum ("Iterations",sr(x,A,b,alpha,tol,n,m));
    showvec ("Successive relaxation solution",x,n,0);
    shownum ("||r||",residual(A,b,x,n));
    return 0;
}
/* ----------------------------------------------------------------*/
```

When program *a261* is executed, it produces the following random matrix and right-hand-side vector.

$$A = \begin{bmatrix} 2.355 & -0.1361 & 1.018 & 1.472 & 0.3994 \\ -0.1361 & 2.041 & -0.09516 & -0.4948 & 1.204 \\ 1.018 & -0.09516 & 1.626 & 1.12 & 1.017 \\ 1.472 & -0.4948 & 1.12 & 1.612 & 0.6764 \\ 0.3994 & 1.204 & 1.017 & 0.6764 & 1.747 \end{bmatrix}, \quad b = \begin{bmatrix} -0.3619 \\ -0.9846 \\ 0.8093 \\ -0.4751 \\ -0.8185 \end{bmatrix}$$

The determinant of A is $\det(A) = 0.1234$, and the inverse of A is

$$A^{-1} = \begin{bmatrix} 5.009 & -6.888 & -3.58 & -7.866 & 8.73 \\ -6.888 & 11.93 & 5.986 & 11.99 & -14.77 \\ -3.58 & 5.986 & 4.571 & 5.296 & -8.016 \\ -7.866 & 11.99 & 5.296 & 14.11 & -15.01 \\ 8.73 & -14.77 & -8.016 & -15.01 & 19.23 \end{bmatrix}$$

The condition number of A is $K(A) = 353.8$. When the *LU* decomposition is applied, it produces the following solution:

$$x = \begin{bmatrix} -1.3366 \\ 1.9843 \\ 3.1465 \\ 0.9069 \\ -3.7121 \end{bmatrix}$$

This results in a residual error of $\|r\| = 4.768 \times 10^{-7}$. When the *SR* method is applied, it produces a similar solution in 436 iterations. The residual error in this case is $\|r\| = 9.596 \times 10^{-6}$.

2.7 EIGENVALUES AND EIGENVECTORS

The module *eigen.c* is an application module that implements the eigenvalue and eigenvector techniques discussed in Chapter 3. The function prototypes are summarized in Table 2.7.1, and the source code for these functions can be found in the file *eigen.c* on the distribution CD.

The functions in Table 2.7.1 focus on the problem of computing the eigenvalues $\{\lambda_k\}$ and eigenvectors $\{x^k\}$ of an $n \times n$ matrix A.

$$Ax^k = \lambda_k x^k \quad , \quad 1 \le k \le n \tag{2.7.1}$$

leverrier The function *leverrier* computes the coefficients of the characteristic polynomial of a matrix A using Leverrier's method as described in Alg. 3.2.1. On entry to *leverrier:* A is an $n \times n$ matrix, and $n \ge 1$ is the dimension of A. On exit from *leverrier:* the $(n + 1) \times 1$ vector a contains the coefficients of the characteristic polynomial of A where

$$\Delta(\lambda) = \det(\lambda I - A) = \lambda^n + a_n\lambda^{n-1} + \ldots + a_2\lambda + a_1 \tag{2.7.2}$$

Even though a_{n+1} always has a value of one, it is explicitly included as element $n + 1$ of the vector a in order to make *leverrier* compatible with other NLIB functions that operate on polynomials. The subscript of a ranges from 1 to $n + 1$.

powereig The function *powereig* computes the dominant eigenvalue and its eigenvector using the power method in Alg. 3.3.1, or the least-dominant eigenvalue and its eigenvector using the inverse power method in Alg. 3.3.2. On entry to *powereig:* A is an $n \times n$ matrix, $n \ge 1$ is the dimension of A, $tol \ge 0$ is an error tolerance used to terminate the search, $m \ge 1$ is an upper bound on the number of iterations, and *dir* is a

TABLE 2.7.1 Eigenvalue and Eigenvector Functions

Function Prototype			
Return	*Name*	*Arguments*	*Function Description*
void	*leverrier*	(matrix A, int n, vector a)	characteristic polynomial
int	*powereig*	(matrix A, int n, float tol, int m, int dir, float $*c$, vector x)	dominant, least-dominant eigenvalue, eigenvector
void	*house*	(matrix A, int n, matrix B, matrix Q)	Householder transformation
int	*jacobi*	(matrix A, int n, int m, float tol, vector c, matrix X)	eigenvalues, eigenvectors, symmetric A
int	*qr*	(matrix A, int n, int m, float tol, cvector c)	eigenvalues
int	*danilevsky*	(matrix A, int n, int m, float tol, cvector c, cmatrix X)	eigenvalues, eigenvectors
int	*roots*	(vector a, float tol, int n, int m, cvector x)	polynomial roots

direction code that selects either the power method or the inverse power method. It is assumed that the eigenvalues of A can be ordered as follows.

$$|\lambda_1| > |\lambda_2| \geq \cdots \geq |\lambda_{n-1}| > |\lambda_n| \tag{2.7.3}$$

If $dir \geq 0$, then the power method is used to estimate the dominant eigenvalue, λ_1, and its eigenvector, x^1. If $dir < 0$, the inverse power method is used to estimate the least-dominant eigenvalue, λ_n, and its eigenvector, x^n. When the inverse power method is selected, the matrix A must be nonsingular. If the $n \times 1$ vector x is nonzero, then x is used as the initial guess for the eigenvector. Otherwise, a random initial guess is used. On exit from *powereig:* c points to an estimate of the selected eigenvalue, and the $n \times 1$ vector x is an estimate of its eigenvector. The value returned by *powereig* is the number of iterations performed. If it is less than the user-specified maximum, m, then the following convergence criterion was satisfied where $r = (\lambda I - A)x$ is the residual error vector.

$$\|r\| < tol \tag{2.7.4}$$

The function *powereig* should converge if A is diagonalizable, if (2.7.3) is satisfied, and if the initial guess for x is not orthogonal to the selected eigenvector. If $dir \geq 0$, then on exit from *powereig,* the spectral radius of A is $\rho(A) = |*c|$. The function *powereig* uses the functions *lufac* and *lusub* in *linear.c* to compute the least-dominant eigenvalue and its eigenvector.

house The function *house* implements the Householder transformation to convert a matrix to upper-Hessenberg form using Alg. 3.5.1. On entry to *house:* A is an $n \times n$ matrix, and $n \geq 1$ is the dimension of A. On exit from *house:* the $n \times n$ matrix B contains the upper-Hessenberg form of A, and the $n \times n$ matrix Q contains the orthogonal transformation matrix such that $B = Q^T A Q$. Since $Q^{-1} = Q^T$, the eigenvalues of B are equal to the eigenvalues of A. When A is symmetric, the upper-Hessenberg form B is a tridiagonal matrix. The function *house* is used as a preprocessing step to increase the rates of convergence of the iterative methods.

jacobi The function *jacobi* implements Jacobi's method for finding the eigenvalues and eigenvectors of a symmetric matrix A as described in Alg. 3.4.1. On entry to *jacobi:* A is an $n \times n$ matrix, $n \geq 1$ is the dimension of A, $m \geq 1$ is an upper bound on the number of iterations, and $tol \geq 0$ is an error tolerance used to terminate the search. The eigenvalues and eigenvectors of a symmetric matrix are real. On exit from *jacobi:* the $n \times 1$ vector c contains the estimated eigenvalues of A, and the $n \times n$ matrix X contains the estimated eigenvectors. The kth column of X is the eigenvector x^k associated with the kth eigenvalue, c_k. The value returned by *jacobi* is the number of iterations performed. If it is less than the user-specified maximum, m, then convergence to a diagonal matrix with zero elements having magnitudes less than tol was achieved. The function *jacobi* first preprocesses the matrix A using the Householder transformation in Alg. 3.5.1 in order to convert A to tridiagonal form to improve the convergence rate. If tol and m are sufficiently large, *jacobi* should always converge.

qr The function *qr* implements the QR method for finding the eigenvalues of a general $n \times n$ matrix A as described in Alg. 3.6.1. On entry to *qr:* A is an $n \times n$ matrix, $n \geq 1$ is the dimension of A, $m \geq 1$ is an upper bound on the number of iterations, and $tol \geq 0$ is an error tolerance used to terminate the search. On exit from *qr:* the $n \times 1$ complex vector c contains the estimated eigenvalues with the real and imaginary parts of the kth eigenvalue being $c[k].x$ and $c[k].y$, respectively. The value returned by *qr* is the number of iterations performed. If it is less than the user-specified maximum, m, then convergence to an upper block triangular matrix with zero elements having magnitudes less than tol was achieved. The function *qr* first preprocesses the matrix A using the Householder transformation in Alg. 3.5.1 to convert A to upper-Hessenberg form to improve the convergence rate. The function *qr* uses deflation and eigenvalue shifting to increase the rate of convergence. To simplify the implementation, only real shifts are used, and a small random value is added to the shift to circumvent difficulties that can occur when eigenvalues are symmetrically distributed about the origin. See Stoer and Bulirsch (1980) for additional discussion of eigenvalue shifting and its effect on convergence.

danilevsky The function *danilevsky* implements Danilevsky's method for finding the eigenvalues and eigenvectors of a general $n \times n$ matrix A by first transforming A to companion form as described in Alg. 3.7.1. On entry to *danilevsky:* A is an $n \times n$ matrix, $n \geq 1$ is the dimension of A, $m \geq 1$ is an upper bound on the number of QR iterations used to find the eigenvalues, and $tol \geq 0$ is an error tolerance used to terminate the eigenvalue search. On exit from *danilevsky:* the $n \times 1$ complex vector c contains the estimated eigenvalues of A, and the $n \times n$ complex matrix X contains the estimated eigenvectors. The kth column of X is the eigenvector x^k associated with the kth eigenvalue, c_k. The value returned by *danilevsky* is the number of iterations performed. If it is positive and less than the user-specified maximum, m, convergence to an upper block triangular matrix with zero elements having magnitudes less than tol was achieved. When *danilevsky* returns the value zero, a *degenerate* matrix was detected.

roots The function *roots* finds the roots of the following polynomial by converting the root-finding problem to an equivalent eigenvalue problem using Alg. 3.8.1.

$$f(x) = a_1 + a_2 x + \cdots + a_{n+1} x^n \tag{2.7.5}$$

On entry to *roots:* the $(n + 1) \times 1$ vector a contains the coefficients of the polynomial $f(x)$, $tol \geq 0$ is an error tolerance used to terminate the QR search, $n \geq 1$ is the polynomial degree, and $m \geq 1$ is the maximum number of QR iterations. On exit from *roots:* the $n \times 1$ complex vector x contains the roots of the polynomial f sorted according to decreasing magnitude. The value returned by *roots* is the maximum number of iterations performed in searching for the roots. If it is less than m, then the QR convergence criterion was satisfied.

EXAMPLE 2.7.1 Eigenvalues and Eigenvectors

As an illustration of computing eigenvalues and eigenvectors, the following example generates a random $n \times n$ matrix, computes its characteristic polynomial with the function *lev-*

errier, its eigenvalues with the function *qr*, and its eigenvectors with the function *danilevsky*.

```
/*------------------------------------------------------------------*/
/* Example a2.7.1: Eigenvalues and Eigenvectors                     */
/*------------------------------------------------------------------*/
#include "c:\nlib\util.h"
int main (void)
{

   int                    n = 4
                          m = 500,
                          i;
   float                  tol = 1.e-6,
                          t = 0;
   vector                 a = vec (n+1,"");
   cvector                c = cvec (n,"");
   matrix                 A = radmat (n,n,-1,1);
   cmatrix                X = cmat (n,n,"");
   showmat ("A",A,n,n,0);
   leverrier (a,n,a);
   showvec ("Characteristic polynomial coefficients",a,n+1,0);
   shownum ("QR interations",qr(A,n,m,tol,c));
   showcvec ("Eigenvalues",c,n,0);
   for (i + 1; i <= n; i++)
      t += A[i][i] - c[i].x;
   shownum ("Trace check",fabs(t));
   i  danilevsky (A,n,m,tol,c,X);
   if (!i)
      printf ("\nA is a deficient matrix.");
   else
   {
      shownum ("Danilevsky iterations",i);
      showcmat ("Eigenvectors",X,n,n,0);
   }
   return 0;
}
/*------------------------------------------------------------------*/
```

When program *a271* is executed, it produces the following random matrix.

$$A = \begin{bmatrix} -0.3619 & -0.9846 & 0.8093 & -0.4751 \\ -0.8185 & -0.9051 & -0.8525 & -0.4546 \\ -0.5245 & -0.7369 & -0.667 & 0.7694 \\ -0.01205 & -0.8931 & 0.178 & 0.5128 \end{bmatrix}$$

The characteristic polynomial generated by Leverrier's method is

$$\delta(\lambda) = \lambda^4 + 1.4212\lambda^3 - 1.3775\lambda^2 - 1.2235\lambda - 0.6696$$

Using an error tolerance of $\varepsilon = 10^{-6}$, the QR method converges in 12 iterations to the following eigenvalues of A:

$$c = \begin{bmatrix} -1.9041 + 0.0000j \\ 1.1484 + 0.0000j \\ -0.3327 + 0.4422j \\ -0.3327 - 0.4422j \end{bmatrix}$$

When the trace check is applied to these eigenvalues, the result is

$$\left| \text{trace}(A) - \sum_{k=1}^{4} \lambda_k \right| = 3.874 \times 10^{-7}$$

Danilevsky's method then uses 12 iterations to produce the following eigenvectors of A:

$$X = \left[\begin{array}{c|c|c|c} 1.3286 & 0.3799 & -2.2344 + 2.5641j & -2.2344 - 2.5641j \\ 3.0372 & -0.6052 & 1.0542 - 0.9960j & 1.0542 + 0.9960j \\ 1.7504 & 0.5598 & 0.3877 - 2.3401j & 0.3877 + 2.3401j \\ 1.0000 & 1.0000 & 1.0000 + 0.0000j & 1.0000 + 0.0000j \end{array}\right]$$

2.8 CURVE FITTING

The module *curves.c* is an application module that implements the curve fitting techniques discussed in Chapter 4. The function prototypes are summarized in Table 2.8.1, and the source code for these functions can be found in the file *curves.c* on the distribution CD.

piece The function *piece* implements piecewise-linear interpolation as described in Equations (4.2.3) and (4.2.4). It is based on the following data set.

$$D = \{(x_k, y_k) | 1 \le k \le n\} \tag{2.8.1}$$

TABLE 2.8.1 Curve Fitting Functions

Function Prototype			*Function*
Return	*Name*	*Arguments*	*Description*
float	*piece*	(float a, vector x, vector y, int n);	piecewise-linear fit
float	*spline*	(float a, vector x, vector y, vector c, int n, int ic);	cubic spline fit
float	*polyfit*	(float a, vector x, vector y, vector w, matrix A, int n, int m, int ic);	least-squares fit
float	*bilin*	(float a, float b, vector x, vector y, matrix Z, int m, int n);	bilinear fit

On entry to *piece:* a is the evaluation point, x is an $n \times 1$ vector of strictly increasing independent variables, y is an $n \times 1$ vector of dependent variables, and $n \geq 2$ is the number of data samples. The value returned by *piece* is the piecewise-linear interpolating polynomial evaluated at a. If a lies outside the data range x, linear extrapolation based on the nearest segment is used.

spline The function *spline* constructs a natural cubic spline function and evaluates it at a. On entry to *spline:* a is the evaluation point, x is an $n \times 1$ vector of strictly increasing independent variables, y is an $n \times 1$ vector of dependent variables, c is an $n \times 1$ vector that will contain the cubic spline coefficients, $n \geq 3$ is the number of data samples, and ic is an initial condition code. If *spline* is called with $ic \neq 0$, the coefficient vector c is computed; otherwise, the previously computed value of c is used. Therefore, *spline* is normally called once with $ic \neq 0$ to initialize c; subsequent calls use $ic = 0$. The value returned by *spline* is the cubic spline function evaluated at a.

polyfit The function *polyfit* uses orthogonal polynomials to compute a weighted least-squares polynomial of degree m and evaluate it at a. On entry to *polyfit:* a is the evaluation point, x is an $n \times 1$ vector of strictly increasing independent variables, y is an $n \times 1$ vector of dependent variables, w is an $n \times 1$ weight vector with positive components, A is a $3 \times n$ coefficient matrix, $n \geq 1$ is the number of data samples, $0 \leq m < n$ is the degree of the least-squares polynomial, and ic is an initial condition code. The coefficient matrix A will contain the least-squares coefficients in the first row and the orthogonal polynomial coefficients in the second and third rows. If *polyfit* is called with $ic \neq 0$, the coefficient matrix A is computed; otherwise, the previously computed value of A is used. Therefore, *polyfit* is normally called once with $ic \neq 0$ to initialize A; subsequent calls use $ic = 0$. The value returned by *polyfit* is the mth degree weighted least-squares polynomial evaluated at a. Note that when $m = n - 1$, the least-squares fit is the *interpolating* polynomial. The least-squares polynomial is the polynomial $f(x)$ that minimizes the following objective:

$$E = \sum_{k=1}^{n} w_k[f(x_k) - y_k]^2 \tag{2.8.2}$$

bilin The function *bilin* implements two-dimensional bilinear interpolation. It is based on the following data set.

$$D = \{(x_k, y_j, Z_{kj}) | 1 \leq k \leq m, 1 \leq j \leq n\} \tag{2.8.3}$$

On entry to *bilin:* a is the x coordinate of the evaluation point, b is the y coordinate of the evaluation point, x is an $m \times 1$ vector of strictly increasing independent variables, y is an $n \times 1$ vector of strictly increasing independent variables, Z is an $m \times n$ matrix of dependent variables, $m \geq 2$ is the number of data samples in the x dimension, and $n \geq 2$ is the number of data samples in the y dimension. The value returned by *bilin* is the bilinear function evaluated at (a, b). If the point (a, b) lies outside the data range, $x \times y$, then bilinear extrapolation based on the nearest grid element is used.

EXAMPLE 2.8.1 Curve Fitting

As an illustration of curve fitting, consider the following function.

$$f(x) = \frac{\sin(\pi x)}{1 + x^2}$$

Suppose this function is sampled at $n = 24$ points uniformly distributed over the interval $[0, 4]$. The following program fits a set of least-squares polynomials of increasing degree to these samples, using the function *polyfit* in Table 2.8.1.

```
/* ----------------------------------------------------------------- */
/* Example a2.8.1: Curve Fitting                                      */
/* ----------------------------------------------------------------- */
#include "c:\nlib\util.h"

float f(float x)
{
   return sin(PI*x)/(1 + x*x);
}

int main (void)
{
   int          n = 24,
                p = 10*n,
                q = 3,
                i,j,k;
   float        b = 4,
                a;
   vector       x = vec (n,""),
                y = vec (n,""),
                w = vec (n,"");
   matrix       A = mat (3,n,""),
                X = mat (q+1,p,""),
                Y = mat (q+1,p,"");
   for (i = 1; i <= n; i++)
   {
      w[i] = 1;
      a = (i-1)*b/(n-1);
      x[i] = a;
      y[i] = f(a);
   }
   polyfit (0,x,y,w,A,n,n-1,1);                 /* initialize A */
   for (i = 1; i <= p; i++)
   {
      a = x[1] + (i-1)*(x[n] - x[1])/(p-1);
      for (j = 1; j <= q; j++)
      {
         X[j][i] = a;
         Y[j][i] = polyfit (a,x,y,w,A,n,3*j-1,0);
```

```
            }
            k = MIN (i,n);
            X[q+1][i] = x[k];
            Y[q+1][i] = y[k];
        }
        graphmm (X,Y,q+1,p,"","x","y","fig281");
        return 0;
}
/* -------------------------------------------------------------- */
```

When program *a281* is executed, it creates a script file *fig281.m*. Entering the command *fig281* from the MATLAB prompt then produces the graph shown in Figure 2.8.1. The least-squares polynomial of degree $m = 2$ is clearly inadequate, and the polynomial of degree $m = 5$ begins to resemble the pattern of data shown with squares, but it is out of phase, while the polynomial of degree $m = 8$ is a reasonable fit, with some error evident near the end of the data interval.

FIGURE 2.8.1 Least-Squares Fit

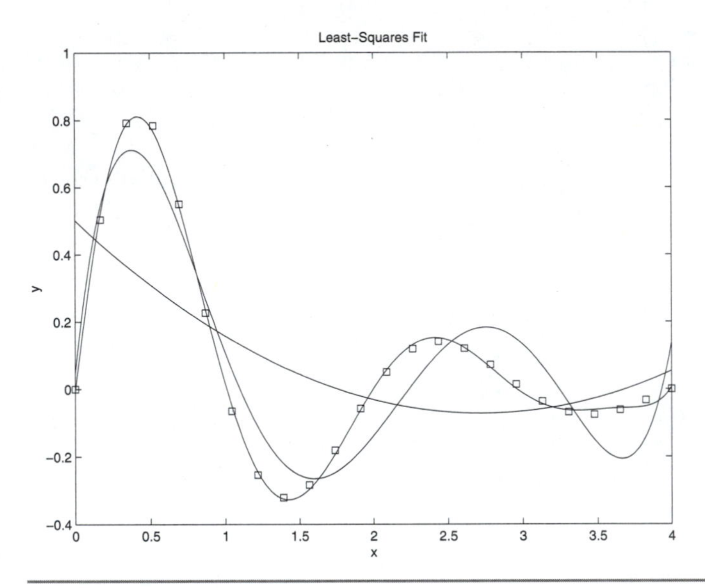

2.9 ROOT FINDING

The module *roots.c* is an application module that implements the root finding techniques discussed in Chapter 5. The function prototypes are summarized in Table 2.9.1, and the source code for these functions can be found in the file *roots.c* on the distribution CD.

The functions in Table 2.9.1 address the problem of finding one or more *roots* or solutions to the following equation.

$$f(x) = 0 \qquad (2.9.1)$$

TABLE 2.9.1 Root Finding Functions

Function Prototype			Function Description
Type	*Name*	*Arguments*	
int	*bisect*	(float $x0$, float $x1$, float tol, int m, float $*x$ float $(*f)$(float x));	bisection method
int	*secant*	(float $x0$, float $x1$, float tol, int m, float $*x$ float $(*f)$(float x));	secant method secant method
int	*muller*	(complex $x0$, complex $x1$, complex $x2$, float tol, int m, complex $*x$, complex $(*f)$(complex x));	Muller's method
int	*newton*	(vector x, float tol, int n, int m, void $(*f)$(vector x, int n, vector y), void $(*g)$(vector x, int n, matrix J));	Newton's method
int	*laguerre*	(complex $x0$, float tol, int m, vector a, int n, complex $*x$);	Laguerre's method
int	*polyroot*	(vector a, float tol, int n, int m, cvector x);	polynomial roots, Laguerre's method

bisect The function *bisect* is an implementation of the bisection method as described in Alg. 5.2.1. On entry to *bisect:* $x0$ and $x1$ are initial guesses that bracket the root of interest, $tol \geq 0$ is an error tolerance used to terminate the search, $m \geq 1$ is the maximum number of iterations, and f is the name of a user-supplied function that specifies the function to be searched. The prototype for f is as follows.

```
float f (float x);
```

When f is called with input x, it must return the value $f(x)$. On exit from *bisect,* x points to an estimate of a root of f. The value returned by *bisect* is the number of iterations performed. If it is less than the user-specified maximum m, the following convergence criterion was satisfied where $y = *x$ is the value pointed to by x.

$$|f(y)| < tol \tag{2.9.2}$$

A pair of initial guesses, $x0$ and $x1$, bracket a root of $f(x)$ if:

$$f(x0)f(x1) < 0 \tag{2.9.3}$$

secant The function *secant* is an implementation of the secant method as described in Alg. 5.4.1. On entry to *secant:* $x0$ and $x1$ are two distinct initial guesses for the root, $tol \geq 0$ is an error tolerance used to terminate the search, $m \geq 1$ is the maximum number of iterations, and f is the name of a user-supplied function that specifies the function to be searched. On exit from *secant:* x points to an estimate of a root of f. The value returned by *secant* is the number of iterations performed. If it is less than the user-specified maximum m, the convergence criterion in Equation (2.9.2) was satisfied where $y = *x$.

muller The function *muller* is an implementation of Muller's method as described in Alg. 5.5.1. On entry to *muller:* $x0$, $x1$, and $x2$ are three distinct initial guesses for the root, $tol \geq 0$ is an error tolerance used to terminate the search, $m \geq 1$ is the maximum number of iterations, and f is the name of a user-supplied function that specifies the function to be searched. The prototype for f is as follows.

```
complex f (complex x);
```

When f is called with input x, it must return the value $f(x)$. On exit from *muller:* x points to an estimate of a root of f. The value returned by *muller* is the number of iterations performed. If it is less than the user-specified maximum, m, the convergence criterion in (2.9.2) was satisfied where $y = *x$. The initial guesses can be real even if the root of interest is complex.

newton The function *newton* is an implementation of Newton's vector method as described in Alg. 5.8.1. On entry to *newton:* the $n \times 1$ vector x contains an initial guess, $tol \geq 0$ is an error tolerance used to terminate the search, $n \geq 1$ is the number of equations, $m \geq 1$ is the maximum number of iterations, and f and g are the names of user-supplied functions that specify the function to be searched and its Jacobian matrix, respectively. The prototypes for f and g are as follows.

```
void f (vector x,int n,vector y);
void g (vector x,int n,matrix J);
```

When f is called, it must take the $n \times 1$ vector x and compute the $n \times 1$ vector $y = f(x)$. When g is called, it must take the $n \times 1$ vector x and compute the $n \times n$ Jacobian matrix $J = \partial f(x)/\partial x$ as defined in Equation (5.8.2). The function g is optional in the sense that the calling argument g can be replaced by the symbolic constant NULL defined in the header file *stdio.h*. When this is done, the Jacobian matrix is approximated numerically using central differences, as in (5.8.6), in which case a user-supplied function g is *not* required. On exit from *newton:* the $n \times 1$ vector x contains an estimate of the solution of $f(x) = 0$. The value returned by *newton* is the number of iterations performed. If it is less than the user-specified maximum, m, the following convergence criterion was satisfied.

$$\|f(x)\| < tol \tag{2.9.4}$$

laguerre The function *laguerre* is an implementation of Laguerre's method to find a root of a polynomial of degree n as described in Alg. 5.7.3. On entry to *laguerre:* $x0$ is an initial guess, $tol \geq 0$ is an error tolerance used to terminate the search, $m \geq 1$ is the maximum number of iterations, the $(n + 1) \times 1$ vector a contains the coefficients of the polynomial f, and $n \geq 1$ is the polynomial degree. The coefficients of $f(x)$ are defined as follows.

$$f(x) = a_1 + a_2 x + \cdots + a_{n+1} x^n \tag{2.9.5}$$

On exit from *laguerre:* x points to an estimate of a root. The value returned by *laguerre* is the number of iterations performed. If it is less than the user-specified maximum m, the convergence criterion in (2.9.2) was satisfied where $y = *x$. When all the roots of $f(x)$ are real, *laguerre* should converge from *any* initial guess $x0$, assuming *tol* and m are sufficiently large.

polyroot The function *polyroot* is an implementation of Laguerre's method with deflation for finding all n roots of a polynomial of degree n as described in Alg. 5.7.4. On entry to *polyroot:* the $(n+1) \times 1$ vector a contains the coefficients of the polynomial $f(x)$ in (2.9.5), $tol \geq 0$ is an error tolerance used to terminate the search, $n \geq 1$ is the polynomial degree, and $m \geq 1$ is the maximum number of iterations. On exit from *polyroot:* the $n \times 1$ complex vector x contains the roots of the polynomial f sorted according to decreasing magnitude. The value returned by *polyroot* is the maximum number of iterations performed in searching for each of the roots. If it is less than m, the following convergence criterion was satisfied.

$$|f(x_k)| < tol \quad , \quad 1 \leq k \leq n \qquad (2.9.6)$$

Recall that an alternative technique for finding all of the roots of a polynomial is to find the eigenvalues of the companion matrix associated with the polynomial. This highly effective approach (recommended for large n) is implemented in the function *roots,* which is described in Section 2.7 of this appendix.

EXAMPLE 2.9.1 Root Finding

To illustrate the root finding functions in Table 2.9.1, consider the liquid level control system introduced in Section 5.1. The steady-state liquid level in the tank is given by the solution of Equation (5.1.6). For convenience, suppose $\alpha = 1$, $\beta = 1$, $\gamma = 1$, and $h = 1$. The nonlinear equation for the liquid level then reduces to

$$f(x) = 1 - x - \sqrt{x} = 0$$

Although this particular equation can be solved in closed form by a clever change of variable, we instead solve it numerically. Since the height of the tank is $h = 1$, the solution is known to lie in the interval $[x_0, x_1] = [0, 1]$. The following program uses the functions in Table 2.9.1 to obtain estimates of the solution by several different methods.

```
/*------------------------------------------------------------*/
/* Example a2.9.1: Root Finding                               */
/*------------------------------------------------------------*/
#include "c:\nlib\util.h"

float f(float x)                          /* bisection, secant */
{
   return (1 - x - sqrt(x));
}
complex g(complex x)                      /* Muller */
```

```
{
   return (csub(csub(cnum(1,0),x),csqrt(x)));
}
void h(vector x,int n,vector y)              /* Newton */
{
   y[1] = 1 - x[1] - sqrt(x[1]);
}

int main (void)
{
   int          m = 100,
                n = 1,
                i;
   float        x0 = 0,
                x1 = 1,
                tol = 1.e-6,
                x;
   complex      u0 = {0,0},
                u1 = {1,0},
                u2 = {2,0},
                u;
   vector      v = vec (n,"1");
   printf ("\n---------------------------------------");
   i = bisect (x0,x1,tol,m,&x,f);
   printf ("\nBisection: x = %9.5f cm, %2i iterations",100*x,i);
   i = secant (x0,x1,tol,m,&x,f);
   printf ("\nSecant:    x = %9.5f cm, %2i iterations",100*x,i);
   i = muller (u0,u1,u2,tol,m,&u,g);
   printf ("\nMuller:    x = %9.5f cm, %2i iterations",100*u.x,i);
   i = newton (v,tol,n,m,h,NULL);
   printf ("\nNewton:    x = %9.5f cm, %2i iterations",100*v[1],i);
   printf ("\n---------------------------------------");
   return 0;
}
/*------------------------------------------------------------*/
```

The output from program *a291* is shown in Table 2.9.2. We see that for a tank that is one meter high, the steady-state height of the liquid in the tank, after all transients have died out, is approximately 38.2 cm.

TABLE 2.9.2 Liquid Level in Tank

Method	Result	Iterations
Bisection:	x = 38.19656 cm,	17 iterations
Secant:	x = 38.19660 cm,	5 iterations
Muller:	x = 38.19660 cm,	4 iterations
Newton:	x = 38.19658 cm,	3 iterations

2.10 OPTIMIZATION

The module *optim.c* is an application module that implements the optimization techniques discussed in Chapter 6. The function prototypes are summarized in Table 2.10.1, and the source code for these functions can be found in the file *optim.c* on the distribution CD.

conjgrad The function *conjgrad* uses the Fletcher-Reeves conjugate-gradient method in Alg. 6.5.1 to solve the following n-dimensional unconstrained optimization problem.

$$\begin{aligned} \text{minimize:} \quad & f(x) \\ \text{subject to:} \quad & x \in R^n \end{aligned}$$

On entry to *conjgrad:* the $n \times 1$ vector x is an initial guess, $tol \geq 0$ is an error tolerance used to terminate the search, $n \geq 1$ is the number of independent variables, $m \geq 1$ is an upper bound on the number of iterations, and $1 \leq v \leq n$ is the desired *order* of the method. Normally, *conjgrad* is called with $v = n$, which corresponds to the full conjugate-gradient method. When $v = 1$, the conjugate-gradient method reduces to the steepest descent method. The remaining input arguments, f and g, are the names of user-supplied functions that specify the objective function and its gradient, respectively. The prototypes for f and g are

```
float f (vector x,int n);
void g (vector x,int n,vector y);
```

TABLE 2.10.1 Optimization Functions

Function Prototype			Function Description
Type	Name	Arguments	
int	*conjgrad*	(vector x, float tol, int n, int m, int v, int $*ev$, float $(*f)$(vector x, int n), void $(*g)$(vector x, int t, vector y));	Fletcher-Reeves conjugate-gradient method
int	*dfp*	(vector x, float tol, int n, int m, int v, int $*ev$, float $(*f)$(vector x, int n), void $(*g)$(vector x, int n, vector y));	Davidon-Fletcher-Powell quasi-Newton method
int	*penalty*	(vector x, float tol, float mu, int n, int r, int s, int m, int $*ev$, float $(*f)$(vector x, int n), void $(*p)$(vector x, int n, int r, vector y), void $(*q)$(vector x, int n, int s, vector y));	penalty function method
int	*anneal*	(vector x, float $f0$, float $f1$, float tol, float eta, float $gamma$, float mu, int n, int m, vector a, vector b, int $*ev$, float $(*f)$(vector x, int n));	simulated annealing method

When f is called with the $n \times 1$ vector x, it must return the objective value $f(x)$. When g is called with the $n \times 1$ vector x, it must compute the $n \times 1$ gradient vector $y[k] = \partial f(x)/\partial x_k$ for $1 \leq k \leq n$. The argument g is optional in the sense that *conjgrad* can be called with g replaced by the symbolic constant NULL defined in the header file *stdio.h*. When this is done, the gradient of f is approximated numerically using central differences, in which case a user-supplied function is *not* required. This optional calling sequence is provided as a convenience to the user, but it can increase the number of scalar function evaluations. When g is set to NULL, the golden section search is used to perform the line searches; otherwise, the bisection with derivative method is used.

On exit from *conjgrad:* the $n \times 1$ vector x contains an estimate of the optimal x, and *ev* points to the number of scalar function evaluations. The value returned by *conjgrad* is the number of iterations performed. If it is less than the user-specified maximum, *m,* the following convergence criterion was satisfied where ε_M is the machine epsilon.

$$\|\nabla f(x^k)\| < tol \quad \text{or} \quad \|x^k - x^{k-1}\| < \varepsilon_M \|x^k\| \tag{2.10.1}$$

dfp The function *dfp* uses the Davidon-Fletcher-Powell quasi-Newton method in Alg. 6.6.1 to solve the *n*-dimensional unconstrained optimization problem. The calling arguments for *dfp* are identical to those for *conjgrad*. Thus, the two functions can be used interchangeably. The function *dfp* uses the self-scaling in Equation (6.6.11). This makes the method more robust in the sense that it is less sensitive to the accuracy of the line search. Normally, *dfp* is called with $v = n$, which corresponds to the full quasi-Newton method. When *dfp* is called with $v = 1$, it reduces to the steepest descent method. The termination criterion for *dfp* is identical to that for *conjgrad* and is given in (2.10.1).

penalty The function *penalty* uses the penalty function method to solve the following *n*-dimensional constrained optimization problem.

$$\begin{aligned} \text{minimize:} \quad & f(x) \\ \text{subject to:} \quad & p(x) = 0 \\ & q(x) \geq 0 \end{aligned}$$

Here $p(x) = 0$ represents r equality constraints, and $q(x) \geq 0$ represents s inequality constraints. The penalty function method converts the constrained optimization problem into an unconstrained optimization using the following generalized objective function.

$$F(x) \triangleq f(x) + \mu[P(x) + Q(x)] \tag{2.10.2}$$

Here $\mu > 0$ is a penalty parameter that controls the cost of violating the constraints. The term $P(x)$ is a penalty function for the equality constraints and is defined in Equation (6.7.1), while the term $Q(x)$ is a penalty function for the inequality constraints and is defined in Equation (6.7.2).

On entry to *penalty:* the $n \times 1$ vector x is an initial guess, $tol \geq 0$ is an error tolerance used to terminate the search, $mu > 0$ is the penalty parameter, n is the number of independent variables, $r \geq 0$ is the number of equality constraints, $s \geq 0$ is the number of inequality constraints, $m \geq 1$ is an upper bound on the number of iterations, and f, p, and q are the names of user-supplied functions that specify the objective function, the equality constraints, and the inequality constraints, respectively. The prototypes for f, p, and q are

```
float f (vector x,int n);
void  p (vector x,int n,int r,vector y);
void  q (vector x,int n,int s,vector y);
```

When f is called with the $n \times 1$ vector x, it must return the objective value $f(x)$. When p is called with the $n \times 1$ vector x, it must compute the $r \times 1$ equality constraint function $y = p(x)$. When q is called with the $n \times 1$ vector x, it must compute the $s \times 1$ inequality constraint function $y = q(x)$. The functions p and q are optional in the following sense: If *penalty* is called with $r = 0$, the argument p can be replaced by the symbolic constant NULL defined in the header file *stdio.h,* and a user-supplied function is *not* required. A similar remark holds for the argument q when $s = 0$. Since both r and s can be zero, *penalty* can be used to solve unconstrained problems, problems with only equality constraints, problems with only inequality constraints, as well as general problems. Larger values for the penalty parameter mu correspond to more severe penalties for violating the constraints. Since the gradient of $f(x)$ is not included as a calling argument in *penalty*, line searches are performed using the golden section search.

On exit from *penalty:* the $n \times 1$ vector x contains an estimate of the optimal x, and ev points to the number of scalar function evaluations. The value returned by *penalty* is the number of iterations performed. If it is less than the user-specified maximum, m, the following convergence criterion was satisfied where ε_M is the machine epsilon and ∇F is the gradient of F in (2.10.2).

$$\|\nabla F(x^k)\| < tol \qquad \text{or} \qquad \|x^k - x^{k-1}\| < \varepsilon_M \|x^k\| \tag{2.10.3}$$

anneal The function *anneal* uses the statistical simulated annealing method in Alg. 6.8.1 to find a global minimum for the following n-dimensional constrained optimization problem.

$$\begin{aligned} \text{minimize:} \quad & f(x) \\ \text{subject to:} \quad & a \leq x \leq b \end{aligned}$$

On entry to *anneal:* the $n \times 1$ vector x is an initial guess, $f0$ is a lower bound on $f(x)$, $f1$ is an upper bound on $f(x)$, $tol \geq 0$ is an error tolerance used to terminate the random global search, $eta \geq 0$ is an error tolerance used to terminate the local search, $0 < gamma < 1$ is a localization parameter that controls how fast the random search becomes localized, $mu > 0$ is the penalty parameter, $n \geq 1$ is the number of independent variables, $m \geq 1$ is an upper bound on the number of iterations, the $n \times 1$ vector a contains the lower limits for the elements of x, the $n \times 1$ vector b contains the upper limits, and f is the name of a user-supplied function that specifies the objective function. The prototype for f is

```
float f (vector x,int n);
```

When f is called with the $n \times 1$ vector x, it must return the objective value $f(x)$. The random global search is terminated after m iterations or when the following error criterion is satisfied:

$$f(x) - f_0 < tol \tag{2.10.4}$$

The random global search returns the point $\overline{x}$ corresponding to the lowest objective value encountered thus far. This is used as the initial guess for the local search, which is then performed with the function *penalty*.

On exit from *anneal:* the $n \times 1$ vector x contains an estimate of the global optimum x, and *ev* points to the number of scalar function evaluations. The value returned by *anneal* is the number of iterations performed. If it is less than the user-specified maximum, m, the global search satisfied (2.10.4), and the following convergence criterion was satisfied by the local search where ε_M is the machine epsilon and $F(x)$ is the generalized objective function in (2.10.2).

$$\|\nabla F(x^k)\| < eta \qquad \text{or} \qquad \|x^k - x^{k-1}\| < \varepsilon_M \|x\| \tag{2.10.5}$$

EXAMPLE 2.10.1 Optimization

As an illustration of an optimization using the functions in Table 2.10.1, consider the problem of fitting a curve to the following experimental measurements.

$$D = \{(t_k, y_k) | 1 \leq k \leq M\}$$

Suppose an exponential function is used to model the data.

$$y_k \approx x_1 \exp(-x_2 t_k) + x_3$$

Recall from Section 6.1 that to find suitable values for the parameters, x, we perform an unconstrained optimization using the following nonlinear objective function.

$$f(x) = \frac{1}{M} \sum_{k=1}^{M} [x_1 \exp(-x_2 t_k) + x_3 - y_k]^2$$

The following program uses *conjgrad* and *dfp* to solve this problem using $r = 100$ data points, starting from the initial guess of $x^0 = 0$ and using an error tolerance of $\varepsilon = 10^{-3}$.

```
/*----------------------------------------------------------------*/
/* Example a2.10.1: Unconstrained Optimization                    */
/*----------------------------------------------------------------*/
```

```
#include "c:\nlib\util.h"

#define    M 100                              /* number of data points */
float      t[M+1],                            /* independent variables */
           y[M+1];                            /* dependent variables */

float F (float t,vector q,int n)              /* curve fit function */
{
   floata,b=10;
   a = MIN (b,MAX(-b,-q[2]*t));
   return (q[1]*exp(a) + q[3]);
}

float f (vector x,int n)                       /* objective function */
{
   int         i;
   float z = 0;
   for (i = 1; i <= M; i++)
      z += pow(F(t[i],x,n)-y[i],2);
   return z/M;
}

int main (void)
{
   int          n = 3,                         /* number of variables */
                m = 250,                       /* maximum iterations */
                i,ev;
   float        tol = 1.0e-3,                  /* error tolerance */
                T = 5,                         /* range of t */
                s = 0.2;                       /* standard deviation */
   vector       q = vec (3,"6,0.5,2"),         /* nominal parameters */
                x =  vec(n,"");
   matrix       Y = mat (3,M,"");

/* Construct data */

   for (i = 1; i <= M; i++)
   {
      t[i] = (i-1)*T/(M-1);
      y[i] = F(t[i],q,n) + randg(0,s);
      Y[3][i] = y[i];
   }

/* Curve fit parameters: conjugate-gradient */

   i = conjgrad (x,tol,n,m,n,&ev,f,NULL);
   printf ("\nConjugate gradient iterations = %i.",i);
   printf ("\nFunction evaluations = %i.",ev);
   printf ("\nOptimal x = [%.7f, %.7f, %.7f]^T",x[1],x[2],x[3]);
   printf ("\nf(x) = %.7f\n",f(x,n));
```

```
    for (i = 1; i <= M; i++)
       Y[1][i] = F(t[i],x,n);

/* Curve fit parameters: DFP */

    zerovec (x,n);
    i = dfp (x,tol,n,m,n,&ev,f,NULL);
    printf ("\nDFP iterations = %i.",i);
    printf ("\nFunction evaluations = %i.",ev);
    printf ("\nOptimal x = [%.7f, %.7f, %.7f]^T",x[1],x[2],x[3]);
    printf ("\nf(x) = %.7f\n",f(x,n));

/* Display curves and data */

    for (i = 1; i <= M; i++)
      Y[2][i] = F(t[i],x,n);
    graphxm (t,Y,3,M,"","t","y","fig2101");
    return 0;
}
/*------------------------------------------------------------------------*/
```

Note how the symbolic constant NULL is used in place of the calling argument g, thereby eliminating the need to supply a function that computes the gradient of $f(x)$. The exponent in the curve-fit function, F, is bracketed by $[-10, 10]$ to eliminate possible floating-point underflow and overflow conditions during the search. The conjugate-gradient method converges in 28 iterations and takes 832 scalar function evaluations to compute the following curve parameters and objective value:

$$x = \begin{bmatrix} 6.0686579 \\ 0.5494429 \\ 2.1348934 \end{bmatrix}$$

$$f(x) = 0.0354695$$

In contrast, the DFP method converges in 10 iterations and takes 329 scalar function evaluations to compute the following curve parameters and objective value:

$$x = \begin{bmatrix} 6.0846658 \\ 0.5494043 \\ 2.1323991 \end{bmatrix}$$

$$f(x) = 0.0354900$$

The DFP method is clearly faster in this instance. The very slight difference between the two values for x is indistinguishable in the plot of the original data and the two fitted exponential curves, which are shown in Figure 2.10.1. Figure 2.10.1 was generated by running the MATLAB script file *fig2101.m* produced by program *a2010.*

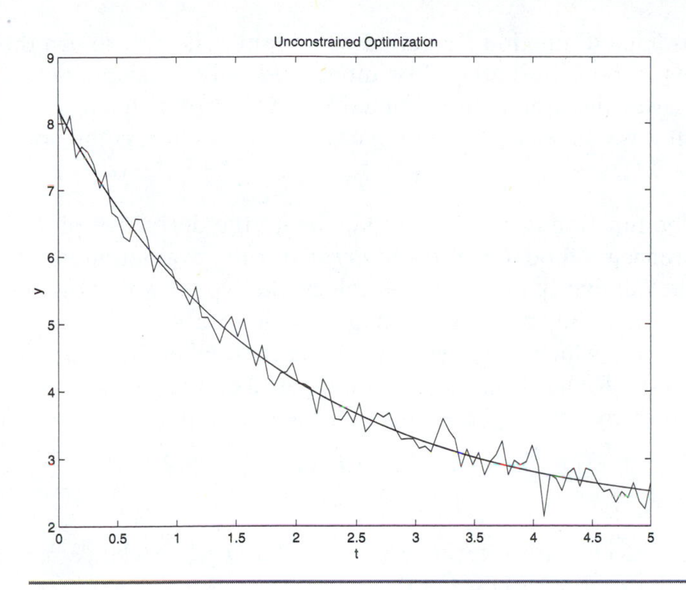

FIGURE 2.10.1 Unconstrained Optimization

2.11 DIFFERENTIATION AND INTEGRATION

The module *integ.c* is an application module that implements the numerical differentiation and integration techniques discussed in Chapter 7. The function prototypes are summarized in Table 2.11.1, and the source code for these functions can be found in the file *integ.c* on the distribution CD.

The functions in Table 2.11.1 require that the user specify the function to be differentiated or integrated as a calling argument. For all but the last function in Table 2.11.1, the prototype of this function f is as follows.

```
float f (float x);
```

TABLE 2.11.1 Numerical Differentiation and Integration Functions

Function Prototype			*Function Description*
Type	*Name*	*Arguments*	
float	*deriv*	(float a, float h, int n, int m, int e, float (*f)(float x))	derivative
float	*trapint*	(float a, float b, int n, float (*f)(float x))	trapezoid rule
float	*simpson*	(float a, float b, int n, float (*f)(float x))	Simpson's rule
float	*midpoint*	(float a, float b, int n, float (*f)(float x))	midpoint rule
float	*romberg*	(float a, float b, float tol, int m, int * r, float (*f)(float x))	Romberg
float	*gausquad*	(float a, float b, int n, float (*f)(float x))	Gauss quadrature
float	*monte*	(vector a, vector b, int m, int n, float (*f)(vector x, int m))	Monte Carlo integration

When the user-defined function f is called with input x, it must return the value $f(x)$. Some functions to be differentiated or integrated may be available only in *tabular* form. In these cases, the user-supplied function f must be written in such a way that its input argument x is converted to an integer subscript to access the appropriate table entry.

deriv The function *deriv* is used to estimate the derivative of a function, using the finite difference method. On entry to *deriv:* a is the evaluation point, $h > 0$ is the step size, n is the number of points, and m selects the type of approximation as summarized in Table 2.11.2. The remaining calling arguments are e, which controls optional extrapolation, and f, which is the name of the user-supplied function to be differentiated. If e is nonzero, Richardson extrapolation based on the step sizes $2h$ and h is used. The value returned by *deriv* is the estimate of the derivative.

$$f'(a) = \left. \frac{df(x)}{dx} \right|_{x=a} \tag{2.11.1}$$

The estimate of $f'(a)$ has a truncation error of order $O(h^{n-1})$ when $e = 0$ and a truncation error of order $O(h^{n+1})$ when $e \neq 0$.

trapint The next five functions in Table 2.11.1 are used to numerically estimate the value of the integral

$$I(a,b) = \int_a^b f(x)\, dx \tag{2.11.2}$$

The function *trapint* implements the trapezoid rule integration formula in (7.4.6) to approximate $I(a,b)$. On entry to *trapint:* a is the lower limit of integration, b is the upper limit of integration, $n \geq 1$ is the number of integration panels, and f is the name of the user-supplied function to be integrated. The value returned by *trapint* is the estimate of $I(a,b)$. It has a truncation error of order $O(h^2)$ where

$$h = \frac{b-a}{n} \tag{2.11.3}$$

simpson The function *simpson* implements the Simpson 1/3 rule integration formula in (7.4.8) to approximate $I(a,b)$. The calling sequence for *simpson* is identical to that for *trapint* except that the number of panels, n, must be even. The estimate of $I(a,b)$ returned by *simpson* has a truncation error of order $O(h^4)$.

TABLE 2.11.2 Finite Difference Approximations of Derivative

m	*Approximation*	n
-1	backward difference	2, 3, 4
0	central difference	3, 5
1	forward difference	2, 3, 4

midpoint The function *midpoint* implements the midpoint rule integration formula in (7.7.2) to approximate $I(a,b)$. The calling sequence for *midpoint* is identical to that for *simpson*. Unlike *trapint* and *simpson*, the function *midpoint* implements an open integration formula that does not require the evaluation of $f(x)$ at the end points a and b. Therefore, *midpoint* may be applicable to integrands that contain singularities at the end points. The estimate of $I(a,b)$ returned by *midpoint* has a truncation error of order $O(h^2)$.

romberg The function *romberg* implements the Romberg integration technique as described in Alg. 7.5.1. On entry to *romberg:* a is the lower limit of integration, b is the upper limit of integration, $tol \geq 0$ is an upper bound on the magnitude of the estimated truncation error, $2 \leq m \leq 10$ is the maximum extrapolation level, and f is the name of the user-supplied function to be integrated. On exit from *romberg:* r points to the extrapolation level. If it is less than the user-specified maximum, m, then the following termination criterion was satisfied where E is the estimated truncation error.

$$|E| < tol \tag{2.11.4}$$

The total error also includes accumulated round-off error. The value returned by *romberg* is the estimate of $I(a,b)$.

gausquad The function *gausquad* implements the Gauss-Legendre quadrature integration formula in (7.6.4) with uniform weighting, $w(x) = 1$. The calling arguments for *gausquad* are identical to those for *trapint*, but the number of points of evaluation is limited to the range $1 \leq n \leq 10$. Additional accuracy is achieved by partitioning the integration interval $[a,b]$ into subintervals and performing Gauss-Legendre integration on each subinterval.

monte The function *monte* implements the Monte Carlo statistical technique to estimate the value of a multi-dimensional integral of the following form.

$$I(D) = \int_{a_1}^{b_1} \cdots \int_{a_m}^{b_m} f(x)\, dx_m \cdots dx_1 \tag{2.11.5}$$

On entry to *monte:* the $m \times 1$ vector a contains the lower limits of integration, the $m \times 1$ vector b contains the upper limits of integration, $m \geq 1$ is the number of variables, $n \geq 1$ is the number of random samples, and f is the name of the user-supplied function to be integrated. The prototype of f is as follows.

```
float f (vector x, int m);
```

When f is called with the $m \times 1$ vector x, it must return the scalar $f(x)$. The domain of integration D is the following rectangular region:

$$D = \{x \in R^m : a_k \leq x_k \leq b_k, 1 \leq k \leq m\} \tag{2.11.6}$$

The value returned by *monte* is an estimate of $I(D)$. The standard deviation of the estimate is proportional to $1/\sqrt{n}$.

EXAMPLE 2.11.1 Integration

Suppose a sequence of random numbers has a Gaussian distribution with mean $\mu = 0$ and standard deviation $\sigma = 1$. Then the probability density function is

$$f(x) = \frac{1}{\sqrt{2\pi}} \exp\left(\frac{-x^2}{2}\right)$$

The probability that a number lies in the interval $[a, b]$ is then

$$I(a,b) = \int_a^b f(x)\, dx$$

There is no closed-form expression for the integral $I(a,b)$, so it must be computed numerically. The following program estimates the value of $I(-1,1)$ using a variety of integration techniques. As a rough check, it also estimates $I(-1,1)$ directly by generating a random sequence and computing the fraction of numbers that lie within $[-1,1]$.

```
/*----------------------------------------------------------------*/
/* Example a2.11.1: Cumulative Probability Distribution*/
/*----------------------------------------------------------------*/
#include "c:\nlib\util.h"

float f (float x)
{
   return exp(-x*x/2)/sqrt(2*PI);
}

int main (void)
{
   int          n[6]={100,100,100,10,10,30000},
                i;
   float        a   = -1,            /* lower limit */
                b   =  1,            /* upper limit */
                tol =  1.e-6,        /* truncation error */
                y = 0,
                x;
   printf ("\ntrapint  = %.7f",trapint(a,b,n[0],f));
   printf ("\nSimpson  = %.7f",simpson(a,b,n[1],f));
   printf ("\nmidpoint = %.7f",midpoint(a,b,n[2],f));
   printf ("\nRomberg  = %.7f",romberg(a,b,tol,n[3],&n[3],f));
   printf ("\nGauss    = %.7f",gausquad(a,b,n[4],f));
   randinit (1000);
   for (i = 0; i < n[5]; i++)
   {
      x = randg (0,1);
      if ((x >= a) && (x <= b)) y++;
   }
   printf ("\nDirect    = %.7f",y/n[5]);
```

```
   shownum ("Extrapolation level",n[3]);
   return 0;
}
/*------------------------------------------------------------------*/
```

The output from program *a2111* is summarized in Table 2.11.3. Notice that Romberg's method converges to a truncation error of less than $\varepsilon = 10^{-6}$ in only $n = 5$ extrapolations in this case. The last row of Table 2.11.3 is a direct estimate of $I(-1,1)$ based on 30,000 random samples.

TABLE 2.11.3 Approximations to $I(-1,1)$

Method	*Points or level*	$I(-1,1)$
Trapezoid	100	0.6826733
Simpson	100	0.6826893
Midpoint	100	0.6826976
Romberg	5	0.6826895
Gauss	10	0.6826847
Direct	30000	0.6828333

2.12 ORDINARY DIFFERENTIAL EQUATIONS

The module *ode.c* is an application module that implements the ordinary differential equations techniques discussed in Chapter 8. The function prototypes are summarized in Table 2.12.1, and the source code for these functions can be found in the file *ode.c* on the distribution CD.

TABLE 2.12.1 Ordinary Differential Equation Functions

Function Prototype			
Type	*Name*	*Arguments*	*Function Description*
long	*rkf*	(vector $x0$, float $t0$, float $t1$, int n, int m, float tol, long q, vector t, matrix X, float $*e$, void $(*f)$(float t, vector x, int n, vector dx));	Runge-Kutta-Fehlberg
long	*bsext*	(vector $x0$, float $t0$, float $t1$, int n, int m, float tol, long q, vector t, matrix X, float $*e$, void $(*f)$(float t, vector x, int n, vector dx));	Bulirsch-Stoer extrapolation
long	*stiff*	(vector $x0$, float $t0$, float $t1$, int n, int m, float tol, long q, vector t, matrix X, float $*e$, void $(*f)$(float t, vector x, int n, vector dx), void $(*fx)$(float t, vector x, int n, matrix J), void $(*ft)$(float t, vector x, int n, vector y));	semi-implicit Bulirsch-Stoer extrapolation
int	*bvp*	(float $alpha$, float $beta$, float a, float b, float tol, int m, int q, vector x, vector y, float $*e$, float $(*g)$(float x, float y, float dy));	finite difference method

All of the functions in Table 2.12.1, except the last one, compute solutions to the following n-dimensional initial value problem:

$$\frac{dx}{dt} = f(t,x) \quad , \quad x(\alpha) = a \tag{2.12.1}$$

The system of differential equations must be defined in a user-supplied function f, whose prototype is as follows.

```
void f (float t,vector x,int n,vector dx);
```

When f is called with the scalar t and the $n \times 1$ vector x, it must evaluate the right-hand-side function $f(t,x)$ in Equation (2.12.1) and return the result in the $n \times 1$ vector dx.

rkf The function *rkf* implements the variable step size Runge-Kutta-Fehlberg method as summarized in Alg. 8.4.1. It computes a solution on a set of m points equally spaced over the interval $[t0, t1]$. On entry to *rkf:* the $n \times 1$ vector $x0$ is the initial state, $t0$ is the initial time, $t1$ is the final time, $n \geq 1$ is the number of equations, $m \geq 2$ is the number of solution points, $tol \geq 0$ is an upper bound on the norm of the local truncation error, $q \geq 1$ is the maximum number of scalar function evaluations, and f is the name of the user-supplied function that defines the equations to be solved.

On exit from *rkf:* the $m \times 1$ vector t contains the equally spaced solution times, the $n \times m$ matrix X contains the solution points, and e points to the norm of the maximum truncation error. The kth column of X contains the solution at time $t[k]$. Thus, the row vector $X[j]$ is the solution to the jth differential equation. The value returned by *rkf* is the number of scalar function evaluations. If it is less than the user-specified maximum, q, the following termination criterion was satisfied.

$$\|E\| < \max\{\|x\|, 1\}tol \tag{2.12.2}$$

Here E is the estimated local truncation error computed in Equation (8.4.7h). Note that when the solution x is large in the sense that $\|x\| > 1$, (2.12.2) becomes a *relative* error bound; otherwise, it is an *absolute* error bound. One evaluation of the $n \times 1$ vector $f(t,x)$ is counted as n scalar function evaluations.

bsext The function *bsext* is similar to *rkf* but uses the Bulirsch-Stoer extrapolation method in Alg. 8.6.1. It computes a solution to (2.12.1) at a set of m points equally spaced over the time interval $[t0, t1]$. The calling sequence for *bsext* is identical to that for *rkf*, so the two functions can be used interchangeably. Typically, *bsext* is used when $f(t,x)$ is relatively smooth and high accuracy is required.

stiff The function *stiff* is designed for stiff systems of differential equations—equations whose solutions contain terms with time scales that differ by more than an order of magnitude. The function *stiff* implements the semi-implicit Bulirsch-Stoer extrapolation method summarized in Alg. 8.7.1. It computes a solution to (2.12.1) at a set of m points equally spaced over the time interval $[t0, t1]$. On entry to *stiff:* the $n \times 1$ vector $x0$ is the initial state, $t0$ is the initial time, $t1$ is the final time, $n \geq 1$ is the number of equations, $m \geq 2$ is the number of solution points, $tol \geq 0$ is an upper bound on the norm of the local truncation error, q is the maximum number of scalar function evaluations, f is the name of the user-supplied function that defines the equations to be solved, fx is the name of an optional user-supplied function that computes the $n \times n$ Jacobian matrix $J(t,x) = \partial f(t,x)/\partial x$ in (8.7.5), and ft is the name of an optional user-supplied function that computes the $n \times 1$ vector $\partial f(t,x)/\partial t$. The prototypes of f, fx, and ft are as follows.

```
void f  (float t,vector x,int n,vector dx);
void fx (float t,vector x,int n,matrix J);
void ft (float t,vector x,int n,vector y);
```

When f is called with the scalar t and the $n \times 1$ vector x, it must evaluate the right-hand-side function $f(t,x)$ in (2.12.1) and return the result in the $n \times 1$ vector dx. When fx is called, it must evaluate the Jacobian matrix defined in Equation (8.7.5) and return the result in the $n \times n$ matrix J. Finally, when ft is called, it must evaluate $\partial f(t,x)/\partial t$ and return the result in the $n \times 1$ vector y. The user functions fx and ft are *optional* in the sense that *stiff* can be called with these arguments set to the symbolic constant NULL defined in the header file *stdio.h*. When this is done, the partial derivatives are evaluated numerically, using central differences. Thus, the user does *not* have to supply the functions fx and ft in this case. Numerical approximations to partial derivatives are provided as a convenience, but they tend to be less accurate, and they can require additional function evaluations.

On exit from *stiff:* the $m \times 1$ vector t contains the equally-spaced solution times, the $n \times m$ matrix X contains the solution points, and e points to the norm of the maximum truncation error. The kth column of X contains the solution at time $t[k]$. Consequently, the row vector $X[j]$ is the solution to the jth differential equation. The value returned by *stiff* is the number of scalar function evaluations. If it is less than the user-specified maximum, q, the termination criterion in (2.12.2) was satisfied.

bvp The function *bvp* uses the finite difference method to solve the following second-order nonlinear two-point boundary value problem:

$$\frac{d^2y}{dx^2} = g\left(x, y, \frac{dy}{dx}\right) \quad , \quad y(\alpha) = a,\ y(\beta) = b \tag{2.12.3}$$

On entry to *bvp: alpha* is the initial value of the independent variable x, *beta* is the final value of x, a is the initial value of the dependent variable y, b is the final value of y, $tol \geq 0$ is an upper bound on the norm of the difference between successive solution estimates, $m \geq 1$ is the number of interior solution points, $q \geq 1$ is the maximum number of iterations, and g is the name of a user-supplied function that describes the equation to be solved. The prototype of g is as follows.

```
float g (float x,float y,float dy);
```

When g is called with scalar inputs x, y, and dy, it must return the value of the right-hand-side function $g(x, y, dy/dt)$ in (2.12.3) where $dy = dy/dx$. On exit from *bvp:* the $(m + 2) \times 1$ vector x contains the equally-spaced values of the independent variable, the $(m + 2) \times 1$ vector y contains the estimated solution with $y[1] = a$ and $y[m + 2] = b$, and e points to the norm of the difference between the last two estimates of the vector y. The value returned by *bvp* is the number of iterations performed. If it is less than the user-specified maximum, q, the infinity norm of the difference between the last two estimates of the vector y was less than *tol*. The function *bvp* uses the function *ludec* in *linear.c*.

EXAMPLE 2.12.1 Differential Equations

Some systems of nonlinear differential equations produce periodic steady-state solutions when started from *any* initial condition. One such system is the *van der Pol* oscillator:

$$\frac{dx_1}{dt} = x_2$$

$$\frac{dx_2}{dt} = -x_1 + \mu(1 - x_1^2)x_2$$

The parameter $\mu \geq 0$ controls the degree of nonlinearity. When $\mu = 0$, this system is a linear harmonic oscillator whose steady-state solution is sinusoidal with the amplitude and phase of the solution dependent on the initial condition $x(0)$. When $\mu > 0$, the steady-state solution is periodic but is no longer a pure sinusoid. Furthermore, the periodic steady-state solution, called a *limit cycle*, is independent of the initial condition when $\mu > 0$ and $x(0) \neq 0$. This can be seen by running the following program, which uses *rkf* and *bsext* to compute a solution to the van der Pol oscillator for two different initial conditions, $x0$ and $x1$.

```
*------------------------------------------------------------------*/
/* Example a2.12.1: The van der Pol oscillator                     */
/*-----------------------------------------------------------------*/
#include "c:\nlib\util.h"

void vanderPol (float t,vector x,int n,vector dx)
{
/* Description: van der Pol oscillator */

   float    mu = 2;
   dx[1]  =  x[2];
   dx[2]  =  -x[1] + mu*(1 - x[1]*x[1])*x[2];
}

int main (void)
```

```
{
   long        q       = 20000,              /* maximum evaluations */
               r;                            /* function evaluations */
   int         n       = 2,                  /* system dimension */
               m       = 500;                /* solution points */
   float       alpha   = 0,                  /* initial time */
               beta    = 30.0,               /* final time */
               tol     = 1.e-5,              /* error bound */
               E;                            /* estimated error */
   vector      t       = vec (m,""),         /* solution times */
               x0      = vec (n,"0.5 0"),    /* initial condition */
               x1      = vec (n,"-1 2");     /* initial condition */
   matrix      X       = mat (2,m,""),       /* plot points x */
               Y       = mat (2,m,""),       /* plot points y */
               Z       = mat (n,m,"");       /* solution points */

/* Solve system from x = x0 */

   r = rkf (x0,alpha,beta,n,m,tol,q,t,Z,&E,vanderPol);
   shownum ("Number of scalar function evaluations",r);
   shownum ("Maximum truncation error",E);
   copyvec (Z[1],m,X[1]);
   copyvec (Z[2],m,Y[1]);

/* Solve system from x = x1 */

   r = bsext (x1,alpha,beta,n,m,tol,q,t,Z,&E,vanderPol);
   shownum ("Number of scalar function evaluations",r);
   shownum ("Maximum truncation error",E);
   copyvec (Z[1],m,X[2]);
   copyvec (Z[2],m,Y[2]);

/* Graph two solution trajectories */

   graphmm (X,Y,2,m,"","x_1","x_2","fig2121",0);
   return 0;
}
/*------------------------------------------------------------------*/
```

The call to *graphmm* in program *a2121* creates a MATLAB script file *fig2121.m*. When *fig2121* is executed from the MATLAB prompt, it produces the plot of x_2 versus x_1 shown in Figure 2.12.1. Notice that one solution starts out outside the limit cycle and spirals in with increasing time, while the other starts out inside the limit cycle and spirals out.

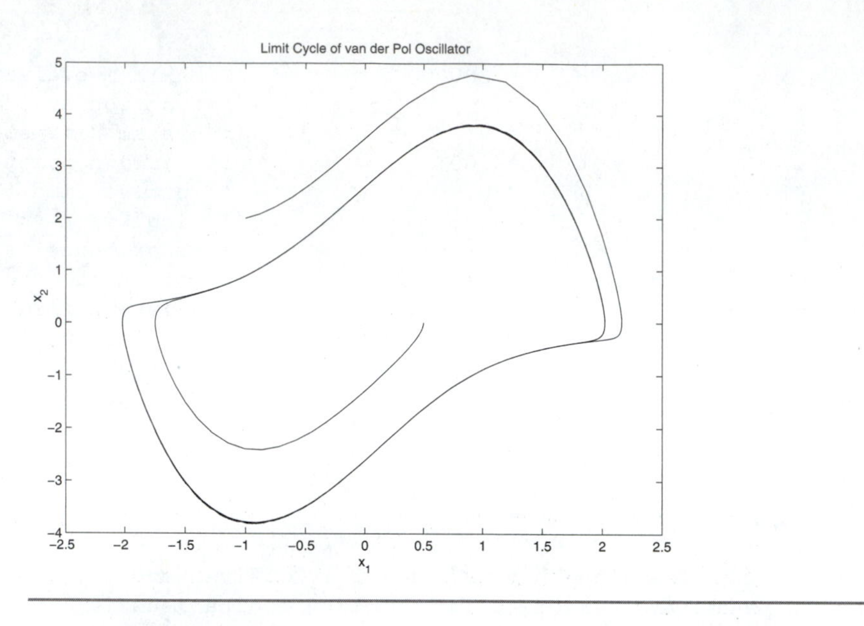

FIGURE 2.12.1 Limit Cycle of the van der Pol Oscillator

2.13 PARTIAL DIFFERENTIAL EQUATIONS

The module *pde.c* is an application module that implements the partial differential equations techniques discussed in Chapter 9. The function prototypes are summarized in Table 2.13.1, and the source code for these functions can be found in the file *pde.c* on the distribution CD. It should be emphasized that it is difficult to formulate general functions for solving partial differential equations given the variety of equations that occur and the different combinations of initial and boundary conditions that are possible. The functions in Table 2.13.1 should be regarded as representative examples that can be used to solve some important classes of partial differential equations subject to common types of initial and boundary conditions. Since the source code for these functions is available, the user is encouraged to use these functions as a starting point to develop additional functions applicable to more general equations and subject to additional types of initial and boundary conditions.

Each of the partial differential equation functions in Table 2.13.1 contains arguments, f and g, which are the names of user-supplied functions. These functions allow the user to specify such things as forcing terms, initial values, and boundary values. In each case, the arguments f and g are *optional* in the sense that either can be replaced by the symbolic constant NULL defined in the header file *stdio.h*. When this is done, the corresponding function is assumed to be zero, which means that a user-supplied function is *not* required in this case.

poisson The function *poisson* uses central difference approximations and the *SOR* method to solve Poisson's equation:

$$\frac{\partial^2 u}{\partial x^2} + \frac{\partial^2 u}{\partial y^2} = f \tag{2.13.1}$$

over the rectangular region

$$R = \{x, y) | 0 < x < a, 0 < y < b\} \tag{2.13.2}$$

subject to the Dirichlet boundary condition, $u(x, y) = g(x, y)$. On entry to *poisson:* $a > 0$ is the maximum value of x, $b > 0$ is the maximum value of y, $m \geq 1$ is the number of steps in x, $n \geq 1$ is the number of steps in y, $q \geq 1$ is the maximum number of iterations, $tol \geq 0$ is an upper bound on the norm of the difference between successive solution estimates, and f and g are the names of user-supplied functions that specify the forcing term and boundary conditions, respectively. The prototypes for f and g are

```
float f (float x,float y);
float g (float x,float y);
```

When f is called with inputs x and y, it must return the value of the forcing term $f(x, y)$ on the right-hand side of Equation (2.13.1). When g is called with (x, y) on the boundary of R, it must return the desired boundary value $u(x, y)$. Note that when the argument f is replaced with the symbolic constant NULL, Poisson's equation reduces to *Laplace's* equation, also called the potential equation. On exit from *poisson: alpha* points to the optimal relaxation parameter computed using (9.2.18), the $m \times 1$ vector x contains the x grid points, the $n \times 1$ vector y contains the y grid points, and the $m \times n$

TABLE 2.13.1 Partial Differential Equation Functions

Function Prototype			*Function*
Type	*Name*	*Arguments*	*Description*
int	*poisson*	(float a, float b, int m, int n, int q, float tol, float $*alpha$, vector x, vector y, matrix U, float $(*f)$(float x, float y) float $(*g)$(float x, float y));	Poisson's equation
void	*heat1*	(float T, float a, int m, int n, float $beta$, float $c1$, float $c2$, vector t, vector x, matrix U, float $(*f)$(float x), float $(*f)$(float x), float $(*g)$(float t, int k));	one-dimensional heat equation
void	*heat2*	(float T, float a, float b, int m, int n, int p, float $beta$,vector x, vector y, matrix U, float $(*f)$(float x, float y), float $(*g)$(float t, float x, float y));	two-dimensional heat equation
void	*wave1*	(float T, float a, int m, int n, float $beta$, vector t, vector x, matrix U, float $(*f)$(float x), float $(*g)$(float x));	one-dimensional wave equation
void	*wave2*	(float T, float a, float b, int m, int n, int p, float $beta$, vector x, vector y, matrix U, float $(*f)$(float x, float y), float $(*g)$(float x, float y));	two-dimensional wave equation

matrix U contains the solution, $u(x, y)$, evaluated at the mn interior grid points. That is, for $1 \le k \le m$ and $1 \le j \le n$,

$$U[k][j] = u(x[k], y[j]) \tag{2.13.3a}$$

$$x[k] = \frac{ka}{m+1} \tag{2.13.3b}$$

$$y[j] = \frac{jb}{n+1} \tag{2.13.3c}$$

The value returned by *poisson* is the number of iterations performed. If it is less than the user-specified maximum, q, the following termination criterion was satisfied where U^i is the ith estimate of the solution.

$$\max_{k=1}^{m} \max_{j=1}^{n} \{|U^i[k][j] - U^{i-1}[k][j]|\} < tol \tag{2.13.4}$$

The estimate, U^0, used to start the *SOR* iterative procedure is the average value of $u(x, y)$ along the boundary of R.

heat1 The next two functions in Table 2.13.1 are used to solve the heat or diffusion equation. The function *heat1* uses the Crank-Nicolson method to solve the one-dimensional heat equation

$$\frac{\partial u}{\partial t} = \beta\left(\frac{\partial^2 u}{\partial x^2}\right) \tag{2.13.5}$$

over the rectangular region

$$R = \{(t, x) | 0 \le x \le T, 0 < x < a\} \tag{2.13.6}$$

subject to a user-selectable combination of Dirichlet and Neumann boundary conditions and the following initial condition.

$$u(0, x) = f(x) \quad , \quad 0 < x < a \tag{2.13.7}$$

On entry to *heat:* $T > 0$ is the maximum value of t, $a > 0$ is the maximum value of x, $m \ge 1$ is the number of steps in t, $n \ge 1$ is the number of steps in x, $beta \ge 0$ is the heat equation coefficient, $c1$ and $c2$ are parameters in the range $[0, 1]$ that select the type of boundary condition, and f and g are the names of user-supplied functions that specify the initial condition and boundary conditions, respectively. The prototypes for f and g are

```
float f (float x);
float g (float x, int k);
```

When f is called with input x, it must return the initial value $u(0, x)$. The function g is used to specify the right-hand side of the following two general boundary conditions.

$$(1 - c1)u(t, 0) - c1u_x(t, 0) = g(t, 1) \tag{2.13.8a}$$

$$(1 - c2)u(t, a) + c2u_x(t, a) = g(t, 2) \tag{2.13.8b}$$

Here $u_x = \partial u/\partial x$. Note that when $c1 = 0$, (2.13.8a) reduces to the Dirichlet boundary condition $u(t,0) = g(t,1)$. Similarly, when $c2 = 1$, (2.13.8b) reduces to the Neumann boundary condition $u_x(t,a) = g(t,2)$. On exit from *heat1:* the $(m + 1) \times 1$ vector t contains the t grid points, the $n \times 1$ vector x contains the x grid points, and the $(m + 1) \times n$ matrix U contains the solution, $u(t, x)$, evaluated at the $(m + 1)n$ interior grid points. That is, for $1 \leq k \leq (m + 1)$ and $1 \leq j \leq n$,

$$U[k][j] = u(t[k], x[j]) \tag{2.13.9a}$$

$$t[k] = \frac{(k - 1)T}{m} \tag{2.13.9b}$$

$$x[j] = \frac{ja}{n + 1} \tag{2.13.9c}$$

The Crank-Nicolson method is unconditionally stable and second-order accurate in both t and x. To obtain good accuracy, the gain parameter γ in Equation (9.3.9) should not be too large. It is recommended that the calling parameters be selected such that $\gamma \leq 1$ where

$$\gamma = \frac{\beta(n + 1)^2 T}{ma^2} \tag{2.13.10}$$

heat2 The function *heat2* uses the alternating direction implicit (ADI) method to solve the two-dimensional heat or diffusion equation

$$\frac{\partial u}{\partial t} = \beta\left(\frac{\partial^2 u}{\partial x^2} + \frac{\partial^2 u}{\partial y^2}\right) \tag{2.13.11}$$

over the rectangular region

$$R = \{(t, x, y) | 0 \leq t \leq T, 0 < x < a, 0 < y < b\} \tag{2.13.12}$$

subject to the Dirichlet boundary condition $u(t, x, y) = g(t, x, y)$ and the following initial condition where $0 < x < a$ and $0 < y < b$.

$$u(0, x, y) = f(x, y) \tag{2.13.13}$$

On entry to *heat2*: $T > 0$ is the maximum value of t, $a > 0$ is the maximum value of x, $b > 0$ is the maximum value of y, $m \geq 2$ is the number of steps in t, $n \geq 1$ is the number of steps in x, $p \geq 1$ is the number of steps in y, $beta \geq 0$ is the heat equation coefficient, and f and g are the names of user-supplied functions that specify the initial conditions and boundary conditions, respectively. The prototypes for f and g are

```
float f (float x,float y);
float g (float t,float x,float y);
```

When f is called with inputs x and y, it must return the initial value $u(0, x, y)$. When g is called with (x, y) on the boundary of R, it must return the desired boundary value $u(t, x, y)$. On exit from *heat2*: the $n \times 1$ vector x contains the x grid points, the $p \times 1$

vector y contains the y grid points, and the $n \times p$ matrix U contains the final solution, $u(T, x, y)$, at the np interior grid points. That is, for $1 \le i \le n$ and $1 \le j \le p$,

$$U[i][j] = u(T, x[i], y[j]) \tag{2.13.14a}$$

$$x[i] = \frac{ia}{n+1} \tag{2.13.14b}$$

$$y[j] = \frac{jb}{p+1} \tag{2.13.14c}$$

The ADI method is an unconditionally stable implicit method. However, to obtain accurate results, the gain parameters γ_x and γ_y in Equation (9.4.10) should not be too large. It is recommended that the calling parameters be selected such that $\gamma_x \le 1$ and $\gamma_y \le 1$ where

$$\gamma_x = \frac{\beta(n+1)^2 T}{ma^2} \tag{2.13.15a}$$

$$\gamma_y = \frac{\beta(p+1)^2 T}{mb^2} \tag{2.13.15b}$$

wave1 The last two functions in Table 2.13.1 are used to solve the wave equation. The function *wave1* uses the explicit central difference method to solve the one-dimensional wave equation

$$\frac{\partial^2 u}{\partial t^2} = \beta\left(\frac{\partial^2 u}{\partial x^2}\right) \tag{2.13.16}$$

over the rectangular region

$$R = \{(t,x) | 0 \le x \le T, 0 < x < a\} \tag{2.13.17}$$

subject to the zero boundary condition $u(t, x) = 0$, and the following initial conditions for $0 < x < a$.

$$u(0,x) = f(x) \tag{2.13.18a}$$

$$u_t(0,x) = g(x) \tag{2.13.18b}$$

On entry to *wave1:* $T > 0$ is the maximum value of t, $a > 0$ is the maximum value of x, $m \ge 1$ is the number of steps in t, $n \ge 1$ is the number of steps in x, $beta > 0$ is the wave equation coefficient, and f and g are the names of user-supplied functions that specify the initial conditions on u and its time derivative, u_t, respectively. The prototypes for f and g are

```
float f (float x);
float g (float x);
```

When f is called with input x, it must return the initial value $u(0, x)$, and when g is called with input x, it must return the initial value $u_t(0, x)$ where $u_t = \partial u/\partial t$. On exit from *wave1:* the $(m + 1) \times 1$ vector t contains the t grid points, the $n \times 1$ vector x contains the x grid points, and the $(m + 1) \times n$ matrix U contains the solution, $u(t, x)$, evaluated at the $(m + 1)n$ interior grid points. That is, for $1 \le k \le (m + 1)$ and $1 \le j \le n$,

$$U[k][j] = u(t[k], x[j]) \tag{2.13.19a}$$

$$t[k] = \frac{(k - 1)T}{m} \tag{2.13.19b}$$

$$x[j] = \frac{ja}{n + 1} \tag{2.13.19c}$$

The function *wave1* is an explicit method that is stable only when the gain parameter γ in Equation (9.5.24) satisfies $\gamma \le 1$. Therefore, it is the *responsibility* of the user to select the calling arguments to *wave1* such that $\gamma \le 1$ where

$$\gamma = \frac{\beta(n + 1)^2 T^2}{m^2 a^2} \tag{2.13.20}$$

wave2 The function *wave2* uses the explicit central difference method to solve the two-dimensional wave equation

$$\frac{\partial^2 u}{\partial t^2} = \beta\left(\frac{\partial^2 u}{\partial x^2} + \frac{\partial^2 u}{\partial y^2}\right) \tag{2.13.21}$$

over the rectangular region

$$R = \{(t, x, y) | 0 \le t \le T, 0 < x < a,\ 0 < y < b\} \tag{2.13.22}$$

subject to the zero boundary condition $u(t, x, y) = 0$, and the following initial conditions where $0 < x < a$ and $0 < y < b$.

$$u(0, x, y) = f(x, y) \tag{2.13.23a}$$

$$u_t(0, x, y) = g(x, y) \tag{2.13.23b}$$

On entry to *wave2*: $T > 0$ is the maximum value of t, $a > 0$ is the maximum value of x, $b > 0$ is the maximum value of y, $m \ge 2$ is the number of steps in t, $n \ge 1$ is the number of steps in x, $p \ge 1$ is the number of steps in y, *beta* > 0 is the wave equation coefficient, and f and g are the names of user-supplied functions that specify the initial conditions on u and its time derivative, u_t, respectively. The prototypes for f and g are

```
float f (float x,float y);
float g (float x,float y);
```

When f is called with inputs x and y, it must return the initial value $u(0, x, y)$, and when g is called with inputs x and y, it must return the initial value $u_t(0, x, y)$ where $u_t = \partial u / \partial t$. On exit from *wave2:* the $n \times 1$ vector x contains the x grid points, the $p \times 1$ vector y contains the y grid points, and the $n \times p$ matrix U contains the final solution, $u(T, x, y)$, at the np interior grid points. That is, for $1 \le i \le n$ and $1 \le j \le p$,

$$U[i][j] = u(T, x[i], y[j]) \tag{2.13.24a}$$

$$x[i] = \frac{ia}{n+1} \tag{2.13.24b}$$

$$y[j] = \frac{jb}{p+1} \tag{2.13.24c}$$

The function *wave2* is an explicit method that is stable only when the criterion in Equation (9.6.12) is satisfied. Therefore, it is the *responsibility* of the user to select the calling arguments to *wave2* such that

$$\frac{T^2}{m^2} \le \frac{1}{4\beta}\left[\frac{a^2}{(n+1)^2} + \frac{b^2}{(p+1)^2}\right] \tag{2.13.25}$$

EXAMPLE 2.13.1 Partial Differential Equation

As an illustration of the use of the functions in Table 2.13.1 to solve a partial differential equation, consider the problem of finding the steady-state temperature distribution in a metal plate of width $a = 2$ and length $b = 2$. This requires a solution to Laplace's equation:

$$\frac{\partial^2 u}{\partial x^2} + \frac{\partial^2 u}{\partial y^2} = 0$$

Suppose the temperature along the edges of the plate satisfies the following boundary condition.

$$u(x, y) = \cos[\pi(x + y)]$$

The following program uses the function *poisson* from Table 2.13.1 to compute the steady-state temperature, using a grid size of $m = n = 39$.

```
/*-----------------------------------------------------------------*/
/* Example a2.13.1: Steady State Temperature                        */
/*-----------------------------------------------------------------*/
#include "c:\nlib\util.h"

float g (float x,float y)                   /* boundary values */
{
   return cos(PI*(x+y));
}

int main (void)
{
   int          m = 39,                     /* x precision */
                n = 39,                     /* y precision */
                q = 500,                    /* maximum iterations */
                i;
   float        a = 2,                      /* maximum x */
                b = 2,                      /* maximum y */
                tol = 1.e-4,                /* error bound */
```

```
                alpha;                  /* relaxation parameter */
   vector       x = vec (m,""),         /* x grid values */
                y = vec (n,"");         /* y grid values */
   matrix       U = mat (m,n,"");       /* solution */

  printf ("Computing solution ... ");
  shownum ("Iterations",poisson (a,b,m,n,q,tol,&alpha,x,y,U,NULL,g));
  shownum ("Optimal relaxation parameter",alpha);
  plotxyz (x,y,U,m,n,"","x","y","u","fig2131");
  return 0;
}
/*----------------------------------------------------------------*/
```

Note how the symbolic constant NULL is used in place of the calling argument *f*, thereby converting Poisson's equation to Laplace's equation. Using an error tolerance of $\varepsilon = 10^{-4}$, the *SOR* method converges in 83 iterations. The optimal value for the relaxation parameter is $\alpha = 1.851$. A plot of the steady-state temperature profile, produced by running the script file *fig2131.m*, is shown in Figure 2.13.1.

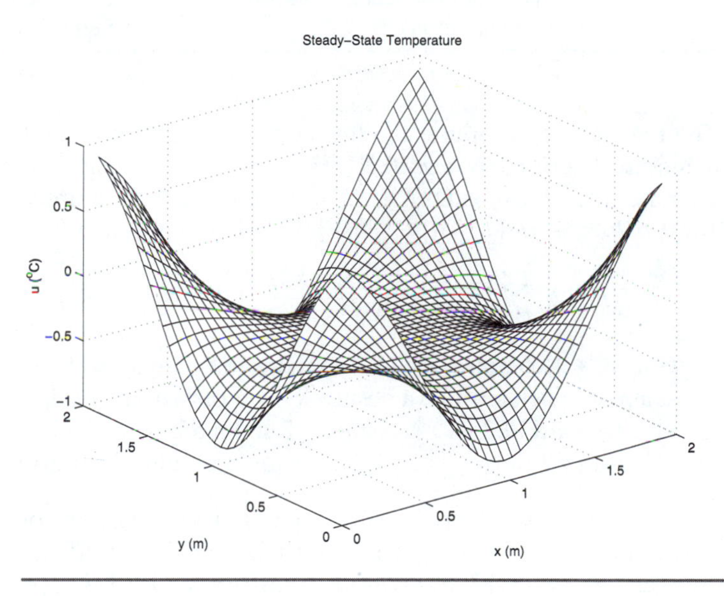

FIGURE 2.13.1 Steady-State Temperature

2.14 DIGITAL SIGNAL PROCESSING

The module *dsp.c* is an application module that implements the digital signal processing techniques discussed in Chapter 10. The function prototypes are summarized in Table 2.14.1, and the source code for these functions can be found in the file *dsp.c* on the distribution CD.

TABLE 2.14.1 Digital Signal Processing Functions

Function Prototype			Function Description
Type	*Name*	*Arguments*	
void	*dft*	(cvector z, int n, int dir)	fast Fourier transform
void	*dft2*	(cmatrix Z, int m, int n, int dir)	2D fast Fourier transform
void	*spectra*	(vector x, int n, float fs, vector a, vector phi, vector f)	magnitude and phase spectra
void	*convolve*	(vector x, vector y, int n, vector z)	convolution
void	*crosscor*	(vector x, vector y, int m, int n, vector rho)	correlation
void	*fir*	(float ($*g$)(float f), int p, vector b)	MA filter
void	*freqrsp*	(vector u, vector y, int n, vector a, vector phi vector f)	frequency response
float	*arma*	(float u, vector $theta$, vector x, int n, int m)	evaluate ARMA model
float	*getarma*	(vector u, vector y, int p, int n, int m, vector $theta$)	least-squares identification: ARMA
float	*lms*	(float u, float y, vector x, vector $theta$, float mu, int m)	adaptive LMS identification: MA

dft The function *dft* computes the fast Fourier transform or FFT using decimation in frequency as described in Alg. 10.3.1. The FFT is an efficient implementation of the discrete Fourier transform or DFT. If $x(k)$ denotes the kth sample of the discrete-time input for $0 \le k < n$ and $i = \sqrt{-1}$, the DFT of $x(k)$ is defined:

$$X(m) \triangleq \sum_{k=0}^{n-1} x(k) \exp\left(\frac{-jmk2\pi}{n}\right) \quad , \quad 0 \le m < n \qquad (2.14.1)$$

On entry to *dft:* the $n \times 1$ complex vector z contains the samples to be transformed; n is the number of samples, which must be a positive power of two; and *dir* is a direction code. If $dir \ge 0$, then the forward FFT of z is computed, and if $dir < 0$, the inverse FFT of z is computed. On exit from *dft:* the $n \times 1$ complex vector z contains the transformed samples.

dft2 The function *dft2* computes the two-dimensional fast Fourier transform or FFT as described in Alg. 10.7.1. The two-dimensional FFT is an efficient implementation of the two-dimensional discrete Fourier transform or DFT. If $z(k,i)$ denotes the (k,i)th sample of the discrete-time input for $0 \le k < m$, and $0 \le i < n$, then the two-dimensional DFT of $x(k,i)$ is defined:

$$X(p,q) \triangleq \sum_{k=0}^{m-1} \sum_{i=0}^{n-1} x(k,i) \exp\left[j2\pi\left(\frac{pk}{m} + \frac{qi}{n}\right)\right] \quad , \quad 0 \le p < m, 0 \le q < n \quad (2.14.2)$$

On entry to *dft2:* the $m \times n$ complex matrix Z contains the samples to be transformed, m is the number of samples in the first dimension, n is the number of samples in the second dimension, and *dir* is a direction code. If $dir \geq 0$, the forward FFT of Z is computed, and if $dir < 0$, the inverse FFT of Z is computed. The number of samples in the first dimension, m, must be a positive power of two, and likewise for the number of samples in the second dimension, n. On exit from *dft2:* the $m \times n$ complex matrix Z contains the transformed samples.

spectra The function *spectra* computes the magnitude and phase spectra of a discrete-time signal $x(k)$. If $X(m) = \text{FFT}\{x(k)\}$ denotes the fast Fourier transform of $x(k)$, the *magnitude* spectrum $A(m)$ and *phase* spectrum $\phi(m)$ are defined by the relationship

$$X(m) = A(m) \exp[j\phi(m)] \quad , \quad 0 \leq m < n \tag{2.14.3}$$

On entry to *spectra:* the $n \times 1$ vector x contains the discrete-time signal; n is the number of samples, which must be a positive power of two; and $fs > 0$ is the sampling frequency in Hz. On exit from *spectra:* the $n \times 1$ vector a contains the magnitude spectrum of x, the $n \times 1$ vector *phi* contains the phase spectrum of x in degrees, and the $n \times 1$ vector f contains the frequencies at which the spectra are evaluated.

$$f[k] = \frac{(k-1)f_s}{n} \quad , \quad 1 \leq k \leq n \tag{2.14.4}$$

Since values of $k > n/2$ correspond to negative frequencies, a and *phi* exhibit symmetry about the point $k = n/2$ because x is real.

convolve The function *convolve* is used to compute the *convolution* of two discrete-time signals. If $x(k)$ and $y(k)$ are the two signals, their convolution is defined:

$$z(i) \triangleq \sum_{k=0}^{i} x(i-k)y(k) \quad , \quad 0 \leq i < n \tag{2.14.5}$$

On entry to *convolve:* the $n \times 1$ vector x contains the samples of the first signal, the $n \times 1$ vector y contains the samples of the second signal, and n is the number of samples. The function *convolve* uses the FFT to efficiently perform a convolution. Consequently, the number samples, n, must be a positive power of two. On exit from *convolve:* the $n \times 1$ vector z contains the convolution of x with y. If x is the pulse response of a linear discrete-time system subject to zero initial conditions, and y is the system input, then z is the system output. An alternative way to find the output of a linear discrete-time system is to use the function *arma*.

crosscor The function *crosscor* is used to compute the *normalized cross correlation* of two discrete-time signals. If $x(k)$ and $y(k)$ denote two discrete-time signals, then the cross correlation of $x(k)$ with $y(k)$ is defined:

$$r_{xy}(i) \triangleq \sum_{k=i}^{n-1} x(k)y(k-i) \quad , \quad 1 \leq i < n \tag{2.14.6}$$

The normalized cross correlation of $x(k)$ with $y(k)$ is found by dividing $r_{xy}(i)$ by its maximum value as follows:

$$\rho_{xy}(i) = \frac{r_{xy}(i)}{\sqrt{r_{xx}(0)r_{yy}(0)}}, \quad 0 \le i < n \tag{2.14.7}$$

On entry to *crosscor:* the $n \times 1$ vector x contains the samples of the first signal, the $m \times 1$ vector y contains the samples of the second signal, n is the number of samples in x, and $m \le n$ is the number of samples in y. The function *crosscor* uses the FFT to efficiently perform the cross correlation. Consequently, n must be a positive power of two. On exit from *crosscor:* the $n \times 1$ vector *rho* contains the normalized cross correlation of x with y. If *crosscor* is called with $y = x$, then *rho* contains the *normalized auto correlation* of x.

fir The function *fir* finds the coefficients of the following linear-phase FIR digital filter, using the Fourier series method described in Alg. 10.6.1.

$$y(k) = \sum_{i=0}^{2p} b_i u(k - i) \tag{2.14.8}$$

On entry to *fir:* g is the name of a user-supplied function that specifies the desired magnitude response of the filter, and p selects the filter order, which is $2p + 1$. The prototype for the function g is

```
float g(float f);
```

When g is called with input frequency f, it must return the desired magnitude response value $g(f)$. The input frequency is expressed as a fraction of the Nyquist frequency f_N, which means that f is normalized to the interval $0 \le f \le 1$. On exit from *fir:* the $(2p + 1)$ vector b contains the coefficients of the filter in (2.14.8). The Hamming window in Table 10.6.2 is used to smooth the ripples in the magnitude response of the filter. Depending on the shape of g, *fir* can be used to design low pass, high pass, band pass, band stop, or more general filters. The filter constructed by *fir* is a linear-phase or constant-delay filter with a delay of p samples. The output of the filter can be computed by calling the function *arma* with $n = 0$, $m = 2p$, and $theta = b$ or by using *convolve*.

freqrsp The function *freqrsp* is used to compute the *frequency response* of a discrete-time linear system. If $u(k)$ denotes the system input and $y(k)$ denotes the system output, the frequency response is $H(m) = \text{FFT}\{y(k)\}/\text{FFT}\{u(k)\}$ or

$$H(m) = A(m) \exp[j\phi(m)] \quad , 0 \le m < n \tag{2.14.9}$$

On entry to *freqrsp:* the $n \times 1$ vector u contains the samples of the input signal, the $n \times 1$ vector y contains the samples of the output signal, n is the number of samples, and $fs > 0$ is the sampling frequency. Since the FFT is used to efficiently compute the frequency response, the number of samples, n, must be a positive power of two. On exit from *freqrsp:* the $n \times 1$ vector a contains the *magnitude response* of the system, the $n \times 1$ vector *phi* contains the phase response of the system in degrees, and the $n \times 1$ vector f contains the frequencies at which the frequency response is evaluated as in (2.14.4).

arma The function *arma* is used to compute the output of the following autoregressive moving average (ARMA) model:

$$y(k) + \sum_{i=1}^{n} \theta_i y(k-i) = \sum_{i=0}^{m} \theta_{n+1+i} u(k-i) \qquad (2.14.10)$$

On entry to *arma:* u is the current value of the system input, the $(n + m + 1) \times 1$ vector *theta* is the parameter vector, and the $(n + m + 1) \times 1$ vector x is the state vector defined as follows.

$$x(k) = [-y(k-1), \dots, -y(k-n), u(k), \dots, u(k-m)]^T \qquad (2.14.11)$$

The remaining arguments are the AR order $n \geq 0$ and the MA order $m \geq 0$. Notice that the current output of the ARMA model, $y(k)$, can be expressed in terms of the parameter vector θ and the current state $x(k)$ using the inner product in (2.3.2) as follows.

$$y(k) = \theta^T x(k) \qquad (2.14.12)$$

On exit from *arma:* the $(n + m + 1) \times 1$ vector x contains the updated state. When *arma* is called, the value of $x[n + 1]$ does not have to be specified because it is replaced by the current input u. The value returned by *arma* is the current value of the system output. The function *arma* computes the output of a linear discrete-time system in real time. If $n = 0$, it evaluates a MA filter, and if $m = 0$, it evaluates an AR filter.

getarma The function *getarma* is used to identify parameter vector, θ, of the ARMA model in (2.14.10) using a least-squares fit. On entry to *getarma:* the $p \times 1$ vector u contains the samples of the input, the $p \times 1$ vector y contains the samples of the output, p is the number of samples, $n \geq 0$ is the AR order, and $m \geq 0$ is the MA order. The number of samples p must satisfy $p \geq n + m + 1$ and should be larger for a reliable fit. On exit from *getarma:* the $(n + m + 1) \times 1$ vector *theta* contains the parameters of the ARMA model. The value returned by *getarma* is the norm of the error in Equation (10.8.7). The function *getarma* uses the function *ludec* from *linear.c* to solve the normal equations.

lms The function *lms* is used to perform real-time identification of the parameters of a MA model, using the least-mean square (LMS) method in Equation (10.8.19). The MA model is described by the following equation, which corresponds to the ARMA model in (2.14.10) with $n = 0$.

$$y(k) = \sum_{i=0}^{m} \theta_{1+i} u(k-i) \qquad (2.14.13)$$

Similarly, the current state of the MA model is an $(m + 1) \times 1$ vector that corresponds to the state vector in (2.14.11), but with $n = 0$.

$$x(k) = [u(k), \dots, u(k-m)]^T \qquad (2.14.14)$$

The output of the MA model is then given by (2.14.12). On entry to *lms:* u is the current input, y is the current output, the $(m+1) \times 1$ vector x is the current state as defined in (2.14.14), the $(m+1) \times 1$ vector *theta* contains the parameters of the MA model, $mu > 0$ is the step size for the steepest descent search, and $m \geq 0$ specifies the order of the MA model. On exit from *lms:* the $(m+1) \times 1$ vector *theta* contains the updated parameter estimates, and the $(m+1) \times 1$ vector x contains the updated state in (2.14.14). When *lms* is called, $x[1]$ does not have to be specified because it is replaced by the current input u. The value returned by *lms* is the current value of the error in Equation (10.8.17). The function *lms* is called each time a new sample becomes available. To ensure convergence, the step size *mu* must satisfy the following constraint where $P(u)$ is the average power of u.

$$0 < mu < \frac{1}{(m+1)P(u)} \tag{2.14.15}$$

EXAMPLE 2.14.1 Digital Signal Processing

Consider the problem of designing a band stop or notch filter with a sampling frequency of $f_s = 1000$ Hz, a pass band gain of one, a low-frequency cutoff of $f_0 = 250$ Hz, and a high-frequency cutoff of $f_1 = 350$ Hz. The following script uses the functions in Table 10.8.2 to perform this task, using an FIR filter of order of $2p + 1 = 61$.

```
/*----------------------------------------------------------------------*/
/* Example a2.14.1: Digital signal processing                           */
/*----------------------------------------------------------------------*/
#include "c:\nlib\util.h"

float g (float f)
{
/* Description: Band stop filter */

   if ((f <= .5) || (f >= .7))
      return 1;
   else
      return 0;
}

int main (void)
{
   int         p = 30,                      /* select order */
               r = 2*p + 1,                 /* number of terms */
               n = 512,                     /* number of samples */
               i;
   float       fs = 1000;                   /* sampling frequencies */
   vector      b   = vec (r,""),            /* filter coefficients */
               x   = vec (r,""),            /* filter state */
               u   = vec (n,"1"),           /* unit pulse input */
               h   = vec (n,"1"),           /* pulse response */
```

```
                    A   = vec (n,""),          /* magnitude response */
                    phi = vec (n,""),          /* phase response */
                    t   = vec (n,""),          /* discrete times */
                    f   = vec (n,"");          /* discrete frequencies */
    fir (g,p,b);
    showvec ("Filter coefficients",b,r,0);
    for (i = 1; i <= n; i++)
    {
       t[i] = (i - 1)/fs;
       h[i] = arma (u[i],b,x,0,2*p);
    }
    graphxy (t,h,3*r/2,"","t (sec)","h","fig2141");
    freqrsp (u,h,n,fs,A,phi,f);
    graphxy (f,A,n/2,"","f (Hz)","A","fig2142");
    return 0;
}
/*------------------------------------------------------------*/
```

A plot of the pulse response of the filter is shown in Figure 2.14.1. Note that it is bounded and goes to zero after 61 samples. All FIR filters are stable and have a pulse response that goes to zero in a finite number of samples. A plot of the magnitude response of the filter is shown in Figure 2.14.2. Observe that the cutoff frequencies are at $f_L = 250$ Hz and $f_H = 350$ Hz per the specifications, the pass band gain is one, and there is very little ripple in the magnitude response.

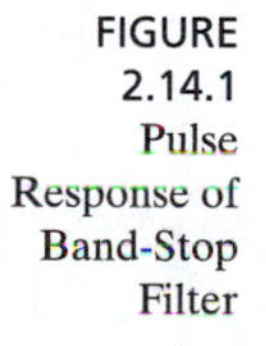

FIGURE 2.14.1 Pulse Response of Band-Stop Filter

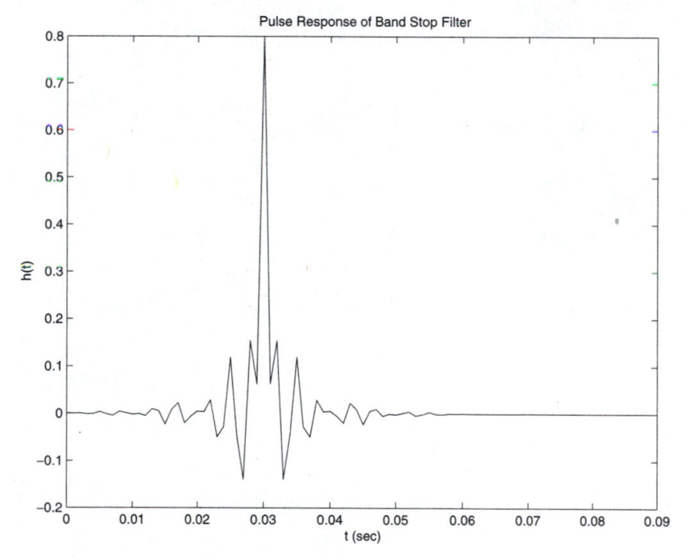

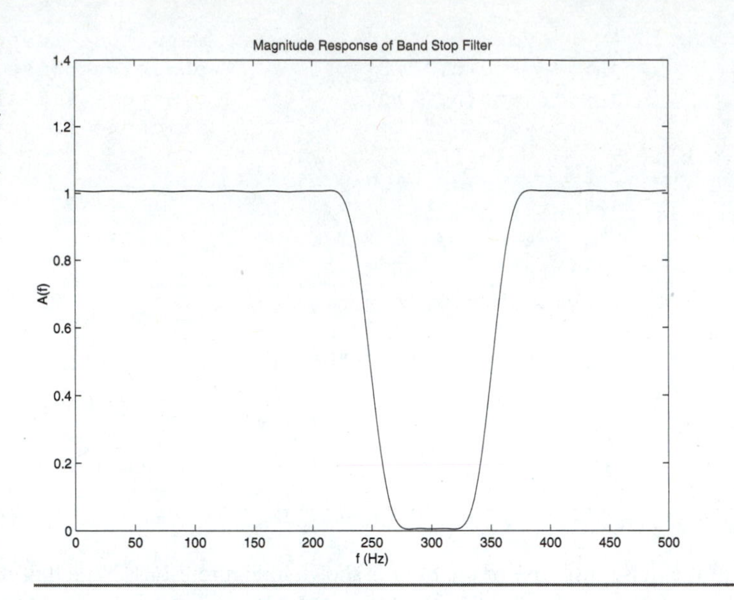

FIGURE 2.14.2 Magnitude Response of Band-Stop Filter

APPENDIX 3

Vectors and Matrices

Vectors and matrices are used extensively in formulating and solving problems in engineering. They provide a natural, elegant, and concise way to describe the essential relationships between problem variables. Indeed, the popular software package MATLAB (short for MATrix LABoratory) uses a matrix as its fundamental underlying data type. Matrix theory is quite elaborate, and there are many books on linear algebra devoted to the topic (see, e.g., Noble, 1969). In this brief appendix, no attempt is made to derive or develop the theory. Instead, only the most essential definitions and basic relationships are summarized. It is *not* necessary for a reader of the text to have taken a course in linear algebra. However, it is *important* that the reader at least be comfortable with the notation of vectors and matrices and be familiar with the basic operations and fundamental definitions.

3.1 VECTOR AND MATRIX NOTATION

Vectors A *vector* is a one-dimensional array of n scalars, which are called the *elements* or components of the vector. The number of elements, n, is the *dimension* of the vector. Vectors are often written by enclosing their elements in square brackets as illustrated by the following examples.

$$x = \begin{bmatrix} x_1 \\ x_2 \\ x_3 \end{bmatrix}, \quad y = \begin{bmatrix} y_1 \\ y_2 \\ \vdots \\ y_n \end{bmatrix}, \quad z = \begin{bmatrix} 4 \\ -2 \\ 5 \\ 1 \end{bmatrix} \tag{3.1.1}$$

The vectors x, y, and z are of dimension 3, n, and 4, respectively. Notice that *single subscripts* are used to denote the elements of the vector. For example, the third element of the constant vector z is $z_3 = 5$. The vectors in (3.1.1) are written as columns of elements and are therefore called *column vectors*. It is also possible to arrange the elements in a row, in which case they are called *row vectors*. To avoid possible ambiguities, *all* vectors are assumed to be column vectors unless specifically noted otherwise. In MATLAB, the kth element of a vector x is denoted $x(k)$, whereas in C, the kth element of a vector x is denoted $x[k]$.

Matrices A *matrix* is a two-dimensional array of scalar elements arranged as m rows by n columns. As with vectors, matrices are often written by enclosing their elements in square brackets as illustrated by the following examples.

$$A = \begin{bmatrix} A_{11} & A_{12} & A_{13} \\ A_{21} & A_{22} & A_{23} \end{bmatrix}, \quad B = \begin{bmatrix} B_{11} & b_{12} \\ B_{21} & B_{22} \\ B_{31} & B_{32} \end{bmatrix}, \quad C = \begin{bmatrix} 7 & -10 \\ 5 & -8 \end{bmatrix} \tag{3.1.2}$$

Here A is a 2×3 matrix, B is a 3×2 matrix, and C is a 2×2 constant matrix. In this case, *double subscripts* are used to denote the elements of the matrix, with the first subscript specifying the row and the second specifying the column. Thus, $C_{21} = 5$ denotes the element of the matrix C appearing in row 2 and column 1. In MATLAB, the element of a matrix A appearing in the kth row and jth column is denoted $A(k, j)$, while in C, it is denoted $A[k][j]$. Notice that a vector can be thought of as a special case of a matrix. Indeed, a column vector of dimension m is an $m \times 1$ matrix, while a row vector of dimension n is a $1 \times n$ matrix.

There are occasions when it is useful to distinguish between members of a sequence of vectors $\{x^1, x^2, \ldots\}$. For example, if the solution to a problem is an $n \times 1$ vector x, then the kth approximation to the solution produced by an iterative method might be represented as x^k. In this case, a *superscript* k is used to denote the kth vector in the sequence. Thus, subscripts are reserved to denote elements of vectors and matrices, while superscripts are used to distinguish between members of a set of vectors. In keeping with this approach, if A is a matrix, then the corresponding lower-case letter with a superscript, a^j, can be used to represent the jth column of A. For example, an alternative way to write the 2×3 matrix A in (3.1.2) in terms of its three columns is

$$A = [a^1, a^2, a^3] \tag{3.1.3}$$

The use of superscripts brings with it the potential for confusion with taking a scalar to an integer power. In those rare instances where the context does not make the meaning of the superscript unambiguous, one can always use parentheses to denote a power as in $(x_1)^2$ to indicate the square of the first component of the vector x. Some authors use bold-faced fonts to distinguish vectors and matrices from scalars. This can be effective for printed material, but it is not a viable approach when writing on a blackboard or in a student notebook.

Transpose The *transpose* of an $m \times n$ matrix A is an $n \times m$ matrix denoted A^T that is obtained from A by interchanging its rows and columns. Therefore, the transpose of the 2×3 matrix A in (3.1.2) is the following 3×2 matrix.

$$A^T \triangleq \begin{bmatrix} A_{11} & A_{21} \\ A_{12} & A_{22} \\ A_{13} & A_{23} \end{bmatrix} \tag{3.1.4}$$

Here the notation, $\triangleq$, means equals by definition. Note that interchanging the rows with the columns is equivalent to interchanging the subscripts, $A^T_{kj} = A_{jk}$. A matrix A is said to be *symmetric* if it is equal to its transpose, $A^T = A$. Since a vector is a special case of a matrix, it is also possible to transpose a vector. For example, a space-saving way to write the 3×1 column vector x in (3.1.1) is as a transposed row vector.

$$x = [x_1, x_2, x_3]^T \tag{3.1.5}$$

Special Matrices An $m \times n$ matrix is *square* if the number of rows equals the number of columns, $m = n$. The elements $\{A_{11}, A_{22}, \ldots, A_{nn}\}$ starting in the upper left corner and ending in the lower right corner are called the *diagonal elements* of the matrix. A matrix A is a *diagonal* matrix if all of its elements are zero except for the diagonal elements. The most important diagonal matrix is the *identity* matrix, denoted I, which has ones along the diagonal. For example, the 4×4 identity matrix is as follows.

$$I \triangleq \begin{bmatrix} 1 & 0 & 0 & 0 \\ 0 & 1 & 0 & 0 \\ 0 & 0 & 1 & 0 \\ 0 & 0 & 0 & 1 \end{bmatrix} \tag{3.1.6}$$

Matrices that have zeros below the diagonal are called *upper-triangular* matrices, while matrices that have zeros above the diagonal are *lower-triangular* matrices. *Sparse* matrices are matrices with a pattern of zeros. Finally, an $m \times n$ matrix of zeros is called the *zero* matrix and is denoted as 0.

Partitioned Matrices Two matrices A and B that have the same number of rows can be stacked side by side to form a third *partitioned* matrix.

$$C = [A, B] \tag{3.1.7}$$

In this case, if A is $m \times n$ and B is $m \times p$, the partitioned matrix C is $m \times (n + p)$ where the first n columns of C are the columns of A, and the last p columns of C are the columns of B. For example, the 3×2 matrix B in (3.1.2) can be stacked next to the 3×1 vector x in (3.1.2) to form the following 3×3 partitioned matrix.

$$[B, x] = \left[\begin{array}{cc|c} B_{11} & B_{12} & x_1 \\ B_{21} & B_{22} & x_2 \\ B_{31} & B_{32} & x_3 \end{array}\right] \tag{3.1.8}$$

It is also possible to stack matrices that have the same number of columns, one on top of the other, to form a partitioned matrix.

3.2 BASIC OPERATIONS

Scalar Multiplication and Division Suppose α is a scalar, x is a vector, and A is a matrix. Then αx is a new vector of the same dimension as x that is obtained by multiplying each element of x by α. A similar remark holds for the matrix αA, which is obtained by multiplying each element of A by α. Thus, for the vector x in (3.1.1) and the matrix A in (3.1.2), we have

$$\alpha x \triangleq \begin{bmatrix} \alpha x_1 \\ \alpha x_2 \\ \alpha x_3 \end{bmatrix}, \quad \alpha A \triangleq \begin{bmatrix} \alpha A_{11} & \alpha A_{12} & \alpha A_{13} \\ \alpha A_{21} & \alpha A_{22} & \alpha A_{23} \end{bmatrix} \tag{3.2.1}$$

In terms of elements, we can write multiplication by a scalar as $(\alpha x)_k = \alpha x_k$ and $(\alpha A)_{kj} = \alpha A_{kj}$. A similar procedure holds for division by α as long as $\alpha \neq 0$.

Matrix Addition and Subtraction Matrices of the *same size* can be added or subtracted from one another to form new matrices. Since vectors are a special case of matrices, this also applies to vectors. For example, if C is the 2×2 constant matrix in (3.1.2) and D is a general 2×2 matrix, then

$$D + C \triangleq \begin{bmatrix} D_{11} + 7 & D_{12} - 10 \\ D_{21} + 5 & D_{22} - 8 \end{bmatrix} \tag{3.2.2}$$

Again, in terms of the elements, we can write matrix addition as $(C + D)_{kj} = C_{kj} + D_{kj}$. Subtraction of matrices is defined in a similar manner with $(C - D)_{kj} = C_{kj} - D_{kj}$.

Matrix Multiplication To examine the notion of matrix multiplication, it is helpful to first consider the special case of multiplying $1 \times n$ row vector x^T times an $n \times 1$ column vector y.

$$x^T y \triangleq x_1 y_1 + x_2 y_2 + \cdots + x_n y_n \tag{3.2.3}$$

The result is a scalar called the *dot product* or inner product of the vector x with the vector y. Thus, the dot product is the sum of the products of the elements of the two vectors. More generally, if A is an $m \times n$ matrix and B is an $n \times p$ matrix, then the matrix product, AB, is an $m \times p$ matrix. The element of AB appearing in the kth row and jth column is the dot product of the kth row of A with the jth column of B. That is,

$$(AB)_{kj} \triangleq A_{k1}B_{1j} + A_{k2}B_{2j} + \cdots + A_{kn}B_{nj} \quad , \quad 1 \le k \le m, 1 \le j \le p \tag{3.2.4}$$

Notice that the matrix A is *compatible* for multiplication with the matrix B only if the number of columns of A is equal to the number of rows of B. That is, the number of columns of the first matrix must equal the number of rows of the second matrix for multiplication to make sense. When two matrices are compatible for multiplication, their product is a matrix that has as many rows as the first matrix and as many columns as the second matrix.

$$\underbrace{A}_{m \times n} \quad \underbrace{B}_{n \times p} = \underbrace{C}_{m \times p} \tag{3.2.5}$$

As a simple illustration, the product of a 2×3 matrix A times a 3×2 matrix B is a 2×2 matrix C computed as follows.

$$\begin{aligned} C &= AB \\ &= \begin{bmatrix} 1 & -2 & 4 \\ 5 & 3 & -6 \end{bmatrix} \begin{bmatrix} 9 & 4 \\ 0 & 7 \\ 6 & -1 \end{bmatrix} \\ &= \left[\begin{array}{c|c} 1(9) - 2(0) + 4(6) & 1(4) - 2(7) + 4(-1) \\ \hline 5(9) + 3(0) - 6(6) & 5(4) + 3(7) - 6(-1) \end{array}\right] \\ &= \begin{bmatrix} 33 & -14 \\ 9 & 47 \end{bmatrix} \end{aligned} \tag{3.2.6}$$

Matrix multiplication is *not* a commutative operation, which means that $AB \neq BA$ in general. Indeed, the product BA may not even make sense due to incompatibility in the sizes of B and A. The transpose of the product of two matrices is related to the transposes of the individual matrices as follows.

$$(AB)^T = B^T A^T \tag{3.2.7}$$

Thus, the transpose of the product is the product of the transposes, but in reverse order.

An important special case of matrix multiplication is multiplication of an $m \times n$ matrix A times an $n \times 1$ column vector x. The result is an $m \times 1$ column vector Ax that can be expressed in terms of the columns of A as follows.

$$Ax = x_1 a^1 + x_2 a^2 + \cdots + x_n a^n \tag{3.2.8}$$

Thus, the product Ax can be thought of as a *linear combination* or weighted sum of the columns of A with the weighting coefficients being the elements of x.

3.3 MATRIX INVERSE

Linear Independence and Rank In certain cases, there is a matrix-equivalent to the reciprocal of scalar called the matrix inverse. To determine when a matrix inverse exists, we must first consider a preliminary notion. A set of vectors $\{x^1, x^2, \ldots, x^n\}$ is a *linearly dependent* set if there exists a nonzero $n \times 1$ vector c such that

$$c_1 x^1 + c_2 x^2 + \cdots + c^n x^n = 0 \tag{3.3.1}$$

Since $c \neq 0$, it follows that at least one of the vectors in a linearly dependent set can be expressed as a linear combination or weighted sum of the other vectors. Any set of vectors that contains the zero vector is linearly dependent. If no nonzero vector c exists satisfying (3.3.1), the set of vectors is said to be *linearly independent.*

The *rank* of an $m \times n$ matrix A is the number of linearly independent columns in the matrix. Since $\text{rank}(A) = \text{rank}(A^T)$, the rank is also equal to the number of linearly independent rows. For a general $m \times n$ matrix A, the rank of A lies in the following range.

$$0 \le \text{rank}(A) \le \min(m, n) \tag{3.3.2}$$

That is, the rank can never be larger than the *minimum* of the number of rows and the number of columns. When $\text{rank}(A) = \min(m, n)$, we say that A is of *full rank*.

Inverse The remainder of this appendix is applicable only to square matrices. For every square $n \times n$ matrix A that is of full rank, there is a unique $n \times n$ matrix A^{-1} called the *inverse* of A that satisfies the following equation where I is the $n \times n$ identity matrix as in (3.1.6).

$$A^{-1}A = AA^{-1} = I \tag{3.3.3}$$

Thus, multiplication of a matrix on the left or the right by its inverse produces the identity matrix. If $\text{rank}(A) < n$, the inverse of A does not exist.

Determinant One way to compute the inverse of a matrix A is in terms of the *determinant,* which is a scalar denoted $\det(A)$. For a 2×2 matrix A, the determinant of A is defined as follows.

$$\det(A) \triangleq A_{11}A_{22} - A_{12}A_{21} \tag{3.3.4}$$

Thus, the determinant of a 2×2 matrix is the product of the diagonal elements minus the product of the off-diagonal elements. For a general $n \times n$ matrix A, the determinant can be defined recursively as follows.

$$\det(A) \triangleq \sum_{j=1}^{n} (-1)^{j-1} A_{1j} \det(B^{1j}) \tag{3.3.5}$$

Here the notation B^{1j} denotes the $(n-1) \times (n-1)$ submatrix of A obtained by eliminating the row one and column j. The operation in (3.3.5) is referred to as expanding the determinant along the first row of A. It is also possible to expand it along other rows or the columns of A. Notice that if the determinant of a scalar is defined as the scalar itself, $\det(\alpha) = \alpha$, when the expansion in (3.3.5) is applied to a 2×2 matrix A, it yields the simplified expression in (3.3.4).

A square matrix A is a *singular* matrix if and only if $\det(A) = 0$. Otherwise, A is said to be a *nonsingular* matrix. A matrix is invertible if and only if it is nonsingular. For the case of a 2×2 matrix A, the inverse can be expressed explicitly in terms of the determinant as follows.

$$\begin{bmatrix} A_{11} & A_{12} \\ A_{21} & A_{22} \end{bmatrix}^{-1} = \frac{1}{\det(A)} \begin{bmatrix} A_{22} & -A_{12} \\ -A_{21} & A_{22} \end{bmatrix} \tag{3.3.6}$$

Here $\det(A)$ is as given in (3.3.4). Thus, to invert a 2×2 matrix, one simply interchanges the diagonal elements, changes the signs of the off-diagonal elements, and divides by the determinant. As an illustration, consider the 2×2 matrix C in (3.1.2). From (3.3.4), the determinant is

$$\det(C) = 7(-8) - 5(-10) = -6 \tag{3.3.7}$$

Since $\det(C) \neq 0$, it follows that C is nonsingular and therefore invertible. Using (3.3.6), the inverse is

$$C^{-1} = \frac{-1}{6} \begin{bmatrix} -8 & 10 \\ -5 & 7 \end{bmatrix} \tag{3.3.8}$$

For some special matrices, the inverse is very simple to compute. A matrix for which $A^{-1} = A^T$ is called an *orthogonal* matrix. For an orthogonal matrix, $A^T A = A A^T = I$. The inverse of the product of two square nonsingular matrices is related to the inverses of the individual matrices as follows.

$$(AB)^{-1} = B^{-1} A^{-1} \tag{3.3.9}$$

As with the transpose operation, the inverse of the product is the product of the inverses, but in reverse order.

3.4 EIGENVALUES AND EIGENVECTORS

Eigenvalues Every $n \times n$ matrix A has associated with it a special polynomial called the *characteristic polynomial,* which is defined as follows.

$$\Delta(\lambda) \triangleq \det(\lambda I - A) \tag{3.4.1}$$

The polynomial $\Delta(\lambda)$ is a polynomial of degree n with a leading coefficient of one. As such, $\Delta(\lambda)$ can be written in factored form as follows.

$$\Delta(\lambda) = (\lambda - \lambda_1)(\lambda - \lambda_2)\cdots(\lambda - \lambda_n) \tag{3.4.2}$$

The roots $\{\lambda_1, \lambda_2, \ldots, \lambda_n\}$ are called the *eigenvalues* of the matrix A. That is, the eigenvalues are the solutions to the *characteristic equation*, $\Delta(\lambda) = 0$. In general, the eigenvalues of A can be complex numbers. However, if the $n \times n$ matrix A is real, then complex eigenvalues always appear in conjugate pairs. If the matrix A is symmetric ($A^T = A$), all the eigenvalues are real. A symmetric matrix is said to be *positive-definite* if for every nonzero vector x, the following scalar is positive.

$$x^T A x > 0 \quad , \quad x \neq 0 \tag{3.4.3}$$

Positive-definite matrices are nonsingular matrices with real positive eigenvalues.

The determinant of an $n \times n$ matrix can be expressed in terms of its eigenvalues as follows.

$$\det(A) = \lambda_1 \lambda_2 \cdots \lambda_n \tag{3.4.4}$$

Thus, the determinant is just the product of the eigenvalues. From this expression, it is clear that a matrix is singular if and only if at least one eigenvalue is zero.

As a simple illustration of computing eigenvalues, consider the 2×2 matrix C in (3.1.2). Using (3.4.1) and (3.3.4), the characteristic polynomial is

$$\begin{aligned}
\Delta(\lambda) &= \det(\lambda I - C) \\
&= \det\left\{\begin{bmatrix} \lambda - 7 & 10 \\ -5 & \lambda + 8 \end{bmatrix}\right\} \\
&= (\lambda - 7)(\lambda + 8) + 5(10) \\
&= \lambda^2 + \lambda - 6 \\
&= (\lambda - 2)(\lambda + 3)
\end{aligned} \tag{3.4.5}$$

Thus, the eigenvalues of C are $\lambda_1 = 2$ and $\lambda_2 = -3$. Notice that $\lambda_1\lambda_2 = -6 = \det(C)$ as expected from (3.4.3). The sum of the diagonal elements of a matrix A is called the *trace* of A.

$$\text{trace}(A) \triangleq A_{11} + A_{22} + \cdots + A_{nn} \tag{3.4.6}$$

The trace of a matrix is simple to compute and it can be used as a partial check on finding the eigenvalues because it is equal to the sum of the eigenvalues

$$\text{trace}(A) = \lambda_1 + \lambda_2 + \cdots + \lambda_n \tag{3.4.7}$$

In the case of the matrix C in (3.1.2), we have $\text{trace}(C) = -1 = \lambda_1 + \lambda_2$.

Eigenvectors For each eigenvalue λ_k, there is a nonzero vector x^k called an *eigenvector* of A that satisfies the following equation.

$$A x^k = \lambda_k x^k \quad , \quad 1 \leq k \leq n \tag{3.4.8}$$

Thus, the eigenvectors of A are special vectors whose directions remain unchanged when multiplied by A; instead, they are simply scaled by the eigenvalue. Let

$X = [x^1, x^2, \ldots, x^n]$ be a matrix whose columns are the eigenvectors of A. If the n eigenvectors of A are linearly independent, X is of full rank, and we can write the following.

$$X^{-1}AX = D \tag{3.4.9}$$

In this case, D is an $n \times n$ diagonal matrix whose diagonal elements are the eigenvalues of A. The n eigenvectors of A are linearly independent when A is symmetric and also when the n eigenvalues of A are distinct.

3.5 VECTOR NORMS

The norm of a vector is a generalization of the absolute value of a scalar. For an $n \times 1$ vector x, the *norm* of x is a measure of the *length* of x denoted $\|x\|$. The following is a family of vector norms called the Holder norms.

$$\|x\|_p \triangleq \left(\sum_{k=1}^{n} |x_k|^p \right)^{1/p} \quad , \quad p \geq 1 \tag{3.5.1}$$

The special case $p = 2$ is called the *Euclidean norm,* which can also be represented in terms of the dot product as $\|x\|_2 = \sqrt{x^T x}$. It can be shown that in the limit as p approaches infinity, $\|x\|_p$ converges to the *infinity norm,* which is defined as follows.

$$\|x\| \triangleq \max_{k=1}^{n} \{|x_k|\} \tag{3.5.2}$$

Every vector norm satisfies a number of import properties. For example, $\|x\| \geq 0$, and $\|x\| = 0$ if and only if $x = 0$. If α is a scalar, then $\|\alpha x\| = |\alpha|\|x\|$. Finally, the *triangle inequality* says that the norm of the sum is bounded from above by the sum of the norms. That is, for two vectors x and y,

$$\|x + y\| \leq \|x\| + \|y\|. \tag{3.5.3}$$

The triangle inequality has a simple and familiar interpretation. Suppose x and y are 2×1 vectors representing the sides of a triangle as shown in Figure 3.5.1. If the Euclidean norm is used, then the triangle inequality simply says that the length of any side of a triangle is less than or equal to the sum of the lengths of the other two sides.

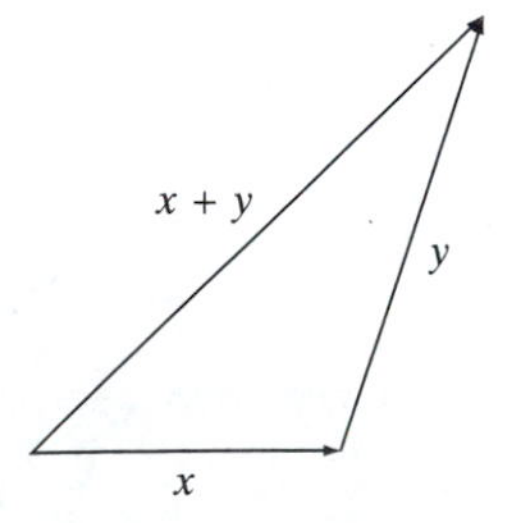

FIGURE 3.5.1
The Triangle Inequality

APPENDIX 4

Answers to Selected Problems

This appendix contains complete solutions to four representative problems from each chapter: two from the Analysis section and two from the Computation section. Files containing the source code for the solutions to the Computation problems can found in the *prob_mat* folder on the distribution CD.

CHAP. 1 NUMERICAL COMPUTATION

1.7 Suppose six bytes are used to represent a floating-point number with $r = 10$ bits used for the exponent.

(a) Find the largest positive number that can be represented. Using Equation (1.2.11) and noting that $2^{r-1} = 512$,

$$\begin{aligned} x_{\max} &= (.11\ldots1)_2 \times 2^{512} \\ &\approx 1.3408 \times 10^{154} \end{aligned}$$

(b) Find the smallest positive number that can be represented. Using (1.2.10),

$$\begin{aligned} x_{\min} &= (.10\ldots0)_2 \times 2^{-511} \\ &= 2^{-512} \\ &\approx 7.4583 \times 10^{-155} \end{aligned}$$

1.13 Derive the following expression for the relative error of the product where $x = x_e + \Delta x$ and $y = y_e + \Delta y$. You can assume that $|r_x| \ll 1$ and $|r_y| \ll 1$.

$$r_{xy} \approx r_x + r_y$$

Using $r_x = \Delta x / x_e$ and $r_y = \Delta y / y_e$,

$$\begin{aligned} r_{xy} &= \frac{xy - x_e y_e}{x_e y_e} \\ &= \frac{(x_e + \Delta x)(y_e + \Delta y) - x_e y_e}{x_e y_e} \\ &= (1 + r_x)(1 + r_y) - 1 \\ &= r_x + r_y + r_x r_y \\ &\approx r_x + r_y \end{aligned}$$

1.18 Write a function called *quadroot* that finds the roots of the polynomial, $p(x) = ax^2 + bx + c$, using the quadratic formula in (1.1.2). The pseudo-prototype is

```
[cvector x] = quadroot (float a,float b,float c)
```

On entry to *quadroot*: a, b, and c are the polynomial coefficients. On exit, the 2×1 complex vector x contains the roots. Include a check for the case $a = 0$. Test *quadroot* by writing a main program that prompts the user for the coefficients and then displays the roots.

```
%-----------------------------------------------------------------
% Problem 1.18
%-----------------------------------------------------------------
clc
clear

% Find roots

fprintf ('Problem 1.18\n');
y = input ('Enter coefficients [a,b,c] (including brackets): ');
roots = quadroot (y(1),y(2),y(3))
wait

function r = quadroot (a,b,c)
%-----------------------------------------------------------------
% Usage:        r = quadroot(a,b,c);
%
% Description: Find the roots of ax^2 + bx + c.
%              Return in the vector r.
%-----------------------------------------------------------------
   if (abs(a) < eps)
      r = -c/b;
   else
      r = zeros (2,1);
      r(1) = (-b + sqrt(b*b - 4*a*c))/(2*a);
      r(2) = (-b - sqrt(b*b - 4*a*c))/(2*a);
   end
%-----------------------------------------------------------------
```

When this script is executed and the user response is [1 2 3], the output produced by the program is $x = -1.000 \pm 1.4142i$.

(P)1.24 Use the technique in (1.6.5) and (1.6.6) to write a function called *randf*, which generates uniformly distributed random numbers in the interval $[0, 1]$. The pseudo-prototype is

```
[int u] = randf (void)
```

When *randf* is called, it should generate the next random number in the sequence. Start the sequence using a seed of one. Test *randf* by writing a main program that computes and displays the first 20 random numbers.

```
%-----------------------------------------------------------------
% Problem 1.24
%-----------------------------------------------------------------
clc
clear
```

```
global random_int
random_int = 1;

% Generate random numbers

fprintf ('Problem 1.24\n');
for k = 1 : 20
   fprintf ('%f\n',randf);
end
wait

function u = randf
%-----------------------------------------------------------------------
% Usage:        u = randf;
%
% Description: Return a random integer in the range 0 to gamma-1
%              where gamma = 2147483646.
%-----------------------------------------------------------------------
   global random_int
   alpha = 16807;
   gamma = 2147483647;
   random_int = rem (alpha*random_int,gamma);
   u = random_int/gamma;
%-----------------------------------------------------------------------
```

When this script is executed, it produces a list of 20 random numbers ranging from 0.00008 to 0.934693.

CHAP. 2 LINEAR ALGEBRAIC SYSTEMS

2.11 Consider the following coefficient matrix.

$$A = \begin{bmatrix} 10 & -1 \\ 0.1 & 0.01 \end{bmatrix}$$

(a) Find the condition number $K(A)$. We use Equation (2.5.8) and Examples 2.5.3 and 2.5.4 as a guide. In this case, $\det(A) = 0.2$ and the inverse of A is

$$A^{-1} = \frac{1}{0.2} \begin{bmatrix} 0.01 & 1 \\ -0.1 & 10 \end{bmatrix}$$

To determine $K(A)$, we use (2.5.5) to compute $\|A\| = \max\{11, 0.11\}$ and then $\|A^{-1}\| = \max\{1.01/0.2, 10.1/0.2\}$. Thus, the condition number of this system is

$$K(A) = 11(50.5) = 555.5$$

(b) Find the condition number $K(A)$ after equilibration. Scaling each row of A by its largest element, the new A and its inverse are as follows.

$$A = \begin{bmatrix} 1 & -0.1 \\ 1 & 0.1 \end{bmatrix}, \quad A^{-1} = \frac{1}{0.2} \begin{bmatrix} 0.1 & 0.1 \\ -1 & 1 \end{bmatrix}$$

The norms of the scaled A and its inverse, are $\|A\| = \max\{1.1, 1.1\}$ and $\|A^{-1}\| = \max\{0.2/0.2, 2/0.2\}$. Thus, the condition number of the scaled system is

$$K(A) = 1.1(10) = 11$$

2.14 Find the inverse of the following matrix A by reducing $C = [A, I]$ to upper-triangular form with elementary row operations and then solving for each column of A^{-1} using back substitution.

$$A = \begin{bmatrix} 2 & -1 & 0 \\ 0 & 4 & 3 \\ 4 & 0 & -1 \end{bmatrix}$$

$$C = [A, I] = \left[\begin{array}{ccc|ccc} 2 & -1 & 0 & 1 & 0 & 0 \\ 0 & 4 & 3 & 0 & 1 & 0 \\ 1 & 0 & -1 & 0 & 0 & 1 \end{array}\right] \rightarrow \begin{bmatrix} 2 & -1 & 0 & 1 & 0 & 0 \\ 0 & 4 & 3 & 0 & 1 & 0 \\ 0 & 2 & -1 & -2 & 0 & 1 \end{bmatrix}$$

$$\rightarrow \begin{bmatrix} 2 & -1 & 0 & 1 & 0 & 0 \\ 0 & 2 & -1 & -2 & 0 & 1 \\ 0 & 4 & 3 & 0 & 1 & 0 \end{bmatrix} \rightarrow \left[\begin{array}{ccc|ccc} 2 & -1 & 0 & 1 & 0 & 0 \\ 0 & 2 & -1 & -2 & 0 & 1 \\ 0 & 0 & 5 & 4 & 1 & -2 \end{array}\right]$$

$$= [U, D]$$

Next, we apply back substitution using each of the columns of D in turn. Column 1:

$$x_3 = 4/5 = 0.8$$
$$x_2 = (-2 + 0.8)/2 = -0.6$$
$$x_1 = (1 - 0.6)/2 = 0.2$$

Column 2:

$$x_3 = 1/5 = 0.2$$
$$x_2 = 0.2/2 = 0.1$$
$$x_1 = 0.1/2 = 0.05$$

Column 3:

$$x_3 = -2/5 = -0.4$$
$$x_2 = (1 - 0.4)/2 = 0.3$$
$$x_1 = 0.3/2 = 0.15$$

Assembling the results into the three columns of A^{-1} then yields

$$A^{-1} = \begin{bmatrix} 0.2 & 0.05 & 0.15 \\ -0.6 & 0.1 & 0.3 \\ 0.8 & 0.2 & -0.4 \end{bmatrix}$$

2.23 Write a program that uses the NLIB functions to compute and print the inverse, determinant, and condition number of the coefficient matrix A in problem 2.22.

```
%-------------------------------------------------------------
% Problem 2.23
%-------------------------------------------------------------
clc
clear
A = [1 -1 4 0 2
```

```
      0 5 -2 7 8
      1 0 5 7 3
      6 -1 2 3 0
      -4 2 0 5 -5];

% Analyze matrix

fprintf ('Problem 2.23\n');
show ('A',A)
show ('inv(A)',inv(A))
show ('det(A)',det(A))
show ('K(A)',condnum(A,5))
%-----------------------------------------------------------------
```

When this script is run, it displays the 5×5 inverse of A and then produces $\det(A) = -3836$ and $K(A) = 35.58$.

(P)2.29 Write a function called *jac* that implements the Jacobi iterative method to solve the system $Ax = b$. The pseudo-prototype for the function is

```
[vector x,int k] = jac (vector x,matrix A,vector b,float tol,int n,int m)
```

On entry to *jac:* the $n \times 1$ vector x is an initial guess, A is an $n \times n$ coefficient matrix, b is an $n \times 1$ right-hand-side vector, $tol \geq 0$ is an error tolerance used to terminate the search, $n \geq 1$ is the number of unknowns, and $m \geq 1$ is an upper bound on the number of iterations. On exit from *jac:* k is the number of iterations performed, and the $n \times 1$ vector x contains the updated estimate of the solution. If $0 < k < m$, the following error criterion must be satisfied where $r = b - Ax$ is the residual error vector:

$$\|r\| < tol$$

If $k = 0$, then one of the diagonal elements of A was found to be zero. Test the function *jac* by writing a program that uses it to solve the following system. Compute and display the norms of the residual error vectors for the first six solution estimates $\{x^1, x^2, \ldots, x^6\}$ and display the final estimate, x^6, using $x^0 = 0$.

$$\begin{bmatrix} 10 & 1 & -1 & 2 & 0 \\ 0 & -12 & 2 & -3 & 1 \\ 2 & -2 & -11 & 0 & 1 \\ 4 & 0 & 1 & 8 & -1 \\ -3 & 1 & 2 & 0 & 9 \end{bmatrix} x = \begin{bmatrix} 19 \\ 2 \\ 13 \\ -7 \\ -9 \end{bmatrix}$$

```
%-----------------------------------------------------------------
% Problem 2.29
%-----------------------------------------------------------------
clc
clear
A = [10 1 -1 2 0
     0 -12 2 -3 1
     2 -2 -11 0 1
     4 0 1 8 -1
     -3 1 2 0 9 ];
```

```
b = [19 2 13 -7 -9]';
n = size (A,1);
m = 6;

% Solve system using Jacobi's method

fprintf ('Problem 2.29\n');
show ('A',A)
show ('b',b)
r = zeros(m,1);
for i = 1 : 6
   x = zeros(n,1);
   [x,k] = jac (x,A,b,0,n,i);
   r(k) = residual (A,b,x);
end
show ('x',x);
show ('||r(x)||',r)

function [x,k] = jac (x,A,b,tol,n,m)
%-----------------------------------------------------------------
% Usage:       [x,k] = jac (x,A,b,tol,n,m)
%
% Description: Solve linear algebraic system, Ax = b, iteratively
%              using the Jacobi's iterative method.
%
% Inputs:      x     = n by 1 vector containing initial guess
%              A     = n by n coefficient matrix (nonzero diagonal
%                      elements)
%              b     = n by 1 right-hand side vector
%              tol   = error tolerance used to terminate search
%              n     = number of components of x
%              m     = maximum number of iterations (m >= 1)
%
% Outputs:     x = n by 1 solution vector
%              k = number of iterations performed.  If 0 < k < m,
%                  then the following convergence criterion was
%                  satisfied where r = b - Ax is the residual error
%                  vector:
%
%                  ||r|| < tol
%
%                  Zero is returned if one of a diagonal element
%                  of A was found to be zero.  In this case, the
%                  equations must be reordered.
%-----------------------------------------------------------------

% Initialize

   k = 0;
   tol   = args (tol,0,tol,4,'jac');
```

```
   n       = args (n,1,n,5,'jac');
   m       = args (m,1,m,6,'jac');
   n = length(x);
   for i = 1 : n
      if abs(A(i,i)) < eps
          return;
      end
   end

% Iterate

   while ((k < m) & (residual(A,b,x) >= tol))
      for i = 1 : n;
         y = b(i);
         for j = 1 : n
            if j ~= i
               y = y - A(i,j)*x(j);
            end
         end
         x(i) = y/A(i,i);
      end
      k = k + 1;
   end
%----------------------------------------------------------------
```

When this script is executed, it produces the following solution estimate and residual errors using six iterations.

$$x = \begin{bmatrix} 2.177 \\ 0.1554 \\ -0.8242 \\ -1.874 \\ -0.1085 \end{bmatrix}, \quad \|r\| = \begin{bmatrix} 3.392 \\ 0.5132 \\ 0.05888 \\ 0.01123 \\ 0.0008854 \\ 0.0002042 \end{bmatrix}$$

CHAP. 3 EIGENVALUES AND EIGENVECTORS

3.4 Suppose A is a triangular $n \times n$ matrix (either upper-triangular or lower-triangular). Show that the eigenvalues of A are the diagonal elements.

Suppose A is lower-triangular. Then, from (3.2.2), the characteristic polynomial of A is

$$\begin{aligned} \Delta(\lambda) &= \det(\lambda I - A) \\ &= \det\left\{\begin{bmatrix} \lambda - A_{11} & 0 & 0 & \cdots & 0 \\ -A_{21} & \lambda - A_{22} & 0 & \cdots & 0 \\ -A_{31} & -A_{32} & \lambda - A_{33} & \cdots & 0 \\ \vdots & \vdots & \vdots & \ddots & \vdots \\ -A_{n1} & -A_{n2} & -A_{n3} & \cdots & \lambda - A_{nn} \end{bmatrix}\right\} \end{aligned}$$

If we expand the determinant about its first row, then

$$\Delta(\lambda) = (\lambda - A_{11}) \det(B)$$

Here B is the $(n - 1) \times (n - 1)$ submatrix of $\lambda I - A$ obtained by eliminating row one and column one. Next, to find the determinant of B, we can expand it along its first row, which yields

$$\det(B) = (\lambda - A_{22}) \det(C)$$

Here C is the $(n - 2) \times (n - 2)$ submatrix of B obtained by eliminating row one and column one. Substituting the expression for $\det(B)$ into the first equation then yields

$$\Delta(\lambda) = (\lambda - A_{11})(\lambda - A_{22}) \det(C)$$

This process can be continued until we are left with the 1×1 submatrix $\lambda - A_{nn}$. Thus, the characteristic polynomial of A can be expressed

$$\Delta(\lambda) = (\lambda - A_{11})(\lambda - A_{22}) \cdots (\lambda - A_{nn})$$

The eigenvalues of A are the roots of the characteristic polynomial. But $\Delta(\lambda)$ is already in factored form. Thus, the eigenvalues of A are the diagonal elements $\lambda_k = A_{kk}$ for $1 \le k \le n$. Finally, from Table 2.3.1, $\det(A) = \det(A^T)$. Thus, for an upper-triangular matrix, it is also the case that the eigenvalues are the diagonal elements.

3.10 Consider the following linear discrete-time system. Determine whether or not this system is stable. Explain.

$$x^{k+1} = \begin{bmatrix} 0 & -.25 \\ 1 & -1 \end{bmatrix} x^k \quad , \quad k \ge 0$$

The eigenvalues of the coefficient matrix are

$$\begin{aligned}
\Delta(\lambda) &= \det(\lambda I - A) \\
&= \det\left\{\begin{bmatrix} \lambda & 0.25 \\ -1 & \lambda + 1 \end{bmatrix}\right\} \\
&= \lambda(\lambda + 1) + 0.25 \\
&= \lambda^2 + \lambda + 0.25 \\
&= (\lambda - \lambda_1)(\lambda - \lambda_2) \\
&= (\lambda + 0.5)^2
\end{aligned}$$

Thus, there is a repeated eigenvalue at $\lambda = -0.5$. From (3.1.14), the spectral radius is

$$\rho(A) = \max\{|-0.5|, |-0.5|\} = 0.5$$

Since $\rho(A) < 1$, it follows from (3.1.13) that the system is stable.

3.13 Write a program that uses the NLIB function *powereig* to find the dominant eigenvalue λ_1 and eigenvector x^1 of the following matrix. Print λ_1, x^1, the spectral radius, and the norm of the residual error vector $r^1 = (\lambda_1 I - A)x^1$.

$$A = \begin{bmatrix} 5 & 4 & 3 & 2 & 1 \\ 4 & 3 & 2 & 1 & 0 \\ 3 & 2 & 1 & 0 & -1 \\ 2 & 1 & 0 & -1 & -2 \\ 1 & 0 & -1 & -2 & -3 \end{bmatrix}$$

```
%----------------------------------------------------------------------
% Problem 3.13
%----------------------------------------------------------------------
clc
clear
A = [5 4 3 2 1
     4 3 2 1 0
     3 2 1 0 -1
     2 1 0 -1 -2
     1 0 -1 -2 -3];
tol = 1.e-6;

% Find eigenvalue and eigenvector

fprintf ('Problem 3.13\n');
show ('A',A)
[c,x,k] = powereig (A,tol,50,1);
show ('Power method iterations',k)
show ('Dominant eigenvalue',c)
show ('Dominant eigenvector',x)
show ('Spectral radius',abs(c))
r = (c*eye(5) - A)*x;
show ('||r(x)||',norm(r,inf))
%----------------------------------------------------------------------
```

When this script is executed, the power method converges in 21 iterations to a dominant eigenvalue of $\lambda_1 = 10$ and a dominant eigenvector of

$$x_1 = \begin{bmatrix} 1 \\ 0.75 \\ 0.5 \\ 0.25 \\ 5.57e - 008 \end{bmatrix}$$

The spectral radius of A is $\rho = 10$, and the infinity norm of the residual error vector is $\|r\| = 8.35498 \times 10^{-7}$.

3.19 Write a program that computes and displays the characteristic polynomial and the spectral radius of the following matrix.

$$A = \begin{bmatrix} 4 & -1 & 3 & 9 & -5 \\ -1 & 2 & 7 & 12 & 7 \\ 3 & 7 & 0 & -11 & 8 \\ 2 & -4 & 14 & 1 & -5 \\ 11 & -8 & 6 & -2 & 13 \end{bmatrix}$$

```
%----------------------------------------------------------------------
% Problem 3.19
%----------------------------------------------------------------------
clc
```

```
clear
A = [4 -1 3 9 -5
     -1 2 7 12 7
     3 7 0 -11 8
     2 -4 14 1 -5
     11 -8 6 -2 13];

% Find characteristic polynomial and spectral radius

fprintf ('Problem 3.13\n');
show ('A',A)
a = poly(A);
show ('Characteristic polynomial coefficients',a')
c = eig(A);
show ('Spectral radius',max(abs(c)))
%------------------------------------------------------------
```

When this script was executed, the characteristic polynomial of A was found to be

$$\Delta(\lambda) = \lambda^5 - 20\lambda^4 + 283\lambda^3 - 4507\lambda^2 + (6.045 \times 10^4)\lambda - (3.53 \times 10^5)$$

The spectral radius was $\rho = 14.6064$.

CHAP. 4 CURVE FITTING

4.3 Find the Lagrange interpolating polynomials $\{Q_1(x), Q_2(x), Q_3(x), Q_4(x)\}$ for the following data set. Use the Lagrange interpolating polynomials to find the interpolating polynomial $f(x)$. Simplify your final answers as much as possible.

$$D = \{(0, -18), (1, -10), (2, 0), (3, 18)\}$$

The Lagrange interpolating polynomials depend only on the x values in D. Using (4.2.7),

$$\begin{aligned}
Q_1(x) &= \frac{(x - x_2)(x - x_3)(x - x_4)}{(x_1 - x_2)(x_1 - x_3)(x_1 - x_4)} \\
&= \frac{(x - 1)(x - 2)(x - 3)}{(0 - 1)(0 - 2)(0 - 3)} \\
&= \frac{x^3 - 6x^2 + 11x - 6}{-6}
\end{aligned}$$

$$\begin{aligned}
Q_2(x) &= \frac{(x - x_1)(x - x_3)(x - x_4)}{(x_2 - x_1)(x_2 - x_3)(x_2 - x_4)} \\
&= \frac{(x - 0)(x - 2)(x - 3)}{(1 - 0)(1 - 2)(1 - 3)} \\
&= \frac{x^3 - 5x^2 + 6x}{2}
\end{aligned}$$

$$Q_3(x) = \frac{(x - x_1)(x - x_2)(x - x_4)}{(x_3 - x_1)(x_3 - x_2)(x_3 - x_4)}$$

$$= \frac{(x-0)(x-1)(x-3)}{(2-0)(2-1)(2-3)}$$

$$= \frac{x^3 - 4x^2 + 3x}{-2}$$

$$Q_4(x) = \frac{(x-x_1)(x-x_2)(x-x_3)}{(x_4-x_1)(x_4-x_2)(x_4-x_3)}$$

$$= \frac{(x-0)(x-1)(x-2)}{(3-0)(3-1)(3-2)}$$

$$= \frac{x^3 - 3x^2 + 2x}{6}$$

Next, from (4.2.9), the interpolating polynomial is

$$\begin{aligned} f(x) &= y_1Q_1(x) + y_2Q_2(x) + y_3Q_3(x) + y_4Q_4(x) \\ &= -18(x^3 - 6x^2 + 11x - 6)/(-6) \\ &\quad -10(x^3 - 5x^2 + 6x)/2 \\ &\quad +18(x^3 - 3x^2 + 2x)/6 \\ &= (3 - 5 + 3)x^3 + (-18 + 25 - 9)x^2 + (33 - 30 + 6)x - 18 \\ &= x^3 - 2x^2 + 9x - 18 \end{aligned}$$

Check: $f(0) = -18$, $f(1) = -10$, $f(2) = 0$, and $f(3) = 18$.

4.8 Some noisy experimental data (e.g., radioactive decay) can be effectively modeled by using an exponential fit of the following form where u and v are unknown parameters.

$$z = u \exp(-vx)$$

This problem can be converted to an equivalent problem of finding a straight-line fit by applying the natural logarithm to both sides of the equation. Let $y = \ln(z)$ and $a = [\ln(u), -v]^T$. Then the above equation can be written as

$$y = a_1 + a_2x$$

This is identical to (4.5.1). Suppose the data samples have uniform weighting as in (4.5.3). Then, from (4.5.5), the optimal least-squares coefficient vector a is the solution of

$$\begin{bmatrix} n & \sum_{k=1}^{n} x_k \\ \sum_{k=1}^{n} x_k & \sum_{k=1}^{n} x_k^2 \end{bmatrix} \begin{bmatrix} a_1 \\ a_2 \end{bmatrix} = \begin{bmatrix} \sum_{k=1}^{n} y_k \\ \sum_{k=1}^{n} x_k y_k \end{bmatrix}$$

Use this result to find the exponential function $f(x) = u \exp(-vx)$ that best fits the following data in a least-squares sense.

$$D = \{(1, 1.84), (2, 0.91), (3, 0.45), (4, 0.26), (5, 0.13)\}$$

The new data set using $y_k = \ln(z_k)$ is

$$\hat{D} = \{(1, 0.61), (2, -0.09), (3, -0.80), (4, -1.35), (5, -2.04)\}$$

The coefficient matrix and right-hand-side vectors are then

$$C_{11} = 5$$

$$C_{12} = 1 + 2 + 3 + 4 + 5 = 15$$

$$C_{21} = C_{12} = 15$$
$$C_{22} = 1 + 4 + 9 + 16 + 25 = 55$$
$$b_1 = 0.61 - 0.09 - 0.80 - 1.35 - 2.04 = -3.67$$
$$b_2 = 0.61 - 2(0.09) - 3(0.80) - 4(1.35) - 5(2.04) = -17.57$$

Thus, the equations to be solved are

$$\begin{bmatrix} 5 & 15 \\ 15 & 55 \end{bmatrix} a = \begin{bmatrix} -3.67 \\ -17.57 \end{bmatrix}$$

Applying Gauss-Jordan elimination yields

$$[C,b] = \left[\begin{array}{cc|c} 5 & 15 & -3.67 \\ 15 & 55 & -17.57 \end{array}\right] \rightarrow \begin{bmatrix} 1 & 3 & -0.73 \\ 15 & 55 & -17.57 \end{bmatrix} \rightarrow \begin{bmatrix} 1 & 3 & -0.73 \\ 0 & 10 & -6.62 \end{bmatrix}$$
$$\rightarrow \begin{bmatrix} 1 & 3 & -0.73 \\ 0 & 1 & -0.66 \end{bmatrix} \rightarrow \begin{bmatrix} 1 & 0 & 1.25 \\ 0 & 1 & -0.66 \end{bmatrix} \rightarrow [I,a]$$

Thus, $a = [1.25, -0.66]^T$. Finally, using $a = [\ln(u), -v]^T$, we have $u = \exp(1.25) = 3.49$ and $v = 0.66$. Thus, the exponential least-squares fit is

$$z = 3.49 \exp(-0.66x)$$

4.13 Write a function called *expfit* that computes the coefficients of the least-squares exponential fit $f(x)$ using the results of problem 4.8. The pseudo-prototype for the function is

```
[float u,float v] = expfit (matrix D,int n)
```

On entry to *expfit:* D is a $2 \times n$ data matrix containing the strictly increasing independent variables in row one and the dependent variables in row two, and $n \geq 2$ is the number of samples. On exit from *expfit:* u is the amplitude and v is the exponent of the least-squares fit $f(x) = u \exp(-vx)$. Test the function with the data set D in problem 4.8. Plot $f(x)$ over $[x_1, x_n]$ and the data points in D on the same graph, and print u and v.

```
%-------------------------------------------------------------------
% Problem 4.13
%-------------------------------------------------------------------
clc
clear
n = 5;
D = [1:n ; 1.84 0.91 0.45 0.26 0.13];

% Find least-squares exponential fit

fprintf ('Problem 4.13\n');
[u,v] = expfit (D,n);
show ('[u,v]',[u,v]);

% Plot exponential curve and data

x = linspace (D(1,1),D(1,n),100);
y = u*exp(-v*x);
```

```
figure
plot (x,y,'-',D(1,:),D(2,:),'s')
xlabel ('x')
ylabel ('y')
title ('A Least-Squares Exponential Fit')
wait

function [u,v] = expfit (D,n)
%-----------------------------------------------------------------
% Usage:        [u,v] = expfit (D,n)
%
% Description: Find the least-squares exponential fit function
%              for the data set D:
%
%              f(x) = u*exp(-v*x)
%
% Inputs:      D = 2 by n matrix with the independent variables
%                  in row 1 and the dependent variables in
%                  row 2.
%              n = number of data points (n >= 2)
%
% Outputs:     u = amplitude of exponential function
%              v = exponent of exponential function
%-----------------------------------------------------------------

% Initialize

n   = args (n,2,n,2,'expfit');

% Form equations

x = D(1,:);
y = log(D(2,:));
C = [n sum(x) ; sum(x) dot(x,x)];
b = [sum(y) ; dot(x,y)];

% Find u and v

a = C\b;
u = exp(a(1));
v = -a(2);
%-----------------------------------------------------------------
```

When this script is executed, it produces parameter values of $u = 3.427$, $v = 0.6553$, or $y = 3.427 \exp(-0.6553x)$. The resulting curve fit to the data is shown in Figure 4.4.1.

4.15 Consider the problem of constructing an interpolating polynomial for the following rational function.

$$f(x) = \frac{1}{1 + x^2}$$

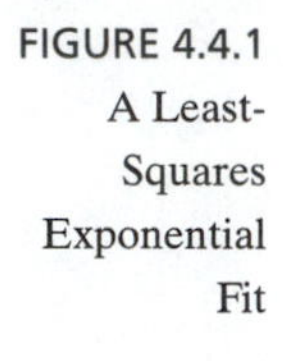

FIGURE 4.4.1 A Least-Squares Exponential Fit

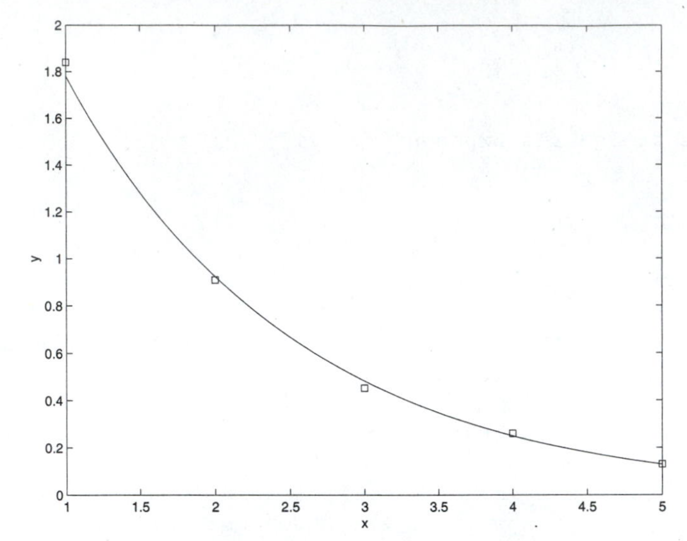

Suppose $n = 9$ samples in the interval $[-4, 4]$ are used. Write a program that uses the roots of the Chebyshev polynomials with appropriate offset and scaling as in (4.4.22) to generate the n samples. Use *polyfit* to compute an interpolating polynomial based on these points. Plot the interpolating polynomial and the data points on the same graph. Compare your results with the case of uniformly distributed samples in Figure 4.4.1 of Chapter 4.

```
%------------------------------------------------------------------
% Problem 4.15
%------------------------------------------------------------------
clc
clear
a = -4;
b = 4;
n = 9;

% Find least squares polynomial

fprintf ('Problem 4.15\n');
k = 1 : n;
r = cos((2*k-1)*pi/(2*n));
x = ((b - a)*r + a + b)/2;
y = 1 ./ (1 + x .* x);
c = polyfit (x,y,n-1);
show ('coefficients',c');

% Plot polynomial and data

x1 = linspace (a,b,100);
y1 = polyval (c,x1);
```

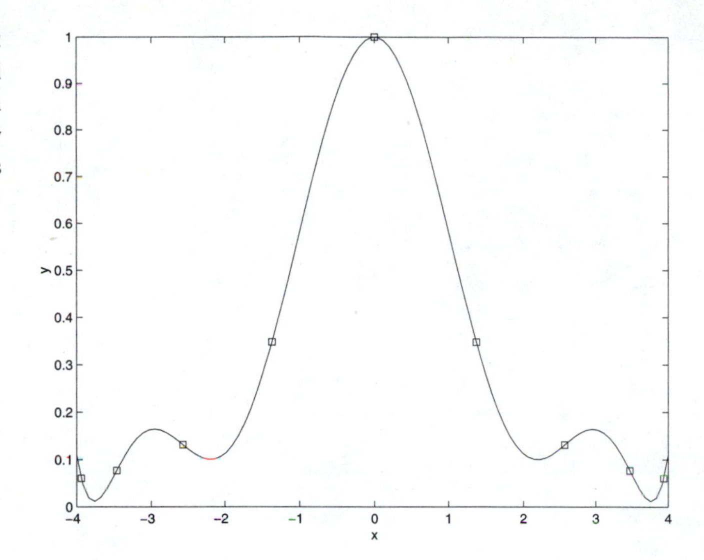

FIGURE 4.4.2 Polynomial Fit with Chebyshev Roots

```
Figure
plot (x1,y1,'-',x,y,'s')
xlabel ('x')
ylabel ('y')
title ('Polynomial Fit with Chebyshev Roots')
wait
%------------------------------------------------------------------
```

When this script is executed, it produces the following coefficient vector.

$$c = \begin{bmatrix} 0.0002131 \\ -3.47e - 018 \\ -0.007884 \\ 9.439e - 017 \\ 0.09994 \\ -7.022e - 016 \\ -0.5091 \\ 1.179e - 015 \\ 1 \end{bmatrix}$$

The resulting interpolating polynomial, based on nonuniform sampling spacing, is shown in Figure 4.4.2. It is clearly a better fit than the uniform case in Figure 4.4.1 of Chapter 4.

CHAP. 5 ROOT FINDING

5.2 Consider the following function.

$$f(x) = \frac{\exp(-x) - x^3}{5}$$

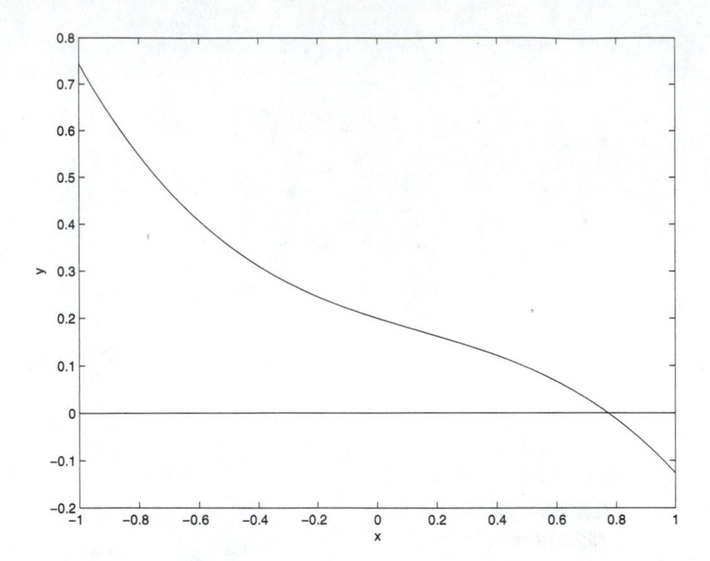

FIGURE 4.5.1 The Function $f(x) = (\exp(-x)/5 - x^3)/5$

(a) Show that this function has at least one root in the interval $[-1,1]$. We check for a sign change:

$$\begin{aligned} f(-1)f(1) &= 0.2(e+1)0.2(1/e-1) \\ &= 0.744(-0.126) \\ &= -0.094 < 0 \end{aligned}$$

(b) Show that f maps the interval $[-1,1]$ into itself by plotting $f(x)$ for $-1 \le x \le 1$. See Figure 4.5.1 where $-0.2 \le f(x) \le 0.8$.

(c) Find bounds γ and δ on the derivative $f'(x)$ such that the following inequality holds.

$$\gamma \le f'(x) \le \delta < 0 \quad , \quad -1 \le x \le 1$$

The derivative is

$$f'(x) = \frac{-(\exp(-x) + 3x^2)}{5}$$

The first term is negative for all x, and the second term is less than or equal to zero. Thus, $f'(x) < 0$ for all x. The first term ranges from a minimum of $-0.2e$ to a maximum of $-0.2/e$. The second term ranges from a minimum of -0.6 to a maximum of 0. Thus, the derivative can be bounded as follows:

$$-0.2(e+3) \le f'(x) \le -0.2/e < 0 \quad , \quad -1 \le x \le 1$$

(d) For what values of b is the following function a contraction mapping?

$$g(x) = x - bf(x) \quad , \quad -1 \le x \le 1$$

From (5.3.11), we require that $0 < bf'(x) < 2$. Since $f'(x) < 0$, this means that $2/f'(x) < b < 0$. From part (c), we can replace $f'(x)$ its minimum value $\delta = -0.2(e+3)$. Thus,

$$\frac{-10}{e+3} < b < 0$$

5.7 Consider the following cubic polynomial. Apply linear synthetic division to verify that $(x + 4)$ is a factor. Find the roots of $f(x)$.

$$f(x) = x^3 + 4x^2 + 9x + 36$$

We apply Alg. 5.7.1 with $n = 3$ and $c = -4$. Using Example 5.7.2 as a guide,

$$b_3 = 1$$
$$b_2 = 4 + (-4)1 = 0$$
$$b_1 = 9 + (-4)0 = 9$$
$$d = 36 + (-4)9 = 0$$

Thus, $d = 0$, which confirms that $x = -4$ is a root. In this case, the deflated polynomial is

$$g(x) = x^2 + 9$$

Thus, the roots of $f(x)$ are $x = -4$, and $x = \pm j3$.

5.11 Write a program that uses the NLIB function *bisect* to find a root of the following function f using $[x_0, x_1] = [0, 10]$ and an error tolerance of $tol = 10^{-6}$. Print the x, $|f(x)|$, and the number of iterations needed for convergence. Verify the result by plotting $f(x)$ over $[0, 10]$ using the NLIB function *graphfun*.

$$f(x) = \ln(2 + x) - \sqrt{x}$$

```
%-----------------------------------------------------------------------
% Problem 5.11
%-----------------------------------------------------------------------
clc
clear
f = inline ('log(2+x) - sqrt(x)','x');
x0 = 0;
x1 = 10;
tol = 1.e-6;
m = 100;

% Find root using bisection search

fprintf ('Problem 5.11\n');
[x,k] = bisect (x0,x1,tol,m,f);
show ('x',x)
show ('|f(x)|',abs(f(x)))
show ('Iterations',k)

% Plot curve

graphfun (x0,x1,'Function f','x','f(x)',f)
%-----------------------------------------------------------------------
```

When this script is executed, the bisection method requires 16 iterations to find $x = 1.73996$. The magnitude of the function at this root is $|f(x)| = 4.85972 \times 10^{-7}$. A graph of the function $f(x)$ is shown in Figure 4.5.2.

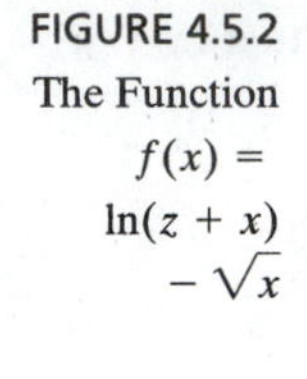

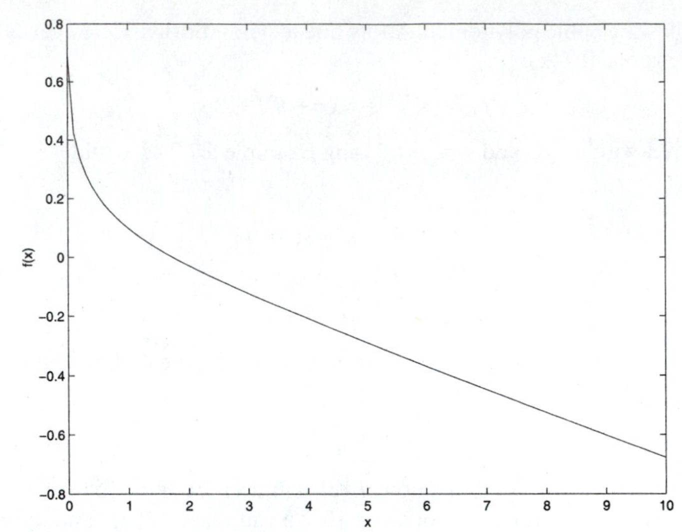

FIGURE 4.5.2 The Function $f(x) = \ln(z + x) - \sqrt{x}$

5.21 Write a program that uses the NLIB function *Newton* to find a root of the nonlinear algebraic system in problem 5.10 using the initial condition $x^0 = [1, -1]^T$ and $tol = 10^{-6}$. Print x, $f(x)$, and the number of iterations needed for convergence.

```
%-------------------------------------------------------------
% Problem 5.21
%-------------------------------------------------------------
clc
clear
x0 = [1 -1]';
tol = 1.e-6;
m = 100;

% Find solution using Newton's method

fprintf ('Problem 5.21\n');
[x,k] = newton (x0,tol,m,'funf521');
show ('x',x)
show ('||f(x)||',norm(funf521(x),inf))
show ('Iterations',k)

function y = funf521 (x)
%-------------------------------------------------------------
% Problem 5.21
%-------------------------------------------------------------
y = [x(1)*cos(pi*x(2)) + x(2)^2 - 1
     x(1)^3 - x(2)*sin(pi*x(1)) + 1];
%-------------------------------------------------------------
```

When this script is executed, it uses 23 iterations to produce the following estimate of a root:

$$x = \begin{bmatrix} 0.5647 \\ 1.205 \end{bmatrix}$$

The infinity norm of the function at this root is $\|f(x)\| = 4.17715 \times 10^{-8}$.

CHAP. 6 OPTIMIZATION

6.4 Suppose the bisection method in Alg. 6.3.3 is used to perform a line search with the derivative approximated numerically using a central difference.

$$F'(y) \approx \frac{F(y + \varepsilon) - F(y - \varepsilon)}{2\varepsilon}$$

Show that the length of the interval of uncertainty is reduced by a factor of approximately 0.707 per function evaluation. How much slower is this (as a percentage) than the golden section search?

Since Alg. 6.3.3 bisects the interval of uncertainty, the length of the interval of uncertainty shrinks by a factor of $\alpha = 0.5$ for each iteration. There is one evaluation of F' per iteration, which corresponds to two evaluations of F using the central difference approximation. Since it takes two function evaluations to shrink by 0.5, the amount it shrinks per evaluation is

$$\alpha = (0.5)^{1/2} \approx 0.707$$

The golden section search shrinks the interval by 0.618. Thus, the bisection method, using central differences to approximate F', is slower by

$$\beta = \frac{100(.707 - .618)}{.618} = 14.4\%$$

6.7 Consider the following constrained optimization problem.

$$\begin{aligned} \text{minimize:} \quad & f(y, z) \\ \text{subject to:} \quad & Ay + Bz = c \\ & q(y, z) \geq 0 \end{aligned}$$

Here A is $r \times r$, B is $r \times (n - r)$, and c is $r \times 1$. Thus, there are r linear equality constraints where $r < n$. Suppose the square matrix A is nonsingular. Convert this n-dimensional optimization problem with r equality constraints and s inequality constraints into a simpler problem of dimension $n - r$ that has no equality constraints. That is, find F, and Q such that the equivalent problem is as follows. Show how to determine y once z is known.

$$\begin{aligned} \text{minimize:} \quad & F(z) \\ \text{subject to:} \quad & Q(z) \geq 0 \end{aligned}$$

We solve the equality constraint for y and then substitute the solution into the objective function and the inequality constraint. Solving $Ay + Bz = c$ for y yields

$$y = A^{-1}(c - Bz)$$

Substituting the result into f and q yields

$$\begin{aligned} F(z) &= f[A^{-1}(c - Bz), z] \\ Q(z) &= q[A^{-1}(c - Bz), z] \end{aligned}$$

6.12 Consider the quadratic objective function in problem 6.6. Write a program that uses the NLIB function *conjgrad* to find the optimal x starting from an initial guess of $x^0 = 0$ and using an error tolerance of $tol = 10^{-3}$. Solve the system twice, once with $v = 1$ corresponding to the steepest descent method and once with $v = n$ corresponding to the full conjugate gradient method. Print the order v, the optimal x, the optimal $f(x)$, the number of iterations, and the number of scalar function evaluations in each case.

From problem 6.6, the objective function is

$$f(x) = 10 + x_1 - 3x_2 + 2x_3 + 8x_4 + (3x_1^2 + 9x_2^2 + 27x_3^2 + 81x_4^2)/2$$

```
%----------------------------------------------------------------
% Problem 6.12
%----------------------------------------------------------------
clc
clear
tol = 1.e-3;
m = 250;
n = 4;
x0 = zeros(n,1);

% Find minimum using steepest descent and conjugate gradient

fprintf ('Problem 6.12\n');
fprintf ('Finding optimal x ...\n')
for v = 1 : n-1 : n
   [x,ev,j] = conjgrad (x0,tol,v,m,'funf612');
   show ('v',v);
   show ('x',x(1),x(2),x(3),x(4));
   show ('f(x)',funf612(x));
   show ('Iterations',j);
   show ('Function evaluations',ev);
end
wait

function y = funf612 (x)
%----------------------------------------------------------------
% Problem 6.12: Objective function
%----------------------------------------------------------------
   d = 10;
   g = [1 -3 2 8]';
   H = diag([3 9 27 81]);
   y = d + g'*x + x'*H*x/2;
%----------------------------------------------------------------
```

When this script is executed, the steepest descent method ($v = 1$) requires 72 iterations and 3823 scalar function evaluations to obtain the following optimal x.

$$x = \begin{bmatrix} -0.333 \\ 0.3333 \\ -0.07407 \\ -0.09875 \end{bmatrix}$$

In this case, the value of the objective at the optimal x was $f(x) = 8.8642$. By contrast, the full conjugate gradient method ($v = 4$) uses five iterations and 270 scalar function evaluations to produce the following estimate for the optimal x.

$$x = \begin{bmatrix} -0.3333 \\ 0.3333 \\ -0.07407 \\ -0.09877 \end{bmatrix}$$

In this case, the value of the objective at the optimal x was again $f(x) = 8.8642$.

6.16 Write a program that uses the NLIB function *dfp* to minimize the following objective function. Use an error tolerance of $tol = 10^{-5}$ and an initial guess of $x^0 = 0$. Print the optimal x, the optimal $f(x)$, the number of iterations, and the number of function evaluations.

$$f(x) = (x_1 + x_2 - 2)^2 + (x_1 - x_2 + 3)^4$$

```
%-------------------------------------------------------------------
% Problem 6.16
%-------------------------------------------------------------------
clc
clear
tol = 1.e-5;
m = 100;
n = 2;
v = n;
x0 = [0 0]';

% Find minimum using Davidon-Fletcher-Powell method

fprintf ('Problem 6.16\n');
[x,ev,j] = dfp (x0,tol,v,m,'funf616');
show ('x',x
show ('f(x)',funf616(x));
show ('Iterations',j);
show ('Function evaluations',ev);
wait

function y = funf616 (x)
%-------------------------------------------------------------------
% Problem 6.16: Objective function
%-------------------------------------------------------------------
   y = (x(1) + x(2) - 2)^2 + (x(1) - x(2) + 3)^4;
%-------------------------------------------------------------------
```

When this script is executed, the DFP method requires four iterations and 219 scalar function evaluations to produce the following estimate to the optimal x.

$$x = \begin{bmatrix} -0.5043 \\ 2.504 \end{bmatrix}$$

The value of the objective at the optimal x is $f(x) = 5.50468 \times 10^{-9}$.

CHAP. 7 DIFFERENTIATION AND INTEGRATION

7.3 Consider the function $f(x) = x \exp(-x)$. Fill in Table 4.7.1 with numerical estimates of the first derivative of $f(x)$ at $x = 0$ using a step size of $h = 0.1$.

The exact value of the derivative is

$$
\begin{aligned}
f'(0) &= [-x \exp(-x) + \exp(-x)]|_{x=0} \\
&= \exp(0) \\
&= 1
\end{aligned}
$$

Using Table 7.2.1, the forward, backward, and central difference approximations using $h = 0.1$ are

$$f'_F(0) = \frac{1}{0.1}\left[\frac{-3}{2}f(0) + 2f(0.1) - \frac{1}{2}f(0.2)\right] = 0.9909$$

$$f'_B(0) = \frac{1}{0.1}\left[\frac{3}{2}f(0) - 2f(-0.1) + \frac{1}{2}f(-0.2)\right] = 0.9889$$

$$f'_C(0) = \frac{1}{0.1}\left[\frac{-1}{2}f(-0.1) + \frac{1}{2}f(0.1)\right] = 1.0050$$

Thus, the percentage errors are

$$
\begin{aligned}
\Delta f'_F(0) &= 100(0.9909 - 1) = -0.91\% \\
\Delta f'_B(0) &= 100(0.9889 - 1) = -1.11\% \\
\Delta f'_C(0) &= 100(0.0050 - 1) = 0.50\%
\end{aligned}
$$

7.6 Suppose an integration formula with a truncation error of order $O(h^q)$ is used to compute

$$I(a,b) = \int_a^b f(x)\, dx$$

The magnitude of the truncation error is 0.01 when the number of integration panels is $n = 10$. What is the minimum value of n needed to ensure that the magnitude of the truncation error is less that 0.0001? What other source of error contributes to the total error?

Let $E(h)$ be the truncation error where $h = (b - a)/n$. Since $E(h)$ is of order $O(h^q)$, there exists a constant α such that

$$
\begin{aligned}
E(h) &\approx \alpha h^q \\
&= \frac{\alpha(b - a)^q}{n^q}, \quad h \ll 1
\end{aligned}
$$

Since $E(h) = 0.01$ when $n = 10$, this yields $0.01 \approx \alpha(b - a)^q/10^q$ or

$$\alpha \approx \frac{10^{q-2}}{(b - a)^q}$$

TABLE 4.7.1 Estimates of First Derivative

Type	*Points*	$f^{(1)}(0)$	*Percent Error*
Forward	3	0.9909	−0.91
Backward	3	0.9989	−1.11
Central	3	1.0050	0.50

Setting $E(h) = 0.0001$ and solving for n then yields $0.0001 \approx 10^{q-2}/n^q$ or

$$n^q \approx 10^{q+2}$$

Thus, the number of panels is the smallest integer n such that

$$n \geq 10^{(q+2)/q}$$

The other major source of error is round-off error associated with finite-precision arithmetic.

7.18 Write a program that uses the NLIB functions *midpoint*, *trapint*, and *simpson* to estimate the value of the following integral. Print the estimates for the number of panels n ranging from 2 to 64 in powers of 2.

$$I(-1,1) = \int_{-1}^{1} \exp[\sin(\pi x)]\, dx$$

```
%----------------------------------------------------------------------
% Problem 7.18
%----------------------------------------------------------------------
clc
clear
a = -1;
b = 1;
f = inline ('exp(sin(pi*x))','x');
g = fopen ('out','w');

% Evaluate integrals

fprintf ('Problem 7.18\n');
fprintf ('     n      midpoint    trapezoid       simpson\n\n');
for n = [2 4 8 16 32 64]
   y(1) = midpoint (a,b,n,f);
   y(2) = trapint  (a,b,n,f);
   y(3) = simpson  (a,b,n,f);
   fprintf ('%6g %12.6f %12.6f %12.6f\n',n,y(1),y(2),y(3));
   fprintf (g,'%6g & %12.6f & %12.6f & %12.6f \\\\ \n ',n,y(1),y(2),
           y(3));
end
fclose(g);
wait
%----------------------------------------------------------------------
```

When this script is executed, it produces the results shown in Table 4.7.2.

7.24 Write a program that uses the NLIB function *romberg* to compute the value of the following integral using a truncation error bound of $tol = 10^{-6}$. The program should print the number of extrapolation levels used and the value of the integral.

$$I(0,1) = \int_{0}^{1} \frac{x^3 \exp(-x^2)}{2 + \cos[\ln(1 + x^2)]}\, dx$$

TABLE 4.7.2 Estimates of Integral

n	*Midpoint*	*Trapezoid*	*Simpson*
2	3.086161	2.000000	2.000000
4	2.521184	2.543081	2.724108
8	2.532131	2.532132	2.528483
16	2.532132	2.532132	2.532132
32	2.532132	2.532132	2.532132
64	2.532132	2.532132	2.532132

```
%-----------------------------------------------------------------------
% Problem 7.24
%-----------------------------------------------------------------------
clc
clear
a = 0;
b = 1;
tol = 1.e-6;
m = 5;
f = inline ('x^3*exp(-x^2)/(2 + cos(log(1 + x^2)))','x');

% Evaluate integral

fprintf ('Problem 7.24\n');
[y,r] = romberg (a,b,tol,m,f);
show ('Extrapolation level',r)
show ('y',y);
%-----------------------------------------------------------------------
```

When this script is executed, it uses four extrapolation levels to produce an estimate of $I(0,1) = 0.04585$.

CHAP. 8 ORDINARY DIFFERENTIAL EQUATIONS

8.2 Consider the DC motor shown in Figure 8.11.1. The motion of this electromechanical device is modeled by a mechanical equation obtained from Newton's second law, an electrical equation obtained from Kirchhoff's voltage law, and an algebraic electromechanical coupling equation.

$$J\frac{d^2\phi}{dt^2} + \mu\frac{d\phi}{dt} = \tau$$

$$L_a\frac{dI_a}{dt} + R_aI_a + k_b\frac{d\phi}{dt} = V$$

$$\tau = k_tI_a$$

The system variables are the motor shaft angle ϕ, the torque developed at the motor shaft τ, the armature winding current I_a, and the applied armature voltage V. The remaining parameters are all positive constants. They include the moment of inertia about the motor shaft J, the coefficient of viscous friction μ, the armature winding inductance L_a, the armature winding resistance R_a, the back emf constant k_b, and the torque constant k_t. Write the equations of motion of this DC motor as a first-order system using the state variables $x = [\phi, d\phi/dt, I_a]^T$. Is this system linear?

We differentiate the elements of x and substitute on the right-hand side using the definition of x and the equations of motion of the motor.

$$\begin{aligned}
\frac{dx_1}{dt} &= \frac{d\phi}{dt} \\
&= x_2 \\
\frac{dx_2}{dt} &= \frac{d^2\phi}{dt^2} \\
&= \frac{1}{J}\left(\tau - \mu\frac{d\phi}{dt}\right) \\
&= \frac{1}{J}(k_t I_a - \mu x_2) \\
&= \frac{1}{J}(k_t x_3 - \mu x_2) \\
\frac{dx_3}{dt} &= \frac{dI_a}{dt} \\
&= \frac{1}{L_a}\left(V - k_b\frac{d\phi}{dt} - R_a I_a\right) \\
&= \frac{1}{L_a}(V - k_b x_2 - R_a x_3)
\end{aligned}$$

These equations are linear and can therefore be represented in matrix form as

$$\frac{dx}{dt} = \begin{bmatrix} 0 & 1 & 0 \\ 0 & -\mu/J & k_t/J \\ 0 & k_b/L_a & -R_a/L_a \end{bmatrix} x + \begin{bmatrix} 0 \\ 0 \\ V/L_a \end{bmatrix}$$

8.4 Suppose we have developed a numerical differential equation solution technique that has a local truncation error of order $O(h^p)$, and suppose $E(h)$ and $E(h/2)$ denote local truncation error estimates using step sizes of length h and $h/2$, respectively. What is the relationship between $E(h)$ and $E(h/2)$ when $|h| \ll 1$?

Since the method is of order $O(h^p)$, there exists a constant α such that

$$E(h) \approx \alpha h^p \quad , \quad h \ll 1$$

It then follows that $E(h/2) \approx 2\alpha(h/2)^p$. The factor two arises because two steps of length $h/2$ are needed to generate one step of length h. Thus, the following relationship exists between successive truncation errors when h is small.

$$E(h/2) \approx \frac{E(h)}{2^{p-1}} \quad , \quad h \ll 1$$

8.12 The motion of a satellite tumbling in space can be modeled by the following nonlinear Euler equations.

$$J_1\frac{dx_1}{dt} = (J_2 - J_1)x_2x_3 + \tau_1$$

$$J_2 \frac{dx_2}{dt} = (J_3 - J_1)x_3 x_1 + \tau_2$$

$$J_3 \frac{dx_3}{dt} = (J_1 - J_2)x_1 x_2 + \tau_3$$

Here x_k denotes the angular velocity, J_k denotes the moment of inertia, and τ_k denotes the applied torque about axis k for $1 \le k \le 3$. Suppose the satellite is non-symmetrical with $J_1 = 1$, $J_2 = 3$, and $J_3 = 2$. Write a program that uses the NLIB functions to solve this system over the time interval $[0, 10]$ when the applied torques are $\tau = \exp(-t)[2, 0, t]^T$. Start from the initial condition $x(0) = [1, 0, -1]$ and plot $x(t)$ versus t.

```
%--------------------------------------------------------------
% Problem 8.12
%--------------------------------------------------------------
clc
clear
n = 3;
x0 = [1 0 -1]';
t0 = 0;
t1 = 10;
m = 150;
tol = 1.e-5;
q = 3000;

% Solve ode

fprintf ('Problem 8.12\n');
[t,X,e,k] = rkf (x0,t0,t1,m,tol,q,'funf812');
graphxy (t,X,'Satellite Motion','t','x');


function dx = funf812 (t,x)
%--------------------------------------------------------------
% Problem 8.12: Satellite equations
%--------------------------------------------------------------

   dx = zeros (3,1);    % column vector
   J   = [1 3 2];
   tau = exp(-t)*[2 0 t];
   dx(1) = ((J(2) - J(1))*x(2)*x(3) + tau(1))/J(1);
   dx(2) = ((J(3) - J(1))*x(3)*x(1) + tau(2))/J(2);
   dx(3) = ((J(1) - J(2))*x(1)*x(2) + tau(3))/J(3);
%--------------------------------------------------------------
```

When this script is executed, it produces the satellite motion shown in Figure 4.8.1.

8.15 Consider the circular membrane shown in Figure 8.11.2. When the membrane is under constant pressure, p, the displacement y (out of the page) as a function of radial position x can be modeled as follows (Nakamura, 1993).

$$\frac{d^2y}{dx^2} + \left(\frac{1}{x}\right)\frac{dy}{dx} + \frac{p}{\tau} = 0$$

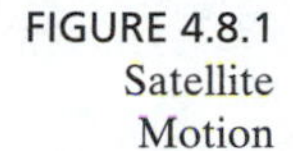

FIGURE 4.8.1 Satellite Motion

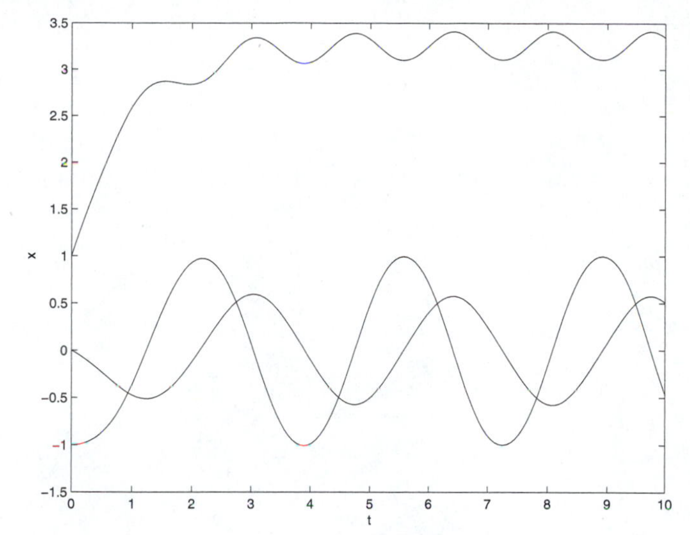

Here the parameter $\tau > 0$ is the membrane tension. The boundary conditions at the inner and outer radius of the annular membrane are $y(\alpha) = 0$ and $y(\beta) = 0$. Write a program that uses the NLIB function *bvp* to find the membrane displacement profile $y(x)$. Use $p = 200$, $\tau = 80$, $\alpha = 0.5$, and $\beta = 1.5$. Solve the system for $m = 19$ points using an error tolerance of $tol = 10^{-4}$. Print the number of iterations and plot the cross-section $y(x)$ versus x.

```
%--------------------------------------------------------------------
% Problem 8.15
%--------------------------------------------------------------------
clc
clear
alpha = 0.5;
beta  = 1.5;
a     = 0;
b     = 0;
tol   = 1.e-4;
m     = 19;
q     = 10;
g     = inline ('-200/80 - y1/x','x','y','y1');

% Solve bvp

fprintf ('Problem 8.15\n');
fprintf ('Solving equations ...\n');
[x,y,r] = bvp (alpha,beta,a,b,tol,m,q,g);
show ('Iterations',r)
graphxy (x,y,'Membrane Displacement','x','y')
%--------------------------------------------------------------------
```

When this script is executed, it requires six iterations to converge to the solution shown in Figure 4.8.2.

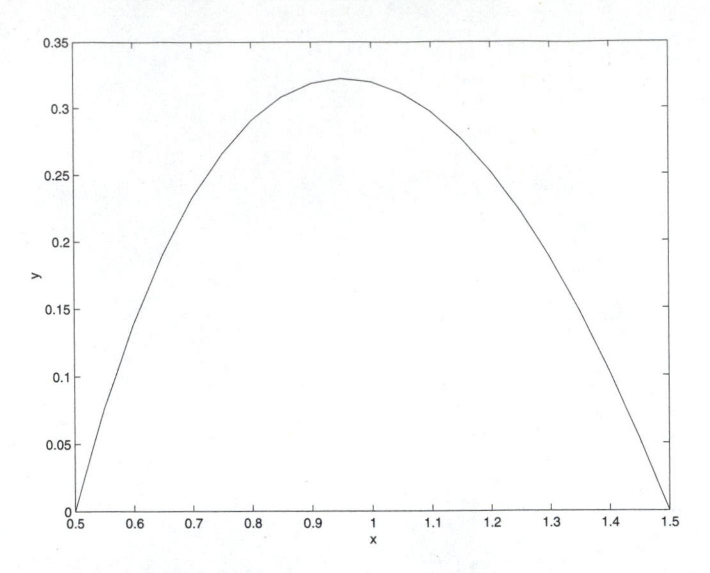

FIGURE 4.8.2 Membrane Displacement

CHAP. 9 PARTIAL DIFFERENTIAL EQUATIONS

9.3 Fisher developed the following model for the spread of genes in a population where p denotes a probability of occurrence of a specific allele and α and β are positive constants (Edelstein-Keshet, 1988).

$$\frac{\partial p}{\partial t} = \beta\left(\frac{\partial^2 p}{\partial x^2}\right) + \alpha p(1 - p)$$

Let p_j^k denote $p(t_k, x_j)$ with $\Delta t = t_{k+1} - t_k$ and $\Delta x = x_{k+1} - x_k$. Develop an explicit difference equation approximation for this *nonlinear* partial differential equation, using a forward difference in time and a central difference in space.

From Table 7.2.1 or (9.3.6a), the two-point forward difference approximation to $p_t = \partial p/\partial t$ is

$$p_t \approx \frac{p_j^{k+1} - p_j^k}{\Delta t}$$

Next, from Table 7.2.2 or (9.3.6b), the three-point central difference approximation to $p_{xx} = \partial^2 p/\partial x^2$ is

$$p_{xx} \approx \frac{p_{j+1}^k - 2p_j^k + p_{j-1}^k}{\Delta x^2}$$

Substituting these approximations into Fisher's model and solving for p_j^{k+1} then yields

$$p_j^{k+1} = p_j^k + \frac{\rho \Delta t}{\Delta x^2}(p_{j+1}^k - 2p_j^k + p_{j-1}^k) + \alpha \Delta t p_j^k(1 - p_j^k)$$

9.8 Find the amplitude factor λ for the FTCS method in problem 9.7, using a candidate solution of the form $u_j^k = (\lambda)^k \exp(ij\pi/p)$ where $i = \sqrt{-1}$ and p is any nonzero integer. For what values of γ is the FTCS method stable?

From problem 9.7, the FTCS equation is

$$u_j^{k+1} = u_j^k - \frac{\gamma(u_{j+1}^k - u_{j-1}^k)}{2}$$

Substituting the candidate solution $u_j^k = (\lambda)^k \exp(ij\pi/p)$ and removing the factor $(\lambda)^k \exp(ij\pi/p)$ from each term yields

$$\lambda = 1 - \gamma\left(\frac{\exp(i\pi/p) - \exp(-i\pi/p)}{2}\right)$$
$$= 1 - i\gamma \sin(\pi/p)$$

Thus,

$$|\lambda|^2 = 1 + \gamma^2 \sin^2(\pi/p)$$

For stability of the method, it is necessary that $|\lambda| \leq 1$ for all nonzero intergers p. Thus, the only γ for which the FTCS method is stable is $\gamma = 0$. From problem 9.7, $\gamma > 0$. Thus, the FTCS method is unconditionally unstable.

9.12 Random motion of particles at the molecular level causes movement of matter from regions of high concentration to regions of low concentration through the process of *diffusion*. Suppose a substance is constrained to move in one dimension. If $c(t, x)$ denotes the concentration of the substance at time t and position x, then, using *Fick's law,* we can model the change in concentration due to diffusion as follows.

$$\frac{\partial c}{\partial t} = D\left(\frac{\partial^2 c}{\partial x^2}\right)$$

Here $D > 0$ is the diffusion coefficient. If the substance is oxygen and the medium is air at 20°C, the diffusion coefficient is $D = 0.201$ (Leyton, 1975). Suppose we are interested in solving this system for $0 \leq t \leq 30$ and $0 < x < 5$. The initial and boundary conditions are as follows:

$$u(0, x) = (x - 2.5)^2$$
$$u(t, 0) = 1$$
$$u_x(t, 5) = 0$$

Write a program that uses the NLIB function *heat1* to solve this system. Print γ and plot the solution as a surface.

```
%-------------------------------------------------------------------
% Problem 9.12
%-------------------------------------------------------------------
   clc
   clear
   m = 50;                      % t precision
   n = 39;                      % x precision
   c = [0 1];                   % boundary condition type
   a = 5;                       % maximum x
   T = 30;                      % maximum t
   D = 0.201;                   % diffusion coefficient
```

```
    f = inline ('(x - 2.5)^2','x');
    g = inline ('1*(k - 2) + 0*(k-1)','t','k');

% Find gamma

    fprintf ('Problem 9.12\n');
    dt = T/m;
    dx = a/(n+1);
    gamma = D*dt/(dx*dx);
    show ('gamma',gamma)

% Solve diffusion equation

    disp ('Solving diffusion equation ...')
    [t,x,U] = heat1 (T,a,m,n,D,c,f,g);
    plotxyz (t,x,U,'Diffusion Equation Solution','t','x','u')
%-------------------------------------------------------------------
```

When this script is executed, it produces $\gamma = 7.718$ and the surface plot shown in Figure 4.9.1.

9.14 Consider the problem of determining the displacement of a taut elastic string of length $a = 4$. Suppose the tension and density of the string are such that the wave equation coefficient is $\beta = 4$. Thus, the string displacement is a solution of

$$\frac{\partial^2 u}{\partial t^2} = 4\left(\frac{\partial^2 u}{\partial x^2}\right)$$

The string is fixed at both ends, and the initial conditions are

$$u(0, x) = \sin(\pi x/4)$$
$$u_t(0, x) = 0$$

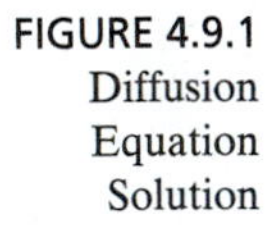
FIGURE 4.9.1 Diffusion Equation Solution

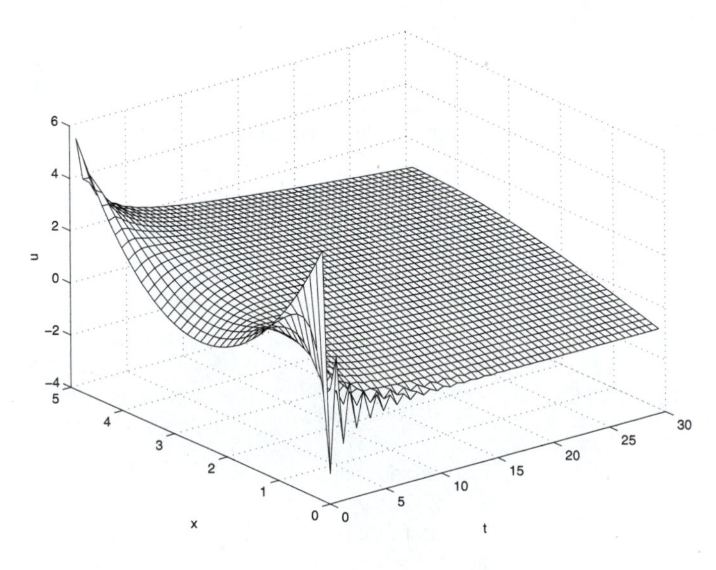

Write a program that uses the NLIB function *wavel* to find the string displacement $u(t, x)$. Print the value of γ and plot the solution as a surface over $0 < t < 4$ and $0 < x < 4$ using a 40×19 grid.

```
%------------------------------------------------------------------------
% Problem 9.14
%------------------------------------------------------------------------
   clc
   clear
   a = 4;                  % string length
   T = 2;                  % maximum t
   m = 40;                 % t precision
   n = 39;                 % x precision
   beta = 4;               % gain
   f = inline ('x*(x - 4)','x');
   g = inline ('0','x');

% Find gamma

   fprintf ('Problem 9.14\n');
   dt = T/m;
   dx = a/(n+1);
   gamma = beta*dt*dt/(dx*dx);
   show ('gamma',gamma);

% Solve wave equation

   disp ('Solving wave equation ...');
   [t,x,U] = wavel (T,a,m,n,beta,f,g);
   plotxyz (t,x,U,'Elastic String','t','x','u');
%------------------------------------------------------------------------
```

When this script is executed, it produces $\gamma = 7.718$ and the surface plot shown in Figure 4.9.2.

CHAP. 10 DIGITIAL SIGNAL PROCESSING

10.2 Consider the analog signal $x_a(t) = \cos(2\pi f_0 t)$. The Fourier transform of this signal is as follows, where $\delta(x)$ denotes the unit impulse, which is zero for $x \neq 0$.

$$X(f) = \frac{\delta(f + f_0) + \delta(f - f_0)}{2}$$

For what range of sampling intervals, T, can this signal be reconstructed from its samples?

A cosine is a pure tone with all of the power concentrated at $f = \pm f_0$. Therefore, the magnitude spectrum satisfies $|X(f)| = 0$ for $|f| > f_0$. That is, the signal is bandlimited to $B = f_0 + \varepsilon$ where $\varepsilon > 0$ can be arbitrarily small. Since we must sample at a rate that is at least twice the bandwidth, this means $f > 2f_0$ or $1/T > 2f_0$. Thus, to avoid aliasing,

$$0 < T < \frac{1}{2f_0}$$

FIGURE 4.9.2
Elastic String

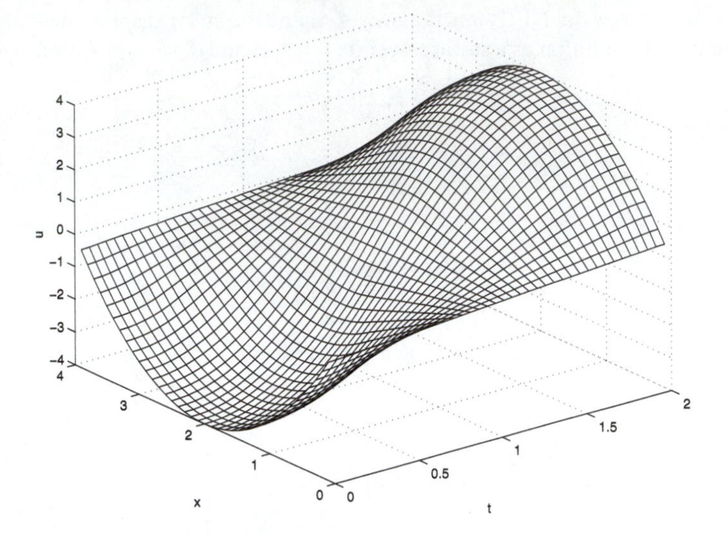

10.6 Consider the following discrete-time system. For what values of α is this system stable?

$$y(k) = u(k) - 1.6y(k-1) - (0.64 + \alpha^2)y(k-2)$$

From (10.5.10), the characteristic polynomial of this system is

$$\begin{aligned}\Delta(z) &= z^2 + a_1 z + a_2 \\ &= z^2 + 1.6z + 0.64 + \alpha^2\end{aligned}$$

Using the quadratic formula, the roots of this polynomial (poles of the system) are

$$\begin{aligned}p_{1,2} &= \frac{-1.6 \pm \sqrt{2.56 - 4(0.64 + \alpha^2}}{2} \\ &= -0.8 \pm j\alpha\end{aligned}$$

From (10.5.11), this system is stable if and only if $|p_k| < 1$ for $1 \le k \le 2$. Thus, $\sqrt{(-0.8)^2 + \alpha^2} < 1$ or $0.64 + \alpha^2 < 1$. Solving for α, the system is stable for the following range:

$$-0.6 < \alpha < 0.6$$

10.11 Consider the following discrete-time linear system. Write a program that uses the NLIB function *arma* to compute and plot the pulse response, $h(k)$, for $0 \le k < N$ where $N = 64$.

$$y(k) = u(k-1) - 0.81y(k-2)$$

```
%-------------------------------------------------------------------
% Problem 10.11
%-------------------------------------------------------------------
  clc
  clear
  N = 64;
```

```
   n = 2;
   m = 1;
   r = n + m + 1;
   theta = [0 0.81 0 1]';
   x = zeros (r,1);
   h = zeros (N,1);
   u = zeros (N,1);

% Find pulse response

   fprintf ('Problem 10.11\n');
   u(1) = 1;
   for i = 1 : N
      [x,h(i)] = arma (u(i),theta,x,n,m);
   end
   graphmat (h,'Pulse Response','k','h(k)')
%----------------------------------------------------------------
```

When this script is executed, it produces the pulse response plot shown in Figure 4.10.1.

10.19 Write a program that uses the NLIB function *arma* to find the first 20 samples of the Fibonacci sequence by computing the pulse response of the system in problem 10.4. The script should graph the ratio $h(k)/h(k-1)$ and print the final estimate of the golden ratio.

```
%------------------------------------------------------------------
% Problem 10.19
%------------------------------------------------------------------
   clc
   clear
   N = 20;
```

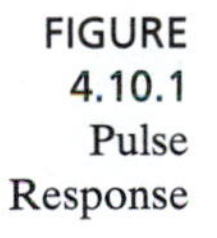

FIGURE 4.10.1 Pulse Response

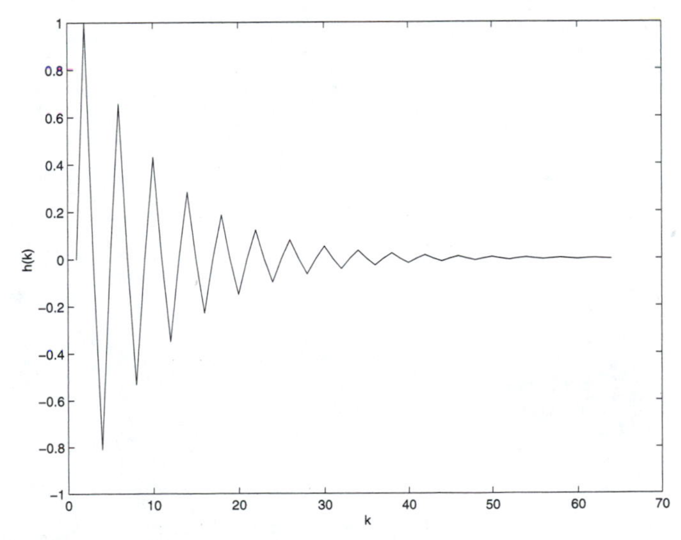

```
    n = 2;
    m = 0;
    r = n + m + 1;
    theta = [-1 -1 1]';
    x = zeros (r,1);
    h = zeros (N,1);
    u = zeros (N,1);
    g = zeros (N-1,1);

% Find pulse response

    fprintf ('Problem 10.19\n');
    u(1) = 1;
    for i = 1 : N
       [x,h(i)] = arma (u(i),theta,x,n,m);
       if i > 1
          g(i-1) = h(i)/h(i-1);
       end
    end
    graphmat (g,'Fibbonaci ratio','k','h(k)/h(k-1)')
    show ('Golden ratio',g(N-1))
%-------------------------------------------------------------------
```

When this script is executed, produces the plot shown in Figure 4.10.2. The final estimate of the golden ratio is $\gamma = 1.61803$.

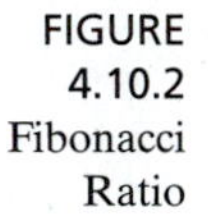

FIGURE 4.10.2 Fibonacci Ratio

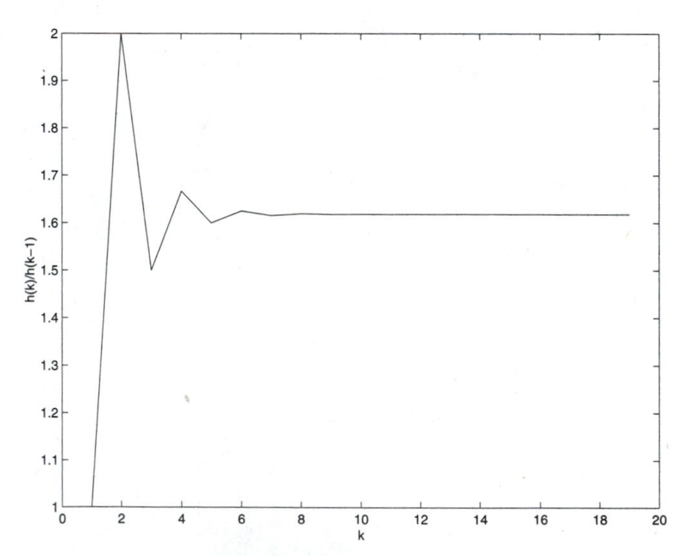

Index

G

H

Q

R

S